DESIGN OF STEEL STRUCTURES

Third Edition

About the Author

S K Duggal is currently Professor and Head in Civil Engineering at Motilal Nehru National Institute of Technology, Allahabad. He obtained his BE, ME and PhD degrees from Allahabad University. He has an experience of more than 32 years of teaching at both undergraduate and postgraduate levels. He is also the author of *Surveying*, Volumes I and II, published by TMH, apart from the numerous research papers that he has presented and published in several national and international conferences and journals respectively.

DESIGN OF STEEL STRUCTURES

Third Edition

S K Duggal
Professor andHead
Department of Civil Engineering
MNNIT
Allahabad

McGraw Hill Education (India) Private Limited

Published by McGraw Hill Education (India) Private Limited
Registered Office: Anjana Complex No: 5/90 A, Butt Road, St. Thomas Mount, Chennai – 600016

Design of Steel Structures, 3/e

36th reprint 2024

ISBN-13: 978-0-07-026068-9
ISBN-10: 0-07-026068-0

37 38 39 40 41 APO 28 27 26 25 24

Typeset at The Composers, 260, C.A. Apt., Paschim Vihar, New Delhi 110 063 and printed and bound in India at A.P. Offset Pvt. Ltd., 25/487, Zulfe Bengal, Dilshad Garden, New Delhi 110095

Cover Printer: A.P. Offset Pvt. Ltd.
Visit us at: www.mheducation.co.in
Write to us at: support.india@mheducation.com
CIN: U80302TN2010PTC111532
Toll free Number: 1800 103 5875

Preface

The main objective of this book is to present a systematic text on the design of steel structures based on the latest Indian Standard codes. The book covers the fundamental aspects of analysis and design, and also discusses practical requirements such as safety, feasibility, and economy of steel structures. Theoretical concepts regarding the strength of materials are applied to the design of steel structures based on given specifications.

Though primarily intended for undergraduate students and practising engineers, it is hoped that the text would also serve postgraduate students as an introductory reference.

In the preparation of the third edition of this volume, the objective was to introduce, wherever necessary, material which embodies the most recent design methodologies and helps to make the text more lucid and clear.

The earlier edition contained numerous illustrative examples, each of which was intended to assist the reader in understanding a certain principle or particular design method. The examples cited in the earlier edition are retained as these were focused on a specific topic detailing the basic design methods, and, subsequently, are related to other design situations. Problems included at the end of chapters are also retained for the same reason.

IS: 1893, Criteria for Earthquake Resistant Design of Structure is revised as IS: in 2002. The equivalent lateral force method for determining lateral loads due to earthquakes is described in Chapter 1 as per the revised code.

High Strength Friction Grip bolts have been detailed in Chapter 2 and examples increased to make the principles of design of High Strength Friction Grip Bolted connections more illustrative. These bolted connections have been also added in the rest of the chapters wherever members along with connections were to be designed.

The first chapter discusses loads on structures, stresses, the concept of factor of safety, methods of design, and gives an introduction to the types of rolled steel sections. Chapter 2 deals with simple connections, rivets, welds, bolts and pins. Discussions on the design of tension, compression, and flexural members are presented in the next seven chapters. Chapter 10 pertains to eccentric connections. The design concepts discussed in Chapter 2–10 are applied to the design of industrial buildings in Chapter 11.

The arrangement of the chapters has been reorganized to keep the sequence logical. Chapter 13 on steel stacks has been introduced after Chapter 12 on water tanks. Chapter 13 deals with the steel stacks used for emitting flue gases from plants and industries. Chapter 14 discusses the theory and design of steel bins for storage of construction materials. All the three chapters discuss elevated structures and structurally behave in a similar fashion. Tubular structures are discussed in Chapter 15.

Chapter 16 discusses the principles of design of light-gauge steel construction. The concept of cold formed steel sections is a relatively new development in the design of steel structures. The practical design expressions are developed for compression and flexural members using the buckling theory of thin plates. The design of connections for members made up with steel sheets is discussed. Chapter 17 introduces the concepts of elementary plastic analysis and design.

The revised code **IS: 800**, based on limit state design, is expected shortly by the Bureau of Indian Standards. Chapter 18 introduces the principles and concepts adopted and major changes in the code. The objective of writing this chapter is to make the teachers and the students familiar with the objectives of the new code.

The book contains a number of illustrative examples for explaining important design principles and methods. Additional references are given at the end of each chapter.

Instructor resources are accessible at https://connect.mheducation.com. Student resources can be accessed using the scratch code at www.mheducation.com.

For Instructors

- PowerPoint lecture slides

For Students

- Sample Chapter on Rigid Frames
- List of Symbols

I am extremely grateful to Dr Mohan Kalani of IIT Bombay, Dr R Vaidyanathan of Government College of Technology, Coimbatore and Dr A R Santhakumar of Anna University, Madras, for reviewing the manuscript and providing their useful suggestions.

I would like to acknowledge the contribution of the following reviewers who provided their valuable feedback for this edition.

S P Singh
Department of Civil Engineering
National Institute of Technology
Jalandhar, Punjab

D C Rai
Department of Civil Engineering
Indian Institute of Technology
Kanpur, UP

Ashok Gupta
Department of Civil Engineering,
Indian Institute of Technology, Delhi

Anurag Mishra
Department of Civil Engineering,
Malaviya National Institute of
Technology, Jaipur, Rajasthan

M K Mondal
Department of Civil Engineering
Bhagalpur College of Engineering,
Bhagalpur, Bihar

G Bairagi
Department of Civil Engineering
Jalpaiguri Govt. Engineering College
Jalpaiguri, West Bengal

Ramesh Ogale
Department of Civil Engineering
SSVPS's B S Deore College of Engineering, Dhule, Maharashtra

L M Gupta
Departmentt of Civil Engineering
National Institute of Technology
Nagpur, Maharashtra

G Sathyavathi
Department of Civil Engineering
SRM Institute of Science and Technology, Chennai, Tamil Nadu

S R Satish Kumar
Structural Engineering Laboratory
Department of Civil Engineering
Indian Institute of Technology
Chennai, Tamil Nadu

D Rama Seshu
Department of Civil Engineering
National Institute of Technology
Warangal, AP

All efforts have been made to keep errors to a minimum, but because of the large amount of design calculations involved, some errors may have crept in inadvertently. I would be grateful if such oversights are pointed out by the readers. Feel free to write to me at *shasikantduggal@yahoo.co.in*.

S K Duggal

FOREWORD

by
Dr C V R Murty
Director
Indian Institute of Technology Hyderabad

This text is a special compilation of the spectrum of topics relevant to steel structures. This book of 18 chapters contains three distinct sets of contents, namely, the *structural components*, *connections between structural components*, and *structural systems*. This text presents in detail the first two sets of contents. It deals with virtually all sub-topics that so far were presented in a scattered way in the various Indian author textbooks on steel structures. The third set of contents has applications in focus. The applications presented should be useful for those seeking information on a variety of industrial structures. Items like steel water tanks, stacks and bins are not common in many texts on the subject. The inclusion of the treatment of steel bridges, tubular steel structures and light-gauge steel structures should be valuable to readers. India now has a new Indian Standard for design of steel structures. In light of this, the all-new chapter on limit state design of steel structures should be very timely. Each chapter is scheduled to have solved examples and exercises; the academic community—teachers and students alike—may find this valuable.

At this point of time, when the Indian Standard on design of steel structures (IS:800) has been abruptly and completely changed in its recently released third edition, the academic and professional communities need to fall back on conceptual level learning to understand the new code. This fact along with the lack of a commentary released by the Bureau of Indian Standards of the new standard IS: 800–2007 will mean that this textbook amongst others should help the knowledge-hungry readership to learn the intent behind the provisions of the code. I hope that this textbook will serve the intended need. Professor Duggal's earlier edition of the textbook is quite popular with the students and teachers. I am hoping that this textbook will follow the suit.

. . . .

FOREWORD

by
Dr Prem Krishna
Former Professor and Head
Department of Civil Engineering
University of Roorkee (Now IIT Roorkee)

I feel greatly honoured to have been asked to write a foreword for the book entitled *Design of Steel Structures* by Prof. S K Duggal, which is now going into its third edition.

Steel as a material has been on our horizons for several centuries, but its history as a construction material may be no more than about a couple of centuries. The advent and development of reinforced and prestressed concrete over the last century provided very versatile structural materials, and particularly with the developments in prestressed concrete, some established areas of application of steel began to be replaced by the former. Steel industry was then, for a small period, being seen as a 'sunset' industry. There has since been a remarkable period of renaissance, not only because of the many advantages of steel but because in many applications it is the only obvious choice. Such applications include tall structures, now looking to reach heights of 1 kilometre, wide span structures, and long span bridges for which spans up to 5 km are being considered. The Indian situation has not been too different.

The burgeoning need for the infrastructural growth in our country has created a big challenge for Civil Engineering in general—a challenge which is being difficult to meet partly because of the slump, for the last few decades, in civil engineering education. Whereas the use of steel as a structural material has the potential to offer effective and quick solutions, problems are compounded further for the construction industry because of the lack of preparedness to utilize steel on all fronts—inadequate production, general preference for the use of concrete, inadequate design and fabrication capabilities. Fortunately, the picture is changing. It is expected that the country will produce by 2020, almost four times the steel it produces today. A new design code (IS 800) has been issued recently by the Bureau of Indian Standards. Civil Engineering education is picking up. The Institute for Steel Development and Growth (INSDAG), set up at Kolkata by the Ministry of Steel, has been playing a commendable catalytic role in pushing steel towards becoming a 'sunrise' industry.

This provides for an interesting backdrop for further growth of literature on steel structures. It is imperative that good quality education is imparted to Civil Engineering graduates to enable them to design steel structures. Professor Duggal has to be complimented for bringing to bear his experience of teaching in producing this book on the Design of Steel Structures with a very comprehensive treatment of the subject. The book has already established itself as a popular text and the author has continued to improve it and widen its scope in the second and the current editions, such that the desired purpose of educating graduates in understanding the subject, as well as to serve the design practitioners, is duly met.

. . . .

FOREWORD

by
Dr Krishen Kumar
Ret Professor
Department of Civil Engineering
Indian Institute of Technology Roorkee

The book is written with greater emphasis on basic principles for structural design of steel structures, much sought after by the civil engineering students, while not losing sight of requirements for professional training. This feature, combined with several others, makes it a near ideal text book for students as well as professionals.

The presentation is comprehensive in the sense that the theory and principles of design are explained in simple, easily understandable language and amply explained with the help of illustrative examples. Many figures and sketches have been drawn in three dimensions for immediate visualization of the details of built-up members and the connections.

There are three aspects that a designer shall always keep in mind while designing a steel structure as these refer to the vulnerability of steel structures and these have been kept in mind by the author in preparing this book. It has been well recognized that greater care is needed in dealing with the following three aspects in the design of steel structures—firstly, the buckling aspect since the members as well as their elements are generally quite slender and tend to buckle (fail prematurely) under stresses such as axial or flexural compression, shear or torsional stresses, and the buckling could either be local or general; secondly, the efficiency of the connections that need to be compact, sufficiently rigid and capable of transferring the design loads besides giving rise to least stress concentrations, the last being especially relevant for fatigue resistance requirement when dynamic loads are significant; and, thirdly, the flexibility aspect, since there could possibly be large deflections leading to high secondary stresses or to instability, making it necessary to limit the deflections as per codal practice/specifications in the design of conventional structures.

The author has been periodically updating the book to improve upon the deficiencies and in this edition three new chapters have also been added, viz., stacks, use of light-gauge sections, and, introduction to limit state design; making the book more comprehensive and useful.

Contents

Visual Walkthrough

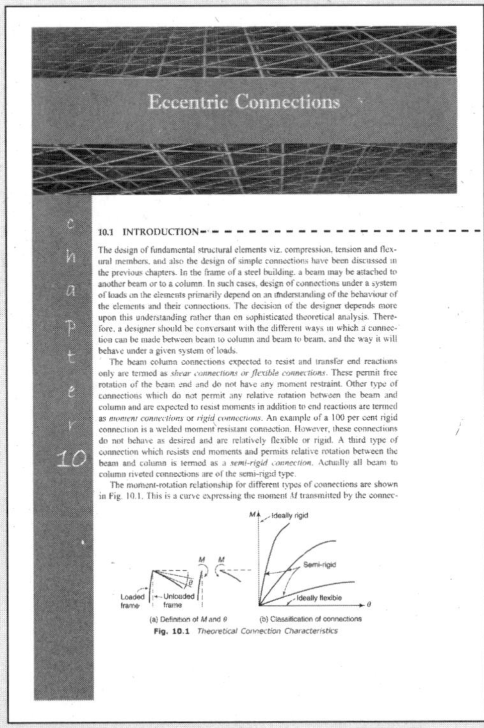

Eccentric Connections

chapter 10

10.1 INTRODUCTION

The design of fundamental structural elements viz. compression, tension and flexural members, and also the design of simple connections have been discussed in the previous chapters. In the frame of a steel building, a beam may be attached to another beam or to a column. In such cases, design of connections under a system of loads on the elements primarily depend on an understanding of the behaviour of the elements and their connections. The decision of the designer depends more upon this understanding rather than on sophisticated theoretical analysis. Therefore, a designer should be conversant with the different ways in which a connection can be made between beam to column and beam to beam, and the way it will behave under a given system of loads.

The beam column connections expected to resist and transfer end reactions only are termed as *shear connections or flexible connections*. These permit free rotation of the beam end and do not have any moment restraint. Other type of connections which do not permit any relative rotation between the beam and column and are expected to resist moments in addition to end reactions are termed as *moment connections* or *rigid connections*. An example of a 100 per cent rigid connection is a welded moment resistant connection. However, these connections do not behave as desired and are relatively flexible or rigid. A third type of connection which resists end moments and permits relative rotation between the beam and column is termed as a *semi-rigid connection*. Actually all beam to column riveted connections are of the semi-rigid type.

The moment-rotation relationship for different types of connections are shown in Fig. 10.1. This is a curve expressing the moment M transmitted by the connec-

(a) Definition of M and θ (b) Classification of connections

Fig. 10.1 *Theoretical Connection Characteristics*

Each chapter begins with an ***Introduction*** that states the purpose and goals of the chapter.

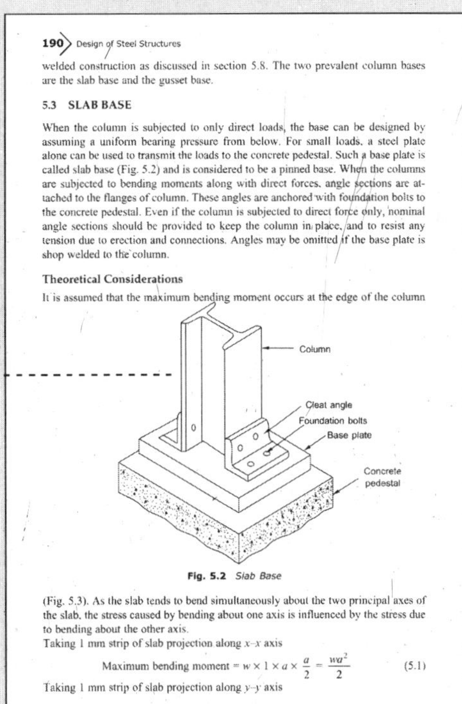

welded construction as discussed in section 5.8. The two prevalent column bases are the slab base and the gusset base.

5.3 SLAB BASE

When the column is subjected to only direct loads, the base can be designed by assuming a uniform bearing pressure from below. For small loads, a steel plate alone can be used to transmit the loads to the concrete pedestal. Such a base plate is called slab base (Fig. 5.2) and is considered to be a pinned base. When the columns are subjected to bending moments along with direct forces, angle sections are attached to the flanges of column. These angles are anchored with foundation bolts to the concrete pedestal. Even if the column is subjected to direct force only, nominal angle sections should be provided to keep the column in place, and to resist any tension due to erection and connections. Angles may be omitted if the base plate is shop welded to the column.

Theoretical Considerations

It is assumed that the maximum bending moment occurs at the edge of the column

Fig. 5.2 *Slab Base*

(Fig. 5.3). As the slab tends to bend simultaneously about the two principal axes of the slab, the stress caused by bending about one axis is influenced by the stress due to bending about the other axis.

Taking 1 mm strip of slab projection along x–x axis

$$\text{Maximum bending moment} = w \times 1 \times a \times \frac{a}{2} = \frac{wa^2}{2} \quad (5.1)$$

Taking 1 mm strip of slab projection along y–y axis

The gradation of ***topics*** provide a quick look into the concepts involved in the analysis of structural elements.

Visual Walkthrough

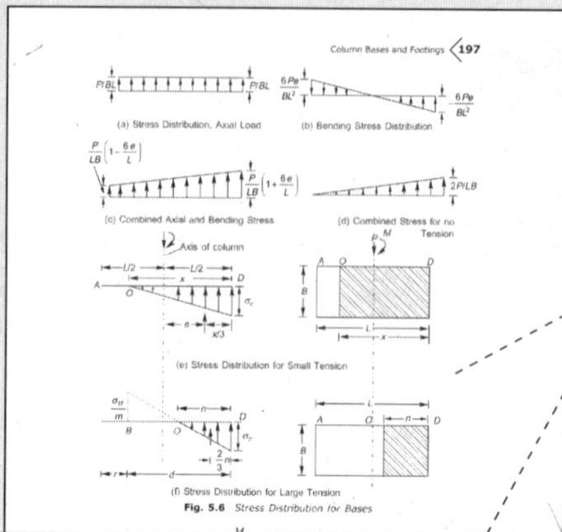

Fig. 5.6 *Stress Distribution for Bases*

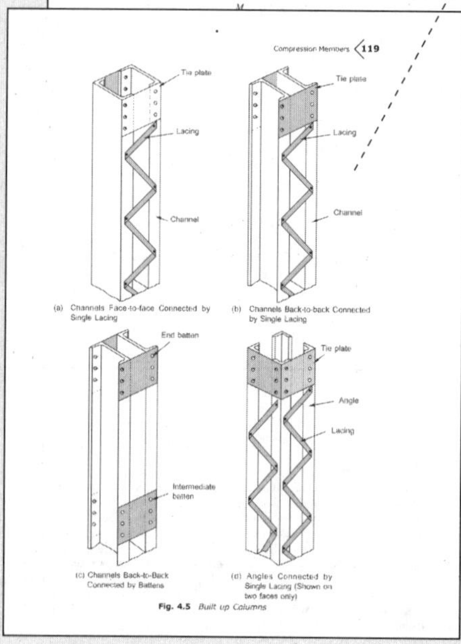

Fig. 4.5 *Built up Columns*

Figures are used to give a pictorial view and to illustrate the arrangement of various elements. ***Diagrams*** are given for visualization of explanation.

A variety of typical ***Solved Examples*** are given to reinforce the concepts.

Try two I.S.A. 75 × 50 × 8 mm @ 145.2 N/m.

Net area = gross area – area of rivet hole

$= 1876 - 2 \times 21.5 \times 8 = 1532 \text{ mm}^2$

Strength of the member $= 1532 \times 150 \times 10^{-3} = 229.800 \text{ kN} > 200 \text{ kN}$

The section can be revised for economy.

Try 2 I.S.A. 70 × 45 × 8 mm @ 131.4 N/m

$$A = 1716 \text{ mm}^2, \quad r = 21.9 \text{ mm}$$

Net area provided = gross area – area of rivet hole.

$= 1716 - 2 \times 21.5 \times 8 = 1372 \text{ mm}^2$

Strength of the member $= 1372 \times 150 \times 10^{-3} = 205.800 \text{ kN} > 200 \text{ kN}$

which is all right.

$$\frac{l}{r} = \frac{3000}{21.9} = 136.98 < 350$$

It is to be noted that this section is the most economical, and stable. However, the cost of making extra holes and of tack rivets will have to be added to do the final estimation and selection.

Example 6.5 The bottom chord tension member of a roof truss is subjected to an axial pull of 300 kN. A differential chain hoist arrangement is attached to the bottom chord and gives a point load of 10 kN at the centre. The length of the member between panel points is 4.0 m. Design the section consisting of two unequal angles with long legs kept back-to-back and turned upwards. Assume f_y = 250 N/mm^2.

Solution

Allowable axial tensile stress, $\sigma_{at} = 0.6 \times 250 = 150 \text{ N/mm}^2$

$$\text{Net area required for axial pull} = \frac{300 \times 10^3}{150} = 2000 \text{ mm}^2$$

Increase the net area by 50% to account for bending moment

$= 1.5 \times 2000 = 3000 \text{ mm}^2$

A tentative gross area required $= 1.4 \times 3000 = 4200 \text{ mm}^2$

Let us assume the section to be weakened by one rivet hole and let the diameter of rivet be 14 mm.

Try 2 I.S.A. 150 × 75 × 10 mm from I.S. Handbook No. 1 (longer legs back-to-back).

The relevant properties of the sections are

$C_{xx} = 53.2$ mm, $A = 4312 \text{ mm}^2$, $I_{xx} = 998.2 \times 10^4 \text{ mm}^4$

Net area provided $= 4312 - 2 \times 15.5 \times 10 = 4002 \text{ mm}^2$

$$\sigma_{at,\,cal} = \frac{300 \times 10^3}{4002} = 74.96 \text{ N/mm}^2$$

$$\sigma_{btx,\,cal} = (M/I_{xx})\, C_{xx}$$

Visual Walkthrough

348 Design of Steel Structures

$$M_t = \frac{2F_a}{L}\left(\frac{L}{2} - \frac{a}{4}\right)^2 = \frac{2 \times 0.75}{5.5} \times \left(\frac{5.5}{2} - \frac{3}{4}\right)^2 = \frac{6}{5.5} = 1.09 \text{ kNm}$$

Try ISLC 150 @ 141.3 N/m

$$Z_{cx} = 93 \times 10^3 \text{ mm}^3,\ Z_{cy} = 20.2 \times 10^3 \text{ mm}^3,\ C_{yy} = 23.8 \text{ mm}$$

$$\sigma_{bc,t} = \frac{M_z}{Z_x} + \frac{M_y}{Z_y}$$

$$= \frac{2.1818 \times 10^6}{93 \times 10^3} + \frac{1.09 \times 10^6}{20.2 \times 10^3} = 77.42 \text{ N/mm}^2 < 0.66 f_y$$

which is safe.

Exercises

Note Assume the yield stress of steel to be 250 N/mm² of not specified in the problem.

8.1 Design a simply supported gantry girder to carry an electric overhead travelling crane for the following data:

Crane capacity	320 kN
Weight of crane and crab	300 kN
Weight of crane	200 kN
Minimum approach of crane hook	1.20 m
Distance between c/c of wheels	3.20 m
Distance between c/c of gantries	16.0 m
Span of gantry girder	4.00 m
Weight of rails	300 N/m
Height of rails	75 mm
Yield stress of steel	280 MPa

8.2 Design a simply supported gantry girder to be used in an industrial building for the following data:

Crane capacity	100 kN
Weight of crab	35 kN
Weight of crane (excluding crab)	160 kN
Minimum clearance between crane hook and gantry girder	1.00 m
Wheel base	3.00 m
Distance between c/c of gantries	20.0 m
Distance between c/c gantry columns	6.0 m
Crane type	M.O.T.

8.3 Two electrically operated overhead travelling cranes are to be used in a bay of an industrial building. Design the gantry girder for the following data:

Crane capacity	250 kN (each)
Bay width	18 m
Spacing of columns	8 m

Exercises for practice are given to prepare students for real-life problems.

Appendix contains important data for design.

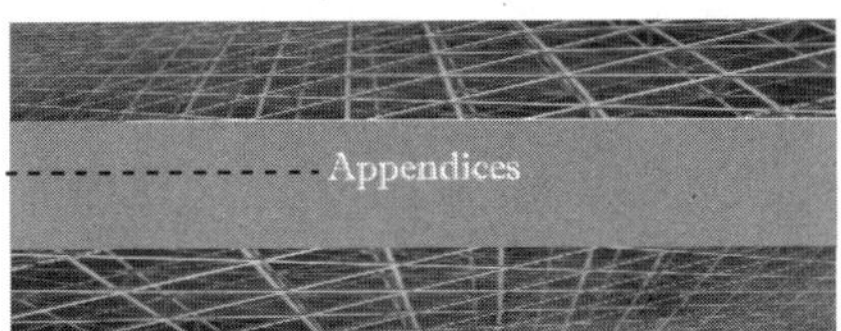

I ABBREVIATIONS

I.S.J.B.	Indian Standard Junior Beam
I.S.L.B.	Indian Standard Light Beam
I.S.M.B.	Indian Standard Medium Beam
I.S.H.B.	Indian Standard Column Section
I.S.W.B	Indian Standard Wide Flange Beam
I.S.J.C.	Indian Standard Junior Channel
I.S.L.C.	Indian Standard Light Channel
I.S.M.C.	Indian Standard Medium Channel
I.S.J.T.	Indian Standard Junior Tee Bar
I.S.S.C.	Indian Standard Special Channel
I.S.N.T.	Indian Standard Normal Tee Bar
I.S.H.T.	Indian Standard Wide Flange Tee Bar
I.S.S.T.	Indian Standard Long Legged Tee Bar
I.S.L.T.	Indian Standard Light Tee Bar
I.S.A.	Indian Standard Equal/Unequal Angle
I.S.B.A.	Indian Standard Bulb Angle
I.S.RO.	Indian Stand Round Bar
I.S.SQ.	Indian Standard Square Bar
I.S.F.	Indian Standard Flat
I.S.PL.	Indian Standard Plate
I.S.SH	Indian Standard Sheet
I.S.ST	Indian Standard Strip

II UNIT WEIGHT OF BUILDING MATERIALS

1. Asbestos cement sheets	120 to 156 N/m²
2. Asphalt	2200 N/m³
3. Brick	21600 N/m³
4. Brick work in cement mortar	19000 N/m³
5. Brick work in lime mortar	18000 N/m³
6. Cement	14400 N/m³
7. Cement mortar	20800 N/m³
8. Earth	18000 N/m³
9. Galvanised steel sheet	53.2 to 138.1 N/m²
10. Gravel	18000 N/m³

General Considerations

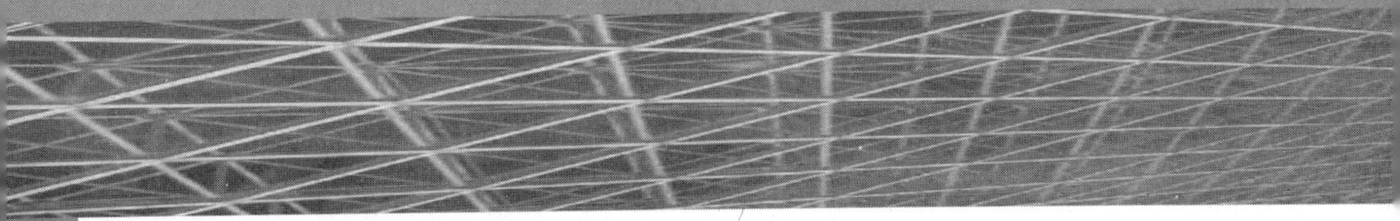

1.1 INTRODUCTION

The design of steel structures involves the planning of the structure for specific purposes, proportioning of members to carry loads in the most economical manner, and considerations for erection at site. First, the structure should serve the purpose for which it is intended and this is achieved by proper functional planning. Secondly, it should have adequate strength to withstand direct and induced forces to which it may be subjected during its life span. An inadequate assessment of forces and their effects on the structure may lead to excessive deformation and its failure. Therefore, the design of structures includes functional planning, acknowledge of the various forces, strength of materials and the design methods. In addition the structure should be economical and easy to erect. An economical structure requires an efficient use of steel and skilled and unskilled labour. Although this objective can usually be accomplished by a design that requires a minimum amount of steel, savings can often be realized by using more steel if it results in a simpler structural form with less fabrication. In fact, as of today, materials account for one-third or less of the cost of a typical steel structure, whereas labour costs can account for 60 per cent or more.

Steel structures are composed of elements which are rolled to a basic cross-section in a mill, and worked to the desired size and form in a fabricating shop or site. A significant difference between steel and concrete constructions is that the designer has more control over the shape of reinforced cement concrete elements. For building a steel structure, the designer is normally compelled to use standard rolled sections. Fortunately, the variety of steel sections available is so great that any desired structural effect can be achieved in steel.

Steel, as a building material has been used extensively in various types of structures. Some of the examples of civil engineering works in steel are high-rise building skeletons, industrial buildings, transmission towers, railway bridges, overhead tanks, chimneys (stacks), towers, bunkers and silos. We shall now discuss the various aspects of design.

Functional Planning

This takes into consideration the overall planning of the structure in accordance to the purpose it is meant for. It also includes the layout and proper arrangement of space. Functional planning involves a lot of imagination and is done with the help of architects.

Design of Structural Elements

The position of elements, viz. beams, columns, trusses, purlins, etc. are marked on the plan provided by architects. Various combinations of possible loads are ascertained and the members are proportioned on the basis of selected design method. A suitable section for the member thus proportioned is selected from I.S. Hand Book No. 1. The section is then checked for the prescribed limits specified in the code of practice. The section designed should use steel with a minimum strength not exceeding 5 per cent than that required. The structure designed should perform satisfactorily for the intended period with permitted deformations, vibrations and comfort to the user.

A designer should take care of the following important points:

1. A section selected should be available in the local market. Unusual sizes and rare rolled sections cost more and should be avoided.
2. In steel structure buildings, the various elements should be compatible at the joints. If a large number of different shapes and sizes of elements are designed, it will be practically difficult to fit the members, and connections will be a problem. For example, in a continuous beam of a framed building the designer may provide different sections in various spans for economy. But this arrangement will generate connection problems and look unattractive.
3. The steel structures in exposed conditions require frequent maintenance. Therefore, the elements of such structures, e.g. bridges, water tanks, etc. should have access to the exposed elements.
4. Transportation facilities for built-up sections should be explored before designing them.

Erection

Steel structure elements should be erected with proper care to avoid secondary stresses. The section should be of a size which can be easily erected.

Standard Specifications and Codes

To facilitate safe and economical design, standard specifications and codes are prepared on the basis of past design practice, experience based on behaviour of existing structures, knowledge gained from failures, and from research data. The information thus gathered is critically examined, assessed and approved by a body of experts and is published in the form of codes and standards. For this purpose the Bureau of Indian Standards has published a number of codes, standards and hand books, some of which will be referred to frequently and are listed below.

1. I.S. Hand Book No. 1: Properties of Structural Steel Rolled Sections
2. I.S. 875–1987: Code of Practice for Design Loads for Building and Structures
3. I.S. 800–1984: Code of Practice for use of Structural Steel in General Building Construction.

A building code has the force of law and is administered by a government entity. Building codes do not give design procedures, but they do specify the design requirements and constraints that must be satisfied. In contrast to building codes, design specifications give more specific guidance for the design of structural members and their connections. They present the guidelines and criteria that enable a structural engineer to achieve the objectives mandated by a building code. Design specifications represent what is considered to be good engineering practice based on the latest research. They are periodically revised and updated by the issuance of supplements or completely new editions. As with building codes, design specifications are written in a legal format by non-profit organisations. They have no legal standing on their own, but by presenting design criteria and limits in the form of legal mandates and prohibitions, they can easily be adopted, by reference, as part of a building code.

1.2 ADVANTAGES OF STEEL AS A STRUCTURAL MATERIAL

Steel has many advantages as a structural material.

1. Steel members have high strength per unit weight. Therefore, a steel member of a small section which has little self weight is able to resist heavy loads.
2. Being light, steel members can be conveniently handled and transported. For this reason, prefabricated members can be frequently provided.
3. Properly maintained steel structures have a long life.
4. The properties of steel mostly do not change with time. This makes steel the most suitable material for a structure.
5. Steel, being a ductile material, does not fail suddenly, but gives visible evidence of impending failure by large deflections.
6. Additions and alterations can be made easily to steel structures.
7. They can be erected at a faster rate.
8. Steel has the highest scrap value amongst all building materials.

1.3 DISADVANTAGES OF STEEL AS A STRUCTURAL MATERIAL

1. Steel structures, when placed in exposed conditions, are subjected to corrosion. Therefore, they require frequent painting.
2. Steel structures need fire proof treatment, which increases cost.

1.4 STRUCTURAL STEEL

Structural steel has been classified by the Bureau of Indian Standards based on its ultimate or yield strength. For example, Fe-410 steel has a minimum tensile strength of 410 N/mm^2. The mechanical properties of steel largely depend on its chemical composition, rolling methods, rolling thickness, heat treatment and stress history. Some of the important mechanical properties of structural steel are given in Table 1.1.

Table 1.1 *Mechanical Properties of Structural Steels*

S. No.	*Property*	*Value*
1.	Yield stress (f_y)	220–540 N/mm^2
2.	Ultimate tensile strength	1.2 f_y
3.	% Elongation (low carbon steel)	20
4.	Modulus of Elasticity (E)	2×10^5 N/mm^2
5.	Shear modulus (G)	0.4 E
6.	Poisson's ratio (μ)	
	(i) elastic range	0.3
	(ii) plastic range	0.5

The chemical composition of some of the steels is given in Table 1.2. Primary elements such as carbon, sulphur, phosphorus, manganese and silicon influence the mechanical properties of steel. Of these, carbon has the maximum influence on the mechanical properties of steel. Iron-carbon alloys containing up to 2 per cent carbon are called *carbon steel* while those having more than 2% are called *cast steel.* With increase in carbon the tensile strength of steel increases but the ductility falls. However, by alloying chromium, nickel, molebdneum, vanadium, etc. the tensile strength can be increased while retaining the desired ductility.

Table 1.2 *Chemical Composition of Structural Steels*

S. No.	*Quality of Steel*	*Designation*	*I.S. Code*	*Maximum Percentage*					*Tensile Strength (N/mm²)*
				C	*S*	*P*	*Mn*	*Si*	
1.	Standard structural C = 0.2–0.35%	Fe-410S	226	0.23	0.055	0.055	—	0.10	410
2.	Structural ordinary	Fe 310	1977	0.23	0.07	0.07	—	—	310
		Fe 410-O	1977	0.23	0.07	0.07	—	—	410
3.	Weldable	Fe 410-H	2062	0.20	0.055	0.055	1.0	—	410
		Fe 440 HT	8500	0.25	0.055	0.055	1.5	—	440
		Fe 490 HT	8500	0.25	0.04	0.04	1.5	—	490
		Fe 590 HT	8500	0.25	0.05	0.05	1.5	—	590

Fe = steel, C = carbon, S = sulphur, P = phosphorus, Mn = manganese, Si = silicon

1.5 STRESS-STRAIN CURVE FOR MILD STEEL

The stress-strain curve for mild steel presents a lot of information necessary to understand how it will behave in a given situation. Therefore, for a steel design method to be satisfactory it is essential to be conversant with the stress-strain relationship of mild steel.

Curve *OABCDEF* in Fig. 1.1 represents the stress-strain curve when a mild steel specimen is subjected to a gradually increasing tensile load. Various elements of the curve are as follows.

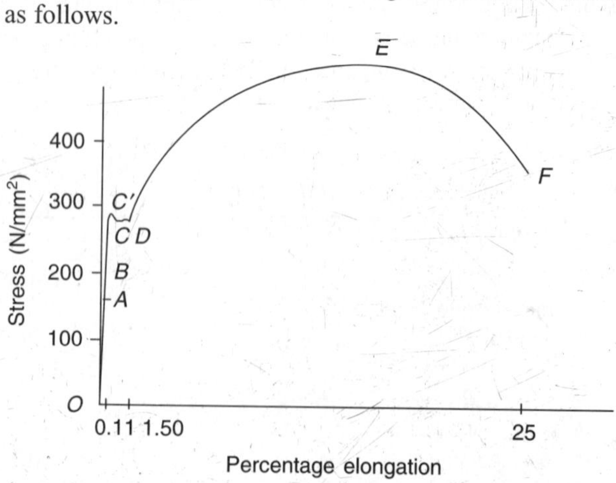

Fig. 1.1 *Stress–strain Curve of Mild Steel*

OAB— represents a straight line curve—the strain produced is proportional to the stress producing it, i.e. it obeys Hook's law.

A— represents the limit of proportionality—the stress beyond which linear variation ceases.

B— represents the elastic limit—the maximum stress up to which a specimen regains its original length on removal of the applied load. This stress, in general, is not measured and *B* is assumed to coincide with *A*.

C′C— represents the upper, lower yield point—the stress at which there is a definite increase in strain without any further increase in stress. The magnitude of stress corresponding to the upper yield point *C′* depends on the cross-sectional shape of the specimen and the type of equipment used to perform the test. In many of the structural steel hot rolled sections, the upper yield point is not obtained due to residual stresses from the hot rolling process. Hence it has no practical significance. The stress corresponding to the lower yield point *C* is the yield stress f_y with a typical magnitude of 250 N/mm^2 for mild steel. The lower yield point is observed if the rate of loading is slow whereas the upper yield point is observed if load is applied rapidly. For the region *OC′*, the material is elastic, and the slope *E*, is the *Young's modulus*. On average *E* is 2×10^5 N/mm^2. The strain at the yield stress is about 0.0012.

CD— represents plastic yielding—it is the strain which occurs after the yield point, with no increase in stress. The point *D* of plastic yield range is somewhat variable but a typical strain is 0.014. The strain in the range *CD* is thus at least ten times the strain at the yield point.

DE— represents strain hardening—it is a range where additional stress produces additional strain. Strain increases fast with stress till ultimate load is reached. As of today this portion of the diagram is not being used for structural design.

E— represents the ultimate stress—the stress corresponding to the ultimate load. The initial slope of this region is about 4 per cent of Young's modulus. At a strain of at least 0.2, the stress reaches its maximum value *E*.

EF— stress falls with rapid increase in strain till the specimen breaks.

F— represents breaking stress—the stress corresponding to the breaking load.

Notes

1. The strain that occurs before the yield point is called *elastic strain*, and that which occurs after the yield point with no increase in stress is called *plastic strain*. The latter strains usually vary from 10 to 15 times the elastic strains.

2. The ideal stress-strain curve for mild steel in compression is identical to the one in tension up to the point of maximum stress. However, the actual behaviour is different and indicates an apparently reduced yield stress in compression. The divergence from the ideal path is called the *Bauchinger effect*.

1.6 ROLLED STEEL SECTIONS

In the design process outlined earlier, one of the objectives—and the primary emphasis of this book—is the selection of the appropriate cross-sections for the individual members of the structure being designed. Most often, this selection will entail choosing a standard cross-sectional shape that is widely available rather than requiring the fabrication of a shape with unique dimensions and properties. The largest category of standard shapes includes those produced by *hot-rolling*. In the manufacturing process, which takes place in a rolling mill, molten steel is taken from the furnace and poured into a *continuous casting* system where the steel solidifies but is never allowed to cool completely. The hot steel passes through a series of rollers that squeeze the material into the desired cross-sectional shape. Rolling the steel while it is still hot allows it to be deformed with no resulting loss in ductility, as would be the case with cold-working. During the rolling process, the member increases in the length and is cut to standard lengths, which are subsequently cut (in fabricating shop) to the length required for particular structure.

Structural steel can be rolled into various shapes and sizes. Usually sections having larger moduli of section in proportion to their cross-sectional areas are preferred. Steel sections are usually designated by their cross-sectional shapes. The shapes of the rolled steel sections available today have been developed to meet structural needs. Cross-section and size are governed by a number of factors: arrangement of material for optimum structural efficiency; functional requirements (surfaces that are easy to connect to, flat surfaces suitable for supporting other materials, etc.); dimensional and weight capacity of rolling mills; and material properties which, for example, inhibit the hot rolling of wide thin elements because of excessive warping or cracking that might occur.

I.S. Hand Book No. 1 published by the Bureau of Indian Standards provides the dimensions, weights and geometrical properties of various sections. The weight of sections in the Hand Book are given in kg/m length. However, for convenience, to use this data in SI units, the weights have been converted in N/m length by multiplying the weights with a factor of 9.81 throughout the text, solved examples and exercises. Structural shapes are abbreviated by a certain system described in the Hand Book for use in drawings, specifications and designs.

Cross-sections of some of the more commonly used hot rolled shapes are shown in Fig. 1.2 and are described as follows.

1. Rolled steel I-Sections — Fig. 1.2(a)
2. Rolled steel channel sections — Fig. 1.2(b)
3. Rolled steel T-Sections — Fig. 1.2(c)
4. Rolled steel angle-sections — Fig. 1.2(d)
5. Rolled steel tube-sections — Fig. 1.2(e)
6. Rolled steel bars — Fig. 1.2(f)
7. Rolled steel flats — Fig. 1.2(g)
8. Rolled steel plates
9. Rolled steel sheets
10. Rolled steel strips

Angle sections were probably the first shapes rolled and produced in 1819 in America. I-beam shape was introduced by Zores of France in 1849. By 1870 Channels and Tees were developed. All these early shapes were made of wrought iron. The first true skeletal frame structure, the Home Insurance Company Building, was built in Chicago in 1884.

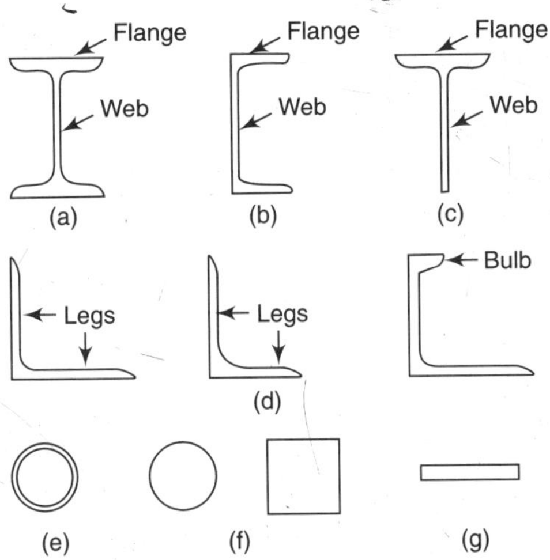

Fig. 1.2 *Rolled Structural Shapes*

The designations of various types of rolled steel sections are listed in Appendix I.

An I-Section is designated by its depth and weight, e.g., I.S.L.B. 500 @ 735.8 N/m means, the I-Section is 500 mm deep and the self weight is 735.8 N per metre length.

A channel section is designated by its depth and weight. For instance, I.S.L.C. 350 @ 380.6 N/m means that the channel section is 350 mm deep and the self weight is 380.6 N per metre length.

Note All standard I-beams and channels have a slope on the inside face of the flange of $16\frac{2}{3}$%.

A T-Section is designated by its depth and weight, e.g., I.S.N.T. 150 @ 223.7 N/m, means the Tee-Section is 150 mm deep and the self weight is 223.7 N per metre length.

An angle-section is designated by its leg lengths and thickness. For example, I.S.A. 40 × 25 × 6 mm means, the section is an unequal angle with legs 40 mm and 25 mm in length and thickness of the legs 6 mm.

Note Bulb angles need a special mention as are generally used in shipbuilding and are uncommon in structures. However these have been common in small depth gantry girders, where the bulb helps to stiffen the outstanding leg, especially when the angle, is under compression along its length. A small depth horizontal girder, whose flanges are the bulbs counteracts the bending action which sometimes occurs when the craneman, instead of using a direct lift and cross travel, drags the load across the shop floor in a direction normal to the girder length.

Steel tubes are designated by their outside diameter and self weight.

Rolled steel bars may be circular or square and are designated by diameter or side respectively, e.g., I.S.SQ. 10 mm means a square bar of 10 mm side and I.S.RO. 10 mm means a round bar of 10 mm diameter.

Steel flats are designated by width and thickness of the section, e.g., 30 I.S.F. 10 mm means the flat is 30 mm wide and 10 mm thick.

Steel plates are designated by length, width and thickness, e.g., I.S.PL. 2000 mm × 1000 mm × 8 mm, means the plate is 2000 mm long, 1000 mm wide and 8 mm thick.

Steel sheets are designated by length, width and thickness, e.g., I.S.SH. 2000 mm × 600 mm × 4 mm, means the sheet is 2000 mm long, 600 mm wide and 4 mm thick.

Steel strips are designated by width and thickness, e.g., I.S.ST. 200 mm × 2 mm, means the strip is 200 mm wide and 2 mm thick.

In most cases, one of the standard shape described above will satisfy design requirements. If the requirements are especially server, then a built-up section (Fig. 1.3) may be needed. Such a situation arises when none of the standard rolled shapes are large enough; that is, the cross-section does not have enough area or moment of inertia.

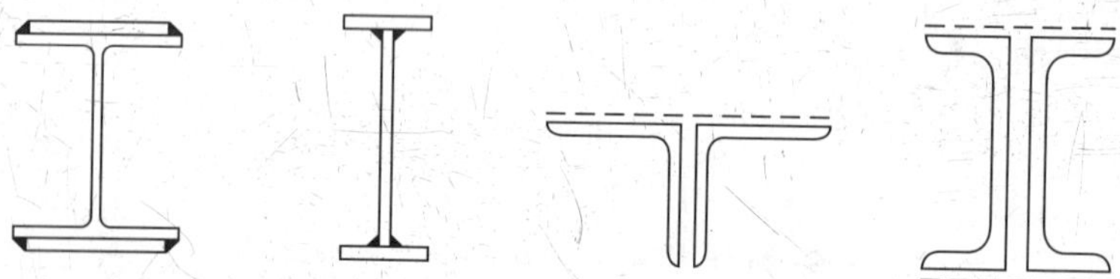

(a) I-Section with cover plates (b) welded plate girder (c) Double angle section (d) Double channel section

Fig. 1.3 *Built-up Sections*

Sometimes a standard shape is augmented by additional cross-sectional elements, as when cover plate(s) is (are) connected to one/both flange(s) of the section (Fig. 1.3(a, b)). Built-up shapes can also be created by attaching two or more standard rolled shapes to each other. Most widely used combinations are double angles or double channels placed back to back and connected at intervals along their length (Fig. 1.3(c, d)). There are many other possibilities, some of which are illustrated throughout this book.

Choice of Section

The design of steel sections is governed by cross-sectional area and section modulus. It has been seen that a variety of steel sections are rolled, but due to the limitations of rolling mills only a few are available. Also, if a section is in demand, it is rolled regularly but one which is in little demand is rolled only on order and hence costs more. Therefore, the design is not only governed by sectional properties but also on availability of the section in the market, which becomes a major consideration. Another factor governing the choice is the ease with which sections can be connected.

1.7 LOADS

For the safe design of a structure it is essential to have a knowledge of various types of loads and their worst combinations to which it may be subjected during its life span.

Types of Loads

The following are the various types of loads.

1. Dead load
6. Earth pressure

2. Live loads
3. Wind load
4. Seismic load
5. Snow load
7. Water current load
8. Impact load
9. Temperature and erection effects

1.8 DEAD LOAD

The dead load of a structure is not known before it is designed. It is assumed or estimated on the basis of experience. After designing, the assumed dead load is compared with the actual dead load. If the difference is significant, the assumed dead load is revised and the structure is redesigned. Appendix II gives the dead loads of some structural materials. Dead load calculations should also include the superimposed loads that are permanently attached to the structure, e.g. partition walls, parapets, etc.

1.9 LIVE LOADS

Live loads are those which may change in position and magnitude. The live loads on roofs are specified in I.S: 875–1987 and are listed in Appendix III. The live loads on floors as specified in Appendix III, may be reduced in the design of columns, piers, walls, and foundations, as given in Table 1.3.

Table 1.3 *Percentage Reduction of Total Live Load*

Number of Floors Carried by Member Under Consideration	*% Reduction of Total Live Load on all Floors Above the Member Under Consideration*
1	0
2	10
3	20
4	30
5 to 10	40
Over 10	50

Note The above reduction is not made for storage buildings and factories designed for live loads less than 5000 N/m^2. If the live load is more than 5000 N/m^2 the reduction may be applied provided that the loading assumed for any column, pier, etc. is not less than what it would have been if all the floors had been designed for 5000 N/m^2 with no reduction.

When a single span of beam or girder supports not less than 50 m^2 of floor area at one general level, the live load may be reduced in the design of beam or girder by 5 per cent for each 50 sq. metres supported, subject to maximum reduction of 25 per cent. This is not applicable to storage buildings.

1.10 WIND FORCES

All exposed structures, irrespective of their heights, are affected by wind forces. As wind blows against a structure, its surface (the outer shell) experiences the effect of wind force. The wind pressure intensity at any height of a structure depends upon the velocity and density of air, shape and height of the structure, topography of the surrounding ground surface and the angle of wind attack.

Design Wind Speed

For the calculation of wind load on structures I.S. 875–1987 relates the intensity of wind pressure to the basic maximum wind speed (V_b), over a short interval of 3 seconds, with a 50 years return period, for different zones of the country. The basic wind speed for any site may be obtained from Appendix IV and is modified to include risk level, terrain roughness, height and size of structure, and local topography, to get the design wind velocity (V_z) at any height for the structure.

$$V_z = V_b\, k_1 k_2 k_3 \tag{1.1}$$

where V_z = design wind speed at any height z in m/s
k_1 = probability factor or risk coefficient
k_2 = terrain, height and structure size factor
k_3 = topography factor

Note The design wind speed up to a 10 m height from mean ground level is considered constant.

Risk Coefficient Since the suggested life of buildings and structures is 50 years, a basic wind speed having a mean return period of 50 years is considered. The values of k_1 for the probable periods are given in Table 1.4.

Table 1.4 *Risk Coefficients for Different Classes of Structures in Different Wind Speed Zones*

Class of Structures	*Mean Probable Design Life of Structure in Years*	*k_1 Factor for Basic Wind Speed (m/s) of*					
		33	*39*	*44*	*47*	*50*	*55*
All general buildings and structures	50	1.0	1.0	1.0	1.0	1.0	1.0
Temporary sheds, structures such as those used during construction operations (for example, form-work and false-work), structures during construction stages and boundary walls	5	0.82	0.76	0.73	0.71	0.70	0.67
Buildings and structures presenting a low degree of hazard to life and property in the event of failure, such as isolated towers in wooded areas, farm buildings other than residential buildings	25	0.94	0.92	0.91	0.90	0.90	0.89
Important buildings and structures such as hospitals, communication buildings/towers, power plant structures	100	1.05	1.06	1.07	1.07	1.08	1.08

For exceptionally important structures as nuclear power reactors, satellite communication towers etc., Eq. (1.2) may be used to estimate k_1 for different periods of exposure and chosen probability of exceedance (risk level). The probability level of 0.63 is normally considered sufficient for the design of structures against wind effects.

$$k_1 = \frac{X_{N,P}}{X_{50,0.63}} = \frac{A - B\left[\log\left\{-\frac{1}{N}\log(1-P_N)\right\}\right]}{A + 4B} \tag{1.2}$$

where N = mean probable design life of structure in years
P_N = risk level in N consecutive years (0.63)
$X_{N,P}$ = extreme wind speed for given values of N and P_N
$X_{50,0.63}$ = extreme wind speed for $N = 50$ years, $P_N = 0.63$
A, B = coefficients as given in Table 1.5

Table 1.5 *Values of Coefficients A and B*

Zone	*A*	*B*
33 m/s	83.2	9.2
39 m/s	84.0	14.0
44 m/s	88.0	18.0
47 m/s	88.0	20.5
50 m/s	88.8	22.8
55 m/s	90.8	27.3

k_2 Factor This depends upon terrain, height and structure size. Terrain has been grouped under four categories (Table 1.6) depending upon the effect of obstructions constituting the ground surface roughness.

Table 1.6 *Terrain Category*

Category	*Height of Object*	*Structures Near*
1.	< 1.5 m	Open sea coasts, flat treeless plains
2.	1.5–10 m (with well scattered obstructions)	Parks, air fields, undeveloped sparsely outskirts of towns and suburbs, open land adjacent to sea coast
3.	Upto 10 m (with closely spaced obstructions)	Wooded areas, towns, industrial areas
4.	> 25 m (with high closely spaced obstructions)	Large city centres, well developed industrial complexes

On the basis of size, structures are grouped as (1) Class A—having maximum dimension less than 20 m (2) Class B—having maximum dimension 20–50 m and (3) Class C—having maximum dimension greater than 50 m. The design wind speed at different heights of the structure can be obtained by multiplying the coefficient k_2, given in Table 1.7, with the basic wind speed.

Table 1.7 k_2 Factors to Obtain Design Wind Speed Variation with Height in Different Terrains for Different Classes of Buildings/Structures

Height (m)	Terrain Category 1 Class			Terrain Category 2 Class			Terrain Category 3 Class			Terrain Category 4 Class		
	A	B	C	A	B	C	A	B	C	A	B	C
10	1.05	1.03	0.99	1.00	0.98	0.93	0.91	0.88	0.82	0.80	0.76	0.67
15	1.09	1.07	1.03	1.05	1.02	0.97	0.97	0.94	0.87	0.80	0.76	0.67
20	1.12	1.10	1.06	1.07	1.05	1.00	1.01	0.98	0.91	0.80	0.76	0.67
30	1.15	1.13	1.09	1.12	1.10	1.04	1.06	1.03	0.96	0.97	0.93	0.83
50	1.20	1.18	1.14	1.17	1.15	1.10	1.12	1.09	1.02	1.10	1.05	0.95
100	1.26	1.24	1.20	1.24	1.22	1.17	1.20	1.17	1.10	1.20	1.15	1.05
150	1.30	1.28	1.24	1.28	1.25	1.21	1.24	1.21	1.15	1.24	1.20	1.10
200	1.32	1.30	1.26	1.30	1.28	1.24	1.27	1.24	1.18	1.27	1.22	1.13
250	1.34	1.32	1.28	1.32	1.31	1.26	1.29	1.26	1.20	1.28	1.24	1.16
300	1.35	1.34	1.30	1.34	1.32	1.28	1.31	1.28	1.22	1.30	1.26	1.17
350	1.37	1.35	1.31	1.36	1.34	1.29	1.32	1.30	1.24	1.31	1.27	1.19
400	1.38	1.36	1.32	1.37	1.35	1.30	1.34	1.31	1.25	1.32	1.28	1.20
450	1.39	1.37	1.33	1.38	1.36	1.31	1.35	1.32	1.26	1.33	1.29	1.21
500	1.40	1.38	1.34	1.39	1.37	1.32	1.36	1.33	1.28	1.34	1.30	1.22

k_3 Factor The basic wind speed (V_b) accounts for general site level above mean sea level. However, local features such as hills, valleys, cliffs, escrapments or ridges affect it. The influence of the topographic feature is considered to extend 1.5 L_e upwind and 2.5 L_e downwind of the summit (Fig. 1.4) of crest of the feature, where L_e is the effective horizontal length of the hill depending on upwind slope (θ). The effect is incorporated by a factor k_3.

The value of k_3 for level ground, where the upwind slope is less than 3°, is unity and that for slopes greater than 3° is confined in the range of 1.0 to 1.36.
For a hill or ridge, $k_3 = 1 + Cs$

where $C = 1.2\left(\frac{z}{L}\right)$ for upwind slope 3°–17°

$= 0.36$ for upwind slope > 17°

z = height of crest of hill

L = projected length of upwind zone from average ground level of crest in wind direction

s = a factor obtained from Fig. 1.5 appropriate to the height H above mean ground level and the distance X from the summit or crest relative to the effective length L_e.

Table 1.8 *Force Coefficients C_f for Clad Buildings of Uniform Section (Acting in the Direction of Wind)*

Plan Shape		$V_d b$ m^2/s	*C_f for Height/Breadth Ratio* Up to 1/2	1	2	5	10	20	∞
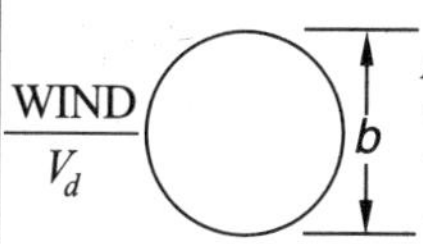	All surfaces rough or with projections	< 6	0.7	0.7	0.7	0.8	0.9	1.0	1.2
		≥ 6							
	Smooth	≥ 6	0.5	0.5	0.5	0.5	0.5	0.6	0.6
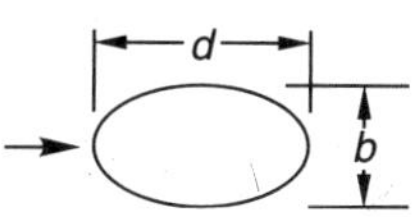	Ellipse $b/d = 1/2$	< 10	0.5	0.5	0.5	0.5	0.6	0.6	0.7
		≥ 10	0.2	0.2	0.2	0.2	0.2	0.2	0.2
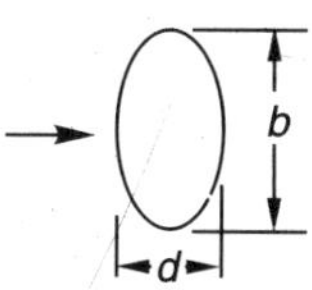	Ellipse $b/d = 2$	< 8	0.8	0.8	0.9	1.0	1.1	1.3	1.7
		≥ 8	0.8	0.8	0.9	1.0	1.1	1.3	1.5
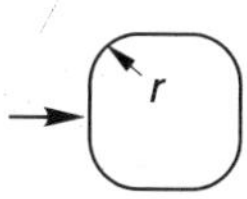	$b/d = 1$ $r/b = 1/3$	< 4	0.6	0.6	0.6	0.7	0.8	0.8	1.0
		≥ 4	0.4	0.4	0.4	0.4	0.5	0.5	0.5
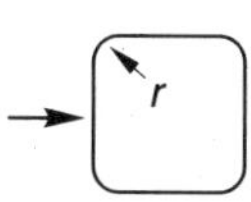	$b/d = 1$ $r/b = 1/6$	< 10	0.7	0.8	0.8	0.9	1.0	1.0	1.3
		≥ 10	0.5	0.5	0.5	0.5	0.6	0.6	0.6

(*Contd.*)

Table 1.8 (*Contd.*)

Plan Shape		$V_d b$ m^2/s	C_f for Height/Breadth Ratio: Up to 1/2	1	2	5	10	20	∞
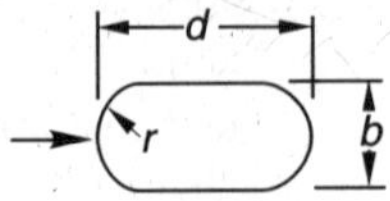	$b/d = 1/2$, $r/b = 1/2$	< 3	0.3	0.3	0.3	0.3	0.3	0.3	0.4
		≥ 3	0.2	0.2	0.2	0.2	0.3	0.3	0.3
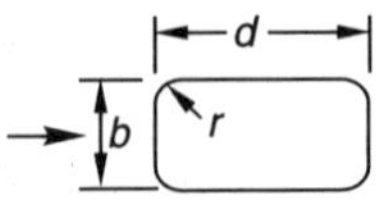	$b/d = 1/2$, $r/b = 1/6$	All values	0.5	0.5	0.5	0.5	0.6	0.6	0.7
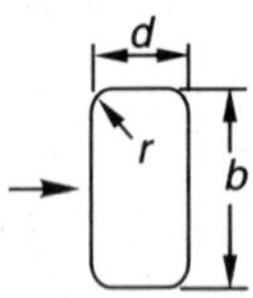	$b/d = 2$, $r/b = 1/12$	All values	0.9	0.9	1.0	1.1	1.2	1.5	1.9
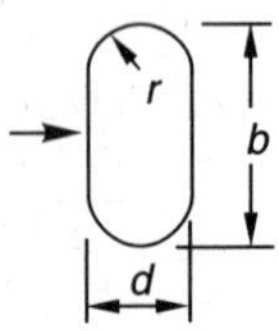	$b/d = 2$, $r/b = 1/4$	< 6	0.7	0.8	0.8	0.9	1.0	1.2	1.6
		≥ 6	0.5	0.5	0.5	0.5	0.5	0.6	0.6
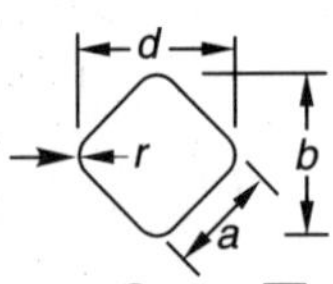	$r/a = 1/3$	< 10	0.8	0.8	0.9	1.0	1.1	1.3	1.5
		≥ 10	0.5	0.5	0.5	0.5	0.5	0.6	0.6
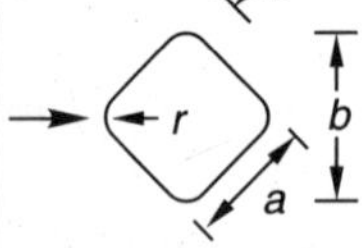	$r/a = 1/12$	All values	0.9	0.9	0.9	1.1	1.2	1.3	1.6
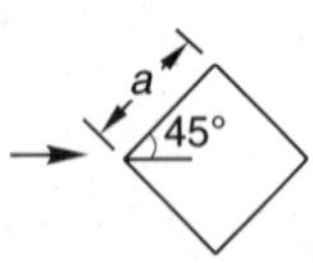	$r/a = 1/48$	All values	0.9	0.9	0.9	1.1	1.2	1.3	1.6
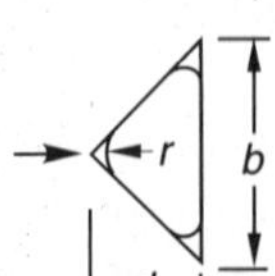	$r/b = 1/4$	< 11	0.7	0.7	0.7	0.8	0.9	1.0	1.2
		≥ 11	0.4	0.4	0.4	0.4	0.5	0.5	0.5
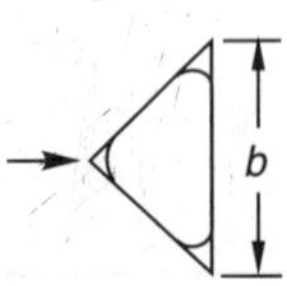	$r/b = 1/12$	All values	0.8	0.8	0.8	1.0	1.1	1.2	1.4

(*Contd.*)

Table 1.8 *(Contd.)*

Plan Shape		$V_d b$ m^2/s	*C_f for Height/Breadth Ratio*						
			Up to 1/2	*1*	*2*	*5*	*10*	*20*	∞
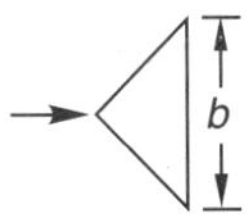	$r/b = 1/48$	All values	0.7	0.7	0.8	0.9	1.0	1.1	1.3
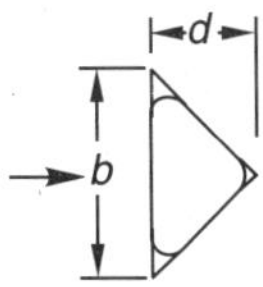	$r/b = 1/4$	< 8 ≥ 8	0.7 0.4	0.7 0.4	0.8 0.4	0.9 0.4	1.0 0.5	1.1 0.5	1.3 0.5
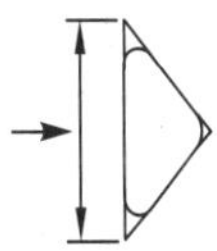	$1/48 < r/b < 1/12$	All values	1.2	1.2	1.2	1.4	1.6	1.7	2.1
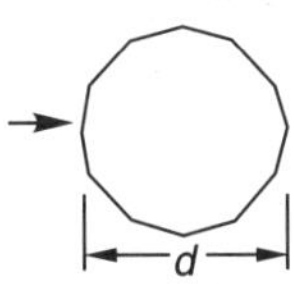	12-sided polygon	< 12 ≥ 12	0.7 0.7	0.7 0.7	0.8 0.7	0.9 0.7	1.0 0.8	1.1 0.9	1.3 1.1
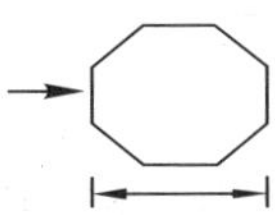	Octagon	All values	1.0	1.0	1.1	1.2	1.2	1.3	1.4
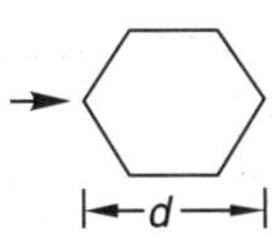	Hexagon	All values	1.0	1.1	1.2	1.3	1.4	1.4	1.5

Notes In this table $v_d\, b$ is an indication of the airflow regime.

Design Wind Pressure This pressure at any height above mean ground level is obtained by

$$p_z = 0.6\, v_z^2$$

where p_z = design wind pressure in N/m^2 at height z

Design Wind Force The total wind load for a building as a whole is given by

$$F = C_f A_e p_z \quad (1.3)$$

where C_f = force coefficient of the structure (Table 1.8)
A_e = effective frontal area
p_z = design wind pressure

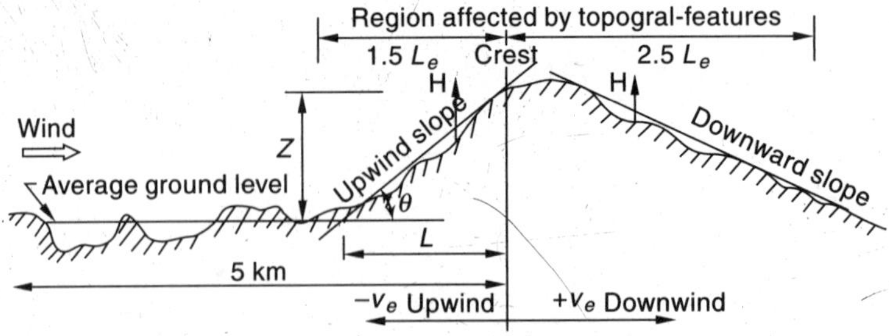

Fig. 1.4 *General Notation*

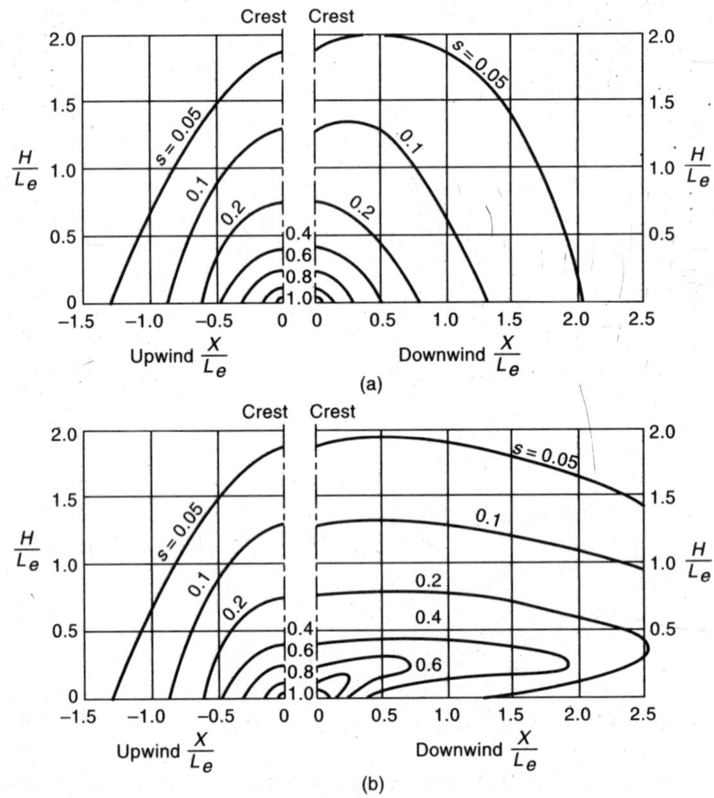

Fig. 1.5 *Factor s for Hill and Ridge*

For flat-sided members such as plates, angles, I-Sections, etc., the force coefficient in two mutually perpendicular directions, normal (C_{fn}) and transverse (C_{ft}) should be considered as given in Table 1.9.

Normal force, $F_n = C_{fn}\, p_z\, k\, l\, b$

Transverse force, $F_t = C_{ft}\, p_z\, k\, l\, b$

where k = a coefficient (Table 1.10) depending upon l/b ratio

l = length of the member

b = width of the member across the direction of wind

Table 1.9 *Force Coefficients (C_f) for Individual Structural Members of Infinite Length*

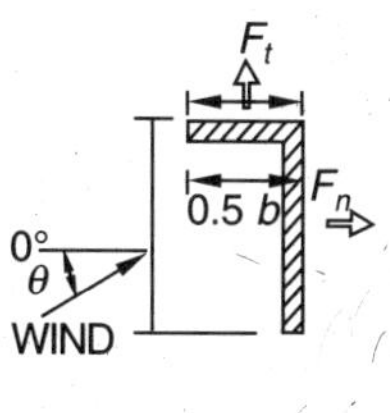

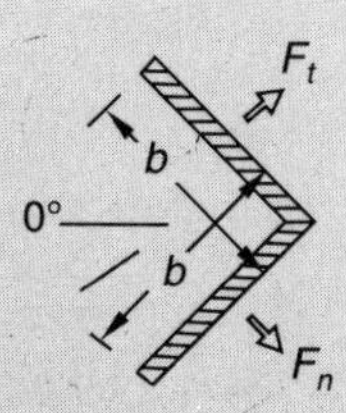

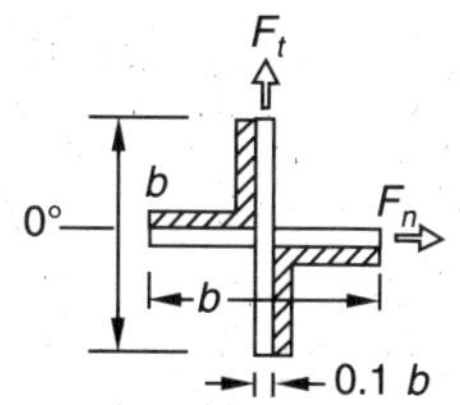

θ	C_{fn}	C_{ft}	C_{fn}	C_{ft}	C_{fn}	C_{ft}
degrees						
0	+1.9	+0.95	+1.8	+1.8	+1.75	+0.1
45	+1.8	+0.8	+2.1	+1.8	+0.85	+0.85
90	+2.0	+1.7	–1.9	–1.0	+0.1	+1.75
135	–1.8	–0.1	–2.0	+0.3	–0.75	+0.75
180	–2.0	+0.1	–1.4	–1.4	–1.75	–0.1

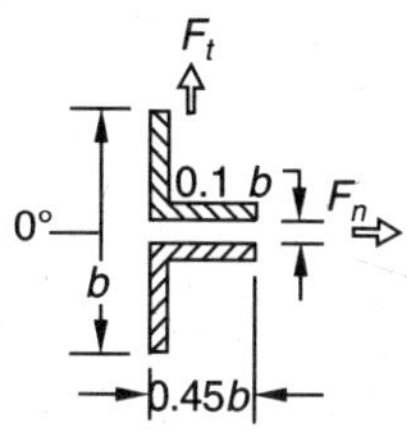

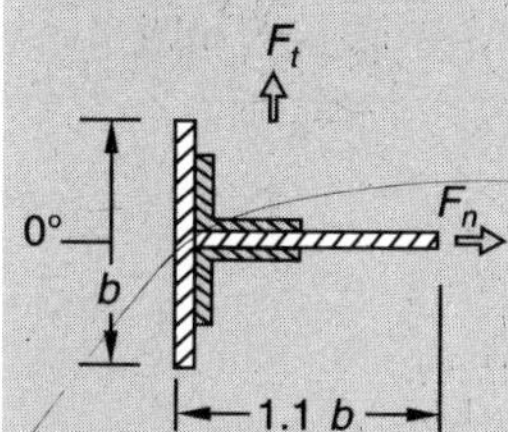

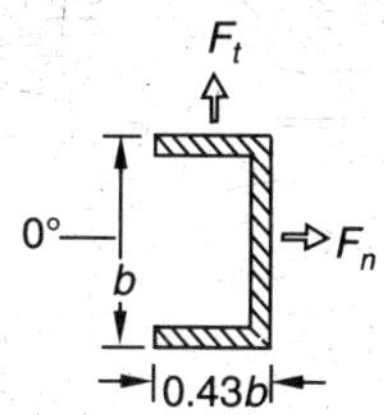

θ	C_{fn}	C_{ft}	C_{fn}	C_{ft}	C_{fn}	C_{ft}
degrees						
0	+1.6	0	+2.0	0	+2.05	0
45	+1.5	–0.1	+1.2	+0.9	+1.85	+0.6
90	–0.95	+0.7	–1.6	+2.15	0	+0.6
135	–0.5	+1.05	–1.1	+2.4	–1.6	+0.4
180	–1.5	0	–1.7	±2.1	–1.8	0

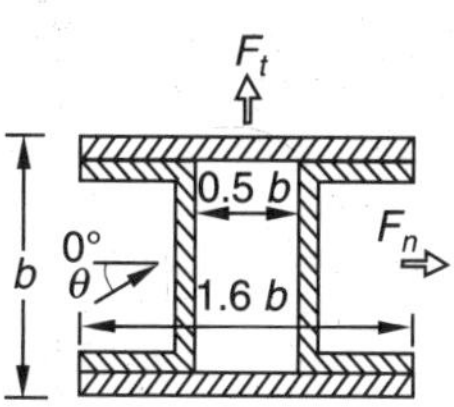

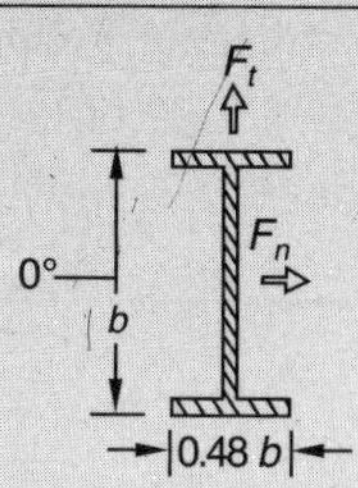

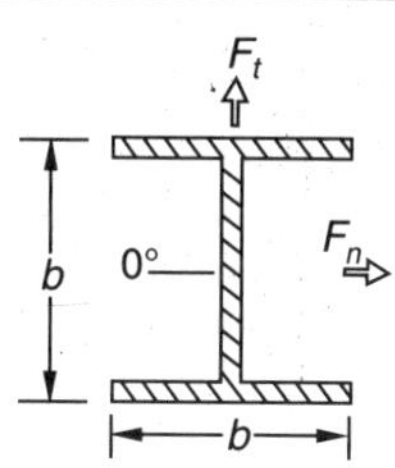

θ	C_{fn}	C_{ft}	C_{fn}	C_{ft}	C_{fn}	C_{ft}
degrees						
0	+1.4	0	+2.05	0	+1.6	0
45	+1.2	+1.6	+1.95	+0.6	+1.5	+1.5
90	0	+2.2	+0.5	+0.9	0	+1.9

(Contd.)

Table 1.9 (*Contd.*)

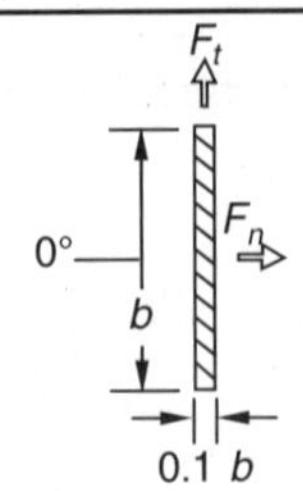

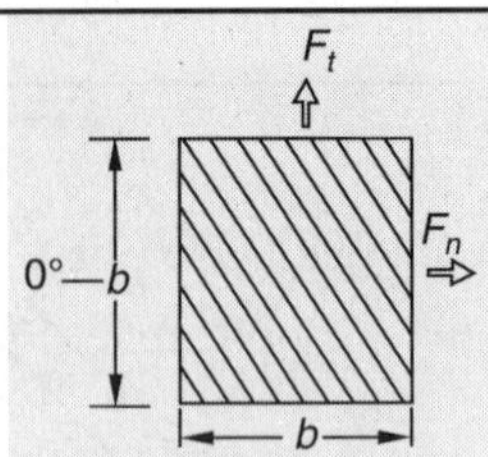

θ	C_{fn}	C_{ft}	C_{fn}	C_{fn}
degrees				
0	+2.0	0	+2.0	0
45	+1.8	+0.1	+1.55	+1.55
90	0	+0.1	0	+2.0

Note In this table, the force coefficient C_f is given in relation to the dimension b and not, as in other cases, in relation to effective frontal area A_e.

Table 1.10 *Reduction Factor k for Individual Members*

l/b or l/D	2	5	10	20	40	50	100	∞
Circular cylinder, subcritical flow ($DV_d \leq 6$ m^2/s)	0.58	0.62	0.68	0.74	0.82	0.87	0.98	1.00
Circular cylinder, supercritical flow ($DV_d \geq 6$ m^2/s)	0.80	0.80	0.82	0.90	0.98	0.99	1.00	1.00
Flat plate perpendicular to wind ($DV_d \geq 6$ m^2/s)	0.62	0.66	0.69	0.81	0.87	0.90	0.95	1.00

Wind force on roofs and walls is given by

$$F = (C_{pe} - C_{pi})\, A\, p_z \tag{1.4}$$

where C_{pe} = external pressure coefficient
C_{pi} = internal pressure coefficient
A = surface area of structural member

Force coefficients for lattice towers are given in Tables 1.11 to 1.13.

Table 1.11 *Overall Force Coefficient for Towers Composed of Flat-sided Members*

Solidity Ratio ϕ	*Force Coefficient for* *Square Towers*	*Equilateral Triangular Towers*
0.1	3.8	3.1
0.2	3.3	2.7
0.3	2.8	2.3
0.4	2.3	1.9
0.5	2.1	1.5

Table 1.12 *Overall Force Coefficient for Square Towers Composed of Rounded Members*

Solidity Ratio of Front Face	Force Coefficient for			
	Subcritical Flow ($DV_d < 6\ m^2/s$)		Supercritical Flow ($DV_d \geq 6\ m^2/s$)	
	Onto face	Onto corner	Onto face	Onto corner
0.05	2.4	2.5	1.1	1.2
0.1	2.2	2.3	1.2	1.3
0.2	1.9	2.1	1.3	1.6
0.3	1.7	1.9	1.4	1.6
0.4	1.6	1.9	1.4	1.6
0.5	1.4	1.9	1.4	1.6

Table 1.13 *Overall Force Coefficient for Equilateral-triangular Towers Composed of Rounded Members*

Solidity Ratio of Front Face ϕ	Force coefficient for	
	Subcritical Flow ($DV_d < 6\ m^2/s$)	Supercritical Flow ($DV_d \geq 6\ m^2/s$)
	All wind directions	All wind directions
0.05	1.8	0.8
0.1	1.7	0.8
0.2	1.6	1.1
0.3	1.5	1.1
0.4	1.5	1.1
0.5	1.4	1.2

1.11 SEISMIC FORCES

When a structure is subjected to ground motions in an earthquake, it responds in a vibratory fashion. The random motion of the ground caused by an earthquake can be resolved in any three mutually perpendicular directions; the two horizontal directions (x and y) and the vertical direction (z). This motion causes the structure to vibrate or shake in all three directions; the predominant direction of shaking is horizontal.

All the structures are primarily designed for gravity loads—force equal to mass times gravity—in the vertical direction. Because of the inherent factor of safety used in the design specifications, most structures tend to be adequate under the vertical shaking. But this should be checked for strength and stability. Generally, however, the inertia forces generated by the horizontal components of ground motion require greater consideration for seismic design. The response of structure to ground motion determines the magnitude of displacements and level of stresses in the elements of the structure. Thus design of a structure requires determination of seismic forces and selection of member sizes of the structure to keep the displacements and stress levels within prescribed limits.

It is important to draw a distinction between force due to wind and those produced by earthquake. Whereas, wind loads are external loads applied and hence proportional, to the exposed surface of the structure. Earthquake loads are essentially inertia forces; result from the distortions produced by both the earthquake motions and inertial resistance of the structure. Their magnitude is then a function of the mass of the structure rather than its exposed surface. Also, in contrast to structural response to essentially static gravity loads or even to wind loads, which can often be validly treated as static loads—the dynamic character of response to earthquake excitation can seldom be ignored. Thus, whereas in designing for static loads one can feel greater assurance about the safety of structure made up of memebers of heavy section, in the case of earthquake loading, the stiffer and heavier structure does not necessarily represent the safe design.

Further, the wind loads actually acting on the buildings are usually somewhat smaller than those usually assumed in design. The situation with regard to earthquake forces is entirely different. The seismic forces specified in the code are quite small, relative to the actual forces expected at least once in the life cycle of the building. In contrast to the wind loads the intensity of earthquake loading depends on properties of the structure. For a structure in a particular earthquake zone, both the wind load and seismic load are computed, and the maximum out of the two is considered for design.

Method for Calculating Seismic Forces (IS: 1893–2002)

There are two methods for finding out seismic forces: seismic coefficient method and dynamic analysis. Only seismic coefficient method is discussed here, as dynamic analysis is beyond the scope of the book. The factors taken into account in assessing lateral design forces are described as follows.

Zone Factor (Z) Seismic zoning assesses the maximum severity of shaking that is anticipated in the region. The factor, thus, is used to obtain the design spectrum depending on the perceived seismic hazard in the zone in which the structure is located. Zone factor as per IS: 1893 (Part 1): 2002 are given in Table 1.14.

Table 1.14 *Zone Factor (Z)*

Seismic Zone	II	III	IV	V
Seismic Intensity	Low	Moderate	Severe	Very Severe
Z	0.10	0.16	0.24	0.36

Importance Factor (I) It is customary to recognise that certain categories of building use should be designed for greater levels of safety and this is achieved by specifying higher lateral design forces. Such categories are the following:

1. Buildings which are essential after an earthquake—hospitals, fire stations, etc.
2. Places of assembly—schools, theatres, etc.
3. Structures whose collapse would endanger the population—nuclear plants, dams, etc.

The importance factors are given in Table 1.15.

Table 1.15 *Importance Factor (I)*

S. No.	Structure	Importance Factor
1.	Important service and community buildings, such as hospitals; schools; monumental structures; emergency buildings like telephone exchanges, television stations, radio stations, railway stations, fire station buildings; large community halls like cinemas, assembly halls and subway stations, power stations	1.5
2.	All other buildings	1.0

Notes

1. The design engineer may choose values of importance factor I greater than those mentioned above.
2. Buildings not covered in S. No. (1) may be designed for higher value of I, depending on economy, strategy considerations like multi-storey buildings having several residential units.
3. This does not apply to temporary structures like excavations, scaffolding etc. of short duration.

Response Reduction Factor (R) For earthquake resistance design a structure is allowed to be damaged in case of severe shaking. Therefore, the structure should be designed for seismic forces much less than what is expected under strong shaking, if the structures were to remain linearly elastic. *Response reduction factor* is the factor by which the actual base shear-force that would be generated if the structure were to remain elastic during its response to the Design Basis Earthquake shaking, should be reduced to obtain the design lateral force. Overstrength, redundancy and ductility together contribute to the fact that an earthquake resistant structure can be designed for much lower force than is implied by the strong shaking. The values of response reduction factor arrived at empirically based on engineering judgement are given in Table 1.16.

Table 1.16 *Response Reducing Factor (R) for Building Systems*

S. No.	Lateral Load Resisting System	R
	Building Frame Systems	
1.	Ordinary RC moment-resisting frame (OMRF)	3.0
2.	Special RC moment-resisting frame (SMRF)	5.0
3.	Steel frame with	
	(a) Concentric braces	4.0
	(b) Eccentric braces	5.0
4.	Steel moment resisting frame designed as per SP 6(6)	5.0
	Building with Shear Walls	
5.	Load bearing masonry wall buildings	
	(a) Unreinforced	1.5
	(b) Reinforced with horizontal RC bands	2.5
	(c) Reinforced with horizontal RC bands and vertical bars at corners of rooms and jambs of openings	3.0

(*Contd.*)

Table 1.16 (*Contd.*)

S. No.	*Lateral Load Resisting System*	*R*
6.	Ordinary reinforced concrete shear walls	3.0
7.	Ductile shear walls	4.0
	Build with Dual Systems	
8.	Ordinary shear wall with OMRF	3.0
9.	Ordinary shear wall with SMRF	4.0
10.	Ductile shear wall with OMRF	4.5
11.	Ductile shear wall with SMRF	5.0

Notes

1. The values of response reduction factors are to be used for buildings with lateral load resisting elements, and not just for the lateral load resisting elements built in isolation.
2. OMRF are those designed and detailed as per IS 456 or IS 800 but not meeting ductile detailing requirement as per IS 13920 or SP 6(6) respectively.
3. SMRF are those designed as OMRF and meeting the ductile detailing requirement.
4. Buildings with shear walls also include buildings having shear walls and frames, but where
   (a) frames are not designed to carry lateral loads, or
   (b) frames are designed to carry lateral loads but do not fulfill the requirements of 'dual systems'.
5. Reinforcement should be as per IS 4326.
6. Prohibited in zones IV and V.
7. Ductile shear walls are those designed and detailed as per IS 13920.
8. Buildings with dual systems consist of shear walls (or braced frames) and moment resisting frames such that:
   (a) the two systems are designed to resist the total design force in proportion to their lateral stiffness considering the interaction of the dual system at all floor levels; and
   (b) the moment resisting frames are designed to independently resist at least 25 per cent of the design seismic base shear.

Fundamental Natural Period (T_a) Because the design loading depends on the building period, and the period cannot be calculated until a design has been prepared, IS 1893 (Part 1): 2002 provides formulae from which T_a may be calculated.

For a moment-resisting frame building without brick infill panels, T_a may be estimated by the empirical expression:

$$T_a = 0.075\, h^{0.75} \quad \text{for RC frame building}$$

$$T_a = 0.085\, h^{0.75} \quad \text{for steel frame building}$$

For all other buildings including moment-resisting frame building with brick infill panels, T_a may be estimated by the empirical expression:

$$T_a = \frac{0.09\,h}{\sqrt{d}}$$

where h is height of building, in metres (this excludes the basement storeys, where basement walls are connected with the ground floor deck or fitted between the building columns. But, it includes the basement storeys, when they are not so connected), and d is the base dimension of the building at the plinth level, in metres, along the considered direction of the lateral force.

Design Response Spectrum Seismic analysis requires that the design spectrum be specified. IS: 1893 (Part 1) 2002 stipulates a design acceleration spectrum or base shear coefficients as a function of natural period. These coefficients are ordinates of acceleration spectrum divided by acceleration of gravity. Relationship holds well in single-degree systems. The spectral ordinates are used for the computation of inertia forces. Appendix V shows the plot between Sa/g and T corresponding to 5 per cent damping for rocky or hard-soil sites. Table 1.17 gives the multiplying factors for obtaining spectral values for various other damping (Note that the multiplication is not to be done for zero period acceleration). The design spectrum ordinates are independent for amounts of damping (multiplication factor 1.0), their variations from one material or one structural solution to another.

Table 1.17 *Multiplying Factor for Obtaining Spectral Values for Damping Other than 5% Damping*

Damping (Per cent)	0	2	5	7	10	15	20	25	30
Factors	3.20	1.40	1.00	0.90	0.80	0.70	0.60	0.55	0.50

Seismic Coefficient Method

This method of finding design lateral forces is also known as *static method* or *equivalent lateral load method.* The procedure does not require dynamic analysis, however, accounts for dynamics of building in approximate manner.

Design Lateral Force Buildings and their elements should be designed and constructed to resist the effects of design lateral force. The design lateral force is first computed for the building as a whole. This design lateral force is then distributed to the various floor levels. The overall design seismic force thus obtained at each floor level is then distributed to individual lateral load resisting elements depending on the floor diaphragm action.

Design Seismic Base Shear The total design lateral force or design seismic base shear (V_B) along any principal direction is determined by

$$V_B = A_h W \tag{1.5}$$

where A_h is the design horizontal acceleration spectrum value, using the fundamental natural period T_a in the considered direction of vibration; and W is the seismic weight of the building. *The design horizontal seismic coefficient* A_h for a structure is determined by the following expression:

$$A_h = \frac{Z}{2}\frac{I}{R}\frac{S_a}{g} \tag{1.6}$$

Note For any structure with $T \le 0.1$ s, the value of A_h will not be taken less than $Z/2$ whatever be the value of I/R.

where Z is the zone factor given in Table 1.14. The factor 2 in the denominator of Z is used so as to reduce the Maximum Considered Earthquake (MCE) zone factor to the factor for Design Basis Earthquake (DBE), I is the importance factor given in Table 1.15. R is the Response reduction factor given in Table 1.16.

S_a/g is the response acceleration coefficient as given in Appendix V for 5 per cent damping based on appropriate natural periods. For other damping values of the structure multiuplying factors given in Table 1.17 should be used. These curves represent free field ground motion.

For rocky or hard-soil sites

$$\frac{S_a}{g} = \begin{cases} 1+15T; & 0.00 \le T \le 0.10 \\ 2.50 & 0.10 \le T \le 0.40 \\ 1.00/T & 0.40 \le T \le 4.00 \end{cases}$$

For medium soil sites

$$\frac{S_a}{g} = \begin{cases} 1+15T; & 0.00 \le T \le 0.10 \\ 2.50 & 0.10 \le T \le 0.55 \\ 1.36/T & 0.55 \le T \le 4.00 \end{cases}$$

For soft soil sites

$$\frac{S_a}{g} = \begin{cases} 1+15T; & 0.00 \le T \le 0.10 \\ 2.50 & 0.10 \le T \le 0.67 \\ 1.67/T & 0.67 \le T \le 4.00 \end{cases}$$

Seismic Weight of Floors The seismic weight of each floor is its full dead load plus appropriate amount of imposed load, the latter being that part of the imposed loads that may reasonably be expected to be attached to the structure at the time of earthquake shaking. It includes the weight of permanent and movable partitions, permanent equipment, part of live load, etc. While computing the seismic weight of each floor, the weight of columns and walls in any storey should be equally distributed to the floors above and below the storey. Any weight supported in between storeys should be distributed to the floors above and below in inverse proportion to its distance from the floors.

Seismic Weight of Building The seismic weight of the whole building is the sum of the seismic weights of all the floors.

Design Imposed Loads for Earthquakes Force Calculation

For various loading classes as specified in I.S. 875 (Part II), the earthquake force should be calculated for the full dead load plus the percentage of imposed load as given in Table 1.18. For calculating the designs seismic forces on the structure, the imposed load on roof need not be considered.

Table 1.18 *Percentage of Imposed Load to be Considered in Seismic Weight Calculation*

Imposed Uniformly Distributed Floor Load (kN/m²)	*Percentage of Imposed Load*
Up to and including 3.0	25
Above 3.0	50

Distribution of Design Force The design base shear (V_B) is distributed to different floor levels, along the height of the building, as per the following expression:

$$Q_i = V_B \frac{W_i h_i^2}{\sum_{i=1}^{n} W_i h_i^2} \tag{1.7}$$

where Q_i is the design lateral force at floor i, W_i is the seismic weight of floor i, h_i is the height of floor i measured from base, and n is the number of storeys in the buildings i.e. the number of levels at which the masses are located.

Notes

1. The method of computing seismic force as discussed above is applicable to buildings not exceeding 40 m in height, in all zones. It can also be used for buildings up to 90 m high in zones I, II and III.
2. For buildings higher than 40 m in zones III, IV and V, modal analysis is done.
3. I.S. 1893 has not been revised for water tanks and bridges. Therefore, the procedure for calculating earthquake forces for these structures have been outlined according to the old code in the respective chapters.

1.12 SNOW LOAD

The snow load depends upon the latitude of the place. The design snow load depends upon the shape of the roof as well as the roofing material. This load acts vertically and may be assumed to be 2.5 N/m² per mm depth of snow.

1.13 EARTH PRESSURE

In the design of structures below ground level, e.g. basement sheet piles, retaining walls, etc. the pressure exerted by soil must be considered. Permissible bearing pressures on subsoil are given in Appendix VI. Many established theories are present, and can be used to compute the earth pressure. Due allowances should be made for possible surcharge from stationary or moving loads.

1.14 WATER CURRENT LOAD

The force exerted due to water current on the piers, abutments and other structures inside water must be taken into consideration, whenever required.

1.15 IMPACT LOAD

Impact load can be computed by multiplying live loads with the impact factors listed in Table 1.19.

Table 1.19 *Impact Factors in Percentage*

S. No.	Type of Structure	Impact Factor in %
1.	Frames supporting lifts and hoists	100
2.	Foundations, footings, piers supporting lifts and hoisting apparatus	40
3.	Light machinery, shaft, motor units	20
4.	Reciprocating machinery or power units	50
5.	Installed machinery	20

Temperature and Erection Effects

Due to fluctuations in temperature, the structural members expand or contract and produce some loading effects in the member, provided the ends are restrained. In a statically determinate system the influence of temperature is basically to change the member lengths and cause secondary stresses. These stresses are sometimes very prominent and may be to the order of 25% of the live loads. Therefore, for structures like bridges, trusses, etc. these must be given due considerations. Erection loads are temporary loads which come into play during construction. These also should be carefully assessed.

1.16 PERMISSIBLE STRESSES

The permissible stresses are some fraction of the yield stress of the material. It may be defined as the ratio of the yield stress to the factor of safety. I.S: 800–1984 specifies the permissible stresses in its various sections and are given in Table 1.20.

Table 1.20 *Permissible Stresses in Steel Structural Members*

S. No.	Types of Stress	Notation	Permissible Stress (MPa)	Factor of Safety
1.	Axial tensile stress	σ_{at}	$0.6\,f_y$	1.67
2.	Maximum axial compressive stress	σ_{ac}	$0.6\,f_y$	1.67
3.	Bending tensile stress	σ_{bt}	$0.66\,f_y$	1.515
4.	Maximum bending compressive stress	σ_{bc}	$0.66\,f_y$	1.515
5.	Average shear stress	τ_{va}	$0.4\,f_y$	2.5
6.	Maximum shear stress	τ_{vm}	$0.45\,f_y$	2.22
7.	Bearing stress	σ_p	$0.75\,f_y$	1.33
8	Stress in slab base	σ_{bs}	185	—

1.17 WORKING STRESSES

The stresses used in practical design are termed as working stresses or safe working stresses. These should never exceed the permissible stresses listed in Table 1.20.

1.18 FACTOR OF SAFETY

It may be defined as the ratio of strength of the member to the expected force. In other terms, when the yield point is well defined, the factor of safety is defined as the ratio of the yield stress to the maximum expected stress. The factor of safety permitted by IS: 800–84 for various stress conditions are as given in Table 1.20.

The concept of introducing a factor of safety is to make the structure safe to account for the following:

1. The analysis methods are based on assumptions and do not give the exact stresses.
2. Structural members may be temporarily overloaded under certain circumstances.
3. The stresses due to fabrication and erection are not considered in the design of ordinary structures.
4. The secondary stresses may be appreciable.
5. Underestimation of the future live loads.
6. Stress concentrations.
7. Unpredictable natural calamities.

1.19 MINIMUM THICKNESS OF STRUCTURAL MEMBERS

The minimum thickness of hot rolled sections should not be less than the limits specified in Table 1.21.

Table 1.21 *Minimum Thickness of Structural Members*

S.No.	*Section*	*Directly Exposed to Weather*	*Accessible for Cleaning and Painting*	*Thickness ∢ (mm)*
1.	Steel	yes	no	8
2.	Steel	no	no	6
3.	Steel	no	yes	6
4.	Tubular steel	yes	yes	4
5.	Tubular steel	no	yes	3.2
6.	Steel in webs of plate girders and crane girders	no		8

Note For steels in contact with water and soil and those subjected to alternate wetting or drying, an additional thickness of 1.5 mm is required. This aggregate thickness should not be less than the appropriate value given in Table 1.21.

1.20 DESIGN METHODS

Design of steel structures consists of the design of steel members and their connections. The design of structural steel elements is based on one or more of the following criteria:

1. Attainment of initial yielding
2. Attainment of full yielding
3. Tensile strength
4. Critical buckling
5. Maximum deflection permitted
6. Stress concentration
7. Fatigue
8. Brittle fracture

A steel structure may be designed by any one of the following methods:

1. Elastic or working stress method.
2. Plastic or limit design method.
3. Limit State method.

In the elastic method of design, the worst combination of loads is ascertained and the members are proportioned on the basis of working stresses. These stresses should never exceed the permissible ones as laid down by the code. This permissible stress will be in the elastic range of steel and will be less than the yield stress, f_y. Steel is a ductile material and from the stress-strain curve it is observed that higher loads than in the elastic method can be applied over the structure. This is due to the fact that a major portion of the curve lies beyond the elastic limit. This extra strength is termed *reserve strength* and forms the basis of plastic design method. This is an aspect of limit design, which confines the structural usefulness up to the plastic strength or ultimate load carrying capacity. In the plastic design method, the working loads are multiplied by the load factor and the members are designed on the basis of the collapse strength. The term plastic is used because, at failures, parts of the member will be subjected to very large strains—large enough to put the member into plastic range.

Limit state method also known as load and resistance factor design (LFRD) is similar to plastic design which considers most critical limit states of strength and serviceability. Load factors are applied to the service loads and then theoretical strength of the member is reduced by the application of a resistance factor. The criterion to be satisfied in the selection of member is factored load ≤ factored strength.

The connection, whether riveted, bolted or welded, can be designed as flexible, semi-rigid or rigid connections. Flexible connections are also known as simple connections. These connections are assumed to resist shear only. Some insignificant bending moment may develop at the connection but a flexible connection does not provide any resistance to moment. Rigid connections, also called the moment connections, resist the shear and bending moments at the connections. Semi-rigid connections resist the bending moment in between the flexible and rigid connections. The moment resistance of such connections is less as compared to rigid connections.

Although the permissible stresses for the main member in a working stress design is based on the initial yielding of the steel, the permissible stresses for fasteners are usually based on the ultimate strength of the connection. The safety factor values of various fasteners are about 2 to 3.

The design of structural elements by elastic and plastic methods and design of connections have been discussed in subsequent chapters in detail. An introduction to limit state method of design is presented in Chapter 18 based on the draft code IS 800 which is expected shortly.

Solved Examples

Example 1.1 The yield strength for a mild steel specimen was found to be 250 N/mm^2. Taking a factor of safety of 2, find out the working stress.

Solution Working stress $= \dfrac{\text{yield strength}}{\text{factor of safety}}$

$$= \frac{250}{2} = 125 \text{ N/mm}^2$$

Example 1.2 A tie bar 50 mm × 8 mm is to carry a load of 80 kN. A specimen of the same quality steel of cross-sectional area 250 mm^2 was tested in the laboratory. The maximum load carried by the specimen was 125 kN. Find the factor of safety in the design.

Solution Yield strength for the steel $= \dfrac{\text{maximum load}}{\text{cross-sectional area}}$

$$= \frac{125 \times 1000}{250} = 500 \text{ N/mm}^2$$

$$\text{Working stress} = \frac{80 \times 1000}{50 \times 8} = 200 \text{ N/mm}^2$$

$$\text{Factor of safety} = \frac{\text{yield strength}}{\text{working stress}}$$

$$= \frac{500}{200} = 2.5$$

Example 1.3 Find out the load on interior column BC on 2nd floor and AB on first floor of a seven storey building as shown in Fig. Ex. 1.3.

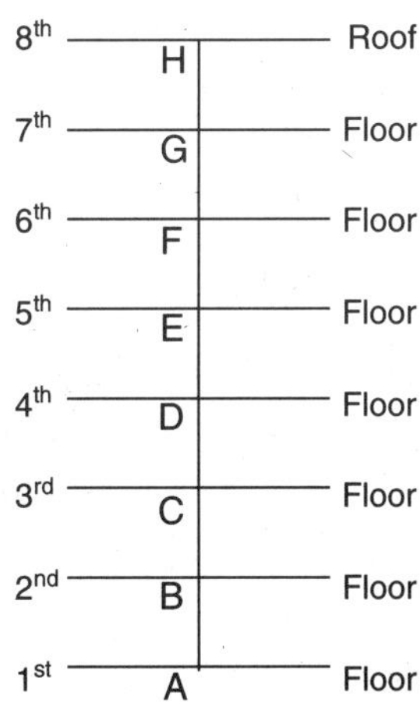

Fig. Ex. 1.3

The live load on each floor	= 4,000 N/m^2
The live load on roof	= 1,500 N/m^2
The dead load of floor finish, slab and beam	= 3,000 N/m^2
Height of each storey	= 3.20 m
The spacing of the columns in both directions	= 4.00 m

Solution Loads from the various floor levels are computed below. The live load is reduced as per the I.S. specifications.

Column	*Floor No.*	*Live Load* N/m^2	*Dead Load* N/m^2	*Total Load from Floor* N/m^2
GH	Roof	1500	3000	4500
FG	7th floor	$0.9 \times 4000 = 3600$	3000	6600
EF	6th floor	$0.8 \times 4000 = 3200$	3000	6200
DE	5th floor	$0.7 \times 4000 = 2800$	3000	5800
CD	4th floor	$0.6 \times 4000 = 2400$	3000	5400
BC	3rd floor	$0.6 \times 4000 = 2400$	3000	5400
AB	2nd floor	$0.6 \times 4000 = 2400$	3000	5400

$$\text{Design load on column BC} = (4500 + 6600 + 6200 + 5800 + 5400 + 5400) \times 4 \times 4$$
$$= 542{,}400 \text{ N} = 542.4 \text{ kN}$$
$$\text{Design load on column AB} = (4500 + 6600 + 6200 + 5800 + 5400 + 5400 + 5400) \times 4 \times 4$$
$$= 628{,}800 \text{ N} = 628.8 \text{ kN}$$

Example 1.4 Estimate the design wind pressure for a 100 m high lattice tower located on the outskirts of Allahabad. Estimate also the wind load in terms of the effective frontal area for the tower.

Solution For Allahabad $V_b = 47$ m/s

$$k_1 = 1.07 \text{ (100 years)}$$
$$k_2 = 1.17 \text{ (category 2, class C)}$$
$$k_3 = 1.0 \text{ (level ground)}$$
$$V_z = V_b\, k_1\, k_2\, k_3$$
$$V_z = 47 \times 1.07 \times 1.17 \times 1.0 = 58.84 \text{ m/s}$$
$$\text{Wind pressure, } p_z = 0.6\ V_z^2$$
$$= 0.6 \times 58.84^2 = 2077.29 \text{ N/m}^2$$
$$\text{Let solidity ratio, } \phi = 0.2$$
$$\text{For } \phi = 0.2,\ C_f = 2.7$$
$$\text{Wind load, } F = C_f\, A_e\, P_d$$
$$= 2.7 \times 2077.29 \times A_e$$
$$= 5608.68\ A_e \text{ Newtons}$$

Example 1.5 Calculate the design wind pressure for a 12 m high water tank located in M.N.N.I.T staff colony, Allahabad.

Solution For Allahabad Zone = 2; $V_b = 47$ m/s

$$k_1 = 1.00 \text{ (50 years)}$$
$$k_2 = 0.934 \text{ (category 3, class A)}$$
$$k_3 = 1.0 \text{ (level ground)}$$
$$V_z = V_b\, k_1\, k_2\, k_3$$
$$V_z = 47 \times 1.0 \times 0.934 \times 1.0 = 43.89 \text{ m/s}$$

Design wind pressure, $P_z = 0.6\ V_z^2$

$$= 0.6 \times 43.89^2 = 1155.79 \text{ N/m}^2$$

Example 1.6 Determine the wind pressure for a bridge (100 years life), 22.5 m span, located about 20 km from Allahabad. Also, find the design wind force in terms of the width b of the bridge.

Solution For Allahabad $V_b = 47$ m/s

$k_1 = 1.07$ (for 100 years)

For 22.5 m span bridge located in the outskirts of Allahabad

$k_2 = 1.15$ (Category 2 Class B)

$k_3 = 1.0$ (level ground)

$$V_z = V_b\, k_1\, k_2\, k_3$$

$$= 47 \times 1.07 \times 1.15 \times 1.0 = 57.83 \text{ m/s}$$

$$\text{Design wind pressure} = 0.6\ V_z^2$$

$$= 0.6 \times 57.83^2 = 2006.58 \text{ N/m}^2$$

For wind normal to bridge

$$F_n = C_{fn}\, p_d\, k\, l\, b \qquad F_t = C_{fl}\, P_d\, k\, l\, b$$

For $\theta = 0$, and assuming double angle sections for the members from Table 1.9;

Force coefficient in the normal direction, $C_{fn} = 1.6$

Force coefficient in the transverse direction, $C_{ft} = 0.0$

Let $l/b = 5.0$; $k = 0.66$

Wind force on the bridge, $F_n = 1.6 \times 2006.58 \times 0.66 \times 22.5 \times b$

$= 47676.34\ b$ Newtons $= 47.67\ b$ kN

Example 1.7 A steel chimney 3.0 m in diameter is situated in a region where the intensity of wind pressure is 1200 N/m^2. Assuming the intensity of wind pressure to be uniform, estimate the shear due to wind load at a level 15 m below the top of the chimney.

Solution Design wind load $= K\, P_1\, A_1$

where

K = shape factor = 0.7

P_1 = intensity of wind pressure = 1200 N/m^2

A_1 = projected area = 3.0 m × 15.0 m

$$P = 0.7 \times 1200 \times (3.0 \times 15.0)$$

$$= 37{,}800 \text{ N} = 37.8 \text{ kN}$$

Example 1.8 The plan and elevation of a three-storey reinforced concrete school building is shown in Fig. Ex. 1.8. The building is located at Kolkata (Zone (III). The type of soil encountered is medium stiff and it is proposed to design the building with special moment resisting frame. The intensity of dead load is 10 kN/m^2 and the floors are to cater imposed load of 3 kN/m^2. Determine the design seismic loads on the structure by static analysis.

Solution

Design parameters:

For seismic zone III, Zone factor, $Z = 0.16$

Importance Factor, $I = 1.5$

Response Reduction factor, $R = 5$

Seismic Weight:

Floor Area = 8 × 8 = 64 m^2

For live load up to and including 3 kN/m^2, percentage of live load to be considered = 25%

Total seismic weight on the floors is given by $W = \Sigma W_i$
where W_i is sum of loads from all the floors which includes dead loads and appropriate percentage of live loads.
Therefore, seismic weight contribution from one floor

$$= 64 \times (10 + 0.25 \times 3) = 668 \text{ kN}$$

Load from roof = 64 × 10 = 640 kN
Hence, total seismic weight of the structure = 2 × 688 + 640 = 2016 kN
Fundamental Natural Period of Vibration (T_a) is given by

$$T_a = \frac{0.09h}{\sqrt{d}}$$

where h is the height of the building in m and d is the base dimension in m at plinth level along the direction of the lateral force.

Since the building is symmetriucal in plan, therefore, fundamental natural period of vibration will be same in both the directions.

Therefore, $$T_a = \frac{0.09 \times 10.5}{\sqrt{8}} = 0.334 \text{ Sec}$$

For medium stiff soil and $T_a = 0.334$; $S_a/g = 2.5$ (Appendix V)
Therefore,

$$A_h = \frac{ZI\,(S_a/g)}{2R} = \frac{0.16 \times 1.5 \times 2.5}{2 \times 5} = 0.06$$

Therefore, design base shear V_B,

$$V_B = A_h\ W = 0.06 \times 2016 = 120.96 \text{ kN}$$

The force distribution with building height is given in Table 1.21.

Table 1.21 *Lateral Load Distribution with Height*

Storey Level	W_i*(kN)*	h_i*(m)*	$W_i h_i^2$	$\frac{W_i h_i^2}{\sum_{i=1}^{n} W_i h_i^2}$	*Lateral Force (kN)* $Q_i = V_b \frac{W_i h_i^2}{\sum_{i=1}^{n} W_i h_i^2}$
3	640	10.5	70560	0.626	75.72
2	688	7	33712	0.299	36.17
1	688	3.5	8428	0.0748	9.05
Σ			112700		120.94

Exercises

1.1 Find out the design load for an interior column of ground floor of an eight storey building for the following data:

Height of each storey = 3.5 m
Spacing of columns c/c in each direction = 3.8 m
Live load on each floor = 2500 N/m^2

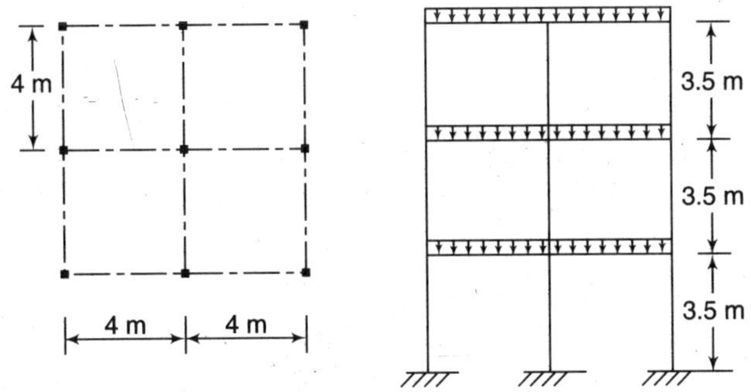

(a) Plan (b) Elevation

Fig. Ex. 1.8 *Building Configuration*

Live load on roof = 1500 N/m^2
Dead load of roof finish, slab and beam = 3000 N/m^2

(*Ans.* 541.5 kN)

1.2 A 90.0 m high chimney is to be constructed at Allahabad. The diameter of the chimney is 3.0 m. Assuming the intensity of wind pressure as 1500 N/m^2, estimate the design wind load at 60 m from the top of the chimney. Assume the intensity of wind pressure to be uniform over the entire height of chimney.

(*Ans.* 189 kN)

1.3 A multistorey building is proposed in Colaba, Mumbai, where a cluster of multistorey buildings exist. Determine the design wind pressure if the return, period is 50 years, the size of the building is 25 m $\times$ 50 m and the height is 150 m. (*Ans.* 1405 N/m^2)

1.4 A three-storey nursing home is to be built in Nainital. The size of the proposed building is 40 m $\times$ 100 m. Compute the design wind pressure on the building if the height of the building is 14 m.

(*Ans.* 1234.52 N/m^2)

1.5 A four-storey RC building is to be analysed using static lateral force method for the following data:

Concrete grade M 25,	Steel grade Fe 415	
Column 300 $\times$ 450 mm,	Beam 300 $\times$ 500 mm	
Slab 100 mm		
RC frame	2 bays of 5 m span each	
Height of each storey	3.3 m	
Imposed load	3.5 kN/m^2	
Strata	Rocky	
Damping	5%	
Density	Masonry	20 kN/m^3
	RCC	25 kN/m^3
Thickness of infill walls		250 mm

(*Ans.* 104.21 kN, 100.05 kN, 44.44 kN, 11.18 kN)

Simple Connections—Riveted, Bolted and Pinned Connections

2.1 INTRODUCTION

The various elements of a steel structure like tension member, compression member and flexural member are connected by fasteners (connectors). Different types of fasteners available are rivets, bolts, pins and welds. The forces exerted by one element on another are transferred through these fasteners, which should therefore be adequate to transmit the forces safely. Often much attention is not given to the design of connections. If the necessary connections are inadequate, the result will be a poor structure in spite of the most efficiently designed member. Therefore the design of connections must be given due importance. The nature of forces and stress distributions also need to be properly evaluated and established.

Of the various types of simple connections used in structures, riveted, bolted and pinned connections behave alike, and are therefore grouped together. These are discussed in the following articles. Welded connections are discussed in Chapter 3.

2.2 RIVETED CONNECTIONS

A rivet is made up of a round ductile steel bar piece (mild or high tensile) called *shank*, with a head at one end. The head can be of different shapes as shown in Fig. 2.1(a–d). The usual form of rivet head employed in structural steel construction is the snap head. The snap heads and pan heads form a projection beyond the plate face, and where this is an objection—as in bearings, where continuity between plate and plate, or between plate and masonry, is necessary—a countersunk head is employed.

The shank is made of the length to extend through the parts to be connected and with sufficient extra length for a second head to be made at the other end. The length of the rivet to be ordered from the market is the grip of the rivet plus the extra length required to form the second head. The grip of the rivet is the distance between the underside of the two heads, i.e., the shank length inside the connection (Fig. 2.1(e)). When the grip of the rivet becomes long, the rivet is subjected to bending in addition to bearing and shearing stresses. The grip has, therefore, an important bearing upon the fixing of the rivet diameter. The grip length should not be more than 4 diameters. If the grip is more, then the diameter should be proportionately larger. The head at the other end can be formed by hand hammering (when the rivet bar is red hot), hydraulic pressure driving or by pneumatic pressure driving.

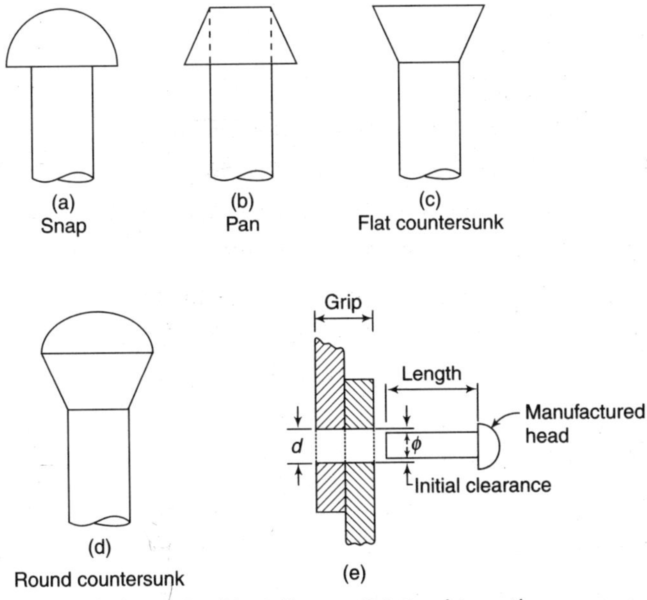

Fig. 2.1 *Rivet Types, Grip and Length*

The diameter of the shank is called the *nominal diameter*. A hole slightly greater than the nominal diameter is drilled in the parts to be connected. The rivet is inserted and the head is formed at the other end. This complete process is called *riveting*. The rivets may be placed in a cold state or may be heated to a red hot state. When the rivets are heated before driving they are called *hot driven field rivets* or *hot driven shop rivets*, depending upon if they are placed in the field or in the workshop. The diameter of the rivets when hot is equal to the diameter of the hole and is called *gross diameter*. The hot rivet becomes plastic, expands and fills the rivet hole completely in the process of forming a head at the other end. On cooling, the rivet shrinks both in length and diameter. Due to the shortening of the rivet shank length, the connected parts become tighter, consequently resulting in the residual tension of an unpredictable amount in the shank and some compression in the plates to be connected. The compression causes friction to slide between the plates and is called *clamping action*. Due to the reduction in diameter of the shank on cooling some space remains between the sides of the hole and the shank.

The use of cold driven rivets is limited as high pressures are required to form the head at room temperature and it may not be feasible to use the equipments for making the head in the field. The strength of the cold driven rivets is more than the hot driven rivets. Their clamping force is however less (as the rivets do not shrink) than that of hot driven rivets. The rivet heads for small diameter rivets can be formed manually with an ordinary hammer. Such rivets are called *hand driven rivets*.

The riveting process is not complete without proper inspection. The rivet placed should be tight and a proper second head concentric with the shank should be made at the other end. A defective rivet can be detected by tapping with a light hammer. If the rivet is loose, it will give a metallic or ringing sound. Such loose rivets should be detected and replaced.

Symbols

The knowledge of rivet symbols is essential for a field engineer to read the drawings. However, in the solved examples, the symbols have not been observed strictly.

Description	*Shop rivets*	*Field rivets*
Round head both sides		
Countersunk near side		
Countersunk far side		
Countersunk both sides		

Rivet Material

The rivets should conform to I.S: 1929–1982 and I.S: 2155–1982 as appropriate. High tensile steel rivets should be manufactured from steel conforming to I.S: 1149–1982.

Types of Riveted Joints

There are two types of riveted joints: lap joint and butt joint.

Lap Joint The two members to be connected are overlapped and connected together. Such a joint is called a lap joint as in Fig. 2.2(a). A single riveted lap joint and a double riveted lap joint are shown in Figs 2.2(b, c) respectively. The load in the lap joint has eccentricity, as the centre of gravity of load in one member and the centre of gravity of load in the second member are not in the same line, as shown in Fig. 2.2(d). Therefore, a couple is formed which causes undesirable bending in the connection and the rivets may fail in tension. To minimise the effect of bending in lap joints at least two rivets in a line should be provided. Also, due to the eccentricity the stresses are distributed unevenly across the contact area between the rivets and members to be connected. This puts a limitation on the use of lap joints.

Butt Joint The two members to be connected are placed end to end. Additional plate/plates provided on either one or both sides, called cover plates, are placed and are connected to the main plates as in Figs 2.2(e) and (h). If the cover plate is provided on one side as in Figs 2.2(e, f) and (g), it is called a single cover butt joint but if the cover plates are provided on both the sides of main plates it is called a double cover butt joint as shown in Figs 2.2.(h, i) and (j). It is more desirable to provide a butt joint than a lap joint for two main reasons:

1. In the case of double cover butt joint the total shear force to be transmitted by the members is split into two parts and the force acts on each half as shown in Fig. 2.2(k). But in the case of lap joint (Fig. 2.2(l)), there is only one plane on which the force acts and therefore the shear carrying capacity of a rivet in a butt joint is double that of a rivet in a lap joint.

2. In the case of a double cover butt joint, eccentricity of force does not exist and hence bending is eliminated, whereas it exists in the case of a lap joint.

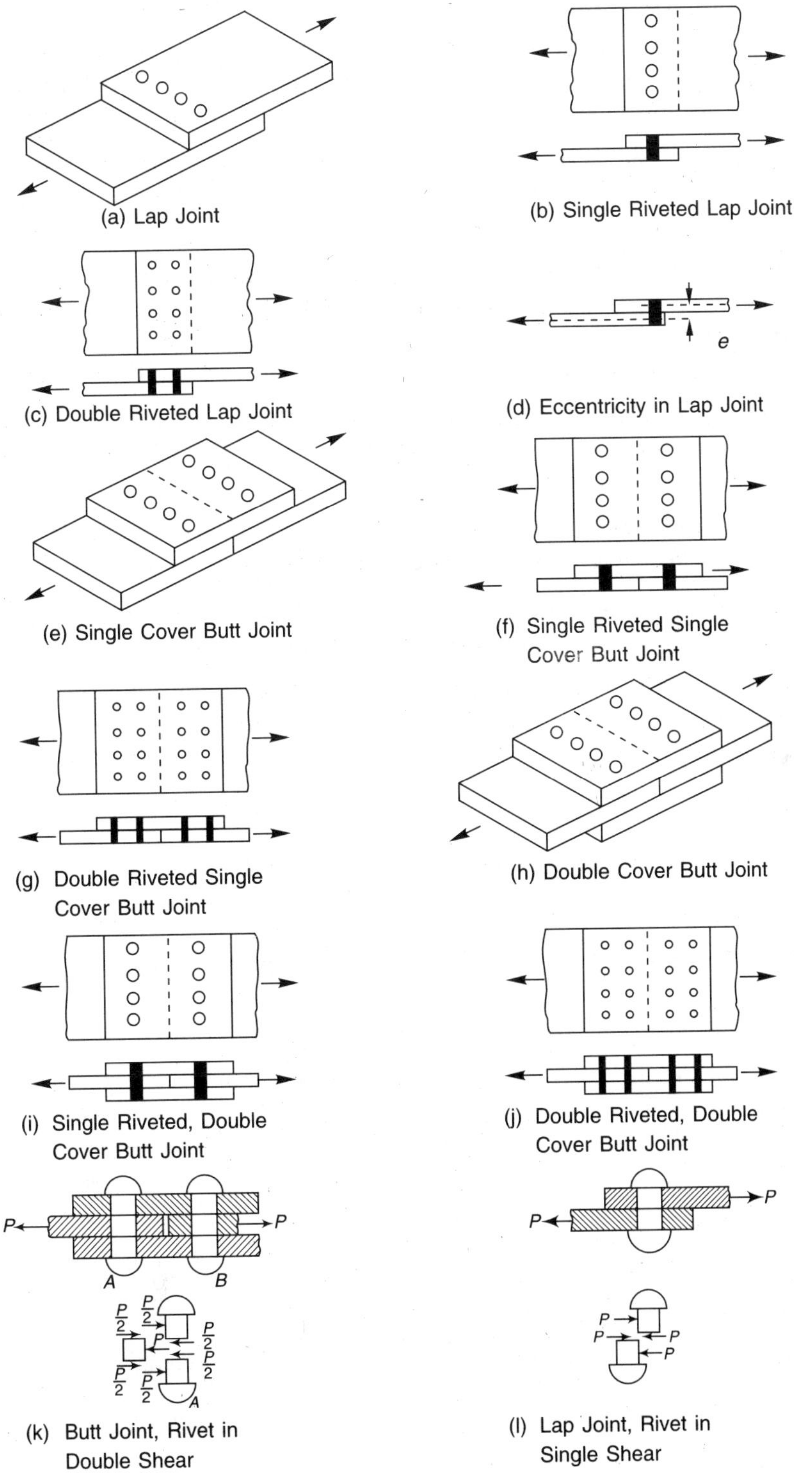

Fig. 2.2 *Types of Riveted Joints*

Failure of Riveted Joints

The riveted joint may fail in any of the following six ways, out of which some failures can be checked by adherence to the specifications of edge distance. Therefore, they are not of much importance, whereas the others require due consideration.

(i) Shear Failure of Rivets (Fig. 2.3(a)) The shear stress in the rivet may exceed the working shear stress in the rivet. Shear stresses are generated because the plates slip due to applied forces.

(ii) Shear Failure of Plates (Fig. 2.3(b)) The internal pressure of over driven (shank length more than the grip) rivets placed at a lesser edge distance than specified causes this failure. This can be checked by providing proper edge distance between the centre of the hole and the end of the plate as specified by I.S. 800.

(iii) Tension or Tearing Failure of Plates (Fig. 2.3(c)) The tensile stress in the plate at the net cross-section may exceed the working tensile stress. Tearing failure occurs when the rivets are stronger than the plates.

(iv) Splitting of Plates (Fig. 2.3(d)) Rivets may have been placed at a lesser edge distance than required causing the plates to split or shear out.

(v) Bearing Failure of Plates (Fig. 2.3(e)) The plate may be crushed when the bearing stress in the plate exceeds the working bearing stress.

(vi) Bearing Failure of Rivets (Fig. 2.3(f)) The rivet is crushed around the half circumference. The plate may be strong in bearing and the heaviest stressed plate may press the rivet.

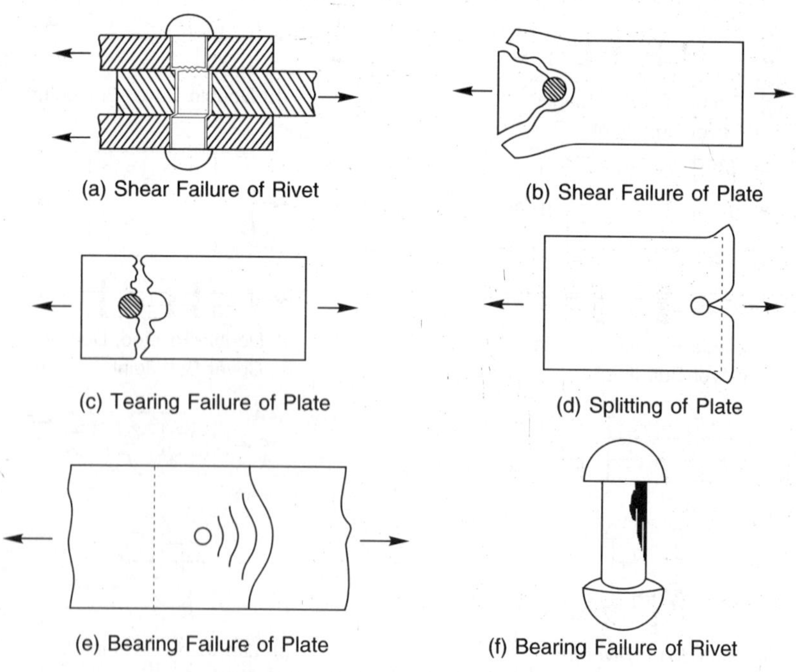

Fig. 2.3 *Failure of Riveted Joints*

Assumptions

Following are the assumptions made in the design of riveted joints. The assumptions are true for static loads approaching ultimate strength.

1. *Rivets are Assumed to be Stressed Equally* All the rivets are assumed to resist the same shear. This is true if the plates are assumed to be perfectly rigid and the rivets perfectly elastic. Actually the outer rivets are subjected to greater shear as compared to the inner rivets and the elastic stress in the members to be jointed does not remain same between two consecutive pair of rivets. Therefore the end rivets *a* and *c* shown in Fig. 2.4(a) deform more than the inner rivets *b*. The elastic stress in the top plate, between rivets *a* and *b*, is twice the elastic stress in the bottom plate and hence the upper plate stretches more between rivets *a* and *b*. Similarly the lower plate stretches more as compared to the upper plate between rivets *b* and *c*. The assumption is approximately valid when the connection is subjected to static loads approaching ultimate strength.

2. *The Rivet Hole is Assumed to be Filled Completely by the Rivet* The assumption is true in the case of cold driven rivets, but in the case of hot driven rivets it is not true as the rivets shrink diametrically on cooling.

3. *Friction Between the Plates is Neglected* The load is transmitted by friction and not by shear or bearing when the load is less than the frictional resistance of the plates. When the load exceeds the frictional resistance it is resisted partly by friction and partly by shear and bearing. Therefore the assumption is true for the kind of action that occurs near failure, when the plates have lost their clamping action because of a number of cycles of alternate loading.

4. *Shear Stress is Assumed to be Uniformly Distributed Over the Gross Cross-sectional Area of the Rivet* The shear stress distribution is not uniform when the loads is within elastic range, but due to the ductility of steel the shear stress distribution tends to be uniform as the load approaches the ultimate load.

5. *Stress in a Plate is Assumed to be Uniform* Due to the presence of holes in the plate the stress distribution is non-uniform. This is because stress concentration takes place near the holes. But as the load approaches the ultimate limit the stress approaches a uniform distribution, as shown in Fig. 2.4(b).

6. *Bending of Rivet is Neglected* For the design of a riveted joint, bending of the rivets is neglected. From Fig. 2.4(c) it is clear that the forces acting on a rivet can never be in a direct line and a bending moment of 3/8 Pt will act on the rivet, if the plates are of the same thickness. For average length rivets, it is probably permissible to neglect bending, but for ones with long grips it cannot be ignored. Bending in rivets should be accounted for if the length to diameter ratio of the rivet is more than 10.

7. *Bearing Stress is Uniform Between Plates and Rivets* The bearing stress is assumed to be uniform over the nominal contact surface of plates and rivets, as shown in Fig. 2.4(d). Since the rivet does not fill the hole completely the actual bearing stress distribution is not uniform. This distribution in lap and butt joints are

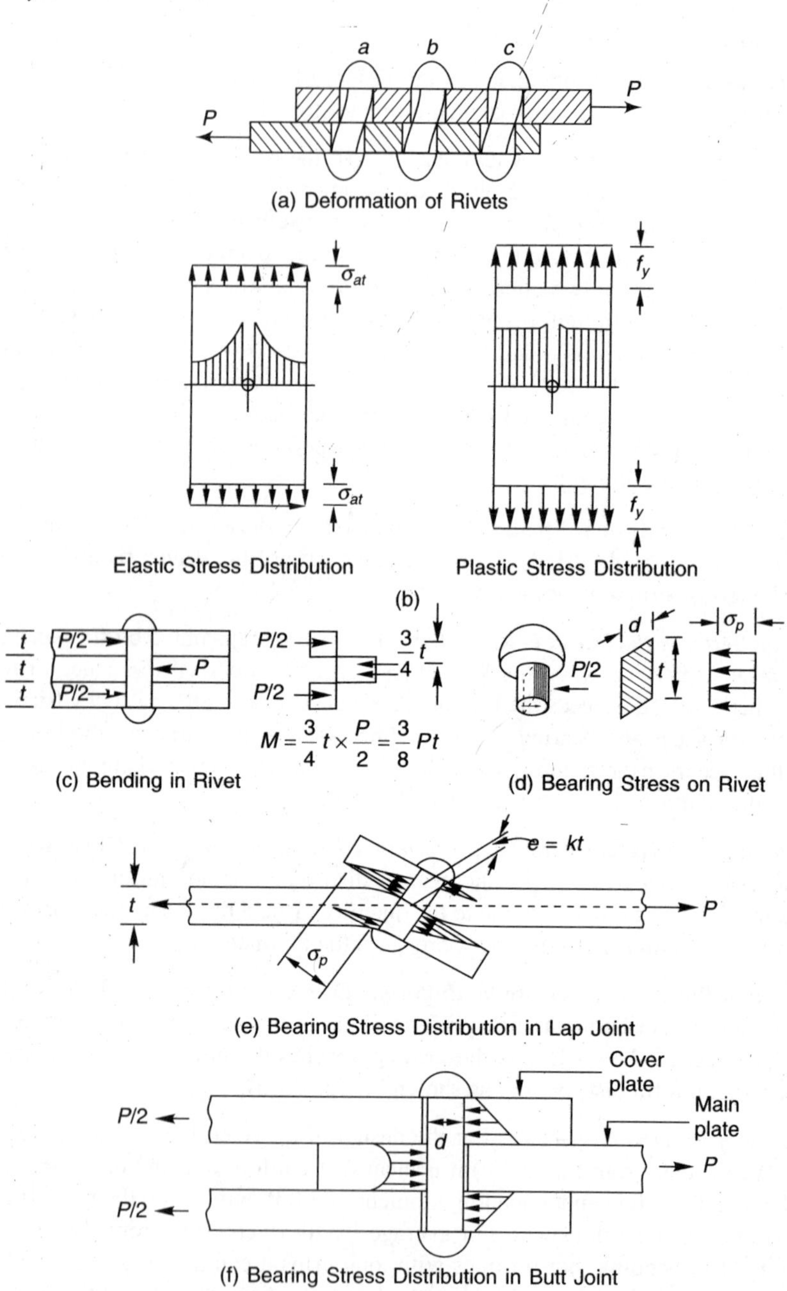

Fig. 2.4 *Deformations of Rivets and Stress Distributions in Joints*

shown in Figs 2.4(e, f) respectively. The bearing stress distribution is non-uniform throughout for a lap joint, whereas in the case of a butt joint it is more or less uniform for main plates, with some variations in cover plates.

Permissible Stresses

As per I.S: 800–1984, permissible stresses for mild steel rivets conforming to I.S: 1367–2002 are stated in Table 2.1. The stresses are in Mega-Pascal units. The permissible stresses in high tensile steel rivets should be the stresses in Table 2.1 multiplied by the ratio of the tensile strength of the rivet material to the tensile strength as specified in I.S: 1148–1973.

Table 2.1 *Permissible Stresses in Rivets*

Type of Rivet	*Axial Tension (MPa)*	*Shear (MPa)*	*Bearing (MPa)*
Power driven	100	100	300
Hand driven	80	80	250

Notes

(i) For field rivets the permissible stresses are reduced by 10%.
(ii) The calculated bearing stress of a rivet on the parts connected by it should not exceed the value of f_y (yield stress) and 1.2 f_y for hand driven and power driven rivets, respectively.

Specifications

To estimate the strength of the riveted joints and to ensure safe design, the specifications for pitch and edge distances as laid down by I.S: 800–1984 must be observed strictly.

(i) Diameter The gross diameter may be found by increasing the nominal diameter (ϕ), by certain allowances as given below.

Gross diameter = nominal diameter + 1.5 mm, $\phi \le 25$ mm
Gross diameter = nominal diameter + 2 mm, $\phi > 25$ mm

(ii) Gauge (g) is the distance between adjacent rivet lines, or the distance between the back of the rolled section and the first rivet line, or centre to centre distance (Fig. 2.5) between two consecutive rivets measured along the width of the member or connection. The gauge lines are specified on the sections tabulated in the I.S. Handbook No. 1 and the rivets should be placed on them in order to facilitate and simplify office and shop work. Gauge lines are also called the rivet lines.

(iii) Pitch (s) is the distance between the centre of two consecutive rivets measured along a row of rivets as shown in Fig. 2.5. A row generally refers to a line of rivets placed parallel to the stress in a member. Since the rivets are staggered the pitch in this case will be referred to as staggered pitch (Fig 2.5(a)). For wide plates (Fig. 2.5(b)) pitch may also be defined as the c/c distance of rivets measured along the length of the member or the connection. Rivets should be placed apart at a sufficient distance for the following reasons:

1. To prevent tension failure of members between the two consecutive rivets, and
2. To permit efficient installation of rivets.

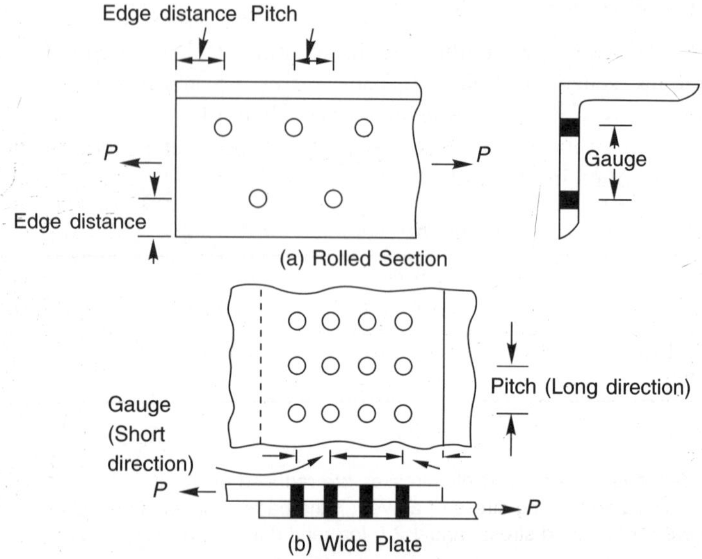

Fig. 2.5 *Gauge and Pitch*

It is also desirable to place the rivets sufficiently close together for the following reasons:

1. To reduce the length of the connection and gusset plate i.e. to have a compact joint, and
2. To have uniform stress in the rivets.

It is assumed that the load on the joint is distributed equally among all the rivets. This is true only when the plates are rigid, but actually the plates are elastic and deformable. If the plates are rigid, all the rivets will be stressed equally, like in Fig. 2.6(a). But as they are deformable, stresses are generated in the plates. The deformations are maximum at the ends and decrease towards the middle, as shown in Fig. 2.6(b). Rivets at the ends will be stressed more as compared to the inner ones, as the slip is more for end rivets. The variation of stress in rivets due to plate deformation will be more if the spacing of rivets i.e. pitch is more. Therefore, a compact joint is desirable to reduce the variation in rivet stress.

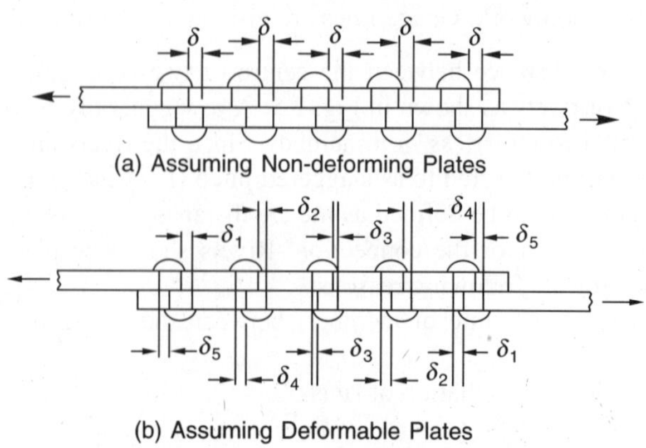

Fig. 2.6 *Deformation of Rivets and Plates*

Minimum Pitch The distance between centre of holes should not be less than 2.5 times the nominal diameter of the rivet. If the rivets are spaced closer than this, very little clearance is left between the last-formed rivet head and the die forming the next rivet head.

Maximum Pitch If the rivets be widely spaced or pitched, the plates, especially if thin, are apt to gap apart in the case of a built-up tension member. The same reasoning applies to a built-up compression member, but with the additional question of buckling. The various plates of a compression member, e.g., a column, strut, or the top flange of an ordinary plate web girder, become little individual columns held to the neighbouring plates by the rivet heads. The rivet heads are under tension, and wider the spacing, the greater is the load on the rivet head, tending to break it away from the shank. Again, the further apart the rivets are, the more flexible become the little columns, so that should one rivet head give way, an individual column of twice the ordinary length occurs. This may finally lead to the entire failure of the built-up member. The limits placed on maximum pitches in various cases are as follows.

1. The distance between the centre of two consecutive rivets in the direction of stress should not exceed $16t$ or 200 mm, whichever is less in tension members, and $12t$ or 200 mm, whichever is less in compression members.

   In the case of compression members in which forces are transmitted through the butting face this distance should not exceed 4.5 times the diameter of the rivet and for a distance from the butting face, equal to 1.5 times the width of the members, where t is the thickness of the thinner outside plate in mm.
2. The distance between the centres of two adjacent rivets, including tacking rivets, should not exceed $32t$ or 300 mm, whichever is less.
3. The distance between the centres of any two consecutive rivets in a line adjacent and parallel to an edge of an outside plate should not exceed (100 mm + $4t$) or 200 mm, whichever is less.
4. Where the rivets are staggered at equal intervals and gauge does not exceed 75 mm, the distance specified in (1) and (2) may be increased by 50%.

(iv) Edge Distance This is the distance from the centre of the rivet hole to the adjacent edge of the member (Fig. 2.5(a)). Rivet holes should not be placed too near the edges for the following reasons:

1. The failure of plate in tension may take place, and
2. The steel of the plate opposite the hole may bulge out and may crack.

The minimum distance from the centre of any hole to the edge of a plate should not be less than that given in Table 2.2. Where two or more parts are connected together a line of rivets should be provided at a distance of not more than (37 mm + $4t$) from the nearest edge. If the joint is not exposed to weather the above stated limit may be increased to $12t$, where t is the thickness of the thinner outside plate.

Table 2.2 *Edge Distance*

Diameter of Hole (mm)	*Distance to Sheared or Hand Flame Cut Edge (mm)*	*Distance to Rolled, Machine Flame Cut, Sawn or Planed Edge (mm)*
13.5 and below	19	17
15.5	25	22
17.5	29	25
19.5	32	29
21.5	32	29
23.5	38	32
25.5	44	38
29.0	51	44
32.0	57	51
35.0	57	51

(v) Tacking Rivets Tacking or stitch rivets are used to make the sections act in unison, and to prevent buckling in compression members, where two or more sections are in contact.

When the distance between the centres of two consecutive rivets in such cases exceeds the maximum specified pitch of $12t$ or 200 mm, whichever is less, in compression members and $16t$ or 200 mm, whichever is less in tension members, additional rivets are provided. These are not subjected to calculated stresses and are called tacking or stitch rivets.

1. Tacking rivets should have a pitch in line not exceeding $32t$ or 300 mm, whichever is less. Where it is exposed to weather, pitch in line should not exceed $16t$ or 200 mm. In both cases, the lines of rivets should not be apart at a distance greater than these pitches.
2. For compression members maximum pitch of tacking rivets should be 600 mm.
3. In case of tension members the tacking rivets should be provided at a pitch in line not exceeding 1000 mm.

(vi) Packings Packings or 'filler plates' are used to make the surfaces of the main plates to be jointed, flush. In Fig. 2.7 a joint has been shown with packings. The left-hand side portion of the joint is shown with rivets which have bent due to flexure. This bending can be overcome by providing some additional rivets as shown on the right-hand side of the joint. By doing so, some of the load from the main plates is picked by the rivets on the main plates and transferred to the lengthened packing plates. This load is distributed approximately evenly throughout this

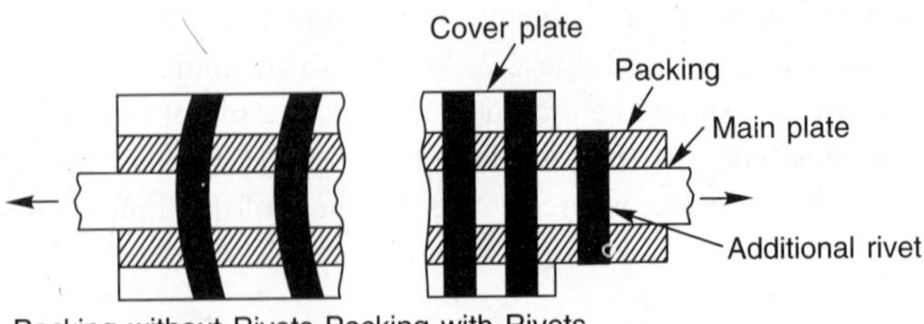

Fig. 2.7 *Bending of Rivets*

extended packing plate portion. The bending action on the rivets over main joint, therefore, is reduced considerably. In fact, if the packers are carried for sufficient length ahead and attached by a corresponding number of extra rivets the bending action on the main rivets will be entirely eliminated. Therefore, extra rivets not subjected to the calculated stresses are provided on extensions of packings.

According to I.S. specifications, rivets or bolts carrying a calculated shear stress through a packing more than 6 mm thick should be increased (from the numbers of rivets require by normal calculation) by 2.5% for each 2 mm thickness of packing. The extra rivets or bolts should be placed on the packing extension. When packings are subjected only to direct compression, the specification for tacking rivets as cited here should not apply.

Strength of a Riveted Joint

Of the failures of riveted joints discussed in Section 2.2.4, the failures of plates in shear, bearing and splitting can be checked simply by providing proper edge distance. Therefore, the strength of a riveted joint is computed by considering the failure of rivets in shear and bearing and of plate in tearing only. The minimum value of these three types of failures is taken as the strength of a riveted joint.

Strength of Rivet in Shear The rivet shank shears along the plane of slip. The number of planes along which the rivet can be sheared indicates the number of shears, i.e. single or double shear (Figs 2.2(l), (k)). The resistance of a rivet to shear depends upon the cross-sectional area of the shank and the allowable unit shear stress of the rivet metal.

Let P_s = strength of riveted joint in shear in Newton

d = gross diameter of rivet in mm

τ_{vf} = permissible shear stress in rivet in MPa

n = number of rivets on each side of joint

n' = number of rivets in one pitch length

Strength of rivet in shear = cross-sectional area × permissible shear stress

Single shear

$$\text{Strength of one rivet} = \frac{\pi}{4} d^2 \tau_{vf} \tag{2.1}$$

$$\text{Strength of riveted joint} = n \frac{\pi}{4} d^2 \tau_{vf} \tag{2.2}$$

$$\text{Strength of riveted joint/pitch length} = n' \frac{\pi}{4} d^2 \tau_{vf} \tag{2.3}$$

Double shear

$$\text{Strength of one rivet} = 2 \frac{\pi}{4} d^2 \tau_{vf} \tag{2.4}$$

$$\text{Strength of riveted joint, } P_s = 2n\frac{\pi}{4}d^2\,\tau_{vf} \tag{2.5}$$

$$\text{Strength of riveted joint/pitch length} = 2n'\frac{\pi}{4}d^2\,\tau_{vf} \tag{2.6}$$

Note The shearing strength of a rivet in double shear is considered to be twice its single shear value. Any further increase of shear value beyond double shear is not permitted for rivets subjected to shear forces on more than two planes.

Strength of Rivet in Bearing The bearing strength of a rivet is the force that can be exerted on it by the section through which it passes. The bearing area of a rivet is cylindrical but for computation purpose, the projected area which is rectangular, is considered as shown in Fig. 2.4(d).

Let P_b = strength of the riveted joint in bearing in Newtons
d = gross diameter of the rivet in mm
t = thickness of thinner section in lap joint or
= thickness of main thinner plate or sum of cover plates thickness, whichever is minimum in butt joint
σ_{pf} = permissible bearing stress in rivet in MPa
n = number of rivets on each side of joint
n' = number of rivets in one pitch length

Strength of rivet in bearing = projected area × permissible bearing stress

$$= dt\,\sigma_{pf} \tag{2.7}$$

$$\text{Strength of riveted joint, } P_b = ndt\,\sigma_{pf} \tag{2.8}$$

$$\text{Strength of riveted joint/pitch length} = n'\,dt\,\sigma_{pf} \tag{2.9}$$

Strength of Plate in Tearing Strength of the plate in tearing depends upon the net section resisting the force. As shown in Fig. 2.8, *x-x* is assumed to be the critical section.

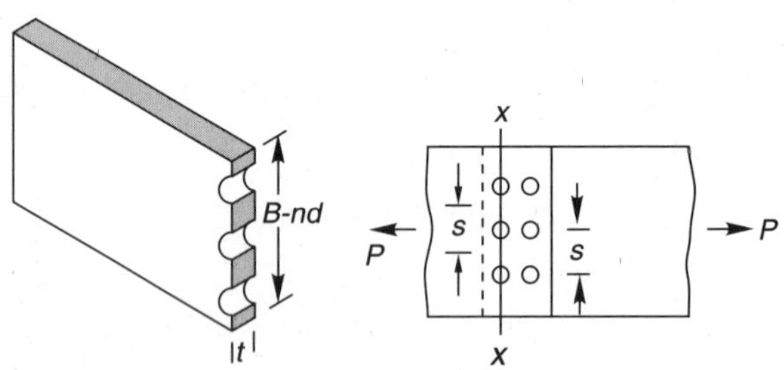

Fig. 2.8 *Net Section of Plate with Chain Riveting*

Let P_t = strength of the riveted joint in tearing in Newton
s = pitch of rivets in mm
B = width of plate in mm
n = number of rivets in section under consideration
d = gross diameter of rivet in mm
σ_{at} = permissible tensile stress in plate in MPa
= $0.6f_y$ (150 MPa for steel with yield stress 250 MPa)

Strength of joint at critical section, P_t = net area along the section × permissible tensile stress in plate

$$P_t = (B - nd)t\,\sigma_{at} \tag{2.10}$$

$$\text{Strength of the joint/pitch length} = (s - d)t\,\sigma_{at} \tag{2.11}$$

Note The net section in Eq. (2.11) is considered to be $(s - d)$ only because the plate tears along the critical transverse section, which is weakened by one rivet hole only. Therefore, only one rivet hole diameter is subtracted contrary to the calculations made for the computation of strength of joint/pitch length in case of shear and bearing, where the number of rivets per pitch were multiplied to the strength of one rivet.

Rivet Value (R_v) The minimum strength of a rivet in shear (Eq. (2.1)) or bearing (Eq. (2.7)) is called the rivet value.

Efficiency of riveted joint (η) This is also called the percentage strength of riveted joint. It is the ratio of the strength of the joint to the strength of the main member expressed as a percentage. The effectiveness of a particular riveted joint is measured by the efficiency.

$$\eta = \frac{\text{strength of riveted joint}}{\text{strength of solid plate}} \times 100$$

Net Section

The section available to resist the stresses, after the holes for the rivets are made, is termed as net section. The rivets may be placed in a variety of patterns, depending upon the space available for connection and the shape of members to be connected. The most common types of rivet patterns are chain riveting (Fig. 2.9(a)), and diamond riveting (Figs 2.9(c), (d)). Staggered pattern shown in Figs 2.9(b) yields more net area of the section and because of this reason this pattern is most suitable for tension members. Staggered and diamond patterns are better as compared to the chain pattern.

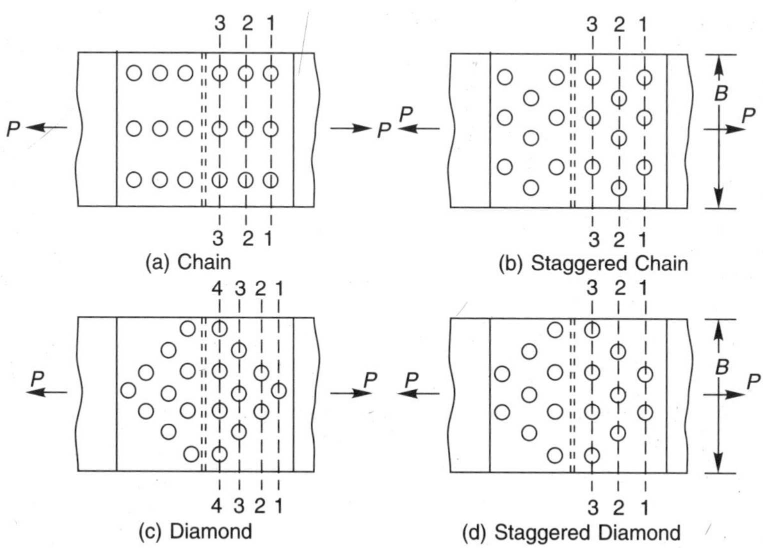

Fig. 2.9 *Rivet Patterns, Shown with the Help of a Single Cover Butt Joint*

Consider a single cover butt joint as shown in Fig. 2.9(c). The plates resist the total load at section 1–1 and successively lesser loads at sections 2–2, 3–3, etc. To understand this consider the butt joint shown in Fig. 2.10(a) wherein the flow of force has been shown from the main plate to the cover plate through the rivets. Force in the main plate before section 1–1 will be equal to P. As soon as it crosses section 1–1, the rivets absorb force equal to their strength and transfer it to the cover plate. Thus the force to be resisted by the main plate after section 1–1 will be less than the force at or before section 1–1. Further as it crosses section 2–2 again, there is a reduction of force in the main plate for the same reason. Therefore, it is desirable to have a less number of rivets providing more net section to resist the load at section 1–1. The number of rivets can be increased in the inward rows 2–2 and 3–3 and consequently a lesser net section may be used as the load on these sections gradually decreases. It will be reverse in the case of the cover plate where the force is maximum at section 3–3. Thus for the main plate the critical section is 1–1 and for the cover plate it is section 3–3. This can also be shown by working out the tearing strength at various sections of the chain (Fig. 2.9(a)) and diamond (Fig. 2.9(c)) patterns.

(i) For chain riveting (Fig. 2.9(a)):

at section 1–1 = $(B - 3d)\, t\, \sigma_{at}$ (2.12)

at section 2–2 = $(B - 3d)\, t\, \sigma_{at} + 3\, R_v$ (2.13)

at section 3–3 = $(B - 3d)\, t\, \sigma_{at} + 6\, R_v$ (2.14)

So the critical section is 1–1 for the main plate and 3–3 for the cover plate.

(ii) For diamond riveting (Fig. 2.9(c)):

at section 1–1 = $(B - d)\, t\, \sigma_{at}$ (2.15)

at section 2–2 = $(B - 2d)\, t\, \sigma_{at} + R_v$ (2.16)

at section 3–3 = $(B - 3d)\, t\, \sigma_{at} + 3R_v$ (2.17)

at section 4–4 = $(B - 4d)\, t\, \sigma_{at} + 6R_v$ (2.18)

So the critical section is 1–1 for the main plate and 4–4 for the cover plate.

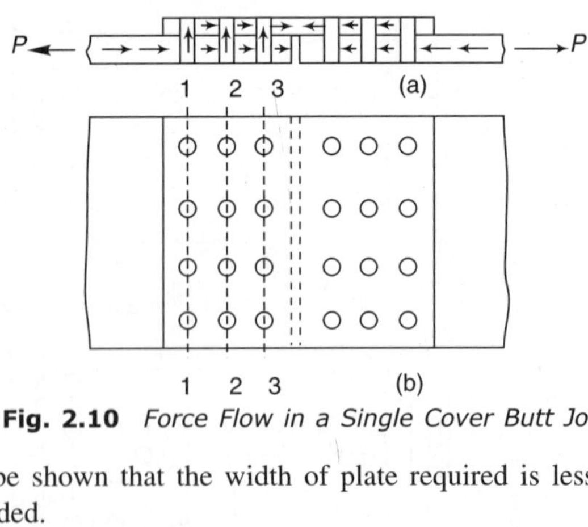

Fig. 2.10 *Force Flow in a Single Cover Butt Joint*

It can also be shown that the width of plate required is less if diamond rivet pattern is provided.

Let n be the number of rivets in chain riveting at the critical section 1–1, Fig. 2.9(a). The required width of the main plate is calculated from

$$P = (B - nd)\, t\, \sigma_{at} \quad \text{or} \quad B = \frac{P}{t\, \sigma_{at}} + nd \tag{2.19}$$

The critical section in diamond riveting will pass through one rivet hole, as in Fig. 2.9(c). The required width of the main plate is calculated from

$$P = (B - d)\, t\, \sigma_{at}$$

or $$B = \frac{P}{t\,\sigma_{at}} + d \qquad (2.20)$$

It is clear from Eqs 2.19 and 2.20, that the width of plate required to accommodate diamond pattern is less than that for chain pattern by $d(n - 1)$. When the members are long, for instance bridge members, this reduction in width affects the economy considerably.

Staggered Pitch (s) This is also known as *alternate* or *reeled pitch*. It is the distance measured along one rivet line, from the centre of a rivet on it to the centre of the adjoining rivet on a lower and parallel rivet line, as shown in Fig. 2.11.

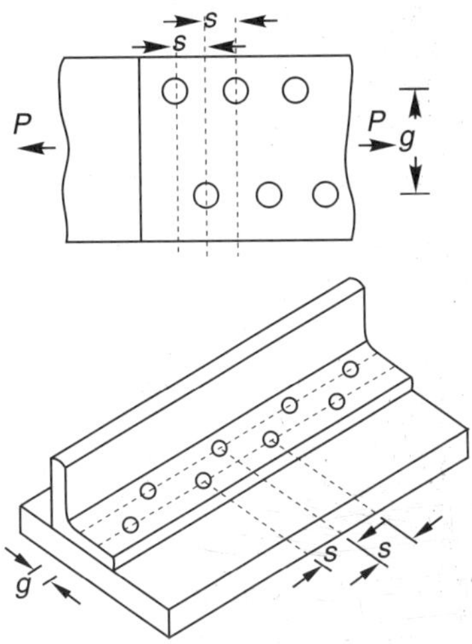

Fig. 2.11 *Staggered Pitch*

Design of Concentric Riveted Joint

When the line of action of a load coincides with the centre of gravity of the rivet group, the joint is said to be concentric. For such joints it is assumed that the stress is equally distributed among all the rivets. A concentric riveted joint design consists in finding out the diameter, the number or pitch of the rivets and the efficiency of the joint.

Diameter of the Rivet The size of a rivet is usually determined by considering the practical aspects of riveting as well as the strength of the structural element. It is economical to use a small number of large diameter rivets rather than a large number with small diameters. This is because for a given plate thickness, in case the shear controls the design, the strength is directly proportional to the square of the diameter of the rivet and large diameter rivets or rivets with double shear may be a choice. Whereas, if bearing controls the design, the strength is directly proportional to the rivet diameter only and in this case a large number of rivets with small diameter are preferred. The nominal diameter of a rivet is assumed to be between 12–25 mm for joining structural elements. As the range is quite large, to have an approximate value *Unwin's formula* may be used.

$$\phi = 6.01\sqrt{t}$$

where ϕ is the nominal diameter of the rivet in mm and t is the minimum thickness of the plates to be jointed, in mm. The limitation of the Unwin's formula is that it gives higher values than required. Therefore, suitable approximation may have to be made for the nominal diameter by the designer.

Number and Pitch of Rivets The number of rivets can be computed by dividing the total load to be transmitted by the rivet value. This value is the minimum strength of one rivet in shear or bearing and can be worked out as discussed in

Section 2.2.8. For a butt joint the calculated rivets are arranged on one side of the joint and the same number of rivets are also provided on the other side of the joint to make a complete joint. The pitch of the rivets should conform to the I.S. specifications, with a minimum to achieve economy and compactness. An optimum pitch of the rivets can be found by equating the tearing strength of the plates to the rivet value.

$$(s - d)t\,\sigma_{at} = R_v \tag{2.21}$$

Efficiency of the Joint After calculating the number of rivets and the pitch, a rivet pattern is decided and the efficiency of the joint computed, which gives an idea of the strength of the joint.

Note When the joints are subjected to reversal of stress, their frictional resistance is lost and the rivets become loose in their holes. In such cases the rivets are subjected to fatigue. To account for this, such joints are overdesigned (i.e. a larger number of rivets is provided than obtained from design) or designed for the sum of tension and compression acting on the member.

Riveted Joint in Shells

Design of riveted joint in shells consists in designing the thickness of shells, diameter of rivets and pitch. The shells can be either cylindrical or spherical. The following expressions are used for calculating stresses in shells.

Cylindrical shells

$$\text{Longitudinal stress} = \frac{pd}{4t}$$

$$\text{Circumferential stress} = \frac{pd}{2t}$$

Spherical shells

$$\text{Longitudinal stress} = \text{circumferential stress} = \frac{pd}{4t}$$

For computing the thickness of the shell the expression giving the maximum stress is used. The riveting of the shell is done in the longitudinal direction. The stress thus computed will be maximum as it is on the net section. To obtain the gross area of the section the above expressions are modified as

$$\text{Longitudinal stress} = \frac{pd}{2\eta t} \qquad \text{(cylindrical shell)}$$

$$= \frac{pd}{4\eta t} \qquad \text{(spherical shell)}$$

where p = pressure inside the shell
d = diameter of the shell
η = efficiency of the joint
t = thickness of the shell

Joints of shells are designed as double cover butt joints. The efficiency of the joint in the beginning is not known; therefore, it is assumed to be between 60 – 75% for a double cover butt joint and the thickness of the main plates forming the shells is calculated. The diameter of the rivet and a suitable pitch conforming to I.S. specifications is assumed. The rivets are placed staggered in two rows to increase the net area. The efficiency of the joint is calculated and if required, the design is revised.

2.3 BOLTED CONNECTIONS

A bolt may be defined as a metal pin with a head at one end and a shank threaded at the other end to receive a nut as in Fig. 2.12(a). Steel washers are usually provided under the bolt as well as under the nut to serve two purposes:

1. to distribute the clamping pressure on the bolted member, and
2. to prevent the threaded portion of the bolt from bearing on the connecting pieces.

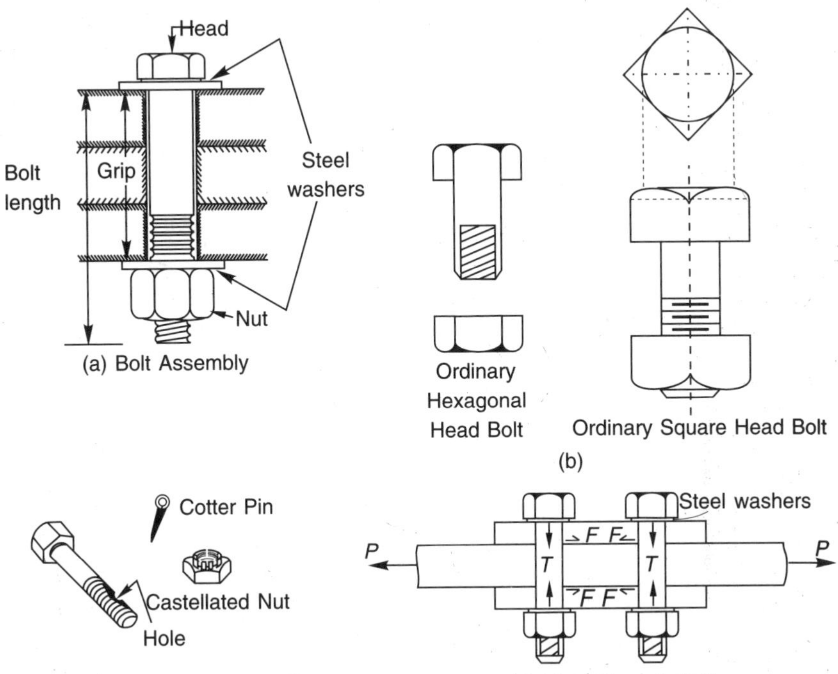

Fig. 2.12 *Bolted Joints*

In order to assure proper functioning of the connection, the parts to be connected must be tightly clamped between the bolt head and nut. If the connection is subjected to vibrations, the nuts must be locked in position. Bolted connections are quite similar to riveted connections in behaviour but have some distinct advantages as follows:

1. The erection of the structure can be speeded up, and
2. Less skilled persons are required.

The general objections to the use of bolts are:

1. Cost of material is high: about double that of rivets.
2. The tensile strength of the bolt is reduced because of area reduction at the root of the thread and also due to stress concentration.

3. Normally these are of a loose fit excepting turned bolts and hence their strength is reduced.
4. When subjected to vibrations or shocks bolts may get loose.

Uses

1. Bolts can be used for making end connections in tension and compression members.
2. Bolts can also be used to hold down column bases in position.
3. They can be used as separators for purlins and beams in foundations, etc.

Types There are several types of bolts used to connect the structural elements. Some of the bolts commonly used are:

(a) Unfinished bolts
(b) Turned bolts
(c) Ribbed bolts
(d) High strength bolts
(e) Interference bolts

Unfinished Bolts

Unfinished bolts are also called ordinary, common, rough or black bolts. These are used for light structures (purlins, bracings, etc.) under static loads. They are not recommended for connections subjected to impact load, vibrations and fatigue. Bolts are forged from low carbon rolled steel circular rods, permitting large tolerances. Ordinary structural bolts are made from mild steel rods with square or hexagonal head, as shown in Fig. 2.12(b). Square heads cost less but hexagonal heads give a better appearance, are easier to hold by wrenches and require less turning space. The bolt hole is punched about 1.6 mm more than the bolt diameter. The nuts on bolts are tightened with spud wrenches, producing little tension. Therefore, no clamping force is induced on the sections jointed. Sometimes a hole is drilled in the bolt and a cotter pin with a castellated nut is used to prevent the nut from turning on the bolt, as shown in Fig. 2.12(c). The connections with unfinished bolts are designed in a similar way as all the riveted connections except that the permissible stresses are reduced to account for tolerance provided on shank and threaded portion of the bolts. The requirements regarding pitch and edge distance are same as that for rivets. The permissible stresses are as given in Table 8.1 of I.S: 800–1984.

Turned Bolts

These are similar to unfinished bolts, with the difference that the shank of these bolts is formed from a hexagonal rod. The surface of the bolts are prepared carefully and are machined to fit in the hole. Tolerances allowed are very small. These bolts have high shear and bearing resistance as compared to unfinished bolts. However, these bolts are obsolete nowadays. The specifications for turned bolts are given in I.S: 2591–1982.

Ribbed Bolts

These are also called fluted bolts. The head of the bolt is like a rivet head. The threads and nut are provided on the other end of the shank. From the shank core longitudinal ribs project making the diameter of the shank more than the diameter of the hole. These ribs cut grooves into the connected members while tightening and ensure a tight fit. These bolts have more resistance to vibrations as compared to ordinary bolts. The permissible stresses for ribbed bolts are same as that for rivets.

High Strength Bolts

Normally, for ordinary bolted joints, the force is transferred through the interlocking and bearing of bolts. However, for high strength bolted joints this transfer of force is accomplished through the friction between the interfaces formed between load carrying elements jointed. This friction is developed by applying a load normal to the joint by tightening these bolts to proof load. That is why these bolts are also known as friciton type bolts. These are made from bars of medium carbon steel. Their high strength is achieved through quenching and tempering processes or by alloying steel. Steel washers of hard steel or carburized steel are provided as shown in Fig. 2.12(d), to evenly distribute the clamping pressure on the bolted member and to prevent the threaded portion of the bolt from bearing on the connecting pieces. If the bolts are tightened by the turn of nut method, the nut is made snug and is tightened a half turn more by hand wrenches, then the washers are not required.

As the bolts are in tension up to proof load, loosening of nut and the washer is checked. Because of this property, the vibrations and impact resistance of the joint is also improved. The nut and head of the bolts are kept sufficiently large to provide an adequate bearing area. The specifications for high strength bolts are laid in I.S: 3757–1985 and I.S: 4000–1992. These bolts have a tensile strength several times that of the ordinary bolts. High strength bolts have replaced rivets and are being used in structures, such as high rise buildings, bridges, machines etc. Due to their distinct advantages and vital use, high strength bolts are discussed below in some detail.

Principle of High Strength Friction Grip Bolts The shank of the high strength bolts does not fully fill the hole. So shear and bearing are not the criteria for load transmission as is in the case of rivets, which fill the hole completely. The nut is tightened to develop a clamping force on the plates which is indicated as the tensile force T in the bolt (Fig. 2.12(d)). This tension should be about 90% of proof load. When a shear load is applied to the joint no slip will occur until the shear load exceeds the frictional resistance between the elements jointed. When the shear load exceeds the frictional resistance a slip occurs. On further increase of this load, the gradual slipping brings the bolt in contact with the plate edges.

The horizontal frictional force F, is induced in the joints which is equal to the tensil force T, as in Fig. 2.12(d), in the bolts multiplied by the coefficient of friction.

$$F = \mu T \tag{2.22}$$

This frictional force F should exceed the applied force P on the member.

μ = coefficient of friction or slip factor, is defined as ratio of the load per effective interface required to produce slip in a pure shear joint to the proof load induced in bolt. The recommended slip factors are given in Table 2.4.

Table 2.4 *Slip Factors*

Surface Condition (1)	*Slip Factor* (2)
Weathered, clear of mill scale and loose rust	0.45
Blasted with shot or grit and loose rust removed	0.50
Sprayed with aluminium	0.50
Sprayed with zinc	0.40
Treated with zinc-salient paints	0.35
Treated with etch primer	0.25
Galvanised	0.16
Dry mill scale	0.35

Table 2.5 *Proof Loads for H.S.F.G. Bolts*

Thread (d)	*Nomimnal Stress Area* mm^2	*Property Class* 8.8	10.9
		Proof Load N	
M10	58	33,700	48,100
M12	84,3	50,700	70,000
M14	115	68,800	95,500
M16	157	94,500	130,000
M18	192	115,000	159,000
M20	245	147,000	203,000
M22	303	182,000	252,000
M24	353	212,000	293,000
M27	459	275,000	381,000
M30	561	337,000	466,000
M33	694	416,000	576,000
M36	817	490,000	678,000
M39	976	586,000	810,000

Design of High Strength Friction Grip Bolts

1. The design of high strength bolts is governed by Eq. (2.22). The strength of H.S.F.G. bolt is measured by the proof load of the bolt as given in Table 2.5. This proof load is the minimum shank tension which must be applied to the bolt in order to ensure that the parts to be connected are adequately drawn together.

Let n = number of interfaces

F_s = factor of safety

= 1.4 for all loads except wind load

= 1.2 for wind loads

T = proof load of one bolt

$$F = \frac{n\mu T}{F_s} \tag{2.23}$$

The diameter of the bolt is assumed and proof load is referred from I.S: 3757–1972 (Table 2.5 of the text). From Eq. (2.23) the frictional force F can be computed, and should be more than the load on the joint.

Note

1. HSFG bolts are classified as 10.9 K and 8.8 G depending upon the mechanical properties of the material from which these are formed.
2. HSFG bolts are designated by size, length, symbol and IS code number. For example, a HSFG bolt 20 mm diameter and length 120 mm, conforming to mechanical properties of 10.9 K is designed as M 20 × 120 IS: 3757–10.9 K.

2. In case the tension acts in the direction of bolt axis the clamping force is reduced. If the external tension exceeds the yield strength of the bolt shank then the plates may separate. Therefore, the external tension in the direction of the bolt shank is limited to 0.6 times the proof load of the bolts. However, if fatigue conditions are involved, the maximum permissible external tension of the bolt shank is limited to 0.5 times the proof load.

3. When connections are subjected to external tension in addition to shear, for instance in brackets, the bolts should be proportioned to satisfy the following expression:

$$\frac{\text{calculated shear}}{\text{slip factor} \times \text{number of effective interfaces}} \le \frac{(\text{proof load} - \text{calculated tension} \times f)}{\text{factor of safety}}$$

where f = load factor, is a numerical value by which the load which would cause slip in a joint is divided to give the permissible working load on the joint. It may be taken as 2 when external tension is repetitive and 1.7 when non-repetitive.

Advantage of High Strength Friction Grip Bolts High strength friction grip bolts have replaced the rivets because of their distinct advantages as discussed below. However, the material cost is about 50% greater than that of ordinary bolts and special workmanship is required in installing and tightening these bolts.

1. These provide a rigid joint. There is no slip between the elements connected.
2. Large tensile stresses are developed in bolts, which in turn provide large clamping force to the elements connected. High frictional resistance is developed providing a high static strength of the joint.
3. Because of the clamping action, load is transmitted by friction only and the bolts are not subjected to shear and bearing.
4. The frictional resistance is effective outside the hole and therefore lesser load is transmitted through the net section. Thus, the possibility of failure at the net section is minimized.
5. There is no stress concentration in the holes; therefore, the fatigue strength is more.
6. The tension in bolts is uniform. Also the bolts are tensioned up to proof load hence, the nuts are prevented from loosening.
7. Few persons are required to make the connections, thus cost is reduced.
8. Noise nuisance is not there as these bolts are tightened with wrenches.
9. The hazard of fire is not there and there is no danger of tossing of the bolt.
10. Alterations can be done easily.
11. For same strength, lesser number of bolts are required as compared to rivets which brings overall economy.

Interference Bolts

These bolts have been developed recently and are improved high strength bolts with ribs on the shank. These have a diameter slightly more than the hole diameter. They can be installed with ordinary spud wrenches and are used for structures like towers, masts, etc. where simple tightening equipments are needed.

2.4 PIN CONNECTIONS

When two structural members are connected by means of a cylindrical shaped pin, the connection is called a pin connection. Pins are manufactured from mild steel bars with diameters ranging from 9 to 330 mm. Pin connections are provided when hinged joints are required, i.e., for the connection where zero moment or free rotation is desired. Introduction of a hinge simplifies the analysis by reducing indeterminacy. These also reduce the secondary stresses. These connections cannot resist longitudinal tension. For satisfactory working it is necessary to minimise the friction between the pin and members connected. High grade machining is done to make the pin and pin hole surface smooth and frictionless. Pins are provided in the following cases:

1. Tie rod connections in water tanks and elevated bins
2. As diagonal bracing connections in beams and columns
3. Truss bridge girders
4. Hinged arches
5. Chain-link cables suspension bridges

Various types of pins used for making the connections are forged steel pin, undrilled pin and drilled pin. To make a pin connection, one end of the bar is forged like a fork and a hole is drilled in this portion. The end of the other bar to be connected is also forged and an eye is made. A hole is drilled into it in such a way that it matches with the hole on the fork end bar. The eye bar is inserted in the jaws of the fork end and a pin is placed. Both the forged ends are made octagonal for a good grip. The pin in the joint is secured by means of a cotter pin or screw, as shown in Fig. 2.13.

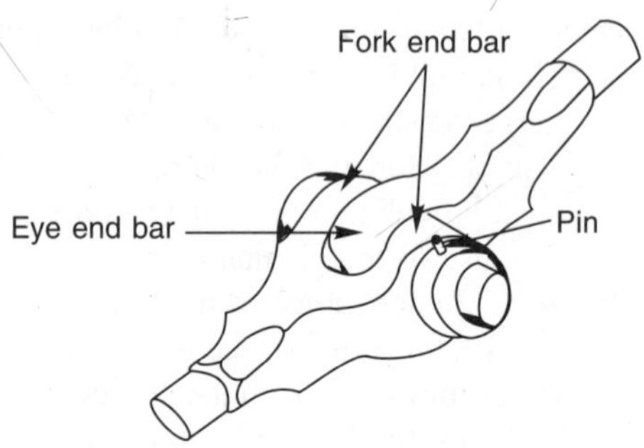

Fig. 2.13 *Pin Connection*

Specifications

Shear Structurally a pin may be treated as a cylindrically shaped beam. Assuming that bending does not exceed yield strength, the shear stress is maximum at the neutral axis of the pin and may be computed by beam-shear formula.

$$\tau_{vf} = \frac{V\,\overline{AY}}{Id}$$

or
$$\tau_{vf} = \frac{V\,\dfrac{\pi d^2}{8}\,\dfrac{2d}{3\pi}}{\dfrac{\pi d^4}{64}\,d}$$

or
$$\tau_{vf} = \frac{16V}{3\pi d^2} \tag{2.24}$$

$$\tau_{vf,\,\text{avg}} = \frac{3}{4}\tau_{vf} \qquad \left(\tau_{vf,\,\text{avg}} = \frac{V}{\pi d^2/4}\right)$$

where τ_{vf} is the maximum shear stress in pin (100 MPa), V is the shear force at the section in Newtons, d is the diameter of the pin in mm, $A\bar{Y}$ is the moment of the area of the cross-section, above the section under consideration, about the neutral axis and I is the moment of inertia of cross-section. Calculations of shear stress from the beam-shear formula give considerable error when the span-to-depth ratio is small, as it usually is for pins. This necessitates the use of nominal shear stress based on the uniform stress distribution over the pin section, in which case $\tau_{vf,\,\text{avg}} = \tau_{vf} = V/A$.

Note The allowable shear stress in the pins is adopted the same as that for power driven rivets.

Bearing A pin in bearing is designed in a similar manner as for rivet. A uniform bearing stress may be assumed for a proper fit between the plates and the pin.

$$\text{Bearing force} = dt\,\sigma_{pf} \tag{2.25}$$

where d is the diameter of the pin in mm, t is the thickness of bar in mm and σ_{pf} is the permissible bearing stress (300 MPa).

Flexure Flexure is most critical in case of pins. The members jointed by pin connections are separated some distance because of the following reasons:

1. to prevent friction
2. to allow for rivet heads, if the member is built up, and
3. to facilitate painting

Large bending moments are generated due to these reasons and the pin diameter is, therefore, generally governed by flexure. Figure 2.14(a) shows a pin connection.

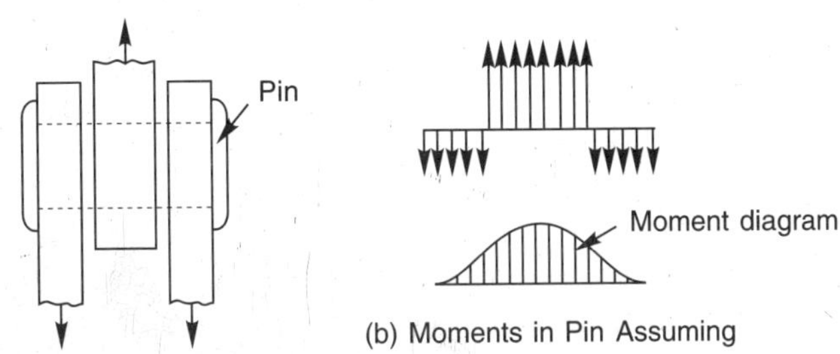

(a) Pin Connection

(b) Moments in Pin Assuming Uniform Load

Fig. 2.14 *Moment in Pin*

For a given bending moment M (Fig. 2.14(b)), assuming the stress to be within the elastic limit, the maximum fibre stress can be obtained by the equation

$$M = \sigma_b Z$$

or

$$M = \sigma_b \frac{\pi d^3}{32}$$

or

$$d = \left(\frac{32M}{\pi \sigma_b}\right)^{1/3} \tag{2.26}$$

where M is the bending moment in N mm and σ_b is the bending stress ($0.66 f_y$).

Design

The design of a pin is similar to that of a rivet. A pin may be assumed to be a large rivet subjected to shear, bearing and flexure. Rivets, as we have seen, are critical in shear and bearing. On the other hand, pins are critical in flexure. Therefore, an efficient design can be obtained by determining the pin size on the basis of its bending strength.

1. The forces are calculated and points on the pin at which the forces from the members are transferred are ascertained.
2. Maximum bending moment on the pin is calculated.
3. The diameter of the pin is calculated by Eq. (2.26).
4. The average shear stress is calculated. It should be less than the allowable shear stress of 100 MPa.
5. The bearing stress is calculated. It should be less than the allowable bearing stress of 300 MPa.

Solved Examples

Example 2.1 Calculate the value of a rivet in a lap joint used to connect two plates 12 mm thick in the following cases:

(a) Power driven rivets, and (b) Hand driven rivets

Solution Nominal diameter of rivet using Unwin's formula,

$$\phi = 6.01\sqrt{t} = 6.01\sqrt{12} = 20.819 \text{ mm} \approx 20 \text{ mm}$$

(Unwin's formula gives results on the conservative side. So let us adopt a rivet diameter 20 mm.)

Gross diameter, $d = 20 + 1.5 = 21.5$ mm

(a) $\tau_{vf} = 100$ N/mm², $\sigma_{pf} = 300$ N/mm².

Strength of rivet in single shear $= \frac{\pi}{4} d^2 \tau_{vf} = \frac{\pi}{4}(21.5)^2 \times 100 \times 10^{-3}$

$= 36.305$ kN

Strength of rivet in bearing $= dt\,\sigma_{pf} = 21.5 \times 12 \times 300 \times 10^{-3} = 77.400$ kN

Hence, the value of rivet, $R_v = 36.305$ kN

(b) $\tau_{vf} = 80$ N/mm^2, $\sigma_{pf} = 250$ N/mm^2.

$$\text{Strength of rivet in single shear} = \frac{\pi}{4} d^2 \tau_{vf} = \frac{\pi}{4} (21.5)^2 \times 80 \times 10^{-3}$$
$$= 29.044 \text{ kN}$$

Strength of rivet in bearing = $dt\ \sigma_{pf} = 21.5 \times 12 \times 250 \times 10^{-3} = 64.5$ kN

Hence, the value of rivet, $R_v = 29.044$ kN

Example 2.2 A 6 mm thick angle section is jointed to a 10 mm thick gusset plate. The angle is supporting a load of 55 kN. Find out the number of 16 mm diameter power driven rivets.

Solution Nominal diameter of rivet, $\phi = 16$ mm

Gross diameter of rivet, $d = 16 + 1.5 = 17.5$ mm

Using Power driven rivets

$$\text{Strength of rivet in single shear} = \frac{\pi}{4} d^2 \tau_{vf} = \frac{\pi}{4} (17.5)^2 \times 100 \times 10^{-3} = 24.052 \text{ kN}$$

For calculating the strength of the rivet in bearing, the minimum thickness of the section to be jointed is considered.

$\therefore$ $t = 6$ mm

Strength of rivet in bearing = $dt\ \sigma_{pf} = (17.5) \times 6 \times 300 \times 10^{-3} = 31.50$ kN

The strength of the rivet is the least of the strength in shear and bearing.

$\therefore$ Strength of rivet, $R_v = 24.052$ kN

$$\text{Number of rivets, } n = \frac{P}{R_v} = \frac{55}{24.052} = 2.2867 \approx 3$$

Provide three rivets to connect angle with gusset plate.

Example 2.3 A single-riveted double-cover butt joint is used to connect two plates 16 mm thick with chain riveting. The rivets used are power driven 20 mm in diameter at a pitch of 60 mm. Find out the safe load per pitch length and efficiency of the joint.

Solution

$\tau_{vf} = 100$ N/mm^2, $\sigma_{pf} = 300$ N/mm^2

$\sigma_{at} = 0.6 f_y = 150$ MPa (for steel with $f_y = 250$ MPa)

Nominal diameter of rivet, $\phi = 20$ mm

Gross diameter of rivet, $d = 20 + 1.5 = 21.5$ mm

Pitch, $s = 60$ mm

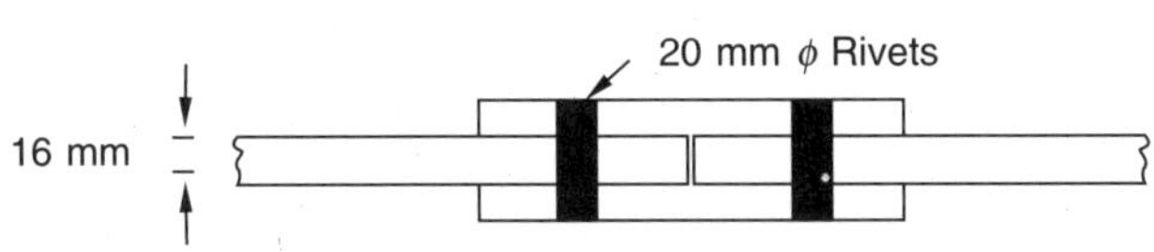

Fig. Ex. 2.3

$$\text{Strength of rivet in double shear} = 2 \times \frac{\pi}{4} d^2 \tau_{vf}$$

$$= 2 \times \frac{\pi}{4}(21.5)^2 \times 100 \times 10^{-3}$$

$$= 72.61 \text{ kN}$$

$$\text{Thickness of cover plate} \nless \frac{5}{8}t$$

$$= \frac{5}{8} \times 16 = 10 \text{ mm}$$

Let us provide two cover plates each 10 mm thick. The thickness to be considered for calculation purposes will be the thickness of the main plate or the sum of thickness of cover plates, whichever is less.

Strength of rivet in bearing = $dt\,\sigma_{pf} = 21.5 \times 16 \times 300 \times 10^{-3} = 103.200$ kN

Strength of plate in tearing = $(s - d)t\,\sigma_{at} = (60 - 21.5) \times 16 \times 150 \times 10^{-3}$

$= 92.400$ kN

Strength of the joint per pitch length in shear is the minimum.

Hence, Strength of joint per pitch length = 72.61 kN

Strength of solid plate per pitch length = $st\,\sigma_{at} = 60 \times 16 \times 150 \times 10^{-3} = 144$ kN

$$\text{Efficiency of joint} = \frac{\text{Strength of joint}}{\text{Strength of solid plate}} \times 100 = \frac{72.61}{144} \times 100 = 50.42\%$$

Example 2.4 Design a riveted joint to connect two plates 14 mm thick. Power driven rivets may be used for making the connection. Assume $f_y = 250$ N/mm^2.

Solution The joint can be designed as a lap joint or as a butt joint. The design of each type of riveted joint is illustrated by this example. Finally the selection of the type of joint is done on the basis of efficiency.

The diameter of the rivet can be found by using Unwin's formula

$$\phi = 6.01\sqrt{t} = 6.01\sqrt{14} = 22.48 \text{ mm} \approx 22 \text{ mm}$$

(As Unwin's formula gives a higher value, provide 22 mm ϕ power driven rivets).

Gross diameter = 22 + 1.5 = 23.5 mm

Lap joint The rivets will be in single shear

$$\text{Strength of rivet in shear} = \frac{\pi}{4} d^2\,\tau_{vf} = \frac{\pi}{4}(23.5)^2 \times 100 \times 10^{-3} = 43.373 \text{ kN}$$

Strength of rivet in bearing = $dt\,\sigma_{pf} = 23.5 \times 14 \times 300 \times 10^{-3} = 98.700$ kN

Strength of rivet/pitch length = 43.373 kN

Let s be the pitch of rivets. For the best design, the rivets should be placed at such a pitch that the strength of rivet per pitch length is equal to the strength of plate in tearing per pitch length.

Strength of plate in tearing = $(s - d)t\,\sigma_{at} = (s - 23.5) \times 14 \times 0.6 \times 250$

or $\quad 43.373 \times 10^3 = (s - 23.5) \times 14 \times 250$

or $\quad s = 44.15 \text{ mm} \approx 45 \text{ mm}$

$$\nless 2.5\phi\ (= 2.5 \times 22 = 55 \text{ mm})$$

Hence, provide rivets at a pitch of 55 mm.

Strength of the plate in tearing = $(55 - 23.5) \times 14 \times 0.6 \times 250 \times 10^{-3} = 66.150$ kN

Strength of joint/pitch length = 43.373 kN

Strength of solid plate/pitch length = $st\,\sigma_{at} = 55 \times 14 \times 0.6 \times 250 \times 10^{-3}$
$= 115.500$ kN

$$\text{Efficiency of the joint} = \frac{\text{strength of joint/pitch length}}{\text{strength of solid plate/pitch length}} \times 100$$

$$= \frac{43.373}{115.500} \times 100 = 37.55\%$$

Butt joint

(i) *Single cover butt joint* The rivets will be in single shear.

$$\text{Strength of rivet in shear} = \frac{\pi}{4} d^2\, \tau_{vf} = \frac{\pi}{4} (23.5)^2 \times 100 \times 10^{-3} = 43.373 \text{ kN}$$

$$\text{Thickness of cover plate} \nless \frac{5}{8} t$$

$$= \frac{5}{8} \times 14 = 8.75 \text{ mm} \approx 10 \text{ mm}$$

The thickness of plate to be considered for bearing of rivets will be = 10 mm.

Strength of rivet in bearing = $dt\,\sigma_{pf} = 23.5 \times 10 \times 300 \times 10^{-3} = 70.500$ kN

Strength of rivet/pitch length = 43.373 kN

Equating the least strength of rivet to the strength of plate per pitch length in tearing,

$$43.373 \times 10^3 = (s - 23.5) \times 10 \times 0.6 \times 250$$

or $$s = 52.415 \text{ mm}$$

$$\nless (2.5\phi = 2.5 \times 22 = 55 \text{ mm})$$

Hence, provide rivets at a pitch of 55 mm.

Strength of the joint/pitch length = 43.373 kN

Strength of solid plate/pitch length = $st\,\sigma_{at}$
$= 55 \times 10 \times 0.6 \times 250 \times 10^{-3} = 82.5$ kN

$$\text{Efficiency} = \frac{\text{strength of joint/pitch length}}{\text{strength of solid plate/pitch length}} \times 100$$

$$= \frac{43.373}{82.5} \times 100 = 52.57\%$$

(ii) *Double cover butt joint*

The rivets will be in double shear.

$$\text{Strength of rivet in shear} = 2 \times \frac{\pi}{4} d^2\, \tau_{vf} = 2 \times \frac{\pi}{4} (23.5)^2 \times 100 \times 10^{-3}$$

$$= 86.747 \text{ kN}$$

The thickness to be considered for the calculation of bearing value and strength in tearing of plate is the main plate thickness or aggregate thickness of cover plates, whichever is less.

$\therefore$ $$t = 14 \text{ mm}$$

Strength of rivet in bearing $= dt\,\sigma_{pf} = 23.5 \times 14 \times 300 \times 10^{-3} = 98.70$ kN

$\therefore$ Strength of rivet per pitch length = 86.747 kN

Equating the strength of rivet per pitch length to the strength of plate per pitch length in tearing.

$$86.747 \times 10^3 = (s - 23.5) \times 14 \times (0.6 \times 250)$$

or $$s = 64.80 \text{ mm} \approx 60 \text{ mm}$$

$$\not< 2.5\phi = 2.5 \times 22 = 55 \text{ mm}$$

Provide rivets at a pitch of 60 mm.

Strength of joint in tearing/pitch length $= (s - d)t\,\sigma_{at}$

$= (60 - 23.5) \times 14 \times 0.6 \times 250 \times 10^{-3}$

$= 76.65$ kN

Strength of solid plate per pitch length $= st\,\sigma_{at}$

$= 60 \times 14 \times 0.6 \times 250 \times 10^{-3}$

$= 126$ kN

$$\text{Efficiency of joint } \eta = \frac{\text{strength of joint per pitch length}}{\text{strength of solid plate per pitch length}} \times 100$$

$$= \frac{76.65}{126} \times 100 = 60.83\%$$

From the above illustration it is clear that the efficiency of a double cover butt joint is maximum and therefore it should be preferred.

Example 2.5 A tie member has to transmit a pull of 300 kN. Design a butt joint to connect it with 12 mm thick gusset plate. Also find the efficiency of the joint. Assume steel of yield stress 250 MPa.

Solution Nominal diameter of the rivet $= 6.01\sqrt{t} = 6.01\sqrt{12} = 20.819$ mm

≈ 20 mm

Provide 20 mm ϕ power driven rivets.

Gross diameter of the rivet = 20 + 1.5 = 21.5 mm

Pitch of the rivets $= 2.5\,\phi = 2.5 \times 20 = 50 \approx 60$ mm

Let us provide a double cover butt joint. The rivets will be in double shear.

$$\text{Shear strength of one rivet} = 2\,\frac{\pi}{4}\,d^2\,\tau_{vf}$$

$$= 2 \times \frac{\pi}{4}(21.5)^2 \times 100 \times 10^{-3} = 72.61 \text{ kN}$$

Bearing strength of one rivet $= dt\,\sigma_{pf}$

$= 21.5 \times 12 \times 300 \times 10^{-3} = 77.4$ kN

Strength of one rivet, $R_v = 72.61$ kN

$$\text{Number of rivets required on each side of the butt joint} = \frac{300}{72.61} = 4.13 \approx 5$$

Arrange the rivets as shown in Fig. Ex. 2.5.

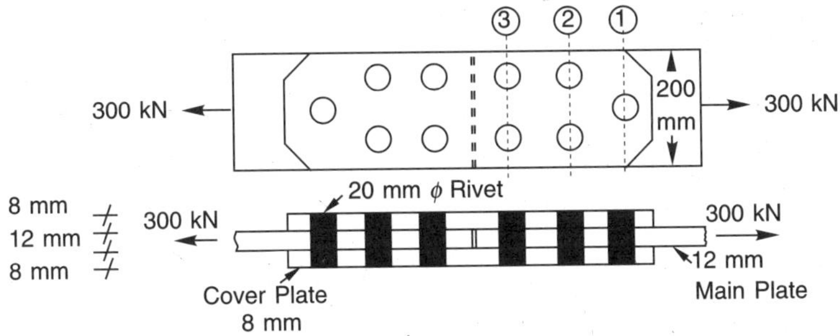

Fig. Ex. 2.5

Plate width The width of the tie plate and the thickness of the cover plate can be computed on the basis of the tearing strength of the plate section. The strength of the joint is minimum at Section (1) – (1). Therefore the width of the plate required will be maximum at Section (1) – (1). The calculations at Sections (2) – (2) and (3) – (3) have also been done to illustrate the above fact.

Let x = minimum width of the tie plate required.

Section (1) – (1)

$$(x - d)t\sigma_{at} = 300 \text{ kN}$$

or $$(x - 21.5) \times 12 \times 0.6 \times 250 = 300 \times 10^3$$

or $$(x - 21.5) = 166.66$$

or $$x = 188.16 \approx 200 \text{ mm}$$

Section (2) – (2)

$$(x - 2d)t\sigma_{at} + nR_v = 300 \times 10^3$$

or $$(x - 2 \times 21.5) \times 12 \times 0.6 \times 250 + 1 \times 72.61 \times 10^3 = 300 \times 10^3$$

or $$(x - 43.0) = 126.32$$

or $$x = 169.32 \text{ mm}$$

Section (3) – (3)

$$(x - 2d)t\sigma_{at} + nR_v = 300 \times 10^3$$

or $$(x - 2 \times 21.5) \times 12 \times 0.6 \times 250 + 3 \times 72.61 \times 10^3 = 300 \times 10^3$$

or $$(x - 43.0) = 45.65$$

or $$x = 88.65 \text{ mm}$$

Hence, the necessary width of the tie plate is 200 mm.

Thickness of cover plate The maximum thickness of the cover plate will be at section (3) – (3) being the critical section.

The thickness t_1 of each cover plate so calculated should not be less than 5/8 times the thickness of the tie plate $\left(\frac{5}{8} \times 12 = 7.5 \text{ mm}\right)$.

Section (3) – (3)

$$(x - 2d) \times 2t_1 \times \sigma_{at} = 300 \times 10^3$$

$$(200 - 43) \times 2t_1 \times 150 = 300 \times 10^3$$

or $$t_1 = 6.369 \text{ mm} \approx 8 \text{ mm}$$

Efficiency of joint

$$\eta = \frac{\text{strength of joint}}{\text{strength of solid plate}} \times 100$$

Tearing strength of plate at Section (1) – (1)

$$= (200 - 21.5) \times 12 \times 0.6 \times 250 \times 10^{-3}$$
$$= 321.3 \text{ kN}$$

Tearing strength of the two cover plates at Section (3) – (3)

$$= (200 - 2 \times 21.5) \times 2 \times 8 \times 0.6 \times 250 \times 10^{-3}$$
$$= 376.8 \text{ kN}$$

Hence the tearing strength of the joint = 321.3 kN

The actual safe load of the joint is the minimum of the strength in shear, bearing and tearing.

Strength of joint in shear = 5 × 72.61 = 363.05 kN

Strength of joint in bearing = 5 × 77.4 = 387.00 kN

Hence, strength of the joint = 321.3 kN

Strength of the solid plate = $xt\sigma_{at} = 200 \times 12 \times 0.6 \times 250 \times 10^{-3} = 360$ kN

$$\text{Efficiency of the joint} = \frac{321.3}{360} \times 100 = 89.25\%$$

Example 2.6 Design the joint *B* of a roof truss as shown in Fig. Ex. 2.6. The members are connected by power driven 20 mm ϕ rivets to a gusset plate 12 mm thick.

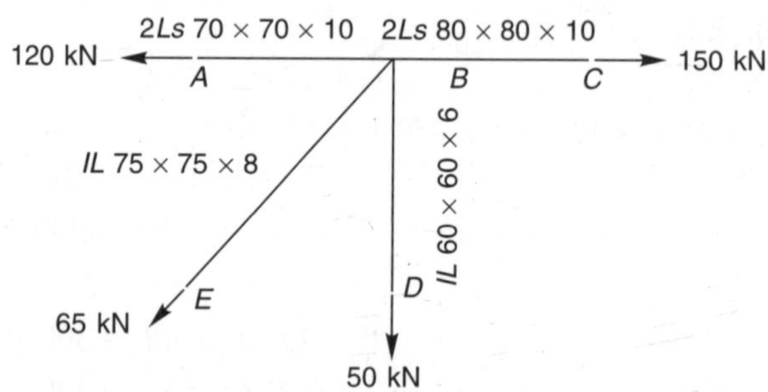

Fig. Ex. 2.6

Solution For power driven rivets

$$\tau_{vf} = 100 \text{ N/mm}^2, \quad \sigma_{pf} = 300 \text{ N/mm}^2$$

Using steel with f_y, 250 N/mm^2

$$\sigma_{at} = 0.6 \times f_y = 0.6 \times 250 = 150 \text{ N/mm}^2$$

Nominal diameter of rivet, ϕ = 20 mm

Gross diameter of rivet, d = 20 + 1.5 = 21.5 mm

Member AB

$$P = 120 \text{ kN}$$

It is built up of 2 I.S.A. 70 mm × 70 mm × 10 mm.
The rivets will be in double shear.

$$\text{Strength of rivet in shear} = 2 \times \frac{\pi}{4} d^2 \tau_{vf}$$

$$= 2 \times \frac{\pi}{4} \times (21.5)^2 \times 100 \times 10^{-3} = 72.61 \text{ kN}$$

$$\text{Strength of rivet in bearing} = dt\sigma_{pf}$$

$$= 21.5 \times 10 \times 300 \times 10^{-3} = 64.500 \text{ kN}$$

Hence, strength of rivet, $R_v = 64.5$ kN

$$\text{Number of rivets} = \frac{120}{64.5} = 1.86 \approx 2$$

Provide 2 rivets.

Member BC

$$P = 150 \text{ kN}$$

It is built up of 2 I.S.A. 80 mm × 80 mm × 10 mm.

The rivets will be in double shear.

$$\text{Strength of rivet in shear} = 72.61 \text{ kN}$$

$$\text{Strength of rivet in bearing} = 64.5 \text{ kN}$$

Hence, strength of rivet, $R_v = 64.5$ kN

$$\text{Number of rivets} = \frac{150}{64.5} = 2.325 \approx 3$$

Provide three rivets.

Member BD

$$P = 50 \text{ kN}$$

It is built up of single I.S.A. 60 mm × 60 mm × 6 mm.

The rivets will be in single shear.

$$\text{Strength of rivet in shear} = \frac{\pi}{4} d^2 \tau_{vf}$$

$$= \frac{\pi}{4} (21.5)^2 \times 100 \times 10^{-3} = 36.305 \text{ kN}$$

The thickness to be considered for bearing will be 6 mm.

$$\text{Strength of rivet in bearing} = dt\ \sigma_{pf}$$

$$= 21.5 \times 6 \times 300 \times 10^{-3} = 38.70 \text{ kN}$$

Hence, strength of rivet, $R_v = 36.305$ kN

$$\text{Number of rivets} = \frac{50}{36.305} = 1.377 \approx 2$$

Provide two rivets.

Member BE

$$P = 65 \text{ kN}$$

It is built up of single I.S.A. 75 mm × 75 mm × 8 mm.
The rivets will be in single shear.

$$\text{Strength of rivet in shear} = \frac{\pi}{4} d^2 \, \tau_{vf} = \frac{\pi}{4}(21.5)^2 \times 100 \times 10^{-3} = 36.305 \text{ kN}$$

The thickness to be considered for bearing will be 8 mm.

$$\text{Strength of rivet in bearing} = dt \, \sigma_{pf} = 21.5 \times 8 \times 300 \times 10^{-3} = 51.60 \text{ kN}$$

Hence, strength of rivet, $R_v = 36.305$ kN

$$\text{Number of rivets} = \frac{65}{36.305} = 1.790 \approx 2$$

Provide two rivets.

Example 2.7 A riveted boiler 2.5 m in diameter is made with a double riveted double cover butt joint. It is subjected to a fluid pressure of 2.0 N/mm^2. Design the longitudinal joints of the boiler. Assume $fy = 250$ N/mm^2.

Solution

$$\text{Pressure } p = 2.0 \text{ N/mm}^2$$

Assume the efficiency of longitudinal joint to be 70% and allowable stress to be 120 N/mm^2.

$$\text{Thickness of plate} = \frac{pd}{2\eta\sigma} = \frac{2 \times 2.5 \times 10^3}{2 \times 0.7 \times 120} = 29.76 \text{ mm} \approx 30 \text{ mm}$$

Provide 30 mm thick plates for making the boiler shell.

$$\text{Actual stress in plate, } \sigma_{\text{cal}} = \frac{pd}{2t\eta} = \frac{2.0 \times 2.5 \times 10^3}{2 \times 30 \times 0.7} = 119.04 \text{ N/mm}^2$$

$$< 120 \text{ N/mm}^2$$

which is safe.

$$\phi = 6.01\sqrt{t} = 6.01\sqrt{30} = 32.918 \text{ mm} \approx 30 \text{ mm}$$

Provide 30 mm ϕ power driven rivets.

$$\text{Gross diameter of rivets, } d = 30 + 2 = 32 \text{ mm}$$

The rivets will be in double shear as a double cover butt joint is provided. It is also double riveted; hence the strength per pitch length of rivet in shear and bearing is doubled.

$$\text{Strength of rivet in shear/pitch length} = 2 \times 2 \times \frac{\pi}{4} d^2 \, \tau_{vf}$$

$$= 2 \times 2 \times \frac{\pi}{4}(32)^2 \times 100 \times 10^{-3}$$

$$= 321.699 \text{ kN}$$

Strength of rivet in bearing/pitch length $= 2dt\sigma_{pf}$

$$= 2 \times 32 \times 30 \times 300 \times 10^{-3}$$

$$= 576.0 \text{ kN}$$

Hence, $R_v = 321.699$ kN

The rivets are staggered so that the section is weakened by one rivet hole only.

Tearing strength of plate/pitch length $= (s - d)t\sigma_{at} = (s - 32) \times 30 \times 150$

Equating the tearing strength of plate/pitch length to the strength of rivet/pitch length,

$$321.699 \times 10^3 = (s - 32) \times 30 \times 0.6 \times 250$$

or $s = 103.488 \approx 100$ mm

$$\nless 2.5\,\phi\ (= 2.5 \times 30 = 75 \text{ mm})$$

which is all right.

Tearing strength of plate/pitch length $= (100 - 32) \times 30 \times 0.6 \times 250 \times 10^{-3}$

$$= 306.0 \text{ kN}$$

Strength of solid plate/pitch length $= 100 \times 30 \times 0.6 \times 250 \times 10^{-3} = 450$ kN

$$\eta = \frac{306}{450} \times 100 = 68\%$$

Example 2.8 A cleat angle 150 mm × 15 mm × 10 mm is connected to a flange of column I.S.H.B. 450 @ 907.4 N/m through three rivets 16 mm in diameter. Find the maximum end reaction of the beam that the connection can safely carry. Assume the rivets to be power driven.

Solution From I.S. Handbook No. 1, the relevant property of I.S.H.B. 450 @ 907.4 N/m, $T = 13.7$ mm

Thickness of angle section = 10 mm

Gross diameter of the rivet = 16 + 1.5 = 17.5 mm

The rivets will be in single shear.

Shear value of one rivet $= \frac{\pi}{4} d^2\, \tau_{vf}$

$$= \frac{\pi}{4}(17.5)^2 \times 100 \times 10^{-3} = 24.05 \text{ kN}$$

Shear strength of the joint = 3 × 24.052 = 72.15 kN

Bearing value of one rivet $= dt\sigma_{pf}$

$$= 17.5 \times 10 \times 300 \times 10^{-3} = 52.5 \text{ kN}$$

Bearing strength of the joint = 3 × 52.5 = 157.5 kN

Hence, strength of the joint = 72.15 kN.

Therefore, the joint can transfer an end reaction of 72.15 kN.

Example 2.9 Two sections 10 mm and 18 mm thick are to be jointed by double cover butt joint. The joint is double riveted with cover plates each 8 mm thick. The load to be transferred by the joint is 500 kN. Design the joint and rivets on packings.

Solution Assume nominal diameter of rivets = 20 mm

Gross diameter = 20 + 1.5 = 21.5 mm

Assume the rivets to be power driven and steel of yield stress, f_y = 250 MPa.

$$\tau_{vf} = 100 \text{ N/mm}^2,\ \sigma_{pf} = 300 \text{ N/mm}^2,\ \sigma_{at} = 0.6 \times 250 = 150 \text{ N/mm}^2$$

The joint is double cover, hence the rivets are in double shear.

$$\text{Value of rivet in double shear} = 2\frac{\pi}{4}d^2\tau_{vf}$$

$$= 2 \times \frac{\pi}{4}(21.5)^2 \times 100 \times 10^{-3} = 72.610 \text{ kN}$$

$$\text{Value of rivet in bearing} = dt\sigma_{pf}$$

$$= 21.5 \times 10 \times 300 \times 10^{-3} = 64.50 \text{ kN}$$

Hence, the rivet value is 64.50 kN.

$$\text{The number of rivets required} = \frac{500}{64.50} = 7.75 \approx 8$$

Since the joint is double riveted, provide four rivets in each row.

$$\text{Strength of rivet/pitch length} = 2 \times 64.5 = 129 \text{ kN}$$

Equating it to the strength/pitch length of the plate in tearing,

$$129 \times 10^3 = (s - 21.5) \times 10 \times 150$$

or

$$s = 107.5 \text{ mm}$$

$$\approx 105 \text{ mm} \nless 2.5\,\phi\ (2.5 \times 20 = 50 \text{ mm})$$

Provide rivets at 105 mm c/c.

Rivets on Packing The two main plates to be jointed are of different thicknesses.

$$\text{The difference in thickness} = 18 - 10 = 8 \text{ mm} > 6 \text{ mm}$$

Therefore additional rivets will have to be provided on the packing

$$\text{Additional rivets} = \frac{2.5}{100} \times \frac{8}{2} \times 7.75 = 0.775$$

Provide 2 rivets 20 mm ϕ on the packing, Fig. Ex. 2.9.

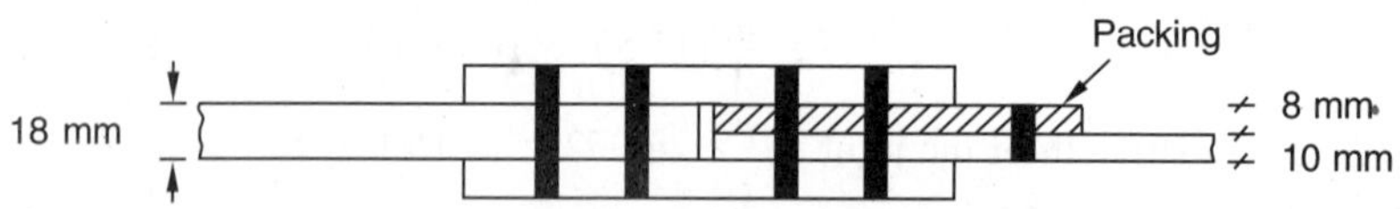

Fig. Ex. 2.9

Example 2.10 Calculate the efficiency of a zigzag double riveted lap joint to connect 10 mm thick plates with 20 mm ϕ power driven shop rivets as shown in Fig. Ex. 2.10 (f_y = 250 N/mm^2)

Solution Consider a 100 mm length of the joint shaded as shown which represents typical conditions. Two rivets will fall in each pitch length formed between lines *a-a* and *b-b*.

$$\phi = 20 \text{ mm},\ d = 20 + 1.5 = 21.5 \text{ mm}$$

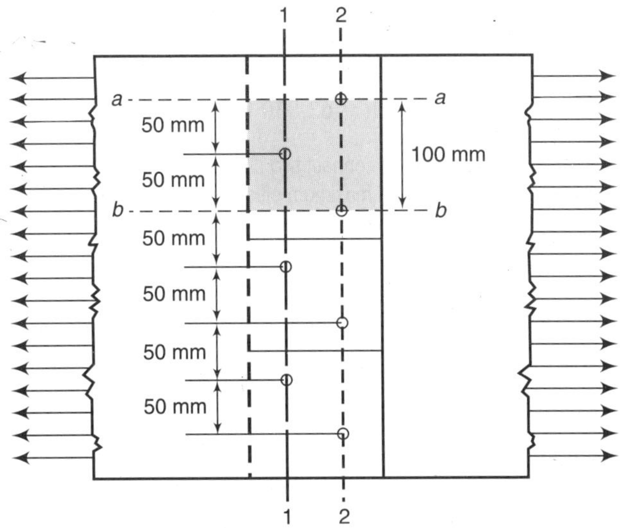

Fig. Ex. 2.10

Strength of rivet per pitch length in

$$\text{Single shear} = 2 \times \frac{\pi}{4} \times 21.5^2 \times 100 \times 10^{-3} = 72.61 \text{ kN}$$

$$\text{Bearing} = 2 \times 21.5 \times 10 \times 300 \times 10^{-3} = 129 \text{ kN}$$

Strength of plate in tearing = $(100 - 21.5) \times 10 \times 0.6 \times 250 \times 10^{-3} = 117.75$ kN

Strength of joint per pitch length = 72.61 kN

Strength of solid plate per pitch length = $100 \times 10 \times 0.6 \times 250 \times 10^{-3} = 150$ kN

$$\text{Efficiency of the joint} = \frac{72.61}{150} \times 100 = 48.41\%$$

Example 2.11 Determine the strength of a 20 mm diameter ordinary turned bolt if it is used to connect:

(a) Two plates each 10 mm thick
(b) Three plates, the outer two plates 8 mm thick each and the intermediate plate 14 mm thick.

$$\tau_{vf} = 100 \text{ N/mm}^2, \qquad \sigma_{pf} = 300 \text{ N/mm}^2$$

Solution

(a) Strength of bolt in single shear $= \frac{\pi}{4} d^2 \tau_{vf}$

$$= \frac{\pi}{4} (20)^2 \times 100 \times 10^{-3} = 31.415 \text{ kN}$$

Strength of bolt hole in bearing $= 20 \times 10 \times 300 \times 10^{-3} = 60$ kN

Hence, strength of the bolt = 31.415 kN.

(b) Strength of bolt in double shear $= 2 \times \frac{\pi}{4} d^2 \tau_{vf}$

$$= 2 \times \frac{\pi}{4} (20)^2 \times 100 \times 10^{-3} = 62.830 \text{ kN}$$

Strength of bolt hole in bearing $= dt\ \sigma_{pf}$

$$= 20 \times 14 \times 300 \times 10^{-3} = 84 \text{ kN}$$

Hence, strength of the bolt = 62.830 kN

Example 2.12 Design a butt joint to connect two plates 240 × 12 mm (f_y = 250 N/mm^2) using ordinary M20 bolts. Arrange the bolts to give maximum efficiency.

Solution Let us provide a double cover butt joint.

$$\text{Thickness of cover plate } = \frac{5}{8} \times 12 = 7.5 \text{ mm} = 8 \text{ mm}$$

The tensile force (T) the main plate can carry

$$= bt\ \sigma_{at} = 240 \times 12 \times 0.6 \times 250/10^3 = 432 \text{ kN}$$

$$\text{Shear strength of bolt} = 2\frac{\pi}{4} d^2 \tau_{vf}$$

$$= 2 \times \frac{\pi}{4} \times 20^2 \times 80/10^{-3} = 50.265 \text{ kN}$$

$$\text{Bearing strength of bolt} = dt\ \sigma_{pf}$$

$$= 20 \times 12 \times 250 \times 10^{-3} = 60 \text{ kN}$$

Hence, strength of the bolt = 50.265 kN

$$\text{Number of bolts required} = \frac{432}{50.265} = 8.594 \approx 9$$

Provide a diamond joint as shown in Fig. Ex. 2.12.

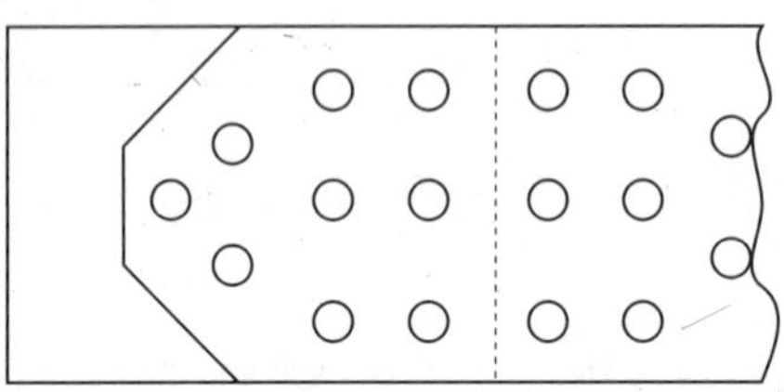

Fig. Ex. 2.12

Example 2.13 An ISA 100 × 100 × 10 mm is subjected to a tension of 66 kN. It is to be jointed to a 12 mm thick gusset plate. Design a high strength bolted joint.

Solution Design load = 66 kN

$$F = \frac{n\,\mu\,t}{F_s}$$

$$66 = \frac{1 \times 0.5 \times T}{1.4}$$

Proof load, T = 184.8 kN

The bolts are under single shear. Let us try High Strength M 14–8.8 G bolts. From Table 2.5, the proof load for the bolt is 68.8 kN.

$$\text{Number of bolts required} = \frac{184.8}{68.8} = 2.68 \approx 3$$

Use 3, High Strength M 14–8.8 G bolts.

Example 2.14 Design a bolted connection using HSFG bolts to connect flange of the bracket-Tee with the column flange as shown in Fig. Ex. 2.14. The double angle tie member is connected to the web of Tee bracket as shown in the Fig. Ex. 2.14 Assume a slip factor of 0.5.

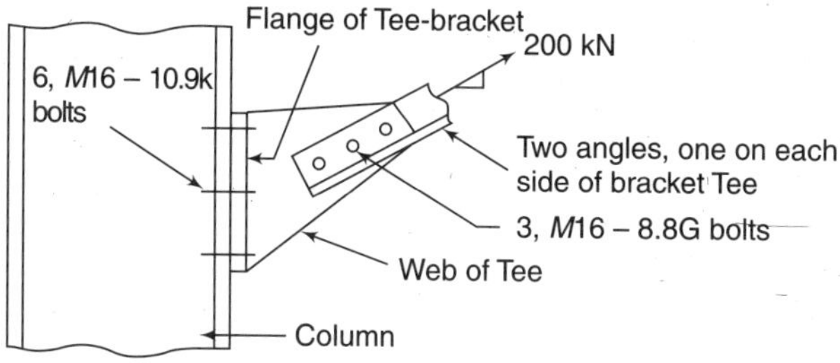

Fig. Ex. 2.14

Solution

Connection of tie member with Tee-bracket

There are two interfaces.

Member force = 200 kN

$$F = \frac{n\mu T}{F_s}$$

$$200 = \frac{2 \times 0.50 \times T}{1.4}$$

Proof load, $T = 280.0$ kN

Let us try M 16–8.8 G bolts

The proof load for the above bolt is 94.5 kN.

$$\text{Number of bolts required} = \frac{280.0}{94.5} = 2.96 \approx 3$$

Provide 3 M 16–8.8 G HSFG bolts for making the connection.

Connection of Tee-bracket with column flange

$$\text{Tension on the connection} = 200 \times \frac{4}{5} = 160 \text{ kN}$$

$$\text{Shear on the connection} = 200 \times \frac{3}{5} = 120 \text{ kN}$$

Let us try total 6 HSFG M 16–10.9 K bolts, three on each part of the Tee-flange on the two sides of the bracket web.

$$\text{Tension per bolt} = \frac{160}{6} = 26.66 \text{ kN}$$

$$\text{Shear per bolt} = \frac{120}{6} = 20 \text{ kN}$$

There is only one interface

$$\frac{\text{calculated shear}}{\text{slip factor} \times \text{number of interfaces}} = \frac{\text{proof load} - \text{calculated tension}}{\text{factor of safety}}$$

$$\frac{20}{0.50} = \frac{\text{proof load} - 26.66 \times 1.7}{1.4}$$

$$\text{Proof load} = 101.32 \text{ kN} \quad < 130 \text{ kN}$$

Provide 6 HSFG bolts M 16–10.9 K

Example 2.15 Design a pin for the tension link shown in Fig. Ex. 2.15. The plate sections (f_y = 250 MPa) are as shown.

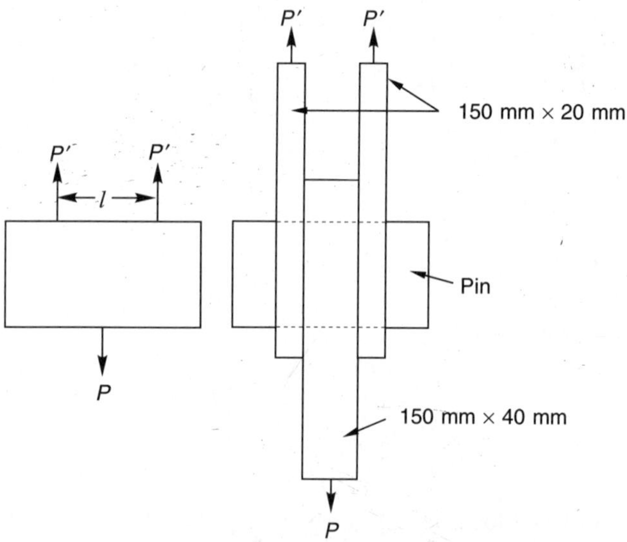

Fig. Ex. 2.15

Solution Pull transmitted by each 20 mm plate:

$$P' = \sigma_{at}\, bt = 0.6 \times 250 \times 150 \times 20 = 450 \times 10^3 \text{ N}$$

$$P = 0.6 \times 250 \times 150 \times 40 \times 10^3 = 900 \text{ kN}$$

$$\text{Maximum bending moment} = \frac{Pl}{4}$$

$$= \frac{900 \times 10^3 \times (40 + 10 + 10)}{4}$$

$$\therefore \qquad = \frac{900 \times 10^3 \times 60}{4} \text{ N mm}$$

$$d = \left(\frac{32\text{ M}}{\pi \sigma_b}\right)^{1/3} = \left(\frac{32 \times 900 \times 10^3 \times 60}{4 \times \pi \times 0.66 \times 250}\right)^{1/3}$$

$$= 94.10 \text{ mm} \approx 100 \text{ mm}$$

Provide a 100 mm diameter pin.

Check

(i) Average shear stress in pin,

$$\tau_{vf} = \frac{450 \times 10^3}{\frac{\pi}{4}(100)^2} = 57.29 \text{ N/mm}^2 < 100 \text{ N/mm}^2$$

which is as it should be.

(ii) Bearing stress in the pin,

$$\sigma_{pf} = \frac{450 \times 10^3}{100 \times 20} = 225 \text{ N/mm}^2 < 300 \text{ N/mm}^2$$

which is all right.

Example 2.16 Design a pin to connect two pairs of parallel 175 mm × 40 mm eye bars (f_y = 250 MPa) as shown in Fig. Ex. 2.16.

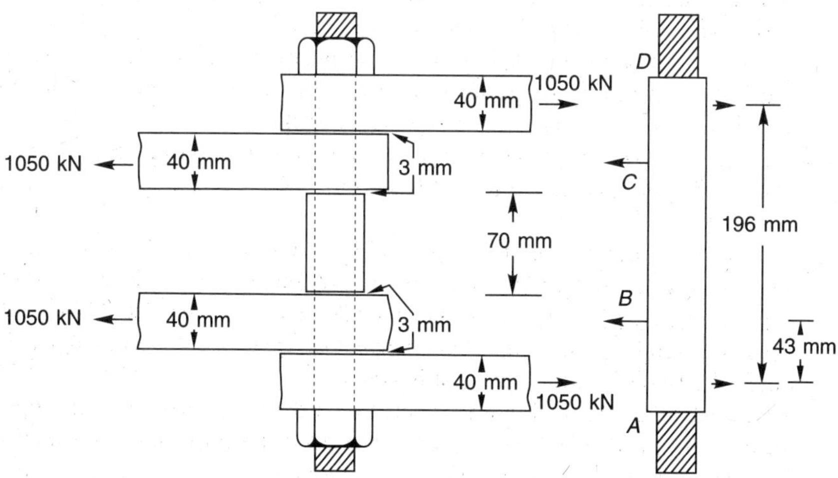

Fig. Ex. 2.16

Solution Maximum tension in eye bar = 175 × 40 × 150 × 10^{-3} = 1050 kN

Maximum bending moment will be at *B* and *C*.

Maximum bending moment = 1050 × 43 = 45,150 kN mm

$$d = \left(\frac{32 \times M}{\pi \times \sigma_{bt}}\right)^{1/3} = \left(\frac{32 \times 45{,}150 \times 10^3}{\pi \times 0.66 \times 250}\right)^{1/3}$$

$$= 140.73 \text{ mm} \approx 150 \text{ mm}$$

Check in shear $\tau_{vf} = \dfrac{1050 \times 10^3}{\frac{\pi}{4} \times 150^2} = 59.4178 \text{ N/mm}^2 < 100 \text{ N/mm}^2$

which is all right.

Check in bearing

$$\sigma_{pf} = \frac{1050 \times 10^3}{150 \times 40} = 175 \text{ N/mm}^2 < 300 \text{ N/mm}^2$$

which is as it should be.

Exercises

Note Use the steel of yield stress 250 MPa if not specified in the problem.

2.1 Compute the value of rivet, 22 mm in diameter, used to connect two plates 12 mm thick in the following cases:
(a) Power driven rivets
(b) Hand driven rivets

2.2 Determine the safe load and the efficiency of a double cover butt joint. The main plates are 14 mm thick connected by 18 mm diameter rivets at a pitch of 100 mm. Design the cover plates also. What is the percentage reduction in the efficiency of the joint if the plates are lap jointed?

2.3 A water tank is made with 10.0 mm thick plates. The plates are jointed by a lap jointed using 18 mm diameter rivets at a pitch of 60 mm. Find the efficiency of the joint if the rivets are power driven.

2.4 Design a riveted joint to connect an angle section tension member 80 mm × 80 mm × 8 mm to a gusset plate 12 mm thick. The member is to carry a load of 100 kN.

2.5 A tension member consists of an I.S.M.C. 200 @ 216.8 N/m. It is to be connected to a gusset plate. Design riveted connections to develop the full tensile strength of the member. Assume yield stress of steel 230 MPa.

2.6 Design a double cover butt joint to connect two plates each 12 mm thick. The load to be transferred by the joint is 400 kN.

2.7 Two plates 12 mm thick are connected by a double riveted lap joint. Power driven 16 mm diameter rivets at 60 mm pitch are provided. Find the load carrying capacity and the efficiency of the joint.

2.8 Design the joints of members of a roof truss with the gusset plate 12 mm thick as shown in Fig. Prob. 2.8.

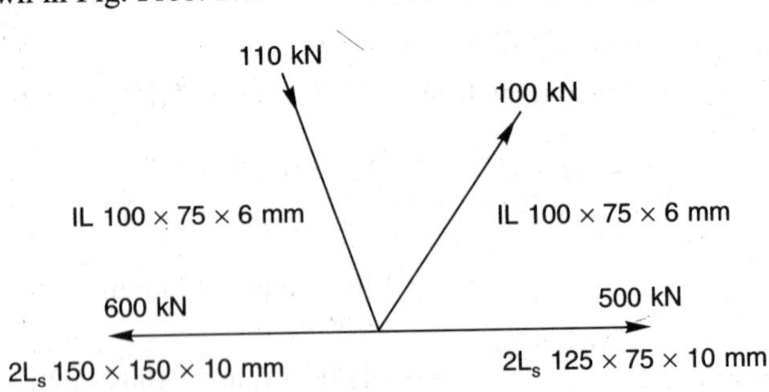

Fig. Prob. 2.8

2.9 A riveted boiler 2400 mm in diameter is subjected to a fluid pressure of 2 N/mm². Design the longitudinal joints of the boiler.

2.10 Design the foundation bolts for a steel chimney. The uplift on the wind-ward side of the 4.0 m diameter chimney is 358 N/mm on its periphery at the base.

2.11 Design the foundation bolts of a gusset base foundation, subjected to a tension of 70 kN.

2.12 A column ISHB 200 @ 392.4 N/m carries a axial compressive load of 380 kN. Design the connections of the gusset material with the column section using HSFG bolts.

2.13 Design a lap joint to connect two plates 16 mm thick using high strength friction grip bolts.

2.14 Design a pin connection to connect two pairs of parallel eye bars 200 mm × 50 mm. The inner pair of eye bars is spaced 80 mm apart.

Simple Connections—Welded Connections

3.1 INTRODUCTION

When two structural members are jointed by means of welds the connection is called a welded connection. A few decades ago designers had a feeling that welded connections were less fatigue resistant and that a good quality welded connection could not be made. These negative feelings had a great impact on the use of welding in structures. But the progress made in welding equipment and electrodes, the advancing art and science of designing for welding, and the increasing trust and acceptance of welding have combined to make it a powerful implement for the expanding construction industry.

The economics inherent in welding are helping to offset increase in the prices of material and cost of labour. In addition, the shortened production cycles made possible by welding, have helped effect a quickening in the pace of new construction. Welding will become increasingly important as more people acquire a greater depth of knowledge and experience that goes with it. Today most of the regulatory agencies and government departments accept welded joints.

There are a number of reasons for using a welded design, but a few basic ones are:

1. Welded designs offer the opportunity to achieve a more efficient use of materials. Welding is the only process that produces a one piece construction.
2. The speed of fabrication and erection helps compress production schedules.

Welding permits architects and structural engineers complete freedom of design. The usage of outstanding design advancements such as open-web expanded (castellated) beams, tapered beams, vierendeel trusses, cellular floor construction, tubular column and trusses are a few examples of welded constructions.

A properly welded joint is stronger than the material jointed. Fused joints create a rigid structure in contrast to the non-rigid structures made using other type of joints. The compactness and greater rigidity of welded joints permits design assumptions to be realised more accurately. Welded joints are better for fatigue loads, impact loads and vibrations.

Welding saves weight and consequently cuts costs. Connecting steel plates are reduced or eliminated since they often are not required. No deductions are there for holes; thus the gross section is effective in carrying loads.

Welding offers the best method for achieving a rigid connection, resulting in reduced beam depth and weight. Thus it noticeably lowers the overall height of a building. The weight of the structure and consequently the static loading is considerably reduced. This saves column steel and reduces foundation requirements. Saving in transportation, handling time and erection is proportional to the weight savings.

3.2 TYPES

The basic types of welding joints can be classified depending upon the *type of weld*, e.g., fillet weld, butt or groove weld, plug weld, slot weld, spot weld, etc., *position*, e.g., flat weld, horizontal weld, vertical weld and overhead weld, etc., and *type of joint*, e.g., butt weld, lap weld, tee weld and corner weld. The various types of welds are shown in Fig. 3.1.

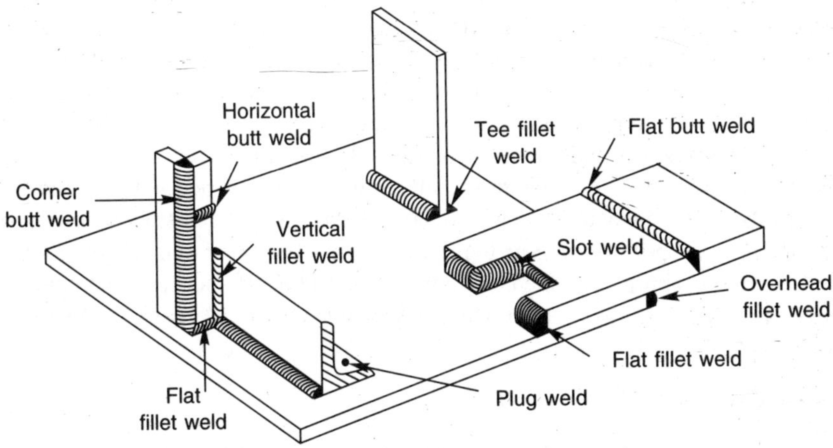

Fig. 3.1 *Types and Positions of Welds*

Fillet and butt welds are most frequently used. Butt welds, are provided when the members to be jointed are lined up. Various types of butt welds are shown in Fig. 3.2. Single V, U, J, etc. are cheaper to form, but require double the weld metal than double grooved joints. The choice between single and double grooves is usually a question of whether the higher cost of preparation is offset by the saving in weld metal. Fillet welds are provided when two members to be jointed are in different planes. This situation is frequently met within structures. Therefore, fillet welds are more common than butt welds. Various types of fillet welds are shown in Fig. 3.3.

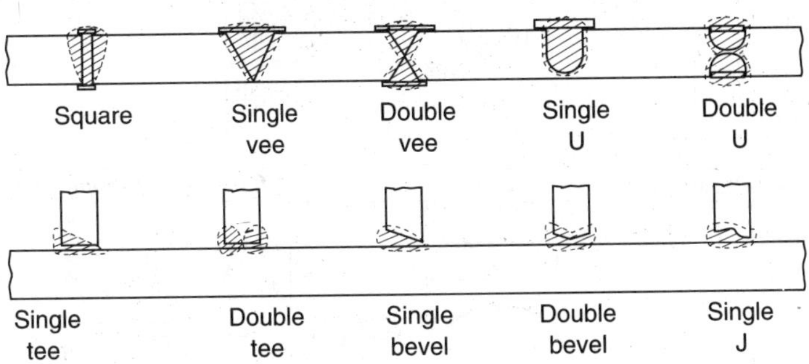

Fig. 3.2 *Types of Butt Welds*

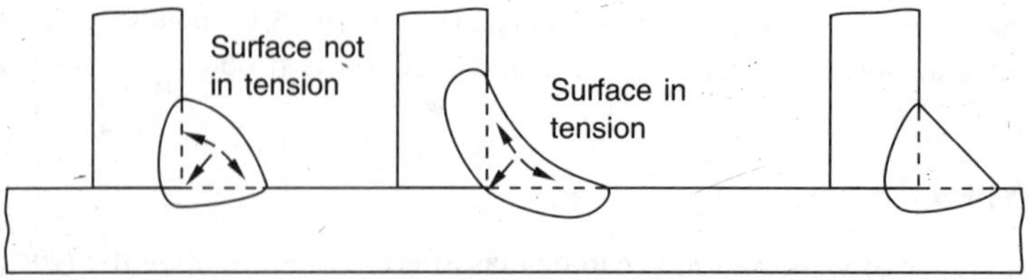

Fig. 3.3 *Types of Fillet Weld*

Fillet welds are normally easier to make, require less material preparation, and are easier to fit than butt welds. On the other hand, for a given amount of weld material, they are not as strong and they cause greater concentration of stress. In lightly stressed structures where stiffness rather than strength controls design and fatigue or brittle fracture is not a problem, fillet welds are entirely adequate and generally more economical. Reliable fillet welds may be, and frequently are, designed for severe stress and service conditions, but, as a rule, butt welds are better in highly stressed structures where smooth flow of stress is necessary. If the butt joint has the same characteristics as the parent metal, is finished smooth with it on both sides (by proper grinding or the like), and has complete penetration with no unweld zones, it approaches the condition of no joint at all and, for most common types of parent metal, may have impact and fatigue resistance equal or superior to that of the base metal.

3.3 SYMBOLS

A knowledge of welding symbols is essential for a site engineer to be able to read the drawings. Symbols save a lot of space as descriptive notes can be omitted.

Table 3.1 depicts the symbols and the method of their representation on the drawings. The drawings need to indicate the side of welding, size, contour and finish, spacing and whether it is field or shop weld. Some examples of symbols and dimensions of welds are shown in Table 3.2 and 3.3, respectively.

Table 3.1 *Description of Type of Weld and its Representation*

Type of weld												
Fillet	*Butt*								*Seam*	*Spot*	*Plug*	*Field weld*
	Square	V	Bevel	U	J	V with broad root face	Bevel with broad root face	Weld with raised edges				

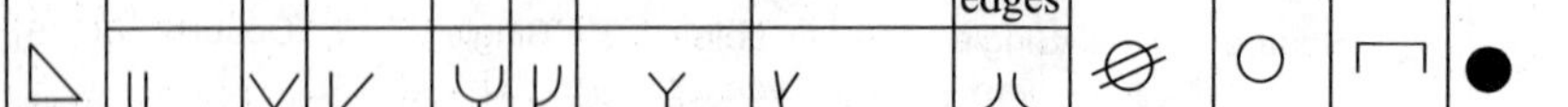

Shape of weld surface	*Symbol*	*Method of representation*
Flat (usually finished flush)	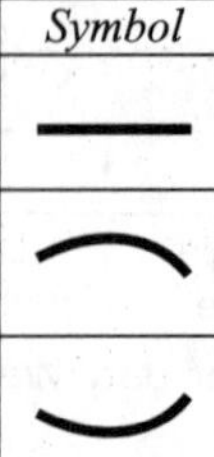	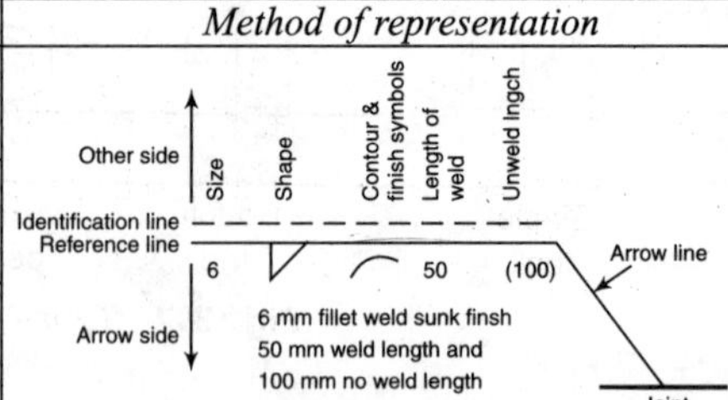
Convex		
Concave		

Notes

1. The position of the arrow line with respect to the weld is generally not significant.
2. The reference line should be a straight line preferably drawn parallel to the bottom edge of the drawing.
3. The symbol is placed on the continuous line side of the reference line if the weld is on the arrow side of the joint.
4. The symbol is placed on the broken line side if the weld is on the other side of the joint.

Table 3.2 *Examples of Use of Main and Supplementary Symbols for Welds*

S. No.	*Designation symbol*	*Illustration*	*Symbolization*
1	Single-V butt weld 	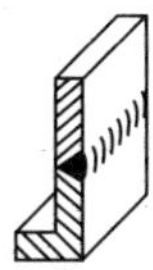	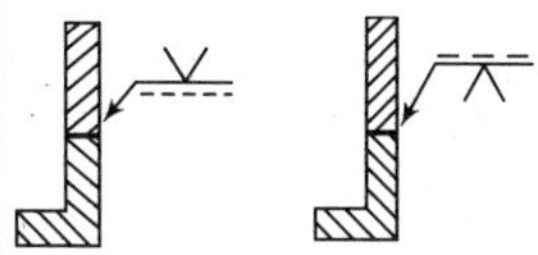
2		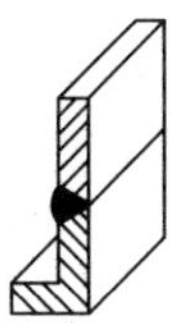	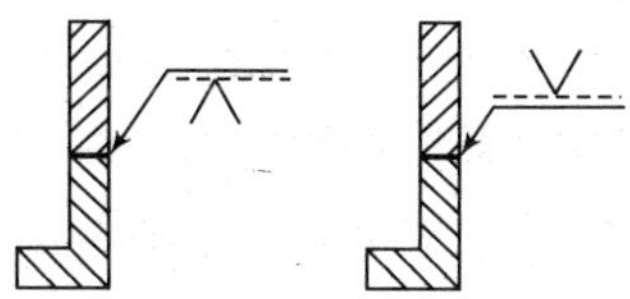
3	Double V butt weld 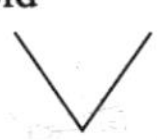	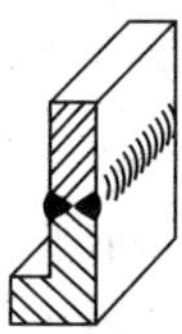	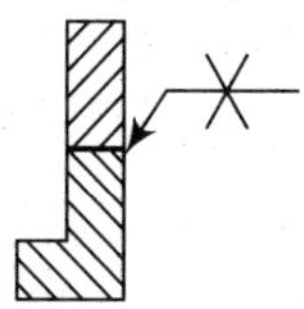
4	Concave fillet weld 	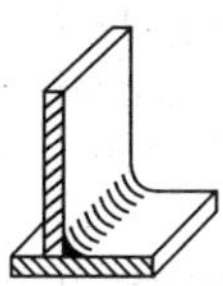	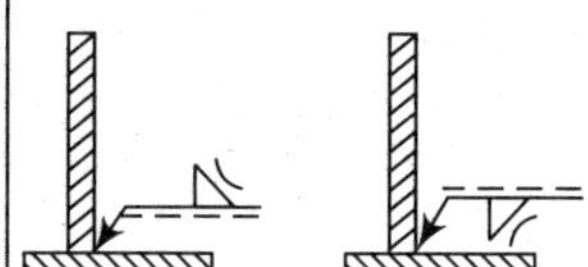

Table 3.3 *Dimensions of Welds*

S. No.	*Designation welds*	*Definition*		*Inscription*
1	Butt weld	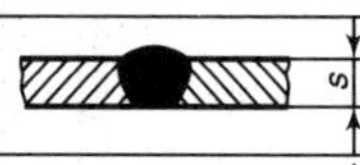	*S*: minimum distance from the surface of the part to the bottom of the penetration, which cannot be greater than the thickness of the thinner part	V
				s ||
				s Y

(*Contd.*)

Table 3.3 *(Contd.)*

S. No.	*Designation welds*	*Definition*		*Inscription*
2	Continuous fillet weld	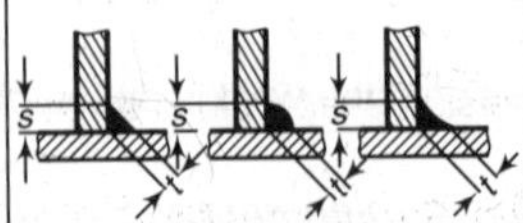	*t*: height of the largest isosceles triangle that can be inscribed in the section *s*: side of the largest isosceles triangle that can be inscribed in the section	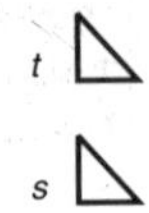
3	Intermittent fillet weld	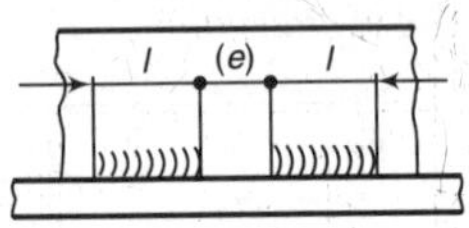	*l*: length of the weld (Without end crates). (*e*): distance between adjacent weld elements. *n*: number of weld elements *t*, *s* } (Same as in 2)	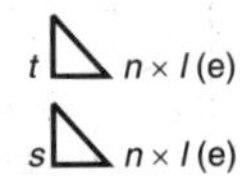
4	Staggered intermittent fillet weld	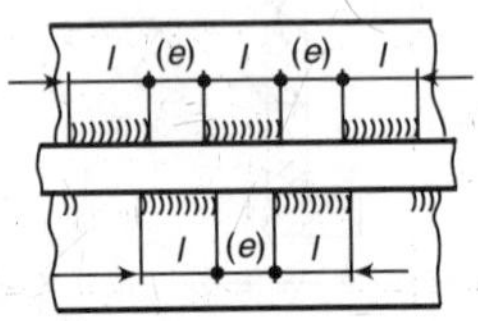	*l*, (*e*), *n* } (Same as in 3) *t*, *s* } (Same as in 2)	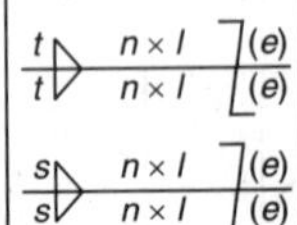

3.4 WELDING PROCESS

Welding consists of joining two steel sections by establishing a metallurgical bond between them through the application of pressure and/or through fusion. The most commonly used process is arc welding—a fusion process. The bond between the metals is produced by reducing the surfaces to be joined to a molten state and then allowing the molten metal to solidify. When the molten metal solidifies, union is completed.

In the arc welding process (Fig. 3.4), the intense heat required to reduce the metal to liquid state is produced by an electric arc. The arc produces a temperature of about 3600 °C between the section to be welded and the electrode. The tremendous

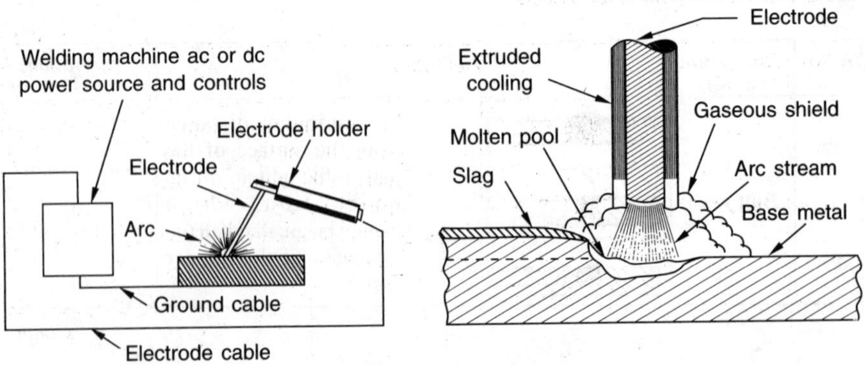

Fig. 3.4 *Metal Arc Welding*

heat at the tip of the electrode melts filler metal and base metal, thus liquifying them in a common pool called a *crater*. As the areas solidify, the metals are joined into one solid homogeneous piece. By moving the electrode along the joint to be welded, the surfaces to be jointed are welded together along their length. In all modern arc welding processes, the arc is shielded for the following reasons:

1. To protect the molten metal from air, either with gas vapour or slag.
2. To add alloying or fluxing ingredients
3. To control the melting of electrode for more effective use of the arc energy.

3.5 WELD DEFECTS

Good welding techniques, standard electrodes and proper joint preparation are the basic tools to achieve a sound weld. However, defects are inevitable and a knowledge of these is essential to minimise them. Some of the common defects in the welds are discussed below and are shown in Fig. 3.5.

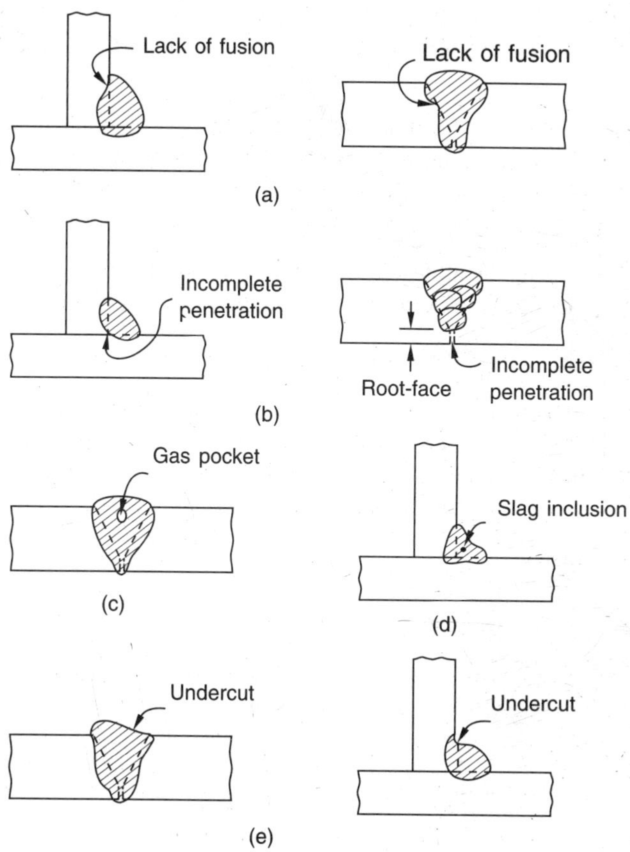

Fig. 3.5 *Weld Defects*

Incomplete Fusion is the failure of the base metal to get completely fused with the weld metal. It is caused by rapid welding and foreign materials on the surfaces to be welded.

Incomplete Penetration is the failure of the weld metal to penetrate the complete depth of joint. It is normally found with single vee and bevel joints and also because of large size electrodes.

Porosity occurs due to voids or gas pockets entrapped in the weld while cooling. It results in stress concentration and reduced ductility of the metal. Normally porosity is not a problem because each void is spherical and not a notch. Even with a slight loss in section because of the voids, their spherical shape may be considered to allow a smooth flow of stress around the void without any measurable loss in strength. Mainly these are caused because of careless use of back up plates, presence of moisture in the electrodes, hydrogen in the steel and excessive current.

Slag Inclusions are metallic oxides and other solid compounds which are sometimes found as elongated or globular inclusions. Being lighter than the molten material these float and rise to the weld surface from where these are removed after cooling of the weld. However, excessive rapid cooling of the weld may cause them to be trapped inside the weld. These present a problem in vertical and overhead welding.

Cracks are divided as hot and cold. Hot cracks occur due to the presence of sulphur, carbon, silicon and hydrogen in the weld metal. Phosphorous and hydrogen trapped in the hollow spaces of the metal structure give rise to the formation of cold cracks. Preheating of the metal to be welded eliminates the formation of cracks.

Undercutting is the local decrease of the thickness of the parent metal at the weld toe. This is caused by the use of excessive current or a very long arc.

3.6 PERMISSIBLE STRESSES

The permissible stresses in joining members of mild steel by welds conforming to I.S: 226–1975 and electrodes conforming to I.S: 814–2004 are given in Table 3.4. The stresses in M.K.S. units are recommended in I.S: 816–1969 (reaffirmed 1998) which is still not revised. The stress values are converted to SI units and presented in Table 3.4.

Table 3.4 *Stress in Welds*

S. No.	*Type of Stress*	*Permissible Stress*
1.	Tension or compression on section through throat of butt weld	150 N/mm^2
2.	Bending stress in tension or compression	165 N/mm^2
3.	Shear on section through throat of butt or fillet weld	108 N/mm^2
4.	Plug weld	108 N/mm^2

Notes

(i) The permissible stresses in shear and tension are reduced to 80% when the welding is done in the field.

(ii) When the effects of wind or earthquake are considered, the permissible stresses are increased by 25%.

(iii) For combination of stresses clause 7.5.2 and 7.5.3 of I.S: 816–1969 (reaffirmed 1998) may be referred to.

3.7 DESIGN OF BUTT WELDS

A butt weld is shown in Fig. 3.6. A square butt weld is provided for sections less than 8 mm. Above this a single U, Vee or double U, Vee, etc. are provided. Butt weld is usually designed for direct tension or compression but if shear is also there then due provision should be made.

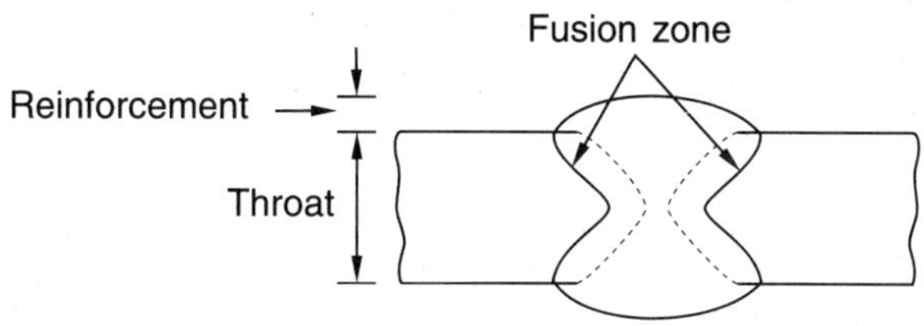

Fig. 3.6 *Butt Weld*

Since a butt weld involves no abrupt change in section at the joint, it is the most suitable form of the weld for transmitting alternating stresses.

Specifications

(i) Reinforcement Reinforcement is the extra weld metal which makes the throat dimension at least 10% greater than the thickness of the welded material. The reason for providing reinforcement is to increase the efficiency of the joint. The provision of reinforcement ensures that the depth through the weld is at least equal to the thickness of the plate and in case of thick plates, the metal within the thickness of the plate is in the annealed condition. However, any reinforcement on the weld is ignored in calculating the stresses. It is also difficult for a welder to make the weld flush with the parent metal; so extra metal is deposited.

Reinforcement makes the butt weld stronger for static loads. But in case of repeated and vibrating loads stress concentration develops in the reinforcement, leading to early failure. Hence, under such type of loads it is undesirable and the weld surface is made flush. Where a flush surface is required, the butt weld is first reinforced and then dressed flush. Subsequent removal of reinforcement is not considered as reducing the strength of the joint. In any case, the reinforcement should not exceed 3 mm.

(ii) Size The size of the butt weld used is specified by the throat dimension. This is also called effective throat thickness. In case of complete penetration of the butt weld the effective throat thickness is taken as the thickness of the thinner member jointed. In case the full penetration of butt weld cannot be achieved, an effective throat thickness of 7/8th of the thickness of the thinner member should be used. But for calculating the stress, the effective throat thickness is assumed to be 5/8th of thickness of the thinner member.

Except for the square edge type made in relatively thin metals, complete penetration welds generally are welded from both the sides. Incomplete penetration butt welds, usually are of the single V or single bevel type. In fact, incomplete penetration is the failure of the base metal and the weld metal to fuse at the root (Fig. 3.5(b)). This defect may be due to faulty design of the groove such as excessive root-face dimension, root gap or groove angle, or it may be due to faulty technique such as the use of excessively large size electrodes, excessive speed or insufficient current. Incomplete penetration is particularly undesirable since it causes stress concentration under load and may cause shrinkage crack.

(iii) Effective Area The effective area of the butt weld is the product of effective throat thickness and the effective length of the butt weld. Effective length is the length of the weld for which the required size of the weld is done.

Design Procedure

1. In case of complete penetration of the butt weld, design calculations are not required as the weld strength at the joint is equal to the strength of the member connected.
2. In case of incomplete penetration of the butt weld, the effective throat thickness is computed and the required effective length is determined to furnish the strength equal to the strength of the members connected.

Let P = strength of the butt weld in Newtons
σ_{tf} = permissible stress in tension or compression in N/mm^2
L = effective length of the weld
t = effective throat thickness

then
$$P = Lt\sigma_{tf} \tag{3.1}$$

3.8 DESIGN OF FILLET WELDS

The fillet weld is done for members which overlap each other. For such joints the critical stress is shear stress. They are also subjected to direct stresses but these are not of much importance. To all external appearances, the concave fillet weld seems to be larger than the convex weld. However, a check of the cross-section may show the concave weld to have less penetration and a smaller throat (Fig. 3.3) than first thought; therefore the convex fillet weld may actually be stronger even though it may have less deposited metal. Originally a concave fillet weld was favoured because it seemed to offer a smoother path for the flow of stress. But experience has shown that single pass fillet welds of this shape have a greater tendency to crack upon cooling, which outweighs the effect of improved stress distribution.

When a concave fillet weld cools and shrinks, its outer face is stressed in tension (Fig. 3.3), whereas a convex fillet weld on cooling shrinks and stresses the outer face in compression. Therefore in concave fillet welds cracks are developed on cooling. However, when concave welds are desired they are made in two or more passes—the first slightly convex, and the other passes are built up to form a concave fillet weld. Concave fillet welds are most suitable under alternating stresses.

Specifications

(i) Size (s) The size of a fillet weld is specified as the minimum leg length of the weld. The leg length is the distance from the root to the toe of the fillet weld (Fig. 3.7(a)). It is measured by the largest right angle triangle which can be inscribed within the weld (Fig. 3.7(a)). This definition would allow an unequal-legged fillet weld (Fig. 3.7(b)). Another definition stipulates the largest inscribed isosceles

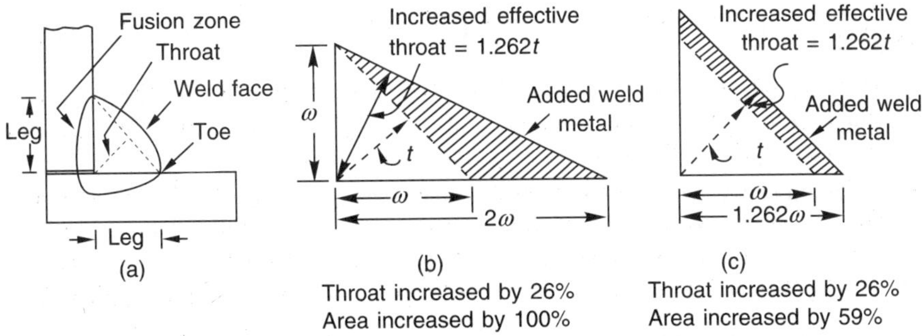

Fig. 3.7 *Leg Length of Fillet Weld*

right triangle and would limit this to an equal-legged fillet weld (Fig. 3.7(c)). Equal-legged fillet welds are preferred since they are easy to make. Unequal-legged fillet welds are sometimes used to get additional throat area, hence strength, when the vertical leg of the weld cannot be increased. An example of this is the attachment of a channel (shear connector) over a beam flange, shown in Fig. 3.8. Here the vertical leg of the fillet weld must be held to the thickness at the outer edge of the channel flange. Additional strength must be obtained by increasing the horizontal leg of the fillet. The weld size for unequal-legged fillet weld is specified by both leg lengths.

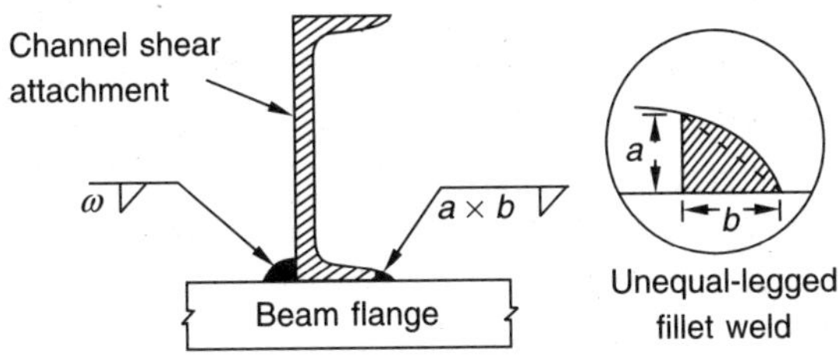

Fig. 3.8 *Welding of Channel Shear Connector*

The maximum size of a fillet weld is obtained by subtracting 1.5 mm from the thickness of the thinner member to be jointed. This specification limits the size of the fillet weld so that total strength may be developed without overstressing the adjacent metal.

A small weld placed on a thick member is undesirable. The heat generated in depositing the small size weld may not be appreciably enough to expand the base metal. Consequently, as the weld cools, it contracts and is prevented from doing so by the stable base metal. This results in initial stresses in the weld metal. Also the welding process will not heat the heavy plate much beyond the immediate vicinity of the weld. The mass of the thick plates will cool the weld too rapidly and tend to make it brittle. To help control this situation, I.S specifies the minimum size of a fillet weld as given in Table 3.5. The minimum size of the weld should not exceed the thickness of the thinner part jointed. In case of welds applied to the round toe of steel sections the maximum size of the weld should not exceed 3/4 of the thickness of the section at the toe.

Usually a weld size closer to the minimum size is selected for the reasons stated as follows:

1. Large size welds require more than one run of welding which means that after the first run of welding chipping and cleaning will be required for proper bond of successive weld runs. This increases the cost.
2. A smaller size weld will be cheaper than a large one for the same strength considering the volume of welding. For example, a 300 mm, 5 mm size fillet weld will have the same strength (115.5 kN) as a 150 mm long and 10 mm size fillet weld. However the volume of weld metal for a 10 mm weld will be $\left[\frac{1}{2}\times 10^2 \times 150 = 7500 \text{ mm}^3\right]$ which is about twice that of 5 mm size weld $\left[\frac{1}{2}\times 5^2 \times 300 = 3750 \text{ mm}^3\right]$.

Table 3.5 *Minimum Size of Weld*

Thickness of thicker member over (mm)	*up to and including (mm)*	*Minimum size (mm)*
0	10	3
10	20	5
20	32	6
32	50	8 first run 10

(ii) Effective Throat Thickness is the shortest distance from the root of the fillet weld to the face of the diagrammatic weld (line joining the toes) as shown in Fig. 3.7(a). The effective throat thickness should not be less than 3 mm.

$$\text{Effective throat thickness} = K \times \text{size of weld} = KS$$

where S is the size of weld in mm and K is a constant. The value of K depends upon the angle between the fusion faces and is given in Table 3.6.

Table 3.6 *Values of Constant K for Different Angles between Fusion Faces*

Angle between fusion faces	*60° – 90°*	*91° – 100°*	*101° – 106°*	*107° – 113°*	*114° – 120°*
K	0.70	0.65	0.60	0.55	0.50

Note Fillet weld is not recommended if the angle between fusion faces is less than 60° or more than 120°.

(iii) Effective Length is the length of the fillet weld for which the specified size and throat thickness of weld exist. It is taken equal to its overall length minus twice the weld size. The deduction is made to allow for craters to be formed at the ends of the welded length. End returns as shown in Fig. 3.9 are made equal to twice the size of the weld to relieve the weld length from high stress concentrations at their ends.

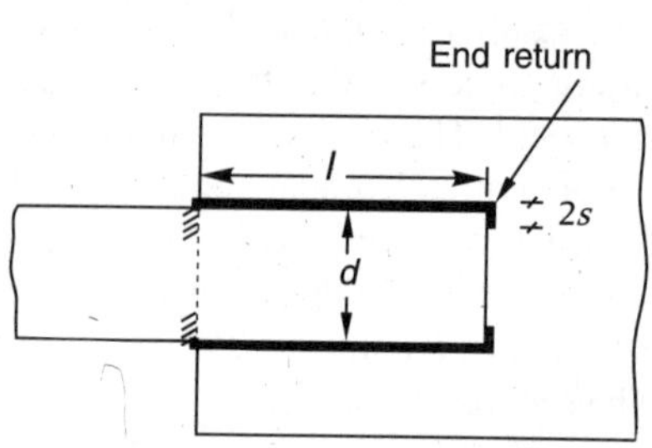

Fig. 3.9 *End Returns*

If the welds are parallel with the lines of stress, as in the case of longitudinal fillet welds, and are placed at the edges of the plates, there is a serious concentration of stress at the edges of the plate. Therefore, the length of the longitudinal (side) fillets on the flat should be not less than the width of the flat. The unevenness of the stress distribution is accentuated as the width of the plate increases. For this reason the perpendicular distance between longitudinal fillet welds is limited to 16 times thickness of the thinner plate jointed. If the plate is wider than this limit, slot or plug welds must be introduced. As a rule the provision of a slot tends to improve the distribution of stress in the plates as shown in Fig. 3.10. Longitudinal fillet welds in slots have the same strength as ordinary longitudinal fillet welds.

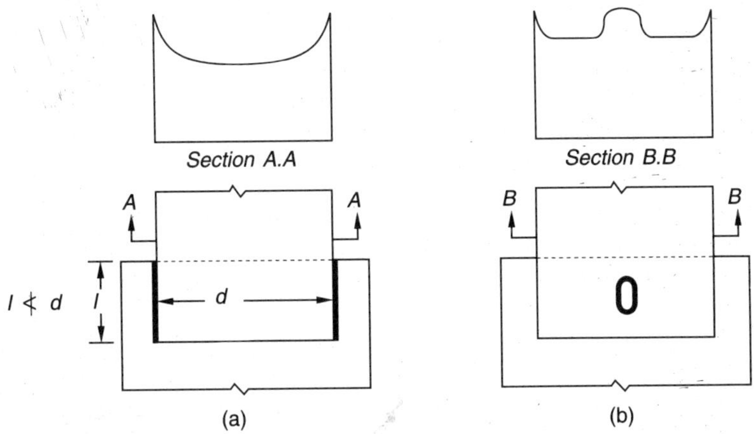

Fig. 3.10 *Distribution of Stress in Side-Welded Tension Bar with and without Slot*

Slots if provided should be set back behind the beginning of the welds on the edges of the plate to ensure that the effective area of plate is not reduced and that the deposition of slots does not lead to serious concentration of stress within the plates. On first sight the joint as shown in Fig. 3.11(b) might appear to be superior in strength to the one in Fig. 3.11(a), but it fails at smaller loads as shown in Fig. 3.11(c) owing to the stress concentration in the plate in the re-entrant angles of the notches.

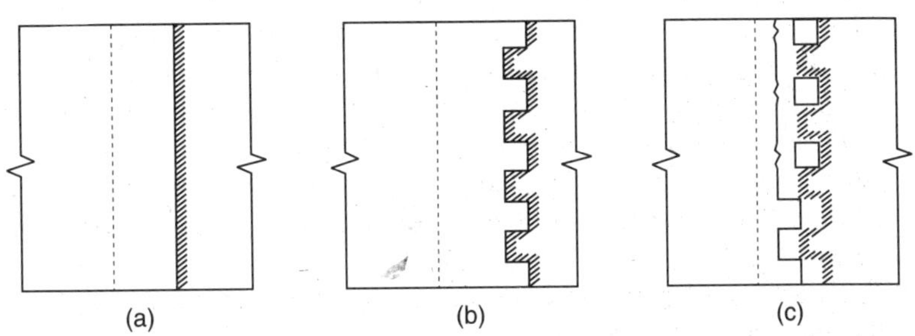

Fig. 3.11 *Unsatisfactory Method of Increasing Length of Fillet Welds by Means of Square Slots*

(iv) Effective Area The effective area of a fillet weld is equal to the effective length of the weld multiplied by the effective throat thickness.

(v) Overlap The overlap of plates to be fillet welded in a lap joint should not be less than 5 times the thickness of the thinner part.

(vi) Transverse Spacing between longitudinal fillet welds should not be more than $32t$ or 300 mm whichever is lesser.

Design

The following assumptions are made in the analysis of welded joints.

1. The welds connecting the various parts are homogeneous, isotropic and elastic elements.
2. The parts connected by the weld are rigid and their deformations are therefore, neglected.
3. Only stresses due to external loads are considered. Effects of residual stresses, stress concentrations and shape of the welds are neglected.

Design Procedure A fillet weld may be subjected to direct, bending and shear stresses but since the permissible stress in shear is always less, it controls the design.

1. The size (S) of the weld is assumed based on the thickness of the members to be jointed.
2. A fillet weld fails by shear at an angle of about 45° through the throat. Its strength is therefore assumed to equal the allowable shearing stress times the theoretical throat area of the weld. The strength of the weld is given by

$$P = Lt\tau_{vf}$$

or

$$P = LKS\,\tau_{vf} \tag{3.2}$$

where L = effective length of the weld in mm
t = throat thickness in mm
S = size of weld in mm
τ_{vf} = shear stress in the weld in N/mm^2
P = strength in newtons.

The strength of the weld/millimetre is calculated from Eq. (3.2).

3. Pull or thrust to be transmitted by the connections is estimated.
4. Effective length of the weld is calculated by dividing the pull or thrust by the strength of weld/mm. The length can either be adjusted as longitudinal fillet welds (parallel to the load axis) or as transverse fillet welds (perpendicular to the load axis) along with longitudinal fillet welds. It is a common practice to treat both the welds as if they are stressed equally.
5. If only a longitudinal fillet weld is made, a check is provided to see that the length of each longitudinal fillet weld is more than the perpendicular distance between them.
6. End returns of length twice the size of the weld are provided at each end of the longitudinal fillet weld.

Notes

1. It is assumed that the strength of the longitudinal and tranverse fillet welds, as shown in Fig. 3.12, is same. Actually the strength of the transverse fillet weld is about 30%

more than the longitudinal fillet weld, because a transverse fillet weld is stressed more uniformly for full length whereas a longitudinal fillet weld is stressed non-uniformly due to varying deformations along the weld length. Another reason for the greater strength of the transverse fillet weld is given by tests which show that failure occurs at an angle other than 45°, giving transverse fillet welds a larger effective throat area.

2. The main reasons for neglecting the greater strength of transverse fillet welds are probably an interest in simplifying design and the fact that in most cases little would be saved by differentiating between transverse and longitudinal welds. Many joints contain combinations of longitudinal, transverse, and oblique welds. Any attempt at a thorough analysis would become complicated and difficult to justify as a design measure.

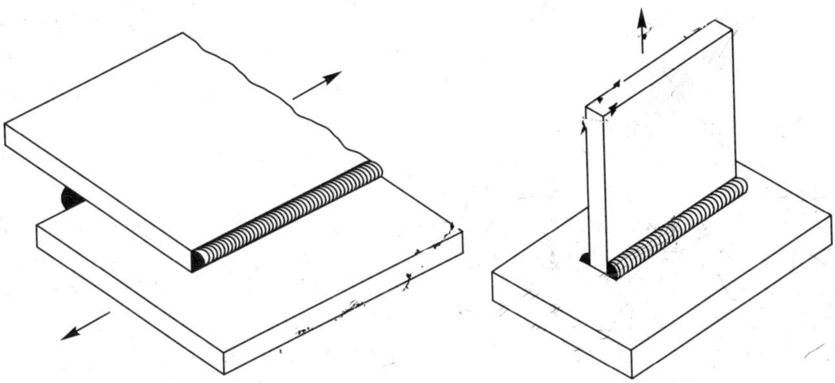

Longitudinal Fillet Weld Transverse Fillet Weld

Fig. 3.12

3.9 DESIGN OF INTERMITTENT FILLET WELDS

Intermittent fillet welds are provided to transfer calculated stress across a joint when the strength required is less than that developed by a continuous fillet weld of the smallest practical size, e.g. in case of connections of stiffeners to the web of plate girders. The fillet weld length required is computed as a continuous fillet weld and a chain of intermittent fillet welds of total length equal to the computed length, with spacing as per I.S. specifications is provided, as shown in Fig. 3.13. Intermittent fillet welds as shown in Fig. 3.13(a) are structurally better than those shown in Fig. 3.13(b).

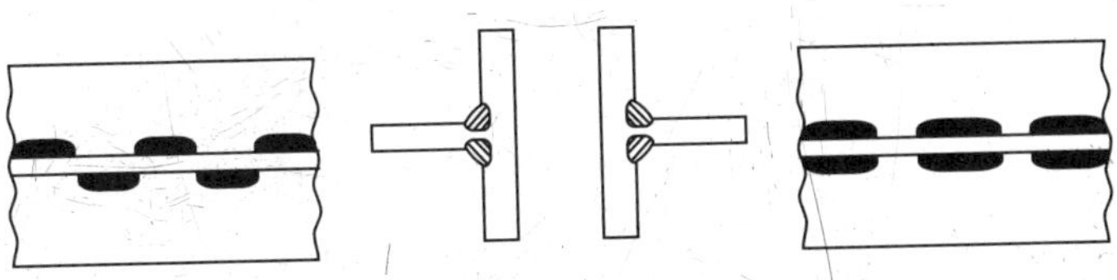

(a) (b)

Fig. 3.13 *Intermittent Fillet Weld*

The question of whether and to what extent intermittent welds should be used involves consideration of the following points:

1. Intermittent fillet welds are not economical unless the weld is of minimum size. A smaller fillet weld of a longer length is usually more economical for

the same strength. This is because the strength of a fillet weld increases directly with size but the weight of the weld metal increases with the square of the size.

2. If automatic welding is to be used the weld should be continuous.
3. If the structure is exposed, the use of continuous welds may be preferable as they are conducive to greater ease of maintenance and longer life of structure.
4. If severe dynamic loads act on the structure, intermittent welds must not be used.

Note Intermittent butt welds may be used to resist shear forces only and in general, are not recommended.

Design

1. The size of weld is assumed and the total effective length of the intermittent fillet weld required is computed.
2. Any intermittent fillet weld section should have a minimum effective length of four times the size of the weld with a minimum of 40 mm, except for plate girders (Clause 11.6 of I.S: 816–1969 (reaffirmed 1998)).
3. The clear spacing between an intermittent fillet weld should not exceed $12t$ for compression, and $16t$ for tension and should in no case be more than 200 mm.
4. At the ends, the longitudinal intermittent fillet weld should be of a length not less than the width of the member or else transverse welds should also be provided. If transverse welds are also provided along with longitudinal intermittent fillet welds, the total weld length at the ends should not be less than twice the width of the member.

3.10 FILLET WELDS FOR TRUSS MEMBERS

Truss members are composed of single angle or double angle sections. The following points should be borne in mind while designing the weld length:

1. The calculated weld length is placed as longitudinal fillet welds either on the two sides parallel to the axis of the load (Fig. 3.14(a)), or on three sides as shown in Fig. 3.14(b), i.e. transverse welds along with longitudinal welds. A longitudinal fillet weld length should never be placed on one side only as there will be a possibility of rotation.

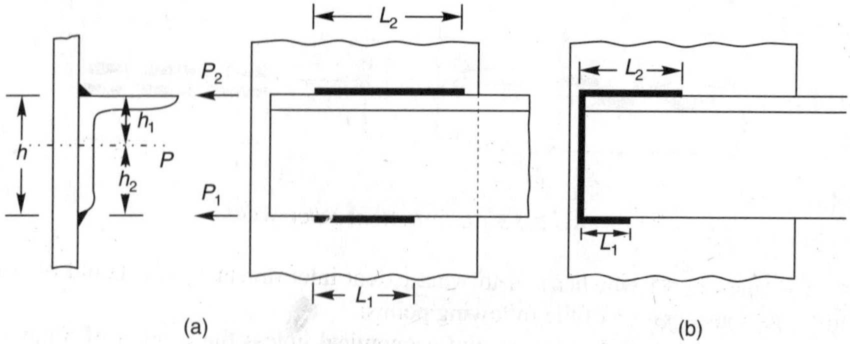

Fig. 3.14 *Fillet Welds for Truss Member*

2. The centre of gravity of the weld should coincide with the centroid of the section used as a truss member. If the member is symmetrical, the welds will be placed symmetrically but if the member is unsymmetrical (angle, channel) as is usually the case of a truss member, the lengths of longitudinal fillet welds are kept different on the two sides, as shown in Fig. 3.14(a), to achieve the above condition.

Let L_1, L_2 = length of longitudinal fillet welds on two sides
P_1, P_2 = force along lengths L_1 and L_2 respectively
P = force acting on the centroid of the section

Taking moment about the line passing through length L_1 (Fig. 3.14(a)),

$$P_2h - Ph_2 = 0$$

or

$$P_2 = \frac{Ph_2}{h} \tag{3.3}$$

Similarly,

$$P_1 = \frac{Ph_1}{h} \tag{3.4}$$

Once the force P_1, P_2 are known the fillet weld lengths can be designed as described in Section 3.8.

The effectiveness of a fillet weld can be increased by connecting the two side welds by an end weld as shown in Fig. 3.14(b). When the weld length to be provided cannot be accommodated over the two parallel sides because of limited overlap, it may be the only choice. This arrangement also reduces the size of gusset plate and results in economy.

Let L_1, L_2 = length of longitudinal fillet welds
h = length of end fillet weld
P = force acting on the centroid of the section
P_3 = force in the end fillet weld

Total length of weld required = $L_1 + L_2 + h$

Taking moment about the line passing through L_1,

$$P_2 h + P_3 \frac{h}{2} - Ph_2 = 0 \tag{3.5}$$

Taking moment about the line passing through L_2,

$$P_1 h + P_3 \frac{h}{2} - Ph_1 = 0 \tag{3.6}$$

Since P_3 (0.7 $hS\ \tau_{vf}$) is known, the above Eqs (3.5) and (3.6) can be solved for P_1 and P_2. Once the forces P_1 and P_2 are known the fillet weld lengths L_1 and L_2 can be designed.

3.11 PLUG AND SLOT WELDS

Plug and slot welds are used most often to tie two parts together and, in particular to reduce the unsupported dimensions of cover plates in compression, which increases the critical stress. They may also be used for shear transmission. Their use is generally reserved for locations where it is impractical to make a fillet weld yet possible to provide a plug or slot. The critical section for either a plug weld or a slot weld is the faying surface between the connected parts. The unit shearing resistance on this section is essentially the same as that of a fillet weld. Most specifications allow the same unit stress. Plug and slot welds should not be used to transmit tension, that is, a force normal to the faying surface. Tensile resistance depends largely upon the degree of penetration of the weld, which is apt to be rather than uncertain in either a plug or a slot weld.

Some engineers distrust plug and slot welds because of the difficulty of inspection. It is rather easy to make a plug or slot weld which appears excellent on the surface yet contains voids at the critical section.

The following specifications should be adhered to while designing plug and slot welds.

1. Width or diameter should be $\geq 3t$ and also ≥ 25 mm.
2. Corner radius in slotted hole should be $\geq 1.5t$ and also ≥ 12 mm.
3. Clear distance between holes should be $\geq 2t$ and also ≥ 25 mm where t is the thickness of plate having a hole or slot.

A combination of plug weld and other types of welds is permissible and the strength of the joint is the sum of the individual capacities of the welds.

3.12 FAILURE OF WELDS

Butt Welds When the butt weld is reinforced on both the sides of the plate, the section through the weld is increased to such an extent that it is unlikely for failure to occur in the weld, and the fracture normally occurs some distance away (Fig. 3.15(a)). The reinforcement acts as a supporting rib which inhibits necking in the immediate vicinity of the weld.

If the weld is ground flush with the surface of the plate, the position of the fracture depends upon the relative strength of the plate and weld metal. If the tensile strength or the yield point is lower for the weld metal than for the plate, failure takes place through the centre of the weld (Fig. 3.15(b)). But if the tensile strength and yield point of the weld metal are higher, failure takes place in the plate away from the weld. Failure in the weld junction is quite unusual.

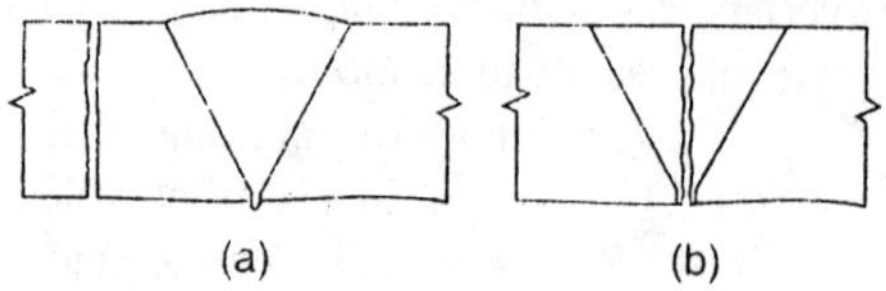

Fig. 3.15 *Failure of Butt Welds*

Under conditions of free bending across a reinforced butt weld, the stiffening effect of the reinforcement inhibits failure at the joint. But since stress concentration occurs at the weld junction, due to change of the section, there is some tendency for a crack to start at the weld junction itself with a flush butt weld, failure under such a condition occurs in the middle of the weld. If there is a wide difference in the tensile strength or yield point between the two metals, the failure may occur at the junction.

End Fillet Weld The plane of the fracture in a normal profile convex end fillet weld (Fig. 3.16(a)) is along the diagonal from the root of the fillet (Fig. 3.16(b)). When the fillet is subjected to shear in addition to the tensile stress, the position of the line of fracture departs from the diagonal according to the relative magnitude of the two stresses (Fig. 3.16(c)). If the fillet legs are unequal, a fracture usually occurs near the shorter leg (Fig. 3.16(d)). If the tensile strength of the weld metal is considerably greater than that of the plate, the fillet may remain intact and be pulled right out of the plate (Fig. 3.16(e)). With all end fillet welds failures occur abruptly, after a small amount of deformation.

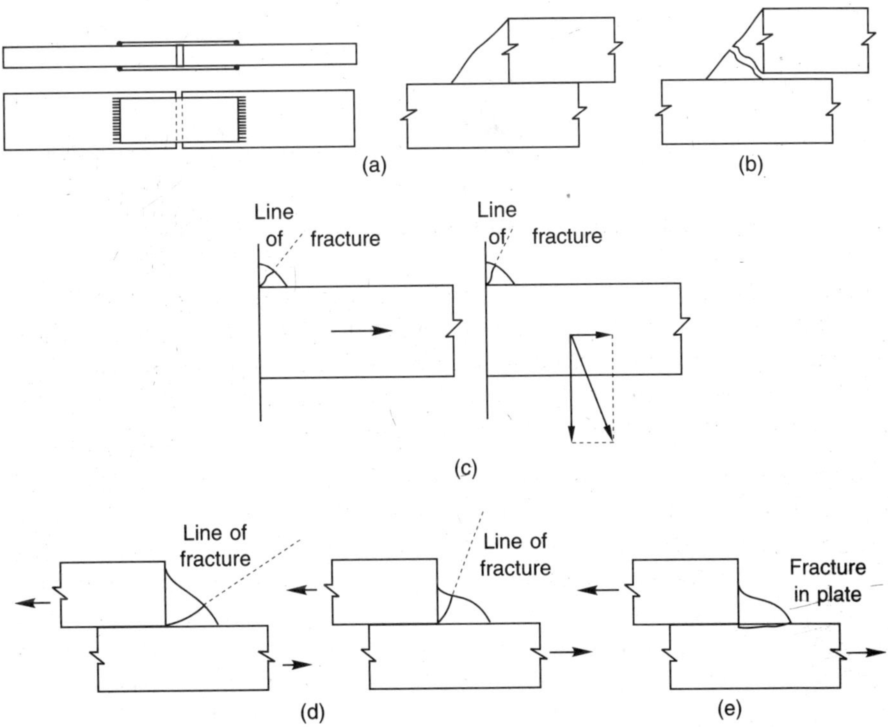

Fig. 3.16 *Position of Fracture in End Fillet Welds*

Side Fillet Weld In a side convex fillet weld subjected to shear stress along the weld, failure occurs down the throat of the weld. The break commences at the toe of the fillet at one or both the ends of the weld and as it progresses, the plane of fracture rotates (Fig. 3.17). The failure is gradual and considerable deformation of the fillet and usually also of the plates takes place before the final fracture.

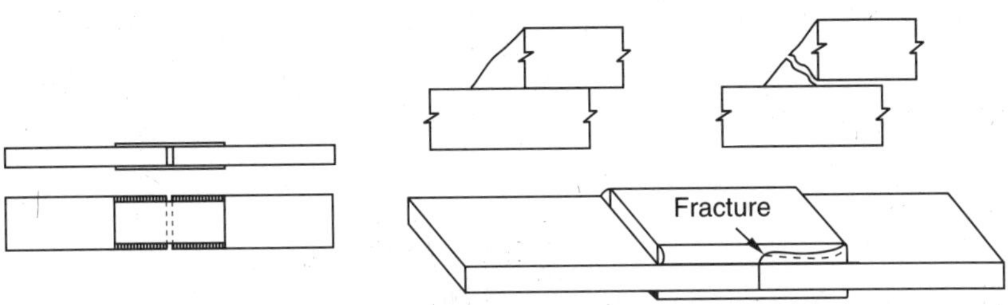

Fig. 3.17 *Position of Fracture in Side Fillet Welds*

3.13 DISTORTION OF WELDED PARTS

Fillet Weld Consider a fillet weld as shown in Fig. 3.18(a). After the weld is cooled, shrinkage is greatest along the face of the weld where large portion of filler metal is deposited. If the vertical plate is free to move (Fig. 3.18(b)), it is pulled over by the shrinkage of the weld without residual stresses due to shrinkage. This is

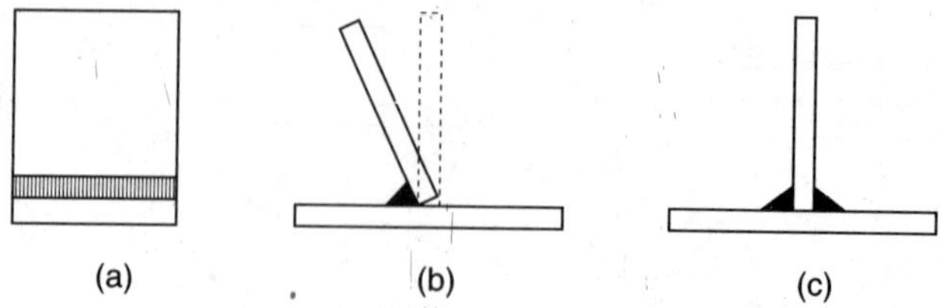

Fig. 3.18 *Distortion and Residual Stresses of Fillet Welds*

so because no resistance is offered to such distortion. Now a fillet weld on other side of the vertical plate (Fig. 3.18(c)), if deposited, will tend to pull the plate back into the original position generating large residual stress in both the welds. This residual stress generated because of asymmetry of welding sequence can be minimized by depositing the two welds simultaneously.

Short-length Butt Weld Consider a butt weld connecting two plates A and A_1 as shown in Fig. 3.19. When molten metal is deposited part A and zone B of plate A_1 are heated rapidly to a very high temperature allowing the steel to behave plastically. Zone C being heated less than the remaining two is assumed to behave elastically. The heated portions tend to expand in proportion to the change in temperature. Zone C, being less heated, tends to expand less and restrains some of expansion of zone B and part A. The distorted shape is shown in Fig. 3.19(b). Since part A and zone B are restrained from expanding while in a plastic state, in cooling they tend to shrink to a length less than their original dimension. Zone C, which did not undergo plastic deformation, tends to shrink to its original dimension and thus will restrain part A and zone B from free shrinkage. When the plates are completely cooled, part A will be shorter than its original length and will present a distorted shape shown in Fig. 3.19(c). Since part A and zone B have been prevented from full shrinkage, they are subjected to tension, whereas zone C is subjected to compression and bending, resulting in essentially compressive stresses at the boundary between zones B and C, and tensile stresses at the free edge of zone C. If the two plates shown in Fig. 3.19 were symmetrical about the weld, the distortions and residual stresses would be reduced considerably.

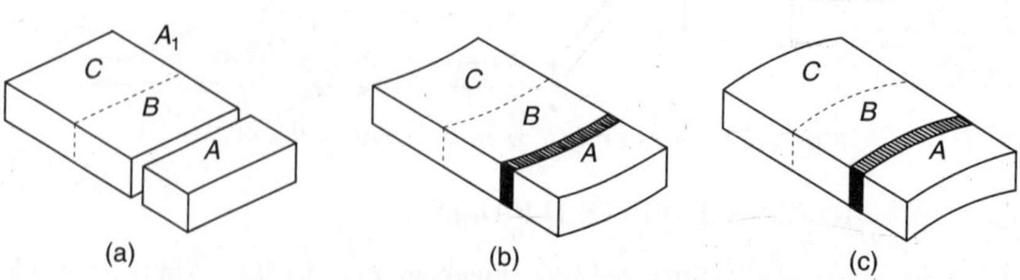

Fig. 3.19 *Distortion and Residual Stresses in Butt Weld*

Long Butt Weld Consider a long butt weld as shown in Fig. 3.20. One part of the plate zone D, is in a heated or expanding state similar to Fig. 3.19(b) while the zone

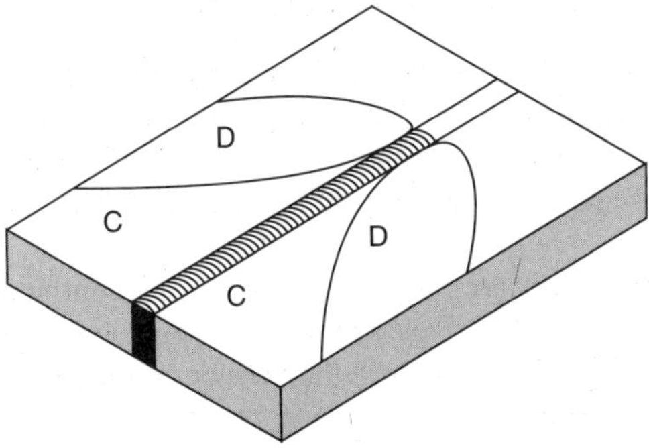

Fig. 3.20 *Heating and Cooling of a Long Butt Weld*

C is in a cooling or contracting state similar to Fig. 3.18(c). The relative areas of two zones, final distortions, and the residual stresses depend on speed of welding and on arrangement and sequence of welds designed to minimize distortions and residual stresses. There are a number of techniques to minimize residual stresses and prevent distortions. Some of which are preheating the parts before welding, hammering the weld to elongate it locally and to relieve the shrinkage forces (peening), etc.

3.14 INSPECTION OF WELDS

A poor weld leads to collapse; therefore proper inspection of the weld is necessary. Some of the methods for inspecting welds are as follows:

(i) Magnetic Particle Method Iron fillings are spread over the weld and it is then subjected to an electric current. The fillings form patterns which are interpreted to locate surface cracks.

(ii) Dye Penetration Method The depth of a crack can be estimated by this method. A dye is applied over the weld surface and then the surplus is removed. A dye absorber is placed over the weld which oozes the dye giving an idea of the depth of the crack.

(iii) Ultrasonic Method In this method, ultrasonic sound waves are sent through the weld. Defects like flaws, blow holes, etc. affect the time interval of sound transmission identifying the defect.

(iv) Radiography X-rays or gamma rays are used to locate defects. This method is used in butt welds only. It cannot be used in fillet welds because the parent material will also form part of the projected picture.

3.15 FILLET WELD VS. BUTT WELD

Fillet weld are preferred in comparison to butt welds due to the following reasons:

1. A fillet weld saves the operation of veeing and finishing the ends of members.
2. In case of a butt weld, members are fabricated slightly long and cut exactly to have a close fit in the field. This process is uneconomical.
3. Butt welds have higher residual stresses.

3.16 WELDED JOINTS VS. RIVETED JOINTS

1. Welded joints are economical. This is because splice plates and rivet materials are eliminated. Also, the gusset plates required are of a smaller size because of the reduced connection length. Labour cost is also less as only one person is required to do the welding whereas at least four persons are required for riveting.
2. Welded structures are more rigid (due to the continuity of the section at the joint) as compared to riveted joints. In riveted joints, cover plates, connecting angles, etc. deflect along with the member during load transfer and make the joint more flexible.
3. Due to the fact that the strength of a welded joint is the same as that of the parent metals, even a smallest piece of the metal which otherwise is a scrap can be used, bringing overall economy.
4. With welding it has become possible to connect tubular sections, which are structurally very economical.
5. Due to the fusion of two metal pieces jointed, a continuous structure is obtained, which gives a better architectural appearance than riveted joints.
6. Alterations can be done with less expenses in case of welding as compared to riveting.
7. The process of welding is quicker in comparison to riveting.
8. The process of welding is silent, whereas in the case of riveting a lot of noise is produced.
9. In welding less safety precautions are required for the public in the vicinity, whereas a hot rivet may toss and injure the persons working.
10. As splice plates, rivets, etc. are not used, the details and drawings of welded structures are easier and less time consuming.
11. The efficiency of a welded joint is more than that of a riveted joint. In fact a proper welded joint may have 100% efficiency.
12. Members to be jointed may distort due to the heat during the welding process, whereas there is no such possibility in riveted joints.
13. The possibility of a brittle fracture is more in the case of welded joints as compared to riveted joints.
14. The inspection of welded joints is difficult and expensive, whereas riveted joints can be inspected simply by tapping the joint with a hammer.
15. A more skilled person is required to make a welded joint as compared to the riveted joint.

3.17 SELECTION OF FASTENERS

After being conversant with various types of fasteners, a designer may choose a particular type based on the following considerations:

(i) Nature of the Connections First, it should be ascertained whether the connection is for a permanent structure or for a temporary one. For permanent structure rivets, welds and high strength bolts are recommended, whereas for temporary structures ordinary bolts and pins may be used.

(ii) Strength and Efficiency Required For high strength and efficiency, welding and HSFG bolts are preferred.

(iii) Availability of Skilled Persons If the joint is a welded or with HSFG bolts, highly skilled people are required, whereas riveting, bolting or pin connections can be made by ordinary technicians.

(iv) Cost The cost of connection influences the overall cost of the structure, e.g., if the connections are to be made in the field, riveted or high strength bolted connections may be economical as compared to welded joints. Also, welding at the site may not be feasible due to lack of power supply. Some of the other factors which should be given due consideration are loading conditions, equipments available and skilled fabricators.

Unfinished bolts, being economical, can be provided for structures subjected to small static loads and temporary structures. The bolting process is rapid and less skilled labour can be employed. High strength friction bolts are costly but are good for fatigue loads. Welding has the biggest advantage of giving rigid (fully moment-resisting) joints. Also, it gives a better appearance and uses metal economically. Rivets are becoming obsolete due to their limitations as discussed earlier. These are preferred only when a simple connection with a small moment-resistance is desired. Normally it has been found economical to use riveted or welded connections in the shop and bolted connections in the field.

Solved Examples

Example 3.1 Two plates of thickness 16 mm and 14 mm are to be jointed by a butt weld, as shown in Fig. Ex. 3.1. The joint is subjected to a tensile force of 280 kN. Due to some reasons the effective length of the weld that could be provided was 175 mm only. Check the safety of the joint if:

1. Single-V butt weld is provided,
2. Double-V butt weld is provided.

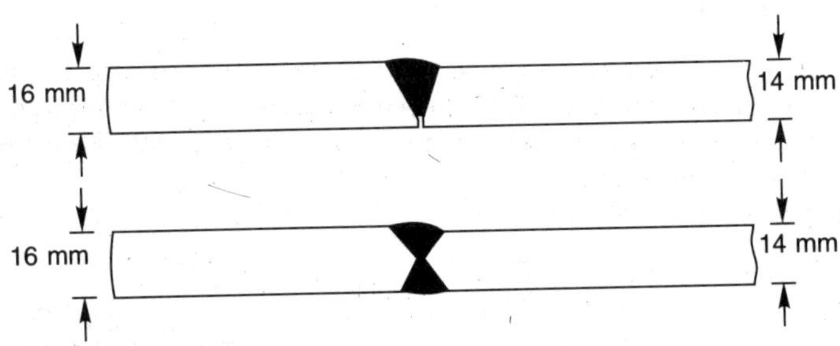

Fig. Ex. 3.1

Solution The allowable stress in butt weld in tension is 150 N/mm^2.

Single-V butt weld In case of single-V butt weld, incomplete penetration of the weld takes place; therefore as per I.S. specifications,

$$\text{Throat thickness} = \frac{5}{8}t$$

$$= \frac{5}{8} \times 14 = 8.75 \text{ mm}$$

Effective length of the weld = 175 mm

Strength of weld = $8.75 \times 175 \times 150 \times 10^3$

= 229.68 kN < 280 kN

which is inadequate.

Double-V butt weld Complete penetration of the butt weld takes place in case of a double-V butt weld. Therefore, as per I.S. specification

Throat thickness = thickness of thinner plate

= 14 mm

Strength of weld = $14 \times 175 \times 150 \times 10^3$

= 367.50 kN > 280 kN

which is adequate and safe.

Example 3.2 A butt weld is used to connect two plates 180 mm × 18 mm each. Find out the stress developed in the weld if it is subjected to a moment of 13000 kN mm.

Solution Assume the butt weld to be double-U type.

$$\text{The section modulus of the butt weld} = \frac{18 \times 180^2}{6} = 97.2 \times 10^3 \text{ mm}^3$$

Permissible bending stress, σ_{bf} = 165 N/mm^2

$$\sigma_{bf,\text{ cal}} = \frac{13000 \times 10^3}{97.2 \times 10^3}$$

$$= 133.74 \text{ N/mm}^2 < 165 \text{ N/mm}^2$$

which is safe.

Example 3.3 A tie member in a truss girder is 250 mm × 14 mm in size. It is welded to a gusset plate 10 mm thick by a fillet weld. The overlap of the member is 300 mm. The weld is 6 mm in size. Find the strength of the joint if the welding is done as shown in Fig. Ex. 3.3. What is the increase in strength if welding is done all around.

Allowable shear stress in weld = 108 N/mm^2.

Solution Effective length of weld

= 2 × 300 + 250

= 850 mm

Effective throat thickness = 0.7 × 6 = 4.2 mm

Strength of the fillet weld = $850 \times 4.2 \times 108 \times 10^{-3}$

= 385.56 kN

When the welding is done all around,

Effective length of weld = 2 × 300 + 2 × 250

= 1100 mm

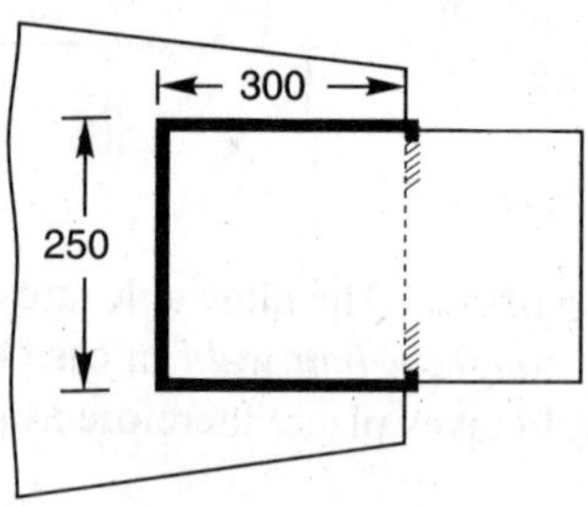

Fig. Ex. 3.3

Strength of the fillet weld $= 1100 \times 4.2 \times 108 \times 10^{-3} = 498.96$ kN

The strength of the weld is increased by

$$498.96 - 385.56 = 113.4 \text{ kN}$$

Example 3.4 A tie member 75 mm × 8 mm is to transmit a load of 90 kN. Design the fillet weld and calculate the necessary overlap.

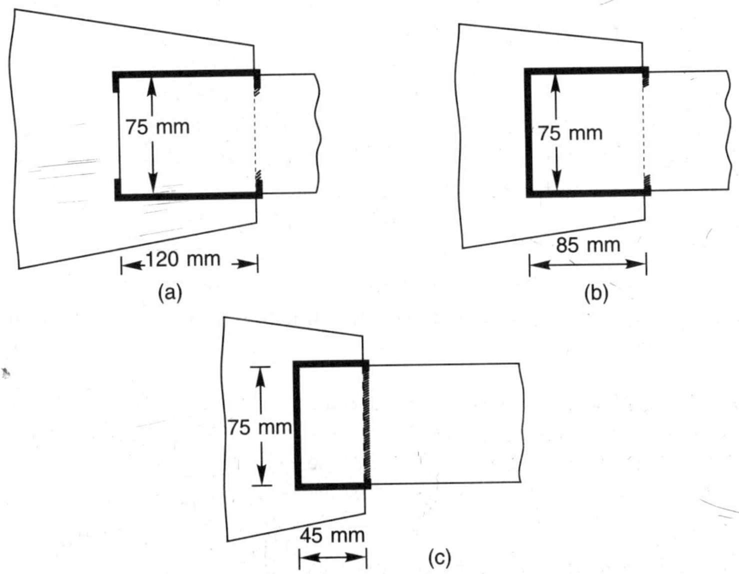

Fig. Ex. 3.4

Solution The tie member can be welded to the gusset plate on two sides parallel to the load axis, or on three sides or on all the four sides. The economy lies in finding out the minimum overlap, keeping in mind the specifications laid down by the I.S. code.

$$\text{Strength of weld} = \text{load over the joint} = 90 \text{ kN} = 90 \times 10^3 \text{ N}$$

Size of weld

Minimum size of weld for 8 mm thick section = 3 mm

Maximum size of weld = 8 – 1.5 = 6.5 mm

Provide a size of weld between 3 mm and 6.5 mm.

Let $S = 5$ mm

Effective throat thickness $= 0.7 \times 5 = 3.5$ mm

(a) *Welds on two sides*

Effective length of the weld

$$= \frac{\text{strength}}{\text{effective throat thickness} \times \text{allowable stress}}$$

$$= \frac{90 \times 10^3}{3.5 \times 108} = 238.09 \approx 240 \text{ mm}$$

$$\text{Length of weld to be provided on each side} = \frac{240}{2} = 120 \text{ mm} \not< 75 \text{ mm}$$

Therefore, provide 120 mm long and 5 mm size fillet welds on the two sides, as shown in Fig. Ex. 3.4(a).

The necessary overlap required is 120 mm. The welds are turned at the ends.

$$\text{Overall length of weld required} = 2(120 + 2 \times 5) = 260 \text{ mm}$$

(b) *Welds on three sides* Length of the weld required is 240 mm. The overlap required can be found by the following relation:

$$240 = 2 \times \text{overlap} + 75$$

or
$$\text{Overlap} = 82.5 \text{ mm} \approx 85 \text{ mm}$$

Provide an overlap of 85 mm as shown in Fig. Ex. 3.4(b).

(c) *Welds on all the four sides* Length of the weld required is 240 mm. The overlap required can be found by the following relation:

$$240 = 2 \times \text{overlap} + 2 \times 75$$

or
$$\text{Overlap} = 45 \text{ mm}$$

Provide an overlap of 45 mm, as shown in Fig. Ex. 3.4(c).

Therefore, it is found that the most economical way to join the member is by welding on all four sides.

Example 3.5 A circular plate, 150 mm in diameter, is welded to another plate by means of a 6 mm fillet weld as shown in Fig. Ex. 3.5. Calculate the greatest twisting moment that can be resisted by the weld.

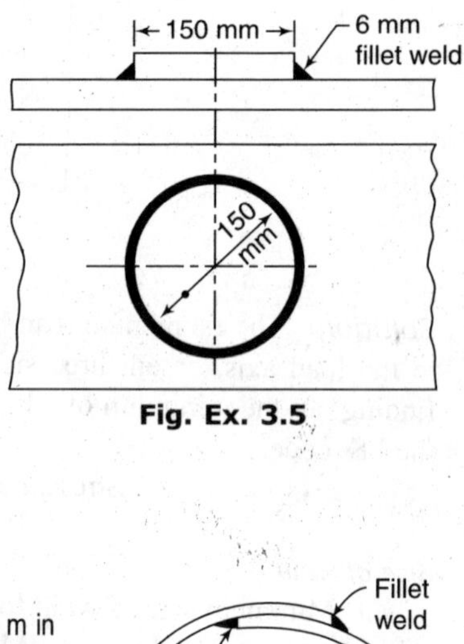

Fig. Ex. 3.5

Solution Safe load that can be resisted per mm length of weld

$= 0.7 \times 6 \times 108$

$= 453.6$ N/mm

Length of the weld $= \pi d = \pi \times 150 = 471$ mm

Greatest twisting moment $= 453.6 \times 471 \times \dfrac{150}{2} \times 10^{-6} = 16.02$ kNm

Example 3.6 A circular penstock of mild steel, 1.2 m in diameter, is fabricated with a 12 mm plate and secured by fillet welds of 8 mm size, provided on the inside and outside of the lapped ends as shown in Fig. Ex. 3.6. Determine the safe internal pressure that can be allowed in the penstock.

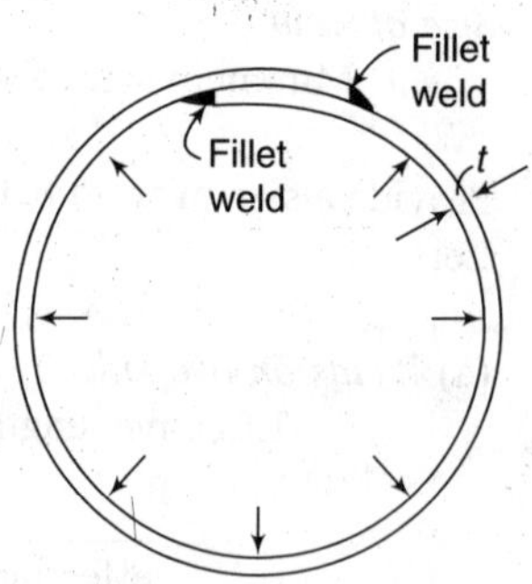

Fig. Ex. 3.6

Solution Throat thickness, $t = 0.7 \times 8 = 5.60$ mm

Internal force per unit length, causing bursting of pipe $= \dfrac{pd}{2}$

where p is the internal pressure and d the diameter of penstock.

Resistance offered by the weld, per unit length = $2t\tau_{vf}$

Hence $\dfrac{pd}{2} = 2t\ \tau_{vf}$

or $p = \dfrac{4t\tau_{vf}}{d} = \dfrac{4 \times 5.6 \times 108}{1200} = 2.016\ \text{N/mm}^2$

Example 3.7 A tie member consisting of angle section I.S.A 80 mm × 50 mm × 8 mm (f_y = 250 MPa) is welded to a 12 mm gusset plate. Design welds to transmit a load equal to the full strength of the member. Refer Fig. Ex. 3.7.

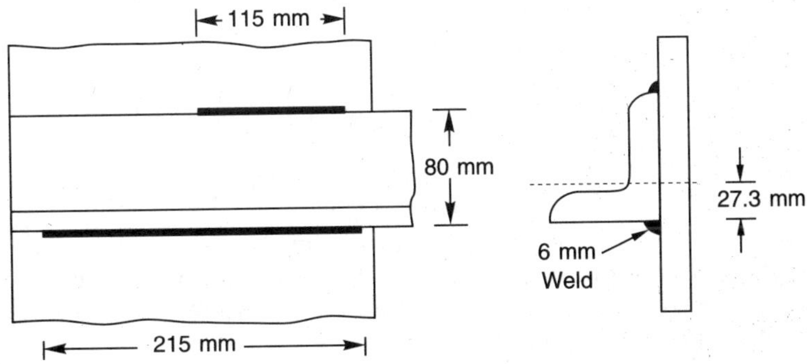

Fig. Ex. 3.7

Solution Let P be the strength of the member.

From I.S. Handbook No. 1, the relevant properties of the angle section are,

$$A = 978\ \text{mm}^2,$$
$$C_{xx} = 27.3\ \text{mm}$$
$$\sigma_{at} = 0.6f_y = 0.6 \times 250 = 150\ \text{N/mm}^2$$
$$P = \text{cross-sectional area} \times \text{permissible stress}$$
$$= 978 \times 150 \times 10^{-3} = 146.700\ \text{kN}$$

The weld will be designed to transmit a force equal to 146.7 kN.

The force resisted by the weld at the lower side of angle,

$$P_1 = \frac{146.7 \times (80 - 27.3)}{80} = 96.638\ \text{kN}$$

The force resisted by the weld at the upper side of angle,

$$P_2 = \frac{146.7 \times 27.3}{80} = 50.061\ \text{kN}$$

Let us assume the size of weld, S= 6 mm (> 3 mm)

Effective throat thickness of weld = 0.7 × 6 = 4.2 mm

$$l_1 = \frac{96.638 \times 10^3}{4.2 \times 108} = 213.04 \approx 215\ \text{mm}$$

$$l_2 = \frac{50.061 \times 10^3}{4.2 \times 108} = 110.36\ \text{mm} \approx 115\ \text{mm}$$

Example 3.8 Design the joint in Example 3.7 if the weld is to be done on the three sides as shown in Fig. Ex. 3.7.

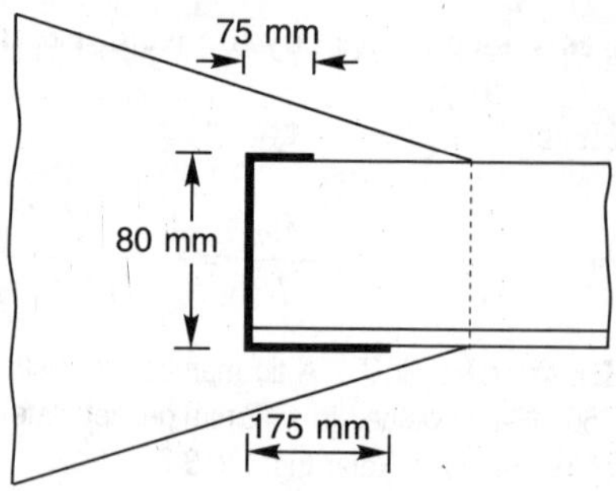

Fig. Ex. 3.8

Solution Total length of weld $= l_1 + l_2 + 80$

Strength of weld/mm length $= 4.2 \times 108$

$= 453.6$ N

Equating the strength of the total weld to the load,

$$(l_1 + l_2 + 80) \times 453.6 = 146.7 \times 10^3$$

$$l_1 + l_2 = 243.41 \text{ mm}$$

Take moment about top edge of angle section,

$$453.6 \times 80 \times \frac{80}{2} + 453.6 \times l_1 \times 80 = 146.7 \times 10^3\ (80 - 27.3)$$

$$l_1 = 173.04 \text{ mm} \approx 175 \text{ mm}$$

Hence, $l_2 = 243.41 - 173.04$

$= 70.37 \text{ mm} \approx 75 \text{ mm}$

Example 3.9 A tie member of a truss consists of double angle section of dimensions 80 mm × 80 mm × 8 mm welded on the opposite side of a 12 mm thick gusset plate. Design a fillet weld to make the joint. Axial tension in the member is 200 kN.

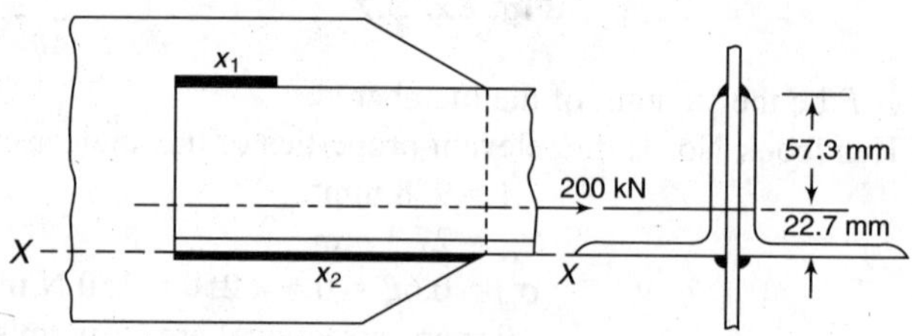

Fig. Ex. 3.9

Solution Refer Fig. Ex. 3.9.

Let the lengths of weld at top and bottom edges be x_1 and x_2, respectively for each angle.

Total length = weld $= 2(x_1 + x_2)$ mm

Maximum size of weld $= 8 - 1.5 = 6.5$ mm

Minimum size of weld = 5 mm

Let us provide 5 mm weld size.

Strength of weld per mm length $= 0.7 \times 5 \times 108 = 378$ N/mm

Equating the strength of the weld to the tension in the member,

$$378 \times 2(x_1 + x_2) = 200 \times 10^3$$

$$(x_1 + x_2) = 264.55 \text{ mm}$$

Taking moment about X-X

$$378 \times 2x_1 \times 80 = 200 \times 10^3 \times 22.7$$

$$x_1 = 75 \text{ mm}$$

$$x_2 = 264.55 - 75 = 189.55 \approx 190 \text{ mm}$$

Provide $x_1 = 75$ mm and $x_2 = 190$ mm

Example 3.10 An I.S.L.C. 300 @ 324.7 N/m is used to transmit a force of 600 kN. The channel section is connected to a gusset plate 10 mm thick. Design a fillet weld if the overlap is limited to 350 mm. Use slot welds if required.

Solution Relevant properties of I.S.L.C. 300 @ 324.7 N/m are,

$$A = 4211 \text{ mm}^2$$

$$T = 11.6 \text{ mm}$$

$$t = 6.7 \text{ mm}$$

The maximum size of weld = 6.7 – 1.5 = 5.2 mm

Provide 5 mm-size weld.

Throat thickness = $0.7 \times 5 = 3.5$ mm

Strength of weld/mm length = $3.5 \times 108 = 378$ N

$$\text{Length of weld required} = \frac{600 \times 1000}{378} = 1587.30 \approx 1590 \text{ mm}$$

The maximum length of the weld that can be provided is

$300 + 2 \times 350 = 1000$ mm (< 1590 mm).

Hence, the length of the weld will be provided with slot welds.

Provide width of slots 25 mm ($3t = 3 \times 6.7 = 20.1$ mm or 25 mm whichever is greater).

Let length of the slot be x,

$$1590 = 2 \times 350 + 300 + 4x$$

$$x = 147.5 \text{ mm} \approx 150 \text{ mm}$$

Provide 150 mm long fillet welds in slots as shown in Fig. Ex. 3.10.

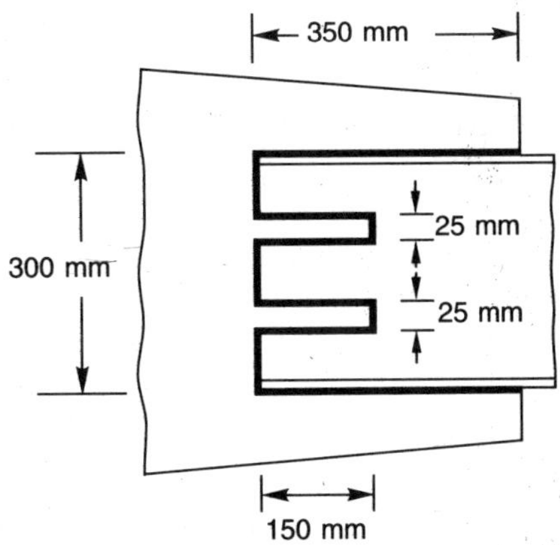

Fig. Ex. 3.10

Example 3.11 Design a suitable fillet weld to connect the web plate to the flange plate and the flange plate to the cover plate of the compression flange of a plate girder as shown in Fig. Ex. 3.11. The web plate is 1000 mm × 12 mm. The flange plate and cover plate are 400 mm × 20 mm and 375 mm × 16 mm respectively. The maximum bending moment $M = 500 \times 10^3$ kN mm and maximum shear force is $V = 1000$ kN.

Solution *Connection of web and flange plate*

Minimum size of weld = 5 mm

Maximum size of weld = 12 – 1.5 = 10.5 mm

Adopt 7 mm size of fillet weld.

$$\tau_{vf,\,\text{cal}} = \frac{V\,\overline{AY}}{I_{XX}\ \Sigma t}$$

$$t = 0.7 \times S = 0.7 \times 7 = 4.9 \text{ mm}$$

$$\Sigma t = 2 \times 4.9 = 9.8 \text{ mm}$$

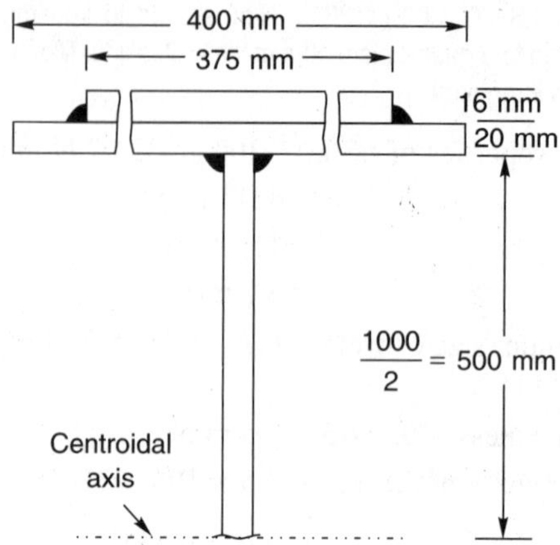

Fig. Ex. 3.11

$$A\bar{Y} = 400 \times 20 \times (500 + 10) + 375 \times 16 \times (500 + 20 + 8)$$
$$= 408 \times 10^4 + 316.8 \times 10^4 = 724.8 \times 10^4 \text{ mm}^3$$

$$I_{XX} = 2 \times \left[\frac{375 \times 16^3}{12} + 375 \times 16 \times 528^2 + \frac{400 \times 20^3}{12} + 400 \times 20 \times 510^2 \right] + \frac{12 \times 1000^3}{12}$$

$$= 8.507 \times 10^9 \text{ mm}^4$$

$$\tau_{vf,\ cal} = \frac{1000 \times 10^3 \times 724.8 \times 10^4}{8.507 \times 10^9 \times 9.8}$$

$$= 86.93 \text{ N/mm}^2 < 108 \text{ N/mm}^2$$

which is all right.

Connection of flange plate to cover plate

Adopt 6 mm size of fillet weld.

$$S = 6 \text{ mm}$$
$$t = 0.7 \times 6 = 4.2 \text{ mm}$$
$$\Sigma t = 2 \times 4.2 = 8.4 \text{ mm}$$
$$A\bar{Y} = 375 \times 16 \times (500 + 20 + 8) = 316.8 \times 10^4 \text{ mm}^3$$

$$\tau_{vf,\ cal} = \frac{1000 \times 10^3 \times 316.8 \times 10^4}{8.507 \times 10^9 \times 8.4}$$

$$= 44.33 \text{ N/mm}^2 < 108 \text{ N/mm}^2$$

which is as should be.

Example 3.12 A 120 mm diameter and 6 mm thick pipe is welded to a 14 mm plate by fillet weld. The pipe is subjected to a vertical load of 3 kN at 1.00 m from the welded end and a twisting moment of 1.2 kNm. Design the joint.

Solution Direct load = 3 kN = 3×10^3 N

$$\text{Bending moment} = 3 \times 10^3 \times 1000 = 3 \times 10^6 \text{ Nmm}$$

$$\text{Twisting moment} = 1200 \times 10^3 \text{ Nmm}$$

Let t = Effective throat thickness of the weld.

$$\text{Polar moment of inertia, } I_{zz} = 2\pi r^3 t$$

$$= 2\pi \times 60^3 \times t = 1357168\, t \text{ mm}^4$$

$$I_{XX} = \frac{I_{zz}}{2}$$

$$= \frac{1357168\, t}{2} = 678584.01\, t \text{ mm}^4$$

$$\text{Shear stress due to direct load} = \frac{3 \times 10^3}{2\pi \times 60 \times t} = \frac{7.957}{t} \text{ N/mm}^2$$

$$\text{Bending stress due to bending moment} = \frac{3 \times 10^6 \times 60}{67.85 \times 10^4 \times t} = \frac{265.29}{t} \text{ N/mm}^2$$

This bending stress shall be treated as shear since, the actual failure will be along the throat of the weld.

$$\text{Shear stress due to twisting moment} = \frac{1200 \times 10^3 \times 60}{135.7168 \times 10^4 \times t} = \frac{53.05}{t} \text{ N/mm}^2$$

$$\text{Combined stress} = \left\{\left(\frac{7.957}{t}\right)^2 + \left(\frac{265.29}{t}\right)^2 + \left(\frac{53.05}{t}\right)^2\right\}^{1/2}$$

or

$$\frac{270.65}{t} = 108$$

$$t = \frac{270.65}{108} = 2.50 \text{ mm} \approx 3 \text{ mm}$$

Hence, provide a 3 mm fillet weld.

Exercises

3.1 Find out the strength of a 6 mm fillet weld per mm length.
A 200 mm wide plate is to be jointed to another plate section 300 mm wide. Find out the strength of the joint if the overlap of the plates is 150 mm. Both longitudinal and end fillet welds are provided. What will be the overlap required if only longitudinal fillet welds are provided? The plates are 8 mm thick.

3.2 Two plates 12 mm thick are jointed by (i) a single V-butt weld (ii) a double V-butt weld. The effective length of the weld is 220 mm. Determine the strength of the welded joints.

3.3 Determine the maximum load which can be applied to the fillet welds for the cases shown in Fig. Prob. 3.3.

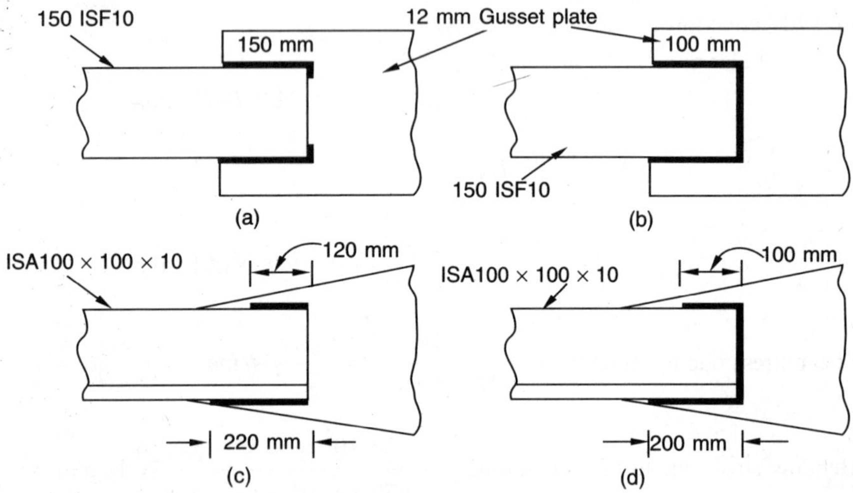

Fig. Prob. 3.3

3.4 An angle section 150 mm × 115 mm × 12 mm is to be connected to a gusset plate 12 mm thick. Design the fillet weld to carry a load equal to the strength of the member.

3.5 Design a fillet weld to join a tension member consisting of 2 I.S.A. 100 mm × 75 mm × 8 mm to a gusset plate 12 mm thick. The tensile load is 280 kN.

3.6 A 120 mm diameter pipe 0.50 m long is welded to 10 mm thick plate at right angles to the plate. A vertical load of 8 kN acts at its free end. The pipe is 6 mm thick. Design the welded connection.

3.7 A 200 mm wide and 16 mm thick plate is to be connected with a 8 mm fillet weld. Determine the length L required to develop the full strength of bar (Fig. Prob. 3.7).

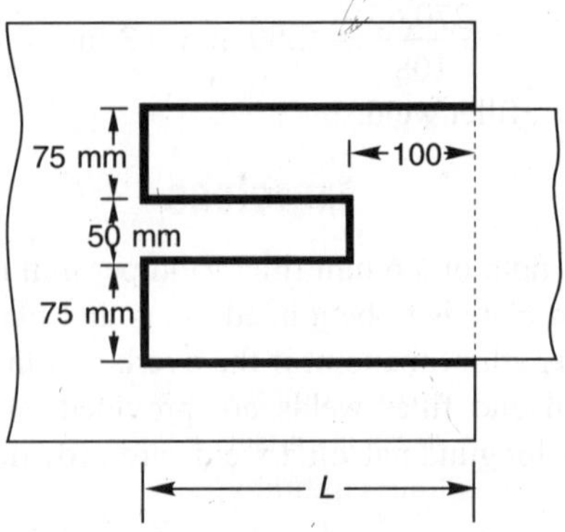

Fig. Prob. 3.7

Compression Members

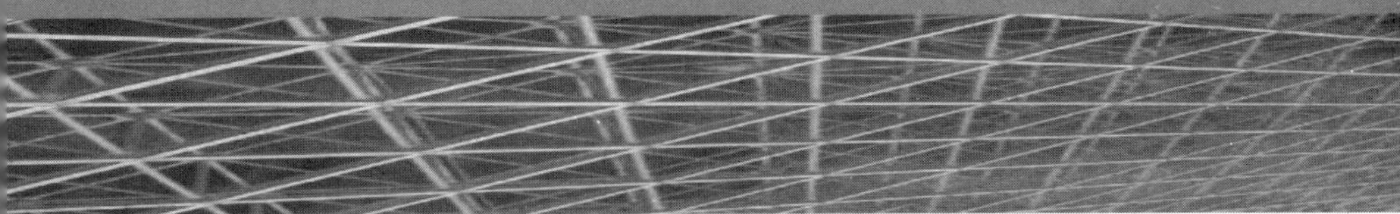

4.1 INTRODUCTION

A *compression member* is a structural member which is straight and subjected to two equal and opposite compressive forces applied at its ends. Different terms are used for a compression member depending upon its position in structures. *Strut* is a compression member used in the roof truss and bracing. They are of a small span and may be vertical or inclined. *Column, stanchion* or *post* is a vertical compression member supporting floors or girders in a building. These compression members are subjected to heavy loads. The *principal rafter* is a top chord member in a roof truss and *boom* is the principal compression member in a crane.

Stability plays an important role in the design of compression members. Ordinary structural analysis is based on the condition of stable equilibrium between internal and external forces, and a linear relationship is assumed to exist between stress and strain. However when buckling is involved, it is necessary to investigate the potentially unstable equilibrium between the external and internal responses that are further complicated by the complex stress-strain relationship of the material extending from elastic to inelastic range. The term unstable used here pertains to a condition in which the slightest increment of deflection results in a further increase, which may lead to collapse of structure.

Excessive compression of long columns may cause yielding or buckling. It can fail due to yielding if it is absolutely straight, has perfectly homogeneous material, concentric loads and no initial residual stresses. These are ideal conditions which may never exist in an actual structure. As compressive loading of a column is increased, it eventually causes some eccentricity. This in turn sets up some bending moment, causing the column to deflect or buckle slightly. This deflection increases the eccentricity and thus the bending moment. This may progress to where the bending moment is increasing at a rate greater than the increase in load, and the column soon fails by buckling. In general long columns fail by elastic buckling, intermediate columns by inelastic buckling, and very short columns usually fail by crushing or yielding. At the point of failure, the stress in a long column will not exceed the proportional limit and it may be much lower than this limit for a very slender column. Failure of the intermediate column occurs after the extreme fibres have reached the yield point. A very short column is not really a column as such but is considered to be a block without buckling.

In any case, some maximum compressive stress can be set as a limit of strength, and an allowable working stress is chosen accordingly. Also it is logical to apply a large factor of safety to a long column and a smaller factor of safety to a shorter one. It should be taken note of that the terms long, intermediate and short columns are only relative. They are defined by the interpretation of their slenderness ratio.

This chapter will discuss the design of individual compression members in order to illustrate the detailed procedure involved in selecting appropriate proportions of such members.

4.2 EFFECTIVE LENGTH

The form of curve into which a compression member tends to deflect depends upon the mode of end fixtures. In each case there is a portion of the length of the compression member which bends as if this part had been a pin-jointed end. The end points of this portion of the compression member are the points of contraflexure.

The effective length, l, of a compression member is the distance between these points. Therefore it should be derived from the actual length and end conditions. The end conditions are accounted for through the use of effective length factors, which when multiplied by the actual length L give the effective length. Since the effective length is based on the end conditions of the compression member a precise determination of the effective length is very difficult and any arbitrary assumption may lead to serious errors. Tables 4.1 and 4.2 provide the effective length of columns and struts respectively, for various end conditions.

An angle strut may be a continuous or a discontinuous strut. A continuous strut is a compression member which is continuous over a number of joints, such as a top chord member of a truss bridge girder, principal rafter of a roof truss, etc. In the chord members of roof truss, double unequal angle sections with their short legs placed back to back on the opposite side of the gusset plate are used. This provides greater overall stiffness to the truss against lateral bending. A discontinuous strut is a compression member which extends between two adjacent joints only, e.g., vertical or inclined compression members in a roof truss.

Table 4.1 *Effective Length of Columns*

End Restraints	*Effective Lengths*	*Figure No.*
(a) Effectively held in position at both ends but not restrained against rotation	$1.00L$	4.1(a)
(b) Effectively held in position at both ends and restrained against rotation at one end.	$0.80L$	4.1(b)
(c) Effectively held in position and restrained against rotation at both ends.	$0.65L$	4.1(c)
(d) Effectively held in position and restrained against rotation at one end, and at the other end restrained against rotation but not held in position.	$1.20L$	4.1(d)
(e) Effectively held in position and restrained against rotation at one end, and at the other end partially restrained against rotation but not held in position.	$1.50L$	4.1(e)
(f) Effectively held in position at one end but not restrained against rotation, and at the other end restrained against rotation but not held in position.	$2.00L$	4.1(f)

(Contd.)

Table 4.1 (*Contd.*)

End Restraints	*Effective Lengths*	*Figure No.*
(g) Effectively held in position and restrained against rotation at one end, but neither held in position nor restrained against rotation at the other end.	2.00L	4.1(g)

End conditions

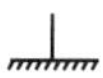	Rotation fixed and translation fixed
	Rotation free and translation fixed
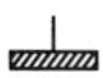	Rotation fixed and translation free
	Rotation free and translation free

Notes

(i) For battened compression members the effective length should be increased by 10%.
(ii) In the case of stepped columns, Appendix D of I.S. 800–1984 may be referred to.

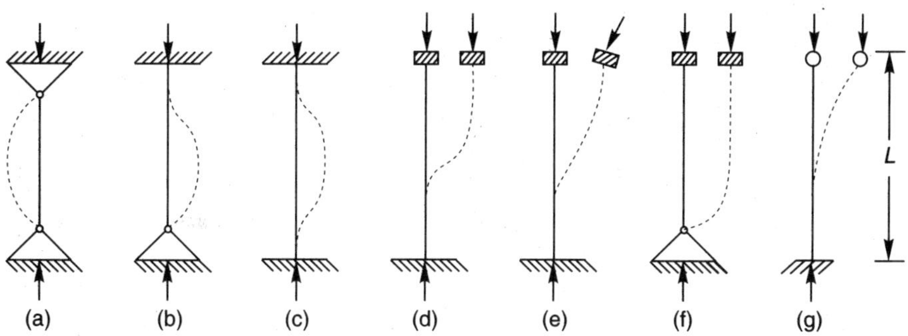

Fig. 4.1 *End Restraint of Compression Member*

Table 4.2 *Effective Length of Struts*

Type	*Sections*	*Effective Lengths*	*Allowable Compressive Stress*	*Figure No.*
1. Continuous	Single or double angle	0.7L to 1.00L	σ_{ac}	
2. Discontinuous	Single angle connected with one rivet	1.00L	0.8σ_{ac}	4.2(a)
3. Discontinuous	Single angle connected with more than one rivet or with weld.	0.85L	σ_{ac}	4.2(b)
4. Discontinuous	Double angles placed back to back on opposite sides of gusset plate	0.70L to 0.85L	σ_{ac}	4.2(c)
5. Discontinuous	Double angles placed back to back on same side of gusset plate	1.00L	0.8σ_{ac}	4.2(d)

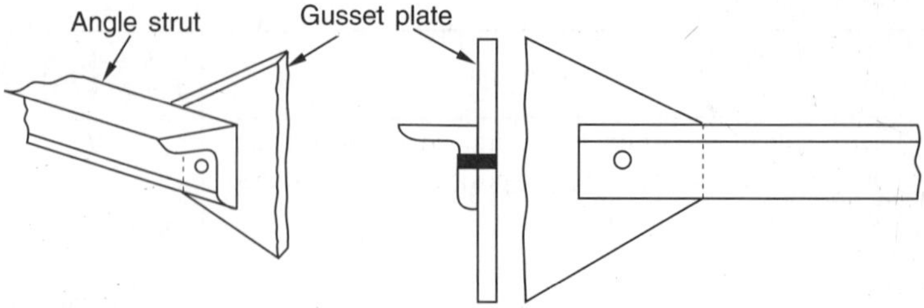

(a) Single angle strut connected by one rivet

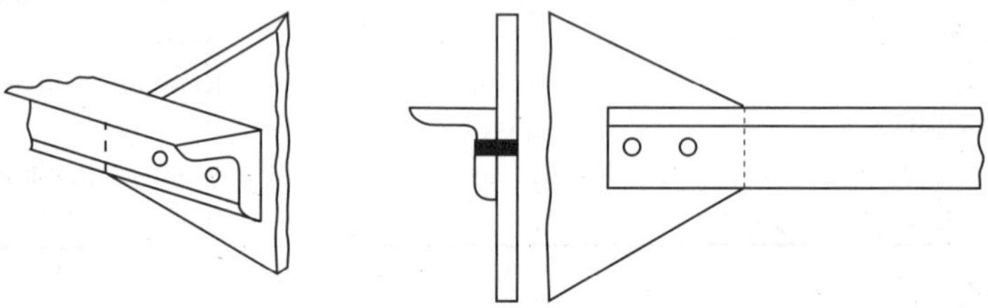

(b) Single angle strut connected by more than one rivet

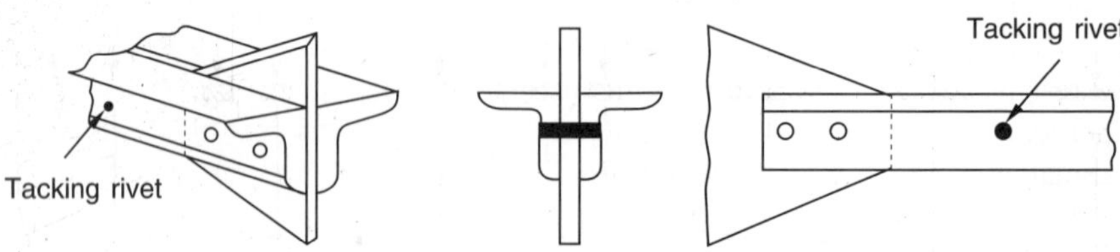

(c) Double angle strut on opposite side of gusset

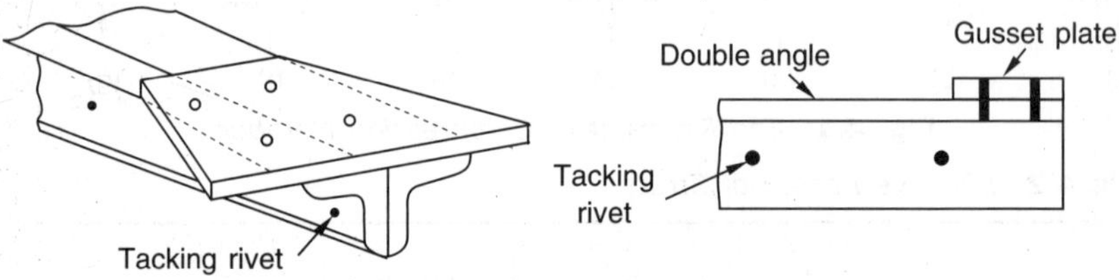

(d) Double angle strut on same side of gusset

Fig. 4.2 *End Restraint of Struts*

4.3 SLENDERNESS RATIO (λ)

The slenderness ratio of a member is the ratio of the effective length to the appropriate radius of gyration $(\lambda = kL/r)$. This is valid only when the column has equal unbraced heights for both axes and end conditions are same for both axes. The appropriate radius of gyration is one which is minimum for a particular section. For example a section asymmetrical about the centroidal axes (angle section, Fig. 4.3) will bend about the principal axis V–V for which the radius of gyration is minimum. On the other hand, a section symmetrical about both the centroidal axes (I-Section) or even with one axis of symmetry (channel section, two angles back to back) will bend about one of the centroidal axis giving lesser radius of gyration. This is because for such sections the principal axes coincide with the centroidal axes.

The slenderness ratios about *XX* and *YY* axes are,

$$\lambda_{XX} = l_{XX}/r_{XX} \qquad r_{XX} = \sqrt{I_{XX}/A}$$

$$\lambda_{YY} = l_{YY}/r_{YY} \qquad r_{YY} = \sqrt{I_{YY}/A}$$

and $$\lambda_{max} = l/r_{min} \qquad r_{min} = \sqrt{I_{min}/A}$$

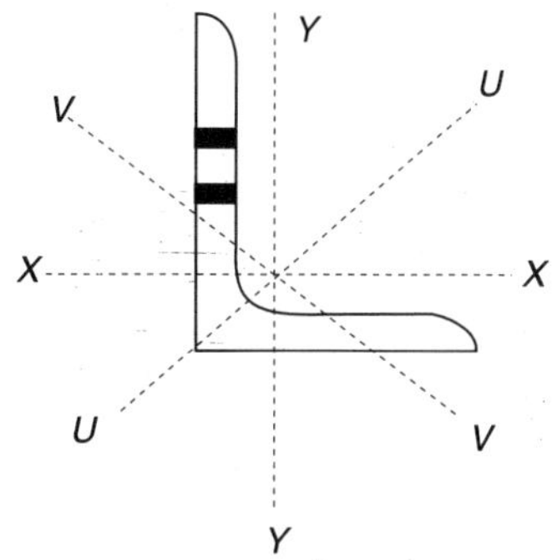

Fig. 4.3 *Axes of Angle Section*

To minimise steel requirements in column design (so as to use the material at the greatest possible stress) the slenderness ratio should be kept as small as possible. The designer can achieve this either by selecting a section which provides the largest value of the minimum radius of gyration without providing more area or by reducing the unsupported length of the column by some means.

Maximum radius of gyration is obtained when material is farthest from the centroid. Thus, for a constant area the material gets thinner as the column size increases and in such a case local buckling limits the column size.

If feasible, the unsupported length can be reduced by furnishing intermediate supports to a compression member by attached construction and thus permitting use of a smaller section at high average stress. If this is possible in only one direction, then the value of the unsupported length will be different for the two directions. Sections with different radii of gyration in two directions may be so chosen as to obtain the values of the two slenderness ratios λ_{xx} and λ_{yy} with least difference (see Example 4.9). The larger of the two would be used for the determination of the nominal axial compressive stress. It would be desirable to orient the section so that the axis having the smallest radius of gyration would be the one braced. Sometimes when the conditions at the column end, about the two axes are different, a balance between the two slenderness ratios (λ_{xx} and λ_{yy}) can be achieved by properly orienting the section and the material can be used to its maximum stress value. An I-Section electric pole fixed at the base and carrying wires parallel to the road is an example wherein the wires provide a hinge condition in the direction of wires (l = 0.8L) and with free end condition in the perpendicular direction (l = 2L). The section is used to its full strength by providing flanges parallel to the direction of wires and thus attempting to equalise λ_{xx} and λ_{yy}.

$$\lambda_{YY} = \frac{0.8L}{r_{YY}}, \quad \lambda_{XX} = \frac{2L}{r_{XX}}$$

The slenderness ratio of compression members is limited because of the following reasons:

1. The effect of accidental and construction (fabrication, transportation and erection) loads are automatically taken care of.
2. The bracing members may be used as a walkway for workmen or to provide temporary support for equipments.
3. To take care of the probability of members being subjected to unexpected vibrations.

The maximum permissible slenderness ratio for compression members are stated in Table 4.3.

Table 4.3 *Maximum Slenderness Ratio (λ) for Compression Members*

S. No.	*Type of Member*	λ
1.	A strut connected by single rivet at each end	180
2.	A member carrying compressive loads resulting from dead loads and imposed loads	180
3.	A member subjected to compressive forces resulting from wind/earthquake forces, provided the deformation of such members does not adversely effect the stress in any part of the structure	250
4.	Compression flange of a beam	300
5.	A member normally acting as a tie in a roof truss or a bracing system not considered effective subjected to possible reversal of stresses resulting from the action of wind or earthquake forces	350

4.4 COLUMN DESIGN FORMULA

The failure of columns excluding the possibility of torsion, is due to bending. The factors that influence the bending behaviour in real columns are, lateral loads, end eccentricity, column curvature and non-homogeneity of material, etc. These factors should be given due consideration. Residual stresses, variation in inelastic stress-strain characteristics, shear strength, local buckling, shape of cross section and end restraints are some of the factors that effect the buckling resistance of columns. It is impractical to incorporate all the factors that affect the strength of the column mathematically in any one column formula. A number of formulae have been suggested by designers. I.S: 800–1984 recommends the use of *Merchant Rankine formula* (with a f.o.s. of 1/0.6).

Merchant Rankine formula in its fundamental form is

$$\frac{1}{(f)^n} = \frac{1}{(f_e)^n} + \frac{1}{(f_y)^n} \tag{4.1}$$

or

$$f = \frac{f_e \times f_y}{[(f_e)^n + (f_y)^n]^{1/n}} \tag{4.2}$$

where f_e is the elastic critical stress (same as f_{cc} used by I.S. code) and f_y is the failure stress.

The direct stress in compression on the cross-sectional area of an axially loaded compression member is limited to 0.6 f_y. Therefore, the formula for permissible compressive stress derived from the Merchant Rankine formula is,

$$\sigma_{ac} = 0.6 \frac{f_{cc} f_y}{[(f_{cc})^n + (f_y)^n]^{1/n}} \tag{4.3}$$

where σ_{ac} = permissible stress in axial compression in MPa

f_y = yield stress of steel, in MPa

f_{cc} = elastic critical stress in compression $= \dfrac{\pi^2 E}{\lambda^2}$

λ = slenderness ratio

n = A factor assumed as 1.4 (ranges between 1-3)

The values of σ_{ac} based on the above formula for some of the I.S. structural steels (for yield stresses from 220 N/mm^2 to 540 N/mm^2) are stated in Table 4.4 of the text. The Merchant Rankine formula gives the average allowable compressive stress for the compression member. Note that it does not give the actual unit stress developed in the compression members by the load. The unit stress resulting from this formula when multiplied by the cross-sectional area of the compression member gives the allowable load which may be supported.

4.5 TYPES OF SECTIONS

Requirements for compression members are more demanding than those for tension members, for here the carrying capacity is a function of the shape as well as the area and also the material properties. The material must be disposed so as to resist effectively any tendency towards general or local instability. Thus the member must be sufficiently rigid to prevent general buckling in any possible direction, and each plate element of the member must be thick enough to prevent local buckling.

The most important property of the section in a compression member is the radius of gyration and thus the moment of inertia and it can be increased by spreading the material of section away from its axis (the material closer to the axis contributes less to it). An ideal section is one which has the same moment of inertia about any axis through its centre of gravity.

Various types of sections with which the students are already conversant may be used as single sections or in combination with either the same or different cross-sections. Of course, the practicability and availability of the sections will have to be ensured. Rolled steel sections cost less than the built-up sections per unit weight and are therefore, preferred.

Rods and bars withstand very little compression when length is more. Hence these are recommended for lengths less than 3 m only.

Tubular sections are most suitable for small loads and lengths. These sections are usually provided for roof trusses and bracings. The use of tubes as compression members was limited over decades due to the difficulty in making connections with rivets/bolts. But with the development of welding techniques its use has become frequent for the following reasons:

1. Tubes have the same radius of gyration in all directions and have a high local buckling strength.
2. Tubes have more torsional resistance.
3. In case of members subjected to wind, round tubes are subjected to less force than flat sections.

Table 4.4 *Permissible Stress σ_{ac} (MPa) in Axial Compression for Steels with Various Yield Stress*

$f_y \rightarrow$ / $\lambda \downarrow$	220	230	240	250	260	280	300	320	340	360	380	400	420	450	480	510	540
10	132	138	144	150	156	168	180	192	204	215	227	239	251	269	287	305	323
20	131	137	142	148	154	166	177	189	201	212	224	235	246	263	280	297	314
30	128	134	140	145	151	162	172	183	194	204	215	225	236	251	266	280	295
40	124	129	134	139	145	154	164	174	183	192	201	210	218	231	243	255	267
50	118	123	127	132	136	145	153	161	168	176	183	190	197	207	216	225	233
60	111	115	118	122	126	133	139	146	152	158	163	168	173	180	187	193	199
70	102	106	109	112	115	120	125	130	135	139	142	147	150	155	160	164	168
80	93	96	98	101	103	107	111	115	118	121	124	127	129	133	136	139	141
90	85	87	88	90	92	95	98	101	103	105	108	109	111	114	116	118	119
100	76	78	79	80	82	84	86	88	90	92	93	94	96	97	99	100	101
110	68	69	71	72	73	74	76	77	79	80	81	82	83	84	85	86	87
120	61	62	63	64	64	66	67	67	69	70	71	71	72	73	73	74	75
130	55	55	56	57	57	58	59	60	61	61	62	62	63	63	64	64	65
140	49	50	50	51	51	52	53	53	54	54	54	55	55	56	56	56	57
150	44	45	45	45	46	46	47	47	48	48	48	49	49	49	49	50	50
160	40	40	41	41	41	42	42	42	43	43	43	43	43	44	44	44	44
170	36	36	37	37	36	37	38	38	38	38	39	39	39	39	39	39	39
180	33	33	33	33	33	34	34	34	34	35	35	35	35	35	35	35	35
190	30	30	30	30	30	30	31	31	31	31	31	31	32	32	32	32	32
200	27	27	28	28	28	28	28	28	28	28	28	28	28	28	28	28	28
210	25	25	25	25	25	25	26	26	26	26	26	26	26	26	26	26	26
220	23	23	23	23	23	23	23	24	24	24	24	24	24	24	24	24	24
230	21	21	21	21	21	21	22	21	22	22	22	22	22	22	22	22	22
240	20	20	20	20	20	20	20	20	20	20	20	20	20	20	20	20	20
250	18	18	18	18	18	18	18	18	18	19	19	19	19	19	19	19	19

The least radius of gyration of a single angle section is small as compared to channels and I-Sections and hence, it is not suitable for long lengths. These are therefore, more commonly used as strut in roof trusses with riveted or welded connections. Single angle sections in general should therefore be avoided, where possible. Equal angles are more desirable and economical than unequal ones, because their least radius of gyration is greater for the same area of steel.

Where compression members are designed for very large structures it may be desirable to use built-up sections. When these are used they must be connected suitably on their open sides to hold the parts together in their proper positions and to assist them in acting together as an integral unit.

Double angles placed back to back (Fig. 4.4(a)) or with legs spread (Fig. 4.4(b)) are most suited for trusses. For a double angle section back-to-back, it is desirable to use unequal angles with the long legs back-to-back to achieve a balance between radius of gyration values about the X and Y axes.

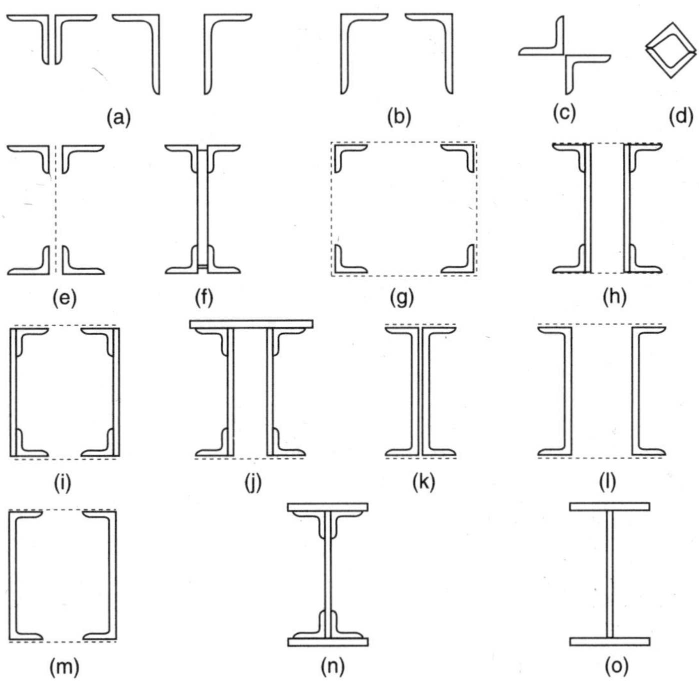

Fig. 4.4 *Shapes of Compression Members*

When the legs are spread, the least radius of gyration is same as that when placed back-to-back. The stiffness of double angle sections with legs spread is greater than when placed back-to-back but this is uneconomical because of the additional cost of lacing and riveting. However, it has no advantage over a double angle placed back-to-back regarding the least radius of gyration. Two angle sections can also be used in the form of a star (Fig. 4.4(c)). This cruciform arrangement of double angles is most effective because of its approximately equal radii of gyration in two directions. The box section can be formed with welded connection (Fig. 4.4(d)). The least radius of gyration of the section shown in Figs 4.4(c) and (d) of two equal angles is same and is more than that of the sections shown in Fig. 4.4(a, b).

Four angle sections can be used as shown in Fig. 4.4(e-g). The built-up section shown in Fig. 4.4(e) is provided when the load is small. For higher loads the section shown in Fig. 4.4(f) is preferred. In order to obtain higher values of least radius of gyration in shapes (e) and (f), the longer legs of unequal angles should be kept horizontal, as indicated. The section shown in Fig. 4.4(g) is used for moderate loads and large length, i.e. when the least radius of gyration required is large but due to excessive lacing it is uneconomical. The connecting systems can be replaced by plates on one side or on more than one side, when a greater cross-sectional area is required or where stiffer bracing between the parts is necessary, as shown in Fig. 4.4(h), (i), (j). Such sections are normally provided in bridge trusses.

Two channels placed back-to-back (Fig. 4.4(l)) result in a small value of radius of gyration about *y-y* axis and are, therefore, seldom recommended. Two channels spaced apart (Fig. 4.4(l)) constitute a good section. Two channels placed face to face (Fig. 4.4(m)) provide a larger value of radius of gyration as compared to channels back-to-back (Fig. 4.4(l)) and separated apart for the same spacing. Thus two channels face to face are the ideal section for compression members as these provide more rigidity. However, when a large radius of gyration is not the criteria for the type required, channels are placed back-to-back as the lacing is minimised. Also the exposed location of both ends of each rivet makes it easier to fabricate and further lower the cost. Channels placed face to face are also called placed toe-to-toe or placed with flanges turned inwards.

The dotted lines in the sections shown in Fig. 4.4 represent connecting systems such as lacings, etc., or discontinuous parts and the solid lines represent parts that are continuous for the full length of members.

Generally rolled I.S.H.B. section with additional intermediate support in the weak direction are used as column sections. For large loads, when rolled I-Section do not suffice, an I-Section can be built up with plates and angles in case riveted connections are done (Figs 4.4(f) and (n)). On the other hand, such a section is built up with plates only if welded connections are done (Fig. 4.4 (o)). Two I-Sections laced together can also be used for higher loads.

4.6 BUCKLING

When an axially loaded compression member becomes unstable overall (that is, not locally unstable), it can buckle in one of the following of three ways.

Flexural Buckling It is a deflection caused by bending, or flexure, about the axis corresponding to the largest slenderness ratio. This is usually the minor principal axis—the one with smallest radius of gyration. Compression members with any type of cross sectional configuration can fail in this way.

Torsional Buckling The flexural buckling considered above is due to bending alone, that is, the sections displace from their original position by translation without rotation. Thin wall members with open cross-sectional shape are sometimes weak in torsion and hence may buckle by twisting rather than bending. Torsional buckling occurs when the torsional rigidity of the member is appreciably smaller than its bending rigidity. This type of failure is caused by twisting about the longitudinal axis of member. It can occur only with doubly symmetrical cross sections with very slender cross-sectional elements. Standard hot-rolled shapes are not susceptible to torsional buckling, but a member built from thin plate elements may be and should be investigated. The cruciform shape shown in Fig. 4.4(c) is particularly vulnerable to this type of buckling.

Flexural-torsional Buckling This type of failure is caused by a combination of flexural buckling and torsional buckling. The member bends and twists simultaneously. This type of failure can occur only with unsymmetrical cross-sections, both those with one axis of symmetry—such as channels, structural tees, double angle shapes, and equal-leg single angles—and those with no axis of symmetry, such as unequal-leg single angles.

Generally, the specifications require an analysis of torsional or flexural-torsional buckling only when appropriate.

4.7 DESIGN OF AXIALLY LOADED COMPRESSION MEMBERS

Following are the assumptions made while designing a column.

1. The ideal column is assumed to be absolutely straight having no crookedness, which never occurs in practice.
2. The modulus of elasticity is assumed to be constant in a built-up column.
3. Secondary stresses (which may be of the order of even 25%-40% of primary stresses) are neglected.

The cross-sectional shape of an axially loaded column depends largely on the length and load on the column. In its design information required will be the length of the member, the end conditions and the loads it has to support. The designer is supposed to select a section which provides a large radius of gyration without providing more area and in which the average compressive stress does not exceed the allowable compressive stress. To compute these stresses, the cross-sectional area and the radius of gyration must be known. So there are two unknowns and it becomes essential to assume one out of the two based on some principles, and compute the other. The section is then checked for safety. The procedure is thus of trial and error and is as follows:

1. Average allowable compressive stress in the section is assumed. It should not be more than the upper limit for the column formula specified by the relevant code.
2. The cross-sectional area required to carry the load at the assumed allowable stress is computed.

$$A = \frac{P}{\text{allowable compressive stress}}$$

   where A is the tentative cross-sectional area required (in mm^2), and P is the load on column in Newtons.
3. A section that provides the estimated required area is selected from I.S. Hand Book No. 1. The section is so chosen that the minimum radius of gyration of the section selected is as large as possible. The appropriate least radius of gyration for the section selected is recorded.

4. The effective length of the column is calculated on the basis of end conditions, from Table 4.1 or Table 4.2 as appropriate, and the slenderness ratio is computed ($\lambda = l/r$), which should be less than the permissible slenderness ratio (Table 4.3).
5. For this estimated value of slenderness ratio, the maximum allowable compressive stress, σ_{ac} is calculated from Table 4.4 of the text. In case of struts this value may have to be reduced to 80 per cent depending upon the case as indicated in Table 4.2.
6. The load carrying capacity of the member is computed by multiplying the maximum compressive stress thus obtained with the cross sectional area provided. This value of the load carrying capacity of the member should be more than the load to be supported by it.

Notes

1. To assume allowable compressive stress in the first step of the design procedure, reference may be made to Table 4.4 of the text. For steel with f_y = 250 MPa a trial value of allowable compressive stress may be assumed between 60-85 MPa for struts and 85–110 MPa for columns.
2. Note that steps 1 and 2 are only for beginners. An experienced designer will select a section straightaway from IS Hand Book No. 1 and will apply check for its safety and economy.
3. If the load carrying capacity of the column exceeds the load over the column by more than 5%, the section needs a revision (being oversafe and uneconomical). However, many-times it may not be possible due to the limitation of sections available.

4.8 BUILT-UP COLUMNS (LATTICED COLUMNS)

The size and shape of standard rolled steel sections are limited because of the limitations of rolling mills. When rolled sections do not furnish the required sectional area or when a special shape or large radius of gyration is required in two different directions a built-up section is fabricated.

To achieve economy in the design of heavily loaded long columns the least radius of gyration of column section is increased to maximum. Rolled sections are kept away from the centroidal axis of the column and are connected by some connecting system. The commonly used lattice are lacing bars (Fig. 4.5(a), (b) and (d)), batten plates (Fig. 4.5(c)), lacing with battens (Fig. 4.7(e)) and, perforated cover plates.

When lacings are provided with battens, the shear component of the axial load tends to expand the column laterally. Because of lacing an accordian-like action of the system exists which means the lateral expansion of the column must accompany shortening under load, if there is to be no shortening of lacing bars. The introduction of battens restrain this action and the lacing system with battens behaves as a truss which may result in large stresses leading to failure. Therefore lacing with battens is not recommended.

When four angles are used as a built-up column, lacing is staggered (Fig. 4.5(d)). If the lacing connections in the two adjacent faces are made to coincide then the angle is fully restrained at that point and tends to buckle between the lacing connections in the weak direction (bending about V–V axis). On the other hand, if the lacing connections are staggered in the two adjacent faces then the angle is not fully restrained at the connections and tends to buckle along an axis other than V–V.

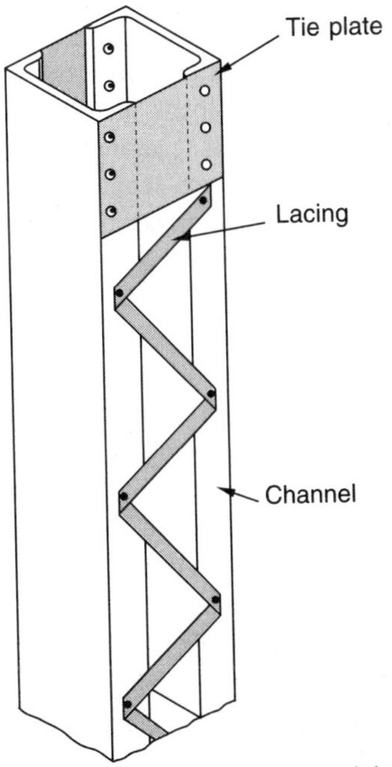

(a) Channels Face-to-face Connected by Single Lacing

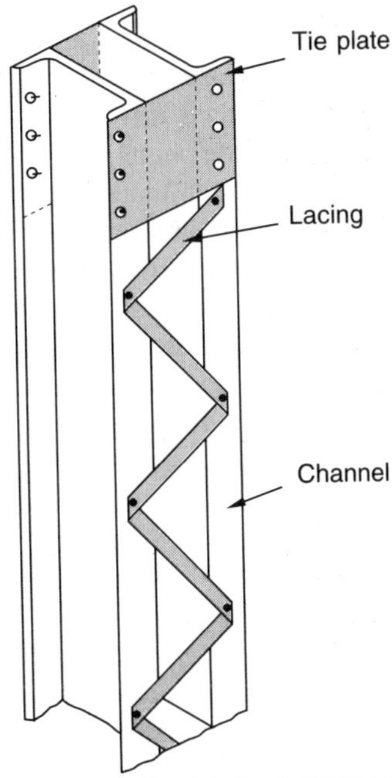

(b) Channels Back-to-back Connected by Single Lacing

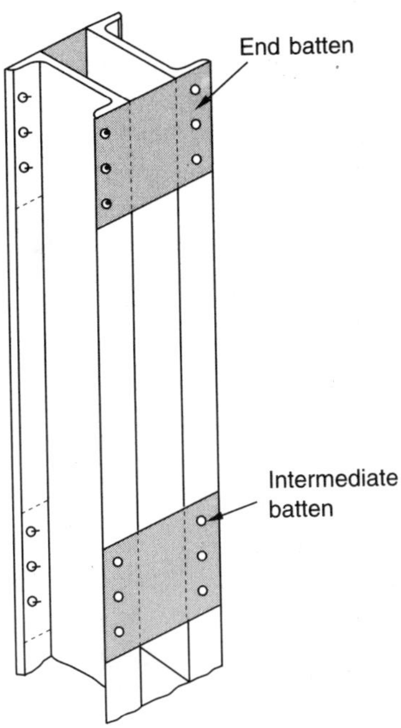

(c) Channels Back-to-Back Connected by Battens

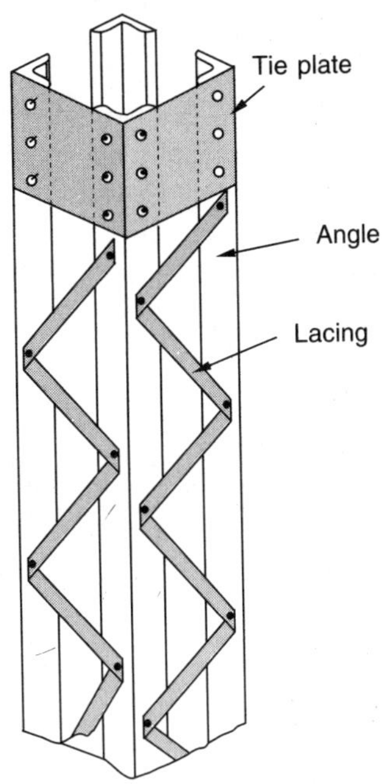

(d) Angles Connected by Single Lacing (Shown on two faces only)

Fig. 4.5 *Built up Columns*

Lattice, the lacing bars and batten plates, are not load carrying elements. Their function is primarily to hold the main component members (different sections) of the column in their relative position and equalize the stress distribution in them. They also provide points of intermediate support for each separate part of a built-up column. Thus for a built-up column lacings or battens are economical only if the increase in permissible stress for the load carrying members permit a greater reduction in weight than what is added by lacing or batten. However a built-up column, designed as an axially loaded column, may be accidentally loaded eccentrically or may have initial croockedness. Variable bending moments will be induced in such columns because of the eccentricity between the centroidal axis of the column and line of action of the applied load. Due to this there will be related shear forces in the plane of the cross-section and in the connecting system. The lattice must be capable of resisting the following loads: stresses due to external transverse loads on the column, stresses due to the deflected shape of the column, stresses due to shortening of main segments and stresses due to crushing induced by the curvature of the column. Usually, the latter two are negligible and are not considered in the design. The codes specify a minimum value of shear component as specified percentage of axial load on column to be considered in design.

Built-up columns are formed by using two or more structural shapes called main segments. These may in turn be assumed to be made up of a series of small columns of length C (Fig. 4.6(b)), formed due to the open lattice connection and are called the component columns. If this length C is large the component columns may buckle alternately in and out between the connections and the buckling would be as of pin-ended columns. The necessity of connecting the component members of a built-up column can be understood better by considering a compression member composed of several main component members, each of which will tend to buckle laterally if not tied together to act as an integral unit to support the load. If there were no such connection, each branch would be subjected under load to buckling with respect to its own axis. When connecting systems are installed, the stiffness of the column as a whole increases considerably. The sections used to form the column

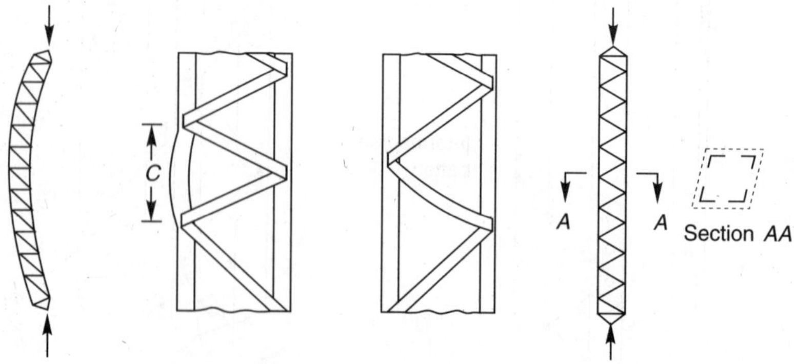

(a) Failure of Column as a Whole

(b) Failure of Main Component Segment

(c) Failure of Lacing

(d) Distortion of Cross-section

Fig. 4.6 *Failure Modes of a Latticed Column*

act as a single section, withstanding buckling with respect to the axis about which the sections were spaced.

The buckling strength of a latticed column is smaller than that of solid column having the same area and the same slenderness ratio provided that the solid column does not buckle locally because of thinness of material. This is because the shearing component of the axial load produces deformation in the lattice which tend to reduce the over-all stiffness of the column and therefore reduce the buckling strength of the column.

In the design of built-up columns the following conditions are considered, which are illustrated by a four angle section lattice column (Figs 4.6(a)–(d)).

1. Buckling of the column as a whole under the axial load (Fig. 4.6(a)).
2. Buckling of the component column (Fig. 4.6(b)).
3. Failure of lattice member (Fig. 4.6(c)).
4. Distortion of the cross-section (Fig. 4.6(d)).

4.9 LACING

Flat or angle sections are normally used as lacings. The purpose of lacing is to hold the various parts of a column straight, parallel, at a correct distance apart and to equalize the stress distribution between its various parts. Various arrangements of lacings with two channels back-to-back are shown in Figs 4.7(a)–(e). The firm lines representing lacings show single lacing on one face and the dotted lines represent lacing on the other face (Fig. 4.7 (a–c)). Double lacing system is shown in Fig. 4.7(d). Lacings can also be used with battens as shown in Fig. 4.7(e), but this arrangement is not preferred as it gives undesirable effects. Lacing bars should not

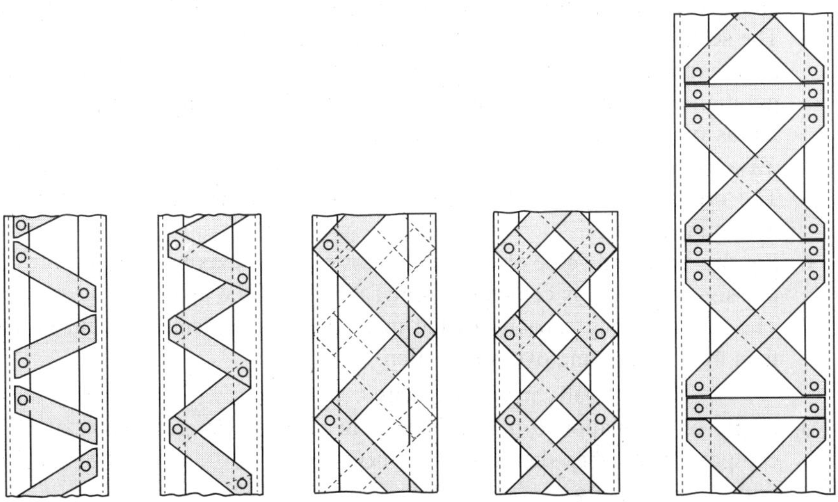

(a) Single Flat Lacing Non-Intersecting (b) Single Flat Lacing Intersecting (c) Single Flat Lacing, Non-Mirror Image (Not Recommended) (d) Double Flat Lacing Intersecting (e) Double Flat Lacing with Batten (Not Recommended)

Fig. 4.7 *Lacing Systems*

project beyond the column section. Usually these are connected with a single rivet at the end but sometimes two rivets are provided assuming that one rivet out of the two may be faulty, though the stresses are normally small and only one rivet may be sufficient. There may be possibility that a rivet connecting the lacing flat at a point fails, then the length C (Fig. 4.6(b)) of the component member will become double. This possibility can be overcome by either lacing the sections as shown in Fig. 4.7 (a) or by lacing the column on the far end as shown in Fig. 4.7(c). The latter is not recommended by the I.S. code. Single or double lacings can be provided depending upon the criteria discussed below and are then designed as compression members. The transverse shear force for which the lacing system is satisfactory if properly designed is usually very small and single lacing system is sufficient, but the designers consider double lacing system to be superior though uneconomical, and recommend it.

Design

The following are the steps in the design of laced columns:

1. Average allowable compressive stress is assumed. For steel with $f_y = 250$ MPa a beginner may assume a trial value between 110–150 MPa.
2. Required cross-sectional area to carry the load at the assumed allowable stress is computed.

$$A = \frac{P}{\text{allowable compressive stress}}$$

   where A = cross-sectional area in mm^2.
3. Select two channels or four angles or two I-sections with or without extra plates as required, from I.S. Handbook No. 1. The area provided is recorded.
4. The sections are so spaced that the radius of gyration of the section about the axis perpendicular to the plane of lacing should not be less than the radius of gyration about the axis in the plane of lacing. This is achieved by making the radius of gyration about the YY-axis equal to or greater than that about the XX-axis.
5. The effective length of the column is estimated and the slenderness ratio determined.
6. For the estimated value of the slenderness ratio the maximum allowable compressive stress σ_{ac} is computed from Table 4.4 of the text.
7. The load carrying capacity of the member is computed. It should be more than the load coming over the section.

$$\text{Load carrying capacity} = \sigma_{ac} A > P$$

8. Angle of inclination θ of the lacing bar with the longitudinal axis of the component member should be kept between 40°–70°, as per I.S. specifications.
9. The maximum spacing of lacing bars C should be such that the minimum slenderness ratio of the component member c/r_{yy} is not greater than 50 or 0.7 times the slenderness ratio of the member as a whole, where C is the length of component member and r_{yy} is the radius of gyration about YY-axis of the component member.

Note In the single lacing system, to satisfy this criteria the angle θ can be increased from 40° up to 70° only. If still the component column is found to be unsafe a double lacing system remains the only choice.

10. The lacing for compression member should be proportioned to resist a total transverse shear V equal to 2.5% of the axial force in the column. This shear V is divided equally in all parallel planes N in which there are shear resisting elements, such as lacing or continuous plates. Hence V/N is the transverse force to which the lacing is subjected (Fig. 4.8).

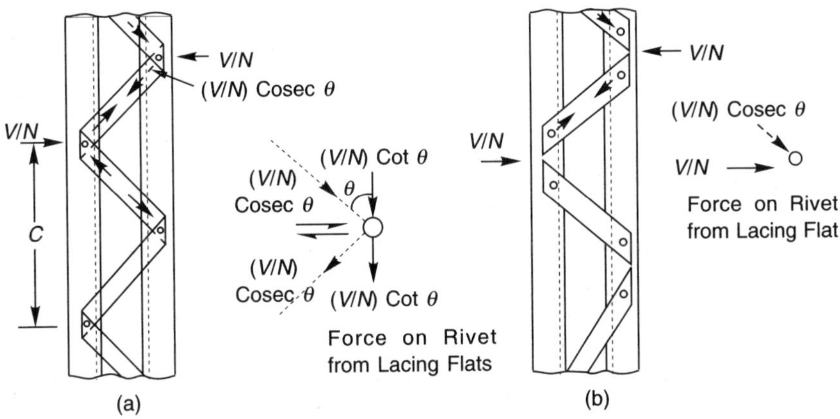

Fig. 4.8 *Force in Lacing and Connector*

Notes

1. $N = 2$ for two channels laced on both faces or four angles laced on all the faces.
2. To the shear V so determined any shear due to the weight of the member or due to other forces, is added and the lacing is proportioned for the combined shear.

11. The compressive force in the lacing bar is computed, which is equal to (V/N) cosecθ for single lacing system and $(V/2N)$ cosecθ, for a double lacing system.
12. The section of the lacing flat is initially assumed and then checked for safety.

Width First of all the diameter of the rivet is assumed. Depending upon this the flat width is selected from Table 4.5.

Table 4.5 *Width of Lacing Flat*

Nominal Diameter of Rivet in mm	*Width of Lacing Flat in mm*
16	50
18	55
20	60
22	65

Thickness If single lacing system is adopted, the thickness t of lacing flat $\nless$ 1/40 of the length between the inner end rivets or welds. If double lacing system is adopted, the thickness t of the lacing flat $\nless$ 1/60 of the length between the inner end rivets or welds.

Note Rolled sections other than flats, e.g. angle sections and tubes can also be used as lacing.

13. Minimum radius of gyration for the lacing flat is computed, $r = t/\sqrt{12}$.
14. The slenderness ratio of the lacing bar is computed which should be < 145. In a riveted construction, the effective length of the lacing bar is the length between the inner end rivets for a single lacing system and 0.7 times of this distance for a double lacing system. In welded constructions, the effective length of the lacing bar is 0.7 times the distance between the inner ends of the welds connecting the lacing bars to the member.
15. The compressive strength of the lacing flat is computed for the calculated value of the slenderness ratio of the lacing flat. This should be more than the force on the flat.
16. The tensile strength of the flat, which is equal to $(b - d)t\ \sigma_{at}$ and should be more than the force on the lacing flat, is calculated.
17. In case of riveted constructions, the rivet value is calculated and should be more than the load coming over the rivet. In case the two lacing flats are riveted at same point the load over the rivet will be $2(V/N)\cot\theta$, whereas for lacing flats connected at different points the load on the rivet will be (V/N) cosec θ. In case of welded construction, the welding of lacing bars to the main members should be sufficient to transmit the load in the bars. Welding should be provided along each side of the lacing bar for the full length of the lap. The overlap of lacing bar should not be less than four times the thickness of the bar or the member, whichever is less.
18. To check the distortion of the column section at the column ends tie plate (batten) is provided at the end of the lacing system and this should be able to resist the forces to which the lacing flat is subjected. The tie plate is designed as a batten plate.

4.10 BATTEN

Battens are plates or any other rolled sections used to connect the main components of compression members. Battens should be placed opposite to each other on the two parallel faces and should be spaced and proportioned uniformly throughout, as shown in Fig. 4.9. The number of battens should be such that the member is divided into at least three bays within its actual length. Battens provided at the ends of columns are called end battens and all other battens are called intermediate battens. Battened columns have the same strength as laced columns but are uneconomical and are therefore used rarely. Battens are not recommended for a column subjected to an eccentric load in the plane of a connecting system.

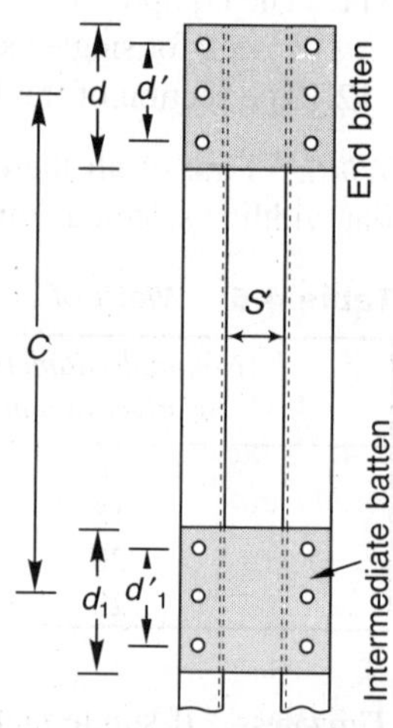

Fig. 4.9 *Batten*

Design

Column sections to be battened are designed in a way similar to a laced column.

The only difference is in the effective length which is increased by 10% in the case of a battened column. Once the column section is derived (steps 1 to 7 section 4.9.1), the battens can be designed as follows:

1. The maximum spacing of the battens C should be such that the minimum slenderness ratio of the component member C/r_{yy} is not greater than 50 or 0.7 times the slenderness ratio of the member as a whole, about an axis parallel to the battens.
2. The section of the batten is selected as follows and is then checked for the forces to which it is subjected.

Depth The effective depth of the end batten should not be less than the distance between the centre of gravity of the component members and this should be more than twice the width of one component member. The depth of the intermediate batten is taken as 3/4 of the effective depth of the end batten and should be more than twice the width of one component member. Refer to Fig. 4.9 where two channels have been shown placed back-to-back, spaced apart by a distance S', and connected by battens.

$$\text{Effective depth of end batten } d' = S' + 2c_{yy}$$

$$\text{Overall depth of end batten } d = d' + 2 \times \text{edge distance}$$

$$\text{Effective depth of intermediate batten } d_1' = \frac{3}{4} d'$$

$$\text{Overall depth of intermediate batten } d_1 = d_1' + 2 \times \text{edge distance}$$

where c_{yy} = the distance taken from I.S. Handbook No. 1 for the section

Thickness The thickness t of battens should not be less than 1/50 of the distance between the innermost connecting lines of the rivets or welds.

$$t = \frac{1}{50}(S' + 2g) \qquad \text{(for rivets in one row)}$$

where g = gauge distance taken from I.S. Handbook No. 1 for that particular section.

Note The above requirements of size and thickness do not apply if rolled sections, e.g., angles, channels, I-Sections, etc. are used as battens.

3. Battens should be designed to carry the bending moment and shear arising from the transverse shear force, V, which is 2.5% of the total axial force on the whole compression member. This transverse shear force is divided equally in all the parallel planes N in which there are shear resisting elements such as battens or continuous plates battens.
4. Battens should be able to resist the longitudinal shear and moment arising from the transverse shear V (Fig. 4.10).

$$V_1 = \frac{VC}{NS}; \qquad M = \frac{VC}{2N}$$

where V_1 is the longitudinal shear force, V is the transverse shear force, C is the centre to centre distance of battens and S is the minimum transverse distance between the centroids of the rivet group/weld.

Moment

The Two Forces V/2N Equal and Opposite at Distance C, Give Rise to Moment,

$$M = \frac{V}{2N} \times C = \frac{VC}{2N}$$

Longitudinal Shear

$$V_1 = \frac{VC/2N}{S/2} = \frac{VC}{NS}$$

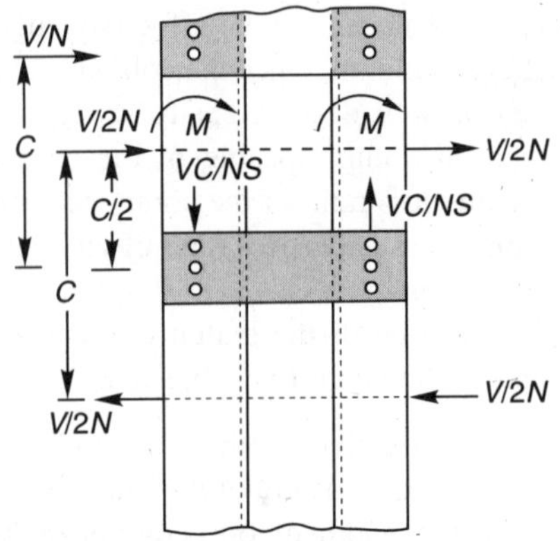

Fig. 4.10 *Forces on Battens*

5. The average shear stress is calculated in the section of battens and it should be less than the allowable average shear stress.
6. The bending stress in the section of the batten is calculated and it should be less than the allowable bending stress ($0.66f_y$).

$$\sigma_{bc,\,\text{cal}} = \frac{M}{Z} = \frac{M}{\frac{1}{6}td^2} = \frac{6M}{td^2}$$

7. The connection (riveted/welded) is designed to resist the longitudinal shear and bending moment. If riveted construction is done; not less than two rivets should be used. If welded construction is done, the aggregate length of weld on each edge (parallel to the depth) should not be less than half the depth of batten plate. At least one third of the weld should be placed at each end of this edge.

4.11 COMPRESSION MEMBERS COMPOSED OF TWO COMPONENTS BACK-TO-BACK

The compression members may also be designed as two rolled sections placed back-to-back and not connected by any connecting systems as has been discussed in Section 4.8. In such a case, the following specifications given by I.S: 800–1984 may be followed;

1. Two rolled sections when placed back-to-back or separated by a small distance should be connected together by rivets or welds so that the slenderness ratio of each member between the two consecutive connections is not greater than 40 or 0.6 times the slenderness ratio of the column as a whole.
2. The ends of the strut should be connected together with not less than two rivets or an equivalent weld and there should not be less than two additional connections spaced equidistant in the length of the strut. When there is small spacing between the two sections, washers and packings should be provided to make connections. When the legs of angles or the table of Tee is ≥ 125 mm or where web of channel is 150 mm wide, not less than two rivets or bolts should be used in each connection.

3. The rivets should not be less than 16 mm in diameter for members ≤ 10 mm thick, 20 mm in diameter for members ≤ 16 mm thick, and 22 mm in diameter for members > 16 mm thick.
4. Such compression members connected by rivets/welds should not be subjected to transverse loading in a plane perpendicular to the washer-riveted or welded surface.
5. When placed back-to-back the spacing of the rivets should not exceed $12t$ or 200 mm and the longitudinal spacing between intermittent weld should not be more than $16t$, where t is thickness of the thinner section.

4.12 ENCASED COLUMN

In tall buildings, designed with skeleton as a steel structure, the columns are encased in concrete to provide flush surfaces from the architectural point of view, (Fig. 4.11). It increases the fire resistance of the column and checks corrosion of the outer columns, which otherwise have to be painted regularly. Another use of this type of column is in the basements. In the design of such a column the following specifications as laid by I.S: 800–1984 must be followed:

1. The member should be a symmetrical I-Section shape or channels back-to-back with or without cover plates.
2. The overall dimensions of the steel sections should not exceed 750 × 450 mm over plating where used, the larger dimension being measured parallel to web.
3. The column should be unpainted and solidly encased in ordinary dense concrete with a 20 mm aggregate and of minimum M–15 grade.
4. The minimum width of solid casing is b_0 + 100 mm, where b_0 is the width of steel flange of the column.
5. The surface and edges of the steel column should have a concrete cover of not less than 50 mm.
6. The casing should be reinforced with steel wires in the form of strirrups which should be at least 5 mm diameter, 150 mm c/c. These stirrups are supported by 10 mm diameter bars at the four corners.
7. Steel core encased columns should be machined accurately at splices.

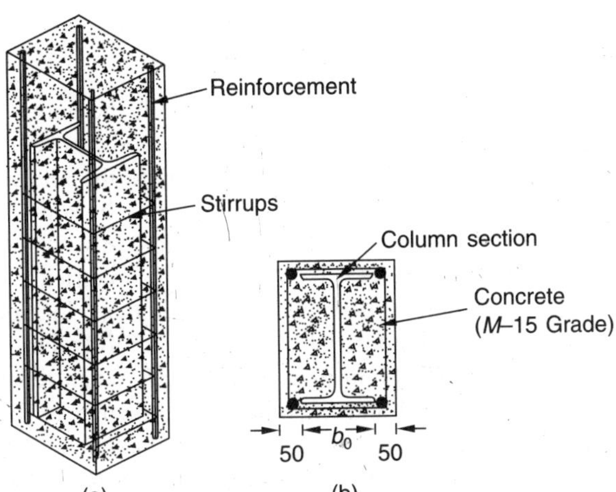

Fig. 4.11 *Cased Column*

Design

The steel column section is assumed to carry the entire load and is designed as a simple column but is checked in a different manner. An encased column is designed as follows:

1. Allowable working stress in the column is assumed and the cross-sectional area required for the load over it is computed.
2. A suitable section is chosen from I.S. Handbook No. 1. A section furnishing a large radius of gyration is preferred. The slenderness ratio of the column is then worked out. It should be less than 250.
3. From Table 4.4 of the text, the allowable axial stress in compression σ_{ac} is found.
4. The load carrying capacity of the column is worked out and is checked.
5. The load carrying capacity of the encased column is computed and it should not exceed two times the load permitted on an uncased column. To compute the load carrying capacity of the encased column the allowable compressive stress is calculated as follows:

   The radius of gyration for the encased column about *YY*-axis is given by $r_{yy} = 0.2\ (b_0 + 100)$ mm, where b_0 = width of the steel flange in mm.

   The slenderness ratio of the encased column l/r_{yy} is computed and the corresponding allowable stress in compression σ_{ac} is found from Table 4.4 of the text.
6. If a column is encased, the concrete is assumed to assist in carrying the load over its rectangular cross-section. The allowable compressive load in such a case is

$$P = A_{sc}\ \sigma_{sc} + A_c\ \sigma_c$$

where A_{sc} = cross-sectional area of steel

A_c = cross-sectional area of concrete,

σ_{sc} = permissible stress in steel in compression, and

σ_c = permissible stress in concrete in compression.

But this does not apply to struts of overall dimension greater than 1000 mm × 500 mm, the dimension of 1000 mm being measured parallel to the web or to the box sections.

4.13 ECCENTRICALLY LOADED COLUMNS

It is seldom that the axis of the load coincides with the geometrical axis of the column. The flexural stresses thus induced in the column have been given due consideration in the design formula used to calculate the allowable compressive stress. But various situations arise where the load has to be kept out of line with the column axis intentionally. Basically there can be two cases:

1. The line of action of the load is parallel to the geometrical axis of the column. Such a type of loading is called eccentric loading, e.g. load on column bracket from a gantry girder of industrial building, beam to column connections, etc. Also, the load on the column may be inclined. When resolved it gives two components, one parallel and the other perpendicular to the axis of the column. Such members are subjected to combined axial and bending stresses.
2. The line of the action of the load is perpendicular to the geometrical axis of the member. Such type of a load is called a lateral load, e.g. columns subjected to wind loads in tall buildings.

Apart from the above conditions, the compression member may be subjected to bending stresses and this should be given due consideration, e.g. struts in a truss bend due to their self weight inducing bending stresses.

Bending is very important in the case of compression members. The bending moment gives rise to thrust in the compression members which increases the lateral deflection. Therefore, loads which are initially axial become eccentric and assist bending. But in the case of tension members, the load tries to pull the member and reduces the lateral deflection of the member, if any. This is why the slenderness ratio of compression members is limited to below the limits for tension members.

Probably a more important point in the foregoing is the eccentricity of the load on the section. There is generally, however unintensional, an eccentricity of loading which causes bending with consequent additional fibre stress. Eccentricity of loading is more dangerous with a compression member than with a tension member, because in the former case an increase in the load is accompanied by an increase in buckling or displacement of the centre line of the section away from the load line, while in the latter case the centre line exhibits a greater tendency to coincide with the load line as explained above.

Eccentricity of the Load

When the load is transferred to the column from some eccentricity (the distance from the assumed point of application of the load to the centroid of column) the loads are assumed to be applied at positions as specified below:

1. *Stiffened seat* mid point of stiffening seat, (Fig. 4.12(a)) (middle of bearing length)
2. *Unstiffened seat* outer face of vertical leg of seat angle (Fig. 4.12(b))
3. *Cleats to web of beam* face of compression member (no eccentricity) (Fig. 4.12(c))

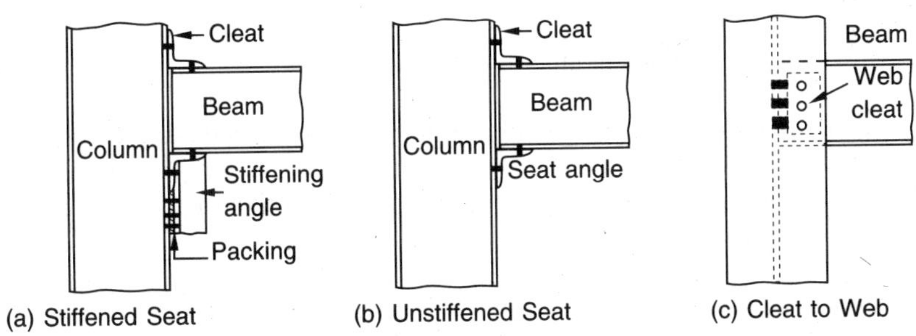

Fig. 4.12 *Eccentricity of Column Loads*

The eccentricity of the loads may be assumed as discussed and the bending moment may be computed by $M = Pe$, where P is the load and e the eccentricity. But for an exact analysis the following procedure should be carried out.

Let us consider a column AD, Fig. 4.13, with both ends hinged, to which a bracket plate CBE is attached. P is the load at an eccentricity of e from the centroidal axis of the section.

Let

$$AB = a$$

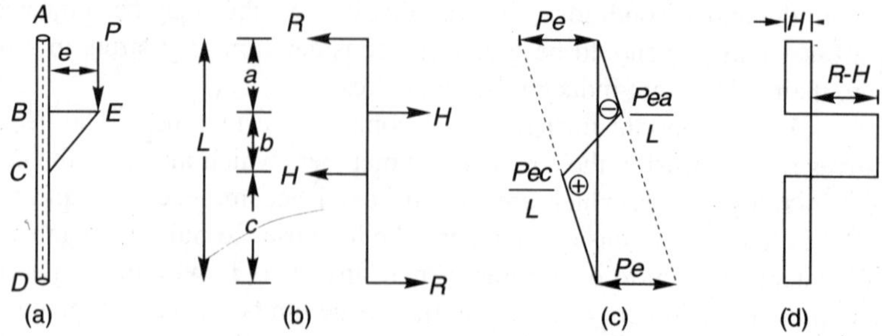

Fig. 4.13 *Shear Force and Bending Moment Diagrams for a Column with Bracket*

$$BC = b$$
$$CD = c$$

Consider the moment about the toe or heel (Fig. 4.13(a)), $M = Pe$. This gives rise to two equal and opposite reactions H at the ends of bracket plate, (Fig. 4.13(b)).

For equilibrium,

$$Hb = Pe$$

or
$$H = \frac{Pe}{b}$$

Let R be the reaction at the two ends A and D. The bending moments at these points are zero,

or
$$RL = Hb = Pe$$

or
$$R = \frac{Pe}{L}$$

$$\text{Bending moment at } B \text{ is} = -Ra = -\frac{Pea}{L}$$

$$\text{Bending moment at } C = +Rc = +\frac{Pec}{L}$$

The bending moment has a straight line variation in the portions AB, CB and CD and is plotted in Fig. 4.13(c).

The maximum value of the bending moment in the column can be when the bracket is at the top or at the bottom of the column.

$$\text{Bending moment} = R(L - b) = Pe\left(1 - \frac{b}{L}\right) \text{ at the top of bracket}$$

$$= R(L - b) = -Pe\left(1 - \frac{b}{L}\right) \text{ at the bottom of bracket}$$

Note It is observed that the maximum bending moment never reaches the value of Pe.

The shear force is plotted in Fig. 4.13(d). The shear force in various portions AB, BC, and CD of the column are R, $R - H$ and R respectively.

Similarly the values of bending moments for various end conditions of the column can be calculated and used for the design.

Combined Permissible Axial and Bending Compressive Stresses

In most of the practical situations as discussed above, columns are subjected to bending moments in addition to axial loads. For such a column section the maximum stress should not exceed the yield stress i.e.

$$\sigma_y = \frac{P}{A} + \frac{M}{z} \tag{4.4}$$

$$= \sigma_{ac,\,\mathrm{cal}} + \sigma_{bc,\,\mathrm{cal}}$$

or

$$\frac{\sigma_{ac,\,\mathrm{cal}}}{\sigma_y} + \frac{\sigma_{bc,\,\mathrm{cal}}}{\sigma_y} \le 1 \tag{4.5}$$

Since the permissible axial compressive stress depends upon the slenderness ratio, the permissible bending compressive stress depends on the lateral instability of the beam and different f.o.s. are used for loads and bending moments, Eq. (4.5) may be modified as follows.

$$\frac{\sigma_{ac,\,\mathrm{cal}}}{\sigma_{ac}} + \frac{\sigma_{bc,\,\mathrm{cal}}}{\sigma_{bc}} \le 1 \tag{4.6}$$

where $\sigma_{ac,\,\mathrm{cal}}$ = calculated average axial compressive stress
σ_{ac} = permissible axial compressive stress
σ_{bc} = permissible bending compressive stress
$\sigma_{bc,\,\mathrm{cal}}$ = maximum bending compressive stress without any consideration of additional moment due to axial load interacting with deflection

The analysis of members subjected to both bending and axial load is satisfactory so long as the axial load is not too large. The presence of axial load produces secondary moments, and unless the axial load is relatively small, these additional moments must be accounted for. This secondary moment is largest where the deflection is largest; the additional moment causes an additional deflection over and above that resulting from the transverse load. The deflections and these secondary moments can be found by second-order methods usually implemented with a computer program and are impractical for manual calculations. IS: 800 permits the use of moment amplification method. This method entails computing the maximum bending moment resulting from flexural loading by a first-order analysis, then multiplying by a moment amplification factor to account for the secondary moment. The expression for amplification factor is as follows.

$$\text{Amplification factor} = \frac{1}{1 - \dfrac{P}{P_E}}$$

Note The above amplification factor assumes the member braced against side sway that is, one who's ends cannot translate with respect to each other.

Euler load, $$P_E = \frac{\pi^2 EI}{l^2} = \frac{\pi^2 EAr^2}{l^2} = \frac{\pi^2 EA}{(l/r)^2} = \frac{\pi^2 EA}{\lambda^2}$$

$$\frac{P}{P_E} = \frac{P}{\pi^2 EA/\lambda^2} = \frac{n\sigma_{ac,\,\text{cal}}}{\pi^2 E/\lambda^2} = \frac{n\sigma_{ac,\,\text{cal}}}{f_{cc}}$$

Thus the maximum bending compressive stress taking into account the additional moment due to axial load interacting with deflections

$$= \frac{\sigma_{bc,\text{cal}}}{1 - \dfrac{P}{P_E}} = \frac{\sigma_{bc,\text{cal}}}{1 - \dfrac{n\sigma_{ac,\text{cal}}}{f_{cc}}}$$

where f_{cc} = elastic critical stress in compression = $\dfrac{\pi^2 E}{\lambda^2}$

n = factor of safety in compression

λ = slenderness ratio = $\dfrac{l}{r}$

P_E = Euler load

Eq. (4.6) may be rewritten as

$$\frac{\sigma_{ac,\,\text{cal}}}{\sigma_{ac}} + \frac{\sigma_{bc,\,\text{cal}}}{\left\{1 - \dfrac{n\sigma_{ac,\,\text{cal}}}{f_{cc}}\right\}\sigma_{bc}} \le 1 \tag{4.7}$$

Note One divided by the term within parentheses is referred to as the *amplification factor* which is based upon equal column end moments causing a single curvature deflection. It takes into account secondary moments. When f_{cc} is very large and the axial stress very small, the amplification factor is negligible. On the other hand, if σ_{ac} is significant the amplification factor increases the ratio $\sigma_{bc,\text{cal}}/\sigma_{bc}$ and accounts for the additional bending stress due to secondary moment caused by an axial load.

In the derivation of Eq. (4.7) it is assumed that the column is braced against side sway and is subjected to equal end moments producing single-curvature bending (Fig. 4.14(a)). Maximum moment amplification occurs at the centre where the deflection is largest. For equal end moments, the moment is constant throughout the length of the member, so the maximum primary moment also occurs at the centre. Thus the maximum primary moments and the maximum secondary moments are additive. However, if applied end moments produce reverse-curvature bending as shown in Fig. 4.14(c), the maximum primary moment will be at one of the ends, and maximum moment amplification occurs between the ends. Depending on the value of the axial load P, the amplified moment can be either larger or smaller than the end moment.

I.S: 800–1984 recommends the above interactive formula with a factor of safety n as 1/0.6 and incorporates a reduction factor C_m, to consider the end conditions (other than equal column end moments) and side sway of the columns in frames, which should be multiplied by the amplified bending stress ratio. Thus, Eq. (4.7) can be rewritten as

$$\frac{\sigma_{ac,\text{cal}}}{\sigma_{ac}} + \frac{C_m\,\sigma_{bc,\text{cal}}}{\left\{1 - \dfrac{\sigma_{ac,\,\text{cal}}}{0.6\,f_{cc}}\right\}\sigma_{bc}} \le 1 \tag{4.8}$$

The interactive Eq. (4.8) can be extended to make it applicable to compression members subjected to axial load and bending moments about both the axes. Thus

$$\frac{\sigma_{ac,\,cal}}{\sigma_{ac}} + \frac{C_{mx}\,\sigma_{bcx,\,cal}}{\left\{1-\frac{\sigma_{ac,\,cal}}{0.6\,f_{ccx}}\right\}\sigma_{bcx}} + \frac{C_{my}\,\sigma_{bcy,\,cal}}{\left\{1-\frac{\sigma_{ac,\,cal}}{0.6\,f_{ccy}}\right\}\sigma_{bcy}} \le 1 \tag{4.9}$$

If the additional moments due to axial loads interacting with deflections are neglected (as per I.S: 800–1984 if the ratio of $\sigma_{ac,\,cal}/\sigma_{ac} \le 0.15$), Eq. (4.9) reduces to

$$\frac{\sigma_{ac,\,cal}}{\sigma_{ac}} + \frac{\sigma_{bcx,\,cal}}{\sigma_{bcx}} + \frac{\sigma_{bcy,\,cal}}{\sigma_{bcy}} \le 1 \tag{4.10}$$

where

$\sigma_{bcx,\,cal}$ = calculated bending compressive stress due to the bending moment about major axis

σ_{bcx} = permissible bending compressive stress about major axis taking into account lateral instability

$\sigma_{bcy,\,cal}$ = calculated bending compressive stress due to bending moment about minor axis

σ_{bcy} = permissible bending compressive stress about minor axis

x–x, y–y = major and minor axis respectively

C_m = a coefficient called reduction factor (to account for end conditions and also in certain combinations of actual bending moments where the amplification factor may overestimate the secondary effect) whose value is established by relative size and direction of the column end moments and is never more than one. It may be taken as follows for members in frames:

(i) side sway not prevented, i.e., no bracing against sidesway buckling is provided, $C_m = 0.85$. It may be larger but not less than 0.85.

(ii) For braced columns, side sway is prevented and not subjected to transverse load between support in the plane of bending, $C_m = 0.6 - 0.4\beta \;\; \ge 0.4$.

where

β = ratio of smaller to larger moment at the ends of the member in the plane of bending. Figure 4.14 depicts the three end conditions frequently encountered in steel buildings.

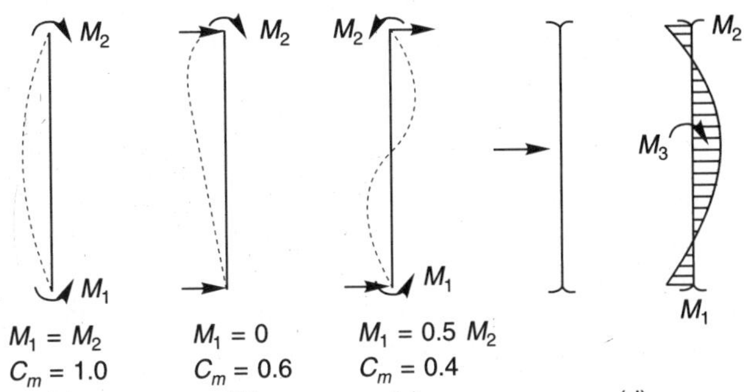

Fig. 4.14 *Reduction Factors for End Braced Columns*

The signs when using the ratio β are as follows:

This is in agreement with the definition of the amplification factor (i.e., single curvature bending is more critical for buckling instability than reversed curvature, and similarly the load-deflection effects will be reduced).

(iii) ends are restrained against rotation, $C_m = 0.85$

(iv) ends are not restrained against rotation, $C_m = 1.00$

When the column has transverse loading as shown in Fig. 4.14(d),

$$C_m = 1.0 + \frac{\psi \sigma_{ac,\,cal}}{\sigma_{ac}'}$$

where $\sigma_{ac,\,cal}$ = actual column stress

σ_{ac}' = Euler stress

ψ = factor depending on end restraint and transverse loading determined from Table 4.6.

Table 4.6 *Values of Factor ψ for Tranverse Loadings and Different End Conditions*

	ψ	ψ	
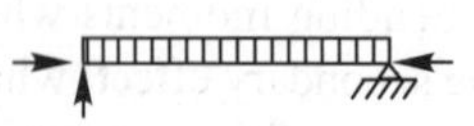	0.0	–0.2	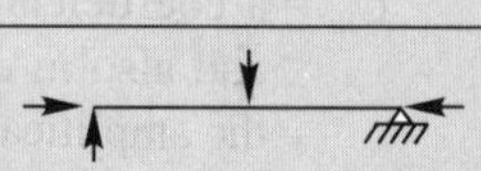
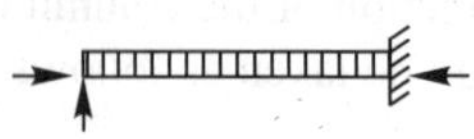	–0.3	–0.4	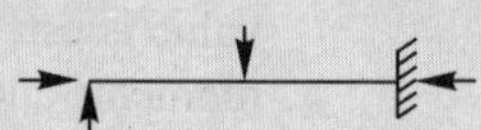
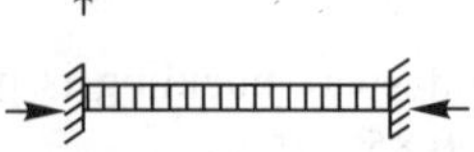	–0.4	–0.6	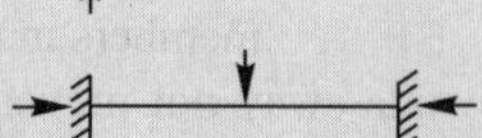

Design of Eccentrically Loaded Columns

Eccentrically loaded columns are subjected to direct and bending stresses. The following procedure may be adopted for its design:

1. The allowable working stress in compression is assumed and the cross-sectional area required for the axial load is computed.
2. The calculated sectional area may be increased by 50–100% to account for the eccentricity of the load or for the bending moment.
3. Suitable section is chosen from I.S. Hand Book No. 1. The slenderness ratio of the column is worked out.
4. From Table 4.4 of the text the allowable compressive stress σ_{ac} is determined for the calculated slenderness ratio.
5. The compressive stress $\sigma_{ac,\,cal}$ in the column is calculated by dividing the load with the sectional area provided $\sigma_{ac,\,cal} = P/A$.
6. The bending stress in the column is calculated by dividing the bending moment with the section modulus $\sigma_{bc,\,cal} = M/Z$.

7. The permissible bending stress is taken to be either $0.66f_y$ if bending is about the minor axis, or obtained from Table 7.1, of the text if bending is about the major axis. In the latter case permissible bending stresses are reduced to account for lateral instability.
8. The column section is checked for combined stresses as follows:

The ratio of $\sigma_{ac,\,cal}$ to σ_{ac} is worked out.

If $\dfrac{\sigma_{ac\,,\,cal}}{\sigma_{ac}} \leq 0.15$,

then
$$\frac{\sigma_{ac\,,\,cal}}{\sigma_{ac}} + \frac{\sigma_{bcx\,,\,cal}}{\sigma_{bcx}} + \frac{\sigma_{bcy\,,\,cal}}{\sigma_{bcy}} \leq 1$$

This is the straight line interaction formula which does not include the influence of secondary moments resulting from axial forces, the various combination of end moments and the possible side sway effect on a beam-column. The above formula in its revised form when $\dfrac{\sigma_{ac\,,cal}}{\sigma_{ac}} > 0.15$ is as follows.

$$\frac{\sigma_{ac\,,\,cal}}{\sigma_{ac}} + \frac{C_{mx}}{\left\{1 - \dfrac{\sigma_{ac\,,\,cal}}{0.60\,f_{ccx}}\right\}} \cdot \frac{\sigma_{bcx\,,\,cal}}{\sigma_{bcx}} + \frac{C_{my}}{\left\{1 - \dfrac{\sigma_{ac\,,\,cal}}{0.60\,f_{ccy}}\right\}} \cdot \frac{\sigma_{bcy\,,\,cal}}{\sigma_{bcy}} \leq 1$$

4.14 SPLICES

A joint when provided in the length of the column is called a splice. Splices for axially loaded compression members require little attention. If a member is loaded concentrically, no splice is required. Compression will be transmitted by direct bearing, and column sections could be rested one on top of each other and function satisfactorily. However, the load is never truly axial and the real column has to resist bending due to this eccentrically applied load. In addition to this the columns may be subjected to bending moments. Also, the bearing surfaces of the adjacent sections can never be machined to perfection. Thus the adjoining pieces must be positively connected. Column sections can be spliced in the following cases:

1. When the length of the column is more than the length of the column section available, a number of pieces are jointed to furnish the full length of the column.
2. In case of multistorey buildings, the section of the column required for the various storeys may be different, as the load goes on increasing for columns of the lower storeys.

Theoretically, a splice plate should be located at the point of contraflexure of the column. If the column ends are restrained in direction and position, this point will be at the middle of the column due to wind stresses. However, due to direct load there will be two points of contraflexure varying from the middle of column to the points above or below the middle, depending upon the amount of wind stresses.

Therefore, it is common practice to design a column of two-storey length and splice it at about 30–150 cm above the floor level. This arrangement also keeps the splice clear of beam or wind brackets.

In splicing compression members it is essential to notice the conditions of bearing surfaces of the sections being spliced, i.e. whether the ends have been faced/milled/machined for complete bearing or not. If the bearing in the two sections is achieved completely a large portion of the load passes down to the lower column directly and the splice is designed for the remaining load. However, if the exact proportion of the load transmitted by the splice plate cannot be determined accurately because of the difficulty to achieve perfect matching of the two column ends, it will be desirable to design the splice for the strength of the member. On the contrary, the tension splices are designed to transmit the full load. Following specifications are followed in the design of splices:

1. Where the ends of compression members are faced for complete bearing over the whole area, these should be spliced to hold the connected members accurately in position, and to resist any tension when bending is present.
2. Where such members are not faced for complete bearing, splices should be designed to transmit all the forces to which these are subjected.
3. Splices are designed as short columns.

The various types of splices used in compression members are shown in Figs 4.15 and 4.16. The splice system shown in Fig. 4.15(a) is used to connect two

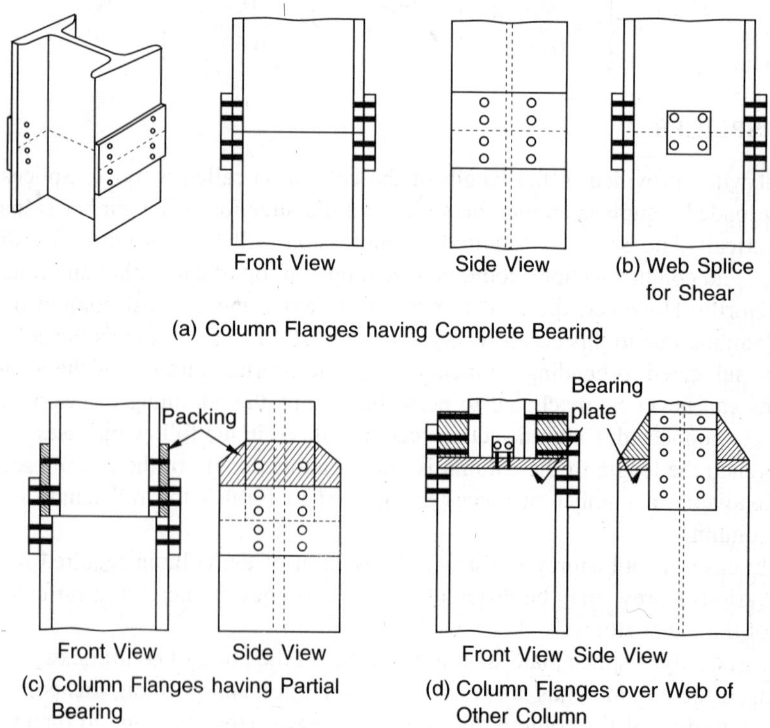

Fig. 4.15 *Column Splices*

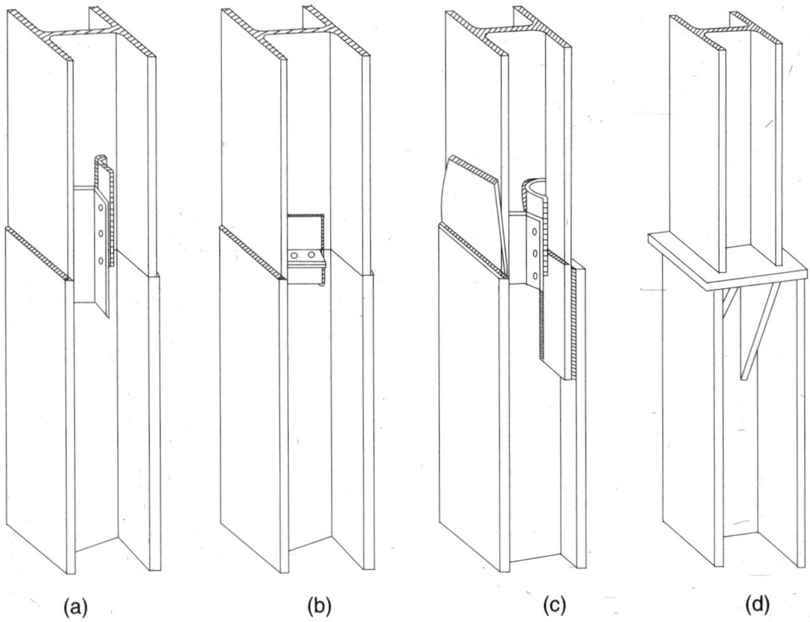

Fig. 4.16 *Weled Column Splices*

column sections having the same cross-section, i.e. the flanges of the upper storey column have full bearing over those of the lower storey column. If the flanges of the upper storey column have partial bearing over those of the lower storey column (i.e. the two column lengths do not have same cross-section) the splice shown in Fig. 4.15(c) is used. Additional packing plates will be required to ensure that the face of the upper column flanges is in the same plane as those of the lower storey column ones. If the flanges of the column of upper storey do not have any bearing over the flanges of the lower storey column (i.e., the flanges of the above storey column rest over the web of the lower storey column) then a bearing plate is provided over lower storey column. The size of the bearing plate is kept equal to the lower storey column cross-section and the next storey column is placed over the bearing plate (Fig. 4.15(d)). Suitably thick packing plates are provided on the two sides of the column of the upper storey and then splice plates are placed and connected.

When column sections are to be spliced by welding (Figs 4.16(a, b)) the ends are first milled for a square bearing surface. Then the two lower erection splice angles are shop-welded on opposite sides of the web of the heavier column section, so as to project past the end of the column. The outstanding legs of these angles are provided with holes for erection bolts to engage the outstanding legs of the other two angles that are shop welded to the upper column section. These erecting angles may be placed horizontally on the web of the column as shown in Fig. 4.16(b). Since they do not project beyond the ends of column, the possibility of damage during transit or erection is overcome. If the columns to be spliced are of different sections, Fig. 4.16(c), the splice plates are first shop-welded (fillet) to the inside face of the flange of the lower column. They are milled with the lower column section. As an alternate to this, splice plates with their lower edges prepared for welding are shop fillet welded to the outside face of the flanges on the upper column. Sometimes when new steel is erected adjacent to an old structure, a combination of this procedure may be used.

If the lower storey column is much deeper than the upper storey column section, stiffeners can be welded as shown in Fig. 4.16(d). These stiffeners will be placed directly below the flanges of the upper column which reduces the thickness of the bearing plate. If the section where splicing is done is subjected to shear force, then the web of the column is spliced with a pair of splice plates (one on each side of the web) as shown in Fig. 4.15(b).

Design

The first step is to ascertain the nature of the loads to which the splice is subjected. They are the axial compressive load, bending moment and occasionally shear.

1. For axial compressive load the splice plates are provided on the flanges of the two column sections to be spliced.

   If the column has machined ends, the splice is designed only to keep the columns in position and to carry tension due to the bending moment to which it may be subjected. The splice plate and the connections should be designed to carry 50% of the axial load and tension (if any due to bending moment). If the ends of the columns are not machined, the splice and connections are designed to resist the total axial load and any tension, if present due to the bending moment.

   The load P_1 for the design of splice and connection due to axial load P on the column will be $P/4$ (machined ends) and $P/2$ (non machined ends).

   The load due to bending moment, $P_2 = \dfrac{M}{\text{lever arm}}$

   where the lever arm is the c/c distance of the two splice plates and P_2 is the load due to the bending moment.
2. Splice plates are assumed to act as short columns with zero slenderness ratio. So these plates will be subjected to full allowable compressive stress ($0.6\, f_y$).
3. The cross-sectional area of the splice plate is calculated by dividing the load coming over the splice by the allowable compressive stress.
4. The width of the splice plate is usually kept equal to the width of the column flange. The thickness of the splice plate is found by dividing the cross-sectional area of the splice plate by the width of the splice plate provided.
5. Nominal diameter of the rivet for connections is assumed and the rivet value is computed. The number of rivets are found by dividing the total load coming on splice by the rivet value.
6. In case a bearing plate is to be designed between two column sections, the length and width of the plate are kept equal to the size of lower storey column and the thickness is computed by equating the moment due to the load to the moment of resistance of plate section.

Notes

1. If the joint is subjected to a bending moment along with the axial load, the splice plate is provided on flanges. The total load to which splice and connections are designed will

be sum of the axial load and the load due to the bending moment ($P_1 + P_2$). Rivets are also checked in tension.

2. When shear force acts in addition to gravity loads, a splice plate is provided on the web (as the web takes the maximum shear stress).
3. The splices should, as far as possible, be so placed that the centroidal axis of the splice coincides with the centroidal axis of the member jointed. This is done to avoid any eccentricity.
4. If packings are provided between the splice plate and the column flange additional rivets as per I.S. specifications are provided on the packings.

Solved Examples

Example 4.1 Calculate the value of the least radius of gyration for a compound column consisting of I.S.H.B. 250 @ 536.6 N/m with one cover plate 300 mm × 20 mm on each flange.

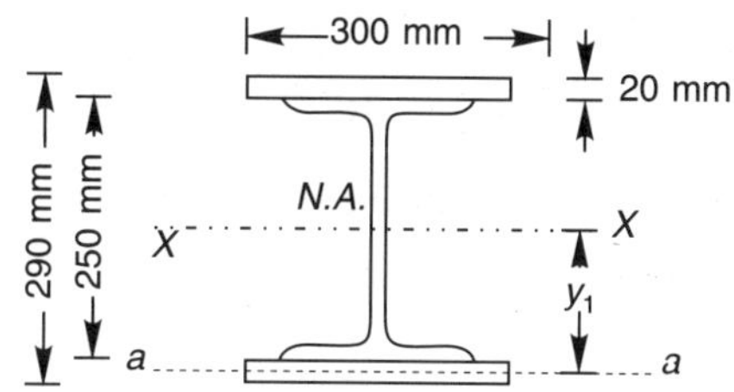

Fig. Ex. 4.1

Solution Refer to Fig. Ex. 4.1

Fom I.S, Handbook No. 1, I_{xx} of I.S.H.B. 250 = 7983.9×10^4 mm^4 and $A = 6971$ mm^2

$$I_{xx} \text{ for plates} = 2[I_{aa} + A_p y_1^2]$$

$$= 2\left[\frac{300 \times 20^3}{12} + 300 \times 20 \times (125 + 10)^2\right]$$

$$= 21910 \times 10^4 \text{ mm}^4$$

Total $I_{xx} = 7983.9 \times 10^4 + 21910 \times 10^4 = 29893.9 \times 10^4$ mm^4

Area of the built up section = $(6971 + 2 \times 300 \times 20) = 18971$ mm^2

$$r_{xx} = \sqrt{\frac{I_{xx}}{A}} = \sqrt{\frac{29,893.9 \times 10^4}{18,971}} = 125.52 \text{ mm}$$

I_{yy} of I.S.H.B. 250 @ 536.6 N/m = 2011.7×10^4 mm^4

$$I_{yy} \text{ of plates} = 2 \times \frac{20 \times (300)^3}{12} = 9000 \times 10^4 \text{ mm}^4$$

Total $I_{yy} = 2011.7 \times 10^4 + 9000 \times 10^4 = 11,011.7 \times 10^4$ mm^4

$$r_{yy} = \sqrt{\frac{I_{yy}}{A}} = \sqrt{\frac{11,011.7 \times 10^4}{18971}} = 76.187 \text{ mm}$$

Example 4.2 An I.S.A. 100 mm × 100 mm × 6 mm (f_y = 250 N/mm^2) is used as a strut in a truss. The length of the strut between the intersections at each end is 3.0 m. Calculate the strength of the strut if,

(a) it is connected by two rivets at each end
(b) it is connected by one rivet at each end
(c) it is welded at each end.

Solution From I.S. Handbook No. 1, the minimum radius of gyration of I.S.A. 100 × 100 × 6 mm, r_{vv} = 19.5 mm, A = 1167 mm^2

(a) L = 3.0 m

From Table 4.2 of the text, the effective length

$$l = 0.85 \times 3000 = 2550 \text{ mm}$$

$$\frac{l}{r} = \frac{2550}{19.5} = 130.769 < 350$$

From Table 4.4. of the text (Table 5.1 of I.S: 800–1984),

for l/r = 130.769 and f_y = 250 N/mm^2; σ_{ac} = 56.538 N/mm^2

Allowable working compressive stress = 56.538 N/mm^2

Strength of the strut = 56.538 × 1167

= 65,980.54 N = 65.98 kN

(b) L = 3.0 m

From Table 4.2 of the text, the effective length,

$$l = 1.0 \times 3000 = 3000 \text{ mm}$$

$$\frac{l}{r} = \frac{3000}{19.5} = 153.846 < 180$$

From Table 4.4 of the text (Table 5.1 of I.S: 800–1984),

for l/r = 153.846 and f_y = 250 N/mm^2; σ_{ac} = 43.46 N/mm^2

Allowable working compressive stress = 0.80 σ_{ac} = 0.80 × 43.46

= 34.76 N/mm^2

Strength of the strut = 34.76 × 1167

= 40,564.92 N = 40.564 kN

(c) When the strut is welded at each end, the specifications as laid by the I.S. code are similar to case (a) and, therefore, the strength of the strut will be 65.98 kN.

Example 4.3 Calculate the strength of a discontinuous strut of length 3.2 m. The strut consists of two unequal angles 100 mm × 75 mm × 8 mm (f_y = 250 N/mm^2), with long legs connected and placed:

(a) on the opposite side of a gusset plate (Fig. Ex. 4.3(a))
(b) on the same side of a gusset plate (Fig. Ex. 4.3(b))

Note The strut is tack riveted and is connected to a 10 mm gusset plate.

Solution From I.S. Handbook No. 1, the properties of two angles back-to-back and spaced 10 mm apart are:

$$A = 2672 \text{ mm}^2$$

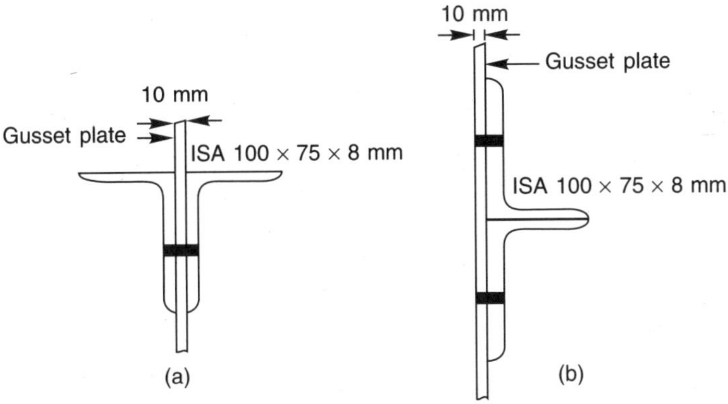

Fig. Ex. 4.3

(a) Angles placed on opposite side of gusset plate:

$$r_{xx} = 31.4 \text{ mm}, r_{yy} = 32.2 \text{ mm}$$

Minimum radius of gyration, $r = 31.4$ mm

$$l = 0.85L$$
$$= 0.85 \times 3.2 \times 10^3 = 2720 \text{ mm}$$

$$\frac{l}{r} = \frac{2720}{31.4} = 86.624$$

From Table 4.4 of the text (Table 5.1 of I.S: 800–1984),
for $l/r = 86.624$ and $f_y = 250$ N/mm^2; $\sigma_{ac} = 93.7136$ N/mm^2

Strength of the strut = 93.7136 × 2672
= 250,402.74 N = 250.40 kN

(b) Angles placed on same side of gusset plate:

$$r_{xx} = 21.8 \text{ mm}, r_{yy} = 44.1 \text{ mm}$$

$$l = L$$
$$= 3.2 \times 10^3 = 3200 \text{ mm}$$

$$\frac{l}{r} = \frac{3200}{21.8} = 146.78$$

From Table 4.4 of the text (Table 5.1 of I.S: 800–1984),
for $l/r = 146.78$ and $f_y = 250$ N/mm^2; $\sigma_{ac} = 46.93$ N/mm^2

Allowable working compressive stress = 0.80 × 46.93 = 37.54 N/mm^2

Strength of the strut = 37.54 × 2672
= 100,317.56 N = 100.317 kN

Example 4.4 Calculate the safe axial load for a stanchion I.S.H.B. 350 @ 710.2 N/m, 3.5 m high. It is to be used as an uncased column in a single storey building. The column is restrained in direction and position at both the ends. $f_y = 250$ N/mm^2

Solution

$$l = 0.65L$$
$$= 0.65 \times 3.5 \times 1000 = 2275 \text{ mm}$$

From I.S. Handbook No. 1, the relevant properties of I.S.H.B. 350 @ N/m are,

$$r_{xx} = 146.5 \text{ mm}, r_{yy} = 52.2 \text{ mm}, A = 9221 \text{ mm}^2$$

Minimum radius of gyration, $r = 52.2$ mm

$$\frac{l}{r} = \frac{2275}{52.2} = 43.58$$

For $\frac{l}{r} = 43.58$, and $f_y = 250$ N/mm^2,

$$\sigma_{ac} = 136.49 \text{ N/mm}^2$$

$\therefore$ Safe axial load $= 136.49 \times 9221$

$$= 1{,}258{,}574.29 \text{ N} = 1258.57 \text{ kN}$$

Example 4.5 Calculate the safe load over a compression member of unsupported length of 4 8 m. The member is to be used in a transmission line tower and is as shown in Fig. Ex. 4.5. The ends are held in position but not restrained in direction. The overall section of the member is 180 mm × 180 mm. Assume f_y = 250 N/mm^2.

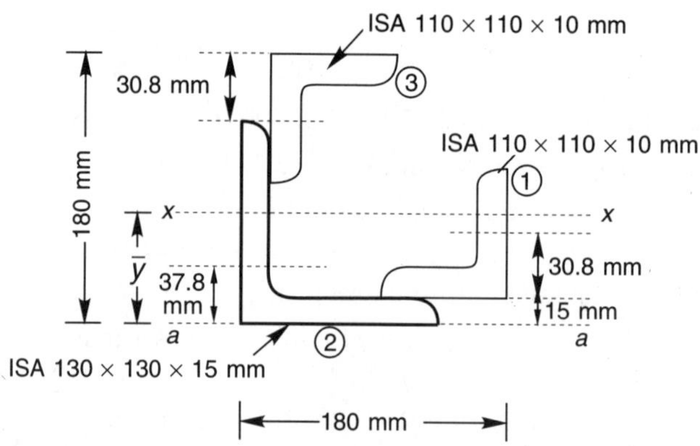

Fig. Ex. 4.5

Solution The relevant properties of the angle sections used from I.S. Handbook No. 1 are as follows:

I.S.A. 110 mm × 110 mm × 10 mm

$A = 2106$ mm^2, $I_{xx} = I_{yy} = 238.4 \times 10^4$ mm^4, $r_{xx} = r_{yy} = 33.6$ mm, $c_{xx} = c_{yy} = 30.8$ mm

I.S.A. 130 mm × 130 mm × 15 mm

$A = 3681$ mm^2, $I_{xx} = I_{yy} = 574.6 \times 10^4$ mm^4, $r_{xx} = r_{yy} = 39.5$ mm, $c_{xx} = c_{yy} = 37.8$ mm

Let the distance of the centroidal axis *xx* from the face *aa* of the section be $\bar{y}$. Taking the moment of the area about the axis *aa*,

$$(2106 + 3681 + 2106)\,\bar{y} = 2106(30.8 + 15) + 3681\,(37.8)$$
$$+ 2106(180 - 30.8)$$

or $$7893\,\bar{y} = 96{,}454.8 + 139{,}141.8 + 314{,}215.2$$

or $$\bar{y} = \frac{549{,}811.8}{7893} = 69.658 \text{ mm}$$

Moment of inertia about xx-axis (I_{xx}) can be found as follows:

$$I_{xx} = I_{xx1} + I_{xx2} + I_{xx3}$$

Moment of inertia of angle section 1 about centroidal axis xx,

$$I_{xx1} = 238.4 \times 10^4 + 2106 \times (69.658 - 30.8 - 15)^2$$
$$= 358.27 \times 10^4 \text{ mm}^4$$

Moment of inertia of angle section 2 about centroidal axis xx,

$$I_{xx2} = 574.6 \times 10^4 + 3681 \times (69.658 - 37.8)^2$$
$$= 948.19 \times 10^4 \text{ mm}^4$$

Moment of inertia of angle section 3 about centroidal axis xx,

$$I_{xx3} = 238.4 \times 10^4 + 2106(180 - 69.658 - 30.8)^2$$
$$= 1570.85 \times 10^4 \text{ mm}^4$$

$$I_{xx} = (358.27 + 948.19 + 1570.85) \times 10^4$$
$$= 2877.31 \times 10^4 \text{ mm}^4$$

The two angle sections (1) and (3) are placed in such a way that the moment of inertia about yy-axis will be same as that about the xx-axis. Hence r_{xx} and r_{yy} will be equal.

$$\text{Minimum radius of gyration, } r = r_{xx} = \sqrt{\frac{I_{xx}}{A}}$$

$$= \sqrt{\frac{2877.31 \times 10^4}{2106 + 3681 + 2106}} = 60.37 \text{ mm}$$

$$l = 4800 \text{ mm}$$

$$\frac{l}{r} = \frac{4800}{60.37} = 79.5$$

For $\frac{l}{r} = 79.5$, and $f_y = 250$ MPa,

$$\sigma_{ac} = 102.04 \text{ N/mm}^2$$
$$\text{Safe load} = 102.04 \times 7893 \times 10^{-3}$$
$$= 805.44 \text{ kN}$$

Example 4.6 Determine the allowable compressive load which the member shown in the Fig. Ex. 4.6 can support, if the member is of 5.5 m effective length. Assume $f_y = 250$ N/mm^2.

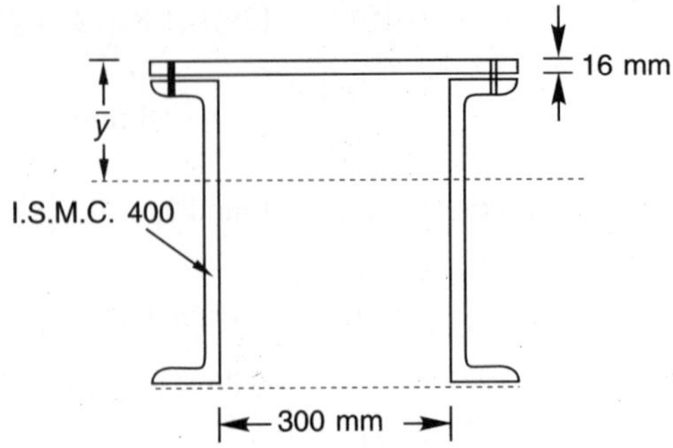

Fig. Ex. 4.6

Solution From I.S. Handbook No. 1, the relevant properties of ISMC 400 @ 484.6 N/m are,

A = 6293 mm^2, B = 100 mm, T = 15.3 mm, t = 8.6 mm, r_{xx} = 154.8 mm, g = 60 mm, I_{xx} = 15,082.8 × 10^4 mm^4, I_{yy} = 504.8 × 10^4 mm^4, C_{yy} = 24.2 mm

$$\text{Area of channel sections} = 2 \times 6293 = 12{,}586 \text{ mm}^2$$

$$\text{Area of plate section} = (300 + 2 \times 100) \times 16 = 8000 \text{ mm}^2$$

$$\text{Total area provided} = 12{,}586 + 8000 = 20{,}586 \text{ mm}^2$$

Let the distance of neutral axis from top be $\bar{y}$.

$$\bar{y} = \frac{2 \times 6293 \times \left(\frac{400}{2} + 16\right) + 8000 \times \frac{16}{2}}{20{,}586}$$

$$= 135.168 \text{ mm}$$

$$I_{xx} = 2[15{,}082.8 \times 10^4 + 6293 \times (216 - 135.168)^2]$$

$$+ \frac{500 \times 16^3}{12} + 500 \times 16 \times (135.168 - 8)^2$$

$$= 51{,}342.82 \times 10^4 \text{ mm}^4$$

$$I_{yy} = 2 \times 504.8 \times 10^4 + 16 \times \frac{500^3}{12} + 2 \times 6293 \times (150 + 24.2)^2$$

$$= 55869.22 \times 10^4 \text{ mm}^4$$

As I_{xx} is less than I_{yy}, the minimum radius of gyration will be r_{xx}.

$$\text{Minimum radius of gyration, } r = \sqrt{\frac{51342.82 \times 10^4}{20{,}586}} = 157.92 \text{ mm}$$

$$\frac{l}{r} = \frac{5.5 \times 10^3}{157.92} = 34.82$$

From I.S: 800–1984, the allowable compressive stress for l/r = 34.82 and f_y = 250 N/mm^2 is

$$\sigma_{ac} = 142.10 \text{ N/mm}^2$$

$$\text{Allowable safe load} = 142.10 \times 20586$$

$$= 2920356 \text{ N} = 2920.356 \text{ kN}$$

Example 4.7 Design a single angle discontinuous strut to carry a load of 47 kN. The length of the strut is 3.0 m between intersections. The strut is connected to 12 mm thick gusset plate with (a) 24 mm ϕ power driven rivet(s) (b) 20 mm ϕ hand driven rivet (s).

Solution Let us assume allowable compressive stress of 60 MPa.

$$\text{Cross-sectional area required} = \frac{47 \times 10^3}{60}$$

$$= 783.33 \text{ mm}^2$$

(a) *Strut connected by 24 mm ϕ rivet(s)*

$$\phi = 24 \text{ mm}$$

$$d = 24 + 1.5 = 25.5 \text{ mm}$$

$$\text{Strength of rivet in single shear} = \frac{\pi}{4} d^2\ \tau_{vf}$$

$$= \frac{\pi}{4} 25.5^2 \times 100$$

$$= 51000 \text{ N} = 51 \text{ kN}$$

$$\text{Strength of rivet in bearing} = d\ t\ \sigma_{pf}$$

$$= 25.5 \times 8 \times 300 = 61200 \text{ N} = 61.2 \text{ kN}$$

$$\text{Hence, rivet value, } R_v = 51 \text{ kN}$$

$$\text{Number of rivets required, } N = \frac{47}{51} = 0.92 \simeq 1$$

Let us provide 24 mm ϕ power driven one rivet at each end of the strut for making the connection.

From I.S. Handbook No. 1 select an angle section giving a value close to the required area (783.33 mm^2). Provide a trial section I.S.A. equal angle $60 \times 60 \times 8$ mm (weight 68.7 N/m).

$$A = 896 \text{ mm}^2$$

$$\text{Minimum radius of gyration, } r = r_{vv} = 11.5 \text{ mm}$$

$$l = 1.0 \times 3 \times 10^3 = 3000 \text{ mm}$$

$$\frac{l}{r} = \frac{3000}{11.5} = 260.86 > 180 \qquad \text{not safe}$$

Since the section is not safe for the slenderness ratio limit, calculations for its load carrying capacity are meaningless and are, therefore, not made.

Select another trial section I.S.A. $100 \times 100 \times 8$ mm (weight = 118.7 N/m).

$$A = 1539 \text{ mm}^2$$

$$\text{Minimum radius of gyration, } r = 19.5 \text{ mm}$$

$$\frac{l}{r} = \frac{3000}{19.5} = 153.84 < 180$$

For $\frac{l}{r}$ = 153.84 and f_y = 250 MPa,

$$\sigma_{ac} = 43.46 \text{ N/mm}^2$$

$$\text{Allowable compressive stress} = 0.8 \times 43.46 = 34.76 \text{ N/mm}^2$$

$$\text{Load carrying capacity} = 34.76 \times 1539$$

$$= 53509.8 \text{ N} = 53.50 \text{ kN}$$

$$> \text{load over strut}$$

(b) *Strut connected by 20 mm ϕ rivet(s)*

$$\phi = 20 \text{ mm}$$

$$d = 20 + 1.5 = 21.5 \text{ mm}$$

Let us provide 20 mm ϕ hand driven 2 rivets at each end of the strut for making connection.

$$\text{Strength of rivet in single shear} = \frac{\pi}{4} d^2 \tau_{vf}$$

$$= \frac{\pi}{4} \times 21.5^2 \times 80 = 29044 \text{ N} = 29.04 \text{ kN}$$

$$\text{Strength of rivet in bearing} = d\, t\, \sigma_{pf}$$

$$= 21.5 \times 10 \times 250 = 53750 \text{ N} = 53.75 \text{ kN}$$

$$\text{Hence, rivet value, } R_v = 29.04 \text{ kN}$$

$$\text{Number of rivets required, } N = \frac{47}{29.04} = 1.618 \simeq 2$$

Select a trial section 80 × 80 × 10 mm.

$$A = 1505 \text{ mm}^2$$

$$\text{Minimum radius of gyration, } r = 15.5 \text{ mm}$$

$$l = 0.85 \times 3000 = 2550 \text{ mm}$$

$$\frac{l}{r} = \frac{2550}{15.5} = 164.516 < 350$$

For $\frac{l}{r}$ = 164.516 and f_y = 250 MPa,

$$\sigma_{ac} = 39.1936 \text{ N/mm}^2$$

$$\text{Load carrying capacity} = 39.1936 \times 1505$$

$$= 58{,}986.36 \text{ N} = 58.98 \text{ kN}$$

$$> \text{load over the strut}$$

The above section is oversafe, therefore, let us select another trial section. Try I.S.A. 100 × 75 × 8 mm (weight 103 N/m).

$$A = 1336 \text{ mm}^2$$

$$\text{Minimum radius of gyration, } r = 15.9 \text{ mm}$$

$$\frac{l}{r} = \frac{2550}{15.9} = 160.37$$

For $\frac{l}{r}$ = 160.37 and f_y = 250 MPa,

$$\sigma_{ac} = 40.852 \text{ N/mm}^2$$

$$\text{Load carrying capacity} = 40.852 \times 1336$$

$$= 54578.272 \text{ N} = 54.57 \text{ kN}$$

$$> \text{load over the strut}$$

It should be noted that by merely increasing one rivet in the connection, considerable economy has been achieved. The weight of section, when two rivets are used for connection reduces from 118.7 N/m to 103 N/m.

Example 4.8 Design a double angle discontinuous strut to carry a load of 90 kN. The length of the strut is 3.0 m between intersections. The two angles are placed back to back (with long legs connected) and are tack riveted.

(a) Angles are placed on opposite sides of 12 mm gusset plate.
(b) Angles are placed on same side of 12 mm gusset plate.

Solution Let us assume allowable compressive stress of 60 MPa.

$$\text{Cross-sectional area required} = \frac{90 \times 10^3}{60} = 1500 \text{ mm}^2$$

(a) *Angles placed on opposite sides of gusset plate* From I.S. Handbook No. 1 select 2 I.S.A. 65 × 65 × 8 mm (weight = 151 N/m).

$$\text{Area provided} = 1952 \text{ mm}^2$$

$$\text{Minimum radius of gyration, } r = r_{xx} = 19.6 \text{ mm}$$

$$l = 0.85L$$

$$= 0.85 \times 3000 = 2550 \text{ mm}$$

$$\frac{l}{r} = \frac{2550}{19.6} = 130.10 < 350$$

For $\frac{l}{r}$ = 130.10 and f_y = 250 MPa,

$$\sigma_{ac} = 56.94 \text{ N/mm}^2$$

$$\text{Safe load} = 56.94 \times 1952$$

$$= 111{,}146.88 \text{ N} = 111.146 \text{ kN}$$

$$> \text{load over the strut}$$

The designer may now try to economise the section.

Try a section having a greater radius of gyration and less weight/m run. Select 2 I.S.A. 70 × 70 × 6 mm (weight = 123.6 N/m).

$$A = 1612 \text{ mm}^2$$

$$\text{Minimum radius of gyration, } r = 21.4 \text{ mm}$$

$$\frac{l}{r} = \frac{2550}{21.4} = 119.158 < 350$$

For $\frac{l}{r}$ = 119.158 and f_y = 250 MPa,

$$\sigma_{ac} = 64.67 \text{ N/mm}^2$$

$$\text{Safe load} = 64.67 \times 1612$$

$$= 104{,}248.04 \text{ N} = 104.248 \text{ kN}$$

$$> \text{load over the strut}$$

Provide 2 I.S.A. 70 × 70 × 6 mm on opposite side of the gusset plate (Fig. Ex. 4.8(a)). Frcm the above trial it is observed that the weight of the section is reduced from 151 N/m to 123.6 N/m and the section is still oversafe. Therefore, if desired another section may be tried to bring further economy.

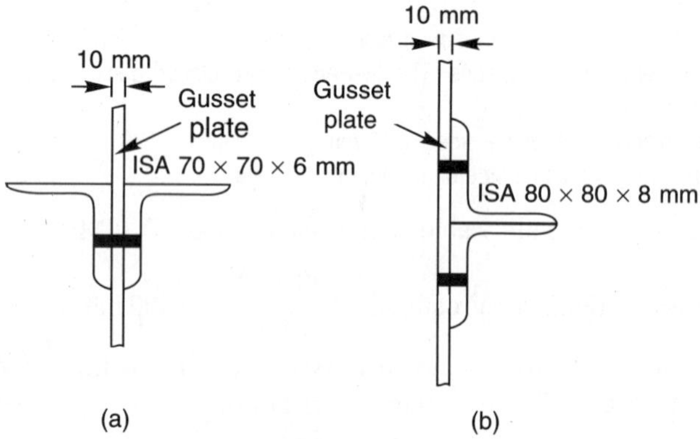

Fig. Ex. 4.8

(b) *Angles placed on same side of gusset plate* Select 2 I.S.A. 65 × 65 × 8 mm

$$A = 1952 \text{ mm}^2$$

$$\text{Minimum radius of gyration, } r = 19.6 \text{ mm}$$

$$l = 3000 \text{ mm}$$

$$\frac{l}{r} = \frac{3000}{19.6} = 153.06 < 350$$

For $\frac{l}{r} = 153.06$ and $f_y = 250$ MPa,

$$\sigma_{ac} = 43.776 \text{ N/mm}^2$$

$$\text{Allowable working compressive stress} = 0.8 \times 43.776 = 35.02 \text{ N/mm}^2$$

$$\text{Safe load} = 35.02 \times 1952$$

$$= 68{,}359 \text{ N} = 68.359 \text{ kN}$$

$$< \text{load over the strut}$$

The design is not satisfactory.

Select 2 I.S.A. 80 × 80 × 6 mm (weight = 143.2 N/m).

$$A = 1858 \text{ mm}^2$$

$$\text{Minimum radius of gyration, } r = r_{xx} = 24.6 \text{ mm}$$

$$\frac{l}{r} = \frac{3000}{24.6} = 121.95 < 350$$

For $\frac{l}{r} = 121.95$ and $f_y = 250$ MPa,

$$\sigma_{ac} = 62.635 \text{ N/mm}^2$$

$$\text{Allowable working compressive stress} = 0.8 \times 62.635 \text{ N/mm}^2 = 50.108 \text{ N/mm}^2$$

$$\text{Safe load} = 50.108 \times 1858 = 93{,}100.664 \text{ N} = 93.100 \text{ kN} > \text{load over strut}$$

This design is safe and economical and, therefore, satisfactory. Provide 2 I.S.A. 80 × 80 × 6 mm on the same side of gusset plate (Fig. Ex. 4.8(b)).

Note From this example it can be concluded that it is economical to provide two angles on the opposite side of the gusset plate.

Example 4.9 Design a column to support an axial load of 700 kN. The column has an effective length of 7 m with respect to the *x*-axis and 5.0 m with respect to the *y*-axis. $f_y = 250$ N/mm^2

Solution Let us assume the allowable axial compressive stress as 85 N/mm^2.

$$A = \frac{700 \times 10^3}{85} = 8235.29 \text{ mm}^2$$

From I.S. Handbook No. 1 try I.S.H.B. 350 @ 661.2 N/m. The relevant properties of the section are,

$$A = 8591 \text{ mm}^2$$

$$r_{xx} = 149.3 \text{ mm}, \ r_{yy} = 53.4 \text{ mm}$$

$$\frac{l_x}{r_{xx}} = \frac{7 \times 10^3}{149.3} = 46.885 < 180$$

which is safe.

$$\frac{l_y}{r_{yy}} = \frac{5 \times 10^3}{53.4} = 93.63 < 180$$

which is safe.

From Table 4.4 of the text for $l_x/r_{xx} = 46.885$ and $f_y = 250$ MPa

$$\sigma_{ac1} = 134.1805 \text{ N/mm}^2$$

$$\text{Load carrying capacity} = 134.1805 \times 8591 = 1152.744 \text{ kN} > 700 \text{ kN}$$

which is all right.

For $l_y/r_{yy} = 93.63$, and $f_y = 250$ MPa,

$$\sigma_{ac2} = 86.37 \text{ N/mm}^2$$

$$\text{Load carrying capacity} = 86.37 \times 8591 = 742004.67 \text{ N} = 742 \text{ kN} > 700 \text{ kN}$$

which is safe.

Hence, provide I.S.H.B. 350 @ 661.2 N/m.

Example 4.10 Design a stanchion 3.5 m long in a building subjected to an axial load of 350 kN. Both the ends of the stanchion are effectively restrained in direction and position. Asume f_y = 250 MPa.

Solution Assume allowable axial compressive stress = 85 N/mm^2

$$\text{Cross-sectional area required} = \frac{350 \times 10^3}{85} = 4117.64 \text{ mm}^2$$

From I.S. Handbook No. 1 select I.S.H.B. 150 @ 339.4 N/m.

The relevant properties of the section are,

$$A = 4408 \text{ mm}^2$$

$$\text{Minimum radius of gyration, } r = r_{yy} = 33.5 \text{ mm}$$

$$l = 0.65\,L$$

$$= 0.65 \times 3.5 \times 10^3 = 2275 \text{ mm}$$

$$\frac{l}{r} = \frac{2275}{33.5} = 67.91 < 180$$

which is as it should be.

For $\frac{l}{r} = 67.91$ and $f_y = 250$ N/mm^2,

$$\sigma_{ac} = 114.09 \text{ N/mm}^2$$

$$\text{Safe load carrying capacity} = 114.09 \times 4408$$

$$= 502{,}908.7 \text{ N} = 502.90 \text{ kN} > 350 \text{ kN}$$

This section is oversafe, therefore let us try another one. Select I.S.H.B. 150 @ 265.9 N/m. The relevant properties are,

$$A = 3448 \text{ mm}^2$$

$$\text{Minimum radius of gyration, } r = r_{yy} = 35.4 \text{ mm}$$

$$\frac{l}{r} = \frac{2275}{35.4} = 64.265 < 180$$

For $\frac{l}{r} = 64.265$ and $f_y = 250$ MPa,

$$\sigma_{ac} = 117.735 \text{ N/mm}^2$$

$$\text{Safe load carrying capacity} = 117.735 \times 3448 \times 10^{-3}$$

$$= 405.950 \text{ kN}$$

$$> 350 \text{ kN}$$

which is all right.

Hence, provide I.S.H.B. 150 @ 265.9 N/m.

Example 4.11 An 8.0 m long column of an industrial building supports a bracket 1.5 m × 1.0 m and 1.8 m below the end supporting truss as shown in Fig. Ex. 4.11. The reaction from the truss is 140 kN. The bracket supports a beam transmitting an end reaction of 75 kN. The eccentricity of the end reaction from the column axis is 1.00 m. The ends of the column are restrained in position but not in the direction. Design the column.

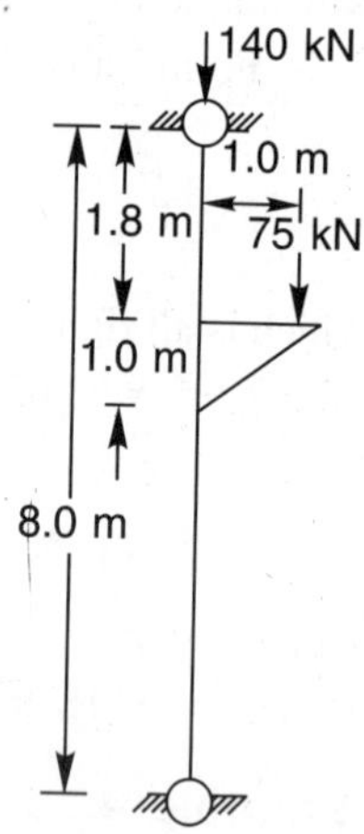

Fig. Ex. 4.11

Solution

$$l = L$$

$$= 8.0 \text{ m} = 8 \times 10^3 \text{ mm}$$

$$e = 1 \times 10^3 \text{ mm}$$
$$a = 1.8 \times 10^3 \text{ mm}$$
$$c = (8 - 1.8 - 1) \times 10^3 = 5.2 \times 10^3 \text{ mm}$$

$$\text{Maximum negative bending moment} = \frac{Pea}{l}$$
$$= \frac{75 \times 10^3 \times 1 \times 10^3 \times 1.8 \times 10^3}{8 \times 10^3}$$
$$= 16{,}875 \times 10^3 \text{ Nmm}$$

$$\text{Maximum positive bending moment} = \frac{Pec}{l}$$
$$= \frac{75 \times 10^3 \times 1 \times 10^3 \times 5.2 \times 10^3}{8 \times 10^3}$$
$$= 48{,}750 \times 10^3 \text{ Nmm}$$

Hence, maximum design moment = 48750×10^3 Nmm

Design axial load = $140 \times 10^3 + 75 \times 10^3 = 215 \times 10^3$ N

Assuming allowable axial compressive stress $\sigma_{ac} = 70$ N/mm^2,

$$A = \frac{215 \times 10^3}{70}$$
$$= 3071.42 \text{ mm}^2$$

Let area thus calculated be increased by 100% to account for bending moment.

$$\text{Area} = 2 \times 3071.42 = 6142.85 \text{ mm}^2$$

From I.S. Handbook No. 1 select a suitable section furnishing cross-sectional area next to the value of area as calculated above. Provide I.S.H.B. 250 @ 500.3 N/m. The relevant properties of the section are,

$$A = 6496 \text{ mm}^2,\ r_{xx} = 109.1 \text{ mm},\ Z_{xx} = 618.9 \times 10^3 \text{ mm}^3$$
$$r_{yy} = \text{Minimum radius of gyration, } r = 54.9 \text{ mm,}$$
$$\frac{l}{r} = \frac{8000}{54.9} = 145.719 < 180$$

For $\frac{l}{r} = 145.719$ and $f_y = 250$ MPa,

$$\sigma_{ac} = 49.84 \text{ N/mm}^2$$

Now,
$$\frac{T}{t} = \frac{9.7}{6.9} = 1.4 < 2$$

and
$$\frac{D}{T} = \frac{250}{9.7} = 25.7$$

For $\frac{D}{T} = 25.7$ and $\frac{l}{r} = 145.719$, from Table 7.1-B of the text,

$$\sigma_{bc} = 99.3 \text{ N/mm}^2$$

$$\sigma_{ac,\ \text{cal}} = \frac{215 \times 10^3}{6496} = 33.097 \text{ N/mm}^2$$

$$\sigma_{bc,\ \text{cal}} = \frac{48,750 \times 10^3}{618.9 \times 10^3} = 78.768 \text{ N/mm}^2$$

$$\frac{\sigma_{ac,\,\text{cal}}}{\sigma_{ac}} = \frac{33.097}{49.84} = 0.664 > 0.5$$

The interaction formula to be used for checking the safety is

$$\frac{\sigma_{ac,\,\text{cal}}}{\sigma_{ac}} + \frac{C_{mx}\ \sigma_{bcx,\text{cal}}}{\left\{1 - \frac{\sigma_{ac,\,\text{cal}}}{0.60\ f_{ccx}}\right\}\sigma_{bcx}} \le 1$$

The bending of the column takes place about x-axis as the eccentricity of the load is about x-axis. Therefore, the slenderness ratio in the plane of bending $= \frac{l}{r_{xx}}$

$$= \frac{8000}{109.1} = 73.32$$

The ends of the column are hinged, therefore,

$$C_{mx} = 1.0$$

$$f_{ccx} = \frac{\pi^2 E}{\lambda^2} = \frac{\pi^2 \times 2 \times 10^5}{73.32^2} = 367.18 \text{ N/mm}^2$$

$$\frac{33.097}{49.84} + \frac{1 \times 78.768}{\left\{1 - \frac{33.097}{0.6 \times 367.18}\right\} \times 99.3} = 1.598 > 1$$

which is unsafe.

Try another section I.S.H.B. 400 @ 759.3 N/m. The relevant properties of the section are,

$A = 9866 \text{ mm}^2$, $r_{yy} = 52.6$ mm, $r_{xx} = 168.7$ mm, $z_{xx} = 1404.2 \times 10^3 \text{ mm}^3$

Minimum radius of gyration, $r = 52.6$ mm

$$\frac{l}{r} = \frac{8000}{52.6} = 152.0913$$

For $\frac{l}{r} = 152.0913$, and $f_y = 250$ MPa,

$$\sigma_{ac} = 44.16 \text{ N/mm}^2$$

Now, $$\frac{T}{t} = \frac{12.7}{9.1} = 1.4$$

and $$\frac{D}{T} = \frac{400}{12.7} = 31.5$$

For $\frac{D}{T} = 31.5$, and $\frac{l}{r_{yy}} = 152.0913$, from Table 7.1-B of the text,

$$\sigma_{bc} = 89.75 \text{ N/mm}^2$$

$$\sigma_{ac,\text{ cal}} = \frac{215 \times 10^3}{9866} = 21.79 \text{ N/mm}^2$$

$$\sigma_{bc,\text{ cal}} = \frac{48750 \times 10^3}{1404.2 \times 10^3} = 34.72 \text{ N/mm}^2$$

$$\frac{\sigma_{ac\text{ , cal}}}{\sigma_{ac}} = \frac{21.79}{44.16} = 0.49 > 0.15$$

Hence, $$\frac{\sigma_{ac\text{ , cal}}}{\sigma_{ac}} + \frac{C_{mx}\ \sigma_{bcx\text{ , cal}}}{\left\{1 - \frac{\sigma_{ac\text{ , cal}}}{0.60\ f_{ccx}}\right\}\sigma_{bcx}} \le 1$$

$$\frac{l}{r_{xx}} = \frac{8000}{168.7} = 47.42$$

$$f_{ccx} = \frac{\pi^2 \times 2.0 \times 10^5}{47.42^2} = 877.82 \text{ N/mm}^2$$

Hence, $$\frac{21.79}{44.16} + \frac{1 \times 34.72}{\left\{1 - \frac{21.79}{0.6 \times 877.82}\right\} \times 89.75} = 0.8969 < 1$$

which is all right.

Example 4.12 Design a continuous strut 9.0 m long used as principal rafter of the roof truss. The reactions from the purlins are 2 kN, 3 kN and 4 kN at the intermediate points as shown in the Fig. Ex. 4.12(a). In addition to the above forces the rafter is subjected to a compressive force of 100 kN.

Solution The problem consists of finding out the maximum bending moment on the rafter due to the applied normal loads on the purlins. The rafter is then designed for the axial compressive force and maximum bending moment.

Analysis for moments There are several methods by which the bending moments in the continuous beam can be found. Let us use the moment distribution method to calculate the moments.

Fixed end moments

$$M_{FAB} = -\left\{\frac{2 \times 0.3 \times 2.7^2}{3^2} + \frac{3 \times 2.1 \times 0.9^2}{3^2}\right\}$$

$$= -1.053 \text{ kNm}$$

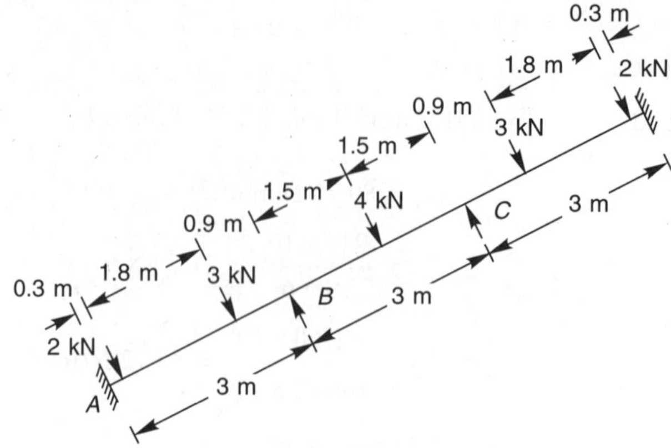

Fig. Ex. 4.12(a)

$$M_{FBA} = +\left\{\frac{2\times 0.3^2 \times 2.7}{3^2} + \frac{3\times 2.1^2 \times 0.9}{3^2}\right\}$$

$$= +1.377 \text{ kNm}$$

$$M_{FBC} = -\frac{WL}{8} = -\frac{4\times 3}{8} = -1.5 \text{ kNm}$$

$$M_{FCB} = +1.5 \text{ kNm}$$

By symmetry, $M_{FCD} = -1.377$ kNm and $M_{FDC} = 1.053$ kNm

Distribution factors Assuming the moment of inertia to be constant throughout the rafter,

$$K_{BA} = \frac{I}{L} = \frac{I}{3}$$

$$K_{BC} = \frac{I}{L} = \frac{I}{3}$$

$$K_{CB} = \frac{I}{L} = \frac{I}{3}$$

$$K_{CD} = \frac{I}{L} = \frac{I}{3}$$

$$\text{Distribution factor for } BA = \frac{K_{BA}}{K_{BA} + K_{BC}} = \frac{\frac{I}{3}}{\frac{I}{3} + \frac{I}{3}} = 0.5$$

Similarly distribution factors, for $BC = 0.5$, $CB = 0.5$, and $CD = 0.5$

Joint	*A*	*B*		*C*		*D*
Distribution factor		0.5	0.5	0.5	0.5	
Member	AB	BA	BC	CB	CD	DC
F.E.M	–1.053	1.377	–1.5	1.5	–1.377	1.053
Balance		0.062	0.062	–0.062	–0.062	
Carry over	0.031		–0.031	0.031		–0.031
Balance		0.016	0.016	–0.016	–0.016	
Carry over	0.008		–0.008	0.008		–0.008
Balance		0.004	0.004	–0.004	–0.004	
Moments	–1.014	1.458	–1.458	1.458	–1.458	+1.014

Maximum positive bending moment = 3.00 – 1.458 = 1.542 kNm

Maximum negative bending moment = 1.458 kNm

∴ Design moment = 1.542 kNm

The bending moment diagram for the rafter is shown in Fig. Ex. 4.12(b).

Forces The strut is subjected to

Axial compressive force = 100 kN

Bending moment = 1.542 kNm

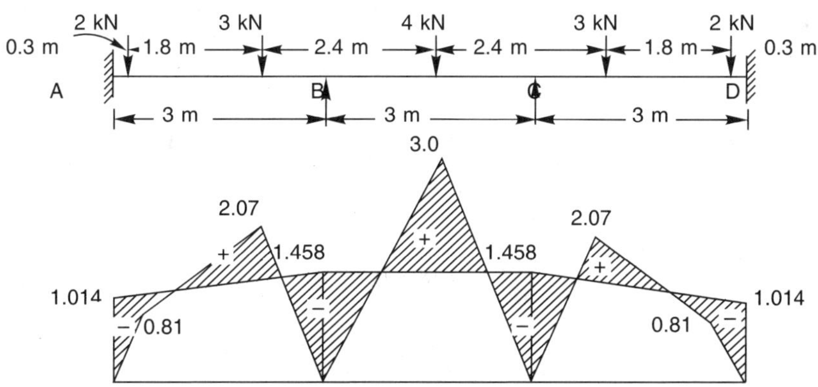

Fig. Ex. 4.12(b)

Assume allowable axial compressive stress = 60 N/mm²

$$A = \frac{100 \times 10^3}{60} = 1666.66 \text{ mm}^2$$

Increase the area by 50% to account for the bending moment

$$\text{Area} = 1.5 \times 1666.66 = 2499.99 \text{ mm}^2$$

If a single angle section is provided it will bend about an axis other than the *xx*-axis and it is difficult and time consuming to determine the axis of bending due to the effect of the gussets at the panel point and connections at the purlin points. If a double angle section is provided, there will be bending about the *xx*-axis (in the vertical plane). Therefore let us provide a double angle section (tack riveted).

Provide 2 I.S.A. 70 × 70 × 10 mm. The properties of the angles are,

$A = 2604\text{ mm}^2$, $W = 200.2\text{ N/m}$, $r_{xx} = 21.0\text{ mm}$, $r_{yy} = 29.7\text{ mm}$, $Z_{xx} = 23.4 \times 10^3\text{ mm}^3$

$$\text{Effective length } l_{xx} = 0.7 \times \text{panel length} = 0.7 \times 3 \times 10^3\text{ mm}$$

$$\text{Effective length } l_{yy} = 1.0 \times \text{distance between purlins} = 1.0 \times 2.4 \times 10^3\text{ mm}$$

$$\frac{l_x}{r_{xx}} = \frac{0.7 \times 3 \times 10^3}{21} = 100$$

$$\frac{l_y}{r_{yy}} = \frac{1 \times 2.4 \times 10^3}{29.7} = 80.8$$

For $\frac{l}{r} = 100$ and $f_y = 250$ MPa,

$$\sigma_{ac} = 80\text{ N/mm}^2$$

$$\sigma_{bc} = 0.66 \times 250 = 165\text{ N/mm}^2$$

(Full allowable bending compressive stress $0.66 f_y$ is considered when the table of the angle section is in compression.)

$$\sigma_{ac,\text{ cal}} = \frac{100 \times 10^3}{2604} = 38.40\text{ N/mm}^2$$

$$\sigma_{bc,\text{ cal}} = \frac{1542 \times 10^3}{23.4 \times 10^3} = 65.89\text{ N/mm}^2$$

$$\frac{\sigma_{ac,\text{ cal}}}{\sigma_{ac}} = \frac{38.4}{80} = 0.48 > 0.15$$

Check

$$\frac{\sigma_{ac,\text{ cal}}}{\sigma_{ac}} + \frac{C_{mx}\ \sigma_{bcx,\text{ cal}}}{\left\{1 - \dfrac{\sigma_{ac,\text{ cal}}}{0.6\ f_{ccx}}\right\}\sigma_{bcx}} \le 1.0$$

$$f_{ccx} = \frac{\pi^2 E}{\lambda^2} = \frac{\pi^2 \times 2 \times 10^5}{(100)^2} = 197.4$$

$$\therefore \quad \frac{38.4}{80} + \frac{0.85 \times 65.89}{\left\{1 - \dfrac{38.4}{0.6 \times 197.4}\right\} \times 165} = 0.98 < 1$$

which is safe.

Example 4.13 A stanchion in a building supports beams from three sides. The reactions from the beams are transferred at the eccentricities as shown in Fig. Ex. 4.13. The effective length of the column is 3.5 m. Design the column section. f_y = 250 N/mm^2.

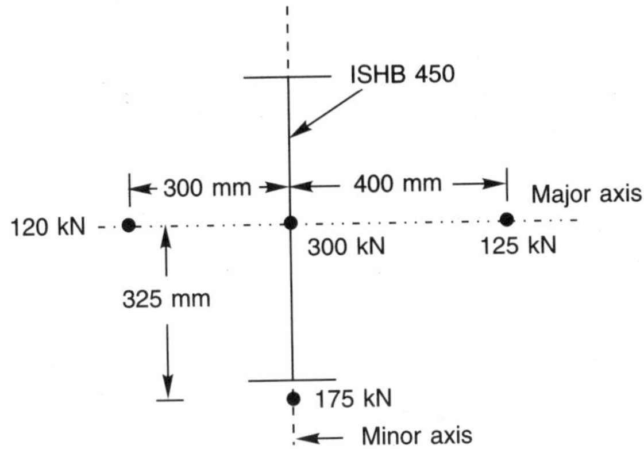

Fig. Ex. 4.13

Solution For the sections having $I_{yy} < I_{xx}$, the bending stresses about the major axis (*xx*-axis) are reduced to account for lateral instability whereas the bending stresses about the minor axis are taken as the maximum bending compressive stresses.

Assume allowable axial compressive stress as 85 N/mm^2.

$$P = 300 + 120 + 175 + 125 = 720 \text{ kN}$$

$$A = \frac{720 \times 10^3}{85} = 8470 \text{ mm}^2$$

As the column is subjected to bending moments, the area computed above is increased, say by 25%, to account for the bending moments.

$$\text{Area} = 1.25 \times 8470 = 10588 \text{ mm}^2$$

From I.S. Handbook No. 1, select I.S.H.B. 450 @ 855.4 N/m.

The relevant properties of this section are

$A = 11114$ mm^2, $Z_{xx} = 1742.7 \times 10^3$ mm^3, $Z_{yy} = 238.8 \times 10^3$ mm^3, $B = 250$ mm, $T = 13.7$ mm, $t = 9.8$ mm, $D = 450$ mm, $r_{yy} = 51.8$ mm, and $r_{xx} = 187.8$ mm

Bending moment about *xx*-axis, $M_{xx} = 175 \times 325 = 56875$ kNmm

Bending moment about *yy*-axis, $M_{yy} = 125 \times 400 - 120 \times 300 = 14{,}000$ kNmm

$$\sigma_{ac,\text{ cal}} = \frac{720 \times 10^3}{11{,}114} = 64.78 \text{ N/mm}^2$$

$$\sigma_{bcx,\text{ cal}} = \frac{M_{xx}}{Z_{xx}} = \frac{56875 \times 10^3}{1742.7 \times 10^3} = 32.6361 \text{ N/mm}^2$$

$$\sigma_{bcy,\text{ cal}} = \frac{M_{yy}}{Z_{yy}} = \frac{14{,}000 \times 10^3}{238.8 \times 10^3} = 58.6264 \text{ N/mm}^2$$

Now, $$\frac{T}{t} = \frac{13.7}{9.8} = 1.397 < 2$$

and $$\frac{d_1}{t} = \frac{450 - (2 \times 13.7)}{9.8} = 43.12 < 85$$

So, $$\frac{T}{t} < 2 \quad \text{and} \quad \frac{d_1}{t} < 85$$

Now, $$\frac{D}{T} = \frac{450}{13.7} = 32.8467$$

and $$\frac{l}{r_{yy}} = \frac{3500}{51.8} = 67.56$$

For $\frac{D}{T} = 32.8467$ and $\frac{l}{r_{yy}} = 67.56$, from Table 7.1-B of the text,

$$\sigma_{bcx} = 144.172 \text{ N/mm}^2$$

and $$\sigma_{bcy} = 0.66 f_y = 0.66 \times 250 = 165 \text{ N/mm}^2$$

For $\frac{l}{r} = 67.56$, and $f_y = 250$ MPa,

$$\sigma_{ac} = 114.44 \text{ N/mm}^2$$

$$\frac{\sigma_{ac,\,\text{cal}}}{\sigma_{ac}} = \frac{64.78}{114.44} = 0.566 > 0.15$$

Check, $$\frac{\sigma_{ac,\,\text{cal}}}{\sigma_{ac}} + \frac{C_{mx}\,\sigma_{bcx,\,\text{cal}}}{\left\{1 - \frac{\sigma_{ac,\,\text{cal}}}{0.6 f_{ccx}}\right\}\sigma_{bcx}} + \frac{C_{my}\,\sigma_{bcy,\,\text{cal}}}{\left\{1 - \frac{\sigma_{ac,\,\text{cal}}}{0.6 f_{ccy}}\right\}\sigma_{bcy}} \le 1$$

Assume $$C_{mx} = 0.85$$

and $$C_{my} = 0.45$$

$$\frac{l}{r_{xx}} = \frac{3500}{187.8} = 18.63$$

$$f_{ccx} = \frac{\pi^2 \times 2 \times 10^5}{18.63^2} = 5681.5 \text{ N/mm}^2$$

$$\frac{l}{r_{yy}} = 67.56$$

$$f_{ccy} = \frac{\pi^2 \times 2 \times 10^5}{67.56^2} = 432.0 \text{ N/mm}^2$$

$$\therefore \quad \frac{64.78}{114.44} + \frac{0.85 \times 32.6361}{\left\{1 - \frac{64.78}{0.6 \times 5681.5}\right\} \times 144.172} + \frac{0.45 \times 58.6264}{\left\{1 - \frac{64.78}{0.6 \times 432}\right\} \times 165}$$

$$= 0.975 < 1$$

which is safe.

Example 4.14 A column I.S.H.B 300 @ 618 N/m in a Hi-rise building supports beams from two sides as shown in Fig. Ex. 4.14. The beams are welded to the column flanges and transfer end reactions of 150 kN (Beam 1) and 250 kN (Beam 2) and an axial compressive force of 450 kN from the top storeys. Check the safety of the column if its effective length is 3.2 m. Assume, C_{mx} = 0.85 and f_y = 250 N/mm^2.

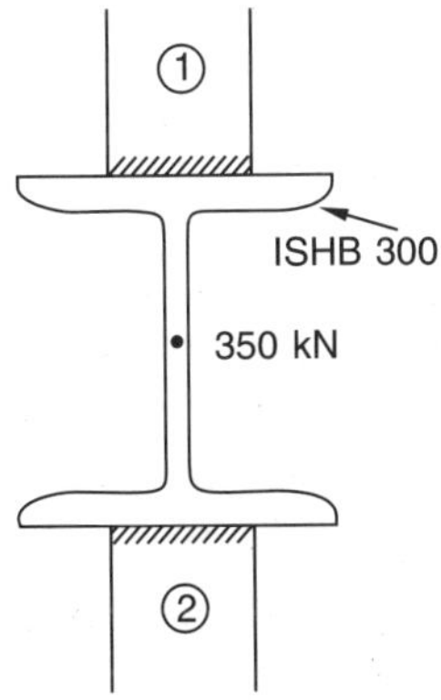

Fig. Ex. 4.14

Solution Load passing through the centre of gravity of the section, $P = 450 + 150 + 250 = 850$ kN

From I.S. Handbook No. 1, the relevant properties of I.S.H.B 300 @ 618 N/m are

$A = 8025$ mm^2, $Z_{xx} = 863.3 \times 10^3$ mm^3, $B = 250$ mm,
$T = 10.6$ mm, $D = 300$ mm, $t = 9.4$ mm,
$r_{xx} = 127.0$ mm, and $r_{yy} = 52.9$ mm

$$M_{xx} = 250 \times \frac{300}{2} - 150 \times \frac{300}{2} = 15000 \text{ kN mm}$$

$$\sigma_{ac,\text{ cal}} = \frac{850 \times 10^3}{8025} = 105.91 \text{ N/mm}^2$$

$$\sigma_{bcx,\text{ cal}} = \frac{M_{xx}}{Z_{xx}} = \frac{15000 \times 10^3}{863.3 \times 10^3} = 17.37 \text{ N/mm}^2$$

Now, $$\frac{T}{t} = \frac{10.6}{9.4} = 1.127$$

and $$\frac{d_1}{t} = \frac{300 - (2 \times 10.6)}{9.4} = 29.66 < 85$$

Now, $$\frac{D}{T} = \frac{300}{10.6} = 28.30$$

and $$\frac{l}{r_{yy}} = \frac{3.2 \times 10^3}{52.9} = 60.49$$

For $\frac{D}{T} = 28.30$ and $\frac{l}{r_{yy}} = 60.49$, from Table 7.1B of the text,

$$\sigma_{bcx} = 149.046 \text{ N/mm}^2$$

For $\frac{l}{r} = 60.49$ and $f_y = 250$ N/mm^2,

$$\sigma_{ac} = 121.51 \text{ N/mm}^2$$

$$f_{ccx} = \frac{\pi^2 \times 2 \times 10^5}{(3200/127)^2} = 3109.12 \text{ N/mm}^2$$

Check

$$\frac{\sigma_{ac,\,cal}}{\sigma_{ac}} + \frac{C_{mx}\ \sigma_{bcx,\,cal}}{\left\{1 - \dfrac{\sigma_{ac,\,cal}}{0.6\ f_{ccx}}\right\}\sigma_{bcx}} \le 1$$

or

$$\frac{105.91}{121.51} + \frac{0.85 \times 17.37}{\left\{1 - \dfrac{105.91}{0.6 \times 3109}\right\} \times 149.046} = 0.9766 < 1$$

which is safe.

Example 4.15 Design a cased column to carry an axial load of 1100 kN. The column is 5 m long and restrained in position and direction at both ends. Use M-15 grade of concrete.

Solution Assume allowable compressive stress 110 MPa,

$$\text{Cross-sectional area required} = \frac{1100 \times 1000}{110} = 10{,}000 \text{ mm}^2$$

Select I.S.H.B. 450 @ 855.4 N/m from I.S. Handbook No. 1.

$$A = 11114 \text{ mm}^2,\ Z_{xx} = 1742.7 \times 10^3 \text{ mm}^3,\ B = 250 \text{ mm}$$

Minimum radius of gyration = 51.8 mm

Now,

$$l = 0.65L$$
$$= 0.65 \times 1000 \times 5 = 3250 \text{ mm}$$

Uncased column

$$\frac{l}{r} = \frac{3250}{51.8} = 62.74 < 250$$

which is all right.

For $\frac{l}{r} = 62.74$ and $f_y = 250$ MPa,

$$\sigma_{ac} = 119.256 \text{ N/mm}^2$$

$$\text{Load carrying capacity} = 119.256 \times 11{,}114$$
$$= 1{,}325{,}411.2 \text{ N} = 1325.411 \text{ kN}$$

which is all right.

Cased column Column size is 550 mm × 350 mm. The column is encased in concrete with a 50 mm cover all around.

$$r_{yy} = 0.2(B + 100)$$
$$= 0.2(250 + 100) = 0.2 \times 350 = 70 \text{ mm}$$

$$\frac{l}{r} = \frac{3250}{70} = 46.428$$

For $\frac{l}{r} = 46.428$ and $f_y = 250$ MPa

$$\sigma_{ac} = 134.500 \text{ N/mm}^2$$

$$\text{Load carrying capacity} = 134.500 \times 11{,}114 = 1{,}494{,}833.0 \text{ N}$$

$$= 1494.833 \text{ kN} < 2 \times 1325.411 \text{ kN}$$

which is safe.

Allowable safe load on column treating it as a reinforced column,

$$\text{Steel area} = 11{,}114 \text{ mm}^2$$

$$\text{Concrete area} = (550 \times 350 - 11{,}114)$$

$$= 181{,}386 \text{ mm}^2$$

Allowable safe load = safe load on uncased column + safe load on concrete

$$= 119.256 \times 11{,}114 + 181{,}386 \times 3.75$$

$$(0.25 f_{ck} = 0.25 \times 15 = 3.75)$$

$$= 2005.60 \text{ kN}$$

> load carrying capacity of cased column

which is safe.

Example 4.16 Design a built up column 10 m long to carry an axial load of 750×10^3 N. The column is restrained in position but not in direction at both the ends. Provide single lacing system with riveted connections. Assume $f_y = 250$ MPa.

(a) Design the column with two channels placed back-to-back
(b) Design the column with two channels placed toe-to-toe
(c) Which of the two systems is economical
(d) Design the lacing system with welded connections for channels back-to-back.

Solution

Design of column

$$P = 750 \times 10^3 \text{ N}$$

$$l = 1.0 \times 10 = 10 \text{ m}$$

Let the allowable axial compressive stress for the column be 110 MPa.

$$\text{Cross-sectional area required} = \frac{P}{\text{allowable axial compressive stress}}$$

$$= \frac{750 \times 10^3}{110} = 6818 \text{ mm}^2$$

Select I.S.M.C. 250 @ 298.2 N/m, 2 in numbers from I.S. Handbook No. 1.

The relevant properties of I.S.M.C. 250 @ 298.2 N/m are,

$$A = 3867 \text{ mm}^2,\ r_{xx} = 99.4 \text{ mm},\ r_{yy} = 23.8 \text{ mm, and } T = 14.1 \text{ mm}$$

$$\text{Area provided} = 2 \times 3867 = 7734 \text{ mm}^2$$

In the design of built-up columns with two sections, the sections are so spaced that the least radius of gyration of the built up section becomes as large a value as possible. Therefore, the radius of gyration about *yy*-axis is increased so that it becomes equal to or more than the radius of gyration about *xx*-axis. This can be achieved by spacing the sections in such a way that r_{xx} becomes r_{min}. Therefore, let us first check the safety of the section after which its spacing can be worked out.

$$\frac{l}{r} = \frac{10 \times 10^3}{99.4} = 100.6036$$

For $\frac{l}{r}$ = 100.6036, and f_y = 250 MPa,

$$\sigma_{ac} = 79.517 \text{ N/mm}^2$$

$$\text{Load carrying capacity} = 79.517 \times 7734 = 614{,}984.47 \text{ N}$$
$$= 614.984 \text{ kN} < \text{load over section}$$

which is not safe.

Try I.S.M.C. 300 @ 351.2 N/m, 2 in numbers. The relevant properties are,

A= 4564 mm^2, r_{xx} = 118.1 mm, r_{yy} = 26.1 mm, C_{yy} = 23.6 mm,
I_{xx} = 6362.6 × 10^4 mm^4, I_{yy} = 310.8 × 10^4 mm^4, g = 50 mm

$$\text{Area provided} = 2 \times 4564 = 9128 \text{ mm}^2$$

$$\frac{l}{r} = \frac{10 \times (10)^3}{118.1} = 84.6740$$

For $\frac{l}{r}$ = 84.6740, and f_y = 250 MPa,

$$\sigma_{ac} = 95.863 \text{ N/mm}^2$$

$$\text{Load carrying capacity} = 95.863 \times 9128 \times 10^{-3}$$
$$= 875.037 \text{ kN} > \text{load over the section}$$

which is safe,

(a) Let us provide the two channels back-to-back and connect them by lacing.

Spacing of channels

$$2I_{xx} = 2\left\{I_{yy} + A\left(\frac{S}{2} + C_{yy}\right)^2\right\}$$

or
$$2 \times 6362.6 \times 10^4 = 2 \times \left\{310.8 \times 10^4 + 4564\left(\frac{S}{2} + 23.6\right)^2\right\}$$

or
$$\left(\frac{S}{2} + 23.6\right)^2 = 13{,}259.859$$

$$S = 183.10 \text{ mm}$$

Let us place the two channels at a spacing of 183.5 mm (Fig. Ex. 4.16(i)).

Lacing system Let us try the single lacing system.

Assume the inclination of lacing bar = 45°

$$\text{Spacing of lacing bars, } C = 2(183.5 + 50 + 50) \cot 45°$$
$$= 2 \times 283.5 \times 1 = 567 \text{ mm}$$

$$\frac{C}{r_{yy}} = \frac{567}{26.1} = 21.724 < 50$$

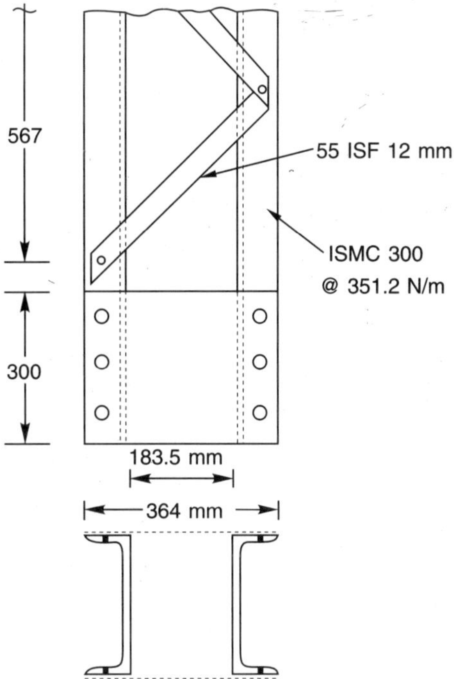

Fig. Ex. 4.16(i)

Also C/r_{yy} should be < 0.7 × 84.674 (= 59.2718), which it is.

$$\text{Maximum shear, } V = \frac{2.5}{100} \times 750 \times 10^3 = 18{,}750 \text{ N}$$

$$\text{Transverse shear in each pannel} = \frac{V}{N} = \frac{1}{2} \times 18{,}750 = 9375 \text{ N}$$

$$\text{Compressive force in lacing bars} = \frac{V}{N} \operatorname{cosec} \theta$$

$$= 9375 \times \operatorname{cosec} 45° = 13{,}258.379 \text{ N}$$

Lacing flats Let us provide 18 mm ϕ power driven rivets.

Gross diameter of rivet = 18 + 1.5 = 19.5 mm

For 18 mm ϕ rivets, width of lacing flat as specified by I.S. Code is 55 mm (Table 4.5 of the text).

$$\text{Minimum thickness of lacing flat} = \frac{1}{40} \times \text{length of flat between inner end rivets.}$$

$$= \frac{1}{40}(183.5 + 50 + 50) \operatorname{cosec} 45°$$

$$= 10.02 \text{ mm} \approx 12 \text{ mm}$$

Provide a flat section 55 I.S.F. 12 mm.

$$\text{Minimum radius of gyration, } r = \frac{t}{\sqrt{12}} = \frac{12}{\sqrt{12}} = 3.464 \text{ mm}$$

$$\frac{l}{r} = \frac{283.5 \times \text{cosec } 45°}{3.464} = 115.72 < 145$$

which is all right.

For $\frac{l}{r}$ = 115.72 and f_y = 250 MPa,

$$\sigma_{ac} = 67.424 \text{ N/mm}^2$$

$$\text{Load carrying capacity} = 67.424 \times 55 \times 12 = 44{,}499.8 \text{ N}$$
$$= 44.49 \text{ kN} > 13.258 \text{ kN}$$

which is sufficient.

$$\text{Tensile strength of lacing flat} = (B - d)t\sigma_{at}$$
$$= (55 - 19.5) \times 12 \times (0.6 \times 250)$$
$$= 63{,}900 \text{ N} = 63.9 \text{ kN} > 13.258 \text{ kN}$$

which is safe.

Connection Assuming that one 18 mm ϕ power rivet is used to connect both the lacing flats with channel at one point as shown in Fig. Ex. 4.16(i).

$$\text{Strength of the rivet in double shear} = 2 \times \frac{\pi}{4} d^2 \tau_{vf}$$

$$= 2 \times \frac{\pi}{4} 19.5^2 \times 100 = 59729 \text{ N}$$

$$\text{Strength of the rivet in bearing} = dt\,\sigma_{pf} = 19.5 \times 12 \times 300$$
$$= 70{,}200 \text{ N}$$

$\therefore$ Strength of the rivet = 59,729 N

$$\text{Force on rivet from flat} = 2\frac{V}{N}\cot\theta$$
$$= 2 \times 9.375 \times 10^3 \cot 45°$$
$$= 18.75 \times 10^3 \text{ N}$$

$$\text{Number of rivets required} = \frac{18.75 \times 10^3}{59729} = 0.31 \approx 1$$

Provide one rivet to connect each end of the flat with channels.

The two lacing flats may be connected at different points with channel section as shown in Fig. 4.8(b). Rivets will be in single shear.

Strength of the rivet in single shear

$$= \frac{\pi}{4} d^2\, \tau_{vf} = \frac{\pi}{4} 19.5^2 \times 100 = 29864.7$$

Strength of rivet in bearing

$$= dt\,\sigma_{pf} = 19.5 \times 12 \times 300 = 70{,}200 \text{ N}$$

Strength of rivet = 29864.7 N

$$\text{Force on rivet from flat} = \frac{V}{N} \text{ cosec } \theta = 13258.379 \text{ N}$$

$$\text{Number of rivets required} = \frac{13258.379}{29864.7} = 0.44 \approx 1$$

Provide 1 rivet to connect each end of the flat with channels.

Note The advantage of connecting each flat separately with the channel has been discussed in Section 4.9. However, this arrangement is costly since additional holes and rivets will be required. Therefore, any one of the above two system of connections may be used, depending upon the descretion of the designer. In the examples that follow the flats are connected at one point to achieve economy.

Tie plate Tie plates are provided at the ends of the laced column,

$$\text{Effective depth} = 183.5 + 2 \times 23.6 = 230.7 \text{ mm} > 2 \times 90 \text{ mm}$$

which is all right.

$$\text{Overall depth of tie plate} = 230.7 + 2 \times 29 = 288.7 = 300 \text{ mm}$$

(minimum edge distance for 18 mm ϕ rivets = 29 mm)

$$\text{Length of the tie plate} = 183.5 + 2 \times 90 = 364 \text{ mm}$$

$$\text{Thickness of the tie plate} = \frac{1}{50}(183.5 + 2 \times 50) = 5.67 \text{ mm} \approx 6 \text{ mm}$$

Provide a tie plate 364 × 300 × 6 mm and connect it to the channels with 18 ϕ rivets.

(b) Let us provide the two channels toe-to-toe and connect them with lacing.

Spacing

$$2I_{xx} = 2\left\{I_{yy} + A\left(\frac{S}{2} - C_{yy}\right)^2\right\}$$

or
$$2 \times 6362.6 \times 10^4 = 2\left\{310.8 \times 10^4 + 4564\left(\frac{S}{2} - 23.6\right)^2\right\}$$

or
$$\left(\frac{S}{2} - 23.6\right)^2 = 13{,}259.859$$

$$S = 277.50 \text{ mm} \approx 278 \text{ mm}$$

Let us place the channels at a spacing of 278 mm (Fig. Ex. 4.16(ii))

Connecting system Provide a single lacing system. Assume the inclination of the lacing bar be 45°.

$$C = 2(278 - 50 - 50) \cot 45° = 356 \text{ mm}$$

$$\frac{C}{r_{yy}} = \frac{356}{26.1} = 13.6398 < 50$$

and should be < 0.7 × 84.674 (= 59.2718), which it is.

$$\text{Compressive force in lacing bar} = 13.258379 \times 10^3 \text{ N}$$

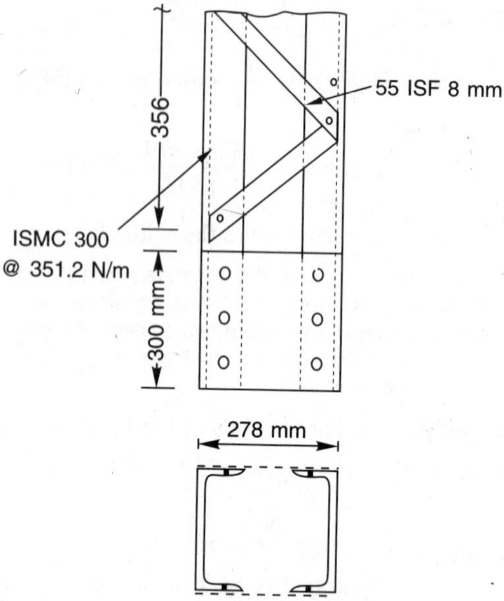

Fig. Ex. 4.16(ii)

Section of lacing flat Let us provide 18 mm ϕ rivets.

$$\text{Gross diameter of rivet} = 18 + 1.5 = 19.5 \text{ mm}$$

$$\text{Width of flat} = 55 \text{ mm}$$

$$\text{Thickness of flat} = \frac{1}{40}(278 - 50 - 50) \times \text{cosec } 45°$$

$$= 6.29 \text{ mm} \approx 8 \text{ mm}$$

Provide 8 mm thick flat.

Check

$$r_{\min} = \frac{t}{\sqrt{12}} = \frac{8}{\sqrt{12}} = 2.309 \text{ mm}$$

$$\frac{l}{r} = \frac{178 \text{ cosec } 45°}{2.309} = 109$$

$$< 145$$

which is safe.

Provide 55 I.S.F. 8 mm flat section.

For $\frac{l}{r}$ = 109.00 and f_y = 250 MPa,

$$\sigma_{ac} = 72.8 \text{ N/mm}^2$$

$$\text{Load carrying capacity} = 72.8 \times 55 \times 8 = 32{,}032 \text{ N} = 32.03 \text{ kN}$$

$$> 13.258 \text{ kN}$$

which is safe.

Tensile strength of lacing flat $= (B - d)t\sigma_{at}$

$= (55 - 19.5) \times 8 \times (0.6 \times 250)$

$= 42{,}600 \text{ N} = 42.600 \text{ kN} > 13.258 \text{ kN}$

which is safe.

Connection

$$\text{Strength of the rivet in double shear} = 2\frac{\pi}{4}d^2\,\tau_{vf} = 2 \times \frac{\pi}{4}19.5^2 \times 100$$

$= 59729 \text{ N}$

Strength of the rivet in bearing $= dt\,\sigma_{pf} = 19.5 \times 8 \times 300 = 46{,}800$ N

Force on the rivet from flat $= 18.75 \times 10^3$ N

$$\text{Number of rivets required} = \frac{18.75 \times 10^3}{46{,}800} = 0.400 \approx 1$$

Provide one rivet 18 mm ϕ for the connection.

Tie plates

Effective depth of tie plate $= S - 2C_{yy} = 278 - 2 \times 23.6 = 230.8$ mm

Overall depth of tie plate $= 230.8 + 2 \times 29 = 288.8 \text{ mm} \approx 300$ mm

Length of the tie plate = 278 mm

$$\text{Thickness of the tie plate} = \frac{1}{50}(278 - 2 \times 50) = 3.56 \text{ mm} \approx 6 \text{ mm}$$

Provide a tie plate 278 × 300 × 6 mm and connect it with 18 mm ϕ rivets.

(c) It is seen that by providing two channels toe-to-toe the lacing size as well as the end tie plate size is reduced. But as spacing is reduced more lacing flats will be required. Thus, overall economy will have to be worked out before choosing between the two possibilities.

(d) From part (a) of this example,

Spacing of the channels = 183.5 mm

Compressive force in the lacing flat = 13.258 kN

Lacing flats Assuming the welding to be done as shown in Fig. Ex. 4.16(iii).

Effective length of the welded lacing flat = 0.7 × 183.5 × cosec 45°

= 181.65 mm

$$\text{Minimum thickness of lacing flat} = \frac{1}{40} \times 183.5 \text{ cosec } 45°$$

$= 6.48$ mm

$\simeq 8$ mm

Let us try flat section 50 I.S.F. 8

$$\text{Minimum radius of gyration, } r = \frac{t}{\sqrt{12}} = \frac{8}{\sqrt{12}} = 2.31 \text{ mm}$$

$$\frac{l}{r} = \frac{181.65}{2.31} = 78.63 < 145$$

which is safe.

For $\frac{l}{r} = 78.63$ and $f_y = 250$ MPa,

$$\sigma_{ac} = 102.50 \text{ N/mm}^2$$

$$\text{Load carrying capacity} = 102.50 \times 50 \times 8$$
$$= 41{,}000 \text{ N} = 41.0 \text{ kN} > 13.258 \text{ kN}$$

which is safe,

$$\text{Overlap of lacing flat} = 50 \text{ mm} > 4 \times 8 \text{ mm}$$

which is safe.

Connection

Flange thickness of I.S.M.C. 300 = 13.6 mm

Minimum size of the weld for 13.6 mm thick member = 5 mm

$$\text{Strength of weld/unit length} = 0.7 \times 5 \times 108 = 378 \text{ N/mm}$$

$$\text{Length of weld required} = \frac{13{,}258}{378} = 35.07 \text{ mm}$$

Adjust the weld length as shown in Fig. Ex. 4.16(iii)

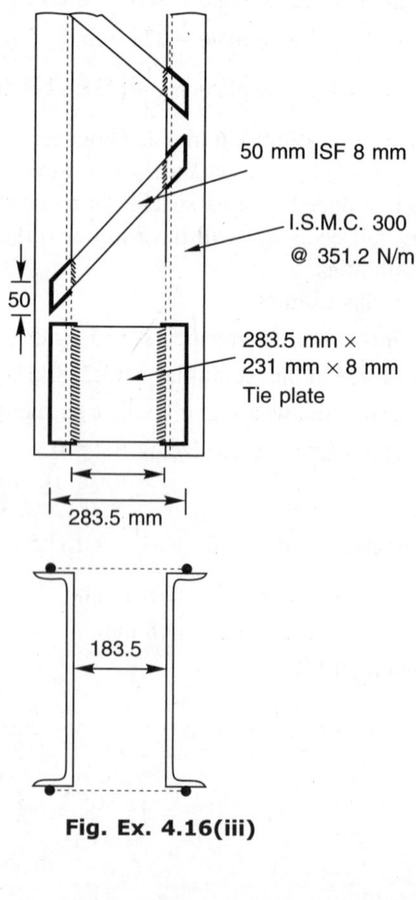

Fig. Ex. 4.16(iii)

Weld length provided = $2 \times 50 + 2 \times 50 = 200$ mm

Tie plate

Overall depth of the plate = effective depth of the tie plate

$= 183.5 + 2 \times 23.6$

$= 230.7 \text{ mm} > 2 \times 90 \text{ mm}$

which is as it should be.

Provide depth of tie plate = 231 mm

Length of tie plate = $183.5 + 2 \times 50 = 283.5$ mm

$$\text{Thickness of the plate} = \frac{1}{50}(183.5 + 2 \times 50) = 5.67 \text{ mm}$$

$\simeq 8$ mm (to accommodate 5 mm weld)

Provide a tie plate $283.5 \times 231 \times 8$ mm and connect it with 5 mm weld as shown in Fig. Ex. 4.16(iii).

Example 4.17 Design a connecting system as battens with riveted connections for the column in Example 4.16. Use two channel sections back-to-back.

Solution

$$P = 750 \times 10^3 \text{ N}$$

$$L = 10.0 \text{ m}$$

$$\text{Effective length} = 1.1 \times 10 = 11 \text{ m}$$

$$\text{Cross-sectional area required} = \frac{750 \times 10^3}{110} = 6818 \text{ mm}^2$$

Provide 2 I.S.M.C. 300 @ 351.2 N/m.

The relevant properties of I.S.M.C. 300 @ 351.2 N/m are,

$A = 4564 \text{ mm}^2$, $r_{xx} = 118.1$ mm, $r_{yy} = 26.1$ mm, $I_{xx} = 6362.6 \times 10^4 \text{ mm}^4$, $I_{yy} = 310.8 \times 10^4 \text{ mm}^4$

Area provided = $2 \times 4564 = 9128 \text{ mm}^2$

$$\frac{l}{r} = \frac{11 \times 10^3}{118.1} = 93.14$$

For $\frac{l}{r} = 93.14$, and $f_y = 250$ MPa,

$$\sigma_{ac} = 86.86 \text{ N/mm}^2$$

Load carrying capacity = $86.86 \times 9128 = 792{,}858.08$ N

$= 792.85 \text{ kN} > 750 \text{ kN}$

which is all right.

The spacing of the channels will be same as calculated in the previous example i.e. 183.5 mm.

In the design of battened column it is desirable to keep $I_{yy} \geq I_{xx}$. Therefore, let us assume a spacing of 200 mm.

Spacing of battens $\frac{C}{r_{yy}}$ should be less than 0.7 times slenderness ratio of column as a whole.

$$\frac{C}{r_{yy}} < 0.7 \times \frac{l}{r}$$

or $$C < 0.7 \times 93.14 \times 26.1$$

or $$C < 1701.66$$

also $$\frac{C}{r_{yy}} < 50$$

or $$C < 50 \times r_{yy}$$

$\therefore$ $$C < 50 \times 26.1 = 1305 \text{ mm} \simeq 1300 \text{ mm}$$

Provide battens at a spacing of 1300 mm (Fig. Ex. 4.17(a)).

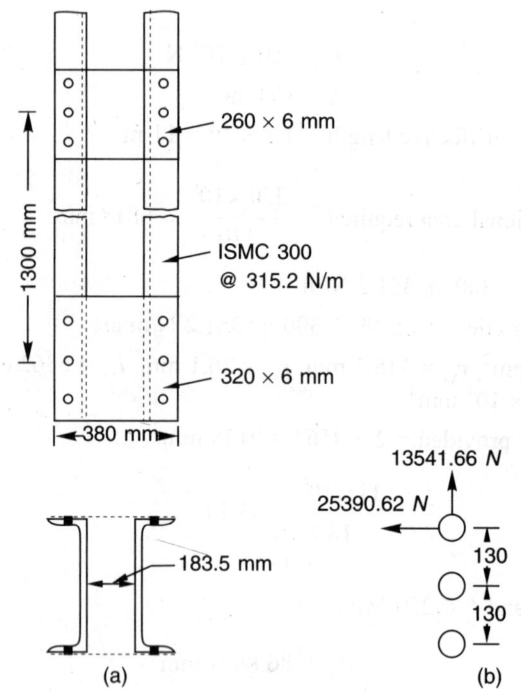

Fig. Ex. 4.17

Size of end battens Provide 20 mm ϕ power driven rivets for connection.

Edge distance = 29 mm ≃ 30 mm

The effective depth $d' = 200 + 2 \times 23.6$

$= 247.2 \text{ mm} > 2 \times 90 \text{ mm}$

which is as it should be.

The overall depth $d = 247.2 + 2 \times 30 = 307.2$ mm $\simeq$ 320 mm

$$\text{Thickness of the batten} = \frac{1}{50}(200 + 2 \times 50) = \frac{300}{50} = 6 \text{ mm}$$

Length of the batten = $200 + 2 \times 90 = 380$ mm

Provide 380 × 320 × 6 mm end batten plates.

Size of intermediate battens Effective depth of the batten plate,

$$d_1' = \frac{3}{4} \times 247.2 = 185.4 \text{ mm} > 2 \times 90 \text{ mm}$$

which is all right.

Overall depth of batten, $d' = 185.4 + 2 \times 30 = 245.4$ mm $\approx$ 260 mm

Thickness = 6 mm

Length = 380 mm

Provide 380 × 260 × 6 mm intermediate batten plates.

Design forces

Transverse shear, V = 2.5% of axial load

$$= \frac{2.5}{100} \times 750 \times 10^3 = 18.75 \times 10^3 \text{ N}$$

$$\text{Longitudinal shear, } V_1 = \frac{18.75 \times 10^3 \times 1300}{2 \times (200 + 2 \times 50)} = 40{,}625 \text{ N}$$

$$\text{Moment, } M = \frac{18.75 \times 10^3 \times 1300}{2 \times 2} = 6{,}093{,}750 \text{ Nmm}$$

Check For end battens,

$$\tau_{va,\text{ cal}} = \frac{40{,}625}{320 \times 6} = 21.158 \text{ N/mm}^2 < 100 \text{ N/mm}^2$$

(0.4 × 250)

which is all right.

$$\sigma_{bc,\text{ cal}} = \frac{6M}{td^2} = \frac{6 \times 6{,}093{,}750.0}{6 \times 320^2}$$

$$= 59.509 \text{ N/mm}^2 < 165 \text{ N/mm}^2 \qquad (0.66 \times 250)$$

which is safe.

For intermediate battens,

$$\tau_{va,\text{ cal}} = \frac{40{,}625}{260 \times 6} = 26.0416 \text{ N/mm}^2 < 100 \text{ N/mm}^2$$

(0.4 × 250)

which is safe.

$$\sigma_{bc,\,cal} = \frac{6M}{td^2} = \frac{6 \times 6{,}093{,}750.0}{6 \times 260^2}$$

$$= 90.144 \text{ N/mm}^2 < 165 \text{ N/mm}^2 \qquad (0.66 \times 250)$$

which is as it should be.

Connection The connections should be designed to transmit both shear and bending moment.

Provide 18 mm ϕ power driven rivets.

$$\text{Gross diameter} = 18 + 1.5 = 19.5 \text{ mm}$$

$$\text{Strength of the rivet in single shear} = \frac{\pi}{4} d^2 \tau_{vf} = \frac{\pi}{4} \times 19.5^2 \times 100$$

$$= 29.8 \times 10^3 \text{ N}$$

$$\text{Strength of the rivet in bearing} = dt\,\sigma_{pf} = 19.5 \times 6 \times 300$$

$$= 35.1 \times 10^3 \text{ N}$$

Hence

$$\text{Rivet value, } R_v = 29.8 \times 10^3 \text{ N}$$

$$\text{Number of rivets required} = \frac{40625}{29.8 \times 10^3} = 1.363$$

Increase the number of rivets to 3 to account for bending moment and provide them in one vertical row.

Check

$$\text{Force in each rivet due to longitudinal shear} = \frac{40625}{3} = 13541.66 \text{ N}$$

Provide a pitch of $\frac{1}{2}(320 - 2 \times 30) = 130$ mm.

$$\text{Force due to moment} = \frac{Mr}{\Sigma r^2}$$

$$= \frac{6{,}093{,}750.0 \times 130}{130^2 + 130^2} = 23{,}437.5 \text{ N}$$

$$\text{Resultant force} = \sqrt{(13{,}541.66)^2 + (23{,}437.5)^2}$$

$$= 27{,}068.3 \text{ N} = 27.068 \times 10^3 \text{ N}$$

$$< 29.8 \times 10^3 \text{ N}$$

which is safe.

$$\text{Number of batten plates on one face} = \frac{\text{length of column}}{\text{spacing of batten}} + 1$$

$$= \frac{10 \times 10^3}{1300} + 1 = 8.69 \approx 9.0$$

$$\text{Revised centre to centre spacing of battens} = \frac{10 \times 10^3}{9 - 1} = 1250 \text{ mm}$$

Example 4.18 Design a column of effective length 5.90 m. It is subjected to an axial load of 1400 kN. Provide two channels back-to-back connected with battens by welded connection. Assume f_y = 250 MPa.

Solution

Assume allowable compressive stress 130 N/mm^2

$$\text{Cross-sectional area required} = \frac{1400 \times 10^3}{130} = 10769.2 \text{ mm}^2$$

Provide 2 I.S.M.C. 350 @ 413 N/m.

The relevant properties of I.S.M.C. 350 are,

$A = 5366$ mm^2, $r_{xx} = 136.6$ mm, $r_{yy} = 28.3$ mm, $I_{xx} = 10008 \times 10^4$ mm^4, $I_{yy} = 430.6 \times 10^4$ mm^4, $C_{yy} = 24.4$ mm

$$\text{Area provided} = 2 \times 5366 = 10,732 \text{ mm}^2$$

$$\lambda = \frac{l}{r} = \frac{1.1 \times 5.90 \times 10^3}{136.6} = 47.51$$

For $\frac{l}{r} = 47.51$, and $f_y = 250$ MPa,

$$\sigma_{ac} = 133.74 \text{ N/mm}^2$$

$$\text{Load carrying capacity} = 133.74 \times 10,732 = 1,435,927 \text{ N}$$

$$= 1435.9 \text{ kN} > 1400 \text{ kN}$$

which is safe.

Spacing of channels

$$2I_{xx} = 2\left\{I_{yy} + A\left(\frac{S'}{2} + C_{yy}\right)^2\right\}$$

or
$$2 \times 10,008 \times 10^4 = 2 \times \left\{430.6 \times 10^4 + 5366\left(\frac{S'}{2} + 24.4\right)^2\right\}$$

or
$$S' = 218.39 \text{ mm} \simeq 220 \text{ mm}$$

Provide 2 I.S.M.C. 350 @ 413 N/m at a spacing of 220 mm back-to-back as shown in Fig. Ex. 4.18.

Spacing of battens $\frac{C}{r_{yy}}$ should be less than 0.7 × slenderness ratio of column as a whole.

$$\frac{C}{r_{yy}} < 0.7 \times \frac{l}{r}$$

or
$$C < 0.7 \times 47.51 \times 28.3$$

or
$$C < 941.17 \text{ mm}$$

Also $\frac{C}{r_{yy}}$ should be less than 50.

or $$\frac{C}{r_{yy}} < 50$$

or $$C < 50 \times 28.3$$

or $$C < 1415 \text{ mm}$$

Provide battens at a spacing of 940 mm.

Size of end battens

$$\text{Overall depth of the batten} = \text{effective depth of the batten}$$
$$= 220 + 2 \times 24.4$$
$$= 268.8 \text{ mm} \simeq 270 \text{ mm}$$

$$\text{Thickness of the batten} = \frac{1}{50} \times 220$$
$$= 4.4 \text{ mm} \simeq 8 \text{ mm}$$

(Overlap $\nless 4 \times 8$ mm)

Provide a 70 mm overlap of battens on channel flange for welding.

$\therefore$ $$\text{Length of batten} = 220 + 2 \times 70 = 360 \text{ mm}$$

Provide 360 × 270 × 8 mm end batten plates.

Size of intermediate battens

$$\text{Overall depth of batten} = \frac{3}{4} \times 270 = 202.5 \simeq 205 \text{ mm}$$

$$\text{Thickness of batten} = 8 \text{ mm}$$

$$\text{Length} = 360 \text{ mm}$$

Provide 360 × 205 × 8 mm intermediate batten plates.

Design forces

$$\text{Transverse shear, } V = 2.5\% \text{ of axial load}$$

$$= \frac{2.5}{100} \times 1400 \times 10^3 = 35 \times 10^3 \text{ N}$$

$$\text{Longitudinal shear, } V_1 = \frac{VC}{NS}$$

$$= \frac{35 \times 10^3 \times 940}{2 \times \left(220 + \frac{2 \times 70}{2}\right)} = 56{,}724 \text{ N}$$

$$\text{Moment, } M = \frac{VC}{2N}$$

$$= \frac{35 \times 10^3 \times 940}{2 \times 2} = 8225 \times 10^3 \text{ Nmm}$$

Check For end battens,

$$\tau_{va,\,cal} = \frac{56{,}724}{270 \times 8}$$

$$= 26.26 \text{ N/mm}^2 < 100 \text{ N/mm}^2 \qquad (0.4 \times 250)$$

which is as it should be.

$$\sigma_{bc,\,cal} = \frac{6 \times 8225 \times 10^3}{8 \times 270^2}$$

$$= 84.6 \text{ N/mm}^2 < 165 \text{ N/mm}^2 \qquad (0.66 \times 250)$$

which is safe.

For intermediate battens,

$$\tau_{va,\,cal} = \frac{56{,}724}{205 \times 8}$$

$$= 34.58 \text{ N/mm}^2 < 100 \text{ N/mm}^2 \qquad (0.4 \times 250)$$

which is all right.

$$\sigma_{bc,\,cal} = \frac{6 \times 8225 \times 10^3}{8 \times 205^2}$$

$$= 146.78 \text{ N/mm}^2 < 165 \text{ N/mm}^2 \qquad (0.66 \times 250)$$

which is all right.

Hence, provide 360 × 205 × 8 mm intermediate batten plates.

Design of weld Welding is done on all the four sides as shown in Fig. Ex. 4.18.

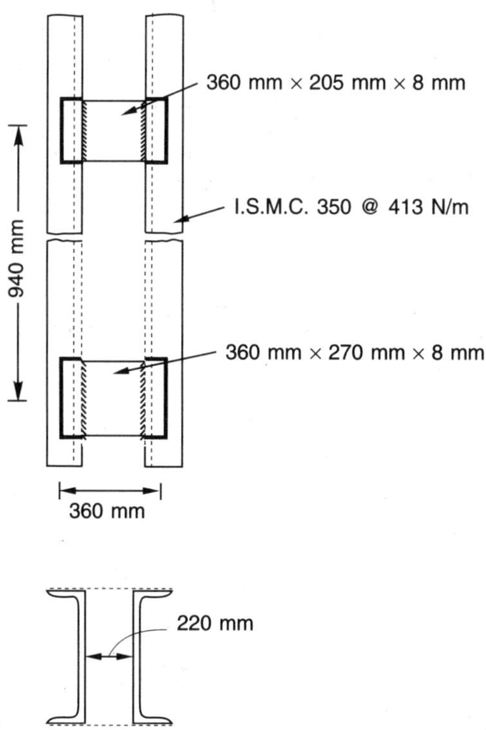

Fig. Ex. 4.18

Let, t = throat thickness of weld.

$$I_{xx} = 2\left[\frac{70\times t^3}{12} + (70\times t)\times\left(\frac{205}{2}\right)^2\right] + \frac{2\times t\times 205^3}{12}$$

Neglecting $\frac{70\times t^3}{12}$ being insignificant,

$$I_{xx} = 290.67 \times 10^4\, t \text{ mm}^4$$

$$I_{yy} = 2\left[\frac{t\times 70^3}{12}\right] + 2\times\frac{205\times t^3}{12} + 2\times 205\times t\times 35^2$$

Neglecting $\frac{2\times 205\times t^3}{12}$ being insignificant

$$I_{yy} = 55.94 \times 10^4\, t \text{ mm}^4$$

$$I_{zz} = I_{xx} + I_{yy}$$

$$= 290.67 \times 10^4 t + 55.94 \times 10^4 t = 346.61 \times 10^4 t \text{ mm}^4$$

$$r = \left\{\left(\frac{205}{2}\right)^2 + \left(\frac{70}{2}\right)^2\right\}^{1/2} = 108.31 \text{ mm}$$

$$\cos\theta = \frac{35}{108.31} = 0.323$$

Direct shear stress, $\tau_{vf_1} = \frac{56724}{(2\times 70 + 2\times 205)\,t} = \frac{103.13}{t}\ \text{N/mm}^2$

Shear stress due to moment, $\tau_{vf_2} = \frac{8225\times 10^3 \times 108.31}{346.61\times 10^4 \times t} = \frac{367.16}{t}\ \text{N/mm}^2$

Combined stress

$$= \left(\left(\frac{103.13}{t}\right)^2 + \left(\frac{367.16}{t}\right)^2 + \frac{2\times 103.13}{t}\times\frac{367.16}{t}\times 0.323\right)^{1/2} = \frac{437.55}{t}$$

$$< 108 \text{ N/mm}^2$$

or $$\frac{437.55}{t} = 108$$

or $$t = \frac{437.55}{108} = 4.05 \text{ mm}$$

$$\text{Size of weld} = \frac{4.05}{0.7}$$

$$= 5.78 \text{ mm } (\not< 5 \text{ mm for } 13.5 \text{ mm thick flange})$$

Provide a 6 mm weld to make the connections.

Example 4.19 Design a built-up column with four angles. The column is 12 m long and supports an axial load of 500 kN. The ends of the column are held in position and restrained against rotation. Design a suitable connecting system. Assume f_y = 250 MPa.

Solution Assume allowable axial compressive stress 120 MPa.

$$\text{Cross-sectional area required} = \frac{500 \times 10^3}{120} = 4167 \text{ mm}^2$$

Let us provide 4 I.S.A. 90 × 90 × 6 mm angle section.

The relevant properties are

$A = 1047 \text{ mm}^2$, $C_{xx} = C_{yy} = 24.2$ mm, $r_{xx} = r_{yy} = 27.7$ mm,
$I_{xx} = I_{yy} = 80.1 \times 10^4 \text{ mm}^4$, Area provided = 4 × 1047 = 4188 mm²

Spacing of angles For allowable compressive stress of 120 MPa and $f_y = 250$ MPa,

$$\text{Allowable } \frac{l}{r} = 64$$

or

$$l = 0.65 \times 12 \times 10^3 = 7800 \text{ mm}$$

$$r = \frac{7800}{64} = 121.875 \text{ mm}$$

Moment of inertia of section required, $I = Ar^2$

$$= 4188 \times 121.875^2$$

$$= 6220.65 \times 10^4 \text{ mm}^4$$

Moment of inertia required = moment of inertia provided

$$6220.65 \times 10^4 = 4 \times 80.1 \times 10^4 + 4188\ \bar{y}^2$$

$$\bar{y} = 118.69 \text{ mm}$$

or

$$\text{Spacing of angles, } S = 2 \times (118.69 + 24.2)$$

$$= 285.78 \approx 286 \text{ mm}$$

$$\text{Load carrying capacity} = 120 \times 4188$$

$$= 502.56 \text{ kN} > 500 \text{ kN}$$

which is safe.

Connecting system Let us provide a double lacing system with the lacing flats inclined at 45°. Rivets are provided at the centre of the leg of the angle.

$$\text{Spacing of lacing bars, } C = (286 - 45 - 45) \cot 45°$$

$$= 196 \text{ mm} \simeq 195 \text{ mm}$$

$$\frac{C}{r_{yy}} = \frac{195}{27.7}$$

$$= 7.03 < 50$$

and should be < 0.7 × 64 (= 44.8), which it is.

$$V = \frac{2.5}{100} \times 500 \times 10^3$$

$$= 12.5 \times 10^3 \text{ N}$$

$$\text{Transverse shear in each plane} = \frac{V}{N}$$

$$= \frac{12.5 \times 10^3}{2} = 6.25 \times 10^3 \text{ N}$$

As double lacing is provided,

$$\text{Compressive force in lacing bar} = \frac{1}{2}\frac{V}{N} \operatorname{cosec}\theta$$

$$= \frac{1}{2} \times 6.25 \times 10^3 \times \operatorname{cosec} 45°$$

$$= 4.42 \times 10^3 \text{ N}$$

Section of lacing flat Let us provide 20 mm ϕ power driven rivets.

Gross diameter = 20 + 1.5 = 21.50 mm

For 20 mm ϕ rivets,

Width of flat = 60 mm

$$\text{Thickness of lacing flat} = \frac{1}{60}(286 - 45 - 45) \operatorname{cosec} 45°$$

$$= 4.619 \text{ mm} \approx 8 \text{ mm}$$

Provide 60 I.S.F. 8 mm flat section.

$$\text{Minimum radius of gyration, } r = \frac{t}{\sqrt{12}}$$

$$= \frac{8}{\sqrt{12}} = 2.31 \text{ mm}$$

$$\frac{l_1}{r} = \frac{0.7 \times (286 - 45 - 45) \times \operatorname{cosec} 45°}{2.31}$$

$$= 83.99 < 145$$

which is safe.

For $\frac{l_1}{r} = 83.99$ and $f_y = 250$ MPa,

$$\sigma_{ac} = 96.61 \text{ N/mm}^2$$

$$\text{Load carrying capacity} = 96.61 \times 60 \times 8 \times 10^{-3}$$

$$= 46.37 \text{ kN} > 4.42 \text{ kN}$$

which is safe.

$$\text{Tensile strength of lacing flat} = (B - d)t\, \sigma_{at}$$

$$= (60 - 21.5) \times 8 \times (0.6 \times 250) \times 10^{-3}$$

$$= 46.200 \text{ kN} > 4.42 \times 10^3 \text{ kN}$$

which is safe.

Connection

$$\text{Strength of the rivet in double shear} = 2 \times \frac{\pi}{4} d^2 \tau_{vf}$$

$$= 2 \times \frac{\pi}{4} \times 21.5^2 \times 100 \times 10^{-3}$$

$$= 72.610 \text{ kN}$$

$$\text{Strength of the rivet in bearing} = dt\, \sigma_{pf}$$

$$= 21.5 \times 6 \times 300 \times 10^{-3} = 38.700 \text{ kN}$$

$$\therefore \quad \text{Strength of the rivet, } R_v = 38.700 \text{ kN}$$

$$\text{Number of rivets} = \frac{2 \times 4.42 \times 10^3 \times \cot 45^\circ}{38.700 \times 10^3}$$

$$= 0.228 \simeq 1$$

Provide one rivet only.

Tie plate Tie plates are provided at each end of the built-up column.

$$\text{Effective depth} = 286 - 2 \times 24.2 = 237.6 \text{ mm} > 2 \times 90 \text{ mm}$$

$$\text{Overall depth of the tie plate} = 237.6 + 2 \times (1.5 \times 20)$$

$$= 297.6 \text{ mm} \simeq 300 \text{ mm}$$

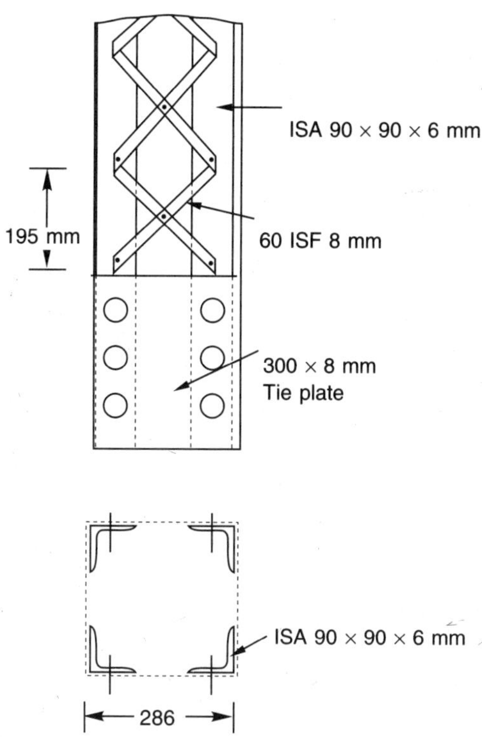

Fig. Ex. 4.19

$$\text{Length of the tie plate} = 286 \text{ mm}$$

$$\text{Thickness of the tie plate} = \frac{1}{50}(286 - 45 - 45)$$
$$= 3.92 \text{ mm} \approx 8 \text{ mm}$$

Provide 286 × 300 × 8 mm tie plate and connect it with rivets as shown in Fig. Ex. 4.19.

Example 4.20 A column I.S.H.B. 300 @ 576.8 N/m is to support a load of 600 kN. The column section is to be spliced at a height of 2.5 m. Design the splice plate. Assume fy = 250 MPa.

Solution Assume the ends of the column sections to be machined for complete bearing. As the column ends are flush it is assumed that 50% of the load is transferred directly and 50% is transferred through the splice and fastenings. Therefore,

$$\text{The load on the splice} = \frac{600 \times 1000}{4} = 150 \times 10^3 \text{ N}$$

Relevant properties of I.S.H.B. 300 @ 576.8 N/m are

$$A = 7485 \text{ mm}^2,\ B = 250 \text{ mm},\ T = 10.6 \text{ mm}$$

$$\text{Sectional area of the splice plate} = \frac{150 \times 10^3}{150} \qquad (\sigma_{ac} = 0.6 \times 250)$$
$$= 1000 \text{ mm}^2$$

Width of the splice plate should be kept equal to the width of the flange.

∴ Width of the splice plate = 250 mm

$$\text{Thickness of the splice plate} = \frac{1000}{250} = 4 \text{ mm} \nless 6 \text{ mm}$$

Provide 6 mm thick splice plate as the column may be exposed to the weather.

Assume nominal diameter of power driven rivet = 14 mm

$$\text{Gross diameter, } d = 14 + 1.5 = 15.5 \text{ mm}$$

$$\text{Strength of the rivet in single shear} = \frac{\pi}{4} d^2 \tau_{vf}$$
$$= \frac{\pi}{4} 15.5^2 \times 100 \times 10^{-3} = 18.87 \text{ kN}$$

$$\text{Strength of the rivet in bearing} = dt\,\sigma_{pf}$$
$$= 15.5 \times 6 \times 300 \times 10^{-3} = 27.9 \text{ kN}$$

$$\text{Strength of the rivet, } R_v = 18.87 \times 10^3 \text{N}$$

$$\text{Number of rivets required} = \frac{150 \times 10^3}{18.87 \times 10^3} = 7.949 \approx 8$$

Provide 14 mm ϕ rivets, 8 in numbers on each side of the splice (joint) to connect the splice plate with the flanges.

$$\text{Minimum pitch} = 2.5\ \phi = 2.5 \times 14 = 35 \text{ mm}$$

Provide a pitch of 50 mm. For 14 mm of rivets, edge distance is 22 mm.

Let us provide an edge distance of 25 mm.

$$\text{Depth of splice plate} = 6 \times 50 + 4 \times 25 = 400 \text{ mm}$$

Provide splice plates 400 mm × 250 mm × 6 mm on column flanges as shown in Fig. Ex. 4.20.

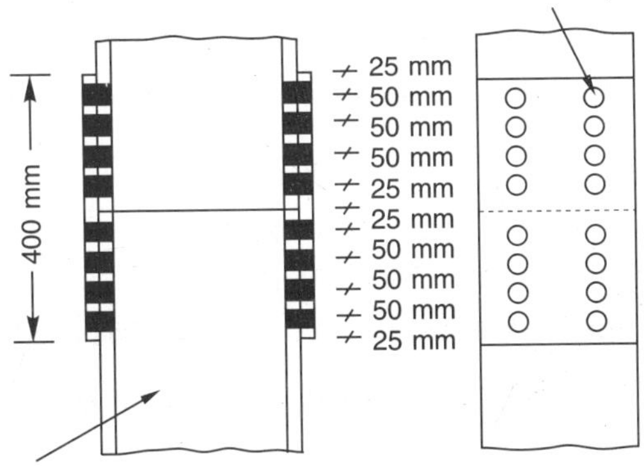

Fig. Ex. 4.20

Example 4.21 A stanchion I.S.H.B. 300 @ 618 N/m in the lower storey of a building is to be jointed to a stanchion I.S.H.B. 200 @ 392.4 N/m of the next upper storey. A load of 650 kN is to be transferred from the top storey stanchion. Design the column splice. The column ends are made flush. Assume the steel with yield stress of 250 MPa.

Solution Relevant properties of I.S.H.B. 300 @ 618 N/m are

$$D_1 = 300 \text{ mm},\ B_1 = 250 \text{ mm},\ T_1 = 10.6 \text{ mm},\ t_1 = 9.4 \text{ mm}$$

Relevant properties of I.S.H.B 200 @ 392.4 N/m are

$$D_2 = 200 \text{ mm},\ B_2 = 200 \text{ mm},\ T_2 = 9.0 \text{ mm},\ t_2 = 7.8 \text{ mm}$$

As the two sections to be spliced are of different depths, filler plates will have to be provided.

$$\text{Thickness of the filler plates} = \frac{300 - 200}{2} = 50 \text{ mm}$$

Bearing plate design A bearing plate is provided between the two column sections.

$$\text{Length of the plate} = D_1 = 300 \text{ mm}$$

$$\text{Width of the plate} = B_1 = 250 \text{ mm}$$

$$\text{Load on each flange} = \frac{650 \times 10^3}{2} = 325 \times 10^3 \text{ N}$$

Distance between the centre of gravity of column flanges of I.S.H.B. 300 @ 618 N/m

$$= 300 - 10.6 = 289.4 \text{ mm}$$

Distance between centre of gravity of column flanges of I.S.H.B. 200 @ 392.4 N/m

$$= 200 - 9.0 = 191 \text{ mm}$$

Distance between the line of action of forces in the flanges of the two column sections

$$= \frac{1}{2}(289.4 - 191) = 49.2 \text{ mm}$$

Moment due to the couple = $325 \times 10^3 \times 49.2 = 15990 \times 10^3$ Nmm

Equating the bending moment to the moment of resistance of the plate section,

$$15990 \times 10^3 = \frac{1}{6} \times 250 \times t^2 \times 185$$

or $$t = 45.54 \text{ mm} \approx 50 \text{ mm}$$

Provide a bearing plate of 300 × 250 × 50 mm.

Splice plate Splice will be designed for 50% of the load on one flange.

$$P = \frac{1}{2} \times 325 \times 10^3 = 162.5 \times 10^3 \text{ N}$$

Cross-sectional area of the splice plate required= $\dfrac{162.5 \times 10^3}{150} = 1083 \text{ mm}^2$

Flange width (B_2) of the top column = 200 mm, and that of the bottom column is 250 mm

Minimum width of splice plate = B_2 = 200 mm

∴ $$\text{Thickness} = \frac{1083}{200} = 5.415 \approx 6 \text{ mm}$$

Provide splice palte 250 × 6 mm in cross-section.

Note The splice plate is made 250 mm wide because the lower column flange is 250 mm wide. But, for calculation of thickness of the splice plate, upper column flange width is considered being less. The width of the splice plate at the extreme top end will be 200 mm only.

Connection Let us provide 18 mm diameter power driven rivets.

Gross diameter = 18.0 + 1.5 = 19.5 mm

Minimum pitch = 2.5 × 18.0 = 45 mm

Provide 60 mm pitch.

For 18 mm ϕ rivets, edge distance = 29 mm.

Let us provide an edge distance of 30 mm.

$$\text{Strength of the rivet in single shear} = \frac{\pi}{4} d^2 \tau_{vf}$$

$$= \frac{\pi}{4} 19.5^2 \times 100 \times 10^{-3} = 29.86 \text{ kN}$$

Strength of the rivet in bearing = $dt\,\sigma_{pf}$

$$= 19.5 \times 6 \times 300 \times 10^{-3} = 35.1 \text{ kN}$$

$$\text{Number of rivets, } n = \frac{162.5 \times 10^3}{29.86 \times 10^3} = 5.44 \approx 6$$

Provide 6 rivets to connect the splice plate on each flange. The rivets are provided in two vertical rows at a pitch of 60 mm and at an edge distance of 30 mm.

$$\text{Depth of splice plate} = 4 \times 60 + 4 \times 30 + 50 = 410 \text{ mm}$$

Provide splice plate 410 × 250 (reducing to 200 mm at the top) × 6 mm as shown in Fig. Ex. 4.21.

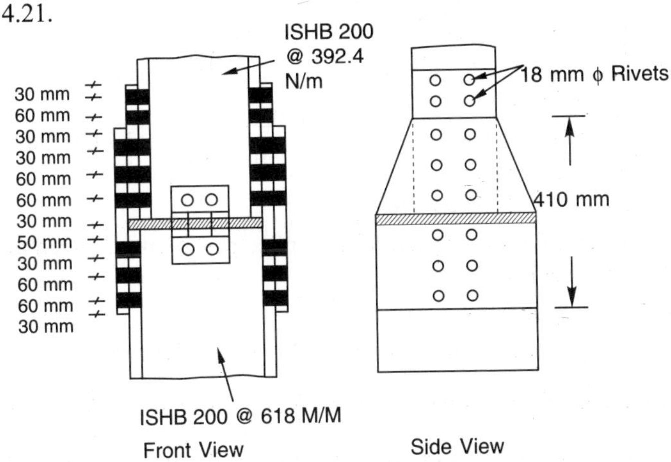

Fig. Ex. 4.21

Number of extra rivets Extra rivets for filler plates are designed as per the I.S. specifications.

$$\text{Thickness of filler plates} = 50 \text{ mm} > 6 \text{ mm}$$

For each 2 mm thickness of packing rivets are increased by 2.5%.

$$\therefore \qquad \text{Increase rivets by } \frac{2.5 \times 50}{2}\% = 62.5\%$$

$$\text{Extra rivets} = 5.44 \times 0.625 = 3.4 \approx 4$$

Provide 4 rivets on packings i.e. 2 in each vertical row on each filler plate.

Note Web cleat angles of nominal size are provided when the column sections to be spliced are not of the same cross-section.

Example 4.22 A column section I.S.H.B. 300 @ 618 N/m (f_y = 250 N/mm^2) is to be spliced. The design loads are:

Axial load over the column	300 kN
Shear force	75 kN
Bending moment	20 kNm

(a) Design the splice plates and riveted connection.
(b) Design HSFG Bolts and revise the splice plate size if possible.

Solution

(a) Assuming the column ends to be machined for complete bearing 50% of the load is transferred directly and 50% by the fasteners. Also, as splicing is done for both the flanges,

$$\text{Direct load on each splice} = \frac{300}{2} \times 0.5 = 75 \text{ kN}$$

$$\text{Load on splice due to moment} = \frac{20 \times 10^6}{300 + 6} = 65.36 \text{ kN}$$

(Assuming 6 mm thick splice plate, the lever arm = 300 + 6 mm)

$$\text{Total design load} = 75 + 65.36 = 140.36 \text{ kN}$$

$$\text{Width of splice plate} = \text{width of the flange} = 250 \text{ mm}$$

$$\text{Thickness of splice plate required} = \frac{140.36 \times 10^3}{250 \times 150} = 3.7429 \text{ mm} \approx 6 \text{ mm}$$

Provide a 250 × 6 mm splice plate.

Length of splice plate The length of the splice plate depends upon the number of rivets in the vertical row.

Let us provide 22 mm ϕ rivets.

$$\text{Gross diameter of rivet} = 22 + 1.5 = 23.5 \text{ mm}$$

$$\text{Strength of rivet in single shear} = \frac{\pi}{4} \times 23.5^2 \times 100 = 43{,}373.6 \text{ N}$$

$$\text{Strength of rivet in bearing} = 23.5 \times 6 \times 300 = 42{,}300 \text{ N}$$

$$\text{The value of one rivet, } R_v = 42{,}300 \text{ N}$$

$$\text{Number of rivets required, } n = \frac{140{,}360}{42{,}300} = 3.318$$

Let us revise the design and provide 20 mm ϕ rivets for economy.

$$\text{The gross diameter of the rivet} = 20 + 1.5 = 21.5 \text{ mm}$$

$$\text{Value of the rivet in single shear} = \frac{\pi}{4} \times 21.5^2 \times 100 = 36{,}305 \text{ N}$$

$$\text{Value of the rivet in bearing} = 21.5 \times 6 \times 300 = 38{,}700 \text{ N}$$

The rivet value is 36305 N

$$n = \frac{140{,}360}{36{,}305} = 3.866 \approx 4$$

Provide four rivets 20 mm ϕ on each side of the splice.

$$\text{Minimum pitch of rivets} = 2.5\,\phi = 2.5 \times 20 = 50 \text{ mm}$$

Provide a pitch of 60 mm.

For 20 mm rivets minimum edge distance is 29 mm

Let us provide edge distance of 30 mm

$$\text{Length of splice plate} = 2 \times 60 + 4 \times 30 = 240 \text{ mm}$$

Provide a splice plate 240 × 250 × 6 mm on column fanges as shown in Fig. Ex. 4.22.

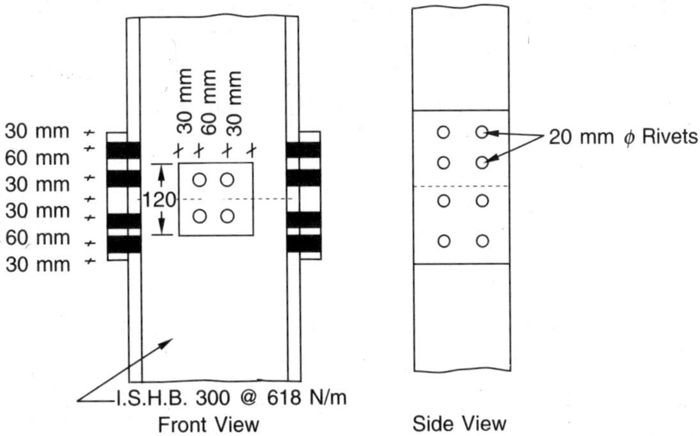

Fig. Ex. 4.22

Splice plate for shear The splice plate for the shear force is provided on the web. A pair of splice plates (one on each side of the web) are provided.

Connections Let us provide 20 mm ϕ rivets.

Gross diameter of the rivet = 20 + 1.5 = 21.5 mm

The rivets will be in double shear

$$\text{Value of the rivet in double shear} = 2 \times \frac{\pi}{4} d^2 \tau_{vf}$$

$$= 2 \times \frac{\pi}{4} \times 21.5^2 \times 100 = 72{,}610 \text{ N}$$

$$\text{Value of rivet in bearing} = dt\, \sigma_{pf}$$

$$= 21.5 \times 9.4 \times 300 = 60{,}630 \text{ N}$$

(With 6 mm shear splice plate 2 Nos.)

The value of rivet = 38,700 N

Maximum shear force in the web = 75 kN

$$\text{Number of rivets required} = \frac{75 \times 10^3}{60{,}630} = 1.237$$

Provide two rivets on each side of the splice,

Minimum pitch = 2.5ϕ = 2.5 × 20 = 50 mm

Provide a pitch of 60 mm

Minimum edge distance for 20 mm ϕ rivets = 29 mm

Let us provide edge distance of 30 mm.

Size of shear splice plate,

Maximum shear force = 1.5 × 75 = 112.5 kN

(since $\tau_{v,\,max} = 1.5\tau_{v\,avg}$)

Maximum permissible shear stress = $0.45 f_y = 0.45 \times 250 = 112.5$ N/mm^2

Let A = cross-sectional area required by one plate.

$$A = \frac{1}{2} \times \frac{112.5 \times 10^3}{112.5} = 500 \text{ mm}^2$$

Provide 140 mm wide splice plate.

Thickness of the splice plate required,

$$t = \frac{500}{140} = 3.57 \text{ mm} \not< 6 \text{ mm}$$

So provide 140 × 6 mm splice plate on each side of the web.

Length of the plate = 60 + 2 × 30 = 120 mm

Provide a pair of 120 × 140 × 6 mm shear splice plates one on each side of the web.

(b) *Connection with HSFG bolts.*

(i) Splice plate connection with flanges of the column selections.
Design load = 140.36 kN
The number of interfaces = 1

$$F = \frac{n\mu T}{F_s}$$

$$140.36 = \frac{1 \times 0.50 \times T}{1.4}$$

Proof load, T = 393 kN

Let us try M20–10.9 k bolts.
Proof load of 2, M20–10.9 k bolts = 2 × 203 = 406 kN > 393 kN.
So provide 2, M20–10.9 k bolts for making the connection.
Edge distance of 20 ϕ bolts = 30 mm ≈ 30 mm.
The depth of splice plate = 4 × edge distance = 4 × 30 = 120 mm.
Hence provide a splice plate 120 × 350 × 6 mm.

(ii) Splice plate connection with web of the column sections.
Shear force = 75 kN.
The number of interfaces = 2.

$$F = \frac{n\mu T}{F_s}$$

$$75 = \frac{2 \times 0.50 \times T}{1.4}$$

Proof load, T = 105 kN

Let us try M16–8.8 G bolts.
Proof load of 2, M16–8.8 G bolts = 2 × 94.5 = 189 kN > 105 kN.
Provide 2, M16–8.8 G bolts in one row.
There will be no change in the size of splice plates on the web.

Example 4.23 Design a butt weld for the column section for Example 4.22.

Solution Let us provide single-V butt weld

$$\text{Effective throat thickness} = \frac{5}{8} \times 10.6 = 6.625 \text{ mm}$$

The relevant properties of I.S.H.B. 300 @ 618 N/m from IS Hand Book No. 1 are:

$D = 300$ mm, $B = 250$ mm, $T = 10.6$ mm, $t = 9.4$ mm, and $h_2 = 25.1$ mm

From Example 4.22, total design load for flange splice = 140.36 kN

$$\text{Required length of weld} = \frac{140.36 \times 10^3}{6.625 \times 150} = 141.24 \text{ mm} < 250 \text{ mm}$$

Hence, provide single-V butt weld for splicing the column sections.
For the shear force of 75 kN, provide double-V butt weld on web.

Effective throat thickness = 9.4 mm

Capacity of weld = 9.4 × 108 = 1015.2 N/mm

$$\text{Required length of weld} = \frac{75 \times 10^3}{1015.2} = 73.87 \text{ mm}$$

Available length of welding = 300 – (2 × 25.1) = 249.8 mm > 79.78 mn

Hence, O.K.

Exercises

Note Use the steel of yield stress 250 MPa if not spcified in the problem.

4.1 A single angle section I.S.A. 60 × 60 × 8 mm, 3.0 m long, is used as a strut. The ends are welded to the gusset plates. Compute the maximum load it can carry.

4.2 A strut consists of a double angle section I.S.A 70 × 70 × 8 mm and is 3.2 m long. The member is connected to the gusset plate by 20 mm ϕ power driven three rivets. Calculate the maximum compressive load the member can transfer.

(a) when the angles are placed on the opposite sides of 12 mm thick gusset plate.

(b) when the angles are placed on the same side of 12 mm thick gusset plate.

4.3 Find the safe compressive axial load for the following sections. (See Fig. Prob. 4.3).

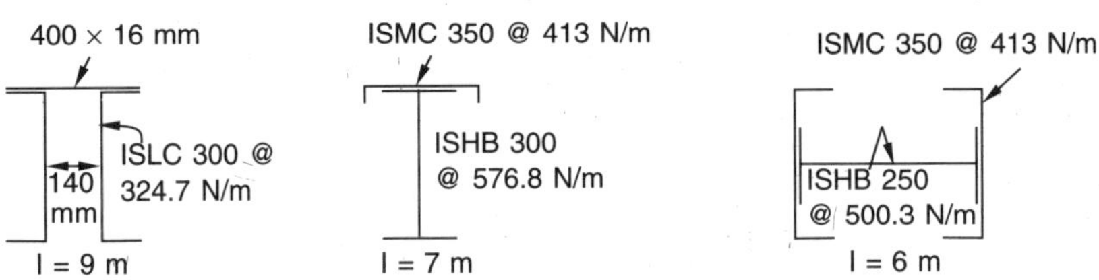

Fig. Prob. 4.3

4.4 A discontinuous strut of 3 m length between the intersections consists of two angles 110 × 110 × 8 mm. The angles are placed back to back on the opposite side of a 10 mm thick gusset plate and are tack riveted. Calculate the percentage change in the load carrying capacity if the two angles are placed on the same side of the gusset plate.

4.5 Design a single discontinuous strut to carry a 100 kN load. The length of the strut between c/c of intersections is 3.40 m. Assume f_y = 280 N/mm^2.

4.6 Design a double angle discontinuous strut to carry a load of 175 kN. The length of the strut between c/c of intersection is 3.85 m. Assume f_y = 280 N/mm^2

4.7 Design a continuous strut to carry a load of 160 kN. The effective length of the strut is 5.8 m. Assume f_y = 230 N/mm^2.

4.8 Design a column section to be used in a public building. Column is 4.80 m long with its ends restrained in direction and position. The column is to support a 1800 kN load.

4.9 Design a built-up column with two channels placed face-to-face. The column is of 6.6 m effective length and supports a load of 1080 kN. Also design the lacing.

4.10 Design a built-up column with four angles laced together. The effective length of the column is 7.20 m and it supports a load of 1200 kN.

4.11 Design a column with a two channel section laced together and placed toe-to-toe and spaced apart to support a load of 3300 kN. The effective length of the column is 5.2 m.

4.12 Design a column of length 6.30 m to carry a load of 700 kN at an eccentricity of 100 mm from the centroidal axis of the column across the web. The ends of the column are pinned.

4.13 Design the above column if the eccentricity of the load is 100 mm along the web and 180 mm across the web.

4.14 Design a column section subjected to an axial load of 520 kN and a moment of 80 kNm about the major axis. The effective length of the column is 4.80 m. How much moment along the minor axis can be allowed along with the above forces for the section designed.

4.15 A column section I.S.H.B. 150 @ 265.9 N/m is to be spliced with another column section I.S.H.B. 150 @ 339.4 N/m. The load on the column is 228 kN. Design the splice. Assume f_y = 220 N/mm^2.

4.16 A column section I.S.H.B. 450 @ 855.4 N/m is to be spliced with a column I.S.H.B. 300 @ 576.8 N/m. The load on the column is 572 kN. Design a suitable splice using HSFG bolts. If a shear force of 110 kN also acts, design the shear splice.

4.17 A column section I.S.H.B. 225 @ 459.1 N/m is to be spliced at floor level. The forces at the section are axial load, transverse shear and bending moment of 325 kN, 90 kN, 31 kN m respectively. Design a suitable splice.

Column Bases and Footings

5.1 INTRODUCTION

Steel columns are normally supported over concrete blocks. However, when the loads supported by these columns are large and the bearing pressure of concrete from below is insufficient to resist the loads, they may fail. Therefore, it is a normal practice to distribute the column loads to steel base plates which are placed over these concrete blocks. In addition to transferring safely the column loads, a base plate also maintains alignment of column in plan, verticality of column and controls column and frames deflections. The main problems in the design of the column bases are to determine the size of the plates and their thicknesses. The size is determined by the required bearing area on foundation; and the required thickness is determined so that bending stress in the plate does not exceed the allowable value. When the soil has very low bearing capacity, large concrete blocks may become uneconomical and in such cases grillage footing is provided.

5.2 TYPES OF COLUMN BASES

The safety of a column and thus of a structure depends mainly upon the stability of the foundation and consequently on the bases, in the case of steel columns. Therefore, the column base should be designed with great care. Column bases can be classified into types: one transmitting direct load only; another carrying an appreciable bending moment in addition to direct load. In the former case, the design of base does not provide a special problem. Usually, a steel plate is used to distribute the column load over a sufficient area to maintain the unit-bending stress within the allowable value for the concrete footing. For an axially loaded column, the pressure distribution between the base plate and concrete pedestal is assumed to be uniform (Fig. 5.1). In actual cases, the pressure distribution is non-uniform and the effect of non-uniformity is accounted for by taking a slightly low permissible bearing pressure of concrete. For column bases transmitting moment, angles anchored to the footing may be used, provided the angles are designed for uplift forces. Stiffeners can be used to advantage for

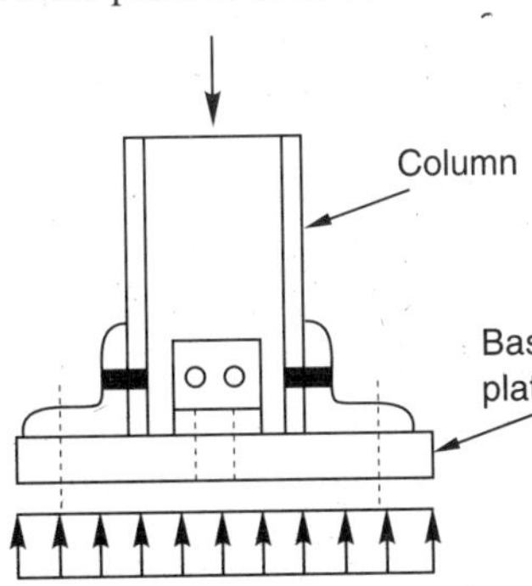

Fig. 5.1 *Assumed Pressure Distribution under Base Plate*

welded construction as discussed in Section 5.8. The two prevalent column bases are the slab base and the gusset base.

5.3 SLAB BASE

When the column is subjected to only direct loads, the base can be designed by assuming a uniform bearing pressure from below. For small loads, a steel plate alone can be used to transmit the loads to the concrete pedestal. Such a base plate is called slab base (Fig. 5.2) and is considered to be a pinned base. When the columns are subjected to bending moments along with direct forces, angle sections are attached to the flanges of column. These angles are anchored with foundation bolts to the concrete pedestal. Even if the column is subjected to direct force only, nominal angle sections should be provided to keep the column in place, and to resist any tension due to erection and connections. Angles may be omitted if the base plate is shop welded to the column.

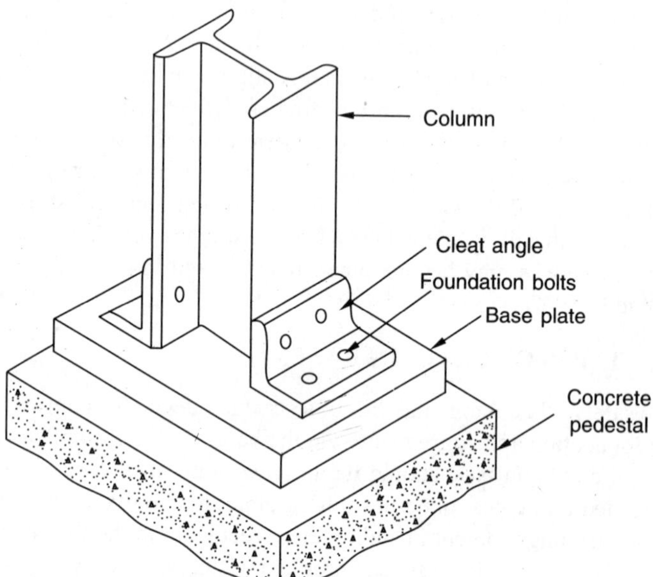

Fig. 5.2 *Slab Base*

Theoretical Considerations

It is assumed that the maximum bending moment occurs at the edge of the column (Fig. 5.3). As the slab tends to bend simultaneously about the two principal axes of the slab, the stress caused by bending about one axis is influenced by the stress due to bending about the other axis.

Taking 1 mm strip of slab projection along x–x axis

$$\text{Maximum bending moment} = w \times 1 \times a \times \frac{a}{2} = \frac{wa^2}{2} \tag{5.1}$$

Taking 1 mm strip of slab projection along y–y axis

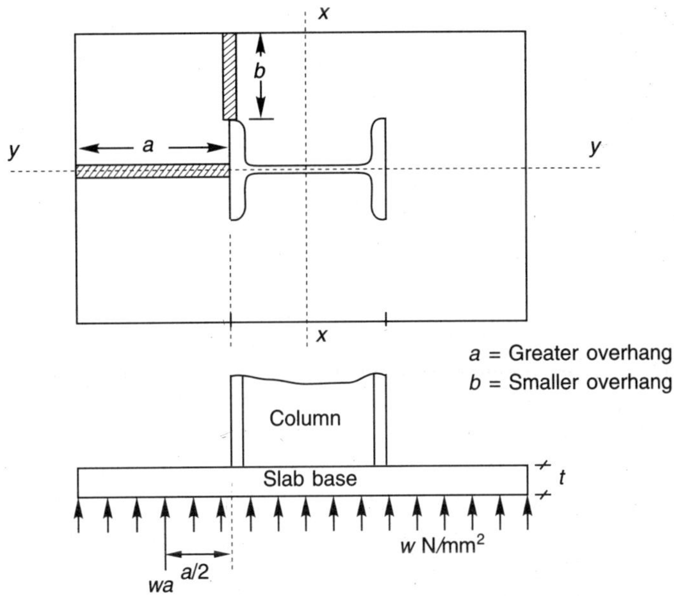

Fig. 5.3 *Slab Base*

$$\text{Maximum bending moment} = w \times 1 \times b \times \frac{b}{2} = \frac{wb^2}{2} \tag{5.2}$$

Let σ_{at_1} and σ_{at_2} be the maximum tensile stresses for overhangs of lengths a and b respectively, and let σ_{bs} be the bending stress in slab base. Then,

Moment of resistance for overhang of length a

$$= \sigma_{at_1} z = \sigma_{at_1} \times \frac{1}{6} \times 1 \times t^2 \tag{5.3}$$

and Moment of resistance for overhang of length b

$$= \sigma_{at_2} z = \sigma_{at_2} \times \frac{1}{6} \times 1 \times t^2 \tag{5.4}$$

For equilibrium, the moment should be equal to moment of resistance. From Eqs (5.1) and (5.3)

$$\frac{\sigma_{at_1} t^2}{6} = \frac{wa^2}{2} \quad \Rightarrow \quad \sigma_{at_1} = \frac{3wa^2}{t^2}$$

and from Eqs (5.2) and (5.4)

$$\frac{\sigma_{at_2} t^2}{6} = \frac{wb^2}{2} \quad \Rightarrow \quad \sigma_{at_2} = \frac{3wb^2}{t^2}$$

But the tensile stress σ_{at_2} in the axial direction x–x causes a compressive stress equal to $\mu\, \sigma_{at_2}$ in the axial direction y–y, where μ is Poisson's ratio, assumed to be 0.25.

The net tensile stress σ_{bs} (axis y-y)

$$= \frac{3\,wa^2}{t^2} - 0.25\left(\frac{3\,wb^2}{t^2}\right)$$

$$\sigma_{bs} = \frac{3\,w}{t^2}\left(a^2 - \frac{b^2}{4}\right) \tag{5.5}$$

or

$$t = \sqrt{\frac{3\,w}{\sigma_{bs}}\left(a^2 - \frac{b^2}{4}\right)} \tag{5.6}$$

where t = thickness of the slab base

σ_{bs} = permissible bending stress in the slab base (185 MPa)

Design Procedure

The design of a base plate consists in finding out its size and thickness. Following are the design steps of base plate:

1. Assume a suitable allowable bearing pressure on the concrete, depending upon the grade of concrete used.
2. The area of a slab base may be computed by

$$A = \frac{P}{\text{allowable bearing pressure on concrete}}$$

where A = area of the base plate required in mm^2

P = direct load

The allowable bearing pressure on the concrete ($0.25\,f_{ck}$) of M–15 grade will be 3.75 N/mm^2.

3. A square base plate is generally provided. The side of the base plate may be worked out by Eq. (5.7).

$$L = B = \sqrt{A} \tag{5.7}$$

Some designers feel that the projections of the base plate beyond column edges a, b as shown in Fig. 5.4 may be kept equal and the sides can be worked out by Eq. (5.8).

$$(D + 2b) \times (b_0 + 2a) = A \tag{5.8}$$

where L = length of base plate in mm

B = width of base plate in mm

a = bigger projection of base plate beyond column in mm

b = smaller projection of base plate beyond column in mm

D = depth of column section in mm

b_0 = width of the flange of column in mm

4. The intensity of pressure w, from the concrete pedestal is worked out by

$$w = \frac{P}{A_1}$$

where w = intensity of pressure from concrete under the slab base in N/mm^2

A_1 = area of base plate provided in mm^2

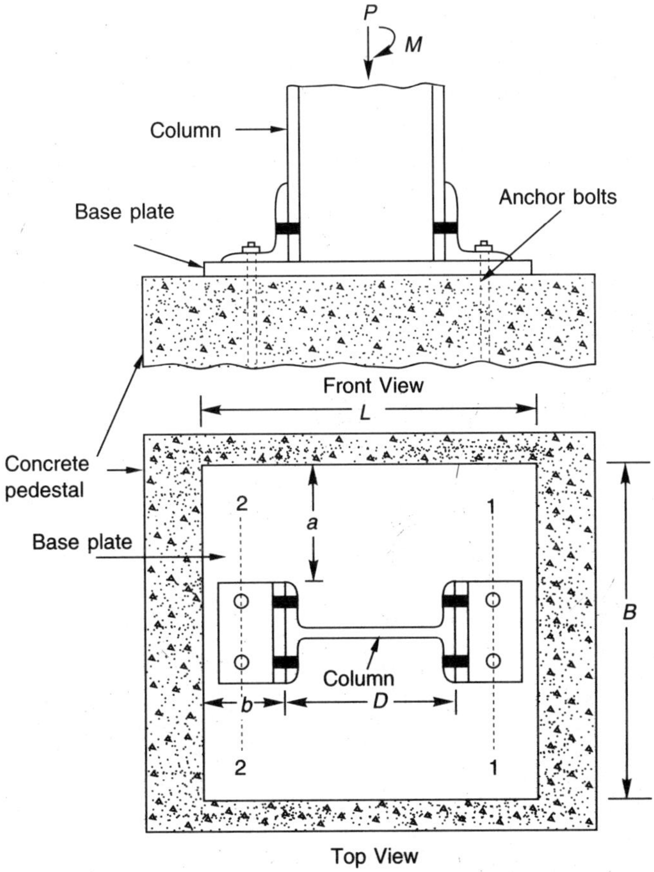

Fig. 5.4 *Projections of Base Plate beyond Column Edges*

5. The minimum thickness of the slab base is calculated from Eq. (5.6).
6. If a solid round steel column is supported over a square base plate, then the thickness of the base plate is calculated from Eq. (5.9).

$$t = 10\left\{\frac{90W}{16\sigma_{bs}} \times \frac{B}{B - d_0}\right\}^{1/2} \tag{5.9}$$

where t = thickness of plate in mm

W = the total axial load in kN

B = the length of the side of the base in mm

$\nless 1.5(d_0 + 75)$mm

d_0 = the diameter of the column in mm

Note It is assumed that the bearing pressures are uniformly distributed over all contact surfaces. The bottom of the column and the supporting surface of the base are machined so that the load is transmitted to the plate by direct bearing.

5.4 GUSSET BASE

When the load over the column section is too large or when the axial load is accompanied with bending moments, a gusset base is provided. It consists of a base plate, two gusset plates and two gusset angles. The gusset plates and angles are placed on flanges as shown in Fig. 5.5. Gusset materials used in the base increase the bearing area, consequently reducing the thickness of the base plate as compared to the slab base. This kind of base may be considered to be rigid. There can be two cases:

(i) Axially loaded columns,
(ii) Eccentrically loaded columns

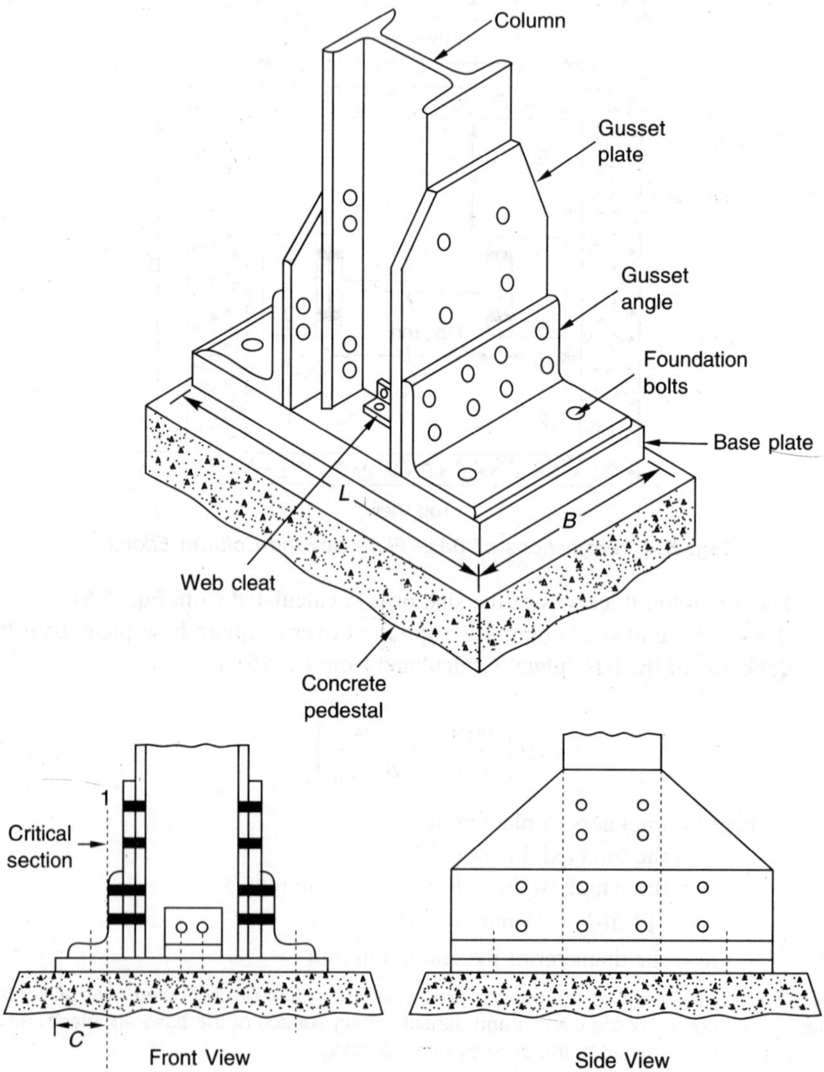

Fig. 5.5 *Gusset Base*

Design Procedure

Following is the procedure adopted for the design of a gusset base:

1. A suitable allowable bearing pressure on concrete depending upon the grade of concrete used is assumed.
2. The area of the base plate is computed by

$$A = \frac{P}{\text{allowable bearing pressure on concrete}}$$

   where A, P and allowable bearing pressure are as already been defined in Section 5.3.
3. The size of the gusset material is assumed. The gusset plate should not be less than 16 mm in thickness. The gusset angle is chosen so as to accommodate two rows of rivets in the vertical leg and one row of bolts in the horizontal leg. Subsequently an unequal angle section is provided. The thickness of the gusset angle should be almost same as that of the gusset plate. The length of the gusset material is normally kept equal to the length of the base plate, parallel to the flange of the column.

   The dimension of base plate parallel to the web,

   L = depth of section + 2(thickness of gusset plate + leg length of angle + overhang)

   (The maximum overhang should be kept as small as possible, say, about 1 cm from the gusset angle toe.)

$$\text{The dimension of base plate parallel to flange, } B = \frac{A}{L}$$

4. The intensity of the bearing pressure w from the concrete below the base is worked out using the following expression.

$$\text{Bearing pressure, } w = \frac{P}{A_1}$$

   where A_1 = area of the base plate provided.
5. The thickness of the base plate is computed by equating the moment at the critical section to the moment of resistance of the gusset at that section. The critical section is assumed to lie at the root of the fillet of the angle section. For unit width of the plate,

$$\text{B.M. at the critical section 1–1, } M = \frac{wc^2}{2} \qquad (5.10)$$

   Moment of resistance at the critical section 1–1,

$$M_R = Z\,\sigma_{bs}$$

$$= \frac{1}{6} \times 1 \times t^2 \times \sigma_{bs} = \frac{t^2}{6}\sigma_{bs}$$

   For equilibrium the moment should be equal to moment of resistance.

$$M = M_R$$

$$\frac{wc^2}{2} = \frac{t^2}{6}\sigma_{bs}$$

or $$t = c\sqrt{3w/\sigma_{bs}} \tag{5.11}$$

$t' = t -$ thickness of gusset angle

where w = intensity of pressure from concrete under the slab base in N/mm^2

c = the portion of the base plate acting as a cantilever in mm

t' = required thickness of base plate in mm

t = thickness of base plate at the critical section in mm

= thickness of base plate + thickness of gusset angle resisting the moment at critical section

σ_{bs} = the permissible bending stress in slab base (185 MPa)

6. The connections are designed for forces to which the column is subjected. If the column ends and gusset materials are not faced/machined for complete bearing, the fasteners are designed for the total forces to be transferred. If they are faced for complete bearing, 50% of the forces are transferred directly by the column and 50% through the fasteners.
7. The diameter of the rivet is assumed and the rivet value is computed in single shear and bearing. The number of rivets are worked out by

$$n = \frac{P'}{R_v} \tag{5.12}$$

where n = number of rivets

P' = the force on the fasteners in Newtons

R_v = rivet value in Newtons

These rivets are provided to connect the gusset plate with the column flange. The height of the gusset plate depends upon the number of rivets to be accommodated. The same number of rivets are provided to connect the gusset plate with the gusset angle.

Nominal size rivets/bolts are provided on the horizontal leg of the gusset angle to keep the column in position.

5.5 DESIGN OF BASES FOR ECCENTRICALLY LOADED COLUMNS

When the base plates are loaded eccentrically the pressure distribution from the concrete does not remain uniform. The compressive stress is maximum near one of the edges of the base plate and decreases towards the other edge.

Let the column be subjected to a load P, at an eccentricity of e, causing a moment $M(= Pe)$ and an axial load P. This gives rise to a combination of axial compressive stress and bending stress.

Due to axial load P, the pressure distribution is uniform and stress diagram is rectangular, as shown in Fig. 5.6(a).

$$\text{The direct stress } = \frac{P}{A_1} = \frac{P}{LB}$$

where A_1 = area of base plate provided.

Due to the bending moment the bending stress developed (Fig. 5.6(b)) is

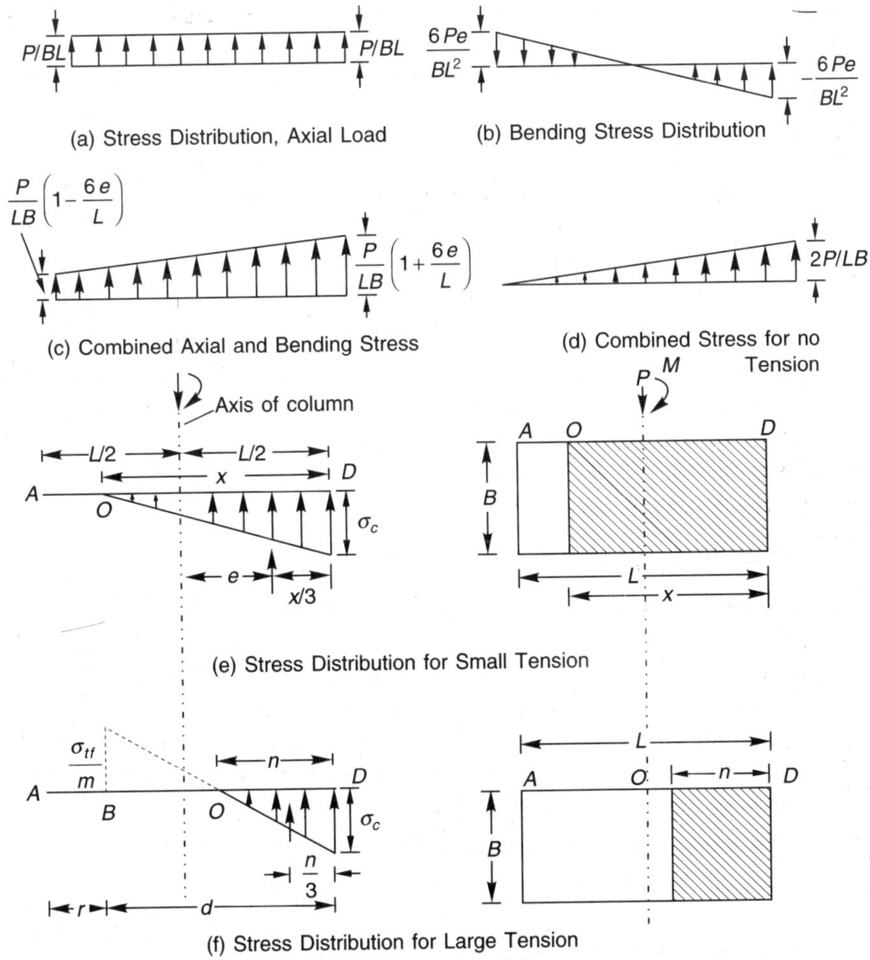

Fig. 5.6 *Stress Distribution for Bases*

$$= \pm \frac{M}{I} y$$

$$= \pm \frac{M}{\frac{1}{12} BL^3} \times \frac{L}{2}$$

$$= \pm \frac{6M}{BL^2} = \pm \frac{6Pe}{BL^2}$$

The combined stress due to axial load and bending moment as shown in Fig. 5.6(c) are,

$$\text{Combined stress, } \sigma = \frac{P}{LB} \pm \frac{6Pe}{BL^2}$$

$$\sigma = \frac{P}{LB}\left(1 \pm \frac{6e}{L}\right) \tag{5.13}$$

It is desirable to have the combined stress compressive because concrete is weak in tension.

Depending upon the eccentricity and length of the base plate, three cases are possible.

From Eq. (5.13),

(i) When there is no tension, $1 - \frac{6e}{L} = 0 \quad \Rightarrow \quad e = \frac{L}{6}$ (Case I)

(ii) When the tension developed is small $e = \frac{L}{6}$ to $\frac{L}{3}$ (Case II)

(iii) When the tension developed is substantial $e > \frac{L}{3}$ (Case III)

The length of base plate required in all the three cases is kept minimum from economy point of view. It is calculated in a way similar to that for an axially loaded column. The width and thickness of the base plate for the above cases are calculated by the procedures given below. The thickness so calculated does not include the thickness of gusset angle.

Case I: When there is no tension $e \leq \frac{L}{6}$

1. Draw the stress diagram (Fig. 5.6(d)) by substituting the value of eccentricity in Eq. (5.13). When $e = \frac{L}{6}$, combined stresses = $\frac{2P}{LB}$, 0 respectively.
2. Calculate the width of base plate using the Eq. (5.13).
3. Calculate the thickness of base plate by equating the bending moment and moment of resistance at the critical section, which is assumed to be at the outer edge of the column section flange.

Case II: When the tension developed is small.

The eccentricity, e lies between $\frac{L}{6}$ to $\frac{L}{3}$

It is assumed that tension in the far side bolts is negligible.

1. For calculating distance x refer to Fig. 5.6(e).

$$\frac{x}{3} + e = \frac{L}{2}$$

$$\Rightarrow \quad x = 3\left(\frac{L}{2} - e\right) \tag{5.14}$$

2. The width of the base plate is determined from Eq. (5.15) as follows.
   Compressive force (C) in concrete = area of stress triangle × width

$$C = \frac{x\,\sigma_c}{2} B$$

This compressive force should be equal to the load acting (P)

$$\frac{x\,\sigma_c}{2} B = P$$

$$\Rightarrow \quad B = \frac{2P}{x\,\sigma_c} \tag{5.15}$$

Substituting the value of x from Eq. 5.14 into Eq. 5.15

$$B = \frac{2P}{3\left(\frac{L}{2} - e\right)\sigma_c} \tag{5.16}$$

3. The thickness of base plate is calculated in a way similar to that in case I.

Case III: When the tension developed is substantial $e > \frac{L}{3}$

The problem is analysed using the concrete beam theory.

1. The depth of the neutral axis is determined as follows.

   From the stress diagram, (Fig. 5.6(f)) $\frac{\sigma_c}{\sigma_{tf}/m} = \frac{n}{d-n}$

$$\Rightarrow \quad n = \frac{d}{1 + \frac{\sigma_{tf}}{m\sigma_c}} \tag{5.17}$$

   where σ_{tf} = tensile stress in the bolts (120 MPa)

   n = depth of neutral axis from the near edge

   m = modular ratio (18)

2. The compressive force C in concrete is calculated by taking moment about the farthest side bolt (i.e. about line of action of tensile force).

$$C\left(d - \frac{n}{3}\right) = M + P\left(\frac{L}{2} - r\right) \tag{5.18}$$

   where d = distance of bolts on the far side from near edge

   r = edge distance of bolts

3. The width of base plate is calculated by using the value of compressive force C and the stress diagram.. The compressive force in the concrete from the stress diagram is given by

$$C = \frac{Bn\sigma_c}{2}$$

$$\Rightarrow \quad B = \frac{2C}{n\sigma_c} \tag{5.19}$$

4. The thickness of base plate is calculated in a way similar to that in case I.
5. The anchor bolts are designed as follows.

   The uplift in bolts, $T = C - P$

   The net area of the bolts at root of threads, $A_s = \frac{T}{\sigma_{tf}}$

Size of the anchor bolts may be chosen between 20 – 36 mm. Length of the bolts should not be less than 24 times the diameter.

5.6 FOUNDATION BOLTS

Foundation bolts also known as *anchor bolts* are generally provided to check the uplift of the base plate. For bases of columns subjected to bending moment in addition to axial load usually, four bolts one at each corner of the base plate will be sufficient. However, the number of bolts can be more but should not exceed eight from construction view point. A minimum two anchor bolts are to be provided even if the column is subjected to only axial load. The number and size of bolts are adjusted so as to generate precompression at least equal to direct compression from the column on the base.

Whenever bending moments are present, foundation bolts of adequate sizes must be designed to hold the column down to the concrete pedestal. Let us say the column is subjected to an axial load P, and a bending moment M (Fig. 5.4). The column will try to overturn either about the line 1–1 or 2–2 passing through the bolts depending upon whether the moment is clockwise or anti-clockwise. The axial load provides a counteracting (stabilizing) moment equal to $Px/2$.

When a clockwise moment (M) acts as shown in Fig. 5.4, bolts A and B will be subjected to uplift, C and D to thrust as shown in Fig. 5.7. These uplifts and thrusts will provide resisting couples = $2P_1x$.

Hence the total stabilizing moment under the action of compressive force P and moment M will be

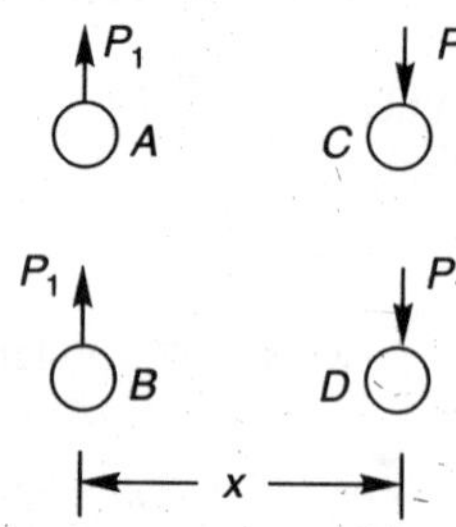

Fig. 5.7 *Tension in Foundation Bolts*

$$P\frac{x}{2} + 2P_1x$$

For equilibrium
$$M = P\frac{x}{2} + 2P_1x$$

$$\Rightarrow P_1 = \frac{M}{2x} - \frac{P}{4} \tag{5.20}$$

where P_1 = uplift force (in N) on bolt
x = distance between bolt lines

Design

1. The uplift force P_1 on the bolts is determined from Eq. (5.20).
2. The area at the root of bolt is given by

$$A_s = \frac{P_1}{\sigma_{tf}}$$

where A_s = area of the bolt at the root of the thread in mm^2
σ_{tf} = permissible tensile stress in bolts
= 120 MPa

Note The diameter of the bolts may be assumed between 20–36 mm (Appendix VII).

5.7 DESIGN OF HOLD-DOWN ANGLES

When there is appreciable uplift on the column, the angle connected to the column flange and base plate is subjected to bending action. The horizontal leg of the angle may be treated as a cantilever beam, fixed at one end by clamping action of the hold down bolts (Fig. 5.8). This is not quite true since there is some restraint offered by the other leg of the angle. Some designers assume the horizontal leg to act as a beam with both the ends fixed. The resulting moment will be half that indicated by the previous approach. Since the vertical leg of the angle is not completely fixed it will increase the moment in the horizontal leg near the bolt. This approach of analyses is probably more correct and is as follows:

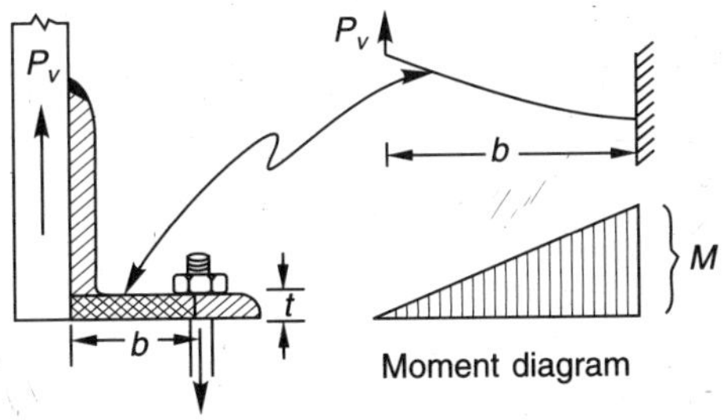

Fig. 5.8

When an uplift force P_v acts, the upper end of the vertical leg if not restrained will tend to move horizontally (Δ_h) as shown in Fig. 5.9.

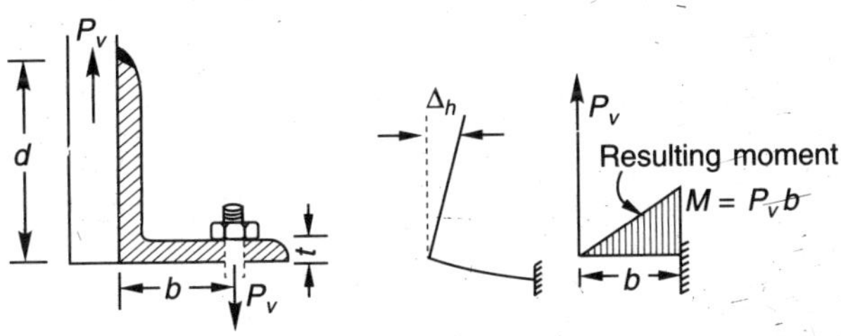

Fig. 5.9

The resulting moment is, $M = P_v\, b$

and $$\Delta_{hv} = \frac{\text{area of moment diagram}}{EI} \times \text{lever arm}$$

$$\Delta_{hv} = \frac{\frac{1}{2}(P_v b)\, bd}{EI} = \frac{P_v b^2\, d}{2\,EI}$$

Since the angle on the other side does provide restraint, a horizontal force P_h is applied to pull the leg back to its support position.

The resulting moment is, $M = P_h\, d$ (Fig. 5.10).

$$\Delta_{hh} = \frac{\text{area 1} \times \text{moment arm 1}}{EI} + \frac{\text{area 2} \times \text{moment arm 2}}{EI}$$

$$= \frac{(P_h\, d)\, bd}{EI} + \frac{\frac{1}{2}(P_h d)\left(\frac{2}{3}d\right)d}{EI} = \frac{P_h\, d^2}{3\,EI}(3b + d)$$

Since $$\Delta_{hh} = \Delta_{hv}$$

$$\frac{P_h\, d^2}{3\,EI}(3b + d) = \frac{P_v b^2\, d}{2\,EI}$$

$$\Rightarrow P_h = \frac{3\,P_v b^2}{2d\,(3b + d)} \qquad (5.21)$$

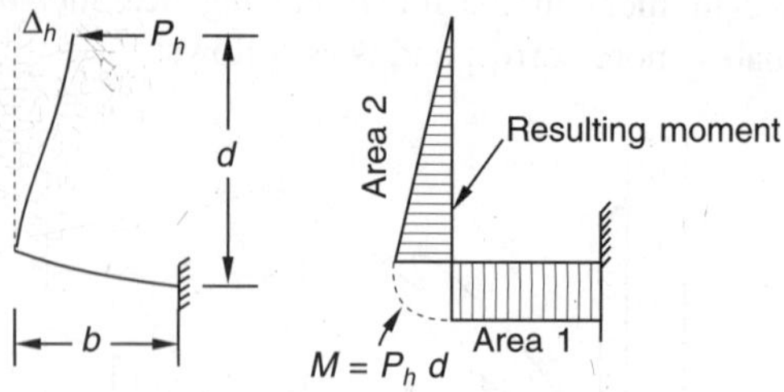

Fig. 5.10

Combining the initial moment resulting from the uplift force (Fig. 5.11(a)) and the secondary moment resulting from the restraint offered by the opposite angle (Fig. 5.11(b)) gives the resulting moment diagram as shown in Fig. 5.11(c).

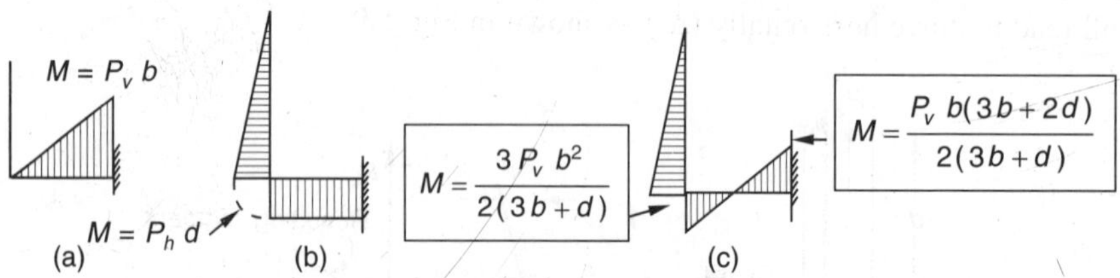

Fig. 5.11

$$\text{Moment at the heel of the angle, } M = \frac{3 P_v b^2}{2(3b+d)}$$

$$\text{and at the hold-down bolts} = \frac{P_v b(3b+2d)}{2(3b+d)}$$

Thickness of the angle

$$z = \frac{M}{\sigma}$$

where σ is the bending stress.

or
$$\frac{Lt^2}{6} = \frac{M}{\sigma}$$

or
$$t = \sqrt{\frac{6M}{L\sigma}} \tag{5.22}$$

When the vertical leg of the angle is welded for its full length as shown in Fig. 5.12(a),

$$t = \sqrt{\frac{3 P_v b}{L\sigma}} \tag{5.23}$$

When the vertical leg is welded only at its toe to the column as shown in Fig. 5.12(b),

$$t = \sqrt{\frac{3 P_v b}{L\sigma}\,\frac{(3b+2d)}{(3b+d)}} \tag{5.24}$$

The design of the base plate is done as described in Sections 5.3, 5.4 and 5.8.

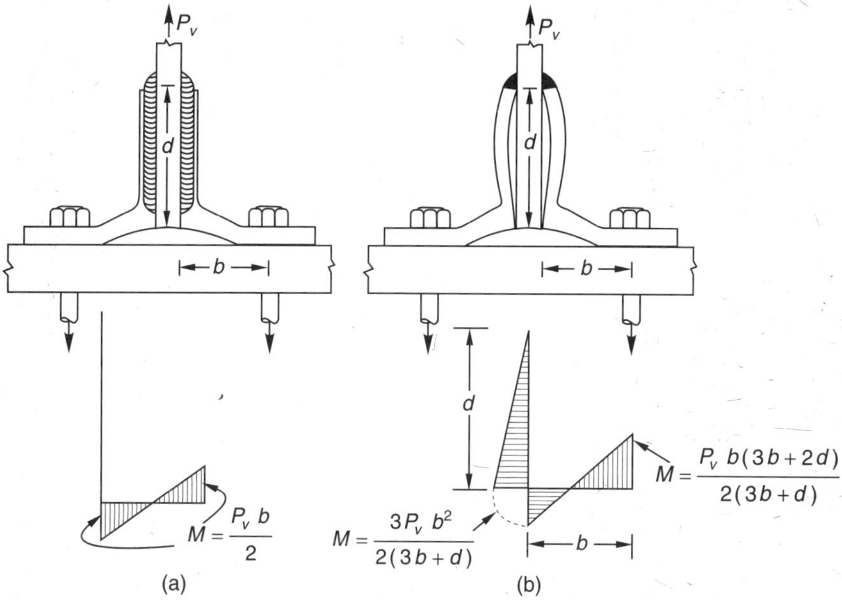

Fig. 5.12

5.8 WELDED COLUMN BASES

Figure 5.13 shows some of the welded column bases for small loads. In Fig. 5.13(a) the column shown is shop welded to the base plate. The others shown in Figs 5.13(b, c) are intended for where the column and base plates are erected separately. The angles are shop welded to the column, and the column field welded to the base plate after erection. The design of these hold down angles is discussed in Section 5.7.

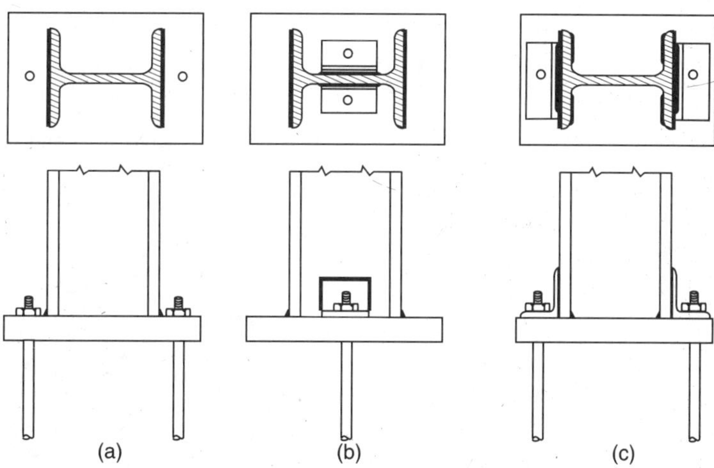

Fig. 5.13 *Welded Column Bases Loaded Lightly*

Some typical welded column bases are shown in Fig. 5.14. In Figs 5.14(a, b) brackets are welded to the flanges to develop greater moment of resistance for attachment to the base plate. The arrangement shown in Fig. 5.14(b) provides an

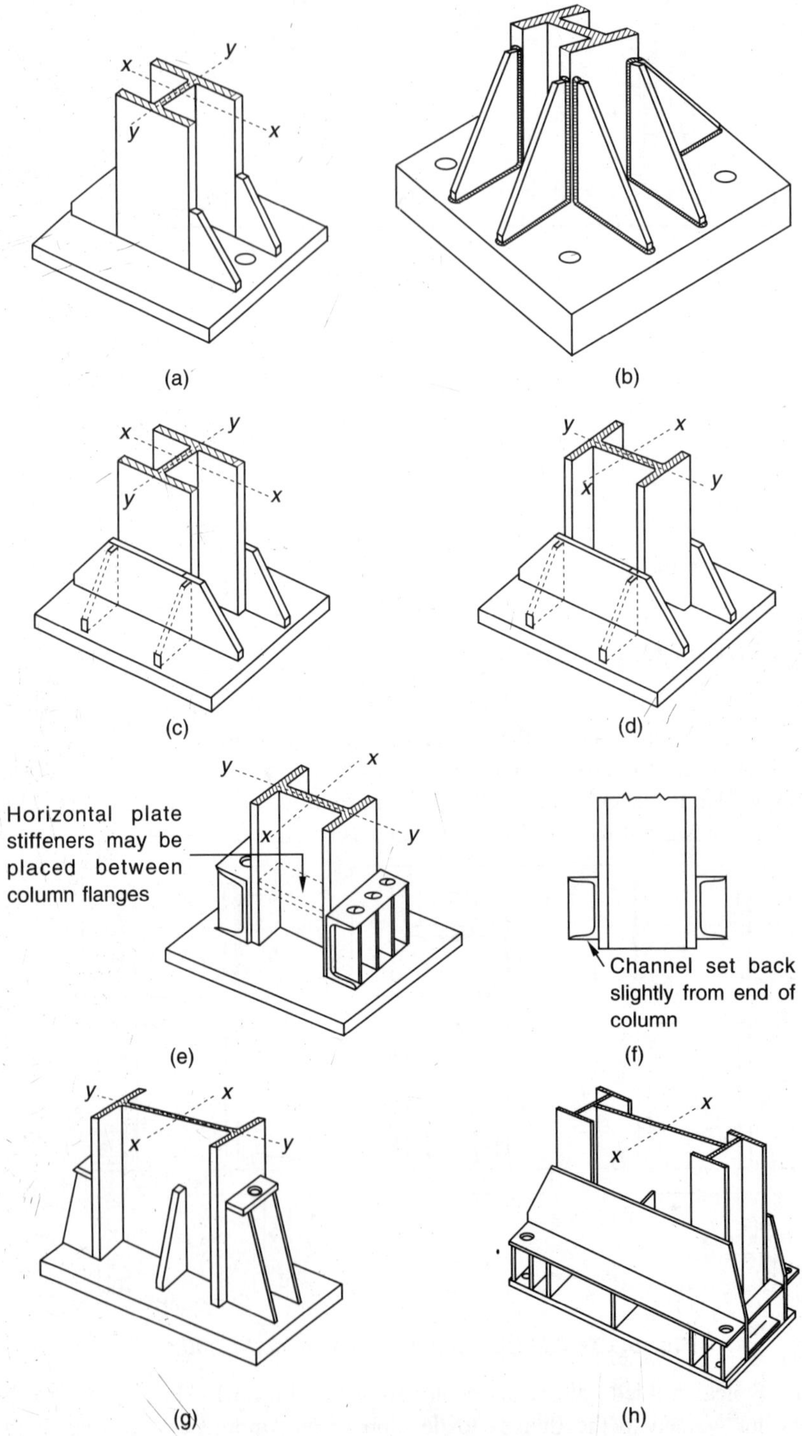

Fig. 5.14 *Typical Welded Column Bases*

excellent arrangement of multistorey buildings. A single bevel or vee joint is prepared by beveling the edge of the bracket; no beveling is done on the column flange.

The arrangement shown in Figs 5.14(c, d) are adopted when a fillet weld is employed. The two bracket plates may be attached to the face of the column flange (Fig. 5.14(c)) or to the outer edge of the column flange (Fig. 5.14(d)) and fillet welded to the base of the column. The former arrangement is a better one since for moments in any rolled section used as a column, greater bending strength and stiffness is obtained about the *x-x* axis. Therefore, if the moment is about *x-x* axis, it would be better to attach the additional plates to the face of the column as in Fig. 5.14(c). This will provide a good transverse fillet across the column flange and good accessibility for welding to the two longitudinal fillet welds along the outer edge of the column flange. The only slight drawback is that the attached plates will not stiffen the overhung portion of the base plate for the bending due to tension in the hold-down bolts, or due to upward bearing pressure of the masonry or concrete support. However, if this is a problem, small brackets shown in dotted lines may be added. Channel sections with stiffeners are shown in Figs 5.14(e, f). These are used for the required moment from the hold down bolts. When the channels are set flush with the column ends (Fig. 5.14(e)) and milled to bear, the additional bearing load must be considered in the design. However, if channels are offset from the column ends (Fig. 5.14(f)), they should be designed for only tension from the hold down bolts. Any vertical tensile load on the channels from the hold down bolts, or vertical bearing load from the base plate will produce a horizontal force at the top which will be applied transverse to the column flange and for this stiffner plates shown by dotted lines are added (Fig. 5.14(e)) between the column flanges.

Figure 5.14(g) shows the hold down bolt supports welded to the column flanges. A typical column base for a crane runway girder column is shown in Fig. 5.14(h). Two large plates are fillet welded to flanges of the column welded to the base plate. Two long narrow plates are welded to this, with small diaphragms. This provides additional strength and stiffeness of the base plate through beam action for the forces from the hold down bolts.

5.9 GRILLAGE FOOTING

A grillage footing, (Fig. 5.15) essentially consists of steel beams encased in concrete. These may be provided when,

1. The loads on the column are extremely heavy, e.g., in tall buildings. The base plate required in such cases will be of a very large dimensions and this may not be feasible.
2. The bearing capacity of the soil on which the concrete block is to be placed may be poor, which results in large dimensions of the concrete pedestal and base plate.

The column in such cases is placed over the base plate which in turn is supported over two or more tiers of beams, and the whole thing is encased in concrete (Fig. 5.15). The successive tiers of beams run at right angles to each other. Concrete in the grillage footing checks the buckling of the beam. As the grillage beams are encased in concrete, most of the requirements of cased beams apply to them. As per

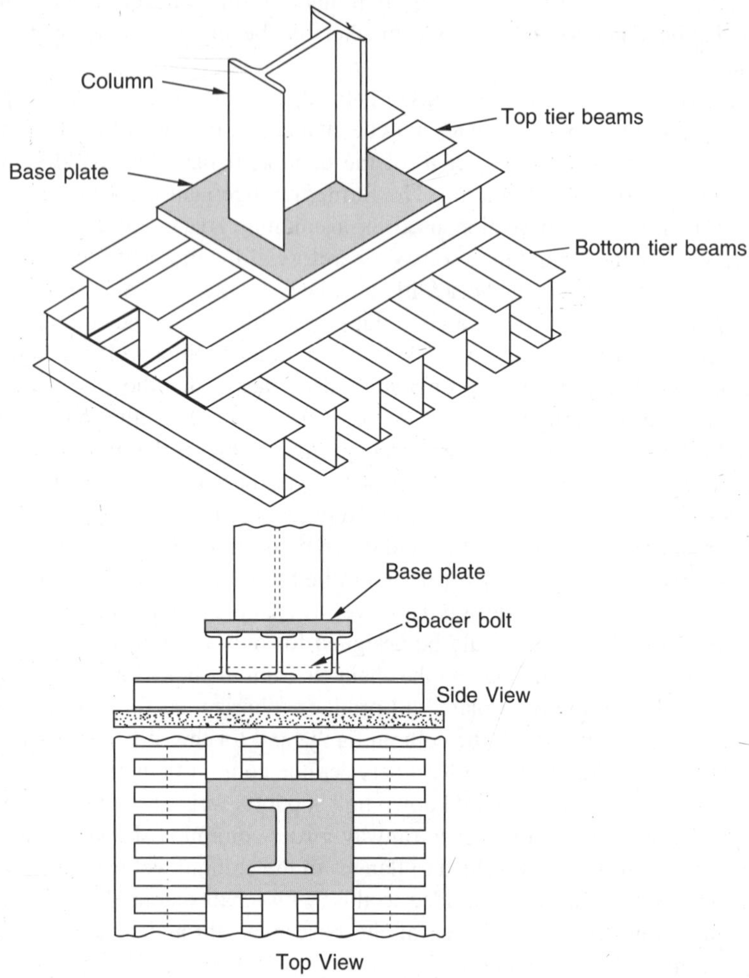

Fig. 5.15 *Grillage Footing*

I.S: 800–1984, the following conditions must be fulfilled:

(a) The beam should be unpainted and solidly encased in concrete with 10 mm aggregate and a working cube strength of not less than 15 N/mm^2 in 28 days. These are never painted as painted beams do not bond with concrete effectively.

(b) The beams in each tier are spaced such that the minimum spacing between the flanges of two consecutive beams is not less than 75 mm. Pipe separators are provided to maintain proper spacing between beams. This is done to place and compact the concrete properly.

(c) A concrete cover of minimum 100 mm is provided all around the beam-tier arrangement.

The size of the base plate is independent of the bearing capacity of the concrete block, since it is to rest directly on the upper tier of the beams. While thickness of the base plate should be sufficient to check punching, its width should cover the upper tier beams completely. The length of the base plate is proportioned to the base connection of the column. The number of upper tier beams depend upon the width of the base plate and as fewer beams can be accommodated, the beam section designed will be quite deep. The length of the upper tier beams depends upon the number of beams to be provided in the lower tier and this in turn depends upon the unit pressure on the foundation and the clearance between the consecutive beams.

Assumptions

1. The load from the column is uniformly distributed over the base plate.
2. Each tier of beams distributes the load uniformly to the lower tier.
3. Intensity of bearing pressure is uniform.

Design Forces

Let P = the load on the column in N

a, b = length and width of the base plate respectively in m

L = length of upper tier beams in m

B = length of lower tier beams in m

The load distribution from the column on the base plate and upper tier beam section is shown in Fig. 5.16. The shear in the beam is zero at its ends and at the centre. Maximum shear occurs at the edge of the base plate.

$$\text{The load/unit length of beam} = \frac{P}{L}$$

$$\text{Maximum shear force} = \frac{P}{L}\left(\frac{L-a}{2}\right)$$

$$V = \frac{P}{2}\left(\frac{L-a}{L}\right) \tag{5.25}$$

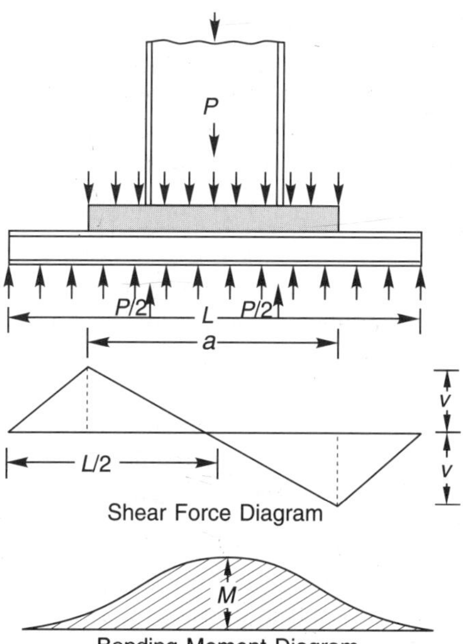

Fig. 5.16 *Shear Force and Bending Moment Diagrams for Grillage Beams*

Similarly, $$V = \frac{P}{2}\left(\frac{B-b}{B}\right) \tag{5.26}$$

The bending moment in the beam is zero at the ends and is maximum at the centre.

$$M = \frac{P}{2}\frac{L}{4} - \frac{P}{2}\frac{a}{4}$$

or $$M = \frac{P}{2}\left(\frac{L}{4} - \frac{a}{4}\right)$$

or $$M = \frac{P}{8}\,(L-a) \tag{5.27}$$

Similarly, $$M = \frac{P}{8}(B-b) \tag{5.28}$$

Design Procedure

As per I.S: 800–1984, the allowable bending stress in grillage beams encased in concrete may be increased by $33\frac{1}{3}$% from that specified in the code.

1. The design load for determining the size of footing is computed. Let load over the column be P. Add 10% of the column load to the load P, to account for the dead load of the column and grillage.
2. The area of the grillage footing is computed by dividing the design load by the bearing pressure on concrete. A square or rectangular footing may be provided in proportion to the size of the base plate.
3. The maximum bending moment for upper tier beams is calculated from Eq. (5.27). The allowable bending stress as per I.S. Specifications will be 1.33×165 MPa ($f_y = 250$ MPa).
4. The section modulus is computed and the number of beams to be provided in the upper tier are fixed by the limitation of the size of the base plate provided. An appropriate section is selected from I.S. Handbook No. 1.
5. The maximum shear force is computed from Eq. (5.25).
$$V = \frac{1}{n}\frac{P}{2}\left(\frac{L-a}{L}\right)$$
where n = number of beams in the upper tier.
6. The beam section provided is checked for shear and crippling.
7. The maximum bending moment for lower tier beams is calculated from Eq. (5.28).
8. The section modulus is computed and the number of beams to be provided in the lower tier beams are fixed based on the length of the upper tier beams.
9. The maximum shear is computed from Eq. (5.26).
$$V = \frac{1}{n}\frac{P}{2}\left(\frac{B-b}{B}\right)$$
where n = number of beams in the lower tier.
10. The beam section provided is checked for shear and crippling.

Solved Examples

Example 5.1 Design a slab base for a column section I.S.H.B. 350 @710.2 N/m subjected to an axial load of 1030 kN for the following conditions:

(a) Load is transferred to the base plate by direct bearing of column flanges.
(b) Load is transferred to the base plate by welded connections.
(c) Design the anchor bolts, if required.

The base rests on a *M*–15 concrete pedestal.

Solution

Safe bearing pressure of concrete = $.25 f_{ck} = 0.25 \times 15 = 3.75\ \text{N/mm}^2 = 3750\ \text{kN/m}^2$

For I.S.H.B. 350; $D = 350$ mm and $B = 250$ mm

$$\text{Required area of the slab base} = \frac{1030}{3750} = 0.2746\ \text{m}^2$$

Provide a square base plate,

$$\text{Side of the base} = \sqrt{0.2746} = 0.524\ \text{m} \approx 525\ \text{mm}$$

Provide a base plate of 525 × 525 mm.

$$\text{Thickness of the base plate, } t = \sqrt{\frac{3w}{\sigma_{bs}}\left(a^2 - \frac{b^2}{4}\right)}$$

where w = pressure on the under side of the base.

$$w = \frac{1030 \times 10^3}{525 \times 525} = 3.736\ \text{N/mm}^2$$

$$a = \text{greater projection} = \frac{525 - 250}{2} = 137.5\ \text{mm}$$

$$b = \text{smaller projection} = \frac{525 - 350}{2} = 87.5\ \text{mm}$$

σ_{bs} = permissible bending stress is slab base = 185 N/mm^2

$$t = \sqrt{\frac{3 \times 3.736}{185}\left((137.5)^2 - \frac{(87.5)^2}{4}\right)} = 32.08\ \text{mm}$$

Provide 36 mm thick plate available from IS Hand Book No. 1.
Provide slab base of 525 × 525 × 36 mm size.

(a) *Riveted slab base* Two cleat angles of nominal size 60 × 60 × 8 mm are provided on the column flanges to keep the column in position as shown in Fig. Ex. 5.1(a).

(b) *Welded slab base* It is desirable to transfer the load to base plate through weld rather than by direct bearing of the column edge. Matching of the bottom section of the column with the top surface of the base plate is not easy. If the direct contact between the two is not perfect, stress concentration occurs. Either fillet welding or butt welding may be done. Since, butt welding requires matching of the edge of the column, fillet weld is generally adopted.

Let us provide fillet welding.
Thickness of base plate is 36 mm.
Thickness of column flange and web are 11.6 mm and 10.1 mm, respectively.
Design load 1030 kN.
Let us try 8 mm weld size.

$$\text{Strength of weld} = 0.7 \times 8 \times 108 = 604.8 \text{ N/mm}$$

$$\text{Required length of weld} = \frac{1030 \times 10^3}{604.8} = 1703 \text{ mm} \approx 1720 \text{ mm}$$

The length available for welding all around the column section

$$= 2 \times 250 + 2 \times (250 - 10.1) + 2 \times (350 - 2 \times 11.6) = 1633.4 \text{ mm} < 1720 \text{ mm}$$

Provide 12 mm thick four bracket plates as shown in Fig. Ex. 5.1(b) to accommodate the weld length along with that can be accommodated on flanges. The vertical sides of the bracket plates will be butt welded to the column flanges.
Provide a weld length as distributed below:

Weld length that can be accommodated on flanges $= 2(250 + 250 - 10.1)$
$= 979.8$ mm

Weld length to be accommodated on each side of bracket plate

$$= \frac{1720 - 979.8}{4 \times 2} = 92.525 \text{ mm} \approx 105 \text{ mm (4 bracket plates welded on the both sides)}$$

Provide welding as shown in Fig. Ex. 5.1(b).

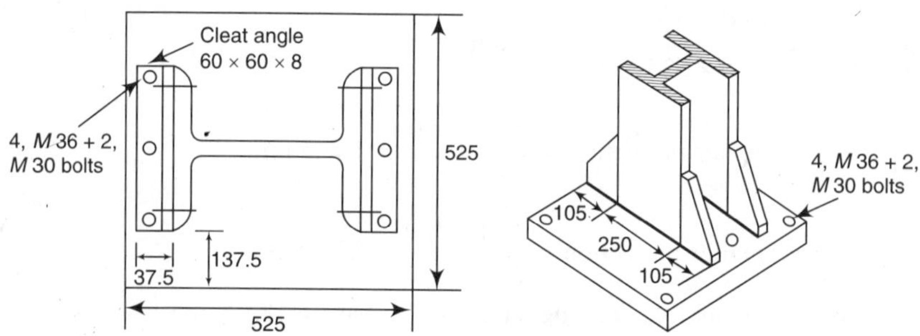

Fig. Ex. 5.1

(c) Since this base is subjected to only axial compressive load of 1030 kN, the anchor bolts will be designed for precompression. Considering the coefficient of friction 0.5 between anchor bolts and concrete, the design force will be 0.5 × 1030 = 515 kN.

$$\text{The stress area required} = \frac{515 \times 10^3}{120} = 4291.66 \text{ mm}^2$$

Provide 4, M 36 and 2 M 30 bolts (stress area 4390 mm^2) as shown in Fig. Ex. 5.1.

Example 5.2 A column section I.S.H.B. 350 @ 661.2 N/m carries an axial load of 1100 kN. Design a suitable gusset base using,

(a) riveted construction (b) HSFG bolts

Allowable bearing pressure on concrete is 4000 kN/m^2.

Solution

$$\text{Allowable bearing pressure on concrete} = 4000 \text{ kN/m}^2 = 4 \text{ N/mm}^2$$

$$\text{Area of the base plate required, } A = \frac{1100 \times 10^3}{4} = 275 \times 10^3 \text{ mm}^2$$

Assume the size of gusset material. Let us provide 16 mm thick gusset plates on the two flanges of column section and two gusset angles I.S.A. 150 × 115 × 15 mm as shown in Fig. Ex. 5.2.

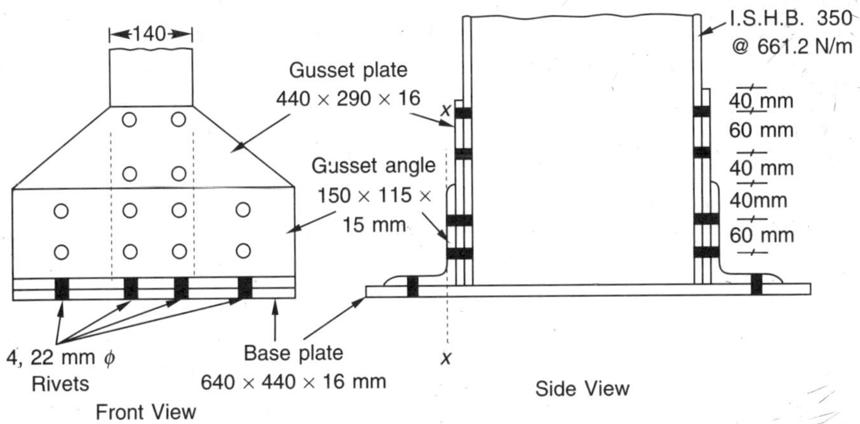

Fig. Ex. 5.2

$$\text{The minimum width required} = 350 + 2 \times 16 + 2 \times 115$$
$$= 612 \text{ mm} \approx 640 \text{ mm}$$

Let us provide width of base plate 640 mm.

$$\text{Projections of base plate beyond gusset angle} = \frac{640 - 612}{2} = 14 \text{ mm}$$

$$\text{Length of base plate} = \frac{275 \times 10^3}{640}$$
$$= 429.7 \text{ mm} \approx 440 \text{ mm}$$

Provide a base plate 640 × 440 mm in size.

$$\text{Average stress under base plate, } w = \frac{1100 \times 10^3}{640 \times 440}$$
$$= 3.906 \text{ N/mm}^2 < 4.0 \text{ N/mm}^2$$

which is safe.

The critical section for the base is x–x, as shown in Fig. Ex. 5.2.

The length of the base acting as cantilever,

$$c = 14 + 115 - 15 = 114 \text{ mm}$$

The combined thickness of base plate and gusset angle at critical section,

$$t = c\sqrt{3w/\sigma_{bs}}$$

$$= 114\sqrt{3 \times 3.906/185} = 28.69 \text{ mm}$$

Thickness of the base plate required = 28.69 – 15 = 13.69 mm

$\nless$ 16 mm (the thickness of gusset plate)

Hence, provide 640 × 440 × 16 mm base plate.

Connections:

(a) Let us provide 20 mm ϕ power driven rivets.

Gross diameter, $d = 20 + 1.5 = 21.5$ mm

$$\text{Strength of the rivet in single shear} = \frac{\pi}{4} d^2 \tau_{vf}$$

$$= \frac{\pi}{4} \times 21.5^2 \times 100 = 36{,}305.03 \text{ N}$$

$$\text{Strength of the rivet in bearing} = dt\, \sigma_{pf}$$

$$= 21.5 \times 11.6 \times 300 = 74820 \text{ N}$$

(The thickness t is taken as the minimum of flange thickness of column section and of gusset plate.)

Hence, rivet value, $R_v = 36{,}305.03$ N

Number of rivets required to connect flange of column to gusset plate assuming the column ends and gusset material to be faced for complete bearing. (50% of the load will pass through fastenings.)

$$n = \frac{0.5\,P}{R_v}$$

$$= \frac{0.5 \times 1100 \times 10^3}{36{,}305.03} = 15.14 \approx 16$$

Therefore, provide 8 rivets on each face in two rows. The number of rivets required to connect the gusset plate with gusset angle are kept same. Therefore, 8 rivets should be used to connect the gusset plate and the gusset angle.

The gusset angles are fastened to the base plate by means of four rivets 22 mm ϕ to keep the column in position. The rivets on the gusset plate will be provided at a pitch as calculated below.

$$R_v = (s - d)\, t\, \sigma_{at}$$

or

$$s = \frac{R_v}{\sigma_{at}\, t} + d$$

$$= \frac{36{,}305.03}{150 \times 11.6} + 21.5 = 42.3 \text{ mm}$$

$$\nless 2.5\,\phi\ (= 2.5 \times 20 = 50 \text{ mm})$$

Therefore, provide a pitch of 60 mm.

Edge distance of 20 mm ϕ rivets is 29 mm. Let us provide an edge distance of 40 mm.

Size of the gusset plate:

$$\text{Height of the gusset plate} = 150 + 2 \times 40 + 60 = 290 \text{ mm}$$

$$\text{Length of the gusset plate} = \text{length of base plate} = 440 \text{ mm}$$

Therefore, provide a gusset plate 290 × 440 × 16 mm. Provide 4 anchor bolts 24 mm ϕ at the corners of the base plate to connect it to the concrete pedestal.

(b) There is only one interface between column flange and gusset plate.

The design force to connect gusset plate with column flange is $\frac{1100}{2} = 550$ kN

$$F = \frac{n\mu T}{F_s}$$

$$550 = \frac{1 \times 0.50 \times T}{1.4}$$

$$T = 1540 \text{ kN}$$

Let us provide M 24–8.8 G bolts.

Proof load of one bolt is 212 kN

$$\text{Number of bolts required} = \frac{1540}{212} = 7.26 \approx 8$$

Provide 8, M 24–8.8 G bolts to connect gusset plate with column flange. Same number of bolts will be required to connect gusset angle with gusset plate. Arrange the bolts in a similar fashion as shown for riveted connection in Fig. Ex. 5.2.

Example 5.3 Design the base plate in Example 5.2, if in addition to the column load it is subjected to a bending moment of:

(1) 60 kNm, and (2) 120 kNm

Design the anchor bolts if required.

Solution

(1)

$$P = 1100 \text{ kN}$$

$$M = 60 \text{ kNm}$$

$$e = \frac{60 \times 10^3}{1100} = 54.54 \text{ mm}$$

Minimum length of base plate, assuming 100 mm projections on each side of column flange,

$$= 350 + 2 \times 100 = 550 \text{ mm}$$

$$\frac{e}{L} = \frac{54.54}{550} = \frac{1}{10.08} < \frac{1}{6}$$

$$\sigma_c = 4 \text{ N/mm}^2$$

Hence,

$$\text{width of base plate} = \frac{P}{\sigma_c L}\left(1 + \frac{6e}{L}\right)$$

$$= \frac{1100 \times 1000}{4 \times 550}\left(1 + \frac{6 \times 54.54}{550}\right) = 797.49 \text{ mm} \approx 810 \text{ mm}$$

Provide a base plate 810 × 550 mm.

Thickness of the base plate: Maximum compressive stress in concrete under near edge of the plate (Fig. Ex. 5.3(a)).

$$= \frac{P}{BL}\left(1+\frac{6e}{L}\right)$$

$$= \frac{1100\times1000}{550\times810}\left(1+6\times\frac{54.54}{550}\right)$$

$$= 3.938 \text{ N/mm}^2$$

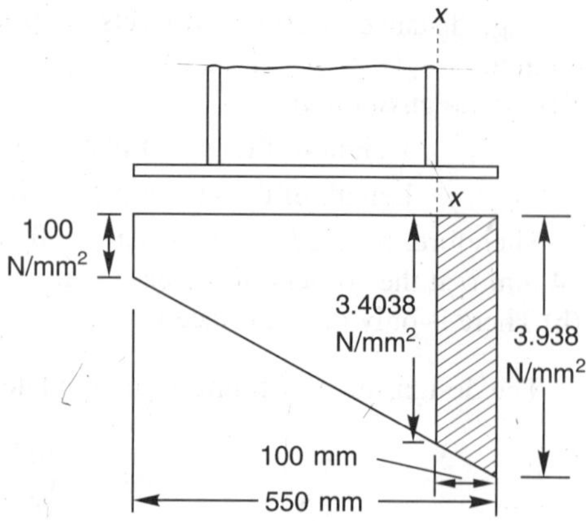

Fig. Ex. 5.3(a)

Minimum compressive stress in concrete under far edge of plate

$$= \frac{P}{BL}\left(1-\frac{6e}{L}\right)$$

$$= \frac{1100\times1000}{550\times810}\left(1-6\times\frac{54.54}{550}\right) = 1.00 \text{ N/mm}^2$$

Stress under critical section x–x $= 1.00 + (3.938 - 1.00) \times \dfrac{450}{550}$

$$= 3.4038 \text{ N/mm}^2$$

Moment at the critical section= moment of hatched area about section x–x

$$= 3.4038 \times 100 \times \frac{100}{2} + \frac{(3.938-3.4038)}{2}\times100\times\left(\frac{2}{3}\times100\right)$$

$$= 18799.66 \text{ N mm}$$

Moment of resistance at critical section x–x

$$= \frac{1}{6}\times1\times t^2\times\sigma_{bs} \quad = \frac{1}{6}t^2\times185 \text{ N mm}$$

Moment of resistance = Moment

$$\frac{1}{6}t^2\times185 = 18,799.66$$

or $$t^2 = 6\times\frac{18799.66}{185}$$

or $$t = 24.69 \text{ mm} \approx 25 \text{ mm}$$

Hence, provide a 25 mm thick plate.

Since the moment is small and the case is of no tensioin, anchor bolts are not required to resist any moment and hence the anchor bolts are not subjected to tension. Therefore the design of anchor bolts is not required. Provide nominal anchor bolts to keep the base in position.

(2) $$P = 1100 \text{ kN}$$

$$M = 120 \text{ kN m}$$

$$e = \frac{M}{P} = \frac{120\times10^3}{1100} = 109.09 \text{ mm}$$

Minimum length of base plate assuming 120 mm projections on each side of column flange (Fig. Ex. 5.3(b)).

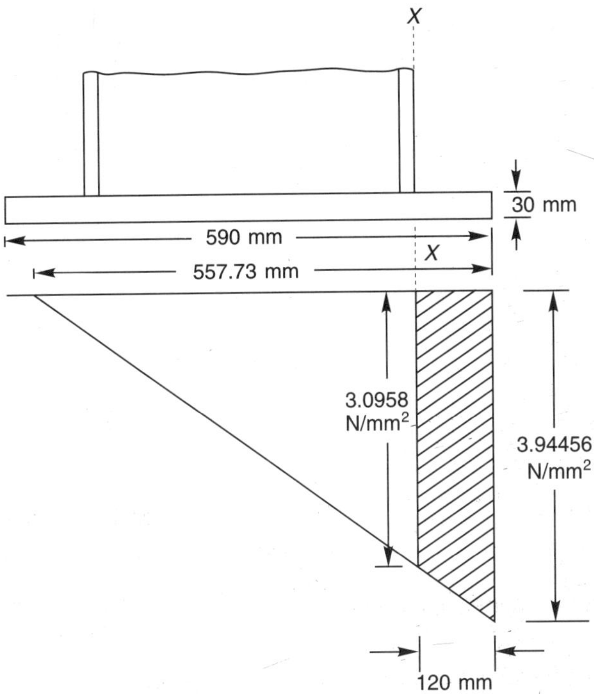

Fig. Ex. 5.3(b)

$$L = 350 + 2 \times 120 = 590 \text{ mm}$$

$$\frac{e}{L} = \frac{109.09}{590} = \frac{1}{5.408}$$

Eccentricity lies between $\frac{L}{3}$ to $\frac{L}{6}$

$\therefore$
$$x = 3\left(\frac{L}{2} - e\right)$$

$$= 3\left(\frac{590}{2} - 109.09\right) = 557.73 \text{ mm}$$

$$B = \frac{2P}{x\sigma_c}$$

$$= \frac{2 \times 1100 \times 10^3}{557.73 \times 4} = 986.14 \text{ mm} \approx 1000 \text{ mm}$$

Provide a base plate 590 × 1000 mm.

Maximum compressive stress in concrete $= 4 \times \frac{986.14}{1000} = 3.94456 \text{ N/mm}^2$

$$\text{Stress at critical section } x\text{–}x = 3.94456 \times \frac{557.73 - 120}{557.73}$$

$$= 3.0958 \text{ N/mm}^2$$

$$\text{Moment at section } x\text{–}x = 3.0958 \times 120 \times \frac{120}{2} + \frac{3.94456 - 3.0958}{2}$$

$$\times \left(\frac{2}{3} \times 120\right) \times 120 = 26{,}363.808 \text{ N mm}$$

$$\text{Moment of resistance} = \frac{1}{6} \times 1 \times t^2 \times 185 \text{ N mm}$$

$$\text{Moment of resistance} = \text{Moment}$$

or $$\frac{1}{6} t^2 \times 185 = 26363.808$$

or $$t = \sqrt{6 \times 26363.808/185}$$

or $$t = 29.24 \text{ mm} \approx 30 \text{ mm}$$

Provide a 1000 × 590 × 30 mm thick base plate. Design of anchor bolts is not required as the tension is assumed to be neglected when eccentricity lies between $\frac{L}{3}$ to $\frac{L}{6}$. Provide nominal anchor bolts to keep the base in position.

Example 5.4 Design a suitable base and anchor bolts for a column subjected to an axial load of 500 kN and a wind moment of 175 kNm. The column section is I.S.H.B. 450 @ 907.4 N/m. Design also welded connection between the column and the base. The safe bearing pressure of concrete may be assumed to be 4000 kN/m^2.

Solution

$$P = 500 \text{ kN}$$

$$M = 175 \text{ kNm}$$

$$e = \frac{M}{P} = \frac{175 \times 10^3}{500} = 350 \text{ mm}$$

Assuming 160 mm projection of the base plate on each side of the flange of the column,

$$\text{the minimum length of base plate} = 450 + 160 + 160 = 770 \text{ mm}$$

$$\frac{e}{L} = \frac{350}{770} = \frac{1}{2.2}$$

This is greater than $\frac{1}{3}$, therefore the design will be done on the basis of the concrete beam theory.

According to the concrete beam theory, the depth of N.A.,

$$n = \frac{d}{1 + \dfrac{\sigma_{tf}}{m\sigma_c}}$$

where $\sigma_{tf} = 120 \times 1.25 = 150 \text{ N/mm}^2$

$\sigma_c = 4 \times 1.33 = 5.32 \text{ N/mm}^2$

(The stresses are increased to take into account the effect of wind.)

Let $m = 18$

If the edge distance for bolts = 75 mm.

$$d = 770 - 75 = 695 \text{ mm}$$

$$n = \frac{695}{1 + \frac{150}{18 \times 5.32}} = 270.80 \text{ mm}$$

$$\text{Lever arm} = d - \frac{n}{3}$$

$$= 695 - \frac{270.80}{3} = 604.73 \text{ mm}$$

Take moment about the line of action of tensile force.

$$C \times 604.73 = 175 \times 10^3 + 500 \times (385 - 75)$$

or $C = 545.69 \text{ kN}$

$$B = \frac{2 \times 545.69 \times 10^3}{270.80 \times 5.32} = 757.5 \text{ mm} \approx 760 \text{ mm}$$

The maximum value of compressive stress will be

$$= 5.32 \times \frac{757.5}{760} = 5.302 \text{ N/mm}^2$$

$$\text{Stress ordinate at the critical section} = 5.302 \times \frac{270.80 - 160}{270.80}$$

$$= 2.17 \text{ N/mm}^2$$

Bending moment at critical section

$$= 3.13 \times 160 \times \frac{160}{2} + \left(\frac{5.302 - 3.13}{2}\right) \times 160 \times \left(\frac{2}{3} \times 160\right)$$

$$= 58{,}598.4 \text{ N mm}$$

$$\text{Moment of resistance} = \frac{1}{6} \times 1 \times t^2 \times 1.33 \times 185 \text{ N mm}$$

or $\frac{1}{6} \times 1 \times t^2 \times 1.33 \times 185 = 58{,}598.4$

or $t^2 = 6 \times \frac{58598.4}{1.33 \times 185}$

or $t = 37.80 \text{ mm} = 40 \text{ mm}$

Anchor bolts:

Tension in windward bolts, $P_1 = C - P$

$= 545.69 - 500 = 45.69 \text{ kN}$

$$\text{Required net area of bolts} = \frac{45.69 \times 10^3}{1.25 \times 120}$$
$$= 304.60 \text{ mm}^2$$

Let us provide two bolts on the tension side.

$$\text{Net area of each bolt} = \frac{304.60}{2} = 152.3 \text{ mm}^2$$

Provide 20 mm diameter bolts each giving a net area of 245 mm^2 on the tension side. Provide 2 bolts of same diameter on the other side also.

$$\text{Length of each bolt} = \frac{\frac{P_1}{2}}{\pi d \times \text{bond stress}}$$

$$= \frac{45.69 \times 10^3}{2 \times \pi \times 20 \times 0.7} = 519.4 \text{ mm} \approx 520 \text{ mm}$$

Hence, provide 4, 20 mm ϕ and 520 mm long bolts.

Example 5.5 Design a two tier grillage foundation to carry an axial load of 1000 kN. A base plate 700 × 700 mm is provided below the stanchion. The concrete is of M-15 grade and bearing pressure of the earth is limited to 150 kN/m^2.

Solution

$$\text{Axial load} = 1000 \text{ kN}$$

Assume the self weight of grillage foundation as 10% of the axial load.

$$\therefore \quad \text{Design load for footing} = 1.1 \times 1000 = 1100 \text{ kN} = 1100 \times 10^3 \text{ N}$$

$$\text{Required area of the grillage foundation} = \frac{1100}{150} = 7.33 \text{ m}^2$$
$$\approx 7.4 \text{ m}^2$$
$$= 740 \times 10^4 \text{ mm}^2$$

Provide a square grillage base.

$$\text{The side of base} = \sqrt{740 \times 10^4}$$
$$= 2720.29 \text{ mm} \approx 2800 \text{ mm}$$

Provide a base of size 2800 × 2800 mm.

Upper tier beams:

$$\text{Maximum bending moment} = \frac{P}{8}(L - a)$$

$$= \frac{1000 \times 10^3}{8}(2800 - 700)$$
$$= 262.5 \times 10^6 \text{ N mm}$$

As the beams are encased in concrete, the allowable bending stresses are increased by $33\frac{1}{3}\%$.

$$\sigma_{bc,\,t\ \text{cal}} = 0.66 f_y \times 1.33$$
$$= 0.66 \times 250 \times 1.33 = 219.45 \text{ N/mm}^2$$

$$Z_{\text{req}} = \frac{262{,}500 \times 10^3}{219.45} = 1196.172 \times 10^3 \text{ mm}^3$$

The minimum clear spacing between the flanges of the beams should be 75 mm as per I.S. specifications.

Let us select 4, I.S.M.B. 225 @ 306.1 N/m.

From I.S. Handbook No. 1, the relevant properties are,

$B = 110$ mm, $t = 6.5$ mm, $h_2 = 25.85$ mm, $Z = 305.9 \times 10^3$ mm^3

$$\text{Clear spacing between the flanges of the beam} = \frac{700 - 4 \times 110}{3}$$
$$= 86.67 \text{ mm} > 75 \text{ mm}$$

which is as it should be.

$$\text{Maximum shear force, } V = \frac{1}{n}\frac{P}{2}\frac{(L-a)}{L}$$

$$= \frac{1}{4} \times \frac{1000}{2}\frac{(2800-700)}{2800} = 93.75 \text{ kN}$$

$$\tau_{va,\ \text{cal}} = \frac{V}{Dt}$$

$$= \frac{93.75 \times 10^3}{225 \times 6.5} = 64.11 \text{ N/mm}^2 < 100 \text{ N/mm}^2$$

which is safe.

Check for crippling:

$$\sigma_{p,\ \text{cal}} = \frac{P}{n\,(a + 2\sqrt{3}h_2\,)t}$$

(where, a = length of base plate)

$$= \frac{1000 \times 10^3}{4(700 + 2\sqrt{3} \times 25.85) \times 6.5} = 48.726 \text{ N/mm}^2$$

$$< 187.5 \text{ N/mm}^2$$

which is safe.

Lower tier beams:

$$\text{Maximum bending moment} = \frac{P}{8}(B-b)$$

$$= \frac{1000 \times 10^3}{8}(2800-700)$$
$$= 262.5 \times 10^6 \text{ N mm}$$

$$Z_{req} = \frac{262.5 \times 10^6}{219.45} = 1196.172 \times 10^3 \text{ mm}^3$$

The length of the footing is 2800 mm. The minimum clear spacing between the flanges should be 75 mm as per I.S. specifications. Let us provide 10, I.S.L.B. 175 @ 163.8 N/m. The relevant properties are,

$$B = 90 \text{ mm}, t = 5.1 \text{ mm}, h_2 = 16.70 \text{ mm}, Z_{xx} = 125.3 \times 10^3 \text{ mm}^3$$

$$\text{Clear spacing between the flanges} = \frac{1}{9}(2800 - 90 \times 10)$$

$$= 211.11 \text{ mm} > 75 \text{ mm}$$

which is as it should be.

$$\text{Section modulus provided} = 10 \times 125.3 \times 10^3$$

$$= 1253 \times 10^3 \text{ mm}^3 > 1196.172 \times 10^3 \text{ mm}^3$$

which is sufficient.

Check for shear:

$$V = \frac{1}{n}\frac{P}{2}\left(\frac{B-b}{B}\right)$$

$$= \frac{1}{10} \times \frac{1000 \times 10^3}{2}\left(\frac{2800 - 700}{2800}\right) = 37.5 \times 10^3 \text{ N}$$

$$\tau_{va,\ cal} = \frac{37.5 \times 10^3}{175 \times 5.1}$$

$$= 42.01 \text{ N/mm}^2 < 100 \text{ N/mm}^2$$

which is safe.

Check for crippling:

$$\sigma_{p,\ cal} = \frac{1000 \times 10^3}{10 \times (700 + 2\sqrt{3} \times 16.7) \times 5.1}$$

$$= 25.875 < 187.5 \text{ N/mm}^2$$

which is safe.

Example 5.6 Two columns I.S.H.B. 350 @ 661.2 N/m and I.S.H.B 400 @ 759.3 N/m are spaced 6 m c/c. The columns carry loads 1100 kN and 1800 kN respectively. Design a combined grillage foundation for the columns. Bearing pressure of the earth is 200 kN/m^2.

Solution The centre of gravity of the loads must be in the same vertical line as the centre of gravity of base of footing.

Let $\bar{x}$ = the distance of C.G. of loads from 1800 kN load.

$$\bar{x} = \frac{1100 \times 6}{1100 + 1800}$$

$$= 2.2758 \text{ m} = 2275.8 \text{ mm}$$

Length of the grillage: A minimum of 500 mm projection will be required for the 1100 kN column base plate.

$$\text{Length of the grillage} = 2[(6000 - 2275.8) + 500]$$
$$= 8448.4 \approx 8450 \text{ mm}$$

Let us provide the grillage length = 8450 mm.

$$\text{The projection for column with 1100 kN load} = \frac{8450}{2} - 3724.2 = 500.8 \text{ mm}$$

Design moment: Assume the grillage to be an inverted beam supported at the columns and subjected to uniform pressure from below the grillage.

$$\text{Intensity of pressure from below the grillage} = \frac{1100 + 1800}{8450}$$
$$= 0.343192 \text{ kN/mm}$$
$$= 343.192 \text{ N/mm}$$

Refer to Fig. Ex. 5.6.

Maximum negative shear below column with 1800 kN load

$$= 343.192 \times 1949.2$$
$$= 668{,}949.85 \text{ N} = 668.9498 \text{ kN}$$

Maximum positive shear below column with 1800 kN load

$$= 1800 - 668.9498$$
$$= 1131.0502 \text{ kN} = 1{,}131{,}050 \text{ N}$$

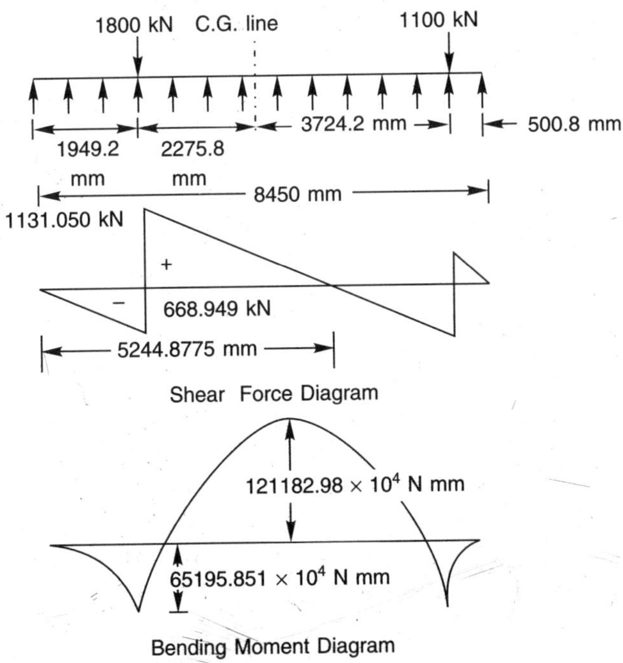

Fig. Ex. 5.6

Shear force is zero at a distance (from the left end of grillage)

$$= 1949.2 + \frac{1,131,050}{343.192} = 5,244.8775 \text{ mm}$$

Bending moment will be maximum at this point as it is the point of inflexion.

The maximum cantilever moment is also calculated. The grillage is designed for the maximum of the two moments.

$$\text{Maximum cantilever moment} = \frac{343.192 \times 1949.2^2}{2}$$

$$= 65,195.851 \times 10^4 \text{ N mm}$$

Maximum bending moment at the point of inflexion

$$= -\frac{343.192 \times 5244.8775^2}{2} + 1800 \times 10^3 \times (5244.8775 - 1949.2)$$

$$= -472,038.97 \times 10^4 + 593221.95 \times 10^4$$

$$= 121,182.98 \times 10^4 \text{ N mm}$$

Therefore, the design moment is $121,182.98 \times 10^4$ N mm.

Section modulus: Allowable bending stresses are increased by $33\frac{1}{3}\%$ as the beams are encased in concrete.

$$Z_{req} = \frac{M}{\sigma_{bc,t}}$$

$$= \frac{121,182.98 \times 10^4}{1.33 \times 165} = 5,522,122.6 \text{ mm}^3$$

Select 3, I.S.L.B. 550 @ 846.6 N/m providing section modulus

$$= 3 \times 1933.2 \times 10^3$$

$$= 5,799,600 \text{ mm}^3 > 5522122.6 \text{ mm}^3$$

which is all right.

Check for shear:

Maximum shear force = 1,131,050 N

$$\tau_{va,\ cal} = \frac{1,131,050}{3 \times (D \times t)}$$

$$= \frac{1,131,050}{3 \times 550 \times 9.9} = 69.24 \text{ N/mm}^2 < 100 \text{ N/mm}^2$$

which is safe.

No bottom tier is required as the area provided by the grillage beam tier brings down the bearing pressure on the foundation below the safe bearing pressure.

Provide a base plate extending over the grillage beams equal to the width as calculated below.

$$\text{Total load on the foundation} = 1100 + 1800 = 2900 \text{ kN}$$

$$\text{Self weight of the foundation} = \frac{10}{100} \times 2900 = 290 \text{ kN}$$

$$\text{Total load} = 2900 + 290 = 3190 \text{ kN}$$

$$\text{Area of the grillage required} = \frac{3190}{200} = 15.95 \text{ m}^2$$

Assuming the bearing pressure from soil 200 kN/m^2.

$$\text{Length of the grillage provided} = 8.450 \text{ m}$$

$$\text{Hence, width of the grillage required} = \frac{15.95}{8.450} = 1.887 \text{ m} \approx 2.0 \text{ m}$$

As the width is quite small, the base plate in the other direction shall run for at least 2.0 m.

Exercises

5.1 Design a suitable slab base for a column section I.S.H.B. 200 @ 365.9 N/m supporting an axial load of 400 kN. The base plate is to rest on a concrete pedestal of *M*–15 grade.

5.2 Design the section of steel column and a suitable base for an axial compressive force of 3500 kN. The effective length of the column is 5.2 m. The safe bearing pressure from concrete may be assumed to be 3.75 N/mm^2.

5.3 A column of 6.0 m effective length is carrying an axial load of 400 kN and a bending moment of 50 kN m. The bearing pressure from the concrete pedestal may be assumed to be 4000 N/m^2. Design a suitable base. Also design the anchor bolts, if required.

5.4 Design a column base for an axial load of 700 kN and a bending moment of 100 kN m. The column section provided is I.S.H.B. 400 @ 806.4 N/m. Design the anchor bolts also, if required. The bearing pressure from concrete may be assumed to be 3750 kN/m^2.

5.5 Design a grillage foundation for a column subjected to an axial load of 1050 kN. The column section provided is I.S.H.B. 450 @ 855.4 N/m with a base plate 700 × 700 mm. The bearing capacity of the soil is 120 kN/m^2.

5.6 Design a grillage foundation for a column subjected to an axial load of 3000 kN. The bearing capacity of the soil is 150 kN/m^2 and the size of base plate is 750 × 750 mm.

5.7 Design a combined grillage footing for the two columns I.S.H.B. 300 @ 576.8 N/m and I.S.H.B. 350 @ 661.2 N/m spaced 3.5 m c/c. The columns carry axial loads of 1200 kN and 2000 kN respectively. The safe bearing pressure from soil is 160 kN/m^2.

Tension Members

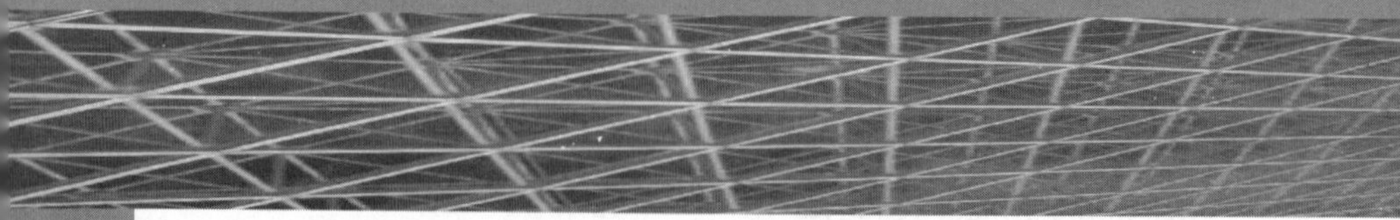

Chapter 6

6.1 INTRODUCTION

A structural member subjected to two pulling (tensile) forces applied at its ends is called a tension member. The member and connections are so arranged that eccentricity in the connection and bending stresses on the member are not developed. The stress in such members is assumed to be uniformly distributed over the net section and hence members subjected only to axial tension are supposed to be the most efficient and economical. However, if some eccentricity exists either due to the member not being perfectly straight or due to eccentricity in connections, either bending stresses are considered in the design or specifications are provided to account for reduction in the net area.

6.2 TYPES OF TENSION MEMBERS

The various forms of tension members are as follows:

Wires and Cables Wire ropes are exclusively used for hoisting purposes and as guy wires in steel stacks and towers. Cables are used in suspension bridges. Since cables are generally long and their flexural stiffness is negligible, initial sag and other geometrical effects must be accounted for in the design. These are not recommended in bracing systems as they cannot resist compression. The advantages of wire rope and cable are flexibility and strength.

Bars and Rods Bars and rods are often used as tension members in bracing systems, as sag rods to support purlins between trusses, to support girts in industrial buildings and where a light structure is desirable, e.g., tower masts, etc. Because of the slenderness of these members, their compression strength is negligible. When rods are used in wind bracing systems, they are pretensioned to reduce the effect of sway. Such members are in general, welded to the gusset plates or may be threaded and bolted. Threading reduces the section and thus, the strength is reduced.

Eye bar tension members are used in truss bridges, chains of suspension bridges and as floor beam hangers for suspension and through arch bridges. Some of the types of connections for rods are shown in Fig. 6.1(a, b). Rods can be threaded and bolted as shown in Fig. 6.1(a), or they can be forged into a loop (Fig. 6.1(b)). Since it is almost impossible to fabricate rods or bars to fit perfectly in structure, turnbuckles (Fig. 6.1(c)) are often required to adjust for the variation in bar length. Flat bars may be riveted bolted, welded, or forged into loop and connected to a pin; an eye bar is shown in Fig. 6.1(d). The larger dimension of the bar is kept vertical to reduce sag.

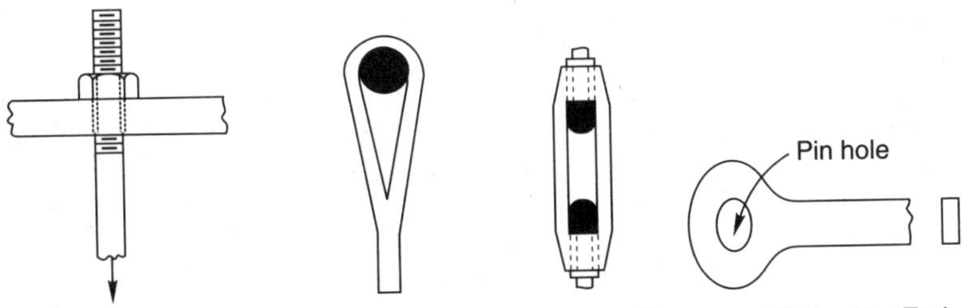

Fig. 6.1 *Connections of Bars and Rods*

The main disadvantage of rods and bars is inadequate stiffness, resulting in noticeable sag under their own weight, especially during erection. Sagging of members built-up of bars and rods may be minimised by limiting the length-diameter ratio, or by fabricating the rod short of its required theoretical length by some arbitrary amount and drawing it into place to provide an initial tension. The same effect may be produced by providing a turn buckle in the rod. Rods are generally connected to supporting members by running them through a hole in the member and attaching a nut on the far end.

The use of rectangular bars is diminishing but they can still be found as hangers in special structures.

Plates and Flats Plates and flat bars are often used as tension members in transmission towers, foot bridges, etc. These are also used in columns to keep the component members in their correct position, e.g. lacing flats, batten plates, end tie plates, etc.

Single and Built-up Structural Shapes Earlier tension members were used to be made always of flat bars, but modern practice uses angles or built-up section where possible. The angle gives a greater rigidity to the structure than the flat bar, and rigidity is of as great importance as strength. In many light bridges flat bar tension member can be seen to vibrate during the passage of the load, while the compression members being necessarily of a section other than a flat bar, can hardly be observed to vibrate at all. Again, the girder may have to act either intentionally as part of a portal frame, and there is thus the possibility of a stress reversal in some members. Angle can take this reversal from a tensile to a compressive stress, whereas flat bars are unfit to carry a compressive load due to their small radius of gyration in one direction. If the tension is very high, channels or I-Sections may be used. It is preferable to select a section which is symmetrical at least about one axis, to avoid eccentricity. It would normally be found that a single angle section will be the most economical. However, it needs close examination for the assumption made is that the section is used for axial tension.

A single angle section with riveted connections produces eccentricity about both the planes, Fig. 6.2(a), while the welded connections may be so designed and adjusted as to reduce eccentricity about one plane only, Fig. 6.2(b). Two angle sections can either be placed back-to-back on the same side of gusset plate, Fig. 6.2(c), or back-to-back on the opposite side of gusset plate, Fig. 6.2(d). When

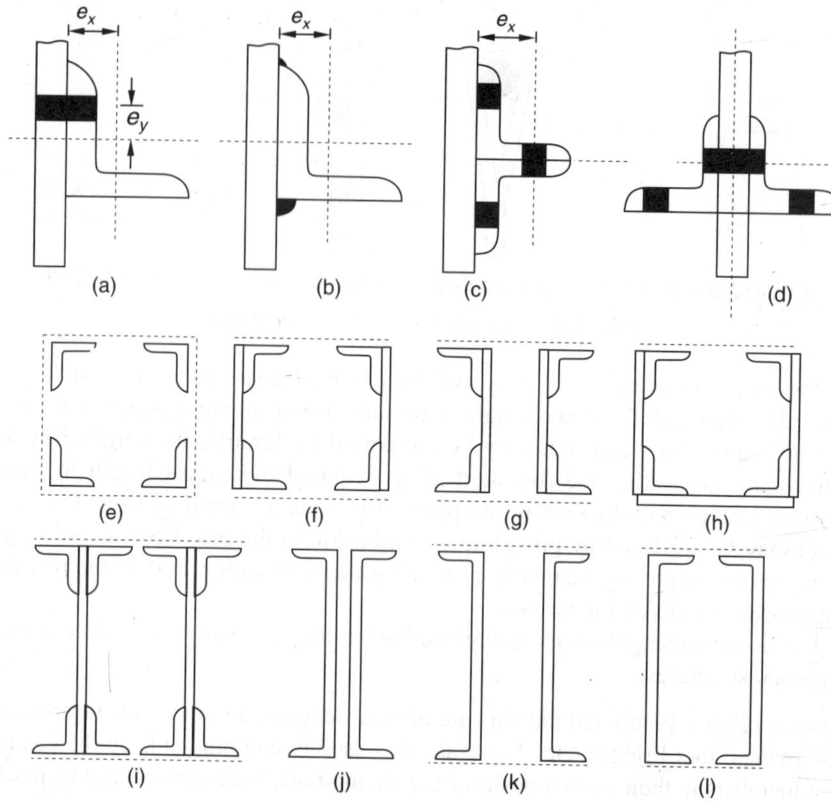

Fig. 6.2 *Shapes of Tension Members*

angles are connected on the same side of the gusset plate, the eccentricity is about one plane (e_x) only which is almost eliminated when the angles are connected on the opposite side of the gusset plate. However, for riveted or bolted connections the eccentricity (e_y) between the connector gauge lines and the centroidal axis of the members, which is very small, may be neglected. Structural Tees make excellent chord members for lightly loaded welded trusses, since the stem may serve as a gusset for attachment for single angle or double angle web members.

Built-up members made up of two or more plates or shapes and connected to act as single member are formed primarily to meet the required area, which cannot be provided by a single rolled shape. Also, built-up members are more rigid because for the same area, much greater moment of inertia can be obtained than with single rolled sections. Another advantage of built-up members is that they can be made sufficiently stiff to carry compression as well as tension and hence they are desirable when stress reversals might occur.

In highway bridge trusses where heavy sections are required, e.g. chord members, angle sections either with connecting system, Figs 6.2(e, f), or with plates, Figs 6.2(g, h, i) are used. Channel sections may also be used effectively when the loads are heavy and more rigidity is required, Figs 6.2(j, k, l). Built-up sections have a great advantage of rigidity over single angle sections. These can also withstand reversal of stresses effectively. I-Sections are occasionally used but are economical over built-up sections shown in Figs 6.2(e to i, k, l).

As there is a variety of sections the designer has ample latitude to select and provide a particular section as a tension member. The choice of cross-section is based on the type of connection to be used. When welded connections are used, plates, tubes, tees are provided, whereas if riveted connections are used angles, channels and I-Sections are preferred.

6.3 PERMISSIBLE STRESSES

Axial Stress

The permissible stress in axial tension on the net effective area of the sections should not exceed 0.60 f_y. The permissible stress in tension has been worked out after applying a factor of safety of 1.65, which is against yielding for tension members for buildings.

$$\sigma_{at} = 0.6 f_y$$

where σ_{at} = stress in axial tension in MPa

f_y = yield stress of steel in MPa

Combined Axial and Bending Tensile Stresses

A member subjected to both axial tension and bending should be proportioned so that the following interaction relationship is satisfied:

$$\frac{\sigma_{at\,,\,\mathrm{cal}}}{0.6 f_y} + \frac{\sigma_{btx\,,\,\mathrm{cal}}}{0.66 f_y} + \frac{\sigma_{bty\,,\,\mathrm{cal}}}{0.66 f_y} \le 1$$

where $\sigma_{at,\,\mathrm{cal}}$ = calculated average axial tensile stress in MPa

$\sigma_{btx,\,\mathrm{cal}}$, $\sigma_{bcy,\,\mathrm{cal}}$ = calculated bending tensile stress in the extreme fibre when bending is about *xx*-axis and *yy*-axis, respectively in MPa

f_y = yield stress of steel in MPa

6.4 SLENDERNESS RATIO (λ)

The slenderness ratio of a tension member is the ratio of its unsupported length (l) to its least radius of gyration (r). Theoretically, there is no limitation on the slenderness ratio of tension members since stability is of little concern. However, they may be subjected to load reversals during transportation, shipping, erection, etc. In order to provide adequate rigidity to prevent undesirable lateral movement or excessive vibrations, design specifications usually contain a limiting slenderness ratio for tension members. IS 800–1984 limits it as listed in Table 6.1.

Table 6.1 *Maximum Slenderness Ratios*

S. No.	*Member*	*Maximum Slenderness Ratio*
1.	A tension member in which a reversal of direct stress due to loads other than wind or seismic forces occurs.	180
2.	A member normally acting as a tie in a roof truss or a bracing system but subjected to possible reversal of stresses resulting from the action of the wind or earthquake forces.	350
3.	Tension members (other than pre-tensioned members)	400

6.5 NET SECTIONAL AREA

The net sectional area of a tension member is the gross-sectional area of the member minus the sectional area of the maximum number of holes. Stresses in a tension member are calculated on the basis of minimum net cross-sectional area available. The reason for considering the net section in the calculation of stresses is the failure of sections with holes. The unit stress in a tension member is increased due to the presence of a hole even if the hole is occupied by a rivet. This is because the area of steel to which load is distributed is reduced and some concentration of stress occurs along the edges of the hole. But for static loading this increase in unit stress is neglected because at yielding, the effect of stress concentration is nullified and tension is therefore assumed to be uniformly distributed over the net section. The net sectional area for various sections can be arrived at as follows:

Threaded Rods

For threaded rods the area at the root of the rod is the net sectional area (Appendix VII).

Bolts, sag rods and tie rods are a few examples of threaded rods. Sag rods are placed between two consecutive purlins in a roof truss system, to achieve economy. Various arrangements of sag rods are shown in Fig. 6.3. Since individual sag rods are placed between successive pairs of purlins, they may be designed individually, each to carry a tangential component from all the purlins below it. Thus sag rods just below and suspended from a ridge purlin will be subjected to maximum force.

The tie rod between ridge purlins at first glance would seem to carry only the horizontal component of the stress in the adjacent sag rod. On the contrary, the stress in the sag rod must be considered a component of that in the tie rod.

Plates

A student is apt to lay great stress upon the fact that a flat bar with a central hole drilled in it fails, under axial tension, at a higher load per net square area than a similar bar without a hole. This is due to the help given by the supposed non-active metal behind the hole, parallel to the direction of pull. Just prior to breaking, the metal tends to narrow at that point where the fracture occurs, and the surplus metal behind the hole restricts this contraction of the strained metal and thus increases the ultimate strength. When there are several holes out of a member they are generally reeled. Now the stress in the plate can hardly be expected to jump from side to side in order to clear the holes without some loss of strength, so that the assumed gain mentioned at the commencement of the paragraph can scarcely be counted upon. Further, it brings in that debatable point as to whether the elastic limit or the ultimate strength should be used when fixing the working stress. The net sectional area of the plate members is obtained by deducting the area of rivet holes from the gross sectional area of plates as follows.

(i) Refer to the plate shown in Fig. 6.4(a) subjected to a pull *P*, and provided with chain riveting. The possibility of failure can be along the section *ABCD*. Net area at the section is equal to the gross area minus the area of rivet holes *B* and *C*.

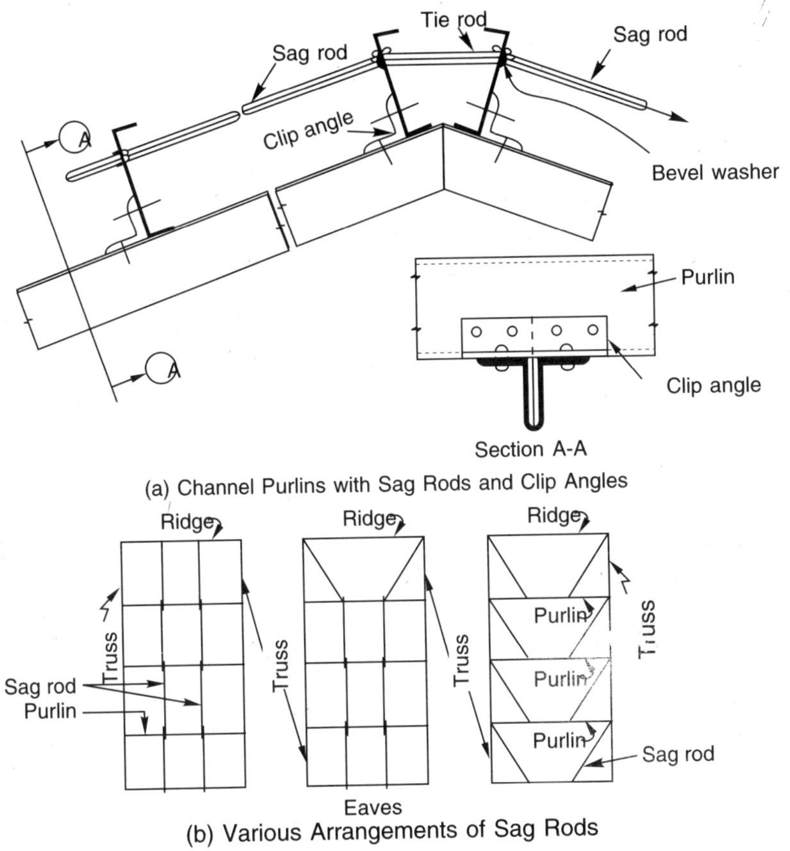

Fig. 6.3 *Channel Purlins Reinforced with Sag Rods*

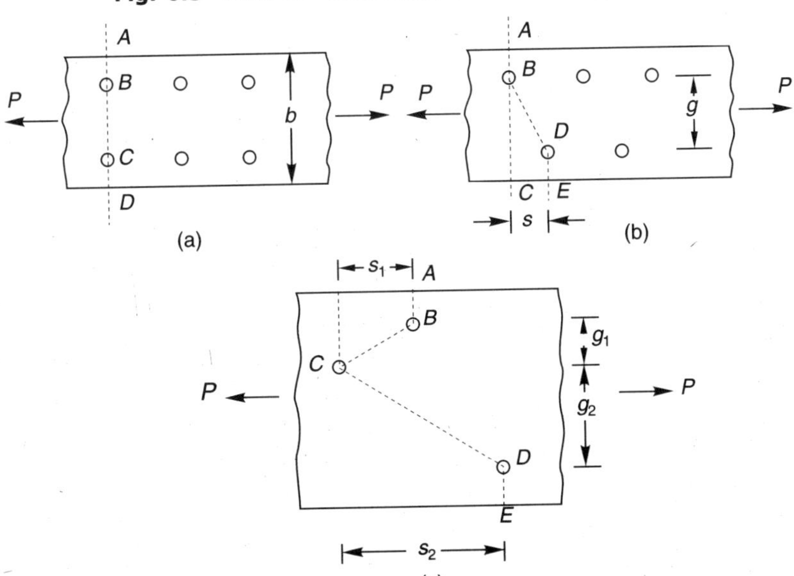

Fig. 6.4 *Net Sectional Area of Riveted Plates*

$$A_{net} = A_{gross} - \text{sectional areas of holes}$$
$$= bt - n(dt)$$
$$= (b - nd)t$$

where A_{net} = net sectional area of plate
A_{gross} = gross sectional area of plates
b = width of plate
n = number of rivets (in the present case = 2)
d = gross diameter of rivet
t = thickness of plates

(ii) Refer to the plate shown in Fig. 6.4(b) subjected to a pull P, and provided with a zig-zag riveting.

To determine the critical net area it might seem logical to compute the area of a section along *ABC* less the area of one rivet hole, and then the area of a section along *ABDE* less two rivet holes. The least of the two values obtained would be the critical value. But this is not correct as along the diagonal line *BD* there is a combination of direct stress and shear and a somewhat smaller area should be used. The strength of the member along section *ABDE* is obviously somewhere between the strength obtained by using the net area along *ABC* and along *ABDE*. I.S. 800 recommends an empirical equation suggested by V.H. Cochrane, to account for staggered holes. Refer Fig. 6.4(c); the area to be deducted is the sum of the sectional area of all holes in a chain of lines extending progressively across the member, less $s^2t/4g$ for each line extending between holes at other than right angles to the direction of stress. The chain of lines should be chosen so as to produce the maximum deductions. For chain of lines *ABCDE*,

$$\text{Deduction} = \text{sectional area of holes} - \left(\frac{s_1^2 t}{4g_1} + \frac{s_2^2 t}{4g_2}\right)$$

$$A_{net} = A_{gross} - \left[\text{sectional area of holes} - \left(\frac{s_1^2 t}{4g_1} + \frac{s_2^2 t}{4g_2}\right)\right]$$

$$= bt - ndt + \frac{s_1^2 t}{4g_1} + \frac{s_2^2 t}{4g_2}$$

$$= \left[b - nd + \frac{s_1^2}{4g_1} + \frac{s_2^2}{4g_2}\right]t$$

If $s_1 = s_2$ and $g_1 = g_2$

$$A_{net} = \left[b - nd + \frac{2s^2}{4g}\right]t$$

$$\text{In general, } A_{net} = \left[b - nd + \frac{n's^2}{4g}\right]t$$

where s = staggered pitch
g = gauge distance
n' = numbers of staggered pitch, and
n = number of holes in the zig-zag line

Angles and Tees

Angle sections are the simplest and most common rolled shape used as tension members in trusses and bracing systems. Normally one leg of the angle section is connected while the other leg is outstanding. In cases where the load is applied to one leg only, it will not cause uniform stress in the section as a whole. The stress distribution is non-uniform near or at the connection because of the eccentricity of load (assumed to be applied at the centroid of sections) w.r.t. the connection. This eccentric force produces bending in the member. Therefore, the connection should be designed so as to reduce the effect of bending to a minimum. In conventional design all effects of eccentricities are often neglected. Sections which are symmetrical about one or both axes are, therefore, an ideal choice. As per I.S. specifications the sectional area of tension members is reduced by making empirical allowances as discussed below:

(i) *In case of single angle connected through one leg (Fig. 6.5(a))*, the net effective sectional area should be taken as

$$A_{net} = A_1 + kA_2$$

where A_1 = effective net cross-sectional area of the connected leg
A_2 = the gross cross-sectional area of the unconnected (outstanding) leg
$k = 3A_1/(3A_1 + A_2)$

Note Where lug angles are used, the effective sectional area of the whole of the angle member should be considered.

(ii) *In case of a pair of angles back-to-back (Fig. 6.5(b)) (or a single Tee) connected by one leg of the angle (or by a flange of the Tee) to the same side of gusset plate*, the net effective area should be taken as

$$A_{net} = A_1 + kA_2$$

where, $k = 5A_1/(5A_1 + A_2)$.

(iii) *For double angles (Fig. 6.5(c)), (or flange of Tees) placed back-to-back and connected to each side of a gusset plate or to each side of part of a rolled steel section*, the net area considered should be taken as,

$$A_{net} = A_{gross} - \text{deduction for rivet holes.}$$

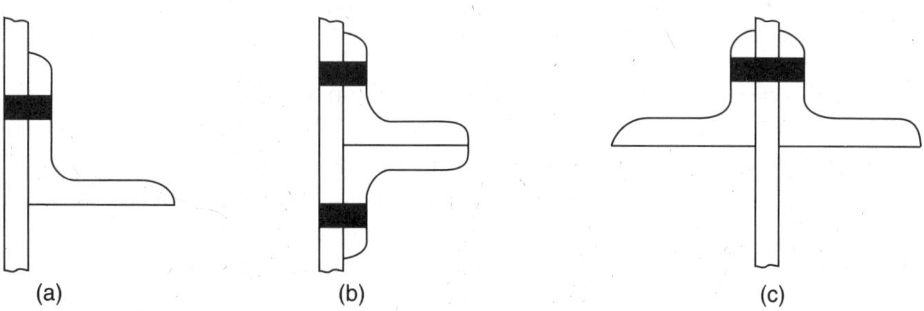

Fig. 6.5 *Net Section of Angles*

Notes

1. In calculating the area for rivet holes the projected area of the hole is considered.
2. In tension members composed of two sections in contact back-to-back or separated back-to-back by a distance not exceeding the aggregate thickness of the connected parts, tacking rivets should be provided at pitch in line not exceeding 1000 mm.
3. When the two sections are not tack riveted the provisions discussed in (ii) and (iii) for calculation of net sectional area do not apply and each section should be designed as a single section (angle or Tee) connected through one leg only.
4. The area of the leg of an angle should be taken as the product of the thickness and the length from the outer corner minus half the thickness and the area of the leg of a Tee as the product of the thickness and the depth minus the thickness of the table (flange).
5. In a built-up member where the chains of holes considered in individual parts do not correspond with the critical chain of holes for the member as a whole, the value of any rivet joining the parts between such chains should be taken into account while calculating the strength of the member.

Channels and I-Sections

The net effective sectional area of a single channel section connected through the web is treated as for double angles connected to one leg each on the gusset. When a rolled or built-up channel- and I-Sections are connected only through flanges, the web is found to be partially ineffective in resisting the applied tensile load. In such cases, the net area may be taken as the total area less half the web area.

6.6 DESIGN OF A TENSION MEMBER

The design of a tension member consists in selecting a section with sufficient area to carry the design loads without exceeding the permissible tensile stress. This selection of a tension member is the most simple and straight forward method because the stability of the tension member is of minor concern. Before designing a tension member the nature of loading and the types of stresses to which the member may be subjected are ascertained. The tension member may be subjected to either pure axial tension or combined axial tension and bending.

Whereas computing the required net area of a tension member is simple enough, the proportioning and arrangement of the member, so that it is appropriate and economical may become quite involved. For instance it is often difficult to arrange a connection without eccentricity; and stress concentration induced by the connected parts can be extremely complicated. Although the designer has considerable freedom in the selection of rolled shapes, the resulting section should be compact; its dimensions should fit into the overall dimensions of the structure, and, in order to minimise stress concentrations, as large a portion as possible should be connected to the gusset plate.

Effect of Staggered Holes In the design of tension members it is desirable to provide as large a net area as possible. Therefore, whenever more than one row of rivets is provided, the rivets are staggered. In case of chain riveting (Fig. 6.4(a)), failure in a section takes place transverse to the line of action of the load, i.e. along *ABCD*. But in the case of staggered rivets (Fig. 6.4(b)), the failure can take place either transversely along *ABC* or along a zig-zag line passing through various rivets *ABDE*. There may be several such paths which may be assumed to be critical. Each such path is checked and the path giving minimum net sectional area is critical.

Tension Member Subjected to Axial Load

The following procedure may be followed in the design of an axially loaded tension member.

1. The net area required (A_{net}) to carry the design load P is obtained by the equation,

$$P = \sigma_{at} A_{\text{net}}$$

2. The net area calculated thus, is increased suitably (25% – 40%) to compute the gross sectional area. From I.S. Handbook No. 1 suitable section/sections providing a cross-sectional area matching with the computed gross sectional area is selected.
3. The number of rivets required to make the connection is calculated. These are arranged in a suitable pattern and the net area of the section provided is calculated. This should be more than the net area calculated in step (1).
4. The slenderness ratio of the member is checked as per the I.S. sepcifications.

Tension Member Subjected to Axial Load and Bending Moment

A tension member may be subjected to bending stresses along with direct tensile stress for the following reasons:

1. It may not be perfectly straight.
2. The member may not be vertical, in which case its self-weight will give rise to bending stresses.
3. The connection may be eccentric. This would induce moments, and thus generate bending stresses.

The bending stresses in the tension members are not of much importance because the direct pull of a tension member tends to reduce the lateral deflection caused by bending.

The following method is useful:

1. The axial tensile force is ascertained and bending moment on the member is computed.
2. The net area is calculated by $P = \sigma_{at} A_{\text{net}}$.
   The area so found may be increased suitably (50% – 100%) to account for bending moment and subsequently by 25% – 40% to compute the gross-sectional area.
3. From I.S. Handbook No. 1 a suitable section furnishing a matching area with that calculated in step (2) is selected and the relevant properties of the section selected are noted.
4. The diameter of rivets is assumed and the rivet value is computed. The number of rivets required for making the connection are calculated. Suitable arrangement of rivets is done and the net area provided by the section is calculated.
5. The actual tensile stress and bending stess in the member are computed.

$$\sigma_{at,\ \text{cal}} = \frac{\text{axial force}}{\text{net area provided}}$$

$$\sigma_{bt,\ \text{cal}} = \frac{\text{bending moment}}{\text{section modulus provided}}$$

6. The safety of the section is checked using the interaction relationship

$$\frac{\sigma_{at,\,\text{cal}}}{0.6\,f_y} + \frac{\sigma_{bt,\,\text{cal}}}{0.66\,f_y} \leq 1$$

6.7 LUG ANGLES

To make the connection at joints, a certain length of the tension member and that of the gusset plate is utilized. If the load is heavy and the number of rivets/length of weld required for making the connection is large, the size of the gusset plate required may be uneconomical. As an alternative to this an additional angle is used along with the tension member for the length required to accommodate the connectors to make the connection. Such an angle is called a lug angle. Lug angles are found to be more effective at the beginning of the connection only due to the non-uniform distribution of the load among the connecting rivets. Lug angles can be eliminated by providing unequal angle sections with a wider leg as a connected leg, and rivets can be provided in two rows. However, in most of the cases the cost of lug angles, their connections and fabrication will be more. This offsets any saving in the gusset plates, thereby defeating the purpose (Ref. Fig. 6.6).

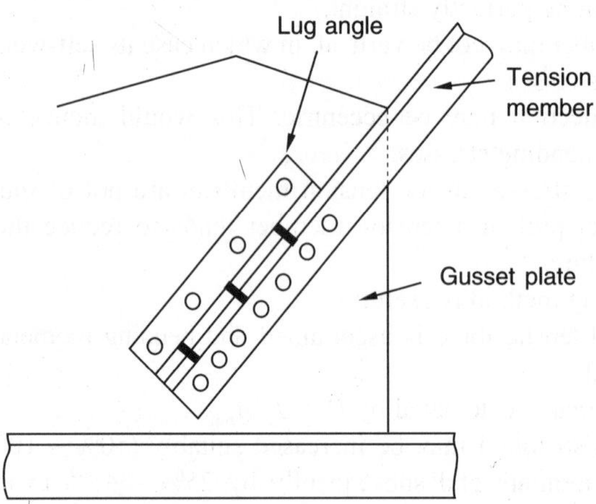

Fig. 6.6 *Lug Angle*

Lug angles may be designed as follows:

1. In the case of angle members, lug angles and their connections to the gussets or other supporting members should be capable of developing a strength not less than 20% in excess of the force in the outstanding leg of the angle. The attachment of the lug angle to the angle member should be capable of developing 40% in excess of that force.
2. Where lug angles are used to connect an angle member, the whole area of the member should be taken as effective. The net area is calculated by deducting the area of the rivet hole from the gross section area.
3. In the case of channel members and the like, the lug angles and their connections to the gusset or other supporting member should be capable of developing a strength of not less than 10% in excess of the force not accounted for by

the direct connection of the member. The attachment of lug angles to the member should be capable of developing 20% in excess of that force.

4 Lug angles connecting a channel shaped member should as far as possible be disposed symmetrically with respect to the section of the member.

6.8 SPLICES

Splices in tension members are used to join two sections when a joint is to be provided, i.e. these replace the members at the joint where it is cut. Where a splice occurs in an angle, channel, Tee or joint section, the force is abstracted from the member through the connections on one side of the joint and is transferred to the splice cover plate/angle section. The force is then carried through these covers across the joint and is transferred to the other portion of the member through the connection.

Splice connection for tension members is usually the riveted or bolted connection. The splice section as well as the rivets are designed for the tensile load to be transmitted by the main tension member. There is always a question whether the splice should be designed for the maximum calculated force or for the actual strength of the member. As per I.S. specification, the splice covers and its connections should be designed to develop the net tensile strength of the main member. The various types of splices that can be provided are shown in Fig. 6.7(a, b, c). If the sections are not of same thickness, packings are introduced, as shown in Fig. 6.7(d), and the extra rivets are determined by 2.5% of the calculated number of rivets for each 2 mm thickness, if packing thickness is more than 6 mm. These additional rivets are provided on extended packing as shown in Fig. 6.7(d).

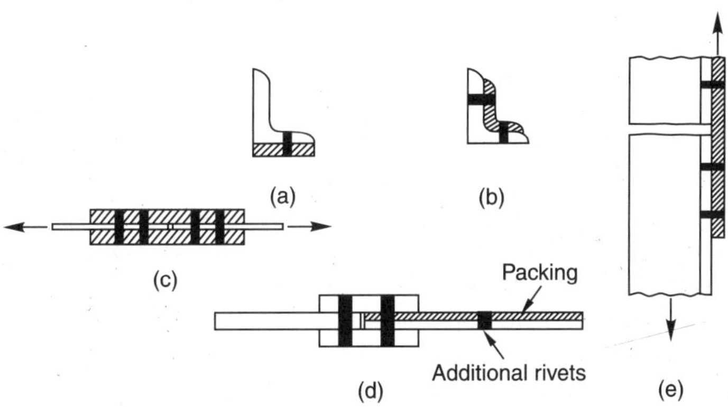

Fig. 6.7

Normally in the design of a tension splice the effect of eccentricity is neglected but as far as possible it should be avoided. Figures 6.7(a, e) show an angle section spliced on one leg by the plate. Such an arrangement causes eccentricity and, thus, moments are created. To overcome this both the legs of the angle should be spliced, as shown in Fig. 6.7(b). Figures 6.7(c, d) show the splicing of plates.

6.9 GUSSET PLATE

A gusset plate is a plate provided to make connections at the place where more than one member is to be jointed, e.g. joints of truss, truss girders, etc. The lines of action of truss members meeting at a joint are assumed to coincide, as shown in Fig. 6.8(a). In case any eccentricity is present, secondary stresses are generated.

The size and shape of the gusset plates are usually decided from the direction of the members meeting at the joint. The plate outlines are fixed so as to meet the specifications of edge distance for the rivets used to connect the various members at that joint.

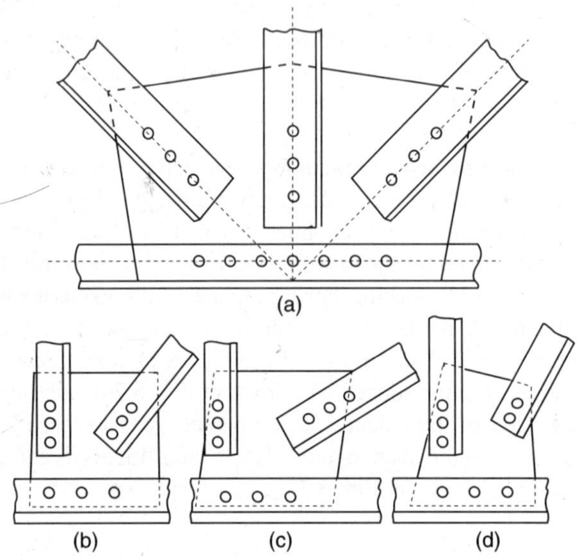

Fig. 6.8 *Shape of Gusset Plate*

The edges of the plates should be kept parallel as far as possible but this causes wastage of material of the gusset plate. Therefore, two sides are kept parallel and the other two are suitably cut. Figure 6.8(b) shows the gusset plate with one side sheared. The opposite side can also be sheared to achieve economy, Fig. 6.8(c). The size of the plate can still be reduced by shearing the top edge of the plate, Fig. 6.8(d), but this gives a bad appearance and generally avoided.

Due to the deflection of the truss, the angle between the truss members reduces. The truss members offer resistance to it, so compressive forces are set up in the portion of the plate between them. This portion of the plate behaves like a column and buckles. Therefore, the length of the top edge of gusset plate is kept minimum. The top edge of the gusset plate is stiffened if it becomes more than 60 times the thickness of the plate.

Design specifications give a little guidance to decide thickness of gusset plates but in any case it should not be less than 12 mm. Structurally a gusset plate is subjected to shear stresses, direct stresses and bending stresses and therefore it should be of ample thickness to resist these at the critical section.

Solved Examples

Example 6.1 A 300 I.S.F. 8 mm is used as a tension member in a lattice girder. It is connected to a 12 mm gusset plate by 18 mm ϕ hand driven rivets. The yield stress of steel is 250 N/mm^2. Calculate the maximum force that the member can carry, if

(a) chain riveting is done, Fig. Ex. 6.1(a)

(b) zig-zag riveting is done, Fig. Ex. 6.1(b)

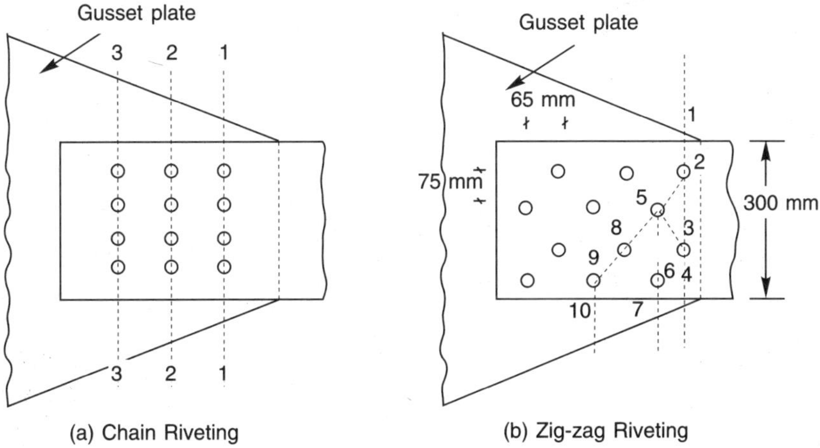

Fig. Ex. 6.1

Solution

$$\phi = 18 \text{ mm}$$

$$\text{Gross diameter, } d = 18 + 1.5 = 19.5 \text{ mm}$$

(a) In chain riveting (Fig. Ex. 6.1(a)) the critical section will be 1–1.

$$\text{Net area} = (b - nd)t$$

$$= (300 - 4 \times 19.5) \times 8 = 1776 \text{ mm}^2$$

$$\text{Strength of member} = 1776 \times 150 \times 10^{-3} = 266.4 \text{ kN}$$

(b) In zig-zag riveting (Fig. Ex. 6.1(b)) the section can fail along (1)–(2)–(3)–(4), (1)–(2)–(5)–(6)–(7), (1)–(2)–(5)–(3)–(4), (1)–(2)–(5)–(8)–(9)–(10). The net area which comes minimum out of this will give the strength of the member.

Section (1)–(2)–(3)–(4)

$$\text{Net area} = (300 - 2 \times 19.5) \times 8 = 2088 \text{ mm}^2$$

Section (1)–(2)–(5)–(6)–(7)

$$\text{Net area} = \left(B - nd + \frac{n' s^2}{4g} \right) t$$

$$= \left(300 - 3 \times 19.5 + \frac{1 \times 65^2}{4 \times 75} \right) \times 8 = 2044.66 \text{ mm}^2$$

The section (1)–(2)–(5)–(3)–(4) should also be checked. But in the present case

$$n = 3, \quad n' = 2$$

If we substitute these values in the equation of the net area, the value will be more than 2044.66 mm^2.

Therefore, the net area at section (1)–(2)–(5)–(3)–(4) has not been worked out.

Section (1)–(2)–(5)–(8)–(9)–(10)

$$\text{Net area} = \left(B - nd + \frac{n' s^2}{4g}\right) t$$

$$= \left(300 - 4 \times 19.5 + \frac{3 \times 65^2}{4 \times 75}\right) \times 8 = 2114 \text{ mm}^2$$

The minimum net area available is at section (1)–(2)–(5)–(6)–(7).

Therefore, the strength of the member will be calculated at this section.

$$\text{Strength} = 2044.66 \times 0.6 \times 250 \times 10^{-3} \qquad (f_y = 250 \text{ N/mm}^2)$$
$$= 306.699 \text{ kN}$$

The strength of the member when chain riveted is 266.4 kN and that when zig-zag riveted is 306.699 kN.

Example 6.2 Determine the tensile strength of a roof truss diagonal 100 × 75 × 10 mm (f_y = 260 N/mm^2) connected to the gusset plate by (Ref. Fig. Ex. 6.2)

(a) 20 mm diameter power driven rivets in one row along the length of the member. The short leg of the angle is kept outstanding.

(b) 5 mm fillet weld.

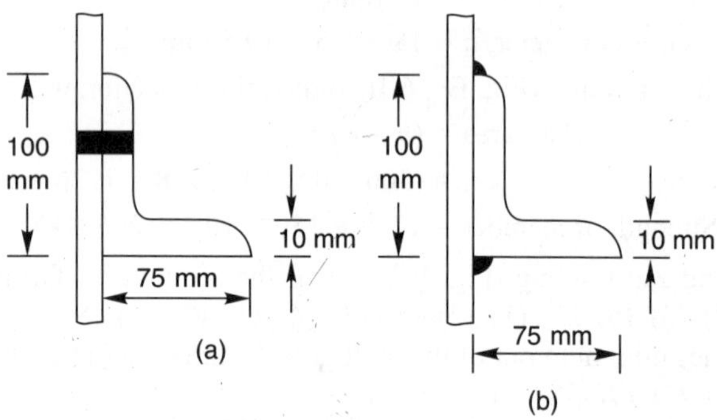

Fig. Ex. 6.2

Solution

(a) Nominal diameter of the rivet, ϕ = 20 mm

Gross diameter of the rivet, d = 20 + 1.5 = 21.5 mm

$$\text{Area of connected leg, } A_1 = \left(100 - 21.5 - \frac{10}{2}\right) \times 10 = 735 \text{ mm}^2$$

$$\text{Area of outstanding leg, } A_2 = \left(75 - \frac{10}{2}\right) \times 10 = 700 \text{ mm}^2$$

$$k = \frac{3A_1}{3A_1 + A_2} = \frac{3 \times 735}{3 \times 735 + 700} = 0.759$$

$$\text{Net area, } A_{net} = A_1 + kA_2 = 735 + 0.759 \,´\, 700 = 1266.32 \text{ mm}^2$$

$$\text{Strength of the diagonal member} = \sigma_{at} \times A_{net}$$
$$= 0.6 \times 260 \times 1266.32 \times 10^{-3} = 197.542 \text{ kN}$$

(b) Area of cross-section of I.S.A. 100 × 75 × 10 mm from I.S. Handbook No. 1,

$$A = 1650 \text{ mm}^2$$

$$\text{Area of connected leg, } A_1 = \left(100 - \frac{10}{2}\right) \times 10 = 950 \text{ mm}^2$$

$$\text{Area of outstanding leg, } A_2 = \left(75 - \frac{10}{2}\right) \times 10 = 700 \text{ mm}^2$$

$$k = \frac{3A_1}{3A_1 + A_2} = \frac{3 \times 950}{3 \times 950 + 700} = 0.8028$$

$$\text{Net area, } A_{net} = A_1 + kA_2$$
$$= 950 + 0.8028 \times 700 = 1511.96 \text{ mm}^2$$

$$\text{Strength of the diagonal} = \sigma_{at} \times A_{net}$$
$$= 0.6 \times 260 \times 1511.96 \times 10^{-3} = 235.865 \text{ kN}$$

Example 6.3 A tie member in a bracing system consists of two angles 150 × 115 × 10 mm (f_y = 250 MPa) with long legs connected to a gusset plate by 18 mm diameter rivets in such a way that each angle section is reduced in section by one rivet hole only. Determine the tensile strength of the member if,

(a) The angles are connected on the same side of the gusset plate, 12 mm thick and are tack riveted.
(b) The angles are connected on the opposite side of the gusset plate, 12 mm thick and are tack riveted.
(c) What is the effect on tensile strength of the member if the angles (in case (b)) are not tack riveted?

Solution Nominal diameter of the rivet, ϕ = 18 mm

Gross diameter of the rivet, d = 18 + 1.5 = 19.5 mm

(a) The relevant properties of 2, I.S.A. 150 × 115 × 10 mm on the same side of gusset plate,

$$A = 5{,}104 \text{ mm}^2$$

$$\text{Area of connected legs} = 2\left(150 - 19.5 - \frac{10}{2}\right) \times 10 = 2510 \text{ mm}^2$$

$$\text{Area of outstanding legs} = 2\left(115 - \frac{10}{2}\right) \times 10 = 2200 \text{ mm}^2$$

$$k = \frac{5A_1}{5A_1 + A_2} = \frac{5 \times 2510}{5 \times 2510 + 2200} = 0.8508$$

$$\text{Net area provided} = A_1 + kA_2$$
$$= 2510 + 0.8508 \times 2200 = 4381.8642 \text{ mm}^2$$
$$\text{Strength of the member} = 4381.8642 \times 150 \times 10^{-3} = 657.2796 \text{ kN}$$

(b) $$\text{Net area provided} = \text{gross area} - \text{area of rivet hole}$$
$$= 5104 - 2 \times 19.5 \times 10 = 4714 \text{ mm}^2$$
$$\text{Strength of the member} = 4714 \times 150 \times 10^{-3} = 707.10 \text{ kN}$$

(c) If the angles are not tack riveted, they behave as single angles connected to a gusset plate.

$$\text{Area of connected legs, } A_1 = 2\left(150 - 19.5 - \frac{10}{2}\right) \times 10 = 2510 \text{ mm}^2$$

$$\text{Area of outstanding legs, } A_2 = 2\left(115 - \frac{10}{2}\right) \times 10 = 2200 \text{ mm}^2$$

$$k = \frac{3A_1}{3A_1 + A_2} = \frac{3 \times 2510}{3 \times 2510 + 2200} = 0.773895$$

$$\text{Net area provided} = 2510 + 0.773895 \times 2200 = 4212.5692 \text{ mm}^2$$
$$\text{Strength of the member} = 4212.5692 \times 150 \times 10^{-3} = 631.885 \text{ kN}$$

Therefore, the load carrying capacity is reduced by 75.215 kN (707.10 – 631.885), if the angles are not tack riveted.

Example 6.4 A bridge truss diagonal carries a pull of 200 kN. The length of the diagonal is 3.0 m. Design a suitable section. The member is connected to a gusset plate 14 mm thick. Assume $f_y = 250$ N/mm^2.

Solution In this example the diagonal member has been designed as a single angle section, a double angle section placed on the opposite side of the gusset plate and a double angle section on the same side of the gusset plate. The relative economy in weight is worked out. However, the additional cost of making holes for rivets and tack rivets has not been considered while working out the economy.

$$\text{Allowable tensile stress, } \sigma_{at} = 0.6f_y = 0.6 \times 250 = 150 \text{ N/mm}^2$$

$$\text{Net cross-sectional area required} = \frac{200 \times 10^3}{150} = 1333.33 \text{ mm}^2$$

Increase the net area obtained by about 40% (to account for rivet holes) to find a tentative gross area required.

$$\text{Gross cross-sectional area required} = 1.4 \times 1333.33 = 1866.67 \text{ mm}^2$$

(a) *Single angle section* It is found from I.S. Handbook No. 1, that as a first trial the angle sections furnishing the required area are I.S.A. equal angle section 100 × 100 × 10 mm (1903 mm^2) and I.S.A. unequal angle section 125 × 75 × 10 mm (1902 mm^2). The weight of each angle is 146.2 N/m. From the economy consideration, both the angles will cost the same. Therefore, the selection will be based on the load carrying capacity of the two angle sections. From the calculations below it is found that the unequal angle section with its long leg connected has a higher load carrying capacity and therefore, should be provided.

Assume nominal diameter of the rivet, $\phi = 20$ mm

$$\text{Gross diameter of the rivet, } d = 20 + 1.5 = 21.5 \text{ mm}$$

Try I.S.A 100 × 100 × 10 mm @ 146.2 N/m:

$$\text{Area of connected leg, } A_1 = \left(100 - 21.5 - \frac{10}{2}\right) \times 10 = 735 \text{ mm}^2$$

$$\text{Area of outstanding leg, } A_2 = \left(100 - \frac{10}{2}\right) \times 10 = 950 \text{ mm}^2$$

$$k = \frac{3A_1}{3A_1 + A_2} = \frac{3 \times 735}{3 \times 735 + 950} = 0.69889$$

$$\text{Net area provided} = 735 + 0.69889 \times 950 = 1398.946 \text{ mm}^2$$

$$\text{Strength of the member} = 1398.946 \times 150 \times 10^{-3} = 209.8419 \text{ kN} > 200 \text{ kN}$$

which is all right.

$$(r_{min} = r_{vv} = 19.4 \text{ mm})$$

$$\frac{l}{r} = \frac{3000}{19.4} = 154.639 < 350$$

which is safe.

Try I.S.A. 125 × 75 × 10 mm @ 146.2 N/m:

$$\text{Area of connected leg, } A_1 = \left(125 - 21.5 - \frac{10}{2}\right) \times 10 = 985 \text{ mm}^2$$

$$\text{Area of outstanding leg, } A_2 = \left(75 - \frac{10}{2}\right) \times 10 = 700 \text{ mm}^2$$

$$k = \frac{3A_1}{3A_1 + A_2} = \frac{3 \times 985}{3 \times 985 + 700} = 0.80848$$

$$\text{Net area provided} = 985 + 0.80848 \times 700 = 1550.937 \text{ mm}^2$$

$$\text{Strength of the member} = 1550.937 \times 150 \times 10^{-3} = 232.6405 \text{ kN} > 200 \text{ kN}$$

which is all right.

$$\frac{l}{r} = \frac{3000}{16.1} = 186.33 < 350$$

which is all right.

It is to be noted that the load carrying capacity of the unequal angle section is substantially large and a still smaller size angle can be provided. Let us try I.S.A. 150 × 75 × 8 mm @ 134.4 N/m:

$$\text{Area of connected leg, } A_1 = \left(150 - 21.5 - \frac{8}{2}\right) \times 8 = 996 \text{ mm}^2$$

$$\text{Area of outstanding leg, } A_2 = \left(75 - \frac{8}{2}\right) \times 8 = 568 \text{ mm}^2$$

$$k = \frac{3A_1}{3A_1 + A_2} = \frac{3 \times 996}{3 \times 996 + 568} = 0.84026$$

$$\text{Net area provided} = 996 + 0.84026 \times 568 = 1473.2733 \text{ mm}^2$$

$$\text{Strength of the member} = 1473.2733 \times 150 \times 10^{-3} = 220.990 \text{ kN}$$

$$> 200 \text{ kN}$$

which is all right

Check for slenderness ratio

$$\frac{l}{r} = \frac{3000}{16.0} = 187.5 < 350$$

which is all right.

Hence, provide I.S.A. 150 mm × 75 mm × 8 mm with long leg connected to the gusset plate.

(b) Let us try a double angle section (tack riveted) placed on the same side of the gusset plate.

Try two I.S.A. 75 × 50 × 8 mm @ 145.2 N/m:

$$A = 1876 \text{ mm}^2, \quad r = 14.0 \text{ mm}$$

Provide 20 mm diameter rivet.

$$\text{Gross diameter, } d = 20 + 1.5 = 21.5 \text{ mm}$$

$$\text{Area of connected legs, } A_1 = 2\left(75 - 21.5 - \frac{8}{2}\right) \times 8 = 792 \text{ mm}^2$$

$$\text{Area of outstanding legs, } A_2 = 2\left(50 - \frac{8}{2}\right) \times 8 = 736 \text{ mm}^2$$

$$k = \frac{5A_1}{5A_1 + A_2} = \frac{5 \times 792}{5 \times 792 + 736} = 0.8432$$

$$\text{Net area provided} = 792 + 0.8432 \times 736 = 1412.6473 \text{ mm}^2$$

$$\text{Strength of the member} = 1412.6473 \times 150 \times 10^{-3} = 211.897 \text{ kN}$$

$$> 200 \text{ kN}$$

which is safe.

$$\frac{l}{r} = \frac{3000}{14.0} = 214.28 < 350$$

which is within the permissible limit.

It is to be noted that by providing the two angles on the same side of the gusset plate the load carrying capacity is reduced as compared to the single angle section. Also, the weight of the double angle section is more as compared to that of the single angle section and thus, uneconomical.

(c) Let us try the double angle section (tack riveted) on the opposite side of gusset plate.

Try two I.S.A. 75 × 50 × 8 mm @ 145.2 N/m.

$$\text{Net area} = \text{gross area} - \text{area of rivet hole}$$
$$= 1876 - 2 \times 21.5 \times 8 = 1532 \text{ mm}^2$$
$$\text{Strength of the member} = 1532 \times 150 \times 10^{-3} = 229.800 \text{ kN} > 200 \text{ kN}$$

The section can be revised for economy.

Try 2 I.S.A. 70 × 45 × 8 mm @ 131.4 N/m

$$A = 1716 \text{ mm}^2, \quad r = 21.9 \text{ mm}$$
$$\text{Net area provided} = \text{gross area} - \text{area of rivet hole.}$$
$$= 1716 - 2 \times 21.5 \times 8 = 1372 \text{ mm}^2$$
$$\text{Strength of the member} = 1372 \times 150 \times 10^{-3} = 205.800 \text{ kN} > 200 \text{ kN}$$

which is all right.

$$\frac{l}{r} = \frac{3000}{21.9} = 136.98 < 350$$

It is to be noted that this section is the most economical, and stable. However, the cost of making extra holes and of tack rivets will have to be added to do the final estimation and selection.

Example 6.5 The bottom chord tension member of a roof truss is subjected to an axial pull of 300 kN. A differential chain hoist arrangement is attached to the bottom chord and gives a point load of 10 kN at the centre. The length of the member between panel points is 4.0 m. Design the section consisting of two unequal angles with long legs kept back-to-back and turned upwards. Assume f_y = 250 N/mm^2.

Solution

Allowable axial tensile stress, $\sigma_{at} = 0.6 \times 250 = 150 \text{ N/mm}^2$

$$\text{Net area required for axial pull} = \frac{300 \times 10^3}{150} = 2000 \text{ mm}^2$$

Increase the net area by 50% to account for bending moment

$$= 1.5 \times 2000 = 3000 \text{ mm}^2$$
$$\text{A tentative gross area required} = 1.4 \times 3000 = 4200 \text{ mm}^2$$

Let us assume the section to be weakened by one rivet hole and let the diameter of rivet be 14 mm.

Try 2 I.S.A. 150 × 75 × 10 mm from I.S. Handbook No. 1 (longer legs back-to-back).

The relevant properties of the sections are

$$C_{xx} = 53.2 \text{ mm}, \; A = 4312 \text{ mm}^2, \; I_{xx} = 998.2 \times 10^4 \text{ mm}^4$$
$$\text{Net area provided} = 4312 - 2 \times 15.5 \times 10 = 4002 \text{ mm}^2$$
$$\sigma_{at,\ \text{cal}} = \frac{300 \times 10^3}{4002} = 74.96 \text{ N/mm}^2$$
$$\sigma_{btx,\ \text{cal}} = (M/I_{xx})\, C_{xx}$$

$$M = \text{B.M. due to thrust} + \text{B.M. due to self weight}$$

$$= \frac{10 \times 10^3 \times 4 \times 10^3}{4} + \frac{331.58 \times 4^2 \times 1000}{8}$$

$$= 1066.3 \times 10^4 \text{ Nmm}$$

$$\sigma_{btx,\text{ cal}} = \frac{1066.3 \times 10^4}{998.2 \times 10^4} \times 53.2 = 56.83 \text{ N/mm}^2$$

Check:

$$\frac{\sigma_{at,\text{ cal}}}{0.6 f_y} + \frac{\sigma_{btx,\text{ cal}}}{0.66 f_y} = \frac{74.96}{0.6 \times 250} + \frac{56.91}{0.66 \times 250} = 0.844 < 1$$

which is safe.

Hence, provide 2 I.S.A. 150 × 75 × 10 mm angle sections with long legs back-to-back as the lower chord member section.

Example 6.6 A diagonal member of a roof carries a maximum axial pull of 300 kN. Design the section and the connections with a 14 mm gusset plate. The length of the gusset plate available for making the connection is 310 mm. Design the lug angle also if required. The steel is of yield stress of 250 N/mm^2.

Solution

Allowable axial tensile stress, $\sigma_{at} = 0.6 \times 250 = 150 \text{ N/mm}^2$

$$\text{Net cross-sectional area required} = \frac{300 \times 10^3}{150} = 2000 \text{ mm}^2$$

The net area is increased by 40% to obtain a tentative gross area.

Gross cross-sectional area required = 1.4 × 2000 = 2800 mm^2

From I.S. Handbook No. 1, select a suitable angle section.

Try I.S.A. 150 × 115 × 12 mm: (See Fig. Ex. 6.6)

The relevant properties are: $A = 3038 \text{ mm}^2$, $r = 24.4$ mm

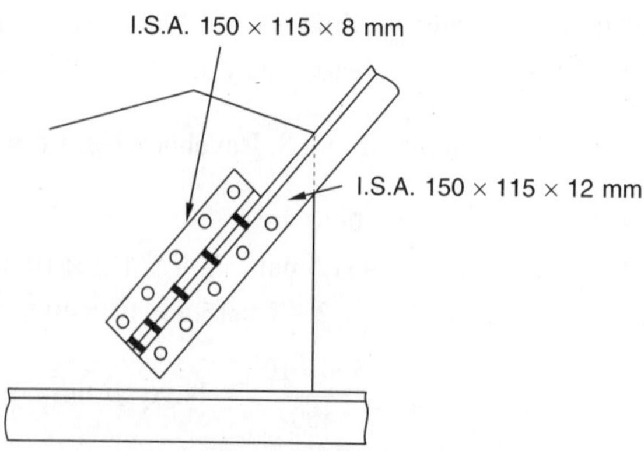

Fig. Ex. 6.6

Let us provide 20 mm diameter power driven rivets for the connection.

Gross diameter of rivet, $d = 20 + 1.5 = 21.5$ mm

$$\text{Area of connected leg, } A_1 = \left(150 - 21.5 - \frac{12}{2}\right) \times 12 = 1470 \text{ mm}^2$$

$$\text{Area of outstanding leg, } A_2 = \left(115 - \frac{12}{2}\right) \times 12 = 1308 \text{ mm}^2$$

$$k = \frac{3A_1}{3A_1 + A_2} = \frac{3 \times 1470}{3 \times 1470 + 1308} = 0.7712486$$

$$\text{Net area provided} = A_1 + kA_2$$
$$= 1470 + 0.7712486 \times 1308 = 2478.79 \text{ mm}^2$$

$$\text{Strength of the member} = 2478.79 \times 150 \times 10^{-3} = 371.818 \text{ kN} > 300 \text{ kN}$$

which is sufficient.

Connection:

$$\text{Strength of rivet in single shear} = \frac{\pi}{4} d^2 \tau_{vf} = \frac{\pi}{4} \times 21.5^2 \times 100 = 36305.03 \text{ N}$$

$$\text{Strength of rivet in bearing} = dt\sigma_{pf} = 21.5 \times 12 \times 300 = 77400 \text{ N}$$

$$\text{Rivet value, } R_v = 36{,}305.03 \text{ N}$$

$$\text{Number of rivets required} = \frac{300 \times 10^3}{36305.03} = 8.263 \approx 10$$

Length of gusset plate required:

Pitch of rivets = $2.5\phi = 2.5 \times 20 = 50$ mm ≈ 60 mm

For 20 mm ϕ rivets, edge distance = 29 mm.

Let us provide a pitch of 60 mm and edge distance of 30 mm.

Minimum length of gusset plate required = $9 \times 60 + 2 \times 30 = 600$ mm

The length of gusset plate available for the connection is 310 mm only, whereas the required connection length is 600 mm. Even if the rivets are staggered and arranged in two rows, the length of the gusset plate available for making the connection will be insufficient. Therefore, provide a lug angle.

Design of lug angle:

Total axial load = 300×10^3 N

Gross area of connected leg of the main angle section

$$= \left(150 - \frac{12}{2}\right) \times 12 = 1728 \text{ mm}^2$$

Gross area of outstanding leg of the main angle section

$$= \left(115 - \frac{12}{2}\right) \times 12 = 1308 \text{ mm}^2$$

In case of unequal angle sections used as tie members, the load gets distributed in the ratio of the gross area of connected and outstanding legs.

Load shared by outstanding leg of main angle

$$= 300 \times 10^3 \times \frac{1308}{1728 + 1308} = 129.249 \times 10^3 \text{ N}$$

$$\text{Load on lug angle} = 1.2 \times 129.249 \times 10^3 = 155.099 \times 10^3 \text{ N}$$

$$\text{Net cross-sectional area required} = \frac{155.099 \times 10^3}{150} = 1033.99 \text{ mm}^2$$

Let us select I.S.A. 150 × 115 × 8 mm angle section from I.S. Handbook No.1.

Cross-sectional area, A= 2058 mm^2

Assume the section to be weakened by one rivet hole only (ϕ = 20 mm)

Net area provided= 2058 – 21.5 × 8 = 1886 mm^2 > 1033.99 mm^2

which is safe.

The lug angle is placed symmetrical to the main angle section. The 115 mm long leg is kept as outstanding leg.

Strength of rivet in single shear = 36305.03 N

Strength of rivet in bearing = 21.5 × 8 × 300 = 51600 N

Rivet value, R_v = 36305.03 N

Number of rivets to connect 150 mm leg of lug angle with 14 mm thick gusset plate

$$= \frac{\text{strength of lug angle}}{R_v} = \frac{155.099 \times 10^3}{36,305.03} = 4.272 \approx 5$$

Provide 5 rivets, 20 mm ϕ to connect lug angle with gusset plate.

∴ Number of rivets to join the main angle leg with gusset plate = 10 – 5 = 5

Check:

$$\text{Load on connected leg} = 300 \times 10^3 \times \frac{1728}{1728 + 1308} = 170.750 \times 10^3 \text{ N}$$

$$\text{Number of rivets required} = \frac{170.750 \times 10^3}{36,305.03} = 4.703 \approx 5$$

which is all right.

Number of rivets required to join outstanding legs of the two angles

$$= \frac{1.4 \times 129.249 \times 10^3}{36,305.03} = 4.984 \approx 5$$

Provide five rivets, 20 mm ϕ to connect the outstanding legs.

Length of gusset plate required = 4 × 60 + 2 × 30 = 300 mm < 310 mm which is all right.

Example 6.7 Design a splice for tension member sections 160 × 10 mm and 250 × 14 mm. The member is subjected to a pull of 200 kN. Assume f_y = 250 N/mm^2.

Solution A pair of splice plates will be provided with one plate on each side of the splice. A suitable arrangement of the connectors designed is assumed and then the width and thickness of the splice plate is computed.

Let nominal diameter of rivet, $\phi = 18$ mm

$\therefore$ Gross diameter = 18 + 1.5 = 19.5 mm

Net tensile strength of the member assuming section to be weakend by one rivet hole (Fig. Ex. 6.7)

$$= (160 - 19.5) \times 10 \times 150/10^3 = 210.75 \text{ kN}$$

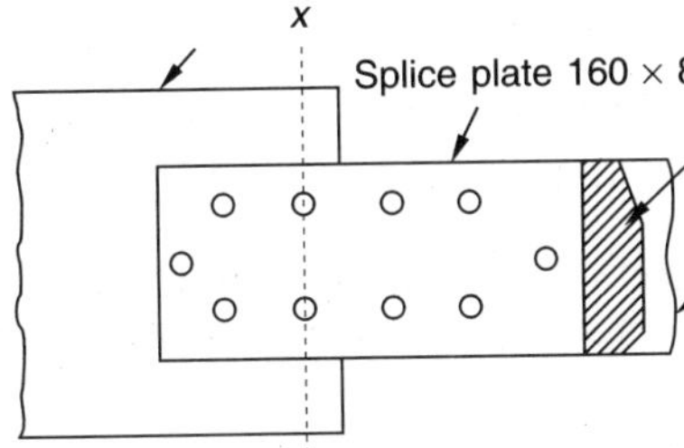

Fig. Ex. 6.7

Let us assume the rivets to be power driven.

$$\tau_{vf} = 100 \text{ N/mm}^2,\ \sigma_{pf} = 300 \text{ N/mm}^2$$

$$\text{Strength of rivet in double shear} = 2 \times \frac{\pi}{4} d^2 \tau_{vf}$$

$$= 2 \times \frac{\pi}{4} 19.5^2 \times 100 = 59{,}729.57 \text{ N}$$

$$\text{Strength of rivet in bearing} = dt\ \sigma_{pf}$$

$$= 19.5 \times 10 \times 300 = 58500 \text{ N}$$

$$\text{Rivet value, } R_v = 58500 \text{ N}$$

$$\text{Number of rivets required on each side of the joint} = \frac{210.75 \times 10^3}{58500}$$

$$= 3.58 \approx 5$$

Provide 18 mm, ϕ rivets, 5 in numbers as shown in Fig. Ex. 6.7.

The two sections are of different thickness. Therefore, packing will be required.

$$\text{Thickness of packing} = 14 - 10 = 4 \text{ mm}$$

As the packing in less than 6 mm, no extra rivets will be required.

The width of splice plate = width of main tie plate = 160 mm

The critical section for the splice plate is, section *x–x*. If *t* is the thickness of splice plates required, then

$$(b - 2d) \times 2t \times \sigma_{at} = \text{net tensile strength of the section}$$

or $$(160 - 2 \times 19.5) \times 2t \times 150 = 210.75 \times 10^3$$

or $$t = \frac{210.75 \times 10^3}{36{,}300}$$

$$= 5.80 \text{ mm} \approx 8 \text{ mm}$$

The sum of the thickness of splice plates should not be less than the maximum thickness of the plates to be connected. Therefore, 8 mm thick plate is sufficient. Provide two 160 mm × 8 mm thick splice plates.

Length of the splice plate

$$\text{Pitch of rivets} = 2.5\ \phi = 2.5 \times 18 = 45 \text{ mm}$$

Provide rivets at a pitch of 60 mm. For 18 mm ϕ rivets edge distance is 29 mm.

Let us provide an edge distance of 30 mm.

$$\therefore \quad \text{Length of splice plate} = 4 \times 60 + 4 \times 30 = 360 \text{ mm}$$

Therefore, provide two 360 × 160 × 8 mm splice plates.

Example 6.8 A tie member of a truss is made up of ISA 75 × 75 × 6 mm and is subjected to a pull of the 90 kN. The length of the member is more than the length of section available. Design the splice using butt weld.

Solution From IS Hand Book No. 1, the cross-sectional area of ISA 75 × 75 × 6 mm is 866 mm^2.

$$\sigma_{at,cal} = \frac{90 \times 10^3}{866} = 103.962 \text{ N/mm}^2$$

Let us provide a single-V butt weld,

Length available for welding = 75 + 75 = 150 mm

$$\text{Effective throat thickness, } t = \frac{5}{8} \times 6 = 3.75 \text{ mm}$$

Capacity of weld, $\sigma_{tf} = 3.75 \times 150 = 562.5$ N/mm

$$\text{Required length of weld} = \frac{90 \times 10^3}{562.5} = 160 \text{ mm} > 150 \text{ mm}$$

Hence, single-V but weld is inadequate.

Let us try double-V butt weld.

Capacity of weld, $\sigma_{tf} = 6 \times 150 = 900$ N/mm

$$\text{Required length of weld} = \frac{90 \times 10^3}{900} = 100 \text{ mm} < 150 \text{ mm}$$

Hence provide double-V butt weld.

Example 6.9 A tension member 10 m long is subjected to an axial tensile load of 1200 kN. Design the section with channels facing each other. The rolled section available is I.S.M.C. 300 @ 351.2 N/m only. The channel sections are weakened by one rivet hole on each flange. Check the adequacy of the section and provide the plates if required.

Solution

$$P = 1200 \text{ kN} = 1200 \times 10^3 \text{ N}$$

$$L = 10 \text{ m} = 10 \times 10^3 \text{ mm}$$

Assume steel with yield stress 250 N/mm^2.

$$\sigma_{at} = 0.60 \times 250 = 150 \text{ N/mm}^2$$

$$\text{Net area required} = \frac{1200 \times 10^3}{150} = 8000 \text{ mm}^2$$

From I.S. Handbook No. 1, note the properties of I.S.M.C. 300 @ 351.2 N/m.

$A = 4564 \text{ mm}^2$, $I_{xx} = 6362.6 \times 10^4 \text{ mm}^4$, $I_{yy} = 310.8 \times 10^4 \text{ mm}^4$, $T = 13.6$ mm

$r_{xx} = 118.1$ mm, $t = 7.6$ mm, $C_{yy} = 23.6$ mm, $g = 50$ mm, $B = 90$ mm

Assume 16 mm ϕ staggered rivets for connection.

$$\text{Gross diameter of rivets} = 16 + 1.5 = 17.5 \text{ mm}$$

Let us provide 2, ISMC 300 @ 351.2 N/m.

Gross area of the channels provided = $2 \times 4564 = 9128 \text{ mm}^2$

Since rivets are staggered, deduction for two rivet holes will be made in the calculation of net area.

$$\text{Net area provided} = \text{gross area} - \text{area of rivet holes} - \frac{1}{2} \text{ web area}$$

$$= 9128 - 2 \times 17.5 \times 13.6 - \frac{1}{2}\,[2 \times (300 - 2 \times 13.6) \times 7.6]$$

$$= 6578.72 \text{ mm}^2 < 8000 \text{ mm}^2$$

which is unsafe.

As the net area provided by two channel sections is less than the net area required additional cover plates are required.

$$\text{Net area to be provided by plate} = 8000 - 6578.72 = 1421.28 \text{ mm}^2$$

Let us provide one additional plate 280 × 6 mm.

Assuming the plate to be connected with flanges of channels by 16 mm ϕ power driven rivets.

Net area provided by plate = $280 \times 6 - 17.5 \times 6 = 1575 \text{ mm}^2 > 1421.28 \text{ mm}^2$ which is all right.

Check: The slenderness ratio of the member should be less than 180.

Channels are placed face to face at spacing 'S' so that

$$r_{yy} \geq r_{xx}$$

or $$I_{yy} \geq I_{xx}$$

$$\bar{y} = \frac{2 \times 4564 \times \left(\frac{300}{2} + 6\right) + 280 \times 6 \times \frac{6}{2}}{2 \times 4564 + 280 \times 6} = 132.22 \text{ mm}$$

$$I_{yy} = I_{xx}$$

or $$2 \times \left[310.8 \times 10^4 + 4564\left(\frac{S}{2} - 23.6\right)^2\right] + 6 \times \frac{280^3}{12}$$

$$= 2 \times 6362.6 \times 10^4 + \frac{280 \times 6^3}{12} + 280 \times 6\left(132.22 - \frac{6}{2}\right)^2$$

$$+ 2 \times 4564\,(156 - 132.22)^2$$

or $$S = 296.7 \text{ mm} \approx 300 \text{ mm}$$

Provide channels at a spacing of 300 mm and provide an additional plate 300 × 6 mm as shown in Fig. Ex. 6.9.

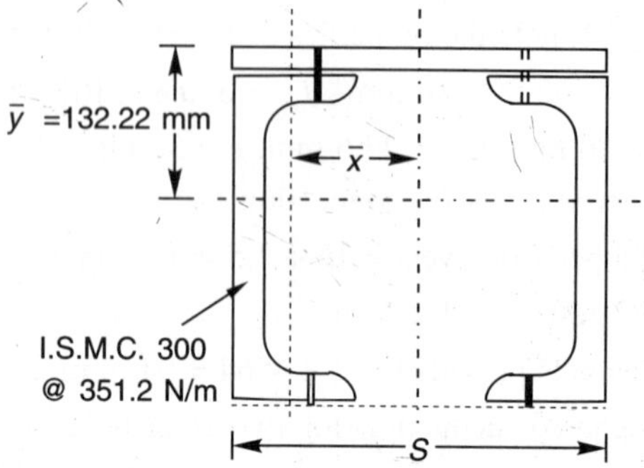

Fig. Ex. 6.9

Since the section of the plate provided is increased from 280 × 6 mm to 300 × 6 mm, the net area provided will be more than net area required and does not require rechecking.

$$I_{xx} = 2 \times 6362.6 \times 10^4 + \frac{300 \times 6^3}{12} + 300 \times 6 \left(132.22 - \frac{6}{2}\right)^2$$

$$+ 2 \times 4564(156 - 132.22)^2$$

$$= 16{,}247.5 \times 10^4 \text{ mm}^4$$

$$A = 300 \times 6 + 2 \times 4564 = 10928 \text{ mm}^2$$

$$r_{xx} = \sqrt{I_{xx}/A} = \sqrt{16247.5 \times 10^4/10928} = 121.9 \text{ mm}$$

$$\lambda = \frac{l}{r_{xx}} = \frac{10 \times 10^3}{121.9} = 82 < 180$$

which is as it should be.

Tie plate:

Effective depth of tie plate = (300 – 2 × 23.6) = 252.8 mm

Overall depth = 252.8 + 2 × 29 = 310.8 mm ≈ 312 mm

(assuming 20 mm ϕ rivets, edge distance = 29 mm)

Length of tie plate = spacing of channels = 300 mm

Thickness of tie plate $= \frac{1}{50}$ (300 – 2 × 50) = 4 mm $\not<$ 6 mm

Provided 312 × 300 × 6 mm tie plates.

Example 6.10 Design sag rods for consecutive purlins near the supported end of a roof truss system as shown in Fig. Ex. 6.10. The purlins are supported at the third points by sag rods. Also design the tie rod between ridge purlins. The data available is

Spacing of truss c/c = 7 m

Spacing of the purlins c/c = 2.0 m
Intensity of the wind pressure = 1500 N/m^2
Slope of the truss = 30°
Self weight of roofing = 200 N/m^2

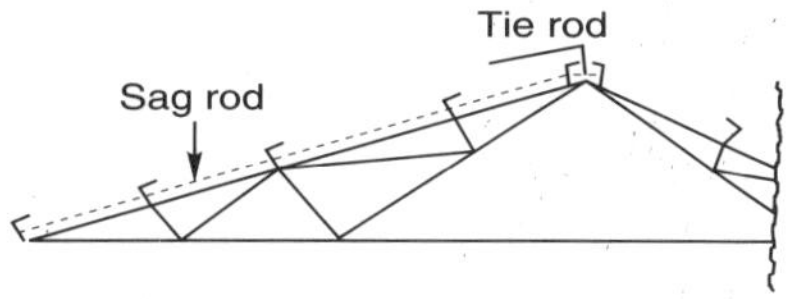

Fig. Ex. 6.10

Solution

Gravity loads:

Dead load from roofing = 200 × 2 = 400 N/m

Let self weight of purlin = 100 N/m

Let there be no access to the roof.

Live load = nil

Total gravity load = 400 + 100 = 500 N/m

Loads normal to roof:

Wind load = 1500 × 2 = 3000 N/m

Component of gravity loads parallel to roof = 500 sin 30° = 500 × 0.5 = 250 N/m

As the sag rods are placed at third points on purlins,

$$\text{Pull on sag rods} = 250 \times \frac{7}{3} = 583.33 \text{ N}$$

$$\text{Allowable axial tensile stress } \sigma_{at} = 0.6 \times 250 = 150 \text{ N/mm}^2$$

$$A_{\text{net}} = \frac{583.33}{150} = 3.888 \text{ mm}^2$$

Assuming the gross area to be about 30% more than the net area.

Tentative area required, $A_{\text{req}} = 1.3 \times 3.888 = 5.055 \text{ mm}^2$

The diameter of the sag rod can be obtained as follows and should not be less than 16 mm.

$$A = \frac{\pi}{4} d^2$$

or

$$d = \sqrt{4A/\pi} = \sqrt{4 \times 5.055/\pi} = 2.536 \text{ mm} \nless 16 \text{ mm}$$

Note As an alternative, the net area of the sag rod may be referred from Appendix VII and the bar size reported.

Provide 16 mm diameter sag rods between purlins.

Tie rods between ridge purlins:

$$\text{Pull in the tie rods} = 4 \times 583.33 \sec 30° = 2694.28 \text{ N}$$

$$A_{\text{net}} = \frac{2694.28}{150} = 17.96 \text{ mm}^2$$

$$A_{\text{gross}} = 1.3 \times 17.96 = 23.35 \text{ mm}^2$$

(increasing net are by 30%)

$$\text{Diameter of the rod} = \sqrt{4 \times 23.35/\pi} = 5.45 \text{ mm} \nless 16 \text{ mm}$$

Provide 16 mm diameter tie rods between ridge purlins.

Exercises

Note Assume the yield stress of steel to be 250 N/mm^2 if not specified in the problem.

6.1 The main tie of a roof truss consists of I.S.A. 150 × 115 × 8 mm and is connected to a gusset plate by 18 mm diameter rivets. Find out the maximum load it can carry.

6.2 A tie member used as a diagonal in a roof truss consists of two angles 75 × 50 × 8 mm (f_y = 280 N/mm^2) placed back-to-back on the same side of the gusset plate. The rivets are 18 mm ϕ provided in one row and the angles are tack riveted. Determine the tensile strength of the member.

6.3 Determine the tensile strength of the member in the above problem if the angle sections are placed on the opposite sides of the gusset plate.

6.4 Determine the net width of the plate shown in Fig. Prob. 6.4. All holes are 20 mm ϕ.

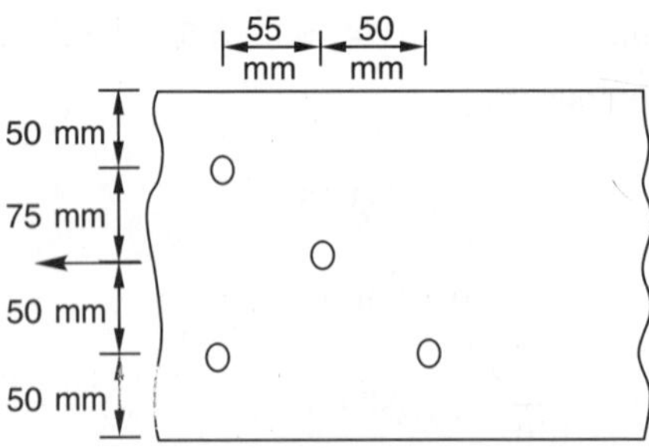

Fig. Prob. 6.4

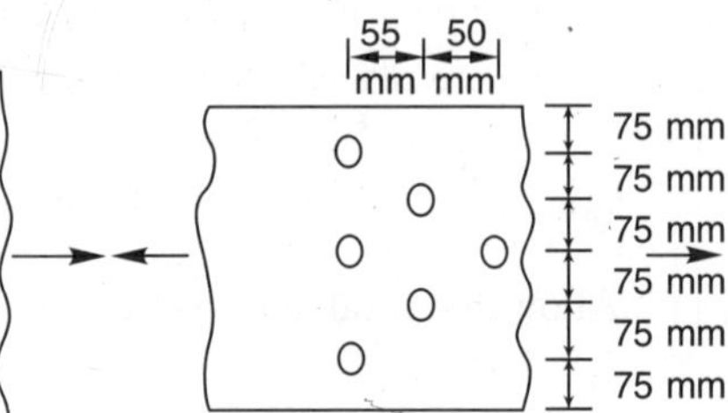

Fig. Prob. 6.5

6.5 Determine the allowable tensile load that can be applied to the plate (f_y = 240 N/mm^2) shown in Fig. Prob. 6.5. Rivets are 20 mm ϕ. The thickness of plate is 25 mm.

6.6 A single angle tension member 200 × 100 × 15 mm is connected by two rows of rivets through its long leg and one row of rivets through the short leg (Fig. Prob. 6.6). Rivets are 20 mm ϕ, power driven. Determine the minimum staggered pitch so that the reduction in area will be only two holes. Also determine the maximum load the member can carry if the staggered pitch is 60 mm.

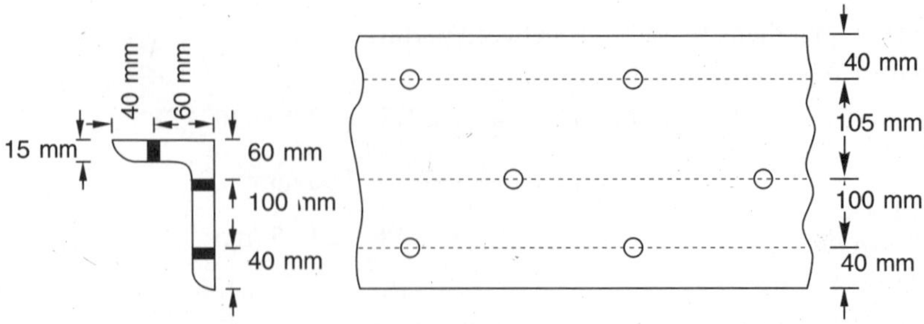

Fig. Prob. 6.6

6.7 A pair of unequal angles made of steel with $f_y = 250$ MPa are chosen to resist a tensile load of 650 kN. The connection requires the arrangement shown in Fig. Prob. 6.7, with two gauges in its long leg and one in the short leg for 20 mm rivets. Select the most economical size of the section.

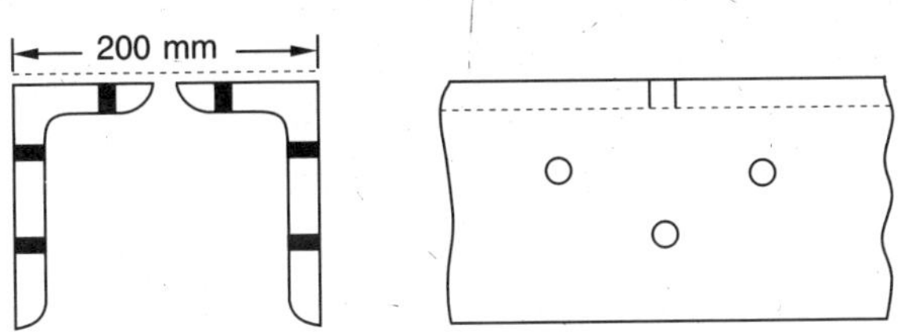

Fig. Prob. 6.7

6.8 Design a tension member to carry a pull of 600 kN. The member is 3.2 m between c/c of intersections.

6.9 Design a tension member 3.4 m between c/c of intersections and carrying a pull of 145 kN. The member is subjected to reversal of stresses.

6.10 Design a tension member to carry a load of 280 kN. The two angles placed back-to-back with long legs outstanding are desirable. The length of the member is 2.9 m.

6.11 A Tee-section 3.8 m long is to be provided to support a tensile load of 220 kN. Design the section.

6.12 A tension member is to consist of four equal angles placed as shown in Fig. Prob. 6.12. The section is to support a load of 1000 kN. The member is 10 m long. The maximum side of the built-up section is restricted to 450 mm. Design the section. Provide additional plates, if required. Each angle may be assumed to have one rivet hole of 20 mm ϕ in the section.

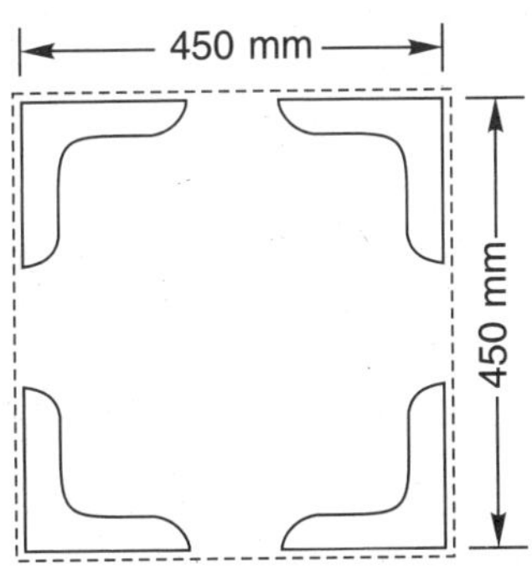

Fig. Prob. 6.12

6.13 Design a suitable splice to connect two mild steel flats 240 I.S.F. 10 mm and 200 I.S.F. 18 mm. The member is subjected to a load equal to the full strength of the member.

6.14 Design a lug angle to be provided with the tie member subjected to a load of 270 kN. The length of the member is 3.65 m.

Beams

7.1 INTRODUCTION

A structural member subjected to transverse loads (loads perpendicular to its longitudinal axis) is called a beam. When provided in buildings they are called *joists*. A large beam supporting a number of joists is called a *girder*. Beam which carry the roof loads in trusses are called *purlins* and those which support the loads from the masonry over the openings are called *lintels*. A roof beam usually supported by purlins is called a *rafter*. A horizontal beam spanning the wall columns of industrial buildings used to support wall coverings is called a *girt*.

The discussion in this chapter will be limited to beams with symmetrical cross-sections. Furthermore, it is assumed that the beam is acted on only by transverse loading and that all the loads and sections lie in the plane of symmetry. It follows that such a beam will be primarily subjected to bending accompanied by shear in the loading plane, with no external torsion and axial force. However, the problem of torsion is not completely avoided for a beam of structural shape, even if the beam shape is symmetrical and loads are in the plane of symmetry. It is because of instability caused by compressive stresses. This instability is called *lateral buckling* when it is of a general nature involving an entire span, or it may be *local buckling*, involving only the local components of a beam. Local buckling is generally a function of the 'width-thickness ratio' of the beam components and can be controlled by following the specifications.

Beams are the most critical members in any structure. Their design should therefore not only be economical but also safe. The main considerations in the design of beams are:

1. They should be proportioned for strength in bending keeping in view the lateral and local stability of the compression flange and the capacity of the selected shape to develop the necessary strength in shear and local bearing.
2. They should be proportioned for stiffness, keeping in mind their deflections and deformations under service conditions.
3. They should be proportioned for economy, paying attention to the size and grade of steel to yield the most economical design.

It is a difficult task for a designer to select a beam size for a given span and load that will satisfy all the above three conditions.

In its simplest form, beam design consists merely of the provision of adequate bending and shear resistance. Though adequate stiffness is also a requirement, but in the majority of cases it is one which is satisfied automatically by members meeting the strength requirements. For optimum bending resistance, as much of the beam's material as possible should be displaced as far as practicable from the neutral axis. The web area must be sufficient to resist shear. Maximum moment and maximum shear usually occur at different cross-sections. However, even when they are found at the same section, their interaction may normally be neglected. Though simple in design, the lateral buckling of beam as a whole, or of its compression flange or its web pose complications. Another problem is of proper depth—an increase in depth may be desirable for moment resistance, it may at the same impair resistance to lateral or web buckling. Before going into depth of design of beams, the suitability of type of rolled steel sections for beams is discussed in the following section.

7.2 TYPES OF SECTIONS

A number of rolled steel sections are used in construction. However, angles and T-Sections are inherently weak in bending while channels can only be used for light loads. A rolled steel I-Section, Fig. 7.1(a), is generally preferred as a beam. Usually the most efficient and economical shapes are I shapes, either rolled or built up. An I-section with cover plates, Fig. 7.1(b) is provided when a large modulus of section (which cannot be furnished by a single rolled I-Section) is required. Generally, I.S.L.B. or I.S.M.B. sections are provided in such cases. When heavy loads are encountered, as may be in the case of a bridge, the I-Section is built up with plates for welded connections (Fig. 7.1(c)) or with plates and angles for riveted connections as shown in Figs 7.1(d) and (e). When the beams are subjected to lateral loads at the compression flange level, an I-Section, preferably ISWB section, along with a channel section (Fig. 7.1(f)) is preferred.

The use of rolled steel channel and angle sections as beams is done with caution except in situations where they have been used and have performed satisfactorily. This is because the load is not likely to be in the plane, which eliminates torsional eccentricities. Also, it is difficult to predict the lateral buckling characteristics of these sections. Such sections are generally used as purlins in roof trusses. However, for higher loads I-Sections may be preferred. Double angles (Fig. 7.1(g)), T-Sections or I.S.J.B. sections (Fig. 7.1(h)) are generally used as lintels. If the head room is restricted for clear height requirements two channels or I-Sections of smaller depths with pipe separators can be provided, as shown in Fig. 7.1(i).

In the case of beams with large spans and light loads, *castellated beams* become the choice. The I-Section is split and rejoined with welding, as shown in Fig. 7.1(j). Such beams are more resistant to bending, as the modulus of the section is increased.

The usual method of selecting a beam section is based upon section modulus. However, the criterion of economy is weight rather than the modulus of section. It is therefore, desirable to choose a light beam furnishing the required modulus of section. Often, it may not be possible to choose the lightest section due to head room limitations, or if adjacent beams are to be matched. Sometimes deflection and occasionally shear may also be the necessary criterion dictating the choice.

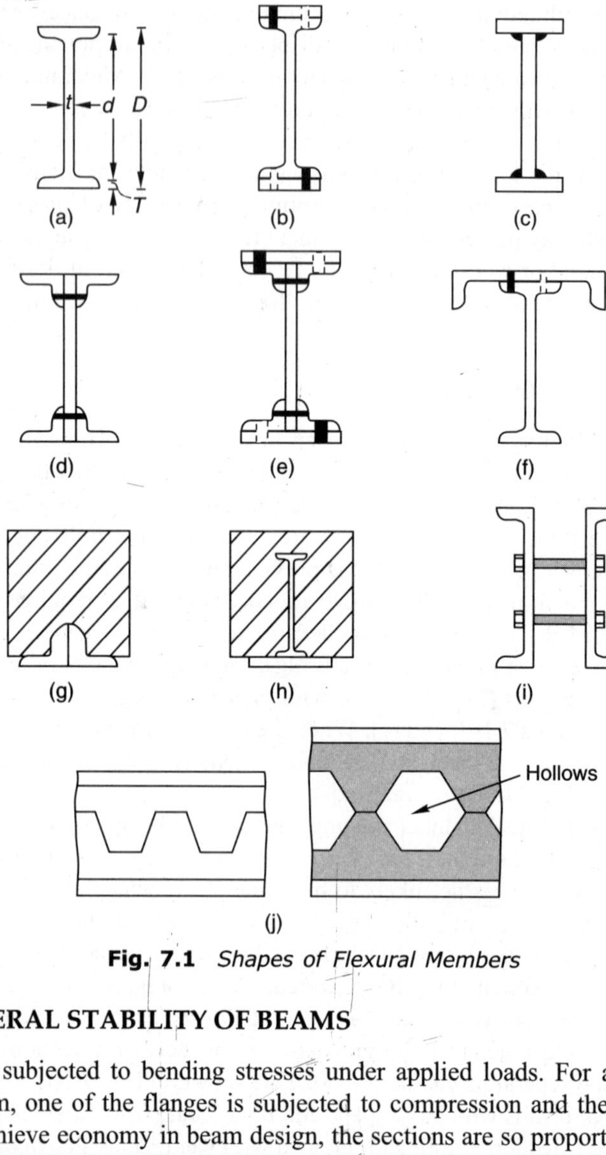

Fig. 7.1 *Shapes of Flexural Members*

7.3 LATERAL STABILITY OF BEAMS

A beam is subjected to bending stresses under applied loads. For a simply supported beam, one of the flanges is subjected to compression and the other to tension. To achieve economy in beam design, the sections are so proportioned that the moment of inertia about the principal axis normal to the web (I_{xx}) is considerably larger than the moment of inertia about the principal axis parallel to the web (I_{yy}). Such a section is relatively weak in bending resistance in the plane normal to the web and, therefore, the compression flange of the beam tends to buckle (Fig. 7.2(a)) in the only direction in which it is free to move, that is, horizontally. This buckling increases with the increase in ratio of moment of inertia about the *xx*-axis to that about the *yy*-axis. On the other hand, the bottom flange, being in tension, tends to remain straight. Since the two flanges and webs actually represent one rigid unit, buckling, if it occurs, can take place only in the manner shown in Fig. 7.2(a). That is, the top flange, tending to buckle, bends farther than the bottom flange, which tends to remain straight, and in consequence the entire cross-section rotates. Thus,

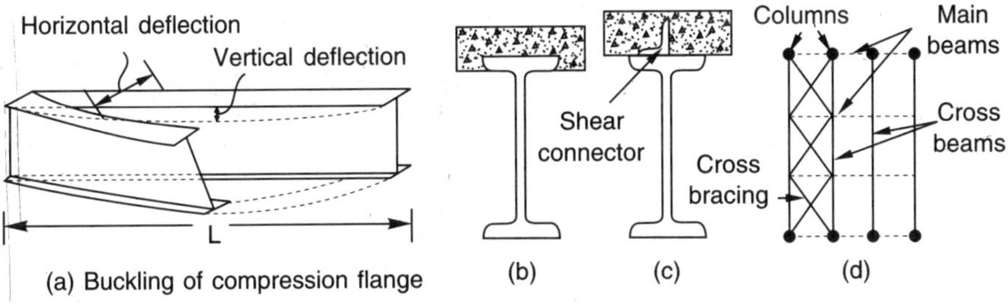

Fig. 7.2 *Beams with Lateral Support*

the lateral buckling of the compression flange is also accompanied by twisting and puts a restriction on the allowable bending stress in compression. The nature of the buckling process for an ideal beam is same as that for an ideal column. That is, the beam is stable with no tendency of lateral-torsional buckling until the bending moment reaches a certain critical magnitude, whereupon the beam becomes unstable and can undergo rotations and lateral deflections of any magnitude, leading to collapse. This necessitates holding of the flange in line, preferably by attached construction. Thus, if the compression flange of a beam can be restrained against lateral buckling by any means, the allowable bending stresses can be utilized to their full value. A beam in which the compression flange is restrained laterally is called a *laterally restrained beam*. Designers have tried different methods to provide effective restraints, some of which are as follows:

(i) By embedding the compression flange of the beam inside the slab concrete (Fig. 7.2(b)). When the compression flange of a beam is fully embedded in a concrete floor, the slab almost certainly furnishes full lateral support. In such a case the question of lateral buckling need not be considered. When the deck or floor-slab merely rests upon the beam, there is the question of whether the potential frictional resistance between the two is capable of developing the force necessary for full lateral support. This must remain largely a matter of judgement as to the reliability of the friction.

(ii) Shear connectors are welded to the compression flange and are embedded in the concrete slab (Fig. 7.2(c)).

(iii) Most non-continuous supports provide adequate lateral support at their points of attachment to the beam if they, in turn, are connected at their far ends to something capable of resisting the transmitted bracing force, that is, to another, stiffer or less heavily loaded beam, a column, a wall, or the ground. When the cross beams join the main beams and are connected to the compression flange, full lateral support should be assumed at the connections. It may happen that the cross beams are connected to the tension flange of the main beam. In such cases, cross bracing as shown in Fig. 7.2(d) is done to achieve the lateral restraint. In practice, however, bracing is often proportioned solely by judgement. Since bracing requirements are usually quite modest, this is generally satisfactory.

There are many other methods also to achieve lateral support but the degree of restraint provided by them is questionable and, therefore, if a doubt arises, it is safer to design the beam as a laterally unsupported beam.

Notes

(i) In case of beams which support concrete slabs over compression flanges with fixed live loads on the slab, full lateral support may be assumed due to the friction between the compression flange and slab. On the contrary, if the loads are moving and vibrations exist, then the friction is reduced and therefore, full lateral support should not be assumed.

(ii) It is difficult to confirm whether a particular beam is laterally restrained or not. It is for the designer to judge if a particular beam has satisfactory lateral support. If a beam is continuous over the support, the compression flange will be the lower flange to some distance on both sides of it and the lower flange will have to be restrained at the supports.

7.4 BENDING STRESS

The important aspects which need consideration in a beam design are moments, shears, deflection, crippling, buckling and lateral support. The choice of a proper rolled section to be used for a beam is often based on its ability to carry the maximum bending moment without exceeding the allowable unit fibre stress. The bending stresses in members subjected to flexure within the elastic range can be determined by using the flexural formula,

$$\frac{M}{I} = \frac{\sigma_{bc,t}}{y}$$

or

$$M = \sigma_{bc,t} Z$$

where M = bending moment

I = moment of inertia

$\sigma_{bc,t}$ = bending stress in compression/tension

y = distance of extreme fibre from N.A.

Z = section modulus

Due to the bending moments in a beam bending stresses are set up in it. For a simply supported beam, the top flange will have a compressive bending stress and the bottom flange will have a tensile bending stress. The maximum bending stress in compression or tension calculated on the section should neither exceed $0.66\ f_y$ (where f_y is the yield stress in MPa) nor the values specified by IS: 800–1984 code, as discussed below.

Laterally Supported Beams

The permissible bending stress in tension, σ_{bt} and maximum permissible bending stress in compression σ_{bc} should not exceed $0.66 f_y$.

Laterally Unsupported Beams

Although the basic permissible stress of $0.66\ f_y$ is based upon the assumption of sufficient lateral support for the compression flange to develop yielding, this support cannot be provided in all cases. When the support is inadequate, lateral buckling occurs before yielding and the maximum permissible bending compressive stress must be reduced accordingly.

Beams and Channels with Equal Flanges The maximum permissible bending compressive stresses for laterally unsupported beams with equal flanges are stated in Tables 7.1A to 7.1F of the text.

Table 7.1(a) *Maximum Permissible Compressive Bending Stresses, σ_{bc} (MPa), in Equal Flange I-Beams or Channels*

$$\text{with } f_y = 250 \text{ MPa}, \ \frac{T}{t} > 2.0 \text{ or } \frac{d_1}{t} > 85$$

$D/T \rightarrow$ $l/r_y \downarrow$	8	10	12	14	16	18	20	25	30	35	40	50	60	80	100
40	160	160	159	159	158	158	158	158	158	157	157	157	157	157	157
45	159	158	157	157	156	156	156	155	155	155	155	155	154	154	154
50	158	157	156	155	154	154	153	153	152	152	152	151	151	151	151
55	157	155	154	153	152	151	150	149	149	148	148	148	148	147	147
60	156	153	152	150	149	148	148	146	145	145	144	144	144	143	143
65	154	152	150	148	147	145	144	143	142	141	140	140	139	139	139
70	153	150	148	146	144	142	141	139	138	137	136	135	135	135	134
75	152	148	145	143	141	139	138	136	134	133	132	131	130	130	129
80	150	147	143	141	138	136	135	132	130	128	128	126	126	125	125
85	149	145	141	138	136	133	132	128	126	124	123	122	121	120	120
90	147	143	139	136	133	130	128	124	122	120	119	117	116	115	115
95	146	141	137	133	130	128	125	121	118	116	114	112	111	110	110
100	145	140	135	131	128	125	122	117	114	112	110	108	107	105	105
110	142	136	131	127	123	119	116	111	107	104	102	99	98	96	95
120	139	133	127	122	118	114	111	104	100	97	94	91	90	88	87
130	137	130	124	118	113	109	106	99	94	90	88	84	82	80	79
140	134	127	120	114	109	105	101	93	88	84	81	78	75	73	72
150	132	124	117	110	105	100	96	88	83	79	76	72	69	67	65
160	129	121	113	107	101	96	92	84	78	74	71	66	64	61	60
170	127	118	110	104	98	93	88	80	74	69	66	62	59	56	55
180	124	115	107	100	94	89	85	76	70	65	62	58	55	52	50
190	122	113	104	97	91	86	82	73	66	62	58	54	51	48	46
200	120	110	102	94	88	83	78	70	63	59	55	50	48	44	43
210	118	108	99	92	86	80	76	67	60	56	52	47	44	41	40
220	116	105	97	89	83	78	73	64	58	53	49	45	42	38	37
230	113	103	94	87	80	75	70	62	55	51	47	42	39	36	34
240	111	101	92	84	78	73	68	59	53	48	45	40	37	34	32
250	109	99	90	82	76	70	66	57	51	46	43	38	35	32	30
260	107	97	88	80	74	68	64	55	49	44	41	36	33	30	28
270	106	95	86	78	72	66	62	53	47	43	39	35	32	28	26
280	104	93	84	76	70	65	60	51	45	41	38	33	30	27	25
290	102	91	82	74	68	63	58	50	44	39	36	32	29	25	24
300	100	89	80	72	66	61	57	48	42	38	35	30	27	24	22

Table 7.1(b) *Maximum Permissible Compressive Bending Stresses, σ_{bc} (MPa), in Equal Flange I-Beams or Channels*

with $f_y = 250$ MPa, $\dfrac{T}{t} \le 2.0$ and $\dfrac{d_1}{t} \le 85$

$D/T \rightarrow$ $l/r_y \downarrow$	8	10	12	14	16	18	20	25	30	35	40	50	60	80	100
40	161	161	160	160	160	160	160	159	159	159	159	159	159	159	159
45	161	160	159	159	158	158	158	157	157	157	157	157	157	157	157
50	160	158	158	157	156	156	156	155	155	155	154	154	154	154	154
55	159	157	156	155	154	154	153	153	152	152	152	151	151	151	151
60	158	156	154	153	152	152	151	150	149	149	149	148	148	148	148
65	156	154	153	151	150	149	148	147	146	146	145	145	144	144	144
70	155	153	151	149	149	147	146	144	143	142	142	141	141	140	140
75	154	152	149	147	146	144	143	141	140	139	138	137	137	136	136
80	153	150	148	145	143	142	140	138	136	135	134	133	132	132	132
85	152	149	146	143	141	139	138	135	133	131	130	129	128	127	127
90	151	147	144	141	139	137	135	131	129	127	126	125	124	123	123
95	150	146	142	139	137	134	132	128	126	124	122	121	120	119	118
100	149	145	141	137	134	132	129	125	122	120	118	116	115	114	113
110	147	142	137	133	130	127	124	119	115	113	111	108	107	105	105
120	144	139	134	129	126	122	119	113	109	106	104	101	99	97	96
130	142	136	131	126	121	118	114	108	103	99	97	94	91	89	88
140	140	133	128	122	118	113	110	103	97	94	91	87	85	82	81
150	138	131	124	119	114	109	105	98	92	88	85	81	78	76	74
160	136	128	121	115	110	106	101	93	87	83	80	75	73	70	68
170	134	126	119	112	107	102	98	89	83	79	75	70	68	64	63
180	131	123	116	109	104	99	94	85	79	74	71	66	63	60	58
190	129	121	113	106	101	95	91	82	75	71	67	62	59	55	54
200	127	118	111	104	98	92	88	79	72	67	63	58	55	51	50
210	125	116	108	101	95	90	85	76	69	64	60	55	52	48	46
220	123	114	106	99	92	87	82	73	66	61	57	52	49	45	43
230	122	112	103	96	90	84	80	70	63	58	55	49	46	42	40
240	120	110	101	94	87	82	77	68	61	56	52	47	43	40	38
250	118	108	99	92	85	80	75	65	59	54	50	44	41	37	35
260	116	106	97	89	83	77	73	63	57	52	48	42	39	35	33
270	114	104	95	87	81	75	71	61	55	50	46	41	37	33	31
280	113	102	93	85	79	73	69	59	53	48	44	39	35	32	30
290	111	100	91	84	77	72	67	58	51	46	42	37	34	30	28
300	109	98	89	82	75	70	65	56	49	45	41	36	32	29	27

Table 7.1(c) *Maximum Permissible Compressive Bending Stresses, σ_{bc} (MPa), in Equal Flange I-Beams or Channels*

with $f_y = 340$ MPa, $\dfrac{T}{t} > 2.0$ or $\dfrac{d_1}{t} > 75$

D/T→ *l/r_y* ↓	*8*	*10*	*12*	*14*	*16*	*18*	*20*	*25*	*30*	*35*	*40*	*50*	*60*	*80*	*100*
40	215	214	212	212	211	211	210	210	209	209	209	209	209	209	209
45	213	211	209	208	207	206	206	205	204	204	204	203	203	203	203
50	210	208	205	204	203	202	201	199	199	198	198	197	197	197	197
55	208	204	202	200	198	197	196	194	193	192	191	191	190	190	190
60	205	201	198	195	193	191	190	188	186	185	185	184	183	183	183
65	203	198	194	191	188	186	185	181	180	178	177	176	176	175	175
70	200	195	190	186	183	181	179	175	173	171	170	169	168	167	167
75	198	192	186	182	179	176	173	169	166	164	163	161	160	159	159
80	195	188	183	178	174	170	168	163	159	157	156	154	153	151	151
85	193	185	179	174	169	165	162	157	153	150	149	146	145	144	143
90	190	182	175	169	165	161	157	151	147	144	142	139	138	136	136
95	188	179	172	165	160	156	152	145	141	137	135	132	131	129	128
100	185	176	168	162	156	151	147	140	135	131	129	126	124	122	121
110	180	170	162	154	148	143	138	130	124	120	117	114	112	109	108
120	176	165	155	147	141	135	130	121	115	110	107	103	101	98	97
130	171	159	149	141	134	128	122	113	106	101	98	93	91	88	87
140	167	154	144	135	127	121	116	105	98	93	90	85	82	79	78
150	163	150	139	129	122	115	110	99	92	87	83	78	75	72	70
160	158	145	134	124	116	110	104	93	86	80	77	72	68	65	63
170	155	141	129	120	111	105	99	88	80	75	71	66	63	59	58
180	151	137	125	115	107	100	94	83	76	70	66	61	58	54	53
190	147	133	121	111	103	96	90	79	72	66	62	57	54	50	48
200	144	129	117	107	99	92	86	75	68	62	58	53	50	46	44
210	140	125	113	103	95	88	83	72	64	59	55	50	46	43	41
220	137	122	110	100	92	85	79	69	61	56	52	47	43	40	38
230	134	119	107	97	89	82	76	66	58	53	49	44	41	37	35
240	131	116	104	94	86	79	74	63	56	51	47	42	38	35	33
250	128	113	101	91	83	76	71	61	53	48	44	39	36	32	31
260	125	110	98	88	80	74	68	58	51	46	42	37	34	31	29
270	122	107	95	86	78	72	66	56	49	44	41	36	32	29	27
280	120	105	93	83	76	69	64	54	47	42	39	34	31	27	25
290	117	102	90	81	73	67	62	52	46	41	37	32	29	26	24
300	115	100	88	79	71	65	60	51	44	39	36	31	28	25	23

Table 7.1(d) *Maximum Permissible Bending Compressive Stresses, σ_{bc} (MPa), in Equal Flange I-Beams or Channels*

with $f_y = 340$ MPa, $\frac{T}{t} \le 2.0$ and $\frac{d_1}{t} \le 75$

$D/T \rightarrow$ $l/r_y \downarrow$	8	10	12	14	16	18	20	25	30	35	40	50	60	80	100
40	217	216	215	214	214	213	213	213	212	212	212	212	212	212	212
45	215	214	212	211	211	210	210	209	208	208	208	208	208	207	207
50	213	211	209	208	207	206	206	205	204	203	203	203	203	202	202
55	211	209	206	205	203	202	201	200	199	198	198	197	197	197	197
60	209	206	203	201	199	198	197	195	193	193	192	191	191	191	190
65	207	203	200	197	195	193	192	189	188	187	186	185	184	184	184
70	205	201	197	194	191	189	187	184	182	181	180	178	178	177	177
75	203	198	194	190	187	184	182	178	176	174	173	172	171	170	169
80	201	195	190	186	183	180	177	173	170	168	167	165	164	163	162
85	199	193	187	183	179	175	173	168	164	162	160	158	157	156	155
90	197	190	184	179	175	171	167	162	158	156	154	151	150	148	148
95	195	187	181	175	171	167	163	157	153	150	148	145	143	142	141
100	193	185	178	172	167	163	159	152	147	144	142	138	137	135	134
110	188	180	172	165	159	155	150	142	137	133	130	126	124	122	121
120	184	175	166	159	152	147	142	133	127	123	120	116	113	110	109
130	180	170	161	153	146	140	135	125	119	114	110	106	103	100	99
140	177	165	156	147	140	134	128	118	111	106	102	97	94	91	89
150	173	161	151	142	134	128	122	112	104	99	95	89	86	83	81
160	169	157	146	137	129	122	117	106	98	92	88	82	79	75	74
170	166	153	142	132	124	117	111	100	92	86	82	76	73	69	67
180	162	149	137	128	120	113	107	95	87	81	77	71	67	63	61
190	159	145	133	124	115	108	102	91	82	76	72	66	63	59	56
200	155	141	130	120	111	104	98	86	78	72	68	62	58	54	52
210	152	138	126	116	108	100	94	83	74	69	64	58	54	50	48
220	149	135	123	113	104	97	91	79	71	65	61	55	51	47	45
230	146	132	119	109	101	94	88	76	68	62	58	52	48	44	42
240	143	128	116	106	98	91	85	73	65	59	55	49	45	41	39
250	141	126	113	103	95	88	82	70	62	57	52	46	43	38	36
260	138	123	110	100	92	85	79	68	60	54	50	44	40	36	34
270	135	120	108	98	89	82	77	65	58	52	48	42	38	35	32
280	133	117	105	95	87	80	74	63	56	50	46	40	36	32	30
290	130	115	103	93	84	78	72	61	54	48	44	38	35	31	29
300	128	112	100	90	82	76	70	59	52	46	42	37	33	29	27

Table 7.1(e) *Maximum Permissible Compressive Bending Stresses, σ_{bc} (MPa), in Equal Flange I-Beams or Channels*

$$\text{with } f_y = 400 \text{ MPa}, \ \frac{T}{t} > 2.0 \text{ or } \frac{d_1}{t} > 67$$

$D/T \rightarrow$ $l/r_y \downarrow$	8	10	12	14	16	18	20	25	30	35	40	50	60	80	100
40	250	248	247	245	245	244	243	243	242	242	242	241	241	241	241
45	247	244	242	240	239	238	237	236	235	235	234	234	234	233	233
50	244	240	237	234	233	231	230	228	227	227	226	226	225	225	225
55	240	235	232	229	226	224	223	221	219	218	217	216	216	216	215
60	236	231	226	223	220	217	216	212	210	209	208	207	206	206	205
65	233	226	221	217	213	210	208	204	202	200	199	197	197	196	195
70	229	222	216	211	207	203	201	196	193	191	189	188	187	186	185
75	226	217	211	205	200	196	193	188	184	182	180	178	177	175	175
80	222	213	206	199	194	190	186	180	176	173	171	168	167	166	165
85	219	209	201	194	188	183	179	172	167	164	162	159	158	156	155
90	216	205	196	188	182	177	173	165	160	156	154	151	149	147	146
95	212	201	191	183	177	171	166	158	152	149	146	142	140	138	137
100	209	197	187	178	171	165	160	151	145	141	138	135	133	130	129
110	203	189	178	169	161	155	149	139	133	128	125	121	118	115	114
120	196	182	170	160	152	145	140	129	121	116	113	108	106	103	101
130	191	176	163	153	144	137	131	119	112	106	103	98	95	92	90
140	185	169	156	146	137	129	123	111	103	98	94	88	85	82	80
150	179	163	150	139	130	122	116	104	96	90	86	81	77	74	72
160	174	158	144	133	124	116	109	97	89	83	79	74	70	67	65
170	169	152	139	127	118	110	104	92	83	78	73	68	64	61	59
180	165	147	134	122	113	105	97	86	78	72	68	63	59	55	54
190	160	143	129	117	108	100	94	82	74	68	64	58	55	51	49
200	156	138	124	113	104	96	90	78	70	64	60	54	51	47	45
210	152	134	120	109	100	92	86	74	66	60	56	51	47	43	41
220	148	130	116	105	96	88	82	71	63	57	53	47	44	40	38
230	144	126	112	101	92	85	79	67	60	54	50	45	41	37	36
240	141	123	109	98	89	82	76	65	57	52	47	42	39	35	33
250	137	119	106	95	86	79	73	62	54	49	45	40	37	33	31
260	134	116	103	92	83	76	70	60	52	47	43	38	35	31	29
270	131	113	100	89	81	74	68	57	50	45	41	36	33	29	27
280	128	110	97	86	78	71	66	55	48	43	39	34	31	27	26
290	125	107	94	84	76	69	64	53	46	41	38	33	30	26	24
300	122	105	92	82	74	67	62	52	45	40	36	31	28	25	23

Table 7.1(f) *Maximum Permissible Bending Compressive Stresses, σ_{bc} (MPa), in Equal Flange I-Beams or Channels*

with $f_y = 400$ MPa, $\frac{T}{t} \le 2.0$ or $\frac{d_1}{t} \le 67$

D/T→ *l/r_y*→	8	10	12	14	16	18	20	25	30	35	40	50	60	80	100
40	253	252	250	249	249	248	248	247	247	246	246	246	246	246	246
45	251	248	246	245	244	243	243	242	241	241	240	240	240	240	239
50	248	245	242	240	239	238	237	235	234	234	233	233	233	232	232
55	245	241	238	236	234	232	231	229	227	227	226	225	225	225	224
60	242	237	234	231	228	226	225	222	220	219	218	217	217	216	216
65	239	234	229	225	222	220	218	215	212	211	210	209	208	207	207
70	236	230	225	220	217	214	212	207	205	203	202	200	199	198	198
75	233	226	220	215	211	208	205	200	197	195	193	191	190	189	188
80	230	223	216	210	206	202	199	193	189	186	185	182	181	180	179
85	227	219	212	205	200	196	192	186	181	178	176	174	172	171	170
90	225	215	207	201	195	190	186	179	174	171	168	165	164	162	161
95	222	212	203	196	190	185	180	172	167	163	161	157	155	153	152
100	219	208	199	191	185	179	175	166	160	156	153	150	148	145	144
110	213	202	191	183	176	169	164	154	148	143	140	135	133	130	129
120	208	195	184	175	167	160	154	144	136	131	127	123	120	117	115
130	203	189	177	167	159	152	146	134	126	121	117	111	108	105	103
140	198	183	171	160	152	144	138	126	117	111	107	102	98	95	93
150	193	178	165	154	145	137	131	118	109	103	99	93	89	86	84
160	188	172	159	148	139	131	124	111	102	96	92	85	82	78	76
170	183	167	154	142	133	125	118	105	96	90	85	79	75	71	69
180	179	162	149	137	127	119	112	99	90	84	79	73	69	65	63
190	175	158	144	132	122	114	108	94	85	79	74	68	64	60	58
200	171	153	139	128	118	110	103	90	81	75	70	63	60	55	53
210	167	149	135	123	114	105	99	86	77	70	66	59	55	51	49
220	163	145	131	119	110	102	95	82	73	67	62	56	52	48	45
230	159	141	127	115	106	98	91	79	70	64	59	53	49	44	42
240	156	138	123	112	102	94	88	75	67	61	56	50	46	42	39
250	152	134	120	108	99	91	85	72	64	58	53	47	43	39	37
260	149	131	117	105	96	88	82	70	61	55	51	45	41	37	34
270	146	128	114	102	93	85	79	67	59	53	49	43	39	35	32
280	143	125	111	99	90	83	77	65	57	51	47	41	37	33	30
290	140	122	108	97	88	80	74	63	55	49	45	39	35	31	29
300	137	119	105	94	85	78	72	61	53	47	43	37	33	29	27

where D = overall depth of beam
d_1 = depth of web
l = effective length of compression flange
r_y = radius of gyration of the section about its axis of minimum strength
T = mean thickness of compression flange
t = thickness of web

Notes

(i) In the case of built-up girders, D should be taken as the over all depth of the girder and T should be taken as $k_1 \times$ mean thickness of the compression flange at the point of maximum bending moment where k_1 is a coefficient depending upon ψ, which is a ratio of the total area of both the flanges at the point of the least bending moment to the area at the point of maximum bending moment.
(ii) The effective length of the compression flange depends upon the end conditions of the beam as elucidated in Table 7.2.

Table 7.2 *Effective Length of Compression Flanges*

Type of Member	*End Conditions of Beam*	*Effective Length, l*
Simply supported beams	*Restrained against torsion and*	
	(i) Unrestrained against lateral bending	L
	(ii) Partially restrained against lateral bending	$0.85L$
	(iii) Fully restrained against lateral bending	$0.7L$
	(iv) Effective lateral bracing at intervals along the length	Distance between intersections of bracings with member
Cantilever beams	*Built in at the support and*	
	(v) Free at the end	$0.85L$
	(vi) Restrained against torsion at the end by continuous construction	$0.75L$
	(vii) Restrained against lateral deflection and torsion at the free end	$0.5L$
	Free at end and continuous at support and	
	(viii) Unrestrained against torsion at support	$3L$
	(ix) Partially restrained against torsion at support	$2L$
	(x) Fully restrained	L

Notes

(i) When the beam ends are not restrained against torsion the effective length in (i), (ii) and (iii) should be increased by 20%.
(ii) If there is a degree of fixity at the free end the effective length should be multiplied by $\frac{0.5}{0.85}$ in (vi), (vii) and by $\frac{0.75}{0.85}$ in (viii), (ix) and (x).

Beams and Plate Girders with Unequal Flanges The maximum permissible bending compressive stress σ_{bc}, for laterally unsupported beams with unequal flanges may be obtained by using the *Merchant Rankine formula*,

$$\sigma_{bc} = 0.66 \frac{f_{cb} f_y}{[(f_{cb})^n + (f_y)^n]^{1/n}} \tag{7.1}$$

where f_y = yield stress of steel
n = a factor assumed as 1.4
f_{cb} = elastic critical stress in bending

$$= k_1(X + k_2Y)\frac{c_2}{c_1} \tag{7.2}$$

$$X = Y\sqrt{1 + (1/20)\,[(lT/r_y D)]^2}\ \text{(MPa)} \tag{7.3}$$

$$Y = \frac{26.5 \times 10^5}{(l/ry)^2}\ \text{(MPa) s} \tag{7.4}$$

k_1 = a coefficient to allow for reduction in thickness or breadth of flanges between points of effective lateral restraint, which depends upon a ratio ψ (Table 7.3 of the text).

ψ = a ratio of the total area of both flanges at the point of least bending moment to the corresponding area at the point of greatest bending moment between such points of restraint.

Table 7.3 *Values of k_1 for Beams with Curtailed Flanges*

ψ	1.0	0.9	0.8	0.7	0.6	0.5	0.4	0.3	0.2	0.1	0.0
k_1	1.0	1.0	1.0	0.9	0.8	0.7	0.6	0.5	0.4	0.3	0.2

Note Flanges should not be reduced in breadth to give a value of ψ lower than 0.25.

k_2 = a coefficient which depends upon a ratio ω, Table 7.4 of the text.

ω = a ratio of the moment of inertia of the compression flange to the sum of the moment of inertia of both the flanges, each calculated about its own axis, parallel to yy-axis of the girder.

Table 7.4 *Values of k_2 for Beams with Unequal Flanges*

ω	1.0	0.9	0.8	0.7	0.6	0.5	0.4	0.3	0.2	0.1	0.0
k_2	0.5	0.4	0.3	0.2	0.1	0	–0.2	–0.4	–0.6	–0.8	–1.0

It may be seen that when the beam is symmetrical, ω becomes 0.5 giving $k_2 = 0$. Thus the additive factor k_2 vanishes for a uniform symmetrical section.

When angle and T-sections are used as beams and the flange (table) is in compression, Fig. 7.3(a), $\omega = 1$ and $k_2 = 0.5$, whereas when the leg is in compression, Fig. 7.3(b), $\omega = 0.0$ and $k_2 = -1.0$. This would suggest that to derive maximum bending strength, sections as shown in Fig. 7.3(a) should be used.

Fig. 7.3

c_1, c_2 = the lesser and greater distances respectively of the extreme fibres from the neutral axis.

Notes

(i) The values of X and Y can either be calculated either from Eqs 7.3 and 7.4 or can be found from Table 7.5 of the text.

(ii) The value of σ_{bc} for calculated value of f_{cb} can be obtained either from Eq. 7.1 or determined from Table 7.6 of the text.

(iii) The values of f_{cb} should be increased by 20% when $\frac{T}{t} \ngtr 2.0$ and $\frac{d_1}{t} \ngtr \frac{1344}{\sqrt{f_y}}$.

(iv) When the beams bent about yy-axis, the maximum permissible bending stress in compression or tension should not exceed 0.66 f_y.

7.5 BEARING STRESS

Bearing stresses (crippling stresses) are set up in the beam at places where concentrated loads or reactions act.

$$\text{The bearing stress, } \sigma_p = \frac{P}{A_e}$$

$$\ngtr 0.75 f_y$$

where P = concentrated load in N

A_e = effective area (b_1 t) at the toe, of the fillet within 30° line drawn from the ends of the bearing (Fig. 7.8)

t = thickness of the web

b_1 = bearing length

f_y = yield stress

If the calculated bearing stress exceeds the permissible value of 0.75 f_y, either the bearing block should be made longer or beams with a thick web or bearing stiffeners should be provided. Design of bearing stiffeners is discussed in Chapter 9.

7.6 SHEAR STRESS

A beam is subjected to two types of shear, i.e. transverse (vertical) shear (Figs 7.4(a) and (b)), and longitudinal shear (Fig. 7.4(c)). Transverse shear failure does not usually occur in a regular steel beam, as the beam will fail due to web crippling prior to the transverse shear failure, but it may occur if the beam is short and heavily loaded or deeply coped as shown in Fig. 7.4(b). Longitudinal or horizontal shear occurs due to the bending of the beam. The fibres above the N.A. shorten in length and those below the N.A. elongate under sagging bending moments. Therefore, the fibres tend to slip over each other and the effect is maximum at the neutral axis. The tendency to slip is resisted by shear strength of the material.

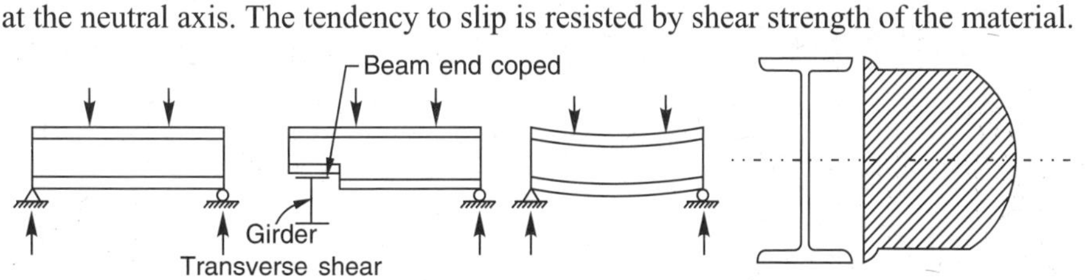

(a) (b) (c) Longitudinal shear (d) Shear in beam

Fig. 7.4

Table 7.5 *Values of X and Y for Calculating f_{cb}*

D/T →	X															Y
l/r_y ↓	8	10	12	14	16	18	20	25	30	35	40	50	60	80	100	
40	2484	2222	2066	1965	1897	1849	1814	1759	1728	1709	1697	1683	1675	1667	1663	1656
45	2103	1856	1708	1612	1546	1499	1465	1411	1380	1362	1349	1335	1327	1319	1315	1309
50	1822	1590	1449	1357	1293	1248	1214	1161	1131	1113	1101	1086	1078	1070	1067	1060
55	1607	1389	1254	1166	1105	1061	1028	976	947	929	917	902	894	886	883	876
60	1437	1232	1104	1020	961	918	886	835	806	788	776	762	754	746	743	736
65	1301	1107	985	904	847	806	775	726	697	679	667	653	645	637	634	627
70	1188	1005	889	811	757	717	687	638	610	592	581	567	559	551	547	541
75	1094	920	810	735	682	644	615	567	540	522	511	497	489	481	478	471
80	1014	849	743	672	621	584	556	509	482	465	454	440	432	424	421	414
85	945	788	687	618	570	533	506	461	434	417	406	392	385	377	373	367
90	886	735	639	573	526	491	464	420	394	377	366	353	345	337	334	327
95	833	689	597	534	488	454	428	385	360	343	332	319	311	304	300	294
100	787	649	560	499	455	423	398	356	331	314	304	290	283	275	272	265
110	708	582	499	443	402	371	347	307	283	268	257	244	237	229	226	219
120	644	527	451	398	359	330	308	270	247	232	222	209	202	194	191	184
130	591	482	411	361	325	298	277	240	218	204	194	181	174	167	163	157
140	546	444	378	331	297	271	251	217	195	181	172	160	153	145	142	135
150	508	412	350	306	274	249	230	197	177	163	154	142	135	145	124	118
160	474	385	326	284	254	230	212	181	161	148	139	127	121	113	110	104
170	445	360	305	265	236	214	197	167	148	135	126	115	109	102	98	92
180	420	339	286	249	221	200	184	155	137	125	116	105	98	92	88	82
190	397	320	270	235	208	188	172	145	127	115	107	96	90	83	80	73
200	376	304	256	222	197	177	162	136	119	107	99	89	83	76	73	66

(Contd.)

Table 7.5 *(Contd.)*

$D/T \rightarrow$	X															Y
$l/r_y \downarrow$	*8*	*10*	*12*	*14*	*16*	*18*	*20*	*25*	*30*	*35*	*40*	*50*	*60*	*80*	*100*	
210	358	288	243	210	186	168	153	128	112	101	93	82	76	70	66	60
220	341	275	231	200	177	159	145	121	105	94	87	77	71	64	61	55
230	326	262	220	191	169	152	138	115	99	89	82	72	66	60	56	50
240	312	251	211	182	161	145	132	109	94	84	77	67	62	55	52	46
250	299	241	202	175	154	138	126	104	90	80	73	64	58	52	49	42
260	288	231	194	167	148	133	121	99	85	76	69	60	55	48	45	39
270	277	222	186	161	142	127	116	95	82	72	66	57	52	46	42	36
280	267	214	180	155	137	122	111	91	78	69	63	54	49	43	40	34
290	257	207	173	149	132	118	107	88	75	66	60	52	46	41	38	32
300	249	200	167	144	127	114	103	84	72	64	57	49	44	38	35	29

Table 7.6 *Values of σ_{bc} Calculated from f_{cb} for Different Values of f_y*

All Units in MPa

$f_y \rightarrow$	220	230	240	250	260	280	300	320	340	360	380	400	420	450	480	510	540
$f_{cb} \downarrow$	1	2	3	4	5	6	7	8	9	10	11	12	13	14	15	16	17
20	13	13	13	13	13	13	13	13	13	13	13	13	13	13	13	13	13
30	19	19	19	19	19	19	19	19	19	19	19	19	19	19	20	20	20
40	25	25	25	25	25	25	25	25	26	26	26	26	26	26	26	26	26
50	30	30	31	31	31	31	31	31	31	32	32	32	32	32	32	32	32
60	36	36	36	36	36	37	37	37	37	37	38	38	38	38	38	38	38
70	41	41	41	41	42	42	42	43	43	43	43	44	44	44	44	44	44
80	45	46	46	46	47	47	48	48	48	49	49	49	49	50	50	50	50
90	50	50	51	51	51	52	53	53	54	54	54	55	55	55	56	56	56
100	54	54	55	55	56	57	57	58	59	59	60	60	60	61	61	62	62
110	58	58	59	60	60	61	62	63	64	64	65	65	66	66	67	67	67
120	61	62	63	64	64	65	67	67	68	69	70	70	71	71	72	72	73
130	65	66	67	67	68	70	71	72	73	74	74	75	76	76	77	78	78
140	68	69	70	71	72	73	75	76	77	78	79	80	80	81	82	83	84
150	71	72	73	74	75	77	79	80	81	82	83	84	85	86	87	88	89
160	74	75	77	78	79	81	82	84	85	87	88	89	90	91	92	93	94
170	77	78	80	81	82	84	86	88	89	91	92	93	94	95	97	98	99
180	79	81	82	84	85	87	89	91	93	94	96	97	98	100	101	102	103
190	82	84	85	87	88	90	93	95	97	98	100	102	102	104	106	107	108
200	84	86	88	89	91	93	96	98	100	102	103	105	106	108	110	111	113
210	86	88	90	92	93	96	99	101	103	105	107	109	110	112	114	116	117
220	89	90	92	94	96	99	102	104	106	109	111	112	114	116	118	120	121
230	90	93	94	96	98	101	104	107	110	112	114	116	118	120	122	124	126

(Contd.)

Table 7.6 (*Contd.*)

All Units in MPa

$f_y \rightarrow$	*220*	*230*	*240*	*250*	*260*	*280*	*300*	*320*	*340*	*360*	*380*	*400*	*420*	*450*	*480*	*510*	*540*
$f_{cb} \downarrow$	*1*	*2*	*3*	*4*	*5*	*6*	*7*	*8*	*9*	*10*	*11*	*12*	*13*	*14*	*15*	*16*	*17*
240	92	94	97	99	100	104	107	110	113	115	117	119	121	124	126	128	130
250	94	96	99	101	103	106	110	113	115	118	120	122	124	127	130	132	134
260	96	98	100	103	105	108	112	115	118	121	123	126	128	131	133	136	138
270	97	100	102	104	107	111	114	118	121	124	126	129	131	134	137	139	142
280	99	101	104	106	108	113	116	120	123	126	129	132	134	137	140	143	145
290	100	103	105	108	110	115	119	122	126	129	132	135	137	141	144	147	149
300	102	104	107	110	112	116	121	125	128	131	135	137	140	144	147	150	153
310	103	106	108	111	114	118	123	127	130	134	137	140	143	147	150	153	156
320	104	107	110	113	115	120	125	129	133	136	140	143	146	150	153	157	160
330	105	108	111	114	117	122	126	131	135	138	142	145	148	152	156	160	163
340	106	110	113	115	118	123	128	133	137	141	144	148	151	155	159	163	166
350	108	111	114	117	120	125	130	134	139	143	147	150	153	158	162	166	169
360	109	112	115	118	121	126	131	136	141	145	149	152	156	161	166	169	172
370	110	113	116	119	122	128	133	138	143	147	151	155	158	163	168	172	175
380	111	114	117	120	123	129	135	140	144	149	153	157	160	166	170	174	178
390	111	115	118	121	125	130	136	141	146	151	155	159	163	168	173	177	181
400	112	116	119	122	126	132	137	143	148	152	157	161	165	170	175	180	184
420	114	118	121	124	128	134	140	146	151	156	160	165	169	175	180	185	189
440	115	119	123	126	130	136	142	148	154	159	164	169	173	179	185	190	195
460	117	121	124	128	132	138	145	151	157	162	167	172	177	183	189	194	200
480	118	122	126	130	133	140	147	153	159	165	170	175	180	187	193	199	204
500	119	123	127	131	135	142	149	155	162	168	173	178	183	190	197	203	209

(*Contd*)

Table 7.6 (*Contd.*)

All Units in MPa

$f_y \rightarrow$	220	230	240	250	260	280	300	320	340	360	380	400	420	450	480	510	540
$f_{cb} \downarrow$	1	2	3	4	5	6	7	8	9	10	11	12	13	14	15	16	17
520	120	125	129	133	136	144	151	158	164	170	176	181	187	194	201	207	213
540	121	126	130	134	138	145	153	160	166	172	178	184	189	197	204	211	217
560	122	127	131	135	139	147	154	161	168	175	181	187	192	200	208	215	221
580	123	128	132	136	140	148	156	163	170	177	183	189	195	203	211	218	225
600	124	129	133	137	141	150	157	165	172	179	185	192	198	206	214	222	229
620	125	129	134	138	143	151	159	166	174	181	187	194	200	209	217	225	232
640	126	130	135	139	144	152	160	168	175	183	189	196	202	211	220	228	235
660	126	131	136	140	145	153	161	169	177	184	191	198	204	214	222	231	238
680	127	132	136	141	145	154	163	171	178	186	193	200	207	216	225	234	242
700	128	132	137	142	146	155	164	172	180	187	195	202	209	218	228	236	244
720	128	133	138	143	147	156	165	173	181	189	196	204	210	220	230	239	247
740	129	134	139	143	148	157	166	174	182	190	198	205	212	222	232	241	250
760	129	134	139	144	149	158	167	175	184	192	199	207	214	224	234	244	253
780	130	135	140	145	149	159	168	176	185	193	201	208	216	226	236	246	255
800	130	135	140	145	150	159	169	177	186	194	202	210	217	228	238	248	257
850	131	137	142	147	152	161	171	180	188	197	205	213	221	232	243	253	263
900	132	138	143	148	153	163	172	182	191	200	208	216	224	236	247	258	268
950	133	138	144	149	154	164	174	183	193	202	211	219	227	240	251	262	273
1000	134	139	145	150	155	165	175	185	195	204	213	222	230	243	255	266	277
1050	135	140	145	151	156	167	177	187	196	206	215	224	233	246	258	270	281
1100	135	141	146	152	157	168	178	188	198	207	217	226	235	248	261	273	285
1150	136	141	147	152	158	168	179	189	199	209	219	228	237	251	263	276	288

(*Contd.*)

Table 7.6 (*Contd.*)

All Units in MPa

$f_y \rightarrow$	*220*	*230*	*240*	*250*	*260*	*280*	*300*	*320*	*340*	*360*	*380*	*400*	*420*	*450*	*480*	*510*	*540*
$f_{cb} \downarrow$	*1*	*2*	*3*	*4*	*5*	*6*	*7*	*8*	*9*	*10*	*11*	*12*	*13*	*14*	*15*	*16*	*17*
1200	136	142	147	153	159	169	180	190	200	210	220	230	239	253	266	279	291
1300	137	143	149	154	160	171	182	192	203	213	223	233	243	257	270	284	297
1400	138	144	149	155	161	172	183	194	205	215	225	236	246	260	274	288	302
1500	139	144	150	156	162	173	184	195	206	217	228	238	248	263	278	292	306
1600	139	145	151	157	163	174	185	197	208	219	229	240	250	266	281	295	309
1700	140	146	151	157	163	175	186	198	209	220	231	242	252	268	283	298	313
1800	140	146	152	158	164	176	187	199	210	221	232	243	254	270	285	301	316
1900	140	146	152	158	164	176	188	200	211	222	234	245	256	272	287	303	318
2000	141	147	153	159	165	177	189	200	212	223	235	246	257	273	289	305	321
2200	141	147	154	160	166	178	190	202	213	225	237	248	259	276	292	309	325
2400	142	148	154	160	166	179	191	203	215	226	238	250	261	278	295	312	328
2600	142	148	154	161	167	179	191	204	216	227	239	251	263	280	297	314	331
2800	142	149	155	161	167	180	192	204	216	228	240	252	264	282	299	316	333
3000	143	149	155	161	168	180	193	205	217	229	241	253	265	283	300	318	335
3500	143	149	156	162	168	181	194	206	218	231	243	255	267	286	303	321	339
4000	143	150	156	163	169	182	194	207	219	232	244	257	269	287	306	324	342
4500	144	150	157	163	169	182	195	208	220	233	245	258	270	289	307	326	344
5000	144	150	157	163	170	183	195	208	221	233	246	259	271	290	309	327	346
5500	144	151	157	163	170	183	196	208	221	234	247	259	272	291	310	328	347
6000	144	151	157	164	170	183	196	209	222	234	247	260	273	291	310	329	348

Some of the other situations where shear determines the design of beams are:

1. when the depth of beam section is small and is loaded uniformly.
2. when large concentrated loads are placed near beam supports. In a tall building the upper storey column may be offset with respect to the lower storey column. The load from the upper storey column will be transferred to the beam end in such a case.
3. In plate girders where the web thickness is small.

A major portion of the shear stress is resisted by the web whereas flanges resist only a minor portion of it (Fig. 7.4(d)). That is why for shear calculations the web size is assumed to be equal to the depth of the girder. The maximum value of the shear stress occurs at the neutral axis. For an I-shapes the average shear stress is only slightly less than the maximum shear stress value. For design purposes therefore, average shear stress is calculated and checked.

The average shear is worked out on the gross area of the web and is taken equal to the web thickness times the overall depth of the beam. Thus, for an I-Section, the shear stress $\tau_{va} = V/tD$, where t = web thickness, and D = overall depth of the beam.

According to I.S: 800–1984 the maximum permissible shear stress, is given by $\tau_{vm} = 0.45 f_y$ and the permissible average shear stress τ_{va} on a web whose depth does not exceed 85 times its thickness, is equal to $0.40 f_y$.

For built-up girders with plates and angles where the depth-thickness ratio of the webs exceeds 85, stiffeners are necessary and the permissible shear stress then depends upon panel dimensions also. For stiffened webs, the permissible average shear stress is determined from Tables 7.7(a)–7.7(c).

Table 7.7(a) *Permissible Average Shear Stress τ_{va} in Stiffened Webs of Steel with f_y = 250 MPa*

d/t	Stress τ_{va} (MPa) for Different Distances c Between Stiffeners												
↓	0.3d	0.4d	0.5d	0.6d	0.7d	0.8d	0.9d	1.0d	1.1d	1.2d	1.3d	1.4d	1.5d
90	100	100	100	100	100	100	100	100	100	100	100	100	100
95	100	100	100	100	100	100	100	100	100	100	100	100	99
100	100	100	100	100	100	100	100	100	100	100	99	99	98
105	100	100	100	100	100	100	100	100	100	99	98	97	96
110	100	100	100	100	100	100	100	100	99	98	96	95	94
115	100	100	100	100	100	100	100	100	98	96	95	94	93
120	100	100	100	100	100	100	100	98	96	95	93	92	91
125	100	100	100	100	100	100	98	97	95	93	92	91	90
130	100	100	100	100	100	99	97	96	94	92	90	89	88
135	100	100	100	100	100	98	96	94	92	90	89	87	86
140	100	100	100	100	99	96	95	93	91	89	87	86	85
150	100	100	100	100	97	94	92	90	88	86	84	83	81
160	100	100	100	98	94	92	89	88	85	83	81	80	78
170	100	100	100	96	92	89	87	85	82	80	78	76	75
180	100	100	98	94	90	87	84	82	80	77	75	73	72
190	100	100	97	92	88	84	82						
200	100	100	95	90	86	82	81						

(Contd.)

Table 7.7(a) (*Contd.*)

d/t ↓	Stress τ_{va} (MPa) for Different Distances c Between Stiffeners												
	0.3d	0.4d	0.5d	0.6d	0.7d	0.8d	0.9d	1.0d	1.1d	1.2d	1.3d	1.4d	1.5d
210	100	99	93	88	83	81							
220	100	98	91	86	81	80							
230	100	96	90	84	79				Non-applicable zone				
240	100	95	88	83	77								
250	100	93	86	82	74								
260	100	92	85	81									
270	99	90	84	81									

Note Intermediate values may be obtained by linear interpolation.

Table 7.7(b) *Permissible Average Shear Stress τ_{va} in Stiffened Webs of Steel with f_y = 340 MPa*

d/t ↓	Stress τ_{va} (MPa) for Different Distances c Between Stiffners												
	0.3d	0.4d	0.5d	0.6d	0.7d	0.8d	0.9d	1.0d	1.1d	1.2d	1.3d	1.4d	1.5d
75	136	136	136	136	136	136	136	136	136	136	136	136	136
80	136	136	136	136	136	136	136	136	136	136	136	136	136
85	136	136	136	136	136	136	136	136	136	136	136	134	133
90	136	136	136	136	136	136	136	136	136	135	133	132	131
95	136	136	136	136	136	136	136	136	135	133	131	129	128
100	136	136	136	136	136	136	136	135	132	130	128	127	126
105	136	136	136	136	136	136	135	133	130	128	126	124	123
110	136	136	136	136	136	135	133	131	128	126	124	122	120
115	136	136	136	136	136	133	131	129	126	123	121	119	118
120	136	136	136	136	135	131	129	127	124	121	119	117	115
125	136	136	136	136	133	129	127	125	121	119	116	114	113
130	136	136	136	135	131	127	125	122	119	116	114	112	110
135	136	136	136	134	129	126	123	120	117	114	111	109	108
140	136	136	136	132	127	124	121	118	115	112	109	107	105
150	136	136	135	129	124	120	117	114	110	107	104	102	100
160	136	136	132	126	120	116	113	110	106	102	99	97	95
170	136	136	129	123	117	112	109	106	101	98	95	92	90
180	136	135	127	119	113	108	105	102	97	93	90	87	84
190	136	133	124	116	110	105	100						
200	136	130	121	113	106	101	96						
210	136	128	118	110	103	97							
220	136	126	116	107	99	93							
230	135	123	113	103	96				Non-applicable zone				
240	134	121	110	100	92								
250	132	119	107	97	89								
260	130	116	104	94									
270	128	114	102	91									

Note Intermediate values may be obtained by linear interpolation.

Table 7.7(c) *Permissible Average Shear Stress τ_{va} in Stiffened Webs of Steel with f_y = 400 MPa*

d/t	*Stress τ_{va} (MPa) for Different Distances c Between Stiffners*												
↓	*0.3d*	*0.4d*	*0.5d*	*0.6d*	*0.7d*	*0.8d*	*0.9d*	*1.0d*	*1.1d*	*1.2d*	*1.3d*	*1.4d*	*1.5d*
70	160	160	160	160	160	160	160	160	160	160	160	160	160
75	160	160	160	160	160	160	160	160	160	160	160	160	159
80	160	160	160	160	160	160	160	160	160	160	159	157	156
85	160	160	160	160	160	160	160	160	160	158	156	154	152
90	160	160	160	160	160	160	160	160	157	155	152	151	149
95	160	160	160	160	160	160	159	157	154	152	149	147	146
100	160	160	160	160	160	160	157	155	151	149	146	144	143
105	160	160	160	160	160	157	154	152	149	146	143	141	139
110	160	160	160	160	159	155	152	149	146	143	140	138	136
115	160	160	160	160	156	152	149	147	143	140	137	135	133
120	160	160	160	159	154	150	147	144	140	137	134	132	129
125	160	160	160	157	152	147	144	141	137	134	131	128	126
130	160	160	160	155	150	145	141	139	134	131	128	125	123
135	160	160	160	153	147	143	139	136	132	128	125	122	120
140	160	160	158	151	145	140	136	133	129	125	122	119	116
150	160	160	155	147	141	135	131	128	123	119	115	112	110
160	160	160	151	143	136	130	126	123	117	113	109	106	103
170	160	158	148	139	132	126	121	117	112	107	103	100	97
180	160	155	144	135	127	121	116	112	106	101	97	93	90
190	160	152	140	131	123	116	111						
200	160	149	137	127	118	111	106						
210	160	146	133	123	114	106							
220	157	143	130	119	109	101							
230	155	140	126	114	105					Non-applicable zone			
240	153	137	123	110	100								
250	151	134	119	106	96								
260	148	131	116	102									
270	146	128	112	98									

Note Intermediate values may be obtained by linear interpolation.

7.7 DEFLECTION

In general, beam deflection is a function of loading, span, modulus of elasticity, and the geometry of cross-section. For regular loading systems, deflections are calculated by the standard formulae. When the loading of a beam or girder is irregular, and of a form not covered by the standard formulae, probably the easiest way to evaluate the deflection is as follows.

Load the beam or girder with the actual moment curve and find the resulting or second bending moment diagram due to this loading. This secondary bending moment curve is the deflection diagram for the beam, and an ordinate at any point represents, to scale, the vertical deflection of the girder at that point. This is known as the method of elastic weights.

Small deflections of the beams do not cause structural problems in general except for discomfort to the users. Excessive deflections may lead to cracks in the plaster or ceilings and may damage the material attached to or supported by the beam. This limits the maximum deflection to be span/325. Deflection limitations may control the proportioning of the cross-section of (1) long lightly loaded spans, (2) beams carrying machinery, cranes, etc., or (3) beams which would vibrate if excessively flexible. Besides this, some of the other reasons for limiting deflections are:

(i) Excessive deflection may create problems for floor or roof drainage.

(ii) In case the beams framed together are of different sections they must deflect alike. Such beams should have identical calculated deflection.

(iii) These may cause undesirable twisting and distortion of connections and connected members and lead to high secondary stresses.

However, a designer can reduce deflections by increasing the depth of the beam section, reducing the span, providing greater end restraints or by other means such as providing camber.

7.8 WEB BUCKLING

The web in a rolled steel section behaves like a column when placed under concentrated loads. The web is quite thin and is, therefore, subjected to buckling. The buckling of the column web is very much influenced by the restraints provided for the flanges. Various types of bucklings that can occur are shown in Fig. 7.5. In all cases, the bottom flange is assumed to be restrained against lateral deflection and rotation. For the top flange, the end restraints and the effective depth of the web to be considered are as below. For the top flange:

(a) restrained against lateral deflection and rotation, the effective depth = $d_1/2$ (Fig. 7.5(a)).

(b) restrained against lateral deflection but not against rotation, the effective depth = (2/3) d_1 (Fig. 7.5(b)).

(c) restrained against rotation but not against lateral deflection, effective depth = d_1 (Fig. 7.5(c)).

(d) not restrained against rotation and lateral deflection, the effective depth = $2d_1$ (Fig. 7.5(d)).

Generally, vertical buckling of webs is not a problem with rolled beam sections. This possibility exists mainly in the rather thin webs of deep plate girders.

To check the buckling of the web bearing stiffeners are provided. As per I.S: 800–1984, bearing stiffeners should be provided where the concentrated loads or reactions exceed the value of $\sigma_{ac}\, tB$.

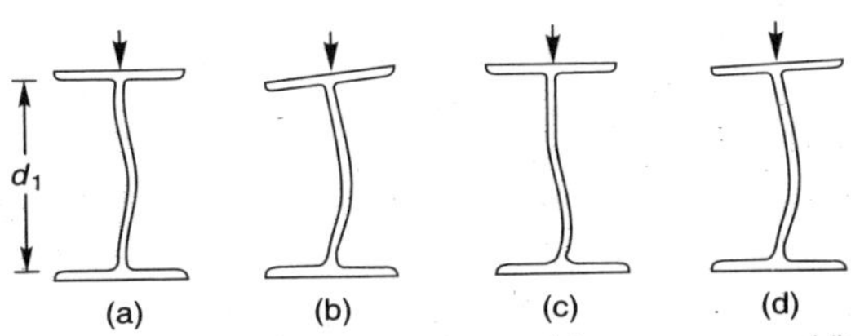

(a) (b) (c) (d)

Fig. 7.5 *Buckling Failure from Vertical Loading*

where

σ_{ac} = the maximum permissible axial stress for columns having a slenderness ratio $d_1\sqrt{3}/t$

t = web thickness

B = length of the stiff portion of the bearing plus the additional length given by dispersion at 45° to the level of the neutral axis. The stiff portion of the bearing should not exceed half the depth of the beam when simply supported, and full depth of the beam when continuous over a bearing (Fig. 7.6).

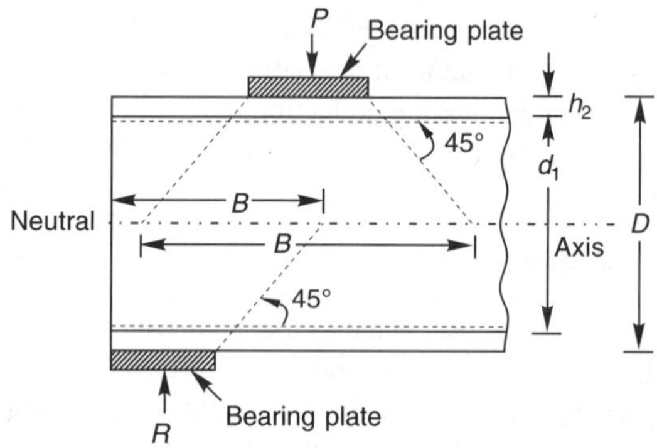

Fig. 7.6 *Bearing Length for Buckling Consideration*

d_1 = clear depth of web between roots of fillet
$= D - 2h_2$

h_2 = depth at root of fillet

D = overall depth of beam

When the flanges are restrained against lateral displacement and rotation, the effective length of the web $l = d_1/2$.

$$\text{Slenderness ratio, } \lambda = \frac{d_1/2}{r_y}$$

$$= \frac{d_1/2}{\sqrt{I_{yy}/A}} = \frac{d_1/2}{\sqrt{\frac{1}{12}Bt^3/(Bt)}} = \frac{d_1}{t}\sqrt{3}$$

7.9 WEB CRIPPLING

Loads and reactions concentrated along a short length of flange of beams are resisted by compressive stresses in the web which vary with distance form the load. The webs of rolled steel sections are, therefore, subjected to a large amount of stresses just below the concentrated loads and above the reactions from the support. Stress concentration occurs at the junction of the web and the flange. As a result, large bearing stresses are developed below the concentrated loads. Consequently, the web near the portion of the stress concentration tends to fold over the flange. This type of local buckling phenomenon is called crippling or crimpling of the web (Fig. 7.7). To keep the bearing stress within permissible limits, the concentrated loads should be transferred from flanges to the web on sufficiently large bearing areas. The root of the fillet is the most critical location for failure because the resisting area has the smallest value there.

$$\text{Bearing stress, } \sigma_p = \frac{P}{\text{bearing area}}$$

where, bearing area = $b_1 t$

Here reference may be made to Fig. 7.8.

Assuming the angle of dispersion of the load to be 30°,

$$\text{Bearing length, } b_1 = b + 2h_2\sqrt{3}$$

(under concentrated loads)

$$b_1 = b + h_2\sqrt{3}$$

(under reactions at support)

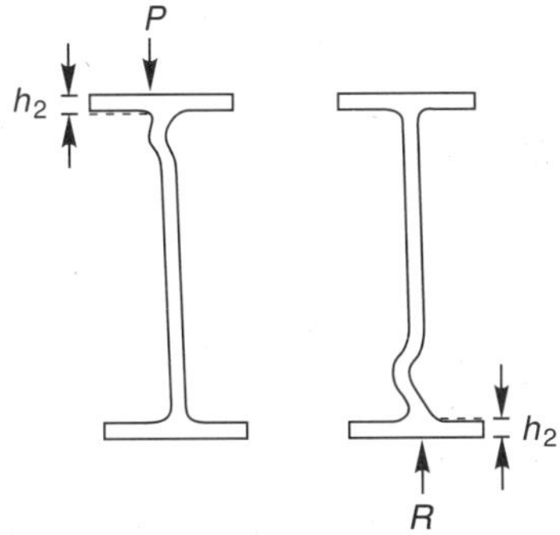

Fig. 7.7 *Web Failure by Local Crippling*

where σ_p = bearing stress in MPa

P = concentrated load or reaction from support in N

t = thickness of web in mm

h_2 = depth of the root of the fillet in mm

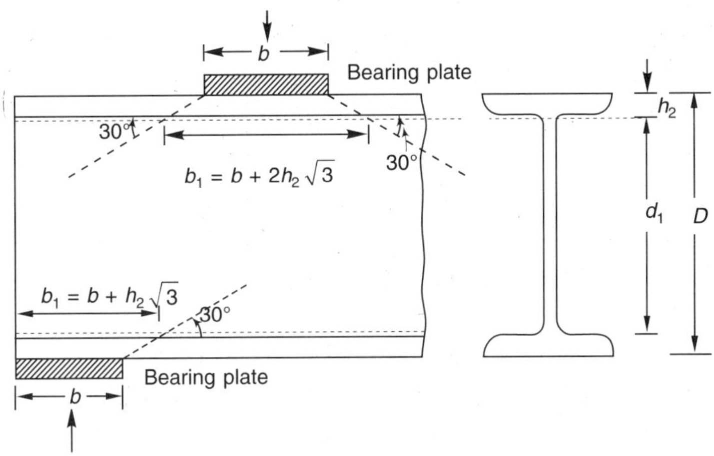

Fig. 7.8 *Bearing Load and Reaction*

In some cases the required bearing length, b_1 may be very small and negligible. But a bearing of minimum 100 mm length should be provided. The value of the calculated stress should be less than the permissible bearing stress. Out of buckling and crippling of the web, it has been found that if the beam section is safe in crippling, it will be safe for buckling also.

Aside from spreading the load over a longer portion of the flange, there are two remedies to the problem. The first is to provide stiffeners which bear against the flanges at the load points and are connected to the web (Section 9.10) so as to transfer the force to it gradually, not abruptly. The other is to make the web thicker. Since web crippling is a problem at only a few sections of a beam, it is economical to provide bearing stiffeners at these sections than to increase the thickness of the entire web.

7.10 DIAGONAL BUCKLING

Diagonal buckling of the web is not at all serious in rolled beam sections. The possibility of diagonal buckling of the web occurs when the ratio, $d_1/t > 85$ where d_1 is the distance between flanges and t is the web thickness. All rolled sections have this ratio less than 70; however, in case of plated girders this ratio may exceed the permissible value and therefore, should be catered for.

Consider an element of plate girder at the neutral axis (Fig. 7.9). The maximum diagonal compression occurs at the neutral axis and will be inclined at 45° to it. Due to horizontal compression above and horizontal tension below the neutral axis, the diagonal compression does not remain inclined at 45° at each and every point in the web except at the N.A. To have a simple solution, we may assume it to be inclined at 45° at each point. The element shown is subjected to vertical shear τ, on its vertical faces. Complementary shear stresses of the same magnitude are developed on the horizontal faces. The stresses produce compression along one diagonal and tension along the other (equal to the shear τ). This compression is called diagonal compression and may cause buckling of the web.

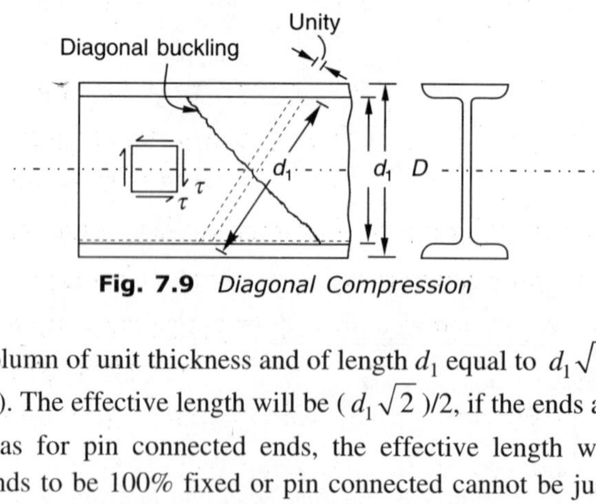

Fig. 7.9 *Diagonal Compression*

Consider a column of unit thickness and of length d_1 equal to $d_1\sqrt{2}$ (as inclined at 45° with N.A.). The effective length will be ($d_1\sqrt{2}$)/2, if the ends are assumed to be fixed, whereas for pin connected ends, the effective length will be $d_1\sqrt{2}$. Assuming the ends to be 100% fixed or pin connected cannot be justified. Therefore, for an exact analysis, some constant will have to be introduced. For a simplified solution, ends are assumed to be fixed.

$$\text{Effective length} = \frac{d_1\sqrt{2}}{2} = \frac{d_1}{\sqrt{2}}$$

$$\text{Minimum radius of gyration} = \frac{t}{\sqrt{12}}$$

$$\text{Slenderness ratio, } \lambda = \frac{d_1/\sqrt{2}}{t/\sqrt{12}} = \frac{d_1\sqrt{6}}{t} = 2.45\frac{d_1}{t}$$

The aveiage compressive stress σ_{ac} for the above calculated slenderness ratio may be found from I.S:800–1984 and should be less than the average allowable shear stress of $0.4 f_y$, or as per the specifications laid in Section 7.6, of the text for built-up girders with plates and angles.

7.11 DESIGN OF LATERALLY SUPPORTED BEAMS

The design of a laterally supported beam consists of selecting a section on the basis of the modulus of section and checking it for shear, deflection and crippling. The design is done in the following manner.

1. The loads that can be expected on the beam are ascertained.
2. The maximum moment and maximum shear force are calculated for the beam.
3. The trial section modulus for the beam is worked out by,

$$Z = \frac{M}{\sigma_{bc,t}}$$

where Z = section modulus
M = maximum bending moment
$\sigma_{bc,t}$ = allowable bending stress in compression/tension ($0.66 f_y$)

4. From I.S. Handbook No. 1 a suitable section having the modulus of section equal to or more than calculated modulus of section is selected; I.S.L.B. or I.S.M.B. sections are preferred.
5. The beam is checked for shear. The average shear stress $\frac{P}{Dt}$ in the beam should not exceed $0.40 f_y$, where P = load or reaction, and t = web thickness.
6. The beam is checked for deflection. The calculated deflections should not exceed span/325.

$$\delta_{\text{cal}} = \frac{5}{384}\frac{wl^4}{EI} \text{ (for uniformly distributed load on simply supported beam)}$$

$$= \frac{1}{48}\frac{Wl^3}{EI} \text{ (for concentrated loads on simply supported beam)}$$

where w = uniformly distributed load in N/mm
W = concentrated load in N
l = effective span of beam c/c of support in mm
I = moment of inertia about xx-axis $(\text{mm})^4$

7. The beam is checked for crippling. The bearing stress calculated should not exceed the value of $0.75 f_y$.

$$\sigma_{p,\text{ cal}} = \frac{P}{(b + 2h_2\sqrt{3})t} \text{ (for concentrated loads)}$$

$$= \frac{P}{(b + h_2\sqrt{3})t} \text{ (for reactions)}$$

7.12 DESIGN OF LATERALLY UNSUPPORTED BEAMS

Most beams are braced by other parts of the structure, such as concrete slabs which are rigidly connected to the supporting beams and girders. Hence, the beams can deflect only vertically and the horizontal or rotational motions are prevented by the floor construction. There are situations, however, where beams or girders have no lateral support or bracing over part or whole of their span, for example, gantry girders. Other such situations occur temporarily during erection, when beams or girders are in place but bracing or floors are not yet erected. For such laterally unbraced beams the compression flange is laterally unsupported and the compression flange may buckle sideways, leading to failure at the elastic limit or even below it.

Beams with Equal Flanges

The design of laterally unsupported beams with equal flanges, e.g., I-Section, channel section, etc. is essentially a matter of trial and error since the section dimensions are not known until a trial section is selected. It is therefore, necessary to assume a section in order to compute the allowable stresses. The design procedure is as follows:

1. The loads which can be expected on the beam are ascertained.
2. The maximum bending moment and maximum shear force in the beam are computed.
3. The modulus of section for the beam is computed by

$$Z = \frac{M}{\sigma_{bc,t}}$$

where Z = section of modulus required
M = maximum bending moment in N/mm
$\sigma_{bc,\,t}$ = reduced permissible bending stress in compression ($< 0.66\,f_y$)

4. From I.S. Handbook No. 1 a suitable section providing the desired modulus of section is selected.
5. The various properties of the selected section are noted. The ratios T/t and d_1/t are calculated and the appropriate table (for steel of given yield stress) is selected from the Table 7.1 A-F of the text, to calculate the permissible bending compressive stress.
6. The permissible bending stress in compression is found corresponding to the calculated values of D/T and l/r_y (D, T, t, d_1, l, r_y have already been defined in Section 7.4 of the text).
7. The moment of resistance M_R, of the section is computed by

$$M_R = \sigma_{bc}\, Z$$

   The moment of resistance thus calculated should be more than the moment over the section.
8. The beam is checked for shear.
9. The beam is checked for deflection.

Beams with Unequal Flanges

Steps (1) to (4) are same as in the design of unsupported beams with equal flanges. The maximum permissible bending compressive stress is calculated as follows.

1. The elastic critical stress, f_{cb}, is calculated using Eq. (7.2) which is:

$$f_{cb} = k_1 (X + k_2 Y) \frac{c_2}{c_1}$$

[where k_1, k_2, X, Y, c_1, c_2 have already been defined in Section 7.4 of the text.]

2. For the calculated value of f_{cb} and the selected value of f_y, the permissible bending stress in compression is found from Table 7.6 of the text.
3. The moment of resistance of the section is computed ($M_R = \sigma_{bc}$ Z) which should be more than the maximum moment.
4. The beam is checked for shear.
5. The beam is checked for deflection.

7.13 BUILT-UP BEAMS (PLATED BEAMS)

For long spans and heavy loads large bending moments are generated. When available rolled-beam sections do not have sufficient strength to resist the external bending moment they may be reinforced along the entire length or part of it. Also, the depth of the beam may be restricted due to the head room requirements. In both the above situations it becomes necessary to provide *built-up* or *compound* beams, as shown in Fig. 7.10.

For flange reinforcement plates are most conveniently used. Such reinforcement also helps to increase the resistance against lateral bending. These additional plates can be connected to the rolled section by riveting, welding or bolting (Figs 7.10(a), (b)). When plates are welded to the flanges, the overhead welding as shown in Fig. 7.10(c) is avoided. One economical application of flange reinforcement is to add plates to portions of beam where high bending moment exists. By reinforcing cantilever or continuous spans in the regions of high bending moment, steel can be used economically. The size and length of the plates can be tailored to suit the moment diagram as shown in Fig. 7.12.

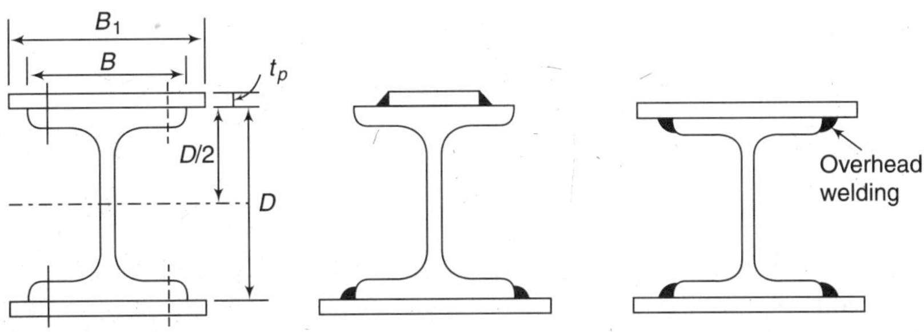

(a) Riveted/bolted (b) Welded (c) Overhead welding is prevented

Fig. 7.10 *Built-up Beams*

The design of a built-up beam is a trial and error procedure. A suitable I-Section is chosen and then the remaining modulus of section required is provided by the plates. The plates are attached to the compression and tension flanges of the beam section. The compound section is then checked for bending stress in compression and tension.

Design of Cover Plates

Let I_{req} = moment of inertia of the section to resist the total bending moment

I = moment of inertia of the desirable beam section available

I_p = moment of inertia of the section required from plate sections

The moment of inertia of the plates can be calculated from the following equation:

$$I_p = I_{\text{req}} - I \tag{7.5}$$

It is assumed that the centre of gravity of the compression flange lies at the junction of flange plate and cover plate.

The above equation can be rewritten as:

$$\frac{I_p}{D/2} = \frac{I_{\text{req}}}{D/2} - \frac{I}{D/2}$$

or
$$Z_p = Z_{\text{req}} - Z \tag{7.6}$$

From Eq. (7.6), the required modulus of section of plates can be worked out.

Now
$$I_p = 2\,[I_p' + A_p\,(D/2)^2\,]$$

Neglecting I_p', being very small as compared to the second term $A_p(D/2)^2$,

$$I_p = 2\,A_p\left(\frac{D}{2}\right)^2$$

or
$$\frac{I_p}{D/2} = 2\,A_p\left(\frac{D}{2}\right)$$

$$Z_p = A_p D$$

or
$$A_p = Z_p/D \tag{7.7}$$

where D = depth of the rolled beam section

I_p' = moment of inertia of the plate about its own *xx*-axis

Z_p = section modulus of plates to be provided

A_p = sectional area of the plates to be provided on each flange.

Knowing the area of the plates A_p, the area of plates on both the flanges A_p' can be calculated by

$$A_p' = 2A_p \tag{7.8}$$

The width B_1 of the flange plates may be kept either equal to or more than the flange width. As per the I.S: 800–1984 specifications, the outstand of the flange plate from the line of connection of plate and flange should not exceed by $20t$, where t is least of the thicknesses of flange plates. The thickness of the flange plate can be computed from Eq. (7.9)

$$t_p = A_p/B_1 \tag{7.9}$$

Bending Stresses

Bending compressive stresses are worked out as described in Section 7.4. If a welded connection is designed the calculated bending stresses in tension and compression are same, but if the connections are riveted the stress in the tension flange is more due to the reduction of area by the rivet holes. Therefore, net and gross sectional areas are calculated and the bending stress in tension is calculated from Eq. (7.10).

$$\sigma_{bt,\text{cal}} = \sigma_{bc,\text{cal}} \frac{\text{gross area of tension flange}}{\text{net area of tension flange}} \tag{7.10}$$

For a laterally supported beam the calculated bending tensile stress will be critical and should be less than the allowable bending tensile stress. However, for laterally unsupported beams the bending compressive stress is likely to be critical.

Design of Connections

The connections may be riveted, bolted or welded. The required strength of the connection is determined by the amount of shear between the connecting elements. In riveted connections, the rivets are staggered to achieve more net area, e.g., if two rows of rivets are provided, only one rivet hole will occur in any of the cross-sections. The nominal diameter of the rivet is assumed and the pitch of the rivets is computed from the Eq. (7.14).

Let τ = shear stress (MPa) at the section
V = shear force (N) at the section
I = moment of inertia
B = width of the flange plate at the section
$A\overline{Y}$ = moment of the area, about N.A., above the section considered
τ_1 = shear force per unit length
s = pitch of the rivets

$$\tau = \frac{VA\overline{Y}}{IB} \tag{7.11}$$

$$\tau_1 = \tau \times B$$

or

$$\tau_1 = \frac{VA\overline{Y}}{IB} \times B$$

or

$$\tau_1 = \frac{VA\overline{Y}}{I}$$

Hence, horizontal shear force per pitch length

$$= \frac{VA\overline{Y}}{I} \times s \tag{7.12}$$

In general, rivets are provided in two rows as shown in Fig. 7.11. Therefore, the horizontal shear force is shared equally by the two rows of rivets and the rivets in each row are subjected to a horizontal shear force per pitch length,

$$\tau_2 = \frac{1}{2}\frac{VA\overline{Y}s}{I} \tag{7.13}$$

The rivet value R_v for the rivet provided is worked out and is equated to this force per pitch length.

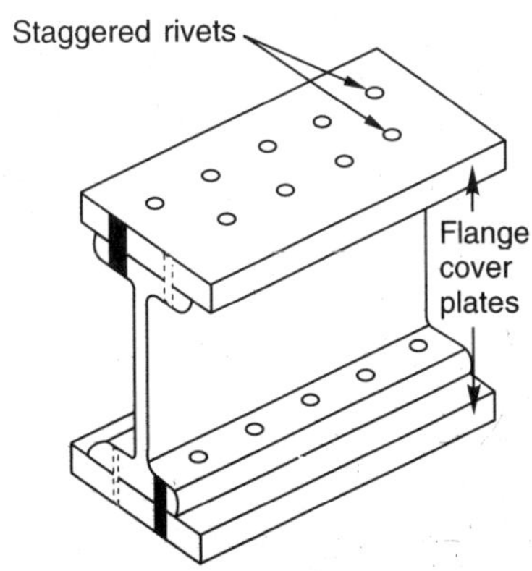

Fig. 7.11 *Staggered Rivets on Flanges of Built up Beams*

$$R_v = \frac{1}{2}\frac{VA\overline{Y}}{I}s \tag{7.14}$$

The pitch calculated from Eq. (7.14) should be checked for the following:

1. Minimum pitch should not be less than 2.5 times the nominal diameter of the rivet.
2. The distance between two adjacent rivets should not exceed (including tacking rivets) 32 t or 300 mm whichever is less.
3. The distance between two adjacent rivets parallel to direction of stress should not exceed 16 t or 200 mm whichever is less on the tension side, or 12 t or 200 mm whichever is less in the compression side.
4. When the rivets are staggered and the gauge distance does not exceed 75 mm the distance specified in (3) can be increased by 50%.

If connections are welded, intermittent fillet welds are provided.

Curtailment of the Cover Plates

A beam is stressed maximum at the point where the bending moment is maximum. At any other point the bending moment and consequently the bending stresses decrease. Therefore, to achieve economy, the maximum section is provided at the point of maximum bending moment and elsewhere it can be reduced. The reduction in the section can be done by cutting the cover plates. The I.S. specification that the first cover plate is carried through the full length of the beam and the remaining plates can be curtailed is applicable for built-up beams fabricated with angle sections for flanges and plate section for web. However, this specification is not applicable to the plated beams. Reference may be made to Fig. 7.12(a) for the following analysis.

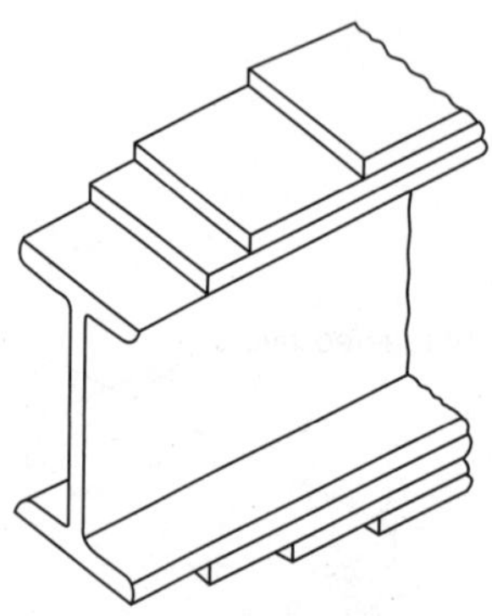

(a) Curtailed Flange Plates

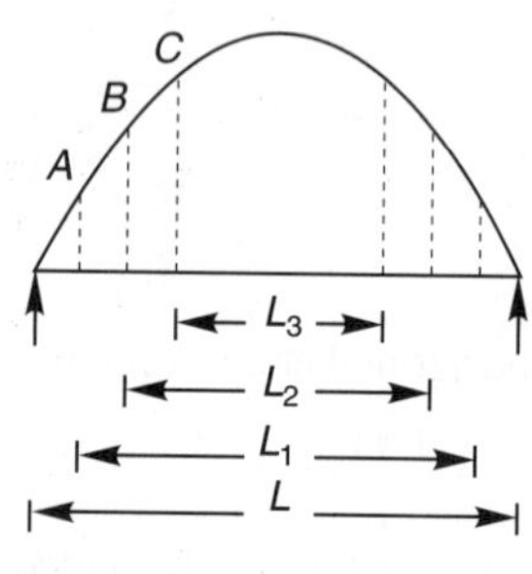

(b) Bending Moment Diagram

Fig. 7.12 *Curtailment of Flange Plates*

Let

x_1, x_2, x_3 = the distance of curtailment of the first, second and third plates from the support

L = effective span of the beam

Z = net modulus of section of the rolled I-Section

Z_1 = net modulus of section of the rolled I-Section with one cover plate

Z_2 = net modulus of section of the rolled I-Section with two cover plates
Z_3 = net modulus of section of the rolled I-Section with three cover plates
w = uniformly distributed load over the beam
L_1, L_2, L_3 = respectively the length of 1st, 2nd and 3rd cover plates to be provided

For a simply supported beam loaded uniformly the maximum bending moment occurs at the centre of the beam. The length of cover plates is so adjusted that the moment of resistance is always more than the bending moment. Equating the bending moment to moment of resistance,

$$M = M_R$$

$$\frac{wL^2}{8} = \sigma_{bt} Z_3 \tag{7.15}$$

At point C,

$$\frac{w(L^2 - L_3^2)}{8} = \sigma_{bt} Z_2 \tag{7.16}$$

At point B,

$$\frac{w(L^2 - L_2^2)}{8} = \sigma_{bt} Z_1 \tag{7.17}$$

At point A,

$$\frac{w(L^2 - L_1^2)}{8} = \sigma_{bt} Z \tag{7.18}$$

From Eqs (7.15) and (7.16),

$$\frac{L^2}{L^2 - L_3^2} = \frac{Z_3}{Z_2}$$

or

$$L_3 = L\sqrt{(Z_3 - Z_2)/Z_3}$$

Similarly,

$$L_2 = L\sqrt{(Z_3 - Z_1)/Z_3}$$

and

$$L_1 = L\sqrt{(Z_3 - Z)/Z_3} \tag{7.19}$$

Some additional length is required for the curtailed plates after the cutoff points. It is supposed that the plates carry the stresses right up to the theoretical point of cut off. This stress at the curtailed section is transferred by means of connectors that provide shear strength equal to the allowable tension or compression in the plate. To accommodate the connectors, the curtailed plate is extended up to the required length. Also, due to curtailment the section is changed abruptly and stress concentration occurs in the vicinity of the curtailed section. This gives rise to fatigue. By extending the cover plate for a small length in the direction of decreasing stress, the stress concentration will be readily absorbed. The curtailment of flange plates on the compression flange is done at the same point as that on the tension flange (Fig. 7.12).

Design Procedure of Built-up Beams

Following are the design steps for a laterally supported built-up beam:

1. The loads which can be expected on the beam are ascertained.

2. The maximum bending moment and maximum shear force in the beam are computed.
3. The trial modulus of section for the compound beam is calculated by,

$$Z_{req} = \frac{M}{\sigma_{bc,t}}$$

where

$\sigma_{bc,t}$ = bending stress in compression and is assumed to be lesser than the permissible bending stress in compression, e.g., for steel with f_y = 250 MPa, it can be assumed to be 150 MPa–165 MPa.

4. From I.S. Handbook No. 1 a suitable section furnishing maximum modulus of section is selected. The remaining modulus of section is provided by the plates using Eq. (7.7).
5. The width of plates is assured and thickness of various flange plates are calculated using Eq. (7.9).
6. The flange cover plates are curtailed using Eq. (7.19).
7. The connections are designed.
8. The maximum bending stress in compression and tension are calculated by Eq. (7.10). The bending stress in tension is checked.
9. The shear stress is calculated and should be less than the permissible shear stress of 0.4 f_y.
10. The beam is checked for crippling, if required.

Note Compound beams are generally kept symmetrical about the axis of bending, as the extreme fibres in the material will be stressed equally in tension and compression. Therefore, if unsymmetrical sections are designed they should be placed in such a way that the neutral axis lies at the mid height of the section so that the top and bottom fibres are stressed equally.

7.14 LINTELS

Beams provided above the openings in walls to support the masonry are called lintels. Some of the sections which can be safely used are shown in Figs 7.1(g), (h). It is desirable that a lintel is built flush from both the sides of the wall. For small openings, flat and plate sections may be used but for openings of moderate size, angles back to back or inverted T-Sections may be the choice. For large openings, channels, I-Sections or built-up sections are preferred. Normally angle sections are used when lateral support from the wall to the angle section can be achieved. If there is any doubt about the lateral support from the wall, I-Section with plates, Fig. 7.1(h), may be used. Masonry is assumed to provide no lateral support even if the angle or channel section are encased in it.

The design loads for lintels are not well defined because it is not certain as to how much load from the masonry will come over the lintel. It is assumed that after setting of the mortar, the load from the masonry is distributed by arch action. As such, designers feel that a small portion of the masonry load above the opening should be considered for the design of lintels.

Case I When the slab over the lintel is above the apex of the equilateral triangle formed on the lintel, the load of the masonry in the triangle thus formed is assumed to act over it, Fig. 7.13(a).

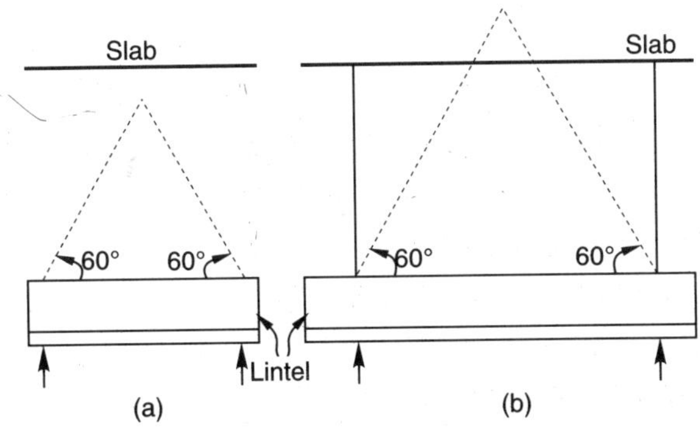

Fig. 7.13 *Loads from Masonry over Lintels*

Case II When the slab over the lintel is below the apex of the equilateral triangle formed on the lintel, the load of the masonry in the rectangle of length equal to the span of lintel, and height equal to the clear height of slab above the lintel, is assumed to act over it, Fig. 7.13(b).

The lintel is either designed as a laterally supported beam and a much bigger section is provided to account for any twisting or it should be designed as a laterally unsupported beam.

Note In Case I, when the design load is from the triangular portion of masonry, the maximum moment will be $Wl/6$, where W = triangular load from masonry, and l = effective span of lintel.

7.15 PURLINS

Purlins are beams provided over trusses to support the roofing between the adjacent trusses, Fig. 7.14(a). These are placed in a tilted position over the principal rafters of the trusses. Channel and angle sections are commonly used as purlins. The two axes u-u and v-v replace the axes x-x and y-y, as shown in Fig. 7.14(b).

The wind force is assumed to act normal to the roof trusses and the gravity loads pass through the centre of gravity of the purlin section. Hence, the purlin section is subjected to twisting in addition to bending. Such bending is called unsymmetrical bending.

Purlins may be designed as simple, continuous or cantilever beams. The simple beam design yields the greatest moments and deflections. It is difficult to ensure continuity through purlin splices, if designed as a continuous beam. Therefore, often designers prefer the cantilever design. If purlins are assumed to be simply supported the moments will be $wl^2/8$. On the other hand, if they are assumed to be continuous, the moments will be slightly less and are taken as $wl^2/10$.

Purlins have a tendency to sag in the direction of sloping roof and sag rods are sometimes provided midway or at third points, between roof trusses, to take up the sag. These rods reduce the moment M_{vv} and result in a smaller purlin section. Sag rods are frequently used with I and channel section purlins, but are rare with angle purlins.

Theoretically, it is desirable to place purlins only at the panel points as otherwise the principal rafter of the roof truss will be subjected to bending. But for large

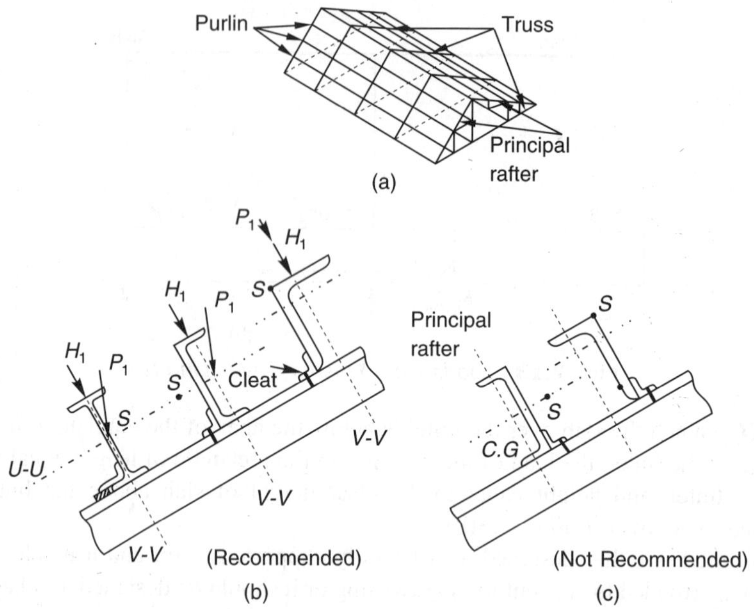

Fig. 7.14 *Purlins*

trusses, if the purlins are placed at panel points their spacing will be large and consequently large section purlins will have to be provided and may be impractical. When purlins are provided at intermediate points the principal rafter is designed for flexure and axial stresses.

A major consideration during erection of purlins is whether to place angle, channel or zee section purlins with their table/flange facing up-slope (Fig. 7.14(b)) or down-slope (Fig. 7.14(c)). To understand this one must have concept of shear centre.

Shear centre for any transverse section of a beam is the point of intersection of the bending axis and the plane of transverse section. The bending axis of the beam is its longitudinal axis through which transverse loads causing bending, must pass in order that bending of the beam is not accompanied by its twisting. For a cross-section with two axes of symmetry, the shear centre will fall at the intersection of the two axes, thus coinciding with the centroid of the section. For sections with one axis of symmetry, the shear centre will fall somewhere on that axis but not necessarily at the centroid of the section. This means that to avoid torsion in some of the beams the line of action of the loads should not pass through the centroid of the section. The shear centre is of particular importance for beams whose cross-sections are composed of thin parts which provide considerable bending resistance but little resistance to torsion. Most of the common structural sections such as channels, angles, etc. made up of thin plates fall into this class.

A pure bending couple acting in a principal plane, produces bending in that plane only and shear stresses do not exist. However, when bending of the beam is produced by transverse loads, shear stresses are generated. Generally, the shear stress on any section will be a force parallel to the plane of loading, which is statically equivalent to the torque created by the eccentric applied load about the shear centre. Hence, if twisting is to be avoided it is necessary to apply loads in the plane of resultant shear stress.

The term *shear flow* is often used when speaking of thin wall members although there is really no flow involved. It refers to shear per mm of cross-section and equals the unit shear stress times the thickness of the member. (The unit shear stress $= \frac{VA\bar{Y}}{Ib}$ and the shear flow $= \frac{VA\bar{Y}}{I}$, if the shear stress is assumed to be constant across the thickness of the section.)

For example in the channel section of Fig. 7.15(a), the shear flow is shown by arrows and in Fig. 7.15(b) the values are totaled for each component of the shape and denoted H and V. The two H values are in equilibrium horizontally and the V value balances the external shear at the section. Although the horizontal and vertical forces are in equilibrium, the same cannot be said for the moment forces unless the line of action of the resultant forces pass through a certain point called the shear centre. The horizontal forces H form a couple. The moment produced by this couple must be opposed by an equal and opposite moment which can only be produced by the two V values. Thus $Ve = Hh$ gives the location of shear centre.

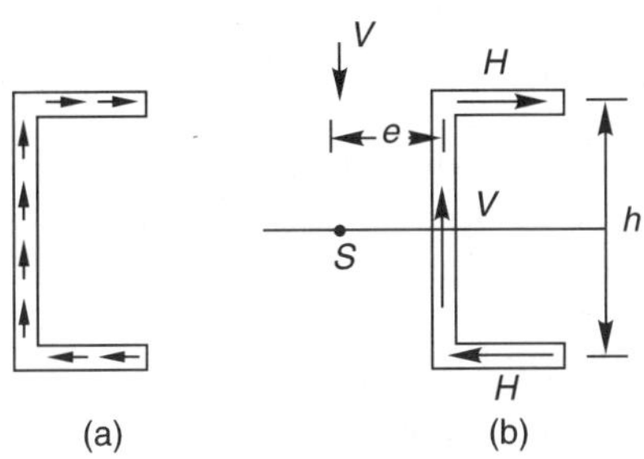

Fig. 7.15 *Shear Centre*

When purlins are placed as shown in Fig. 7.14(b) with table/flange facing up-slope the twisting moment does not cause instability. However, if placed with table/flange facing down-slope the twisting moment causes instability and hence such a placement is not recommended. The location of shear centre for some thin walled sections is given in Appendix VIII.

Design of Channel/I-Section Purlin

The design of a purlin is a trial and error procedure and is done as follows:

1. The span of purlin is the centre to centre distance of adjacent trusses.
2. The gravity loads P, due to sheeting and live load, etc. and the load H due to wind are computed. The loads act as shown in Fig. 7.14(b). The components of the loads along *uu*- and *vv*-axes are computed.
3. The maximum bending moments M_{uu}, M_{vv} along these axes are calculated by

$$M_{uu} = \frac{Pl}{10} \quad \text{and} \quad M_{vv} = \frac{Hl}{10}$$

where P = loads along *vv*-axis (in Newtons)

H = loads along uu-axis (in Newtons)
l = span of the purlin

4. Since purlins are subjected to biaxial bending, the maximum bending stress is given by

$$\sigma_b = \frac{M_{uu}}{I_{uu}} v + \frac{M_{vv}}{I_{vv}} u$$

$$= \frac{M_{uu}}{Z_{uu}} + \frac{M_{vv}}{Z_{vv}}$$

or $$\sigma_b = \frac{M_{uu}}{Z_{uu}} \left(1 + \frac{M_{vv}}{M_{uu}} \frac{Z_{uu}}{Z_{vv}}\right) \quad (7.20)$$

or $$Z_{uu} = \frac{M_{uu}}{\sigma_b} \left(1 + \frac{M_{vv}}{M_{uu}} \frac{Z_{uu}}{Z_{vv}}\right) \quad (7.21)$$

A suitable value of Z_{uu}/Z_{vv} for the I-Section (6-10), and for the channel section (7–9) is assumed. The value of Z_{uu} is determined from Eq. (7.21) and an appropriate section is selected from I.S. Hand Book No. 1. The bending stresses are increased by $33\frac{1}{3}$% to account for wind.

5. The value of maximum bending stress is computed from Eq. (7.20).
6. The calculated value of bending stress should be less than the permissible value of bending stress ($1.33 \times 0.66 \times f_y$)

Design of an Angle Section Purlin

Angle sections are unsymmetrical about both the axes. I.S: 800–1984 permits the use of the angle section as a purlin, provided the slope of the roof truss is less than 30°. The following procedure is adopted for the design:

1. Vertical loads and wind loads are determined. Both the loads are assumed to be normal to roof truss.
2. The maximum bending moment is computed by $\frac{wl^2}{10}$ or by $\frac{Wl}{10}$

   where w = uniformly distributed load
   W = concentrated load at centroid
   l = span of purlin

3. The modulus of section required is calculated by

$$Z = \frac{M}{1.33\sigma_b}$$

   where σ_b = permissible bending stress ($0.66 f_y$).

4. A trial section of angle purlins is arrived at by assuming the depth of the angle section as 1/45 of the span and width of the angle section as 1/60 of the span. The depth and width must not be less than the specified values to ensure that the deflections are not excessive.

5. A suitable angle section is selected from I.S. Hand Book No. 1 for the calculated leg lengths of angle section. The modulus of section provided should be more than modulus of section calculated in step 3.

7.16 ENCASED BEAM

A steel beam totally surrounded by concrete is referred to as encased beam. The natural bond existing between concrete and steel becomes the medium of interaction of the two materials. A steel beam is encased in concrete from an architectural point of view (Fig. 7.16). Also, encasing the beam in concrete increases its fire resistance and reduces deflections. As per I.S: 800–1984, a beam section may be encased if following conditions are fulfilled:

1. The section is of single web I-Section or of double open channel form, with the webs atleast 40 mm apart.
2. The beam is unpainted and solidly encased in concrete of M–15 grade.
3. The minimum width of solid casing is (b_0 + 100) mm, where b_0 is the width of the flange. A cover of 50 mm on each side of section is provided, as shown in Fig. 7.16.
4. 5 mm diameter stirrups are provided at 150 mm c/c spacing and four 10 mm diameter bars are provided on the four corners to keep stirrups in position.

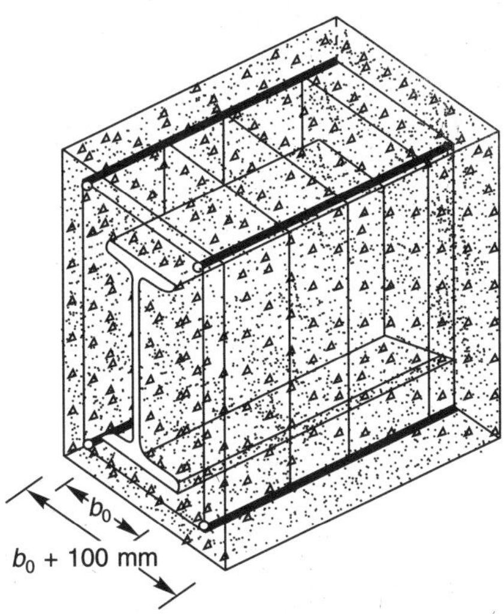

Fig. 7.16 *Encased Beam*

Design

The composite encased beam is considered to be homogeneous by transforming the concrete compression area into an equivalent steel area. The analysis and design of encased beams are based upon the elastic theories of flexure, and the usual assumptions of beam behaviour prevail. The steel beam is proportional to support all the dead loads applied before the concrete hardens, and the encased beam, steel and concrete acting together support all the dead and live loads applied after the concrete has hardened. The bending stresses for loads acting on the encased beam are computed on the basis of the moment of inertia of the composite section. Concrete in tension is neglected.

The compression flange of the beam is assumed to be laterally unsupported. A tentative I shape section is designed as discussed in Section 7.12. The concrete around the compression flange provides a little lateral support. To account for this, the moment of inertia about *yy*-axis is taken as $I_{yy} = Ar_y^2$, where $r_y = 0.2\ (b_0 + 100)$ for encased beams and the bending stresses are calculated again. The permissible bending stresses so calculated should not exceed 1.5 times permitted for the uncased section.

7.17 BEAM BEARING PLATES

In residential buildings where the beams rest on masonry, the pressure developed on the masonry may be excessive due to the high value of reactions at the supports. To check crushing of the supporting masonry, a bearing plate of adequate area is provided. For a simplified analysis, the bearing stress from the masonry below is assumed to be uniform. In fact, stresses are maximum at the edges of the plate near the centre of the beam rather than at the ends. At the other two edges of the plate the stresses are relieved as the plate bends normal to the beam. Therefore, the bearing pressure is non-uniform but to simplify the analysis it is assumed to be uniform. Reference may be made to Fig. 7.17(a).

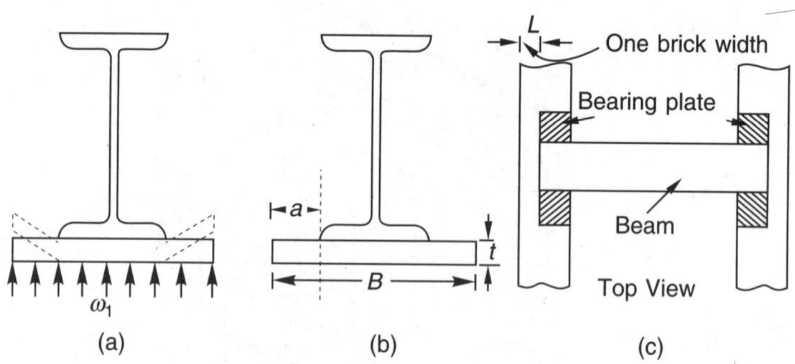

Fig. 7.17 *Bearing Plate*

Design

1. The area of the plate required is calculated by dividing the bearing pressure from the masonry to the reaction. Permissible bearing pressure from brick masonry may be assumed to be 4 N/mm^2.
2. The bearing plate should have enough length along the beam so that the vertical compression of the beam web is under control and that failure by crippling does not occur. The maximum length of the plate is limited by the wall thickness and will be equal to thickness of wall minus one brick width (Fig. 7.17(c)). This width is subtracted to keep the continuation of masonry on the outer face.
3. The width of the bearing plate will be the area of the plate divided by the length.

4. The thickness of the bearing plate can be computed in a similar way as that for a gusset base. The critical section is assumed to be at the edge of the beam flange (Fig. 7.17(b)). Hence,

$$\frac{w_1 a^2}{2} = \sigma_{bs} \times \frac{1}{6} \times 1 \times t^2$$

or

$$t = \sqrt{3 w_1 a^2 / \sigma_{bs}} \tag{7.22}$$

where w_1 = unit pressure from masonry
σ_{bs} = bending stress in slab base (185 MPa)
t = thickness of bearing plate
a = length of cantilever portion of bearing plate

7.18 CASTELLATED BEAM

A beam with a perforated web is called a castellated beam. It is an open web beam but made up of a single rolled wide flange beam section and is formed by flame cutting the beam section in a predetermined pattern and rejoining the segments by welding to produce a regular pattern of holes in the web (Fig. 7.1(j)). The beam section obtained in such a way can be even 50% deeper than the original section. By increasing the depth, the section modulus is increased by about 2.25 times the section modulus of the original beam section. Thus, the load carrying capacity of the beam is increased considerably. A castellated beam has some limitations also, viz. stress concentration occurs near the perforations and the shear carrying capacity is reduced. Stress concentrations may be reduced by making perforations near the neutral axis where the stresses are small and making the cut in a zig-zag way. The shear carrying capacity can be increased by stiffening the web at points of concentrated loads and reaction.

Design

1. The angle of cut is selected to be 45°. For a good design the depth of the stem of the T-Section at the minimum beam cross-section should not be less than 1/4 of the original beam section.
2. The loads over the section from the roof are ascertained and the maximum bending moments are computed.
3. The cross-sectional area of the T-Section at the open throat is calculated. The neutral axis of the section is determined and moment of inertia about the neutral axis is calculated.
4. The moment of resistance of the castellated section which is the product of the resultant tensile or compressive force and the distance between centroids of T-Sections is calculated.

$$M_R = A\, \sigma_{at}\, d \tag{7.23}$$

where A = area of the T-Section at open throat
d = distance between the centroids of T-Sections

The moment of resistance of the castellated beam should be more than the maximum moment.

5. The spacing of castellated beams should not exceed the spacing determined by Eq. (7.24).

$$S = \frac{P'}{W'l} \tag{7.24}$$

where S = c/c distance between the castellated beams in m

P' = net load carrying capacity in N (total load carrying capacity – dead weight of castellated beam)

W' = design load in N/m^2

l = span of the beam in m

6. Stiffeners are designed at the supports and below the concentrated loads.
7. The beam is checked in shear. The average shear at ends is calculated from Eq. 7.25.

$$\tau_{va} = \frac{R}{d't} \tag{7.25}$$

$$< 0.4 f_y$$

where R = end reaction in Newtons

d' = depth of the stem of T-Section

t = thickness of the stem

8. The maximum combined local bending stress and direct stress in T-Segments is also worked out and should be less than the permissible bending stress.
9. The maximum deflection of the T-Segment is calculated. This occurs at the mid span and is due to the net load carrying capacity and local effects.

Let, δ_1 = deflection due to net load carrying capacity

δ_2 = deflection due to local effects

I = average moment of inertia of the section

= 1/2 (moment of inertia of solid section + moment of inertia of hollow section)

I_T = moment of inertia of T-Section

p = number of perforation panels in half span

m, n = distances as shown in Fig. 7.18.

$$\delta_1 = \frac{5}{384}\frac{Wl^3}{EI} \tag{7.26}$$

$$\delta_2 = \frac{V_{avg}\, p(m+n)^3}{24\, EI_T} \tag{7.27}$$

$$\delta = \delta_1 + \delta_2 < l/325$$

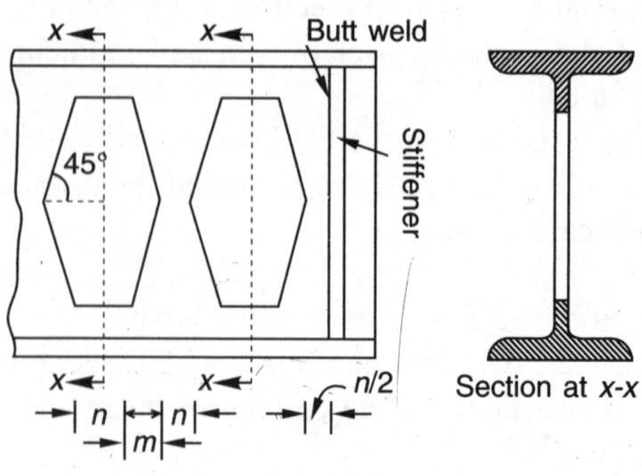

Fig. 7.18 *Castellated Beam*

7.19 EFFECT OF HOLES IN BEAM

Holes are made in the beams for the following reasons:

1. to make riveted/bolted connections and
2. to provide pipes, conduits, etc. in high-rise buildings.

Rivet holes reduce the section modulus and thus affect the bending stresses. Some of the codes do not allow any deduction of holes in flanges until it is more than 15%. Other codes consider the effect of holes in the tension flange only and require calculations of gross and net moments of inertia. The neutral axis in such a case will shift towards the compression flange. But to simplify the analysis, the neutral axis is assumed to be unaffected by the holes. As regards the rivet holes, the normal practice is to deduct all the holes whether they are on the compression flange or on the tension flange.

In case if holes are made for pipes and conduits, extra plates (about 20% more of the area removed) are provided all around the holes to stiffen the cut portion. When holes are made in the web, these do not reduce the section modulus much, but the shear strength is reduced appreciably. If the hole is located at midspan, the shear is minimum and may have little effect on the strength of the girder. If the hole is located near the support in the region of high shear, the additional bending stresses produced by this shear must be added to the conventional bending stresses from the applied beam load.

7.20 COMPOSITE BEAM AND SHEAR CONNECTORS

A composite beam is one whose strength depends upon the mechanical interaction between two or more materials. Reinforced concrete beams are actually composite members but are not designed as such. The term composite beam is applied to steel beams securely bonded with roof floor through shear connectors. These shear connectors check the relative slip that may occur at the interface of the two materials. A shear connector is a mechanical device by which interaction of a concrete slab with a steel beam is accompalished in a composite beam construction. The concrete slab becomes the compression flange of the composite beam and the steel section resists the tensile stresses. Shear connectors serve the following purposes:

1. They allow the concrete slab to become a part of the compression flange of the girder, forming a composite element having greater strength and rigidity.
2. They transfer the longitudinal shear from the concrete to the steel and hold the concrete from uplifting.
3. They laterally support the steel beams.

The shear connectors shown in Fig. 7.19 may be **rigid connectors**—short lengths of bars, angles, Tees, channels welded to the girder flange, deriving their resistance to horizontal shear from the bearing pressure of concrete; **flexible connectors**—such as studs deriving resistance through their bending; **bond or anchorage connectors**—in the form of vertical or inclined mild steel bars or spirals.

Assumptions

1. The concrete slab is continuously connected throughout the entire length of the beam.
2. The slab and the beam do not become separated vertically at any point along the beam.

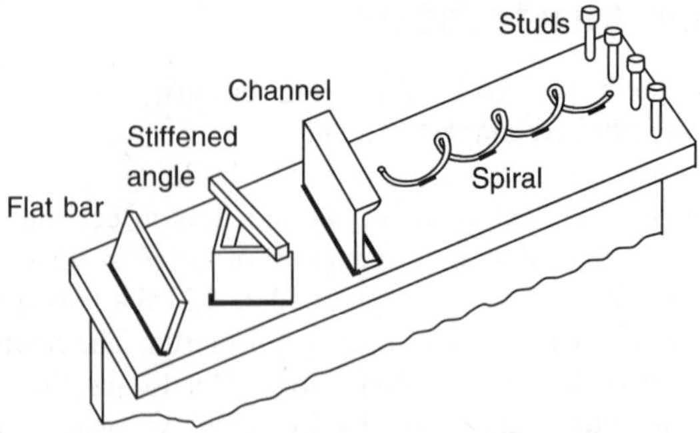

Fig. 7.19 *Typical Shear Connectors*

3. The slip of the shear connector is directly proportional to the load on the connection.
4. The strain distribution is linear over the depth of the member.

Design Requirements Shear connectors transmit horizontal shear forces from the slab into the beam making both the beam and slab act in unison. In addition, they provide anchorage for the slab, thus preventing any tendency for separation of the slab from the beam.

Horizontal Shear Force The size of shear connector is selected and spaced according to the principles of elastic behaviour and longitudinal shear distribution. Thus the spacing of connectors will be a function of shear transfer according to the usual expression for horizontal (longitudinal) shear is

$$V_h = \frac{VA\bar{Y}}{I}$$

where V_h = horizontal shear per unit length at the plane of contact of the RCC slab and beam flange

V = The total vertical shear due to the external loads

$A\bar{Y}$ = The static moment of the transformed area on the slab side of contact surface about the neutral axis of the composite section or the static moment of the area of reinforcement in the RCC slab for negative moment

I = Moment or inertia of the transformed composite section.

However, if the design is based on the shear connectors allowing the composite beam to reach its ultimate load, the force is resisted by the compressive area of the concrete slab as shown in Fig. 7.20.

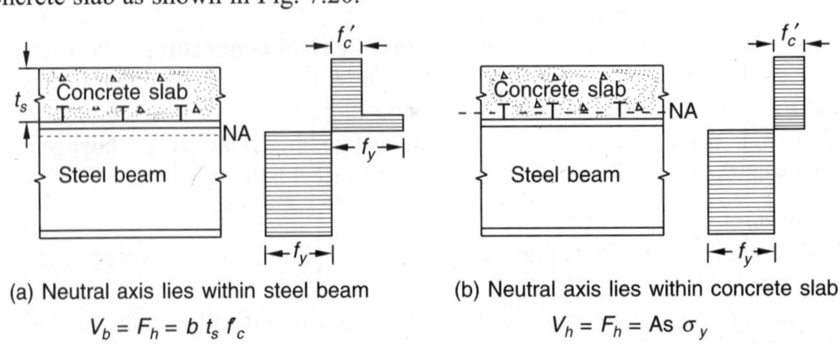

Fig. 7.20 *Composite Action of Steel Beam and Concrete Slab*

For neutral axis lying within the beam (Fig. 7.20(a))

$$V_h = b f_c' t_s = f_c' A_c \tag{7.28}$$

and for neutral axis lying within concrete slab (Fig. 7.20 (b))

$$V_h = A_s f_y \tag{7.29}$$

The horizontal ultimate forces are reduced by a safety factor of 2, and concrete is taken at 85 per cent of its strength. The Eqs (7.28) and (7.29) reduce to

$$V_h = \frac{0.85 f_c' A_c}{2} \tag{7.30}$$

and

$$V_h = \frac{A_s f_y}{2} \tag{7.31}$$

where f_c' = compressive strength of concrete
A_c = cross-sectional area of effective concrete slab
A_s = cross-sectional area of beam
t_s = thickness of concrete slab
b = effective width of slab

The smaller of the two values given by Eqs (7.30) and (7.31) is taken as the total horizontal shear to be carried by all the connectors between the point of maximum bending moment and the ends of the beam, or between the point of maximum moment and a point of contraflexure in a continuous beam.

The number of connectors required on each side of maximum moment based on beam behaviour at the ultimate moment, is determined by dividing the nominal shear load V_h by the allowable shear load Q of the connector to be used.

Allowable Load Q

Rigid connectors

$$Q = F_b A_b \tag{7.32}$$

where F_b = permissible bearing pressure on the concrete

$$= 0.25 f_{ck} \sqrt[3]{\frac{A}{A_b}}$$

A = area to which bearing force is transmitted
$= B t_s$
A_b = bearing area of connector
B = width of flange of beam
f_{ck} = 28 day crushing strength of 150 mm cubes

Anchor connectors

$$Q = K \sigma_{st} + A_t \tag{7.33}$$

where σ_{st} = permissible tensile stress in anchor bars
A_t = cross-sectional area of anchor bars
k = a coefficient (Table 7.8)

Table 7.8 *Value of Coefficient k for Anchor Connectors*

S. No.	Type of Anchor	k
1.	Anchor with bond length not less than 40 times the diameter of the bar	1.0
2.	Looped anchor	
	(i) having loop diameter greater than 15 times the diameter of the bar	1.0
	(ii) having loop diameter less than 15 times the diameter of the bar and used in combination with rigid connectors	0.7
3.	Hooked anchors used in combination with rigid connectors	0.5

Flexible connectors

Studs should have a minimum stud head diameter of $d + 12$ mm, stud head height of 12 mm, ultimate strength of 460 N/mm^2, yield stress of 350 N/mm^2 and elongation of 20 per cent.

$$Q = 1.518\, Hd \sqrt{f_{ck}} \qquad (H/D \geq 4.2) \qquad (7.34)$$

and

$$Q = 6.198\, d^2 \sqrt{f_{ck}} \qquad (H/D < 4.2) \qquad (7.35)$$

where d = diameter of stud

H = height of stud

Channels

$$Q = 0.338(T + 0.5t)\, \frac{L}{\sqrt{f_{ck}}} \qquad (7.36)$$

where T = thickness of flange of channel

t = thickness of web of channel

L = length (in mm) of channel shear connector

Spirals

$$Q = 560\, d \sqrt[4]{f_{ck}} \qquad (7.37)$$

where d = diameter of bar forming spiral

Spacing of Shear Connectors

Rigid connectors The spacing and size of shear connectors should satisfy the following requirements.

1. The bearing pressure on the face of the connector should not exceed the permissible value Q.
2. The longitudinal shear stress along the shearing surface between two consecutive connectors should not be greater than 2.5 times the permissible shear stress for concrete. This condition is deemed to be satisfied, if

$$V_h = 2.5\, b\, q$$

where b = width of the rigid connector flange at the contact surface

q = permissible shear stress measured as indirect tension in the concrete.

3. The projected area of one rigid connector to another, at a slope of 1 in 5 should be at least 3 times the area of the face of the connector.

Anchor connectors The spacing should not be less than 0.7 times the depth of the slab and should not be greater than two times the depth of slab. When combined with rigid connectors the spacing should not be greater than 2.5 times the depth of slab.

Flexible connectors Transverse spacing of the connectors at any section is given by $4d$, where d is the diameter of the stud. In case a single row of studs is used it should be placed over the centre of the beam web.

The maximum longitudinal spacing of the connectors should not exceed 600 mm nor four times the slab thickness.

The minimum spacing of the studs is governed by the welding gun and range approximately 45 mm for 12 mm studs to 56 mm for 25 mm studs.

Effective Width of Concrete Slab The composite beam behaves as a T-beam. The theoretical effective width of the flange is a function of the span of the beam, Poisson's ratio and the shape of moment diagram. When slab is on both sides of the beam. The effective width of the concrete slab should be smallest of the following:

1. Not more than 1/4 of the span.
2. Projection beyond steel flange to be not more than 1/2 the clear distance to the adjacent beam.
3. Not more than 8 times the slab thickness.

For beams on only one side, the projection beyond the beam should be

1. Not more than 1/12 of the beam span.
2. Not more than 1/2 the clear distance to the adjacent beam.
3. Not more than 6 times the slab thickness.

Solved Examples

Example 7.1 A simply supported steel joist with a 4.0 m effective span carries a uniformly distributed load of 40 kN over its span inclusive of self weight. The beam is supported laterally throughout. Select a suitable section (f_y = 250 N/mm^2) and check its safety.

Solution

$$\text{Maximum bending moment, } M = \frac{wl^2}{8} = \frac{40 \times 10^3 \times 4 \times 10^3}{8} = 20 \times 10^6 \text{ Nmm}$$

$$\text{Maximum shear force, } V = \frac{wl}{2} = \frac{40 \times 10^3}{2} = 20 \times 10^3 \text{ N}$$

$$\text{Modulus of section required, } Z = \frac{M}{\sigma_{bc,t}} = \frac{20 \times 10^6}{165} = 121212 \text{ mm}^3$$

The required section modulus is referred from I.S. Hand Book No. 1. An I.S.L.B. 175 @ 163.8 N/m (Z_{xx} = 125,300 mm^3) was selected as a trial. It failed in deflection. Next, a higher section I.S.M.B. 175 @ 189.3 N/m (Z_{xx} = 145,400 mm^3) was checked and was also found to fail in deflection.

Let us try I.S.L.B. 200 @ 194.2 N/m. The relevant properties of the section are:

$$D = 200 \text{ mm}, T = 7.3 \text{ mm}, t = 5.4 \text{ mm}$$

$$I_{xx} = 1696.6 \times 10^4 \text{ mm}^4, Z_{xx} = 169.7 \times 10^3 \text{ mm}^3, h_2 = 17.15 \text{ mm}$$

Check for shear

$$\tau_{va} = \frac{20 \times 10^3}{200 \times 5.4} = 18.518 \text{ N/mm}^2$$

$$< 0.40 f_y \text{ (100 N/mm}^2 \text{ for steel with } f_y = 250 \text{ N/mm}^2)$$

Which is safe.

Check for deflection

$$\delta_{\text{allowable}} = \frac{4 \times 10^3}{325} = 12.30 \text{ mm}$$

$$\delta_{\text{cal}} = \frac{5}{384} \times \frac{wl^4}{EI}$$

$$= \frac{5}{384} \times \frac{40 \times 10^3 \times (4 \times 10^3)^3}{2.00 \times 10^5 \times 1696.6 \times 10^4} = 9.82 \text{ mm} < 12.30 \text{ mm}$$

Which is all right.

Check for crippling Assume a bearing length of 100 mm

$$\sigma_p = \frac{\text{reaction}}{(b + h_2\sqrt{3})\,t} = \frac{20 \times 10^3}{(100 + 17.15\sqrt{3})\,5.4}$$

$$= 28.55 \text{ N/mm}^2$$

$$< 0.75 f_y \text{ (187.5 N/mm}^2 \text{ for steel with } f_y = 250 \text{ N/mm}^2)$$

which is as it should be.

Example 7.2 Design a beam of 5 m effective span, carrying a uniform load of 20 kN/m if the compression flange is laterally unsupported. (Assume $f_y = 250$ N/mm^2)

Solution

$$\text{Effective span } l = 5 \text{ m}$$

$$\text{Live load} = 20 \times 10^3 \text{ N/m}$$

$$\text{Let dead load} = 600 \text{ N/m}$$

$$\text{Design load, } w = 20 \times 10^3 + 600 = 20.6 \times 10^3 \text{ N/m}$$

$$\text{Maximum bending moment, } M = \frac{wl^2}{8} = \frac{20.6 \times 10^3 \times (5 \times 10^3)^2}{10^3 \times 8}$$

$$= 64375 \times 10^3 \text{ Nmm}$$

$$\text{Maximum shear force, } V = \frac{wl}{2} = \frac{20.6 \times 10^3 \times 5}{2} = 51.5 \times 10^3 \text{ N}$$

$$\text{Section modulus required, } Z = \frac{M}{\sigma_{bc}} = \frac{64{,}375 \times 10^3}{110} = 585.227 \times 10^3 \text{ mm}^3$$

From I.S. Handbook No. 1 select a beam section having a section modulus Z_{xx} more than the above calculated value. Provide I.S.M.B. 350 @ 514 N/m. The relevant properties of the section are,

$$Z_{xx} = 778.9 \times 10^3 \text{ mm}^3, I_{xx} = 13{,}630.3 \times 10^4 \text{ mm}^4, T = 14.2 \text{ mm}$$

$$t = 8.1 \text{ mm}, D = 350 \text{ mm}, h_2 = 31.00 \text{ mm}, r_{yy} = 28.4 \text{ mm}$$

$$d_1 = D - 2h_2 = 350 - 2 \times 31 = 288 \text{ mm}$$

Check for section modulus The beam has equal flanges. Therefore, Table 7.1 of the text can be used to check the section. It can also be checked by finding out the elastic critical stress. Both the methods of check have been illustrated.

(i) Method using Table 7.1.

$$\frac{D}{T} = \frac{350}{14.2} = 24.64, \quad \frac{l}{r_{yy}} = \frac{5 \times 10^3}{28.4} = 176.05$$

$$\frac{T}{t} = \frac{14.2}{8.1} = 1.75 < 2, \quad \frac{d_1}{t} = \frac{288}{8.1} = 35.55 < 85$$

As $\frac{T}{t} < 2$ and $\frac{d_1}{t} < 85$, refer to Table 7.1(b) of the text.

For $f_y = 250$ MPa, $D/T = 24.64$, $\frac{l}{r_{yy}} = 176.05$,

Bending compressive stress, $\sigma_{bc} = 87.32 \text{ N/mm}^2$

Moment of resistance, $M_R = 87.32 \times 778.9 \times 10^3 = 68{,}013.548 \times 10^3$ Nmm

$> 64{,}375 \times 10^3$ Nmm

which is safe.

(ii) Method using elastic critical stress

$$\text{The elastic critical stress, } f_{cb} = k_1 (X + k_2 Y) \frac{c_2}{c_1}$$

$\psi = 1$, as the areas of both the flanges at the points of least and maximum bending moments are same.

From Table 7.3 of the text, for $\psi = 1$, $k_1 = 1$

$$\omega = \frac{1}{2} = 0.5$$

From Table 7.4, for $\omega = 0.5$

$$k_2 = 0$$

From Table 7.5, for $\frac{D}{T} = 24.64$, and $l/r_{yy} = 176.05$,

$$X = 161.856 \text{ N/mm}^2, Y = 85.95 \text{ N/mm}^2$$

$$\therefore \quad f_{cb} = 1 \times (161.856 + 0 \times 85.95) \times 1 = 161.856 \text{ N/mm}^2$$

The value of f_{cb} is increased by 20%, if $\frac{d_1}{t} < 85$ and $\frac{T}{t} < 2$.

$$f_{cb} = 1.2 \times 161.856 = 194.227 \text{ N/mm}^2$$

From Table 7.6, of the text, for $f_y = 250$ N/mm², and $f_{cb} = 194.227$ N/mm²,

$$\sigma_{bc} = 87.84 \text{ N/mm}^2$$

$$\text{Moment of resistance} = 87.84 \times 778.9 \times 10^3 = 68{,}418.576 \times 10^3 \text{ Nmm}$$
$$> 64{,}375 \times 10^3 \text{ Nmm}$$

which is as it should be.

Check for shear

$$\tau_{va,\text{cal}} = \frac{51.5 \times 10^3}{350 \times 8.1} = 18.165 \text{ N/mm}^2$$
$$< 0.4 f_y \text{ (for } f_y = 250 \text{ N/mm}^2,\ \tau_{va} = 100 \text{ N/mm}^2)$$

which is safe.

Check for deflection

$$\delta_{\text{allowable}} = \frac{l}{325} = \frac{5 \times 10^3}{325} = 15.3846 \text{ mm}$$

$$\delta_{\text{cal}} = \frac{5}{384} \times \frac{wl^4}{EI}$$

$$= \frac{5}{384} \times \frac{20.6 \times 10^3 \times (5 \times 10^3)^4}{2.00 \times 10^5 \times 13{,}630.3 \times 10^4 \times 10^3} = 6.149 \text{ mm}$$
$$< 15.3846 \text{ mm}$$

which is safe.

Example 7.3 A hall in a building is 15 m × 18 m. The beam arrangement supporting the reinforced concrete slab is shown in Fig. Ex. 7.3(a). Design the beams (f_y = 250 MPa) if the unit weight of concrete is 24 kN/m³.

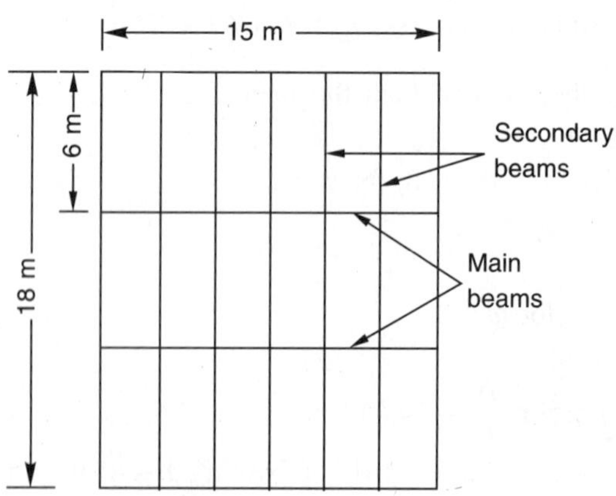

Fig. Ex. 7.3(a)

Solution The beams parallel to the long span will be termed as secondary beams and those parallel to short span, as main beams. Secondary beams are designed to support live loads, dead load of slab and self weight of beams. Main beams are designed for the reactions from secondary beams and their self weight.

Secondary beams

Let, Self weight of beam/m = 0.5 kN

Live load on floor = 4 kN/m^2

Dead weight of 2.5 m strip of concrete slab/m length

$$= 2.5 \times 24 \times \frac{150}{1000} \times 1 = 9 \text{ kN}$$

Live load from roof/m = $4 \times 1 \times 2.5 = 10$ kN

Design load/m length = $10 + 9 + 0.5 = 19.5$ kN

Effective span, $l = 6.0$ m

$$\text{Maximum bending moment, } M = \frac{wl^2}{8} = \frac{(19.5\times10^3 \times 6)\times 6\times10^3}{8}$$

$$= 87.75 \times 10^6 \text{ Nmm}$$

$$\text{Maximum shear force, } V = \frac{wl}{2} = \frac{19.5\times10^3 \times 6}{2} = 58.5 \times 10^3 \text{ N}$$

The compression flange of the secondary beam is embedded in the concrete slab and is therefore designed as a laterally supported beam.

The bending stress $\sigma_{bc,\,t} = 0.66 f_y = 0.66 \times 250 = 165$ N/mm^2

$$\text{Section modulus, } Z = \frac{87.75\times10^6}{165} = 531818.18 \text{ mm}^3$$

From I.S. Handbook No. 1, it is found that I.S.M.B. 300 @ 433.6 N/m has a value of section modulus 573.6×10^3 mm^3 closer to the required section modulus as compared to I.S.L.B. 325 @ 422.8 N/m having a section modulus of 607.7×10^3 mm^3. But the self weight of I.S.L.B. 325 is less than the self weight of I.S.M.B. 300. Therefore, let us try I.S.L.B. 325 @ 422.8 N/m. The relevant properties of the section are

$$D = 325 \text{ mm}, t = 7.0 \text{ mm}, I = 9874.6 \times 10^4 \text{ mm}^4$$

Check for shear

$$\tau_{va,\,\text{cal}} = \frac{58.5\times10^3}{325\times 7.0} = 25.84 \text{ N/mm}^2$$

$$< 0.4 f_y \text{ (for } f_y = 250 \text{ N/mm}^2, \tau_{va} = 100 \text{ N/mm}^2)$$

which is safe.

Check for deflection

$$\delta_{\text{allowable}} = \frac{6\times10^3}{325} = 18.46 \text{ mm}$$

$$\delta_{cal} = \frac{5}{384} \times \frac{(19.5 \times 10^3 \times 6) \times (6 \times 10^3)^3}{2.00 \times 10^5 \times 9874.6 \times 10^4} = 16.66 \text{ mm}$$

$$< 18.46 \text{ mm}$$

which is as it should be.

Main beams Let us assume the compression flanges of the main beams to be properly braced so that complete lateral support may be achieved. Therefore, these will be designed as laterally supported beams.

$$\text{Reaction on the main beam} = 19.5 \times 10^3 \times 6 = 117{,}000 \text{ N} = 117.0 \text{ kN}$$

$$\text{Effective span} = 15 + 0.3 = 15.3 \text{ m}$$

Let us assume the self weight of beam to be 1000 N/m.

$$\text{Maximum bending moment due to self weight} = \frac{(1000 \times 15.3) \times 15.3 \times 10^3}{8}$$

$$= 29.26 \times 10^6 \text{ Nmm}$$

$$= 29.26 \times 10^3 \text{ kNmm}$$

Maximum bending moment due to reactions from secondary beams

$$\text{Maximum reaction from secondary beams} = \frac{2 \times 19.5 \times 6}{2} = 117.0 \text{ kN}$$

$$R_A = R_B = \frac{117.0 \times 5}{2} = 292.5 \text{ kN}$$

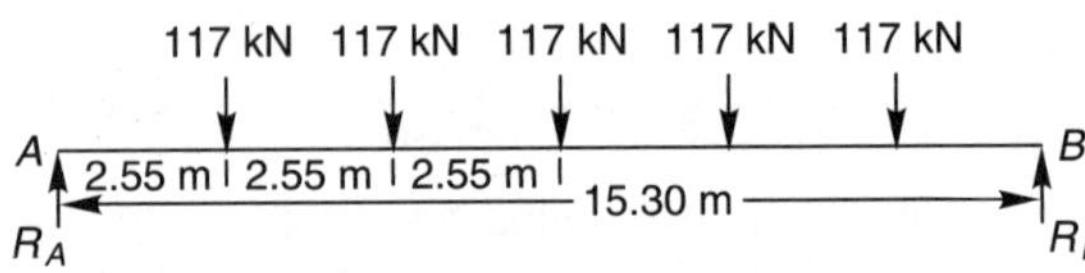

Fig. Ex. 7.3(b)

Maximum bending moment will be at the mid of the main beam (Fig. Ex. 7.3(b))

$$= 292.5 \times 7.65 - 117.0 \times 5.1 - 117.0 \times 2.55$$

$$= 1342.57 \text{ kNm} = 1342.57 \times 10^3 \text{ kNmm}$$

$$\text{Design bending moment} = 1342.57 \times 10^3 + 29.26 \times 10^3$$

$$= 1371.83 \times 10^3 \text{ kNmm} = 1371.83 \times 10^6 \text{ Nmm}$$

$$\text{Section modulus} = \frac{1371.83 \times 10^6}{165} = 8314.15 \times 10^3 \text{ mm}^3$$

Let us try I.S.W.B. 600 with a cover plate 400 mm × 32 mm on each flange. From Table XIV of I.S. Handbook No. 1, the relevant properties of the section are,

$D = 600$ mm, $t = 11.2$ mm, $Z_{xx} = 10{,}905.1 \times 10^3$ mm^3, $I_{xx} = 362{,}048.3 \times 10^4$ mm^4

Check for shear

$$\tau_{va,\,cal} = \frac{292.5 \times 10^3 + 1000 \times 15.3/2}{600 \times 11.2} = 44.66 \text{ N/mm}^2$$

$$< 0.4\, f_y \text{ (for } f_y = 250 \text{ N/mm}^2, \tau_{va} = 100 \text{ N/mm}^2\text{)}$$

which is all right.

Check for deflection

$$\delta_{\text{allowable}} = \frac{15.30 \times 10^3}{325} = 47.07 \text{ mm}$$

$$\text{Deflection due to point loads, } \delta_1 = \frac{Wl^3}{192\, EI}\, n\left[3 - \frac{1}{2}\left(1 + \frac{4}{n^2}\right)\right]$$

$$n - 1 = \text{number of concentrated loads} = 5$$

$$\therefore \quad n = 6$$

$$\delta_1 = \frac{Wl^3}{192\, EI} \times 6 \times \left[3 - \frac{1}{2}\left(1 + \frac{4}{(6)^2}\right)\right]$$

$$= \frac{14.66}{192}\frac{Wl^3}{EI}$$

$$= \frac{14.66 \times 117.0 \times 10^3 \times (15.30 \times 10^3)^3}{192 \times 2.00 \times 10^5 \times 362{,}048.3 \times 10^4} = 44.18 \text{ mm}$$

$$\text{Deflection due to self weight} = \frac{5}{384} \times \frac{(1000 \times 15.3) \times (15.3 \times 10^3)^3}{2.00 \times 10^5 \times 362{,}048.3 \times 10^4}$$

$$= 0.985 \text{ mm}$$

$$\text{Total deflection, } \delta_{\text{cal}} = 44.18 + 0.985 = 45.17 \text{ mm} < 47.07 \text{ mm}$$

which is all right.

Example 7.4 Design a beam of effective span 6.0 m and subjected to a bending moment of 105.3×10^6 Nmm for the following conditions:

(i) The compression flange is laterally unsupported throughout,

(ii) The beam is encased in concrete.

Checks for deflection and shear are not required. Assume f_y = 250 MPa.

Solution

(i) Section modulus, $Z = \dfrac{105.3 \times 10^6}{110} = 957.27 \times 10^3 \text{ mm}^3$

From I.S. Handbook No. 1, it is found that the I.S.M.B. 400 @ 604.3 N/m (Z_{xx} = 1022.9×10^3 mm^3) provides the required section modulus but it was found to fail in bending. Therefore, next higher section I.S.M.B. 450 @ 710.2 N/m is checked.

The relevant properties of the section are,

$$D = 450 \text{ mm}, T = 17.4 \text{ mm}, t = 9.4 \text{ mm}$$

$$I_{xx} = 30,\, 390.8 \times 10^4 \text{ mm}^4,\; Z_{xx} = 1350.7 \times 10^3 \text{ mm}^3,$$

$$r_{yy} = 30.1 \text{ mm}, h_2 = 35.40 \text{ mm}$$

$$d_1 = D - 2h_2 = 450 - 2 \times 35.40 = 379.2 \text{ mm}$$

$$\frac{d_1}{t} = \frac{379.2}{9.4} = 40.34 < 85$$

$$\frac{T}{t} = \frac{17.4}{9.4} = 1.851 < 2$$

as $\frac{d_1}{t} < 85$ and $\frac{T}{t} < 2$ refer to Table 7.1(b),

$$\frac{D}{T} = \frac{450}{17.4} = 25.86$$

$$\frac{l}{r_{yy}} = \frac{6 \times 10^3}{30.1} = 199.33$$

for $\frac{D}{T} = 25.86$ and $\frac{l}{r_{yy}} = 199.33$,

$$\sigma_{bc} = 77.997 \text{ N/mm}^2$$

$$\text{Moment of resistance} = 77.997 \times 1350.7 \times 10^3 = 105.350 \times 10^6$$

$$> 105.3 \times 10^6 \text{ Nmm}$$

which is safe.

(ii) Beam encased in concrete The beam section required is I.S.M.B. 450 @ 710.2 N/m. For the encased beam,

$$r_y = 0.2(b_0 + 100) \text{ mm} = 0.2(150 + 100) = 50 \text{ mm}$$

$$\frac{D}{T} = \frac{450}{17.4} = 25.86, \quad \frac{l}{r_y} = \frac{6 \times 10^3}{50} = 120$$

For $\frac{D}{T} = 25.86$ and $\frac{l}{r_y} = 120$, from Table 7.1(b) of the text,

$$\sigma_{bc} = 112.3 \text{ N/mm}^2 < 116.99 \text{ N/mm}^2 \ (1.5 \times 77.997)$$

which is safe.

Example 7.5 Find the shear centre for the channel section shown in Fig. Ex. 7.5. The channel is subjected to a transverse shear of 30 kN.

Solution

$$I_{xx} = \frac{100 \times 300^3}{12} - \frac{90 \times 280^3}{12} = 6036 \times 10^4 \text{ mm}^4$$

$$\text{Shear at } B = \frac{V A \bar{Y}}{I} = \frac{30 \times 10^3 \times \left\{\left(100 - \frac{10}{2}\right) \times 10\right\} \times 145}{6036 \times 10^4} = 68.46 \text{ N/mm}$$

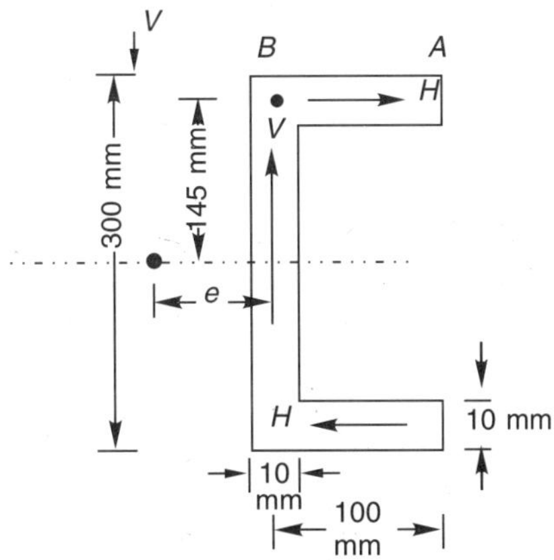

Fig. Ex. 7.5

$$\text{Total } H = \frac{1}{2} \times \left(100 - \frac{10}{2}\right) \times 68.46 \;= 3251.85 \text{ N}$$

$$Ve = H\,h$$

$$30 \times 10^3 \times e = 3251.85 \times (300 - 10)$$

$$\Rightarrow \quad e = 31.43 \text{ mm from the centre of the web.}$$

Example 7.6 For a roof truss system of an industrial building, following data is given

Span of truss = 6.0 m
Spacing of truss = 3.0 m centre to centre
Inclination of roof = 30°
Spacing of purlins = 2.0 m centre to centre
Wind pressure = 1.5 kN/m^2
Roof coverage — A.C. sheets weighing 200 N/m^2

Design:

(i) channel section purlins.
(ii) angle section purlins.
(iii) channel section purlins with sag rods midway.

Solution

(i) Assume dead load of purlins = 100 N/m

Dead load of roof = 2 × 200 = 400 N/m

Total dead load = 400 + 100 = 500 N/m

The dead load acts vertically.

The component of dead load parallel to the roof,

$$H = 500 \sin 30° = 250 \text{ N/m}$$

$$\text{Wind load} = 2 \times 1.5 \times 10^3 = 3000 \text{ N/m}$$

Component of dead load normal to roof = 500 cos 30° = 433.01 N/m

Total load normal to the roof, P = 3000 + 433.01 = 3433.01 N/m

$$M_{uu} = \frac{Pl}{10} = \frac{(3433.01 \times 3) \times 3 \times 10^3}{10} = 3089.70 \times 10^3 \text{ Nmm}$$

$$M_{vv} = \frac{Hl}{10} = \frac{250 \times 3 \times 3 \times 10^3}{10} = 225 \times 10^3 \text{ Nmm}$$

$$Z_{uu} = \frac{M_{uu}}{\sigma_{bc}}\left(1 + \frac{M_{vv}}{M_{uu}} \times \frac{Z_{uu}}{Z_{vv}}\right)$$

Assume $Z_{uu}/Z_{vv} = 7$. Increase the stresses by $33\frac{1}{3}\%$ to account for wind.

$$\therefore \quad Z_{uu} = \frac{3089.70 \times 10^3}{1.33 \times 165} \times \left(1 + \frac{225 \times 10^3}{3089.70 \times 10^3} \times 7\right) = 21{,}256.323 \text{ mm}^3$$

Select I.S.J.C. 100 @ 56.9 N/m from I.S. Hankbook No. 1. The relevant properties are,

$$Z_{xx} = Z_{uu} = 24.8 \times 10^3 \text{ mm}^3,\ Z_{yy} = Z_{vv} = 4.8 \times 10^3 \text{ mm}^3$$

$$\frac{Z_{uu}}{Z_{vv}} = 5.16$$

With the above value of Z_{uu}/Z_{vv},

$$\sigma_{bc} = \frac{3089.70 \times 10^3}{24.8 \times 10^3} \times \left(1 + \frac{225 \times 10^3}{3089.70 \times 10^3} \times 5.16\right) = 171.39 \text{ N/mm}^2$$

$$< 1.33 \times 165 = 219.45 \text{ N/mm}^2$$

which is safe.

Therefore, provide I.S.J.C. 100 @ 56.9 N/m.

(ii) Assume dead load of the purlin as 100 N/m.

Dead load of roof = 2 × 200 = 400 N/m

Total dead load = 400 + 100 = 500 N/m

The dead load is assumed to act normal to the roof.

Wind load = $1.5 \times 10^3 \times 2$ = 3000 N/m

Total load normal to roof surface = 3000 + 500 = 3500 N/m

$$M = \frac{WL}{10} = \frac{(3500 \times 3) \times 3 \times 10^3}{10} = 315 \times 10^4 \text{ Nmm}$$

$$\text{Section modulus required, } Z = \frac{M}{1.33 \times \sigma_{bc}} = \frac{315 \times 10^4}{1.33 \times 165} = 14{,}354.067 \text{ mm}^3$$

$$\text{Depth of angle purlin} = \frac{L}{45} = \frac{3 \times 10^3}{45} = 66.66 \text{ mm} \approx 80 \text{ mm}$$

$$\text{Width of angle purlin} = \frac{L}{60} = \frac{3 \times 10^3}{60} = 50 \text{ mm}$$

Let us try angle section 80 mm × 50 mm × 10 mm.

The section modulus provided = 14400.0 mm^3

> 14354.067 mm^3

which is sufficient.

$$\sigma_{bc,\ \text{cal}} = \frac{315 \times 10^4}{14400} = 218.75 \text{ N/mm}^2$$

$$< 1.33 \times 165 = 219.45 \text{ N/mm}^2$$

which is safe.

(iii) Since the channel section purlins are provided with sag rods midway, the span for calculation of the moment M_{vv} will be half the span of the purlin. However, for moment M_{uu} the span of the purlin will remain equal to the spacing between the trusses.

M_{uu} = 3089.70 Nm   (Refer part (i))

$$M_{vv} = \frac{H\dfrac{l}{2}}{10} = \frac{(250 \times 3) \times \dfrac{3}{2}}{10} = 112.5 \text{ Nm}$$

$$Z_{uu} = \frac{M_{uu}}{\sigma_{bc}}\left(1 + \frac{M_{vv}}{M_{uu}} \times \frac{Z_{uu}}{Z_{vv}}\right) = \frac{3089.70 \times 10^3}{1.33 \times 165} \times \left(1 + \frac{112.5 \times 10^3}{3089.70 \times 10^3} \times 7\right)$$

$$= 1766.70 \text{ mm}^3$$

Select I.S.L.C. 75 @ 55.9 N/m from I.S. Handbook No.1. The relevant properties are

$$Z_{xx} = Z_{uu} = 17.6 \times 10^3 \text{ mm}^3;\ Z_{yy} = Z_{vv} = 4.3 \times 10^3 \text{ mm}^3$$

$$\sigma_{bct,\ \text{cal}} = \frac{3089.70 \times 10^3}{17.6 \times 10^3} \times \left(1 + \frac{112.5 \times 10^3}{3089.70 \times 10^3} \times \frac{17.6 \times 10^3}{4.3 \times 10^3}\right)$$

$$= 201.71 \text{ N/mm}^2 < 1.33 \times 165$$

which is safe.

Example 7.7 Design an I-section purlin, for an industrial building situated in the outskirts of Allahabad, to support a galvanised corrugated iron sheet roof the following data:

Spacing of the truss c/c = 6.0 m
Span of truss = 12.0 m
Slope of roof = 30°
Spacing of purlins c/c = 1.5 m
Intensity of wind pressure = 2 kN/m^2
Weight of galvanised sheets = 130 N/m^2
Yield stress of steel = 250 MPa

Solution

Weight of galvanised corrugated iron sheets = 130 × 1.5 = 195 N/m

Assume dead load of purlin = 100 N/m

Total dead load = 195 + 100 = 295 N/m

The dead load acts vertically downwards.

The component of dead load parallel to the roof = 295 sin 30° = 147.5 N/m

The component of dead load normal to the roof = 295 cos 30° = 255.477 N/m

The wind load is assumed to act normal to the roof.

Wind load = $2 \times 10^3 \times 1.5 \times 1$ = 3000 N/m

Total load on purlin normal to the roof,

$$P' = 3000 + 255.477 = 3255.477 \text{ N/m}$$

The load on purlin parallel to roof,

$$H' = 147.5 \text{ N/m}$$

$$M_{uu} = \frac{P'L}{10} = \frac{(3255.477 \times 6) \times 6 \times 10^3}{10} = 11{,}719{,}717 \text{ Nmm}$$

$$M_{vv} = \frac{H'L}{10} = \frac{(147.5 \times 6) \times 6 \times 10^3}{10} = 531{,}000 \text{ Nmm}$$

Assume $Z_{uu}/Z_{vv} = 7$. Increase the stress by $33\frac{1}{3}\%$ to account for wind.

$$Z_{uu} = \frac{M_{uu}}{\sigma_{bc}}\left(1 + \frac{M_{vv}}{M_{uu}} \times \frac{Z_{uu}}{Z_{vv}}\right) = \frac{11{,}719{,}717}{1.33 \times 165} \times \left(1 + \frac{531{,}000}{11{,}719{,}717} \times 7\right)$$

$$= 70342.75 \text{ mm}^3$$

The nearest value of section modulus from I.S. Handbook No. 1 is furnished by I.S.J.B. 200 @ 97.1 N/m and may be considered to be economical, but the ratio of $Z_{uu}/Z_{vv} = 13.4 > 7$ (assumed), so it will not suit.

Let us try another section I.S.M.B. 125 @ 127.5 N/m.

$$\frac{Z_{uu}}{Z_{vv}} = \frac{Z_{xx}}{Z_{yy}} = \frac{71.8 \times 10^3}{11.7 \times 10^3} = 6.136 \approx 7$$

which is all right.

$$\sigma_{bc,\,cal} = \frac{M_{uu}}{Z_{uu}}\left(1 + \frac{M_{vv}}{M_{uu}} \times \frac{Z_{uu}}{Z_{vv}}\right)$$

$$= \frac{11{,}719{,}717}{71.8 \times 10^3}\left(1 + \frac{531{,}000}{11{,}719{,}717} \times 6.136\right) = 208.60 \text{ N/mm}^2$$

$$< 1.33 \times 165 \text{ N/mm}^2 \ (219.45 \text{ N/mm}^2)$$

which is as it should be.

Note In the Examples 7.6 and 7.7, the intensity of wind pressure is given. However, in actual problems, the design wind pressure and wind load per unit length of purlin should be calculated in accordance to IS 875 (Part 3)-1987. The calculation of wind loads on roof trusses and purlins has been described in Chapter II. Also for calculation of the actual wind load on purlins, reference may be made to Example 11.3.

Example 7.8 Design a lintel over an opening of 3.5 m. The lintel is made in a wall 400 mm thick. The lintel has to support a uniform load of 50 kN in addition to the masonry. The weight of the masonry may be assumed to be 19.2 kN/m^3 and the height of brick work above the lintel is 4.0 m. The yield stress of the steel may be assumed 250 N/mm^2.

Solution Assume 200 mm bearing for the lintel.

Effective span of the lintel = $3.5 \times 10^3 + 2 \times 100 = 3700$ mm

The weight on the lintel due to masonry will be the load of the equilateral triangle formed over the opening. The weight of brickwork in the equilateral triangle,

$$W = \frac{1}{2} \times 3.70 \times 3.70 \sin 60° \times 0.400 \times 19{,}200$$

$$= 45{,}526.6 \text{ N}$$

Uniform load over lintel = 50 kN

Let self weight of beam section = 500 N/m.

$$\text{B.M. due to load } (W) = \frac{Wl}{6} = \frac{45{,}526.6 \times 3700}{6} = 28{,}074{,}737 \text{ Nmm}$$

Total uniform load = $50 \times 1000 + 500 \times 3.7 = 50000 + 1850 = 51{,}850$ N

$$\text{B.M. due to uniform load} = \frac{50{,}740 \times 3700}{8} = 23{,}980{,}625 \text{ Nmm}$$

Total bending moment = 28,074,737 + 23,980,625 = 52,055,362 Nmm

When a beam is embedded in the masonry wall, the latter does not provide any lateral support to it. Therefore, a lintel is designed as a laterally unsupported beam. Assume a reduced value of permissible bending stress in compression, say 100 N/mm^2.

$$Z = \frac{52{,}055{,}362}{100} = 520553.62 \text{ mm}^3$$

From I.S. Handbook No. 1, let us try I.S.M.B. 300 @ 433.6 N/m furnishing section modulus equal to 573.6×10^3 mm^3. The relevant properties of I.S.M.B. 300 @ 433.6 N/m are:

$D = 300$ mm, $T = 12.4$ mm, $t = 7.5$ mm $r_{yy} = 28.4$ mm, $h_2 = 29.25$ mm

$d_1 = 300 - 2 \times 29.25 = 241.5$ mm

$$\frac{d_1}{t} = \frac{241.5}{7.5} = 32.2 < 85$$

$$\frac{T}{t} = \frac{12.4}{7.5} = 1.653 < 2$$

as $\frac{d_1}{t} < 85$ and $\frac{T}{t} < 2$ and $f_y = 250$ N/mm^2, refer to Table 7.1(b).

Now,
$$\frac{D}{T} = \frac{300}{12.4} = 24.1935$$

$$\frac{l}{r_{yy}} = \frac{1.2 \times 3.7 \times 10^3}{28.4} = 156.338$$

Since the lintel is embedded in masonry and provision to restrain the beam ends against torsion cannot be made, the effective length of the compression flange of the beam will be increased by 20% (Table 7.2).

For $\frac{D}{T}$ = 24.1935, $\frac{l}{r_{yy}}$ = 156.338 and f_y = 250 N/mm², from Table 7.1(b),

$$\sigma_{bc} = 96.06 \text{ N/mm}^2$$

$$\text{Moment of resistance} = 96.06 \times 573.6 \times 10^3 = 55101.54 \times 10^3 \text{ Nmm}$$

$$> 52{,}0553.62 \times 10^3 \text{ Nmm}$$

which is safe.

Check for shear

$$\text{Maximum shear force, } V = \frac{45{,}526.6 + 50{,}740}{2} = 48{,}133 \text{ N}$$

$$\tau_{va,\,cal} = \frac{48{,}133}{300 \times 7.5} = 21.39 \text{ N/mm}^2 < 100 \text{ N/mm}^2 \ (0.4\,f_y)$$

which is safe.

Check for deflection

$$\delta_{cal} = \frac{Wl^3}{60\,EI} + \frac{5}{384} \times \frac{wl^4}{EI}$$

$$= \frac{45.526 \times 10^3 \times 3700^3}{60 \times 2.0 \times 10^5 \times 7332.9 \times 10^4} + \frac{5}{384} \times \frac{50.74 \times 10^3 \times 3700^3}{2.0 \times 10^5 \times 7332.9 \times 10^4}$$

$$= 3.058 \text{ mm}$$

$$\delta_{all} = \frac{l}{325} = \frac{3700}{325} = 11.3846 \text{ mm}$$

which is safe.

Example 7.9 Design a beam with 6.5 m effective span, carrying a uniform load of 40.0 kN/m inclusive of self weight over the entire span. The overall depth of the beam is restricted to 350 mm. The compression flange of the beam is laterally supported throughout. (f_y = 250 N/mm²).

Solution

$$\text{Maximum bending moment, } M = \frac{40.0 \times 6.5^2}{8} = 211.25 \text{ kNm}$$

$$\text{Maximum shear force, } V = \frac{40.0 \times 6.5}{2} = 130 \text{ kN}$$

$$\text{Section modulus, } Z = \frac{211.25 \times 10^6}{155} = 1362.9 \times 10^3 \text{ mm}^3$$

Since a plated beam will be provided, a reduced allowable bending stress of 155 N/mm² is assumed to account for rivet holes in the tension flange.

Let us provide I.S.M.B. 225 @ 306.1 N/m. The relevant properties of the section are:

$$A = 3972 \text{ mm}^2,\ B = 110 \text{ mm},\ T = 11.8 \text{ mm},\ t = 6.5 \text{ mm}$$

$$I_{xx} = 3441.8 \times 10^4 \text{ mm}^4,\ Z_{xx} = 305.9 \times 10^3 \text{ mm}^3,\ h_2 = 25.85 \text{ mm},\ g = 60 \text{ mm}$$

$$Z_p = Z_{\text{req}} - Z_{xx} = 1362.9 \times 10^3 - 305.9 \times 10^3 = 1057 \times 10^3 \text{ mm}^3$$

$$A_p = \frac{Z_p}{D} = \frac{1057 \times 10^3}{225} = 4697.7 \text{ mm}^2 \approx 4700 \text{ mm}^2$$

As per clause 3.5.2.1 of I.S. 800–1984, the maximum outstand of flange from the connection should not exceed $16T_1$ in the case of compression flange, where T_1 = aggregate thickness of flange cover plates.

∴ Maximum outstand from connection $= 16\ T_1 = 16 \times 24 = 384$ mm

Provide two plates 12 mm thick on each flange.

$$\text{Width of the cover plate} = \frac{4700}{2 \times 12} = 195.83 \text{ mm} \approx 200 \text{ mm}$$

$$\text{Outstand of flange cover plate from connection} = \frac{1}{2}(200 - 60) = 70 \text{ mm}$$

$$< 384 \text{ mm}$$

Hence, provide two plates 200 mm × 12 mm on each flange.

$$I_{xx,\text{ gross}} = 3441.8 \times 10^4 + \left[2 \times 200 \times \frac{24^3}{12} + 2 \times 24 \times 200 \times \left(\frac{225}{2} + \frac{24}{2}\right)^2\right]$$

$$= 18{,}368.12 \times 10^4 \text{ mm}^4$$

$$\sigma_{bc,\text{ cal}} = \frac{211.25 \times 10^6}{18368.12 \times 10^4} \times \left(\frac{225}{2} + 24\right) = 156.98 \text{ N/mm}^2$$

Check for bending stress in tension Let us connect the cover plates with the flange by 20 mm diameter rivets. The rivets are provided staggered in two rows.

$$\text{Gross area of the tension flange} = 110 \times 11.8 + 2 \times 12 \times 200 = 6098 \text{ mm}^2$$

$$\text{Net area of the tension flange} = 6098 - 21.5 \times (11.8 + 2 \times 12) = 5328.3 \text{ mm}^2$$

$$\sigma_{bt,\text{ cal}} = 156.98 \times 6098/5328.3 = 179.657 \text{ N/mm}^2$$

$$> 165 \text{ N/mm}^2, \text{ which is unsafe.}$$

Try two 14 mm thick plates on each flange.

$$I_{xx,\text{ gross}} = 3441.8 \times 10^4 + 2 \times 200 \times \frac{28^3}{12} + 2 \times 200 \times 28 \times \left(\frac{225}{2} + \frac{28}{2}\right)^2$$

$$= 21{,}437.49 \times 10^4 \text{ mm}^4$$

$$\sigma_{bc,\text{ cal}} = \frac{211.25 \times 10^6}{21437.49 \times 10^4} \times \left(\frac{225}{2} + 28\right) = 138.45 \text{ N/mm}^2$$

$$\text{Gross area of tension flange} = 110 \times 11.8 + 2 \times 14 \times 200 = 6898 \text{ mm}^2$$

$$\text{Net area of tension flange} = 6898 - 21.5 \times (11.8 + 2 \times 14) = 6042.3 \text{ mm}^2$$

$$\sigma_{bt,\,cal} = 138.45 \times 6898/6042.3 = 158.05 \text{ N/mm}^2 < 165 \text{ N/mm}^2$$

which is all right.

Check for shear

$$\tau_{va} = \frac{130 \times 10^3}{225 \times 6.5} = 88.89 \text{ N/mm}^2 < 100 \text{ N/mm}^2$$

which is safe.

Check for deflection

$$\delta_{all} = \frac{l}{325} = \frac{6.5 \times 10^3}{325} = 20 \text{ mm}$$

$$\delta_{cal} = \frac{5}{384} \times \frac{40 \times (6.5 \times 10^3)^4}{2.0 \times 10^5 \times 21,437.49 \times 10^4}$$

$$= 21.68 \text{ mm} > 20 \text{ mm}$$

Since the calculated deflection is slightly more than the allowable one and is not serious, a camber of 5 mm may be provided.

Design of connections 20 mm diameter rivets have been provided in two rows and are staggered to connect the cover plates with the flange.

Gross diameter of the rivets = 20 + 1.5 = 21.5 mm

Maximum shear force, $V = 130 \times 10^3$ N

$$I_{xx} = 21437.49 \times 10^4 \text{ mm}^4$$

$$A\bar{Y} = (200 \times 28 - 21.5 \times 28) \times \left(\frac{225}{2} + 14\right)$$

$$= 632.247 \times 10^3 \text{ mm}^3$$

The rivets will be in single shear as the two cover plates are tied together.

$$\text{Rivet value in single shear} = \frac{\pi}{4} d^2 \tau_{vf} = \frac{\pi}{4} 21.5^2 \times 100 = 36,305.03 \text{ N}$$

$$\text{Rivet value in bearing} = dt\,\sigma_{pf} = 21.5 \times 11.8 \times 300 = 76,110 \text{ N}$$

$$\text{Rivet value, } R_v = 36,305.03 \text{ N}$$

$$s = \frac{2I}{VA\bar{Y}} \times R_v$$

$$= \frac{2 \times 21437.49 \times 10^4}{130 \times 10^3 \times 632.247 \times 10^3} \times 36,305.03$$

$$= 189.38 \text{ mm} \nless 50 \text{ mm } (2.5 \times 20)$$

Hence, provide rivets at a pitch of 160 mm.

Curtailment of cover plate As per I.S. specifications the first cover plate is not curtailed. Since the rolled I-Section flange will extend up to the support, both the cover plates will be curtailed as shown in Fig. Ex. 7.9.

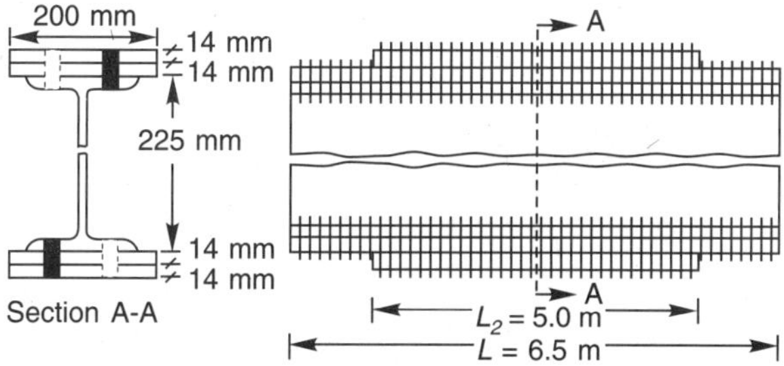

Fig. Ex. 7.9

Net Z_{xx_2} (with two cover plates)

$$I_{xx\text{ gross}} = 21437.49 \times 10^4 \text{ mm}^4$$

$$I_{xx\text{ net}} = I_{xx\text{ gross}} - \text{moment of inertia of rivet holes}$$

$$= 21437.49 \times 10^4$$

$$-2\left[21.5 \times (14 + 14 + 11.8) \times \left\{\frac{225}{2} - 11.8 + \left(\frac{14 + 14 + 11.8}{2}\right)\right\}^2\right]$$

$$= 21437.49 \times 10^4 - 2489.1218 \times 10^4 \text{ mm}^4 = 18948.36 \times 10^4 \text{ mm}^4$$

$$\text{Net } Z_{xx_2} = \frac{18948.36 \times 10^4}{\frac{225}{2} + 14 + 14} = 1348.63 \times 10^3 \text{ mm}^3$$

Net Z_{xx_1} (with one cover plate)

$$\text{Net } I_{xx} = 3441.8 \times 10^4 + 2 \times \left[14 \times 200 \times \left(\frac{225}{2} - 7\right)^2\right]$$

$$-2 \times \left[21.5 \times (14 + 11.8) \times \left\{\frac{225}{2} - 11.8 + \left(\frac{14 + 11.8}{2}\right)\right\}^2\right]$$

$$= 3441.8 \times 10^4 + 6232.9400 \times 10^4 - 1431.6763 \times 10^4$$

$$= 8243.0637 \times 10^4 \text{ mm}^4$$

$$\text{Net } Z_{xx_1} = \frac{8243.0637 \times 10^4}{\frac{225}{2} + 14} = 651.625 \times 10^3 \text{ mm}^3$$

$$L_2 = L\sqrt{(Z_{xx_2} - Z_{xx_1})/Z_{xx_2}}$$

$$= 6.5 \times \sqrt{(1348.63 \times 10^3 - 651.625 \times 10^3)/(1348.63 \times 10^3)}$$

$$= 4.67 \text{ mm}$$

$$x_2 = \frac{L - L_2}{2} = \frac{6.5 - 4.67}{2} = 0.915 \text{ m}$$

$$L_1 = L\sqrt{\frac{(Z_{xx_2} - Z)}{Z_{xx_2}}} = 6.5 \times \sqrt{\frac{(1348.63 \times 10^3 - 305.9 \times 10^3)}{1348.63 \times 10^3}} = 5.715 \text{ m}$$

Since the theoretical length of the first curtailed cover plate (L_1) is quite close to the length of the beam flange (the actual length to be provided will be still more to accommodate additional rivets to dissipate stress at the curtailed section), it is not desirable to curtail it. Hence, provide the first cover plate for the full length of the beam and curtail the top cover plate only.

Moment of resistance of the section with one cover plate

$$= \sigma_{bt} \times Z_{xx_1}$$
$$= 165 \times 651.625 \times 10^3 = 107{,}518 \times 10^3 \text{ Nmm}$$

Average stress in curtailed plate

$$= \frac{107{,}518 \times 10^3}{1348.63 \times 10^3} \times \frac{\left(\frac{225}{2} + 14 + 14\right) - \frac{14}{2}}{\frac{225}{2} + 14 + 14} = 75.75 \text{ N/mm}^2$$

Force at the curtailed section = 75.75 × 200 × 14 = 212100 N

Let us provide 20 mm ϕ power driven rivets.

Gross diameter = 20 + 1.5 = 21.5 mm

$$\text{Value of rivets in single shear} = \frac{\pi}{4} \times 21.5^2 \times 100 = 36305.03 \text{ N}$$

Value of rivet in bearing = 21.5 × 11.8 × 300 = 76,110 N

Value of rivet, R_v = 36305.03 N

$$\text{Number of additional rivets} = \frac{212100}{36305.03} = 5.84 \approx 6$$

Provide total six additional rivets, 20 mm ϕ, in two rows.

Minimum pitch = 2.5ϕ = 2.5 × 20 = 50 mm

For 20 mm ϕ rivets minimum edge distance is 29 mm.

Let us provide a 30 mm edge distance.

Extra length of cover plate = (2 × 50 + 2 × 30) × 2 = 320 mm

Total length of top cover plate = 4.67 + 0.320 = 4.99 m ≈ 5.0 m

Example 7.10 Design an open web (castellated) beam for a span of 17 m. The dead weight of the roofings is 750 N/m^2 and the live load on the roof is estimated to be 2000 N/m^2. Cut the selected I-Section at 45° and adjust the section such that the overall depth of section should not exceed 900 mm. Assume f_y = 250 MPa.

Solution

Dead load = 750 N/m^2, Live load = 2000 N/m^2

Total load, $W' = 2000 + 750 = 2750\ \text{N/m}^2 \approx 3000\ \text{N/m}^2$

Let us try I.S.M.B. 600 @ 1202.7 N/m.

The relevant properties of the section are

$A = 15621\ \text{mm}^2$, $D = 600$ mm, $B = 210$ mm, $T = 20.8$ mm, $t = 12$ mm

$I_{xx} = 91{,}813.0 \times 10^4\ \text{mm}^4$, $I_{yy} = 2651.0 \times 10^4\ \text{mm}^4$, $Z_{xx} = 3060.4 \times 10^3\ \text{mm}^3$, $Z_{yy} = 252.5 \times 10^3\ \text{mm}^3$

Properties of Tee at open throat

Area of flange = $210 \times 20.8 = 4368\ \text{mm}^2$

Area of web = $(150 - 20.8) \times 12$

= $1550.4\ \text{mm}^2$

Total area = $4368 + 1550.4$

= $5918.4\ \text{mm}^2$

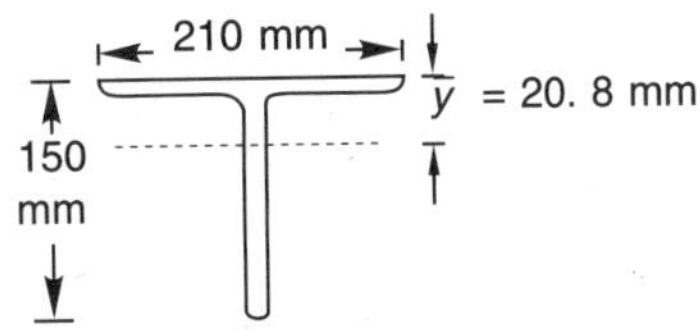

Fig. Ex. 7.10(a)

Take moment of the area about the top edge of the T-Section to find the neutral axis.

$$\bar{Y} = \frac{4368 \times \dfrac{20.8}{2} + 1550.4 \times \left(\dfrac{150 - 20.8}{2} + 20.8\right)}{5918.4} = 30.047\ \text{mm}$$

Moment of inertia of T-Section about N.A.

$$I_T = \left[210 \times \frac{20.8^3}{12} + 4368 \times \left(30.047 - \frac{20.8}{2}\right)^2\right]$$

$$+ \left[\frac{12}{12} \times (150 - 20.8)^3 + 12 \times (150 - 20.8) \times (85.4 - 30.047)^2\right]$$

$$= 87{,}50{,}927.2\ \text{mm}^4$$

Moment of resistance, $M_R = 5918.4 \times (900 - 2 \times 30.047) \times 165 = 820{,}198$ kNmm

Let W be the load that can be applied over the section.

$$W = 8 \times \frac{820{,}198}{17 \times 10^3} = 385.97\ \text{kN}$$

$$\text{Self weight of the beam} = \frac{1202.7 \times 17}{10^3} = 20.445\ \text{kN}$$

Safe distributed load over entire span, $P' = 385.97 - 20.445 = 365.525$ kN

$$\text{Spacing of the beam, } S = \frac{P'}{W'L} = \frac{365.525 \times 10^3}{3000 \times 17} = 7.167\ \text{m}$$

Provide ISMB 600 @ 1202.7 N/m at 7.00 m c/c.

Check for shear

Load/m length of the beam = $3000 \times 7 = 21000$ N

Self weight of beam/m length = 1202.7 N

Total load/m length = $22202.7 \simeq 22203$ N

$$\text{Maximum end shear} = 22{,}203 \times \frac{17}{2} = 188{,}725.5 \text{ N} = 188.725 \text{ kN}$$

Average shear stress,
$$\tau_{va} = \frac{188.725 \times 10^3}{300 \times 12} = 52{,}426 \text{ N/mm}^2$$
$$< 0.40 f_y \ (100 \text{ N/mm}^2)$$

which is safe.

Check for deflection Moment of inertia of perforated section,

$$I_p = 2 \times 5918.4 \times (450 - 30.047)^2 + 2 \times 8{,}750{,}927.2 = 210{,}504.61 \times 10^4 \text{ mm}^4$$

Moment of inertia of solid section,

$$I_s = 210{,}504.61 \times 10^4 + \frac{12 \times 600^3}{12} = 232104.61 \times 10^4 \text{ mm}^4$$

$$\text{Average moment of inertia, } I = \frac{210{,}504.61 \times 10^4 + 232{,}104.61 \times 10^4}{2}$$
$$= 221{,}304.61 \times 10^4 \text{ mm}^4$$

$$\delta_1 = \frac{5}{384} \times \frac{(2 \times 188.725) \times 10^3 \times (17 \times 10^3)^3}{2.00 \times 10^5 \times 221{,}304.61 \times 10^4} = 54.55 \text{ mm}$$

Shear at ends = 188.725 kN

Shear at centre = 0

$$\text{Average shear, } V_{\text{avg}} = \left(\frac{188.725 + 0}{2}\right) = 96.36 \approx 96.5 \text{ kN}$$

Let us provide twenty perforations in the entire span. The dimensions of the perforations will be as shown in Fig. Ex. 7.10(b).

$$p = 10, m = 250 \text{ mm}, n = 200 \text{ mm}, I_T = 8{,}750{,}927.2 \text{ mm}^4$$

$$\delta_2 = \frac{V_{\text{avg}}\, p\, (m + n)^3}{24\, EI_T} = \frac{96.5 \times 10^3 \times 10 \times (250 + 200)^3}{24 \times 2.00 \times 10^5 \times 8{,}750{,}927.2} = 2.093 \text{ mm}$$

$$\delta = \delta_1 + \delta_2 = 55.78 + 2.093 = 57.87 \text{ mm}$$

$$\delta_{\text{all}} = \frac{l}{325} = \frac{17 \times 10^3}{325} = 52.3076 \text{ mm} < 57.87 \text{ mm}$$

The deflection is more than the permissible deflection. To make the beam safe in deflection, a camber in the beam is required equal to 57.87 – 52.307 = 5.566 mm.

Provide a camber of about 20 mm to avoid a feeling of unsafety for the user.

Check for combined stress The beam is also checked for combined local bending and direct stresses at various perforations between 1/4 to 1/8 span.

Maximum end shear = 188.725 kN

Effective span, $l = 17 - 0.85 = 16.15$ m

$$\frac{l}{2} = \frac{16.15}{2} \times 10^3 = 8075 \text{ mm}$$

$$V = \frac{188.725 \times 8 \times 900}{8075} = 168.27 \text{ kN}$$

Taking moments about A (Fig. Ex. 7.10(b)),

$$V_h \times (450 - 30.047) = 168.27 \times \frac{900}{2}$$

or $$V_h = 180.3 \text{ kN}$$

$$\text{Shear stress} = \frac{180.3 \times 10^3}{12 \times 200}$$

$$= 75.1 \text{ N/mm}^2$$

$$< 0.40 f_y \; (100 \text{ N/mm}^2)$$

which is as it should be.

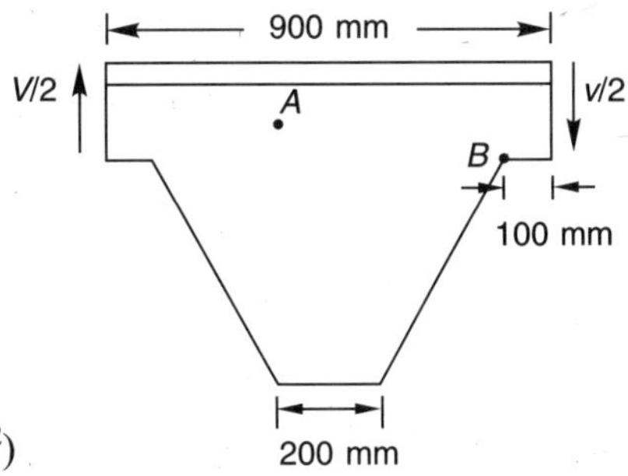

Fig. Ex. 7.10(b)

Combined stress at section 2–2 (at about 1/4, i.e. at about 4.0 m)

The point B of perforation falls 4.6 m from the middle of the span.

Variation of shear is linear.

$$\text{Shear} = \frac{188.725 \times 4.6}{8.075} = 107.50 \text{ kN}$$

There are two flanges, so

$$\text{Shear on each flange} = \frac{107.50}{2} = 53.75 \text{ kN}$$

Variation of bending moment is parabolic.

$$\text{Maximum bending moment at centre} = \frac{188.725 \times 17 \times 10^3}{4} = 802.08 \times 10^3 \text{ Nm}$$

$$\text{Moment at section 1–1} = 802.08 \times 10^3 \times \left[1 - \left(\frac{4.6}{8.075}\right)^2\right]$$

$$= 541.795 \times 10^3 \text{ Nm}$$

$$\text{Direct stress due to moment} = \frac{541.795 \times 10^3 \times 10^3}{5918.4 \times (900 - 2 \times 30.047)} = 109 \text{ N/mm}^2$$

$$\text{Bending stress due to shear} = \frac{53.75 \times 10^3 \times 100 \times (150 - 30.047)}{87.509 \times 10^5}$$

$$= 73.67 \text{ N/mm}^2$$

$$\text{Combined stress} = 109.00 + 73.67 = 182.67 \text{ N/mm}^2$$

Combined stress at section 3–3, i.e. at 3.7 m from the centre

$$\text{Shear} = 188.725 \times \frac{3.70}{8.075} = 86.47 \text{ kN}$$

$$\text{Shear at each flange} = \frac{86.47}{2} = 43.23 \text{ kN}$$

$$\text{Bending moment} = 802.08 \times 10^3 \left[1 - \left(\frac{3.7}{8.075}\right)^2\right] \times 10^3$$

$$= 633.68 \times 10^3 \text{ Nm}$$

$$\text{Direct stress due to moment} = \frac{633.68 \times 10^3 \times 10^3}{5918.4(900 - 2 \times 30.047)} = 127.4 \text{ N/mm}^2$$

$$\text{Bending stress due to shear} = \frac{43.23 \times 10^3 \times 100 \times (150 - 30.047)}{87.509 \times 10^5}$$

$$= 59.25 \text{ N/mm}^2$$

$$\text{Combined stress} = 127.4 + 59.25 = 186.65 \text{ N/mm}^2$$

Combined stress at section 4–4, i.e at 2.80 m from the centre

$$\text{Shear} = \frac{188.725 \times 2.80}{8.075} = 65.44 \text{ kN}$$

$$\text{Shear at each flange} = \frac{65.44}{2} = 32.72 \text{ kN}$$

$$\text{Bending moment} = 802.08 \times 10^3 \left[1 - \left(\frac{2.800}{8.075}\right)^2\right]$$

$$= 705.64 \times 10^3 \text{ Nm}$$

$$\text{Direct stress due to moment} = \frac{705.64 \times 10^3 \times 10^3}{5918.4\,(900 - 2 \times 30.047)} = 141.95 \text{ N/mm}^2$$

$$\text{Bending stress due to shear} = \frac{32.72 \times 10^3 \times 100 \times (150 - 30.047)}{87.509 \times 10^5} = 44.85 \text{ N/mm}^2$$

$$\text{Combined stress} = 141.95 + 44.85 = 186.80 \text{ N/mm}^2$$

Combined stress at section 5–5 i.e. at 1900 mm from centre

$$\text{Shear} = \frac{188.725 \times 1.90}{8.075} = 44.40 \text{ kN}$$

$$\text{Shear at each flange} = \frac{44.40}{2} = 22.20 \text{ kN}$$

$$\text{Bending moment} = 802.08 \times 10^3 \left[1 - \left(\frac{1.9}{8.075}\right)^2\right] = 757.67 \text{ kNm}$$

Direct stress due to bending moment

$$= \frac{757.67 \times 10^3 \times 10^3}{5918.4 \times (900 - 2 \times 30.047)} = 152.42 \text{ N/mm}^2$$

$$\text{Bending stress due to shear} = \frac{22.20 \times 10^3 \times 100 \times (150 - 30.047)}{87.509 \times 10^5} = 30.43 \text{ N/mm}^2$$

$$\text{Combined stress} = 152.42 + 30.43 = 182.85 \text{ N/mm}^2$$

The maximum combined stress is 186.65 N/mm^2, therefore the load coming over the beams will have to be reduced as the permissible stress is 165 N/mm^2. This may be achieved by reducing the spacing of the beams.

Spacing

$$\text{Reduced moment capacity} = \frac{165 \times 802.08 \times 10^3}{186.65} = 709.04 \times 10^3 \text{ Nm}$$

Maximum load that the beam can support per m length

$$= 709.04 \times 10^3 \times \frac{8}{17} = 333.67 \text{ kN}$$

$$\text{Self weight of beam} = 20.445 \text{ kN}$$

$$\text{Load carrying capacity} = 333.67 - 20.445 = 313.225 \text{ kN}$$

$$\text{Spacing, } S = \frac{313.225 \times 1000}{3000 \times 17} = 6.14 \text{ m}$$

Hence, provide beams at 6 m c/c.

End stiffener

$$\text{Maximum width of the end bearing stiffener} = \frac{210 - 12}{2} = 99 \text{ mm}$$

Provide a 95 mm wide stiffener.

Thickness of the stiffener plate $\nless$ 1/12 of the outstand

$$= \frac{1}{12} \times 95 = 7.916 \text{ mm} \approx 10 \text{ mm}$$

Provide an end bearing stiffener plate 95 mm × 10 mm (Fig. Ex. 7.10(c)) on both sides of the web of the castellated beam.

As per I.S. specifications the bearing stiffener should be designed as column, with a distance of web = 40 t, as a part of the stiffener.

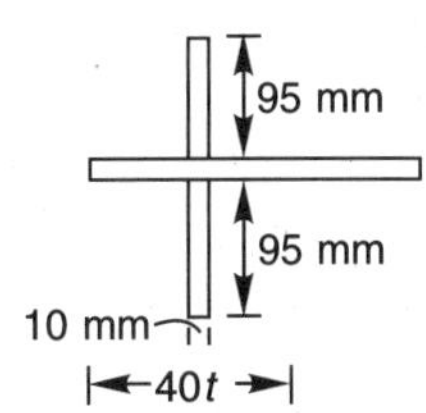

Fig. Ex. 7.10(c)

$$\text{Length of web as stiffener} = 40 \times 12 = 480 \text{ mm}$$

$$\text{Effective area of stiffener} = 2 \times 95 \times 10 + 480 \times 12$$
$$= 7660 \text{ mm}^2$$

Let each stiffener plate be splayed by 5 mm to fit on the weld.

$$\text{Bearing area of stiffener} = 2 \times (95 - 5) \times 10 = 1800 \text{ mm}^2$$

$$\sigma_{p,\text{ cal}} = \frac{188.725 \times 10^3}{1800} = 104.847 \text{ N/mm}^2$$
$$< 0.75 f_y \text{ (185 N/mm}^2)$$

which is safe.

$$\text{Moment of inertia of stiffener} = \frac{10\times(2\times 95+12)^3}{12} + \frac{(480-10)\times 12^3}{12}$$

$$= 6868{,}673.3 + 67680 = 6936{,}353 \text{ mm}^4$$

$$\text{Radius of gyration, } r = \sqrt{I/A} = \sqrt{6936353/7660} = 30.09 \text{ mm}$$

$$\text{Effective length of stiffener} = 0.7 \times \text{height of stiffener}$$

$$= 0.7 \times (900 - 2 \times 20.8) = 600.88 \text{ mm}$$

$$\frac{l}{r} = \frac{600.88}{30.09} = 19.969$$

For $\frac{l}{r} = 19.969$ and $f_y = 250$ N/mm^2,

$$\sigma_{ac} = 148 \text{ N/mm}^2$$

$$\text{Load carrying capacity} = 148 \times 7660 = 1133680$$

$$= 1133.68 \text{ kN} > 377.45 \text{ kN}$$

which is sufficient.

Connection of stiffener Let us provide intermittent fillet welds.

$$\text{Vertical shear/mm length of web} = \frac{377.45\times 10^3}{900 - 2\times 20.8} = 439.71 \text{ N/mm}$$

Let the length of each intermittent fillet weld be 100 mm and the size of weld be 4 mm.

$$\text{Strength of weld per mm length} = 0.7 \times 4 \times 108 = 302.4 \text{ N}$$

The stiffener plate will be welded on both the sides.

$$\therefore \quad \text{Clear spacing of welds} = \frac{4\times 302.4\times 100}{439.71} - 100$$

$$= 175.09 \text{ mm} \not> 10\,t\,(10\times 10 = 100 \text{ mm})$$

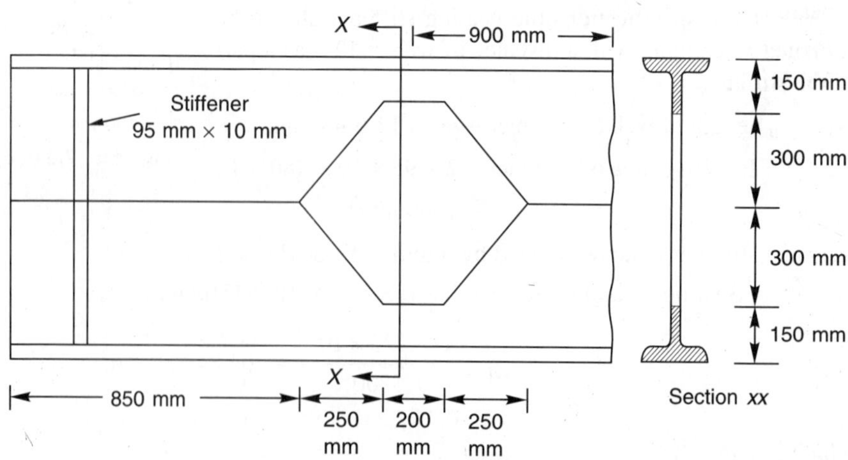

Fig. Ex. 7.10(d)

where t = least thickness of web plate or stiffener plate.
Hence, provide a 4 mm fillet weld 100 mm long at a clear spacing of 100 mm.
The castellated beam section along with stiffener is shown in Fig. Ex. 7.10(d).

Example 7.11 Design a composite beam with flexible shear connectors for the following data:

Span of the beam	5 m
c/c distance of beams supporting slab	2.5 m
Total uniform load over the beam	20 kN/m
Grade of concrete	M–15
Thickness of slab	120 mm
Modular ratio (m)	18

Solution

$$\text{Maximum moment} = \frac{20 \times 10^3 \times 5^2}{8} = 62500 \text{ Nm}$$

$$\text{Section modulus } \frac{62500 \times 10^3}{165} = 378{,}787.8 \text{ mm}^3$$

From I.S. Handbook No. 1 select I.S.L.B. 275 @ 323.7 N/m. The relevant properties of the section are,

$$I = 5375.3 \times 10^4 \text{ mm}^4,\ A = 4002 \text{ mm}^2,\ D = 275 \text{ mm}$$

$$B = 140 \text{ mm},\ T = 8.8 \text{ mm},\ t = 6.4 \text{ mm}$$

Width of composite beam flange

It is minimum of the following

$$b_1 = \frac{L}{4} = \frac{5 \times 10^3}{4} = 1250 \text{ mm}$$

$$b_2 = \frac{1}{2}(2.5 \times 10^3 - 140) = 1180 \text{ mm}$$

$$b_3 = 8 \times 120 = 960 \text{ mm}$$

So the effective width of the concrete slab is 960 mm. The section of composite beam is shown in Fig. Ex. 7.11.

Properties of the composite section
Neutral axis

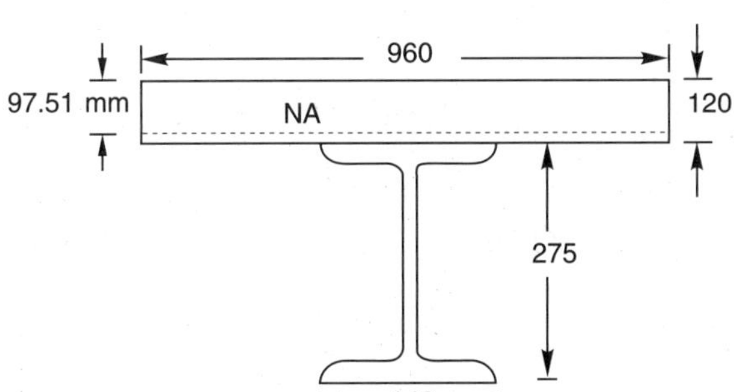

Fig. Ex. 7.11

$$n = \frac{\left(\frac{960}{18}\times 120\right)\times 60 + 4002\times\left(120+\frac{275}{2}\right)}{\frac{960\times 120}{18}+4002} = 97.51 \text{ mm}$$

$$I_{xx} = 5375.3\times 10^4 + 4002\times\left(\frac{275}{2}+120-97.51\right)^2+\frac{960\times 120^3}{18\times 12}$$

$$+\frac{960\times 120}{18}\times 97.51^2$$

$$= 22473.67\times 10^4 \text{ mm}^4$$

$$Z_{top} = \frac{22473.67\times 10^4}{97.51} = 2304.7\times 10^3 \text{ mm}^4$$

$$Z_{bottom} = \frac{22473.67\times 10^4}{(275+120-97.51)} = 755.44\times 10^3 \text{ mm}^3 > 378.78\times 10^3 \text{ mm}^3$$

which is all right.

$$V_h = \frac{A_s f_y}{2} = \frac{4002\times 250}{2} = 500250 \text{ N}$$

Let us provide 16 mm diameter studs. Let the height of studs be 100 mm.

$$\frac{H}{D} = \frac{100}{16} = 6.25 > 4.2$$

Hence, $Q = 1.518\times 100\times 16\sqrt{15} = 9406.70$ N

Number of studs $\frac{500250}{9406.70} = 53.18 \cong 54$

Use 2 studs in each section at a spacing of $4\,d = 4\times 16 = 64$ mm.

The longitudinal spacing of studs

$$= \frac{5\times 10^3}{2}\times\frac{1}{54/2} = 92.59 \text{ mm} < 480 \text{ mm (4 ts)}$$

which is all right.

Provide 16 mm studs at a spacing of 92 mm c/c in two rows.

Exercises

Note Assume the yield stress of steel to be 250 N/mm^2 if not specified in the problem.

7.1 Obtain shear centre for following sections. All dimensions are in mm. (See Fig. Prob. 7.1).

7.2 A steel floor beam in a building has a span of 6.0 m. It is simply supported over supports and carries a uniform load of 40.0 kN/m inclusive of self weight. Design the beam (f_y = 250 MPa) if,

(i) the compression flange is restrained against lateral bending.

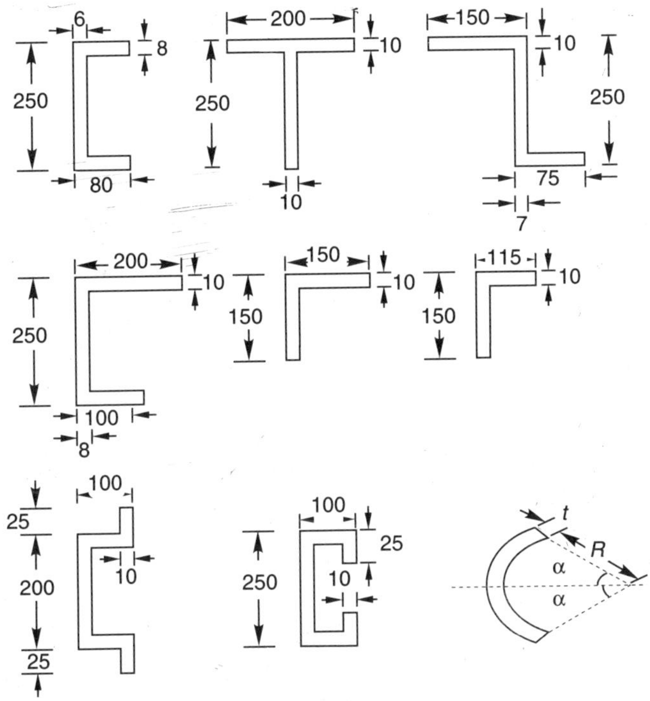

Fig. Prob. 7.1

(ii) the lateral supports for the compression flange are provided only at the ends.

7.3 A conference hall 8 m × 18 m is provided with a 120 mm R.C.C. slab over rolled steel beams spaced 3 m c/c. A wearing coat of 100 mm average thickness is provided. The compression flange of the beams is laterally supported throughout. Design the beam section. Assume f_y = 260 N/mm^2.

7.4 An I.S.L.B. 600 @ 976.1 N/m has been used as a simply supported beam over 4.20 m span. The ends of the beam are restrained against torsion but not against lateral bending. Determine the safe uniform load that the beam can carry. Assume f_y = 340 N/mm^2.

7.5 Design a beam of span 6.50 m to carry a uniformly distributed load of 15 kN/m and two concentrated loads of 75 kN each at 2 m, from either ends of the beam. The only rolled steel I-section available is I.S.M.B. 350 @ 514 N/m (f_y = 250 MPa). Design the beam section if the compression flange is supported laterally.

7.6 A beam is to span an opening of 9 m. It carries a uniform load of 12 kN/m. The depth of the beam is limited to 450 mm from clear head room requirements. Design the cross-section of the beam (f_y = 250 N/mm^2).

7.7 Design a beam 11.0 m in clear span and subjected to a system of loads as shown in Fig. Prob. 7.7. Yield stress may be assumed 250 MPa.

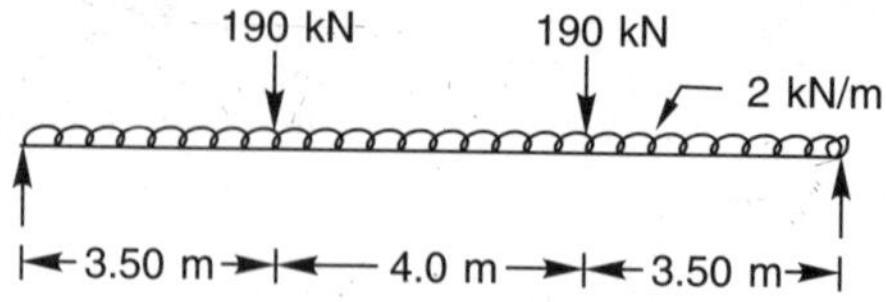

Fig. Prob. 7.7

7.8 Design a two span continuous beam 8 m long, each span being 4.0 m. It supports a design uniform load of 50 kN/m. Use the steel with yield stress of 250 MPa.

7.9 A two span continuous beam (Fig. Prob. 7.9) is built in (fixed) horizontally at the two ends. Design the beam. Assume yield stress 250 MPa.

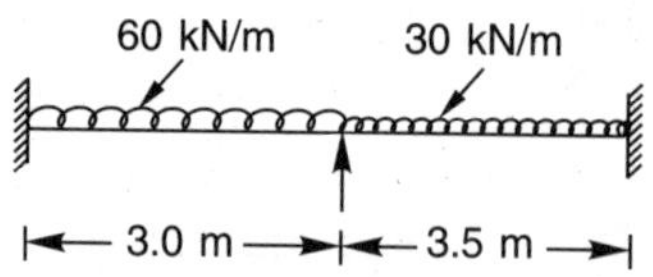

Fig. Prob. 7.9

7.10 Design a fixed ended beam 5.0 m in span, if loaded with a uniform load of 24 kN/m on the left half span. Use the steel with yield stress of 250 Mpa.

7.11 An I.S.W.B. 600 @ 1311.6 N/m (f_y = 250 MPa) transmits an end reaction of 180 kN on the masonry walls 250 mm thick. Design a suitable bearing plate. Assume the allowable bearing stress in masonry to be 5.5 N/mm^2.

7.12 Design a lintel for a wall opening of 2.50 m provided for a window. The data available are

Wall thickness = 225 mm
Wall height above the opening = 1.50 m
Additional load from slab = 15 kN/m
Weight of brick masonry = 20.2 kN/m^3
f_y = 250 N/mm^2

7.13 Design channel section purlins for an industrial building roof for the following data,

Distance between c/c of trusses = 5.0 m
Distance between c/c of purlins = 1.60 m
Inclination of the roof surface to the horizontal = 30°
Weight of G.I. sheets = 133.1 N/mm^2
Wind load normal to the roof = 1500 N/m^2
f_y = 250 N/mm^2

7.14 Design angle section purlins for a truss roof system to support sheeting weighing 160 N/m^2. The trusses are spaced 4.0 m c/c and the purlins are provided 2.0 m c/c. The building is situated in an area where the wind pressure is 1500 N/m^2. Use steel with yield stress 260 N/mm^2.

7.15 An open web roof beam system is provided with perforated beams of clear span 18.0 m. The estimated weight of roofing, ceiling and purlins is 800 N/m^2 and live load is 1500 N/m^2. Design and check the safety (f_y = 250 MPa).

Plate 1

Plate 2

7.16 An encased beam carries a total uniformly distributed load of 50.0 kN/m. The effective span of the beam is 7.0 m. The ends of the beam are restrained against torsion but not against lateral bending. Design the encased beam. Use the steel with yield stress 250 N/mm^2.

7.17 A 150 mm thick balcony slab of R.C.C. is projected from masonry wall with R.B. columns at 2.4 m c/c and 450 × 450 mm in size. The cantilever balcony is supported over beams at the columns. The balcony is 1.5 m wide and 12.0 m long. Design the joist section. An additional load of 1 m high and 100 mm thick masonry parapet may be assumed on the outer ends of the balcony. On the inner balcony end masonry walls and columns are extended for the storey height with a door opening of 2 m × 2.5 m. Use steel with yield stress of 260 N/mm^2.

7.18 Design a composite beam with rigid shear connectors. The beam is of effective span 6.2 m and is subjected to a total load of 36 kN/m.

chapter 8

Gantry Girders

8.1 INTRODUCTION

Overhead travelling cranes are used in industrial buildings to lift and transport heavy jobs, machines, and so on, from one place to another. The crane may be a manually (hand) operated overhead travelling crane (M.O.T.) or an electrically operated overhead travelling crane (E.O.T.). A typical arrangement of a crane system is shown in Fig. 8.1. Usually, the crane consists of a bridge made up of two truss girders. The bridge is termed a *crane bridge, crane girder* or *crab girder*. It spans the bay of the shop and moves in a longitudinal direction. To facilitate movement, wheels are attached to the ends of crane girders. These wheels move over rails placed centrally over the girders which are called *gantry* or *crane gantry girders*. These girders are designed as laterally unsupported beams until the compression flange is laterally supported by either a catwalk or by additional member. Figure 8.2 shows the front view and top view of the typical arrangement of a crane girder, gantry girder and column in a workshop.

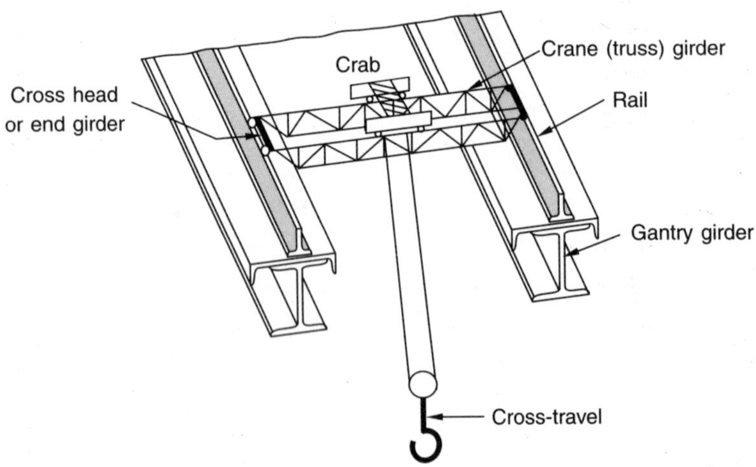

Fig. 8.1 *Gantry and Truss Girder Arrangement*

A trolley (crab) with wheels and a suspended hook is placed over the crane bridge and this arrangement can move in the transverse direction. However, it should be noted that the two movements of the crane, the longitudinal and the transverse cannot be had simultaneously. Bolted clamps or hook bolts are, used to keep the rails in position (Figs 8.3(a) and (b)).

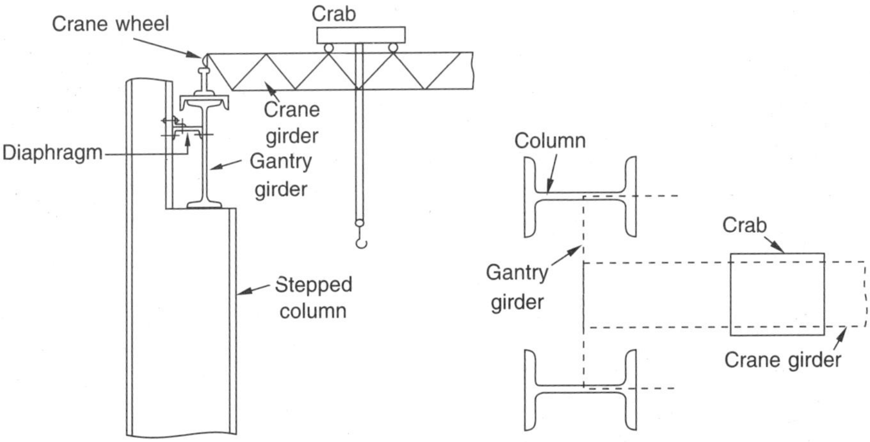

(a) Front View (b) Top View

Fig. 8.2 *Typical Arrangement of Gantry Girder and Crane Girder*

The gantry girders, because of the necessary clearances, must be placed a distance away from the face of columns. A direct connection with the column therefore is impossible. Because of lateral loads, however, some types of connection is mandatory. This is provided by diaphragm as shown in Fig. 8.2(a). In gantry girders, even when deflection is held within limits, there is some rotation at the ends under moving crane. Accordingly some provision for longitudinal movement of the top flange at the ends must be made. This is usually taken care by using bolts with slotted holes (Fig. 8.2(a), 8.3(c)).

Some of the types of sections used for gantry girders are shown in Fig. 8.3. An I-Section is provided for a manually operated overhead crane (Fig. 8.3(a)). The top flange of the I-Section is reinforced with a channel with its flanges turned down on the compression flange (Fig. 8.3(b)) or a channel and a bracket plate are attached to the web of the I-Section (Fig. 8.3(c)) to increase the lateral resistance against the horizontal surge from electrically operated overhead travelling cranes as well as torsional rigidity. If required, still heavier sections as shown in Fig. 8.3(d), may be provided.

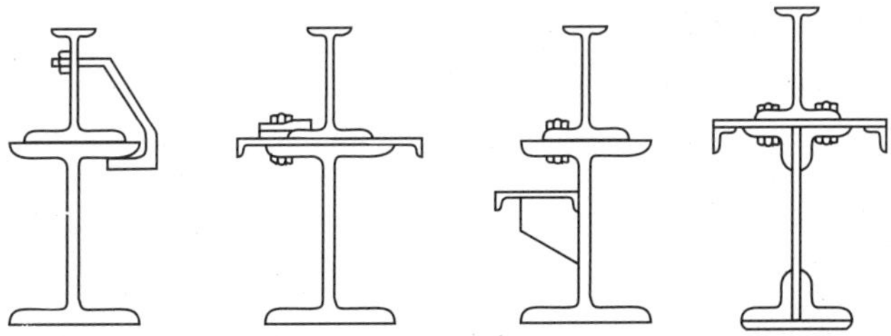

(a) (b) (c) (d)

Fig. 8.3 *Different Shapes of a Gantry Girder*

8.2 LOADS

Gantry girders are unique in themselves. First, it is different from usual beams in buildings. It is laterally unsupported except at the columns. Second, it is one of the very few girders in the buildings that is subjected to impact. Third, it must be analyzed for unsymmetrical bending because of lateral thrust from the starting and stopping of crab. Fourth, it is subjected to longitudinal load due to starting and stopping of the crane bridge itself; and the fifth, these are always simply supported. These are subjected to the forces as follows.

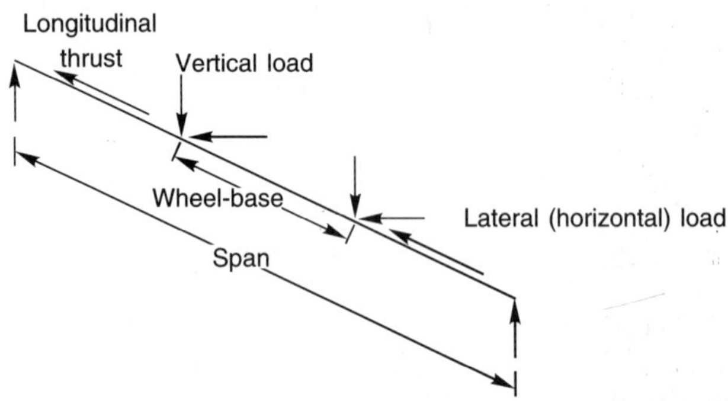

Fig. 8.4 *Loads on Gantry Girder*

1. The reaction from the crane girder, acting vertically downwards (Ref. Fig. 8.4).
2. The longitudinal thrust, due to starting or stopping of crane, acting in the longitudinal direction.
3. The lateral thrust, due to starting/stopping of the crab acting horizontally, normal to the gantry girder.

Vertical Loads

A vertical load acting over the gantry girder is the reaction from the crane girder and consists of the self-weight of the crane, self-weight of the crab and the crane capacity (the maximum load that can be hoisted). To calculate the reaction the maximum wheel load is computed. It occurs when the crab is nearest to the gantry girder. In addition to the reaction from a crane girder, the self-weight of the rail should also be considered.

Lateral Loads

Lateral forces on crane girders may be caused by the

(a) thrust due to the sudden stopping of the crab and load when traversing the crab girders
(b) crab dragging weights across the shop floor.

The foregoing cases cannot occur simultaneously.

In (a), as with the longitudinal gantry girders, the frictional resistance of the rail is transferred into the crab girders and from them into the crosshead girders, thence, as point loads through the main wheels, into the top or compression flanges of the gantry girders.

The positions of the main wheels when maximum lateral bending and shear take place on the gantry girder will be the same as those when maximum vertical bending moment and shear occurs.

(b) The crane is often requisitioned to drag weights across the shop floor. If the load is extremely massive, it is usually mounted on roughly fashioned rollers, probably running on a timber plank track. The lateral thrust and pull on the compressive flanges of the gantry girders then become a matter of conjecture. The resisting forces are, firstly, the friction of the main wheel treads upon the gantry rails and, secondly, the forces offered by the flanges of the main wheels bearing against the gantry rails.

The lateral thrust is assumed to act in the plane of the centre of gravity of the upper flange. Acting as it does at rail level, it has really a lever arm producing torque. This small lever arm and, therefore, the torque are neglected. No help is assumed to be afforded by the lower or tensile flange in resisting lateral thrust. However, should this help be considered, then the torque—due to the thrust multiplied by the distance from the line of action of the thrust to the N.A. (*xx* in this case)—should also be taken into consideration.

Longitudinal Loads

Longitudinal loads are caused due to the stopping or starting of the crane girders, and produce a thrust along the rails. The largest of these, especially in quick-acting electric overhead travelling cranes, is due to the sudden application of the brakes. The frictional resistance to the sliding of the locked wheels upon the rails is supplied by the crane girder. This element in turn distributes it amongst all the crane column shafts.

The lateral and longitudinal thrusts are transferred at the rail level. Therefore, gantry girders are also subjected to bending moments due to these loads.

Impact Loads

The stresses produced in gantry girders due to the above loads are more than those caused by gradually applied loads. This is due to the forces set up by the sudden application of brakes to the rapidly moving loaded cranes acceleration, retardation, vibration, possible slip of slings, etc. The steelwork which carries these quick-acting cranes must be heavier than the steelwork which supports slow-moving cranes. With quick-acting electric overhead travelling (E.O.T) cranes, the stresses in the gantry girders are produced almost instantaneously, whereas with slow-moving hand-operated cranes, the bending stresses in the girder are induced gradually from zero up to their maximum values, as the cranes traverses the girder from the end towards the centre. To account for this, suitable impact factors are introduced as and when applicable. As per I.S: 875–1964, additional loads as listed in Table 8.1 should be considered when structures are subjected to impact loads in addition to live loads.

Table 8.1 *Additional Loads for Structures Subjected to Impact Loads*

Type of Load	*Additional Load*
(a) *Vertical forces transferred to the rails*	
(i) For electric overhead cranes	25% of maximum static wheel load
(ii) For hand operated cranes	10% of maximum static wheel load
(b) *Horizontal forces transverse to the rails:*	
(i) For electric overhead cranes	10% of the weight of the crab and the weight lifted on the crane
(ii) For hand operated cranes	5% of the weight of the crab and the weight lifted on the crane
(c) *Horizontal forces along the rails*	5% of the static wheel load

8.3 SPECIFICATIONS

1. A gantry girder section is subjected to vertical loads and horizontal thrust (surge loads) simultaneously. Therefore, the allowable stresses are increased by 10%. This increase in the allowable stress is not in addition to that allowed for erection loads with or without wind or seismic forces.
2. Either of the two horizontal forces specified in Table 8.1 should be considered to act along with the vertical loads at a time.
3. The vertical deflection of a gantry girder should not exceed the values specified below:
   (i) Where the cranes are manually operated $\frac{L}{500}$
   (ii) Where the cranes are travelling overhead and operated electrically up to 500 kN $\frac{L}{750}$
   (iii) Where the cranes are travelling overhead and operated electrically over 500 kN $\frac{L}{1000}$
   (iv) Other moving loads, such as charging cars, etc. $\frac{L}{600}$

   Where, L = span of the gantry girder.

8.4 DESIGN

The design of rolled sections subjected to lateral loads is a matter of trial and error, but for a built-up girder the following procedure may be adopted. In the design of gantry girders it is assumed that the lateral load is resisted entirely by the compression flange (when applied at the compression flange level). The assumption is fairly justified because if the tension flange is to assist in resisting a lateral load applied to the compression flange it must be pushed sideways by the web. But the lateral stiffness of the web is quite small as compared to that of the compression flange. Therefore, the assistance from the tension flange must be considered negligible.

The gantry girder is designed as follows:

1. Find the maximum wheel load. The worst case of loading occurs when the crane and its load are drawn to the column. Therefore, the maximum vertical load is when the crab is closest to the gantry girder. This distance is the

minimum approach of the crane hook to the gantry girder. The crab in such a position on the crane girder gives the maximum reaction on the gantry girder. The vertical reaction of the crane girder is transferred through its two wheels on to the gantry girder. Therefore, the maximum wheel load is half of this reaction. This maximum wheel load is increased for the impact as specified in Table 8.1 of the text. The lateral thrust due to the cross travel of the crab can be neglected, because the crab, if carrying full load, must go very slowly when in such close proximity to the end of its travel.

2. The maximum bending moment in the gantry girder due to vertical loads is computed. This consists of the bending moment due to the above found maximum wheel loads (with impact) and the bending moment due to the dead load of the girder and rails. The wheel load bending moment is maximum when the two wheels are in such a position that the centre of gravity of the wheel loads and one of the wheel loads are equidistant from the centre of the gantry girder (if only one crane is running over the girder). In other words, the quarter point of the span of the two wheels must coincide with the central line of the girder. If after such a placement one of the wheel load position falls on the adjacent gantry girder, and the gantry girder in question finds only one wheel load, then the maximum bending moment will be when this load is at the mid span.

   When two or more cranes are placed on one gantry, the gantry girder should be designed with the assumption that the cranes will be running buffer to buffer, lifting or carrying their maximum loads simultaneously.

   The bending moment due to dead loads is maximum at the centre of the span. At the point where the bending moment due to the wheel load is maximum (below one of the wheels of the crane girder) the dead load bending moment is slightly less than the maximum dead load bending moment, but for simplifying the calculations the maximum bending moment due to dead loads may directly be added to the maximum wheel load moment.

3. The maximum shear force is computed. This consists of shear force due to wheel loads and dead loads from the gantry girder and rails. The shear due to the wheel load is maximum when one of the wheels is at the support (if only one crane is running over the girder).

4. The section modulus, of the gantry girder is found by the equation

$$Z = \frac{M}{\sigma_{bc,t}} \tag{8.1}$$

   Where M is the maximum bending moment due to vertical loads, and $\sigma_{bc,t}$ is the permissible bending stress ($0.66\, f_y$). A suitable section is selected from I.S. Handbook No. 1. Generally an I-Section with a channel section is provided. As the gantry girders are laterally unsupported the trial section modulus provided (I-Section with channel section on compression flange) is usually 20 to 40% more than the calculated section modulus. Generally, the economical depth of a gantry girder is not less than 1/12 of the span except when the loads are small in relation to the span. The width of the flange should be between 1/40 to 1/30 of the span to prevent excessive lateral deflection.

5. Calculate the moment of inertia of the girder section about both the axes. The maximum bending stress in compression ($\sigma_{bc,\ cal}$) and in tension ($\sigma_{bt,\ cal}$) is calculated by the following equations:

$$\sigma_{bc,\ cal} = \frac{M}{I_{xx\ gross}} Y_1 \tag{8.2}$$

$$\sigma_{bt,\ cal} = \frac{M}{I_{xx\ gross}} Y_2 \tag{8.3}$$

where Y_1 and Y_2 are the distances of extreme fibre of compression and tension flanges from the neutral axis respectively.

6. The bending stress in tension should be less than the permissible bending stress ($1.1 \times 0.66\ f_y$).

Note The beam section here has been checked as that for a laterally supported beam. It is so because only gravity loads have been considered to derive the section wherein the compression flange has been strengthened by providing a channel section which will check the lateral bending of the compression flange. Further, since lateral/longitudinal loads will also be there alongwith gravity loads, the section will be checked for lateral instability. Infact, the forgoing procedure has an advantage. The bending stress value obtained indicates to the designer whether the selected section will be safe for the combination of forces or not. It saves repetition of the design procedure for a laterally unsupported beam.

7. The lateral forces on the girder and the maximum bending moments and shear due to these are calculated. The position of the wheels should be same as that in step (2) and (3).
8. The maximum compressive stress due to lateral loads is calculated. The compression flange is assumed to take lateral loads. Hence the moment of inertia of the compression flange required about its *yy*-axis is calculated. This can be determined by adding I_{xx} of channel section to half the I_{yy} of I-Section. Though approximate but this simplified method gives fairly close and acceptable value.
9. The longitudinal forces along the rails and the stresses (direct and bending) in longitudinal direction are calculated.
10. The total bending compressive stress is found by adding bending stresses, as calculated in step (5) and step (8 or 9, whichever is more). It should not exceed the allowable bending compressive stress.
11. The allowable bending compressive stress σ_{bc} is calculated, as explained in Section 7.12 of the text, and is increased by 10%.
12. Rivets connecting the channel to the I-Section are designed.
13. The maximum shear stress in the gantry girder is checked.
14. The bracket and its connection with column are designed. A pair of bracket plates, one on each flange of I-Section column, connected with a diaphragm is provided to make a seat for gantry girder.

The question now arises as to how much load comes on to each bracket plate. If each bracket plate was independent of its neighbour, then either plate would, in turn, have to support the entire reaction due to the crane and its load. By using a diaphragm one side plate cannot deflect without taking its neighbour with it, and its neighbour cannot deflect unless it also participates in the carrying of the load, since deflection is proportional to load. There is another factor which is sometimes considered, and that is the frequency of this particular case of loading. It will be seldom that the crane wheels will be called upon to carry the maximum load with the crab drawn in tightly against the bracket. The gantry girder is designed for the worst possible cases of loading without consideration as to the laws of chance, and to be consistent the vertical brackets should be designed to meet the maximum load. Conservative designers make each bracket plate strong enough to carry all the reaction, others, in competitive work, often assume that the unloaded plate relieves the loaded plate of a certain amount of load, the maximum load on the loaded plate being vaguely estimated at one-half, three-quarters and seven-eights of the maximum girder reaction. The half is decidedly risky and is not recommended.

Note The computation of design forces and deflection in the procedure laid above, is for one crane running over the gantry girders. Often the industrial buildings require more than one crane at the same level or at different levels in the same aisle, in addition to one or more cranes supported by the same column in the adjacent aisle. If more than one crane is used, the design forces should be suitably computed. It is also desirable to make the girders entirely self-supporting in the adjacent aisles.

Solved Examples

Example 8.1 Design a gantry girder to be used in an industrial building carrying an electric overhead travelling crane, for the following data:

Crane capacity	200 kN
Self-weight of the crane girder excluding trolley	200 kN
Self-weight of the trolley, electric motor, hook, etc.	40 kN
Approximate minimum approach of the crane hook to the gantry girder	1.20 m
Wheel base	3.5 m
c/c distance between gantry rails	16 m
c/c distance between columns (span of gantry girder)	8 m
Self-weight of rail section	300 N/m
Yield stress of steel	250 N/mm^2

Solution

Maximum wheel load

Maximum concentrated load on crane = 200 + 40 = 240 kN

The crane will carry the self-weight as a uniformly distributed load

$$= \frac{200}{16} = 12.5 \text{ kN/m}$$

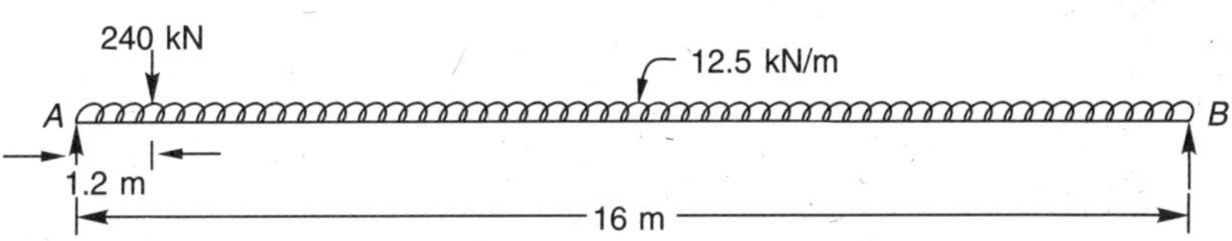

Fig. Ex. 8.1(a)

For maximum reaction on the gantry girder the loads are placed on the crane girder as shown in Fig. Ex. 8.1(a).

Taking moment about B,

$$R_A \times 16 = 240 \times (16 - 1.2) + \frac{12.5 \times 16 \times 16}{2}$$

or $$R_A = 322 \text{ kN}$$

Similarly $$R_B = 118 \text{ kN}$$

The reaction from the crane girder is distributed equally on the two wheels at the end of the crane girder.

$\therefore$ Maximum wheel load on each wheel of crane $= \frac{322}{2} = 161$ kN

Maximum bending moment It consists of maximum moments caused by the moving wheel loads on the gantry girder and self weight of the gantry girder. For maximum bending moment, the wheel loads shall be placed as shown in Fig. Ex. 8.1(b). The calculation of maximum bending moments due to wheel loads and self weight of gantry girder has been done separately because calculation of impact load and bending moment due to it involve live load only.

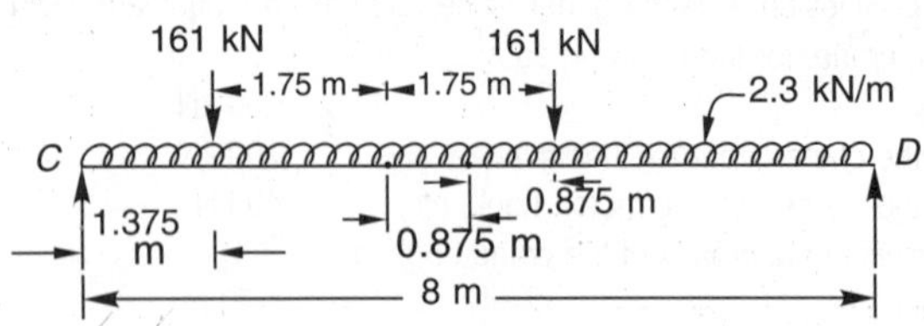

Fig. Ex. 8.1(b)

Assume self weight of gantry girder as 2 kN/m

Total dead load, $w = 2000 + 300 = 2300$ N/m $= 2.3$ kN/m

The position of one wheel load from the mid-point of span

$$= \frac{\text{wheel base}}{4} = \frac{3.5}{4} = 0.875 \text{ m}$$

Bending moment due to live load only:

Taking moment about D,

$$R_C \times 8 = 161 \times (8 - 1.375) + 161 \times 3.125$$

$$R_C = 196.22 \text{ kN}$$

Taking moment about C,

$$R_D \times 8 = 161 \times 1.375 + 161 \times 4.875$$

$$R_D = 125.78 \text{ kN}$$

Maximum bending moment due to live load $= 125.78 \times 3.125 = 393.06$ kN m

Bending moment due to impact $= 0.25 \times 393.06 = 98.26$ kNm

Total bending moment due to live and impact loads $= 393.06 + 98.26$

$= 491.32$ kNm

$$\text{Bending moment due to dead load} = \frac{wl^2}{8} = \frac{2.3 \times 8^2}{8} = 18.4 \text{ kNm}$$

$$\therefore \quad \text{Maximum bending moment} = 491.32 + 18.4 = 509.72 \text{ kNm}$$
$$= 509.72 \times 10^6 \text{ Nmm}$$

Maximum shear force It consists of maximum shear due to moving wheel loads and self weight of the girder. For maximum shear force the wheel loads shall be placed as shown in Fig. Ex. 8.1(c), i.e. one of the wheel loads should be at the support.

Taking moment about D,

$$R_C \times 8 = 161 \times 8 + 161 \times 4.5$$
$$R_C = 251.56 \text{ kN}$$

Hence maximum shear force due to wheel loads = 251.56 kN

Note The value of maximum shear due to self weight and the total maximum shear will be calculated after deriving the section. Once the section is checked in flexure and found safe, the actual self weight will be known, which can be used in shear calculations.

Cross-section of gantry girder

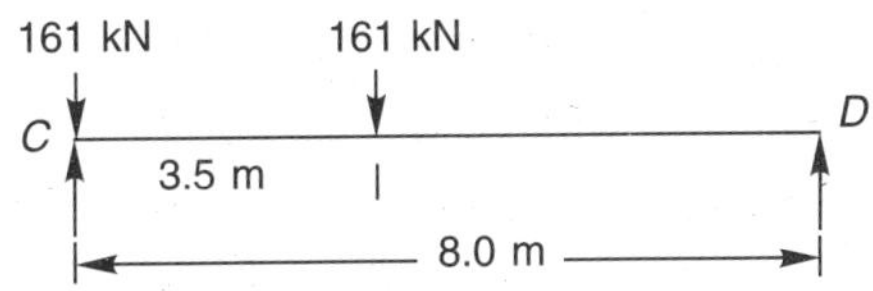

Fig. Ex. 8.1(c)

$$Z = \frac{M}{\sigma_{bc,t}} = \frac{509.72 \times 10^6}{165}$$
$$= 3089.2121 \times 10^3 \text{ mm}^3$$

$$Z_{\text{req}} = 1.25 \times 3089.2121 \times 10^3 = 3861.5151 \times 10^3 \text{ mm}^3$$

The section modulus has been increased by 25 per cent to account for buckling of compression flange.

Select I.S.W.B. 600 @ 1311.6 N/m and I.S.M.C. 300 @ 351.2 N/m from I.S. Hand Book No. 1. The relevant properties of the I-Section and channel sections are,

	I-Section	Channel section
A	17038 mm^2	4564 mm^2
T	21.3 mm	13.6 mm
t	11.2 mm	7.6 mm
B	250 mm	90 mm
I_{xx}	106198.5×10^4 mm^4	6362.6×10^4 mm^4
I_{yy}	4702.5×10^4 mm^4	310.8×10^4 mm^4
C_{yy}	—	23.6 mm

Moment of inertia of gantry girder

The distance of N.A. of built-up section from the extreme fibre of compression flange

$$\bar{Y} = \frac{\Sigma AY}{\Sigma A}$$

$$= \frac{17038 \times (300 + 7.6) + 4564 \times 23.6}{4564 + 17,038} = 247.59 \text{ mm}$$

Gross moment of inertia of the built-up section

$$I_{xx\ gross} = I_{xx\ beam} + I_{xx\ channel}$$

(The channel is placed over the I-section in such a manner that its *yy*-axis becomes the *xx*-axis. Therefore, in calculating I_{xx} of the total section, I_{yy} of the channel will become I_{xx} of the channel.)

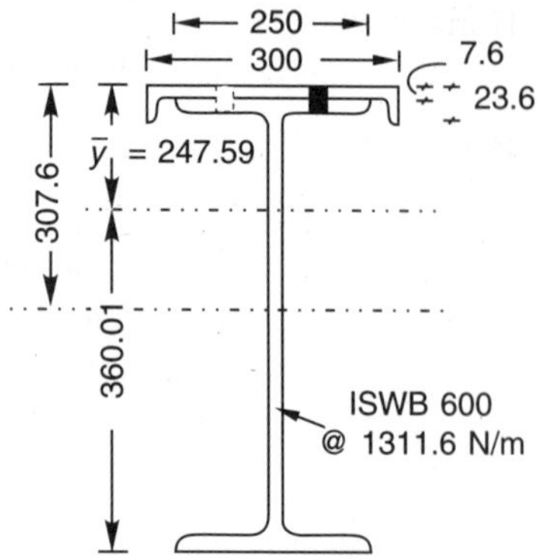

Fig. Ex. 8.1(d)

$$I_{xx}\ \text{gross} = [106{,}198.5 \times 10^4 + 17{,}038 \times (307.6 - 247.59)^2] + [310.8 \times 10^4 + 4564 \times (247.59 - 23.6)^2]$$

$$= 135{,}543 \times 10^4\ \text{mm}^4$$

$$I_{yy\ gross} = I_{yy\ beam} + I_{yy\ channel}$$

$$= 4702.5 \times 10^4 + 6362.6 \times 10^4$$

$$= 11{,}065.1 \times 10^4\ \text{mm}^4$$

Check of bending stress Allowable bending stress, $\sigma_{bct} = 0.66 f_y$

For steel with yield stress 250 N/mm² $\sigma_{bc,\,t} = 165$ N/mm²

Increase the allowable stress by 10% as per I.S. specifications

$$\sigma_{bc,\,t} = 1.1 \times 165 = 181.5\ \text{N/mm}^2$$

$$\sigma_{bc,\ cal} = \frac{509.72 \times 10^6 \times 247.59}{135{,}543 \times 10^4} = 93.108\ \text{N/mm}^2$$

$$\sigma_{bt,\ cal} = \frac{509.72 \times 10^6 \times 360.01}{135{,}543 \times 10^4} = 135.38\ \text{N/mm}^2$$

$$< 181.5\ \text{N/mm}^2$$

$$\tau_{at,\ cal} = \frac{259.56 \times 10^3}{600 \times 11.2} = 38.625\ \text{N/mm}^2$$

$$< 100\ \text{N/mm}^2$$

Lateral forces

Lateral force transverse to the rails = 10% of weight of crab and weight lifted

$$= 0.1 \times 240 = 24\ \text{kN}$$

$$\text{Lateral force on each wheel} = \frac{24}{2} = 12\ \text{kN}$$

Maximum horizontal reaction due to the lateral force by proportion at *C*,

$$= \frac{\text{lateral force} \times \text{reaction at } C \text{ due to vertical load}}{\text{maximum wheel load due to vertical load}}$$

$$= \frac{12 \times 196.22}{161} = 14.625\ \text{kN}$$

Horizontal reaction due to lateral force at D,

$$= 24 - 14.625 = 9.375 \text{ kN}$$

Bending moment due to lateral load by proportion

$$= \frac{12}{161} \times 393.06 \times 10^6 = 29.296 \times 10^6 \text{ Nmm}$$

Maximum shear force due to lateral load by proportion,

$$= \frac{251.56}{161} \times 12 = 18.75 \text{ kN}$$

Bending stress due to lateral loads The moment of inertia of the compression flange about the yy-axis,

= moment of inertia of beam flange + moment of inertia of channel

$$= (4702.5/2) \times 10^4 + 6362.6 \times 10^4 = 8713.85 \times 10^4 \text{ mm}^4$$

$$\text{Bending compressive stress, } \sigma_{bcy} = \frac{29.296 \times 10^6}{8713.85 \times 10^4} \times \frac{300}{2} = 50.43 \text{ N/mm}^2$$

Total bending compressive stress = 93.108 + 50.43 = 143.538 N/mm^2 should be < σ_{bc} (The allowable bending compressive stress)

Allowable bending compressive stress

$$\psi = 1 \quad \therefore \quad k_1 = 1$$

$$\omega = \frac{\text{moment of inertia of compression flange about } yy\text{-axis}}{\text{moment of inertia of built-up section about } yy\text{-axis}}$$

$$= \frac{87{,}135.5 \times 10^3}{110{,}651 \times 10^3} = 0.78748$$

From Table 6.4 of I.S: 800–1984 (Table 7.4 of the text),

$$k_2 = 0.28748$$

Also, $l = 8 \times 10^3$ mm

$$r_{yy} = \sqrt{I_{yy}/A} = \sqrt{11065.1 \times 10^4 / 21602} = 71.57 \text{ mm}$$

$$\therefore \quad \frac{l}{r_{yy}} = \frac{8 \times 10^3}{71.57} = 111.778$$

Mean thickness of flange, T = 21.3 + 7.6 = 28.9 mm

Overall depth of girder, D = 600 + 7.6 = 607.6 mm

$$\frac{D}{T} = \frac{607.6}{28.9} = 21.024$$

For l/r_{yy} = 111.778 and D/T = 21.024, from Table 6.5 of I.S: 800–1984 (Table 7.5 of the text)

$$X = 331.946 \text{ N/mm}^2 \quad \text{and} \quad Y = 212.777 \text{ N/mm}^2$$

Elastic critical stress, $f_{cb} = k_1(X + k_2Y)\ \dfrac{c_2}{c_1}$

$$= 1 \times (331.946 + 0.28748 \times 212.777) \times \frac{360.01}{247.59}$$
$$= 571.6118 \text{ N/mm}^2$$

$$\frac{T}{t} = \frac{28.9}{11.2} = 2.58 > 2$$

Hence, f_{cb} is not increased by 20% as per I.S: 800–1984.

For f_{cb} = 571.6118, and f_y = 250 N/mm², from Table 6.2 of I.S: 800–1984 (Table 7.6 of the text),

$$\sigma_{bc} = 135.580 \text{ N/mm}^2$$

Increase the stress so determined by 10%.

∴ Permissible bending stress in compression = 1.1 × 135.580 = 149.138 N/mm²
> 143.538 N/mm²

which is as it should be.

Longitudinal forces Let the height of rails = 75 mm

Horizontal (longitudinal) force along the rails = 5% of static wheel load
= 0.05 × 161 × 2 = 16.1 kN

Bending moment due to longitudinal force
= 16.1 × (247.59 + 75) = 5193.699 kNmm

Stress in longitudinal direction = $P/A + M/Z$

$$= \frac{16.1 \times 10^3}{21{,}602} + \frac{5193.699 \times 10^3}{5474.273 \times 10^3} = 1.694 \text{ N/mm}^2$$

(Very small, hence negligible).

Connection of flange of ISWB 600 and web of ISMC 300

Maximum shear force due to wheel load = 251.56 kN

Shear due to impact = 0.25 × 251.56 = 62.89 kN

$$\text{Shear due to self-weight} = \frac{(1311.6 + 351.2)}{2} \times 8 = 6651.2 \text{ N} = 6.65 \text{ kN}$$

Total shear force, V = 251.56 + 62.89 + 6.65 = 321.10 kN

Horizontal (longitudinal) shear/mm length

$$\therefore \quad = \frac{VA\bar{Y}}{I}$$

$$= \frac{321.10 \times (7.6 \times 300) \times (247.6 - 7.6/2)}{135{,}543 \times 10^4}$$

= 0.1316835 kN/mm = 131.683 N/mm

Let us provide 22 mm ϕ power driven rivets.

$$\text{Gross diameter} = 22 + 1.5 = 23.5 \text{ mm}$$

$$\text{Strength of rivets in single shear} = \frac{\pi}{4} \times 23.5^2 \times 100 = 43373.61 \text{ N}$$

$$\text{Strength of rivet in bearing} = 23.5 \times 7.6 \times 300 = 53{,}580 \text{ N}$$

$$\text{Rivet value, } R_v = 43373.6 \text{ N}$$

$$\text{Pitch of rivets} = \frac{43{,}373.6}{131.683} = 329.378 \text{ mm}$$

The rivets are provided in two rows and are staggered so that the section is reduced by one rivet hole only.

$$\therefore \quad s = 2 \times 329.378$$

$$= 658.757 \text{ mm}$$

$$\ngtr 12 \times 7.6 = 91.2 \text{ mm} \approx 90 \text{ mm}$$

Provide 22 mm diameter rivets at a pitch of 90 mm.

Check for shear Average shear stress in web of I-Section.

$$\tau_{va} = \frac{321.10 \times 10^3}{600 \times 11.2}$$

$$= 47.78 \text{ N/mm}^2 < 1.1 \times 100 \text{ N/mm}^2$$

which is safe.

$$\text{Average shear stress in web of channel, } \tau_{va} = \frac{14.625 \times 10^3}{300 \times 7.6}$$

$$= 6.414 \text{ N/mm}^2$$

$$< 100 \text{ N/mm}^2$$

which is safe.

Crane Bracket

$$\text{Reaction due to wheel load} = 251.56 \text{ kN}$$

$$\text{Impact load} = 0.25 \times 251.56 = 62.89 \text{ kN}$$

$$\text{Maximum wheel load reaction} = 251.56 + 62.89 = 314.45 \text{ kN}$$

$$\text{Reaction due to self weight of rail girder} = 8 \times (0.3 + 1.3116 + 0.3512)$$

$$= 15.70 \text{ kN}$$

$$\text{Total reaction on bracket} = 314.45 + 15.70 = 330.15 \text{ kN}$$

The end reaction is transferred on to the two bracket plates, each attached to the column flanges.

$\therefore$ Theoretical design load for connection = 330.41/2 = 165.20 kN

To be neither too conservative nor too venturesome, let the active bracket plate carries 75% of the end reaction. Design load for bracket plate and connection = 0.75 × 330.15 = 247.61 kN

Let us provide 22 mm ϕ rivets for connection. Assuming the rivet value to be in shear,

$$\text{Rivet value} = \frac{\pi}{4} \times 23.5^2 \times 100 = 43{,}373.6 \text{ N}$$

The minimum thickness of bracket plate is

$$43{,}373.6 = 23.5 \times t \times 300$$

$$t = 6.15 \text{ mm}$$

Provide an 8 mm thick bracket plate,

$$\text{Minimum pitch of rivets} = 2.5\,\phi = 2.5 \times 22 = 55 \text{ mm}$$

$$\approx 60 \text{ mm}$$

Let us provide rivets in four vertical rows. Assume eccentricity of reaction = 500 mm

$$n = \sqrt{\frac{6\,M}{sn'R_v}} = \left(\frac{6 \times 247.61 \times 10^3 \times 500}{60 \times 4 \times 43{,}373.6}\right)^{1/2} = 8.45 \approx 8$$

Provide 32 rivets, 22 mm ϕ, on each bracket plate with 8 rivets in each row as shown in Fig. Ex. 8.1(e). As the column section is not yet designed, the gauge distance may be assumed to be 60 mm.

Force on extreme critical rivet

Force due to direct load,

$$F_1 = \frac{247.61 \times 10^3}{32} = 7737.81 \text{ N}$$

Force due to moment,

$$F_2 = \frac{Per_n}{\Sigma r^2}$$

$$= 247.61 \times 10^3 \times 500 \times \frac{228.47}{748{,}800}$$

$$= 37774.74 \text{ N}$$

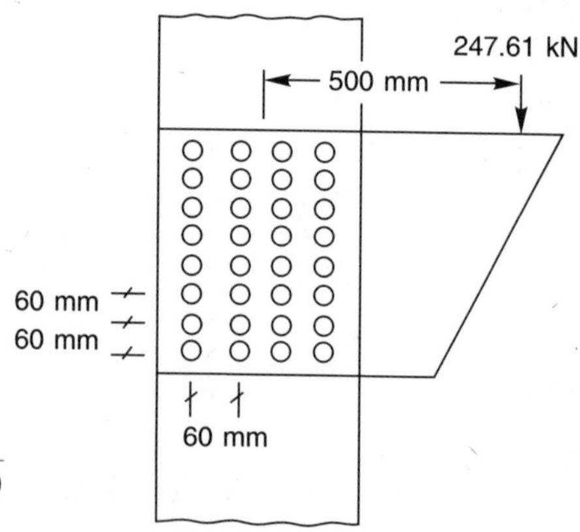

Fig. Ex. 8.1(e)

$\cos\theta = 90/228.47 = 0.3939$

Resultant force

$$= \sqrt{(7737.81)^2 + (37774.74)^2 + 2 \times 7737.81 \times 37774.74 \times 0.3939}$$

$$= 41437.6 \text{ N} < 43{,}373.6 \text{ N}$$

which is safe.

Example 8.2 Design a gantry girder as laterally supported beam to be used in an industrial building carrying a hand operated traveling crane, for the following data:

Crane capacity	= 50 kN
Self-weight of crane girder excluding trolley	= 40 kN
Weight of trolley car	= 10 kN
Approximate minimum approach of the crane hook to gantry girder	= 1 m
Wheel base	= 3 m
c/c distance between gantry rails	= 14 m
c/c distance between column	= 5.5 m
Self-weight of rail section	= 300 N/m
Yield stress of steel	= 250 N/m^2

Solution

Maximum wheel load

Maximum concentrated load on crane = 50 + 10 = 60 kN

Self-weight of crane will act as uniformly distributed load of intensity

$$= \frac{40}{14} = 2.86 \text{ kNm}$$

For maximum reaction of gantry girder, the loads are placed on the girder as shown in Fig. Ex. 8.2(a).

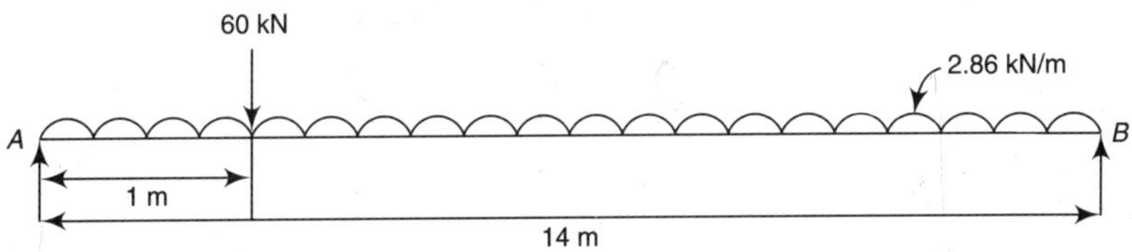

Fig. Ex. 8.2(a)

Taking moment about B

$$R_A \times 14 - 60 \times 13 - 2.86 \times 14 \times 7 = 0$$

or $\quad R_A = 75.73 \text{ kN} = 76 \text{ kN}$

Similarly, $\quad R_B = 100 - 76 = 24 \text{ kN}$

The reaction from the crane girder is distributed equally on two wheels at the end of the crane girder.

$$\text{Maximum wheel load on each wheel of crane} = \frac{76}{2} = 38 \text{ kN}$$

Maximum bending moment

For maximum bending moment, the wheel loads shall be placed as shown in Fig. Ex. 8.2(b).

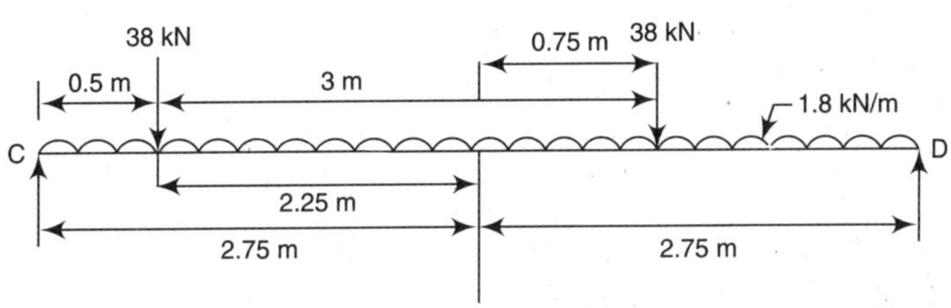

Fig. Ex. 8.2(b)

Assume self-weight of gantry girder as 1.5 kN/m.

Total dead load = 300 + 1500 = 1800 N/m

Taking moment about D,

$$R_C \times 5.5 - 38 \times (2.75 + 2.25) - 38 \times (2.75 - 0.75) = 0$$

$\Rightarrow$ $R_C = 48.36$ kN

and $R_D = 76 - 48.36 = 27.64$ kN

Bending moment under wheel due to live load = $R_D \times 2 = 27.64 \times 2$
= 55.28 kNm

Bending moment due to impact = 0.10×55.28 (10% increase due to M.O.T)
= 5.528 kNm

Total bending moment due to live load and impact loads = 55.28 + 5.528
= 60.808 kNm

$$\text{Bending moment due to dead load} = \frac{wl^2}{8} = \frac{1.800 \times 5.5^2}{8} = 6.806 \text{ kNm}$$

Maximum shear force

For maximum shear force, one of the wheel load should be at support as shown in Fig. Ex. 8.2(c).

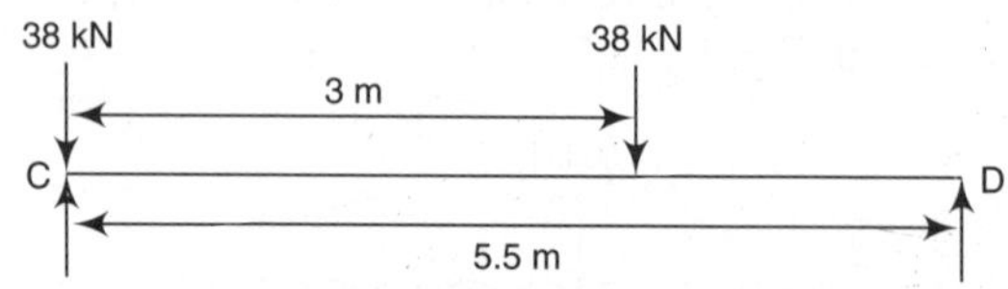

Fig. Ex. 8.2 (c)

$(R_C - 38) \times 5.5 - 38 \times 2.5 = 0$

$\Rightarrow R_C = 55.27$ kN

Hence maximum shear force due to wheel load is 55.27 kN.

Cross section of gantry girder

Gantry girder will be designed as a laterally supported beam for gravity loads. A channel section will be designed for lateral load and welded to the compression flange of gantry girder to check the lateral buckling as shown in Fig. Ex. 8.2(d). The channel section will also be welded to the columns.

$$Z = \frac{M}{\sigma_{bct}} = \frac{67.614 \times 10^6}{165} = 409.781 \text{ cm}^3$$

Let us try ISLB 300 @ 369.8 N/m from IS Hand Book No. 1. The properties are: $A = 48.08$ cm^2, $T = 9.4$ mm, $t = 6.7$ mm, $B = 150$ mm, $I_{xx} = 7332.9 \times 10^4$ mm^4, $I_{yy} = 376.2 \times 10^4$ mm^4

Lateral Forces

Lateral force transverse to rails = 5% of weight of crab and weight lifted

$$= \frac{5}{100} \times 60 = 3 \text{ kN}$$

Lateral force on each wheel, $F_l = 1.5$ kN

Maximum horizontal reaction due to lateral force by proportion at C

$$= \frac{1.5}{38} \times 48.36 = 1.90 \text{ kN}$$

Horizontal reaction due to lateral force at $D = 3 - 1.90 = 1.10$ kN

Bending moment due to lateral load by proportion = $\frac{1.5}{38} \times 55.27 = 2.18$ kNm

Design of channel section

A channel may be connected to I section as shown in Fig. Ex. 8.2(d).

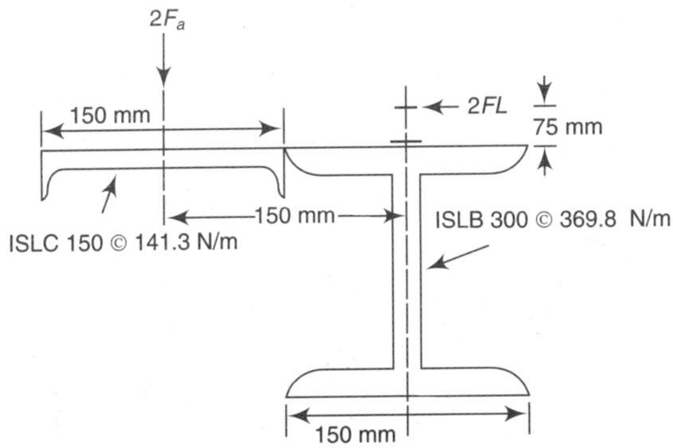

Fig. Ex. 8.2(d)

The section modulus of channel along the *yy* axis will be

$$Z_{yy} = \frac{2.18 \times 10^6}{\sigma_{bc,t}} = \frac{2.18 \times 10^6}{165} = 13212.12 \text{ mm}^3$$

Check for allowable bending stresses

$$\sigma_{bc,t} = 0.66 f_y = 165 \text{ N/mm}^2 \qquad \text{(Fe-250)}$$

The lateral force on head of rails is = 1.5 kN

This force is transferred at C.G. of the channel. Thus at the C.G. of the channel, there will be a force F_l of 1.5 kN and a lateral moment.

$$= 2 \times 1.5 \times e$$

$$e = \text{depth of rail section} + C_{yy} \text{ of channel section}$$

Since the depth of rail section is not known and the channel section is yet to be designed, the lever arm *e* for the moment calculation is assumed as 75 mm.

$$\text{Moment} = 2 \times 1.5 \times 75 = 225 \text{ kNmm}$$

Assuming depth of channel = 150 mm

Distance between centre of gravities of channel section and I-Section,

$$C = \frac{150}{2} + \frac{150}{2} = 150 \text{ mm}$$

$$\text{Axial force, } F_a = \frac{M}{2C} = \frac{225}{2 \times 150} = 0.75 \text{ kN} \qquad \text{(Fig. Ex. 8.2(d))}$$

The channel is subjected to biaxial moments

$$M_x = \frac{2F_l}{L}\left(\frac{L}{2} - \frac{a}{4}\right)^2 = \frac{2 \times 1.5}{5.5} \times \left(\frac{5.5}{2} - \frac{3}{4}\right)^2 = \frac{12}{5.5} = 2.1818 \text{ kNm}$$

$$M_y = \frac{2F_a}{L}\left(\frac{L}{2} - \frac{a}{4}\right)^2 = \frac{2 \times 0.75}{5.5} \times \left(\frac{5.5}{2} - \frac{3}{4}\right)^2 = \frac{6}{5.5} = 1.09 \text{ kNm}$$

Try ISLC 150 @ 141.3 N/m

$$Z_{xx} = 93 \times 10^3 \text{ mm}^3,\ Z_{yy} = 20.2 \times 10^3 \text{ mm}^3,\ C_{yy} = 23.8 \text{ mm}$$

$$\sigma_{bc,\,t} = \frac{M_z}{Z_x} + \frac{M_y}{Z_y}$$

$$= \frac{2.1818 \times 10^6}{93 \times 10^3} + \frac{1.09 \times 10^6}{20.2 \times 10^3} = 77.42 \text{ N/mm}^2 < 0.66 f_y$$

which is safe.

Exercises

Note Assume the yield stress of steel to be 250 N/mm^2 if not specified in the problem.

8.1 Design a simply supported gantry girder to carry an electric overhead travelling crane for the following data:

Crane capacity	320 kN
Weight of crane and crab	300 kN
Weight of crane	200 kN
Minimum approach of crane hook	1.20 m
Distance between c/c of wheels	3.20 m
Distance between c/c of gantries	16.0 m
Span of gantry girder	4.00 m
Weight of rails	300 N/m
Height of rails	75 mm
Yield stress of steel	280 MPa

8.2 Design a simply supported gantry girder to be used in an industrial building for the following data:

Crane capacity	100 kN
Weight of crab	35 kN
Weight of crane (excluding crab)	160 kN
Minimum clearance between crane hook and gantry girder	1.00 m
Wheel base	3.00 m
Distance between c/c of gantries	20.0 m
Distance between c/c gantry columns	6.0 m
Crane type	M.O.T.

8.3 Two electrically operated overhead travelling cranes are to be used in a bay of an industrial building. Design the gantry girder for the following data:

Crane capacity	250 kN (each)
Bay width	18 m
Spacing of columns	8 m

Weight of each crane and crab	250 kN
Minimum approach of crane hook	1.20 m
Minimum distance between cranes	2.0 m
Wheel base	3.2 m
Yield stress of steel	340 N/mm^2

8.4 Design gantry girder in an industrial building for two moving cranes for the following data:

Crane capacity	200 kN
Weight of each crane	150 kN
Weight of each crab	10 kN
Minimum distance of crane hook	1.1 m
Minimum distance between cranes	2 m
Wheel base	3.4 m
Bay width	16 m
Spacing of columns	7 m
Yield stress of steel	250 N/mm^2

Plate Girder

chapter 9

9.1 INTRODUCTION

A flexural member is designed usually for bending, and occasionally for shear. When spans are large and the loads are heavy, even an I-Section with cover plates may be insufficient to support the anticipated loads. In such a case any of the following alternatives may be provided:

(i) two or more I-Sections, connected appropriately
(ii) plate girder
(iii) truss girder

Of the above alternatives, the first is usually uneconomical. Therefore the choice lies between a plate girder and a truss girder only. Plate girders have a moment risisting capacity between rolled I-Sections and truss girders. Plate girders can be built to any desired proportion to suit the particular requirements, and of a cross-section with exactly the properties needed. The disadvantages of the truss girder are higher cost of fabrication and erection, problem of vibration and impact, requirements of higher vertical clearance and costly maintenance. For light loads and short spans (< 10 m) plate girders are uneconomical due to increase in the connection cost and rolled I-Sections are preferred.

Plate girders are built-up flexural members. Their bending resistance can be increased by increasing the distance between the flanges. This in turn also increases the shear resistance as the web area increases. Plate girders are primarily provided in bridges but are also very common in buildings where heavy concentrated loads (from upper storey columns, etc.) act on a long span beam, e.g. a dining hall floor beam of a restaurant requiring a clear space all through.

At the first instance, a plate girder is usually thought as a flexural member whose cross-section is composed of plate elements, which of course is true in case of welded plate girders. Before the widespread use of welding, connecting the components of the cross section was a major consideration in the design of plate girders. All of the connections were made by riveting, so there was no way to attach the flange directly to the web, and additional cross-sectional elements were introduced for the specific purpose of transmitting the load from one component to the other. The usual technique was to use a pair of angles, placed back to back on either side of the web, to attach the flange to the web. One pair of legs was attached to the web, and the other pair attached to the flange, as shown in Fig. 9.1(a).

Riveted plate girders are economically used between 15 m to 30 m span, whereas the welded plate girders may be used up to 100 m span (though a truss girder may be more economical for large spans). A welded plate girder is economical as compared to the riveted plate girder because of the considerable reduction in its self-weight it should be evident that the welded plate girder is far superior to the riveted or bolted girder in terms of simplicity and efficiency.

The design of a plate girder is essentially an exercise in ordinary beam design: the proportioning of a member with a section modulus adequate to resist bending, a web capable of resisting shear in the normal fashion, and sufficient stiffness. The only real difference between it and the design of a rolled beam is that, since the designer has greater control over the dimensions of the section and may make the web thinner in proportion to its depth than in any of the rolled shapes, he must pay more attention to web buckling. The designer has the choice of either making the web thick enough by itself to avoid premature buckling—and this often defeats the purpose of plate girder construction—or placing intermediate stiffeners along its length and thus, by dividing it into smaller panels and reducing the plate span, increasing the web buckling resistance.

9.2 TYPES OF SECTIONS

Several possible plate girder arrangements are shown in Fig. 9.1. When the loads and the span are large, plate girder sections either with riveted connections (Figs 9.1(a) and (b)) or welded connections (Fig. 9.1(c)) may be provided. The number of flange plates can be increased depending upon the moment to be resisted. If the depth is limited and the loads are too large, a box welded plate girder (Fig. 9.1(d)) is provided. A box plate girder with riveted connections can also be made but it is costly as compared to the welded one. Box girders have the advantage of great resistance to lateral buckling and torsion.

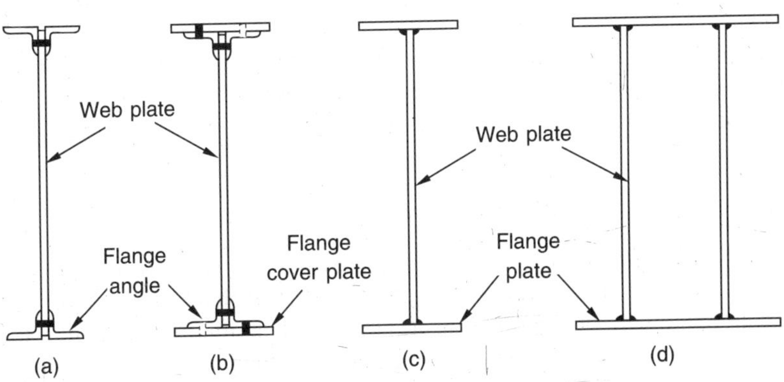

Fig. 9.1 *Plate Girder Arrangements*

9.3 ELEMENTS OF PLATE GIRDER

The various elements of a riveted plate girder are depicted in Fig. 9.2 and are as listed below:

1. Web plate
2. Flange angles with or without flange cover plates for riveted/bolted plate girder and only flange plates for welded plate girders

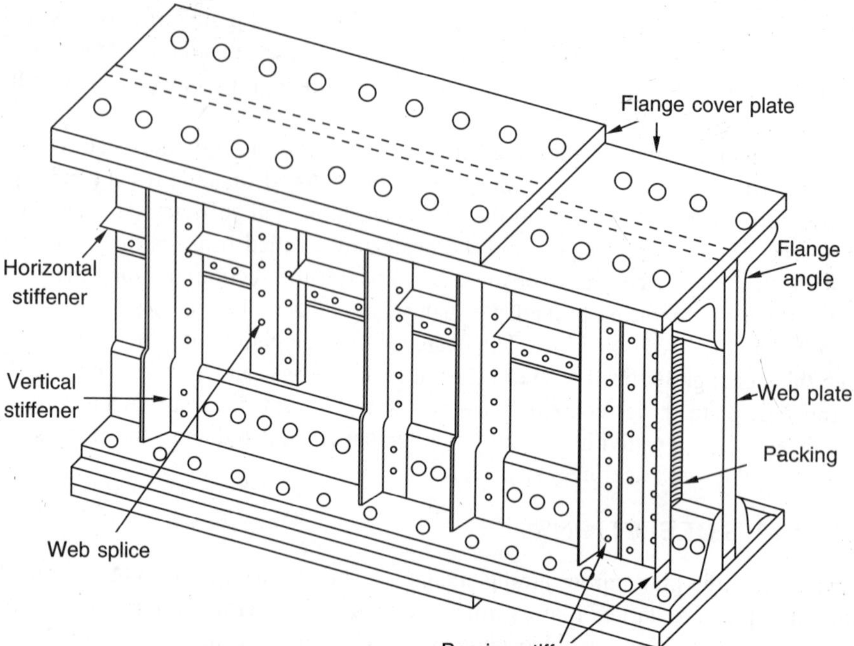

Fig. 9.2 *Elements of Riveted Plate Girder*

3. Bearing stiffeners
4. Vertical stiffeners
5. Horizontal stiffeners
6. Web splices
7. Flange splices

In the following articles the analysis and design for each element of a plate girder is dealt with separately and then an overall design procedure is developed. Special problems of diagonal building of web and local buckling of flange is also addressed.

9.4 GENERAL CONSIDERATIONS

Structural steel design is largely a matter of providing for stability, either locally or in an overall sense. Many standard hot-rolled structural shapes are proportioned so that local stability problems have been eliminated or minimized. When a plate girder is used, however, the designer must account for factors that in many cases would not be a problem with a hot-rolled shape. Most of the flexural strength in plate girders will come from flanges. This necessitates most of the steel to be concentrated in flanges and as far as possible away from the neutral axis of the girder and consequently resulting in deep, thin web. Such webs account for many of the special problems associated with plate girders, including local instability. The thin webs have a possibility of buckling before yielding. The three principal types of buckling to be considered are: shear or diagonal buckling, bending or longitudinal buckling, and bearing or vertical buckling. A thorough understanding of the basis of the IS: 800 provisions for plate girders requires a background in stability theory, particularly plate stability. Such a treatment is beyond the scope of this book, however, and the emphasis in this chapter is on the qualitative basis of the specification requirements and their application.

At a location of high shear in a girder web, usually near the support and at or near the neutral axis, the principal planes will be inclined with respect to the longitudinal axis of the member, and the principal stresses will be diagonal tension and diagonal compression. The diagonal tension poses no particular problem, but the diagonal compression causes the web to buckle. This problem can be addressed in one of three ways (1) the depth-to-thickness ratio of the web can be made small enough that the problem is eliminated, (2) web stiffeners can be used to form panels with increased shear strength, or (3) web stiffeners can be used to form panels that resist the diagonal compression through tension-field action.

Figure 9.3 illustrates the concept of tension-field action. At the point of impending buckling, the web loses its ability to support the diagonal compression, and this stress is shifted to the transverse stiffeners and the flanges. The stiffener resists the vertical component of the diagonal compression, and the flanges resist the horizontal component. The web will need to resist only the diagonal tension, hence the term tension-field action. After initial buckling a plate girder thus acts much like a pratt truss, in which the vertical web members carry compression and the diagonals carry tension, as shown in Fig. 9.3(b), and is able to resist additional shear. Sine the tension field does not actually exist until the web begins to buckle, its contribution to the web shear strength will not exist until the web buckles. The total strength will consist of the strength prior to buckling plus the post-buckling strength derived from tension-field action. The intermediate stiffeners which before initial buckling were assumed to resist no load will after buckling resist compressive load due to diagonal tension. The scope of the book limits the discussion and treatment of a post buckled web since plate girders proportioned in accordance with the provisions of IS 800 are safe against vertical buckling.

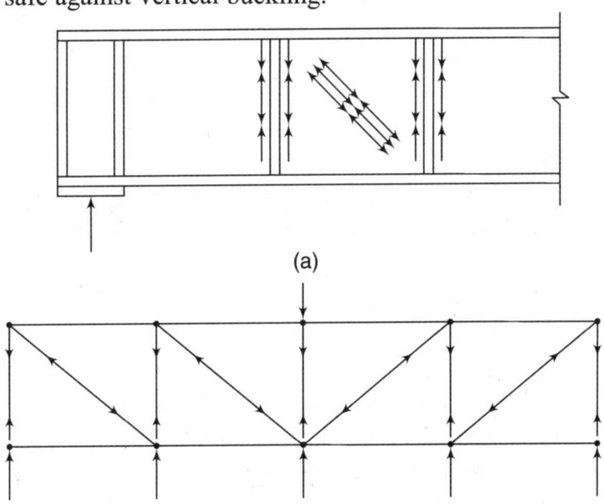

(a)

Note: The arrows indicate forces acting on members, rather than on joints

(b)

Fig. 9.3 *Tension-field Action*

If an unstiffened web is incapable of resisting the applied shear, appropriately spaced stiffeners are used. Cross-sectional requirements for these stiffeners, called intermediate stiffeners, are minimal because their primary purpose is to provide stiffness rather that resist directly applied loads.

Additional stiffeners may be required at points of concentrated loads for the purpose of protecting the web from direct compressive load. These members are called bearing stiffeners, and they must be proportioned to resist the applied loads. They can also simultaneously serve as intermediate stiffeners. In addition, the bearing stiffeners, together with a short length of web, is treated as a column with an effective length less than the web depth and is investigated for compliance with the same specification/provisions as any other compression member. The compressive strength should always be based on the radius of gyration about an axis in the plane of the web, as instability about the other principal axis is prevented by the web itself.

The rivets (bolts)/welds for connecting the components of a plate girder are designed in much the same way as for other riveted (bolted)/welded connections. The flange-to-web welds must resist the horizontal shear at the interface between the two components. This applied shear, called the shear flow, is usually expressed as a force per unit length of girder to be resisted by the weld. The shear flow, based on elastic behaviour, is given by

$$f = \frac{VA\bar{Y}}{I_x}$$

where, $A\bar{Y}$ is the moment, about the neutral axis, of the area between the horizontal shear plane and the outside face of the section.

9.5 PROPORTIONING OF WEB

It is impossible to layout rules for the depth of a plate girder as it is influenced by many factors. Primarily, it depends upon the loads but the head room clearance for high water level and transportation of the fabricated girders to the site may limit it. The depth of the plate girder usually varies between 1/8 to 1/12th of its span, the former being for short spans. The thickness of the web plate is tentatively assumed in the beginning (in no case should it be less than 6 mm if painted and 8 mm if unpainted where the steel is directly exposed to weather and for the main members not directly exposed to weather it should not be less than 6 mm) with a presumption that the stiffeners shall be placed so as to render the web safe in shear.

Depth: The depth of the plate girder for which the weight is minimum is called *economical depth.* A girder giving the least weight is not necessarily economical, because if the depth is more the transportation and erection cost will be more. In any case, it is desirable to know the optimum depth for economy so that even if it cannot be adopted, it will serve as a guide to be considered along with other requirements. Normally a depth less than the depth giving the minimum weight is provided. If the thickness of the plate girder is presumed, an expression can be developed for the economical depth of plate girder.

Let W = total weight of the girder

l = length of the girder

w = weight per unit length of the girder

W_f, W_w, W_s = weight of flanges, web and stiffeners respectively

Weight of the girder = weight of the flanges + weight of the web + weight of the stiffeners

or $$W = W_f + W_w + W_s$$

Let M = maximum design moment

d_1 = depth of the web

t = thickness of the web

A'_f = net flange area

ρ = unit weight of the material

σ_{bt} = bending tensile stress

Weight of Flanges As the cover plates are curtailed the average weight of the flanges is reduced to 80%. The net area of the flange is increased by 10% to account for the additional weight of rivets in the holes.

Hence, average flange area = $1.1 \times 0.80 \times A'_f$

$$= 0.88\ A'_f$$

$$W_f = 0.88 \times 2 \times A'_f \times l \times \rho$$

$$= 1.76\ A'_f \times l \times \rho$$

$$= 1.76 \left[\frac{M}{\sigma_{bt}\, d_1} - \frac{d_1 t}{8} \right] l\rho$$

Weight of Web

$$W_w = d_1 t\, l\rho$$

Weight of Stiffeners The weight of stiffeners is assumed to be 60% of the weight of the web.

$$W_s = 0.60\ d_1\ t\ l\rho$$

Total weight, $$W = 1.76 \left[\frac{M}{\sigma_{bt}\ d_1} - \frac{d_1 t}{8} \right] l\rho + d_1\ t\ l\rho + 0.60\ d_1\ tl\rho$$

$$W = 1.76 \left(\frac{M}{\sigma_{bt}\ d_1} - \frac{d_1 t}{8} \right) l\rho + 1.60\ d_1\ t\ l\rho$$

or $$w = 1.76 \left(\frac{M}{\sigma_{bt}\ d_1} - \frac{d_1 t}{8} \right) \rho + 1.60\ d_1\ t\rho$$

$$= 1.76\ \frac{M}{\sigma_{bt}\ d_1} \rho - \frac{1.76 d_1 t}{8} \rho\ + 1.60\ d_1\ t\rho$$

$$= 1.76\ \frac{M}{\sigma_{bt}\ d_1}\ \rho + 1.38\ d_1\ t\rho$$

For the minimum weight of the plate girder,

$$\frac{\partial w}{\partial d_1} = 0$$

or
$$\frac{\partial w}{\partial d_1} = -1.76 \frac{M}{\sigma_{bt}\, d_1^2} \rho + 1.38\, t\rho$$

or
$$1.38\, t\rho = 1.76 \frac{M}{\sigma_{bt}\, d_1^2} \rho$$

or
$$d_1^2 = \frac{1.76}{1.38} \frac{M}{\sigma_{bt}\, t}$$

or
$$d_1 = 1.1 \sqrt{M/(\sigma_{bt}\, t)} \tag{9.1}$$

The depth given by Eq. (9.1) is called the economical depth of a plate girder.

Notes

1. When the cover plates are not curtailed, the reduction factor of 0.8 is not applicable and the economical depth is given by the equation,

$$d_1 = 1.21 \sqrt{M/(\sigma_{bt}\, t)} \tag{9.2}$$

2. A reduction in the depth of web calculated by Eq. 9.1 gives an increase in weight of about 2 per cent, and the normal depths in practice are usually considerably below those given by Eq. (9.1).

Thickness: As per I.S: 800–1984, the thickness of the web plate should not be less than the following as appropriate:

(i) $\dfrac{d_1 \sqrt{\tau_{va,\,cal}}}{816}$ and $\dfrac{d_1 \sqrt{f_y}}{1344}$, but not less than $d_1/85$, for unstiffened web.

(ii) 1/180 of smallest clear panel dimension and $\dfrac{d_2 \sqrt{f_y}}{3200}$, but not less than $d_2/200$ for vertically stiffened web.

(iii) 1/180 of the smaller dimension in each panel and $\dfrac{d_2 \sqrt{f_y}}{4000}$, but not less than $d_2/250$ for webs stiffened both vertically and horizontally with a horizontal stiffener at a distance from the compression flange equal to 2/5th of the distance from the compression flange to the neutral axis.

(iv) 1/180 of smaller clear panel dimension and $\frac{d_2 \sqrt{f_y}}{6400}$, but not less than $\frac{d_2}{400}$

when there is also a horizontal stiffener at the neutral axis, where d_2 is twice the clear distance from the compression flange angles, or plate or tongue plate to the neutral axis.

Notes

1. To simplify the calculations, Table 6.7 of I.S: 800–1984 can directly be used for finding out minimum thickness of web.
2. The greater clear panel dimension should never exceed 270 *t* and the lesser clear panel dimension 180 *t*, where *t* is the thickness of the web plate.

9.6 PROPORTIONING OF FLANGES

When designing a plate girder to resist the given bending moment, it is desirable to maximize the lever arm of the internal forces, so that the material required for flanges is minimized. Flanges of a riveted plate girder consist of pairs of angles with or without cover plates. In a riveted plate girder, flange angles should form as large a part of the area of the flange as practicable and preferably not less than 1/3rd of the calculated flange area (to keep the centre of gravity of the flange within the back of angles and not in the flange cover plates otherwise the stability is affected). Unequal flange angles with long horizontal legs are provided to serve the following purposes:

1. To increase the moment of inertia of the section, and
2. A large length is available for making the connection with the flange plate.

Sometimes when the shear is heavy, a large number of rivets may be required to connect the flange angle with the web. In such a case, equal angle sections may be preferred, if unequal angles are used, the longer leg may be placed parallel to the web.

When the moment resisting capacity of the plate girder is to be increased, flange cover plates are provided over the flange angles. The moment of inertia and, consequently, the moment resisting capacity of the girder is increased considerably as the flange cover plates are at the greatest distance from the neutral axis. When flange plates are provided above flange angles, these plates should not be extended beyond the specified value of 256 $T/\sqrt{f_y}$ or 16 T, which ever is least, from the stiffening edge, (where T is the thickness of the flange, or the aggregate thickness of the flange plates). The flange plates should not extend more than 20 T beyond the outer line of connection in any case (e.g., flanges and plates in compression with stiffened edges). This is done to safeguard the plates against local buckling.

The effective cross-sectional area of compression flanges is calculated by subtracting any holes left unfilled by rivets or due to manufacturing flaws in the cross-section. The effective sectional area of a tension flange is calculated by subtracting the area of rivet holes from the gross sectional areas of the tension flange and the area of flange plate in excess of 20 T beyond the outer line of connection.

The flange plates should not be thicker than the flange angles in a riveted plate girder. Hence, more than one plate may be required. It is preferred that all the cover plates should have the same thickness but may be of different thicknesses also. In the latter case, outer plates should not be thicker than the inner plates. In such a case, the shape of flange will resemble the bending moment diagram exactly, the wastage of flange steel will be the least and the design will be economical. Further, economy can be achieved by curtailing the length or width of the subsequent flange plates, since the bending moment varies along the span of the plate girder. It is not desirable to reduce the width of the plates as rain water may stand over the projections of the lower plate, causing corrosion.

In a welded plate girder the flange generally consists of one plate only and the variation in the flange area is achieved either by varying the width of the flange plate or its thickness. If more than one flange plate is provided they are connected together by longitudinal welds.

A riveted plate girder flange without flange cover plates is economical, as making holes in plates and outstandings of angles, edge plaining, riveting, etc. are eliminated. On the other hand, an addition of a flange cover plate on the compression flange adds considerably to the flange width reducing its buckling tendency, thus allowing more permissible bending compressive stresses. This addition of the plate also increases the moment of inertia of the girder section as more material is placed away from the neutral axis. It is therefore advantageous to use lighter angles plus a plate for the compression flange, and thicker angles without cover plates for the tension flange. However, as per I.S. specifications, a symmetrical section for girder is recommended.

Plate girders designed as symmetrical sections are proportioned on the basis of the moment of inertia of the gross cross-section with the neutral axis at the centroid of the section. Since it is easy to proportion the flanges on the basis of area rather than by moment of inertia, a tentative flange area is arrived at by the flange area method and is finally checked by the moment of inertia method. Before discussing a procedure to proportion the flange area, it is desirable to understand the neutral axis of the plate girder.

The moment of inertia of a plate girder section is calculated with respect to the neutral axis determined by the gross-section of the girder. Some designers feel that this is unjustified because if the rivet holes on the tension side are given due consideration, i.e., if a fully reduced net section is considered, the neutral axis cannot remain a straight line. It must shift upward at each vertical row of rivets in stiffeners and splices. However, the net sections at the rivets form a very small portion of the total length of the girder and it seems impossible that there can be sudden breaks in the neutral axis and consequent sudden changes in fibre stresses. Since the deflection and distortion of a girder are functions of the gross area of the section, it is reasonable to determine the neutral axis and moment of inertia on the basis of the gross-section and the neutral axis may thus be assumed to be a straight line. The fibre stress calculated on the basis of the gross-section may then be corrected to determine the fibre stress on the net section.

Flange Area Method

This is an approximate method and a trial section of the plate girder can be obtained by this method. The bending stress distribution in the tension and compression flange is assumed to be uniform. The area required for the tension flange is computed and the same area is provided for the compression flange.

$$M = Td_1$$
$$= A_f' \sigma_{bt} d_1$$

or
$$A_f' = \frac{M}{\sigma_{bt} d_1}$$

where M is the bending moment, A_f' is the net flange area, σ_{bt} is the bending tensile stress, T is the force in tension flange and d_1 is the effective depth of the plate girder.

For the calculation of flange area the effective depth d_1 is assumed equal to the depth of the web. This is because the exact value of the effective depth can be found only after the flange section has been derived.

Moment of Inertia Method

This method of proportioning flanges of a plate girder usually yields a better estimate of the required flange area. Here reference may be made to Fig. 9.4.

Let D = overall depth of the girder

d = depth of girder between inner toes of flange angles

d_1 = clear distance between outstanding toes of flange angles

l = span of plate girder

σ_{bc} = allowable maximum bending compressive stress

σ_{bt} = allowable bending tensile stress

I_f = moment of inertia of flanges about their centroidal axes

I_{xx} = moment of inertia of girder about its neutral axis

A_f = gross area of one flange

A_{w1} = area of web between flange angles (as per I.S: 800–1984), i.e., equivalent web area on compression side

t = thickness of web

A_w = area of the web

A_f' = net area of tension flange

ρ = unit weight of steel

A'_{w_1} = net area of web between flange angles (as per I.S: 800–1984), i.e., equivalent web area on tension side

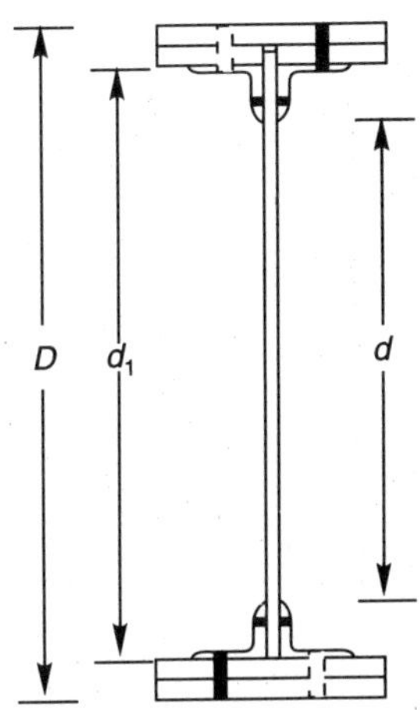

Fig. 9.4 *Cross-section of a Riveted Plate Girder*

The moment of inertia of flanges about their own axis is very small as compared to the moment of inertia of flanges about the centroidal axis. Therefore, the moment of inertia of flanges about their own axis may be neglected.

$$I_{xx} = 2A_f\left(\frac{d_1}{2}\right)^2 + \frac{td_1^3}{12}$$

$$= \frac{2A_f d_1^2}{4} + \frac{td_1^3}{12} \tag{9.4}$$

We know that

$$\frac{\sigma_{bc}}{y} = \frac{M}{I_{xx}}$$

or

$$M = \sigma_{bc}\frac{I_{xx}}{y} \tag{9.5}$$

Substituting the value of I_{xx} in Eq. (9.5),

$$M = \frac{\sigma_{bc}}{d_1/2}\left[\frac{2A_f d_1^2}{4} + \frac{td_1^3}{12}\right] = \sigma_{bc}\left[\frac{2A_f d_1^2}{4d_1/2} + \frac{td_1^3}{12d_1/2}\right]$$

$$= \sigma_{bc}\left[A_f d_1 + \frac{td_1^2}{6}\right] = \sigma_{bc}\left[A_f + \frac{td_1}{6}\right]d_1$$

$$M = \sigma_{bc}\left(A_f + \frac{A_w}{6}\right)d_1 \tag{9.6}$$

or

$$M = \sigma_{bc}\left(A_f + A_{w_1}\right)d_1 \tag{9.7}$$

From Eq. (9.6)

$$A_f + \frac{A_w}{6} = \frac{M}{\sigma_{bc}d_1}$$

or

$$A_f = \frac{M}{\sigma_{bc}d_1} - \frac{A_w}{6} \tag{9.8}$$

Plate 3

Plate 4

Plate 5

or $$A_f = \frac{M}{\sigma_{bc}\, d_1} - A_{w_1} \tag{9.9}$$

To account for rivet holes of the flange, the net area may be assumed to be 3/4th of the gross area. If the rivet size is known at this stage the exact net area may be computed and used. Hence the net area of tension flange

$$A_f' = \frac{M}{\sigma_{bt}\, d_1} - \frac{3}{4}\frac{A_w}{6}$$

or $$A_f' = \frac{M}{\sigma_{bt}\, d_1} - \frac{A_w}{8} \tag{9.10}$$

or $$A_f' = \frac{M}{\sigma_{bt}\, d_1} - A_{w_1}' \tag{9.11}$$

and $$M = \sigma_{bt}\left(A_f' + \frac{A_w'}{8}\right) d_1 \tag{9.12}$$

or $$M = \sigma_{bt}\left(A_f' + A_{w_1}'\right) d_1 \tag{9.13}$$

Note $\frac{A_w}{6}$ and $\frac{A_w}{8}$ are the equivalent web areas for compression and tension flanges respectively. I.S: 800–1984, however, specifies the actual portion of the web between the flange angles to be used as the equivalent web area. This has been represented by A_{w_1}, and A_{w_1}' respectively.

Equation (9.11) usually yields a close estimate of the required net flange area, provided a rough first approximation of the equivalent web area is made. Therefore, it is usual to proportion the flanges initially by the flange area method and later on checked by the moment of inertia method. To have symmetry, the same area as obtained by Eq. (9.11) is provided on the compression flange also. A clearance of 5 mm–10 mm is provided between the top of the web plate and bottom of the flange plate to prevent the load from being distributed directly over the web and causing its crushing.

9.7 SELF-WEIGHT OF PLATE GIRDER

The self-weight of a plate girder is assumed in the beginning (*Wl*/300, *Wl*/400 for riveted and welded plate girders respectively) and the section is proportioned. After designing the plate girder Eq. (9.15) as derived below is used to check the self-weight. A general expression for the self-weight (Section 9.5) of a plate girder is given by

$$w = 1.76\frac{M}{\sigma_{bt}\, d_1}\rho + 1.38 d_1 t \rho$$

For the least weight of plate girder $\frac{\partial w}{\partial d_1} = 0$, and thus

$$1.38\, td_1 = \frac{1.76M}{\sigma_{bt}\, d_1}$$

or
$$t = \frac{1.76}{1.38} \times \frac{M}{\sigma_{bt}\, d_1^2}$$

or
$$t = 1.27 \frac{M}{\sigma_{bt}\, d_1^2} \tag{9.14}$$

Substituting the values of t

$$w = \frac{1.76M}{\sigma_{bt}\, d_1}\rho + 1.38 d_1 \rho \frac{1.27M}{\sigma_{bt}\, d_1^2}$$

$$= \frac{1.76M}{\sigma_{bt}\, d_1}\rho + \frac{1.75M}{\sigma_{bt}\, d_1}\rho = 3.51 \frac{M}{\sigma_{bt}\, d_1}\rho$$

Let $\rho = 7.85 \times 10^{-5}$ N/mm^3

Then $w = 3.51 \times 7.85 \times 10^{-5} \frac{M}{\sigma_{bt}\, d_1}$

or
$$w = 275.535 \times 10^{-6} \frac{M}{\sigma_{bt}\, d_1} \tag{9.15}$$

where w = weight of the plate girder in N/mm
M = bending moment in N mm
σ_{bt} = bending stress in N/mm^2.

9.8 CURTAILMENT OF FLANGE PLATES

The girder section proportion as discussed in the previous sections is required only for a short distance near the point of maximum bending moment. This moment for example in a simple plate girder under uniformly distributed loads occurs near the centre of the span and reduces towards the supports. The bending stresses thus generated are resisted by the flange area. It is therefore possible to use a smaller flange area towards the supports. This reduction in the section may be achieved by cutting of the cover plates one by one until only the web plate and flange angles remain near the supports. But some specifications allow the first flange cover plate to be carried through the length of the girder to serve the following purposes:

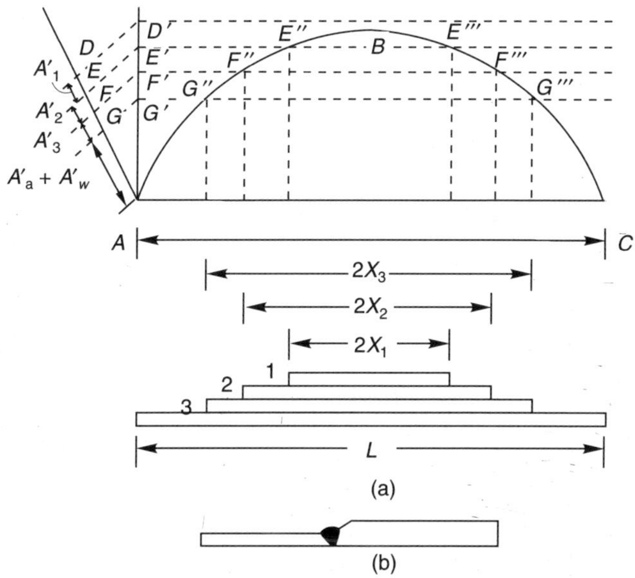

Fig. 9.5 *Curtailment of Flange Plates*

1. It ties together the outstanding legs of the flange angles, and
2. A top flange with the first cover plate curtailed may allow rain water to collect between the flange angles and the web, resulting in corrosion.

The length of curtailment of plates can be found conveniently by the algebraic method or graphical method. The algebraic method of curtailment of flange plates may be used if the plate girder supports only uniformly distributed loads, whereas the graphical method of curtailment of flange plates is applicable to any system of applied loads.

Algebraic Method

Let A'_f = net area of flange (flange angle + flange plates)

A'_1, A'_2, A'_n = net areas of 1st, 2nd, nth cover plates respectively, counted from outside

As the effective depth of the plate girder is approximately same even after curtailment of flange plates, it is therefore assumed to be same throughout the span. If the girder is subjected to a uniform load over the entire span the bending moment diagram is parabolic

or
$$\frac{wl^2}{8} = \sigma_{bt}\,(A'_f + A'_{w_1})\,d_1 \tag{9.16}$$

and
$$\frac{wl^2}{8} - \frac{w(2X_n)^2}{8} = \sigma_{bt}\left[\left(A'_f + A'_{w_1}\right) - \left(A'_1 + A'_2 + \ldots + A'_n\right)\right] d_1 \tag{9.17}$$

Dividing Eq. (9.17) by Eq. (9.16)

$$1 - \frac{(2X_n)^2}{l^2} = 1 - \frac{A'_1 + A'_2 + \ldots + A'_n}{A'_f + A'_{w_1}}$$

or
$$X_n = \frac{l}{2}\frac{\sqrt{A_1' + A_2' + \ldots + A_n'}}{A_f' + A_{w_1}'} \tag{9.18}$$

$\therefore$
$$X_1 = \frac{l}{2}\left(\frac{A_1'}{A_f' + A_{w_1}'}\right)^{1/2}$$

$$X_2 = \frac{l}{2}\left(\frac{A_1' + A_2'}{A_f' + A_{w_1}'}\right)^{1/2}$$

and so on, where $X_1, X_2 \ldots, X_n$ are the distances of the cut-off points from the centre of the plate girder as shown in Fig. 9.5(a).

Graphical Method

If the load is not uniform then the graphical method of curtailment is most suited. First, the bending moment diagram is plotted. The moment of resistance diagram with all cover plates is then plotted (one by one) on the bending moment diagram. The points of intersection of the bending moment diagram with moment of resistance diagram give the cut-off points of the cover plates.

Procedure

1. The maximum bending moment diagram ABC is drawn as shown in Fig. 9.5.
2. The moment of resistance due to various flange elements is calculated.

   Moment of resistance due to the first plate $= A'_1\, \sigma_{bt}\, d_1$

   Moment of resistance due to the nth plate $= A'_n\, \sigma_{bt}\, d_1$

   Moment of resistance due to flange angles and web $= (A'_a + A'_{w_1})\sigma_{bt}\, d_1$

   where A'_a = net area of flange angles
3. From A, a line AD is drawn at a convenient angle and on this the values of areas of different flange elements such as $A'_a + A'_{w_1}$, A'_n, A'_{n-1}, ..., A'_1 are plotted to a suitable scale.
4. From A, a line AD' (equal to the net moment of resistance of the full section) is plotted perpendicular to the line AC.
5. D and D' are joined and from the various plotted points E, F, G, etc., lines EE', FF', GG', etc., are plotted parallel to DD'.
6. Now from E', F', G', etc., lines parallel to AC are drawn. These intercept the maximum bending moment diagrams at E'', E''', F'', F''', G'', G''', etc. These intercepts $E''E'''$, $F''F'''$, $G''G'''$, etc., represent the required theoretical length of the flange elements.

Extension of Curtailed Cover Plates

The curtailed plates are extended for some length beyond the theoretical point of cut-off. Though it is supposed that the plates carry the stresses right up to the theoretical point of cut-off, the stress at the curtailed section has to be transferred.

This is achieved by means of connectors which provide a shear strength equal to the allowable tension or compression in the plate. To accommodate connectors the curtailed plate is extended up to the required length. Another reason for extending the cover plates is to overcome the stress concentration occuring in the vicinity of the curtailed section; the girder section changes abruptly and gives rise to fatigue. By extending the cover plates for a small length in the direction of decreasing stress, the stress concentrations are readily absorbed. The curtailment of flange plates on the compression flange are done at same point as that on the tension flange.

9.9 CONNECTIONS

For Riveted Plate Girder

Flange Angles to Web The rivets connecting flange angle legs to web pass through the angles and web of the plate girder. The rivets are in double shear and in bearing on the web thickness. These rivets are designed to transmit the maximum horizontal shear force resulting from the bending moments and vertical shear associated due to any live load. Reference may be made to Fig. 9.6(a). It is assumed that the maximum web shear is equal to the average shear on the web. Therefore, the horizontal shear per unit length between the web and flange will be same as the horizontal shear per unit length in the web itself near the connection line.

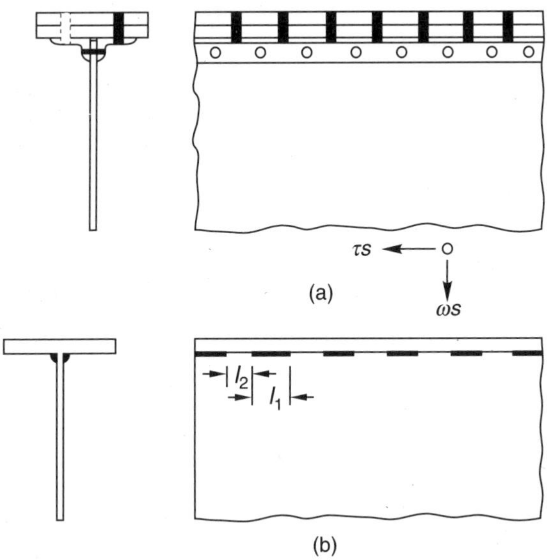

Fig. 9.6 *Flange Angle to Web Connection*

Let R_v = rivet value

τ_1 = shear stress

τ = horizontal shear force/unit length

w = uniformly distributed load over the girder

s = pitch of the rivets

$$\text{Horizontal shear per pitch length} = \tau s$$
$$\text{Vertical shear per pitch length} = ws$$

The resultant force = $\sqrt{(\tau s)^2 + (ws)^2}$

(as the two forces are at a right angle).

The resultant force on a rivet should not exceed the rivet value. Hence,

$$R_v = \sqrt{(\tau s)^2 + (ws)^2}$$

$$R_v = s\sqrt{(\tau^2 + w^2)}$$

or
$$s = \frac{R_v}{\sqrt{(\tau^2 + w^2)}} \tag{9.19}$$

We know that shear stress is given by the equation,

$$\tau_1 = \frac{VA\bar{Y}}{IB}$$

∴ Horizontal shear/unit length, $\tau = \dfrac{VA\bar{Y}}{I_{xx}}$

where V = vertical shear force at the section

$A\bar{Y}$ = statical moment of the flange about N.A. = $A_f \dfrac{d_1}{2}$

$$I_{xx} = 2A_f \frac{d_1^2}{4} + \frac{td_1^3}{12} = \left(A_f + \frac{A_w}{6}\right)\frac{d_1^2}{2}$$

∴
$$\tau = \frac{VA_f \dfrac{d_1}{2}}{\left(A_f + \dfrac{A_w}{6}\right)\dfrac{d_1^2}{2}} = \frac{VA_f}{\left(A_f + \dfrac{A_w}{6}\right)d_1}$$

Substituting the value of τ in Eq. (9.19),

$$s = \frac{R_v}{\left[\left(\dfrac{VA_f}{d_1\left(A_f + \dfrac{A_w}{6}\right)}\right)^2 + w^2\right]^{1/2}} \tag{9.20}$$

or

$$s = \frac{R_v}{\dfrac{VA_f}{\left(A_f + \dfrac{A_w}{6}\right) d_1}} \qquad \text{(if } w = 0\text{)}$$

or

$$s = \frac{R_v \left(A_f + \dfrac{A_w}{6}\right)}{VA_f} d_1 \tag{9.21}$$

Equation (9.20) or (9.21) should be used for computing the pitch of the rivets on the compression flange.

The pitch of the rivets on the tension flange can be computed by replacing the gross flange area $\left(A_f + \dfrac{A_w}{6}\right)$, by $\left(A_f' + \dfrac{A_w}{8}\right)$ and A_f by A_f' in Eq. (9.20) or (9.21). Also, as per I.S. specification $\dfrac{A_w}{6}, \dfrac{A_w}{8}$ (the equivalent web areas) should be replaced by A_{w_1}, A_{w_1}' (the actual web areas between flange angles) respectively. Equations (9.20) and (9.21) can be rewritten as

$$s = \frac{R_v}{\left[\left(\dfrac{VA_f}{d_1 (A_f + A_{w1})}\right)^2 + w^2\right]^{1/2}} \tag{9.22}$$

and

$$s = \frac{R_v (A_f + A_{w_1})}{VA_f} d_1 \tag{9.23}$$

Connection of Flange Angles to Cover Plates The rivets conecting flange angle legs to cover plates, pass through cover plates and flange angles, and are subjected to longitudinal shear only. The rivets are in single shear and bearing. We know that

$$\tau = \frac{VA\bar{Y}}{IB}$$

where τ = shear stress

Let s = the pitch of rivets

$$\text{Horizontal shear per unit length} = \frac{VA\bar{Y}}{I}$$

The horizontal shear in pitch s will be τs. Hence

$$\tau s = \frac{V\overline{AY}}{I_{xx}} s$$

or
$$\tau s = \frac{VA_f \dfrac{d_1}{2}}{\left(A_f + \dfrac{A_w}{6}\right)\dfrac{d_1^2}{2}} s$$

The value of τs should be equal to the rivet value,

or
$$R_v = \frac{VA_f \dfrac{d_1}{2}}{\left(A_f + \dfrac{A_w}{6}\right)\dfrac{d_1^2}{2}} s$$

or
$$s = \frac{R_v\left(A_f + \dfrac{A_w}{6}\right)\dfrac{d_1^2}{2}}{VA_f \dfrac{d_1}{2}}$$

or
$$s = R_v \frac{\left(A_f + \dfrac{A_w}{6}\right) d_1}{VA_f} \tag{9.24}$$

The pitch of the rivets can be computed by Eq. (9.24) for the compression flange. On the tension flange the pitch of the rivets can be computed by replacing the gross area $\left(A_f + \dfrac{A_w}{6}\right)$ by the net area $\left(A_f' + \dfrac{A_w}{8}\right)$ and A_f by A_f' in Eq. (9.24).

Also $\dfrac{A_w}{6}$ and $\dfrac{A_w}{8}$ should be replaced by A_{w1} and A_{w_1}' as per I.S. specification. Equation (9.24) can be rewritten as

$$s = R_v \frac{(A_f + A_{w_1})\, d_1}{VA_f} \tag{9.25}$$

For Welded Plate Girder

Intermittent or continuous fillet welds are provided to connect the flange plates with the web plate. If hand welding is employed, intermittent fillet welding is done and if automatic welding is employed continuous fillet weld may be the choice. The welds are designed for horizontal shear.

$$\text{The horizontal shear per unit length} = \frac{V\,\overline{AY}}{I_{xx}}$$

where V, $A\overline{Y}$ and I_{xx} have already been defined.

In case intermittent fillet welding is done, the strength of the weld when divided by the horizontal shear per unit length gives the pitch of the weld.

The effective length of intermittent welds can be assumed and should comply with the following I.S. specifications:

1. One half of the total intermittent fillet weld is provided on one side of the web and the remaining half on the other side of the web.
2. The actual length of the intermittent weld should be the effective length of each intermittent weld + 2 times the size of the weld (one size of the weld is added on both free ends).
3. Intermittent welds should not be less than 4 times the weld size or 40 mm whichever is more.
4. The clear distance between the intermittent fillet welds should not exceed 12 t or 200 mm, whichever is least for a compression flange and 16 t or 200 mm, whichever is least for a tension flange, where t is the thickness of thinner plate (web or flange).

The actual length l_1, and the spacing l_2, of the intermittent weld can be computed assuming the length according to specifications and then correct dimensions can be reached by trial and error. Reference may be made to Fig. 9.6(b).

9.10 STIFFENERS

The web of the plate girder being relatively tall and thin is a poor compression member and hence, the possibility of vertical and diagonal buckling is always there. Either the web is stiffened vertically as well as horizontally or the compressive stress in the web is kept low enough to prevent buckling.

The elements provided to stiffen the web against buckling are called *intermediate stiffeners*; vertical stiffener and horizontal stiffener. Vertical stiffeners alone or along with horizontal stiffeners divide the web into panels. A single horizontal stiffener along the length of the girder placed in the compression zone may be adequate as a replacement for closely spaced vertical stiffeners. But its effective length and thus buckling is reduced when it is provided along with vertical stiffeners placed on the opposite sides of the web. The vertical stiffeners effectively safeguard the web against shearing buckling stress but have little effect on bending buckling stress, whereas horizontal stiffeners safeguard the web against bending buckling stress. Therefore, when shear stresses predominate, vertical stiffeners result in economic design and when bending stresses predominate, horizontal stiffeners result in economic design. However, when both shear and bending stresses are of equal importance, a combination of vertical and horizontal stiffeners becomes the choice.

The web is assumed to be supported over the edges of the stiffeners. Angle sections are used if the connections are riveted, Fig. 9.7(a), and flat sections are used if the connections are welded, Fig. 9.7(b). Another type of stiffener used to prevent the buckling under the concentrated load or end reaction is called bearing stiffener. These are described in the sections to follow.

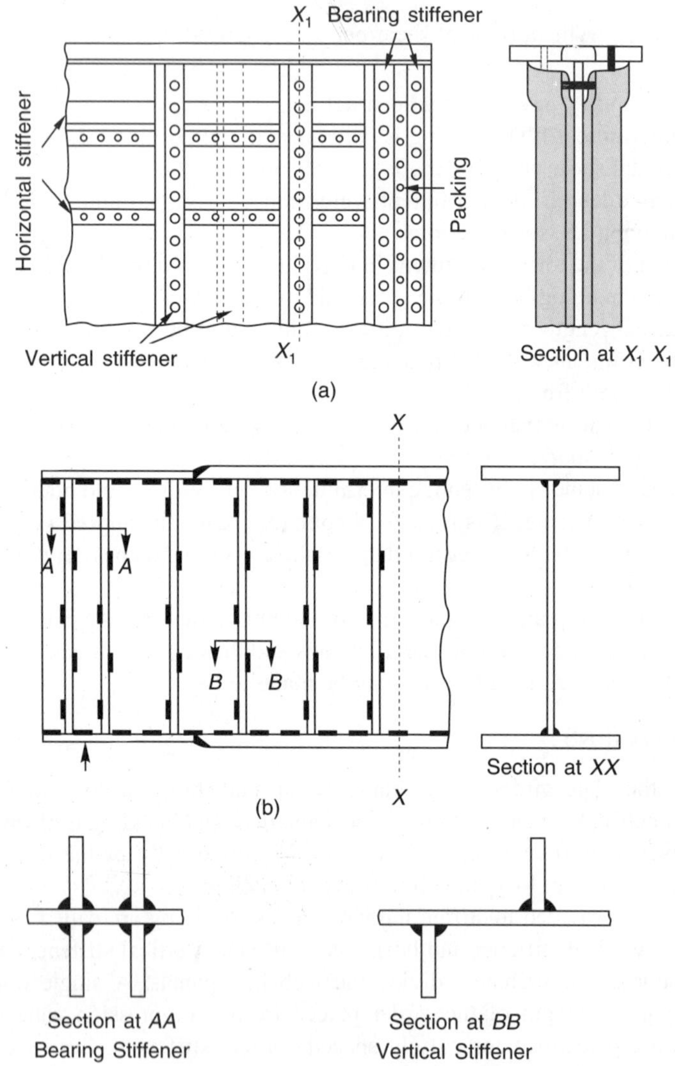

Fig. 9.7 *Arrangement of Stiffeners*

9.11 BEARING STIFFENER

Bearing stiffeners are used to transfer concentrated loads on the girder and heavy reactions at supports to the full depth of the web. A bearing stiffener at the support is called an end bearing stiffener. Bearing stiffeners should be placed:

(i) tight with the web,

(ii) straight (it is not crimped or joggled), Fig. 9.8(a), and

(iii) in pairs of two or four angles, symmetrically placed on both sides of the web (Figs 9.8(b), (c) and (d)) for a riveted plate girder and two or four plates (Fig. 9.8(e)) for a welded plate girder.

As bearing stiffeners are provided straight, the back of the connected leg of the stiffener angle will be at some distance (equal to the thickness of flange angle) from the face of web. To achieve tightness suitable packing plates (equal to the thickness of flange angle) are inserted between the web and bearing stiffener as shown in Fig. 9.8. Additional rivets required because of packing plates are calculated as per I.S. specifications and placed (Fig. 9.8(c)). The bearing stiffener angles extend between the top and bottom flange angles. The ends of the outstanding legs of the stiffener angles are machined to have a close fit and to bear against the legs of the loaded flange angles to relieve the flange connectors from being overstressed. Bearing stiffeners should extend out towards the edges of the flange plates and flange angles as far as possible.

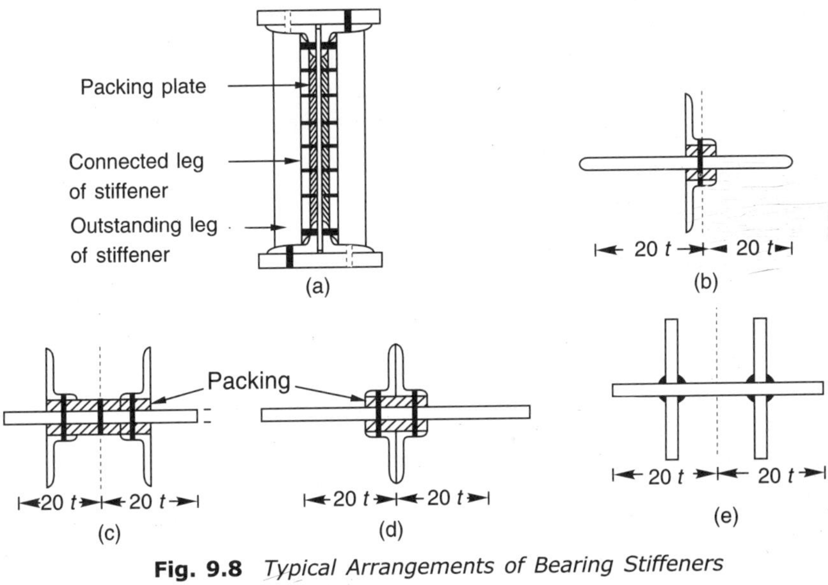

Fig. 9.8 *Typical Arrangements of Bearing Stiffeners*

Design steps

A bearing stiffener must have sufficient contact area between the flange and the stiffener to transfer load in bearing. They are somewhat similar to columns. But because the bearing stiffeners support loads in conjunction with the web which causes variation of load, these are difficult to analyze exactly.

A bearing stiffener is designed as follows:

1. The bearing area required, A = reaction/σ_p

   where, σ_p = permissible bearing stress ($0.75f_y$).

2. In the case of riveted connections angle sections are used for bearing stiffener. Only the length of outstanding leg is considered to provide bearing area and is counted outside the fillets of the flange angle. Therefore, the bearing length is calculated by subracting the length of the root of fillet of flange angle from the length of the outstanding leg of stiffener angle. The length of the outstanding leg for this reason should be more as compared to the connected leg and therefore unequal angle sections are preferred. In the case of welded connections plate or flat sections are used, as shown in Fig. 9.8(e).
3. The bearing stiffeners are designed as columns with the length of the web 20 times the thickness of the web on both sides as shown in Figs 9.8(b), (c), (d) and (e).
4. Since bearing stiffeners act as compression members they are provided straight. For this reason filler plates are provided between the stiffener angles and the web plate. The thickness of filter plate equals the thickness of flange angle.
5. The slenderness ratio of the bearing stiffener is calculated by

$$\lambda = \frac{d_{1\,\text{eff}}}{r}$$

where d_1 = the actual length of stiffener provided (depth of the web)

$d_{1\,\text{eff}} = 0.7\, d_1$ (the effective depth is taken as 0.7 times the depth of web to allow for the fact that the compressive stress is entirely transferred out of the stiffener within the depth of the web)

r = radius of gyration perpendicular to the plane of web of the grider,

$= \sqrt{I/A}$, where I is the moment of inertia and A is the cross-sectional area of the bearing stiffener

6. Allowable compressive stress σ_{ac} corresponding to the calculated value of λ is found from I.S: 800–1984.
7. The load carrying capacity of the stiffener is the product of the allowable compressive stress and bearing area provided and should be more than the concentrated load (or the reaction) to be resisted.
8. Plate girders in which lateral restraint against torsion is provided by the bearing stiffeners at supports, and the compression flange is not supported laterally, the moment of inertia of such stiffeners about centre line of the web plate should not be less than

$$\frac{D^3 T}{250} \times \frac{R}{W}$$

where D = overall depth of the girder

T = maximum thickness of compression flange

R = reaction at the support

W = total load on girder between supports.

9. In case of welded plate girders, the thickness of the plate provided as a bearing stiffener should not be less than 1/12 of the outstand. The outstand of the bearing stiffener is assumed and the thickness of the outstand (t_1) of the stiffener plate will be

$$t_1 = \frac{\text{bearing area required}}{\text{outstand of the stiffener plate}}$$

10. (i) If a riveted joint is provided, the rivet value in double shear and bearing is calculated. The number of rivets required to resist the load are determined.

(ii) If welded connections are used, intermittent fillet welds are designed to transmit the loads. The following specifications should be strictly observed for welding stiffeners by intermittent fillet welds.

(a) The distance between the effective lengths of any two welds, even if staggered on opposite sides of the stiffeners, should not exceed 16 t_1 or 300 mm, whichever is less.

(b) Where intermittent welds are placed on one side of the stiffener or on both sides but staggered the weld length should not be less than 10 t_1.

(c) Where intermittent welds are placed in pairs, one weld on each side of the stiffeners, the effective length of each weld should not be less than 4 t_1.

where t_1 = thickness of the stiffener.

9.12 INTERMEDIATE STIFFENERS

These are also called *stability* stiffeners and are provided to check the diagonal buckling of the web. Depending upon the requirements as discussed in Section 9.5 of the text, vertical (transverse) and horizontal (longitudinal) stiffeners may be provided.

Vertical Stiffeners

These are also called *transverse* stiffeners. Angle sections are provided for riveted plate girders and plate sections for welded plate girders. Angle sections can be in pairs or may be one in number. When a single angle section is used as a vertical stiffener, the angles are placed alternately on the opposite sides of the web plate (Fig. 9.9(a)). It is assumed that the vertical stiffener is not subjected to any load and is selected to provide necessary lateral stiffness only and can therefore, be crimped or joggled for tight fittings as shown in Figs 9.9(b) and (c). Such stiffeners increase the buckling resistance of the web caused by shear.

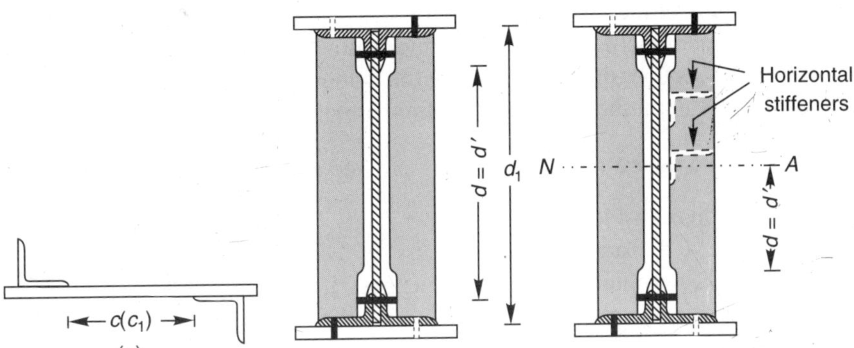

(a)

(b) Vertical Stiffener without Horizontal Stiffeners

(c) Vertical Stiffener with Horizontal Stiffeners

Fig. 9.9 *Intermediate Stiffeners*

Design steps

1. The average shear stress τ_{va} in the web plate is given by

$$\tau_{va} = \frac{V}{d_1 t}$$

where V = maximum shear force at the section
d_1 = effective depth of the web in mm
t = thickness of the web in mm.

2. Vertical stiffeners are provided at a spacing of 0.33 d to 1.5 d, where d is the distance between the flanges ignoring fillets and if horizontal stiffeners are also provided d is the maximum clear depth of the web. Spacing can be reduced near the supports where the shear force is large compared to the middle of the girder. The maximum clear dimension of the panels formed by intermediate stiffeners should not be greater than 270 t and the lesser dimension of the panel should not be greater than 180 t, where t is the thickness of the web.

3. The ratio d/t is determined. The spacing c is calculated from the maximum limit of 180 t.
where c = clear spacing of vertical stiffeners
d = for vertically stiffened webs without horizontal stiffeners is the clear distance between flange angles or where there are no flange angles, the clear distance between the flanges ignoring fillets.
= for vertically stiffened webs with horizontal stiffeners is the clear distance between the tension flange and the horizontal stiffener.

4. The value of the allowable shear stress in the web is found from Table 6.6 of I.S. Code, corresponding to the d/t ratio and maximum spacing c. This value should be greater than $\tau_{va,\,cal}$ in step (1).

5. The length of the outstanding leg of the stiffener should not be greater than $256\ t/\sqrt{f_y}$ for angle sections and 12 t for flats, where t is the thickness of section or flat. On this basis a suitable unequal angle section (riveted connection), or a plate section (welded connection) is selected.

6. The moment of inertia of the stiffener is calculated. If the stiffener is in a pair, the moment of inertia is found about the centre line of the web. If the stiffener is a single section, the moment of inertia is found about the face of the web.

7. The moment of inertia of the section provided should not be less than $1.5\frac{d^3 t^3}{c^2}$ to assure that the stiffeners provide adequate rigidity so that each panel may be considered acting independently from another.
where c = spacing of the stiffeners
t = minimum required thickness of the web.

$$= \frac{V}{\tau_{va}\, d_1}$$

τ_{va} = maximum permissible shear stress
d_1 = distance between the inner toes of the flange angles.

8. (i) For a riveted plate girder the diameter of the rivet is assumed and the rivet value in single shear (single angle stiffener) or in double shear (pair of angles) and in bearing is calculated. The connections are designed to resist a shear of 125 t^2/h kN/m where h is the length of outstand of the stiffener. The pitch of the rivets can be calculated by dividing the rivet value by this value of shear force/m run. It should not exceed the specified values of pitch of rivets as per I.S. specifications.

   (ii) If welded connections are designed staggered intermittent fillet weld is the choice except for bridges where a continuous fillet weld is preferred. The required total length of the weld is calculated by dividing the strength of the weld by the calculated value of shear/m run. The length and spacing of the intermittent fillet weld is then designed for the total required length of the weld.

Notes

1. If the vertical intermediate stiffeners are subjected to bending moments due to eccentricity of the applied loads, the moment of inertia of vertical stiffeners should be increased by $\dfrac{150\,MD^2}{Et} \times 10^4$ mm^4.
2. If the vertical intermediate stiffeners are subjected to lateral loads then the moment of inertia of the stiffener should be increased by

$$\frac{0.3\,VD^3}{Et} \times 10^4 \text{ mm}^4.$$

where M = bending moment, due to eccentricity, in kNm
D = overall depth of girder, in mm
E = Young's modulus of elasticity, 2×10^5 MPa
t = thickness of web in mm
V = transverse force applied at compression flange to be resisted by stiffener, in kN.

Horizontal Stiffeners

Horizontal stiffeners are also called longitudinal stiffeners. These increase the buckling resistance considerably as compared to transverse stiffeners when the web is subjected to bending. The horizontal stiffener consists of angle section for a riveted plate girder and plate section for a welded plate girder and are provided in the compression zone of the web. The first horizontal stiffener is provided at one fifth of the distance from the compression flange to the tension flange. If required another stiffener is provided at the neutral axis. Horizontal stiffeners are not continuous and are provided between vertical stiffeners.

Design steps

1. The moment of inertia of the first horizontal stiffener is calculated by

$$I = 4c_1 t^3$$

where c_1 = actual distance between vertical stiffeners, Fig. 9.9(a), and
t = minimum required thickness of the web.

The moment of inertia of another stiffener, if provided, is calculated by

$$I = d_2\, t^3$$

where d_2 = twice the clear distance from the compression flange angle (or plate in case of welded plate girder) to the neutral axis.

2. Horizontal stiffeners may be provided either as single angle sections or in pairs for a riveted plate girder and plate sections for a welded plate girder. The outstand of the stiffener should not be greater than $256\ t/\sqrt{f_y}$ for angle sections and 12 t for flats.
3. Connections of horizontal stiffeners to web are designed similar to that as for vertical stiffeners.

9.13 WEB SPLICES

The depth of the web plate of a plate girder is generally about four times the width of the flange plates, and, therefore, web plates cannot be obtained in such large widths and long lengths from the rolling mills. Even if they could be obtained they would prove to be too heavy and awkward to handle in the shops. The web plate, if over 10 m or thereby in length, is usually built up of two or more lengths butted together and the joint covered with a splice plate on each face of the web. Further, the presence of a splice permits camber to be given to the girder which would otherwise be difficult to obtain if the plate were in one piece.

The plate girder web may be spliced for one or more of the following reasons:

1. The length of the plate girder may be large whereas the plate are available in limited lengths.
2. Large sized plates are difficult to handle, erect and place as they may get twisted. It is also difficult to transport such plates.

Web splice plates are required only when plate girders are riveted. In the case of welded plate girders the web plates may be butt-welded. Splice plates can be placed anywhere on the web and there is practically no restrictions upon the position of the web splices for girders in multistorey buildings and industrial buildings and the same may be said for a deck span bridge. In a through span bridge with cross girders, the splice must not occur at the junction of the cross to the main girder. Its position is therefore fixed by symmetry, midway between two cross girders. Further, splice should always be placed so that a symmetrical appearance is obtained. The reason is partly aesthetic and partly economic. The latter reason will be apparent as templates, etc., are reduced in number and, therefore, also the cost of fabrication. Primarily the girder web is subjected to shear stresses and hence the splice should be placed at sections where the shear is small.

Some specifications limit the use of splices at the mid span, probably because of large centre line moments, and recommend that the splices should be placed at 1/3rd span, if required. But it is very clear that the moment at 1/3rd span is also appreciable and is not substantially reduced. Some of the specifications recommend splice plates on both sides of the web, below vertical stiffeners near 1/3rd span, with the view that the stiffeners will provide additional strength to the splice. Usually a pair of splices is provided but even a single splice plate if provided, will be satisfactory from the strength point of view. As a precautionary measure an angle is placed over the web splice plate in order to prevent damage to the web while driving the splice rivets. Designers when given a uniform load on a girder often conclude that the maximum shear at a splice is that obtained from the usual shear diagram. This is true if the load is constant and if when first applied to the girder it is distributed evenly throughout. If one end is loaded before the other, this shear curve no longer holds. In this case it is equivalent to a uniform live load advancing on to the span from either end. There must therefore be a shear, some time or other, at the girder centre line. To allow for all contingencies which may arise, it is better—whether the load is dead or live—to find the maximum possible shear at any splice by loading the girder between the splice and the more distant support, and to design the web splices for this shearing force.

A riveted web splice is complicated as the entire depth of the web is not exposed. Several possible splices are shown in Fig. 9.10. The simplest type of a splice consists of a pair of splice plates, one on each side of the web and is called *shear splice* (Fig. 9.10(a)). Its performance is satisfactory but it is not as good as other types of splices shown in Fig. 9.10(b, c), because the rivets in the shear splice away from the neutral axis are subjected to higher loads than the rivets near the neutral axis. Also, the depth of the splice plate is substantially less than the depth of the web plate; which necessitates, more rivets to be placed near the top and bottom edges of the splice plate. This shortcoming can be overcome by providing a moment splice. Reference may be made to Fig. 9.10(b). Moment splice-type I is much stronger than a shear splice.

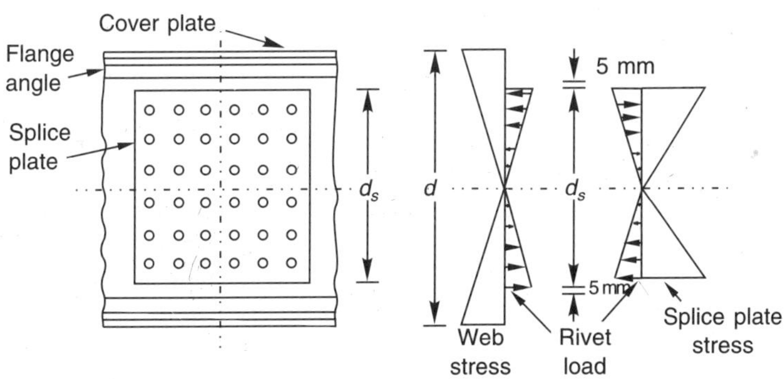

Fig. 9.10(a) *Shear Splice*

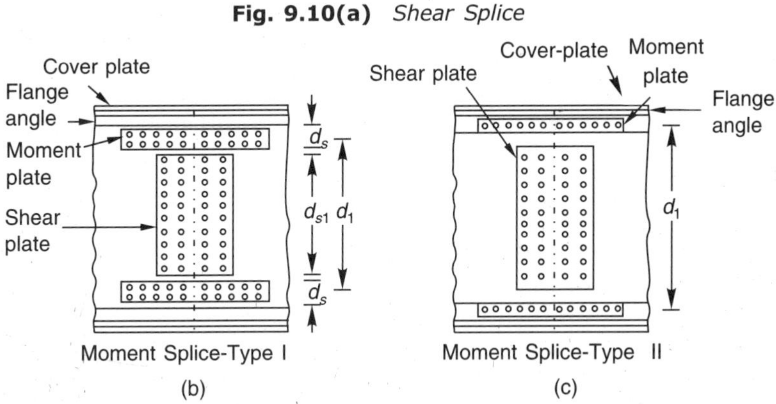

Fig. 9.10(b) and (c) *Moment Splices*

Both the splices discussed above do not run for the full depth of the web plate because the flange angle covers some part of the web. The stress in the web for the lengths covered by the flange angles is probably transferred into the flange angles rather than going to the splice. To overcome the above limitation, the moment splice-type II shown in Fig. 9.10(c) is provided. It consists of a pair of splice plates, one on each side of the web and other pairs of splice plates over flange angles. This is probably the best arrangement and is preferred. The splice plate should not project beyond the flange angle leg because such an arrangement requires filler plates and thus extra connections. This makes the splice arrangement uneconomical.

In all the splice arrangements discussed above, at least two rows of rivets on each side of the web joint should be provided although the calculations may only ask for one. The vertical rivet pitch can be widened out, but must not exceed the maximum permissible pitch of twelve (or sixteen) times the thinnest plate. Two rows of rivets give a fixity or rigidity to the splice plates, and therefore to the whole joint, in a transverse direction, that is lacking where there is only one row of rivets.

In agreement with the assumption that the web takes shear only, the minimum pitch is limited to 2.5 ϕ, and not to any consideration of lowering the web plate's section modulus, as happens in the case of the web which takes bending moment in addition to shear.

Shear Splice

The section of shear splice is so chosen that the moment of resistance of the shear splice is equal to the sum of the moment of resistance of the web plate and the moment due to the maximum actual external shear at the section. For the following derivation reference may be made to Fig. 9.10(a).

Let σ_b = bending stress at the extreme fibre of the web plate

σ_{b1} = bending stress at the extreme fibre of the splice plate

A_w = area of the web plate

A_s = area of the splice plate

d = depth of the web plate to be spliced

d_s = depth of the splice plate.

To simplify the computation of the splice plate area, the moment due to shear is neglected and after the design the splice plate is checked for the total moment on the splice. Hence, moment resisted by the splice plate equals moment resisted by the web plate.

$$M_s = M$$

$$\sigma_{b_1} \times \frac{1}{6} t_s (d_s)^2 = \sigma_b \frac{1}{6} td^2$$

or

$$\sigma_{b_1} = \sigma_b \frac{td}{t_s d_s} \times \frac{d}{d_s}$$

$$\sigma_{b_1} = \sigma_b \frac{A_w}{A_s} \times \frac{d}{d_s} \tag{9.26}$$

We know that $$\frac{\sigma_{b_1}}{\sigma_b} = \frac{d_s}{d}$$

or $$\sigma_{b_1} = \sigma_b \frac{d_s}{d} \tag{9.27}$$

Substituting the value of σ_{b_1} in Eq. (9.26),

$$\sigma_b \frac{d_s}{d} = \sigma_b \frac{A_w}{A_s} \times \frac{d}{d_s}$$

or $$A_s = A_w \left(\frac{d}{d_s}\right)^2 \tag{9.28}$$

The bending moment resisted by the web plate will be

$$M_w = M \frac{I_w}{I} \tag{9.29}$$

where M_w, I_w = moment of resistance and gross moment of inertia of the girder at the web respectively

M, I = bending moment and gross moment of inertia of the girder at the web splice section

Design steps

1. The depth of a splice plate is found by subtracting the leg lengths of the flange angles and a clearance (5 mm) is provided between the splice and flange angles.
2. The area of the splice plate is calculated by using Eq. (9.28). Two splice plates (one on each side of the web) are provided.
3. The thickness of the splice plate is found by dividing the splice plate area of one plate with the depth of the splice plate.
4. The bending moment resisted by the splice plate is computed from,

   $$M_s = M_w + M_v$$

   where M_s = bending moment resisted by the splice plate

   M_w = bending moment resisted by the web plate Eq. (9.29)

   M_v = bending moment due to shear force (= Ve)

   V = shear force at the splice section

   e = eccentricity of vertical shear force from C.G. of rivet group of the splice.
5. The diameter and pitch of rivets are assumed and the rivet value is calculated. The number of rivets in each row is determined by the equation,

   $$n = \sqrt{6M_s / n' s R_v}$$

   where n = number of rivets in one row

   n' = number of rows on one side of the splice

s = pitch of the rivets

R_v = rivet value (rivets will be in double shear).

These rivets will be subjected to direct shear and moment in the plane of the splice plate. The resultant of the two forces on the rivet is calculated and should be less than the rivet value.

6. The minimum width of the plate to accommodate these rivets is computed.

Moment Splice Type I

The number of splice plates required will be six (Fig. 9.10(b)). Four plates (two on each side of web) are placed on the web near the flange angles. These are designed to resist the web moment and are called moment plates. Two plates (one on each side of the web) are placed on the web in between moment plates to resist the shear and are called shear plates.

Let A_s = net area of the two moment plates

d_1 = distance between centre to centre of moment plates

σ_{b1} = bending stress at the centre of the moment plate (assumed to be uniform throughout the moment plate)

σ_b = bending stress at the extreme fibre of the web plate

d = depth of the web plate

d_s = depth of the moment splice plate

d_{s1} = depth of the shear splice plate

n_1 = number of rivets in one vertical row

d_2 = diameter of rivet in millimetres

Moment of resistance of the moment plate should be equal to the bending moment resisted by the web plate,

$$M_w = \sigma_{b1} A_s d_1$$

or

$$M_w = \sigma_b \frac{d_1}{d} A_s d_1 a$$

or

$$A_s = \frac{M_w}{\sigma_b} \times \frac{d}{d_1^2} \tag{9.30}$$

The number of rivets in one vertical row are assumed and the net area of the section is found. If t_1 is the thickness of the moment plate,

$$\text{Net area, } A_s = 2(d_s - n_1 d_2)t_1$$

or

$$t_1 = \frac{1}{2} \times \frac{A_s}{d_s - n_1 d_2} \tag{9.31}$$

Design steps

1. The moment M_w resisted by the web is computed using Eq. (9.29).
2. The bending stress, σ_b at the extreme fibre of the web plate.

$$\sigma_b = \frac{M}{A'_f d}$$

where A'_f = net flange area.

3. The area of two moment splice plates is computed by using Eq. (9.30).
4. The thickness of the moment plate is computed by using Eq. (9.31). The depth of the moment plate is selected such as to accommodate two to three rivets in a vertical row.
5. The horizontal force in the moment plate is the product of σ_{b_1} and the net area of the splice plate A_s.
6. The diameter of the rivet is assumed and the rivet value (in double shear) is calculated. The horizontal force calculated in step (5) is divided by the rivet value to find the number of rivets. If n_1 is the number of rivets in each vertical row, the number or vertical rows can be found to accommodate these rivets. This decides the width of the moment plate.
7. The depth of the shear plate,

$$d_{s_1} = d - 2[a - d_s - b - c]$$

where d = depth of the girder between outer toes of flange angles
a = leg length of the flange angle connected to the web
d_s = depth of the moment plate
b = clearance between flange angle and moment plate
c = clearance between moment and shear plates.
8. The thickness of shear plates (two in numbers) is calculated using:

$$t_2 = \frac{V}{d_{s_1} \tau_{va}}$$

Half the thickness thus calculated, is provided for each shear plate with a minimum of 6 mm.
where τ_{va} = average shear stress ($0.40 f_y$).
9. The diameter of the rivets is assumed and the rivet value (in double shear) is computed. The number of rivets required are found by dividing the shear force with the rivet value. Then the number of vertical rows are decided on the basis of the number of rivets to be accommodated in one vertical row.

Moment Splice Type II

The number of splice plates required will be six. Four plates (two on each side of the web) are placed over the flange angles and two plates (one on each side of the web) are placed on the web (Fig. 9.10(c)). This arrangement is the best splice arrangement and is therefore, sometimes called *rational splice*. Splice plates are designed for the shear and moment to which the web section is subjected, if not spliced. Rivets are designed for the vertical shear due to the shear at the section and the horizontal shear due to the bending moment at the section.

Splice plates on the web

Let M = bending moment at the splice section
V = vertical shear force at the splice section
d_1 = centre to centre distance between the C.G.s of the flanges
Y_1 = distance of extreme fibre of gross area from neutral axis
I = moment of inertia of the plate girder

s = pitch of the rivets

R_v = rivet value

σ_b = bending stress at the C.G. of the rivet group.

$$\text{Vertical shear per unit depth} = \frac{V}{d_1}$$

$$\text{Horizontal shear per unit length} = \sigma_b\, t \qquad \left(\sigma_b = \frac{M}{I} Y_1\right)$$

$$\text{Resultant shear per unit length} = \sqrt{(\sigma_b\, t)^2 + (V/d_1)^2}$$

$$\text{Resultant shear per pitch length} = s\sqrt{(\sigma_b\, t)^2 + (V/d_1)^2}$$

This should be equal to rivet value, i.e.

$$s\sqrt{(\sigma_b\, t)^2 + (V/d_1)^2} = R_v$$

or

$$s = \frac{R_v}{\sqrt{(\sigma_b\, t)^2 + (V/d_1)^2}} \tag{9.32}$$

If n' is the number of vertical rows, then

$$s = \frac{n' R_v}{\sqrt{(\sigma_b\, t)^2 + (V/d_1)^2}} \tag{9.33}$$

Splice plates on flange angles

Let A_f = gross area of the flange angles excluding the web equivalent

A_{w1} = area of the web between the flange angle

σ_{b2} = bending stress at the level of rivets connecting the splice over the flange angle (= MY/I)

Y = distance of the rivet on the splice over flange angle, from the neutral axis

n = number of rivets on one side of splice.

The rivets are subjected to horizontal shears due to bending and vertical shear force.

Horizontal shear force per pitch due to vertical shear force

$$= \frac{V}{d_1} \times \left(\frac{A_f}{A_f + A_{w_1}}\right) s$$

$$\text{Horizontal shear force to be resisted by rivets} = \frac{V}{d_1} \times \left(\frac{A_f}{A_f + A_{w_1}}\right) sn$$

$$\text{Horizontal shear force due to moment} = \sigma_{b_2} A_{w_1}$$

The strength of joint should be equal to total horizontal shear force on splice.

$$n R_v = \frac{V}{d_1} \times \left(\frac{A_f}{A_f + A_{w_1}} \right) sn + \sigma_{b_2} A_{w_1}$$

or

$$n = \frac{\sigma_{b_2} A_{w_1}}{R_v - \frac{V}{d_1} \left(\frac{A_f}{A_f + A_{w_1}} \right) s} \tag{9.34}$$

Design steps

1. The depth of the splice plate on the web will be equal to the depth of the web plate minus the depth of the vertical legs of flange angles. The thickness of the splice plate is kept equal to half the web thickness and should not be less than 6 mm.
2. The number of vertical rows of rivets (two or three) is assumed. The pitch of the rivets is calculated from Eq. (9.33).
3. The number of rows of rivets to be provided is decided and the width of the splice plate is computed.
4. The design of the splice plate over flange angles (for the web) consists in calculating the length and thickness. The area of the web section below the flange angle to be spliced is computed. The area of splice plate is kept equal to the area of the web below the flange angles.
5. The number of rivets required on one side of splice plate are found from Eq. (9.34). From the number of rivets the pitch, the edge distance and the length of the splice plate is decided.
6. The thickness of the splice plate is found by dividing the area calculated in step (4) with the length computed in step (5) and it should not be less than 6 mm.

9.14 FLANGE ANGLE SPLICE

Flange angles are spliced when angles of the required length are not available. It is preferred not to splice both the flange angles at the same section. However, if done so, it is not questionable. Some designers feel that one of the flange angles may be spliced within the first half span and the other one in the second half span. There are several ways to splice flange angles. Some of them are as follows:

1. An angle splice on one side of the flange, (Fig. 9.11(a)).
2. One of the splice angles on the flange angle to be spliced and the other angle (furnishing the remaining splice area) on the other flange angle (Fig. 9.11(b)).
3. One angle splice on the flange angle to be spliced and one splice plate (furnishing remaining splice area) on the other flange angle (Fig. 9.11(c)).

Splice angles are cut to shape to fit the flange angles tightly. When one splice angle is provided, the load from the flange angle is transferred to the web on the same side through the splice angle. The rivets connecting the flange angle with the

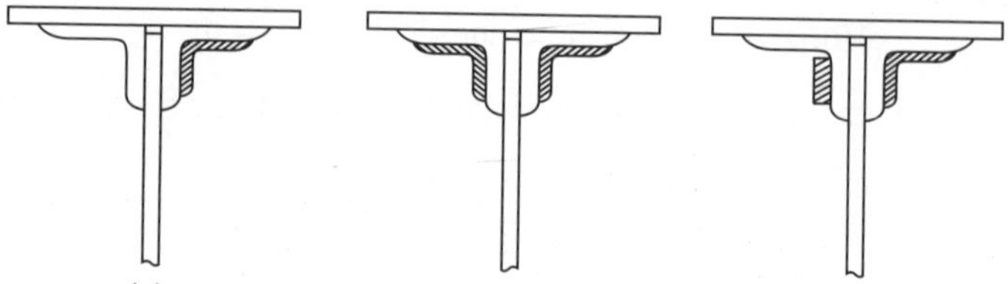

Fig. 9.11 *Flange Angle Splice*

web are sufficient and are provided at the same pitch from over the splice angle. Therefore, no extra rivets are required to connect the splice angle. The edges of the splice angle are trimmed so that these do not project beyond the flange angle (to protect from corrosion). The area of splice should be increased by 5% of the area to be spliced. The load for which the connectors are to be designed is also increased by 5% of the actual force at the splice section.

When two splice angles (Fig. 9.11(b)) or one splice angle and one splice plate (Fig. 9.11(c)) are provided, one of the splice angles or splice plate are not in direct contact with the flange angle to be spliced. A part of the horizontal shear force is assumed to shift from the flange angle to be spliced to the other side of the web, through the web. Therefore, extra rivets are required to resist this force. The force to which the splice angle or splice plate on other side of the web is subjected depends upon their effective cross-sectional areas.

Let P = force in splice plate or splice angle on uncut flange angle
V = shear force at the splice
R_v = strength of the rivet in single shear
n = number of rivets to resist shear force between web and flange
s = pitch of rivets
d_1 = effective depth of plate girder.

The force to be transmitted by the rivets should be equal to the strength of splice angle/splice plate plus the horizontal shear force, i.e.

$$nR_v = P + \text{horizontal shear force} \tag{9.35}$$

$$\text{Horizontal shear force/unit length} = \frac{V}{d_1}$$

$$\text{Horizontal shear force/unit length on one side of web} = \frac{1}{2} \times \frac{V}{d_1}$$

$$\text{Total horizontal shear in one plane} = \frac{1}{2} \times \frac{V}{d_1} ns$$

Equation (9.35) can be rewritten as

$$nR_v = P + \frac{V}{2d_1} \times ns$$

or

$$nR_v - \frac{V}{2d_1} \times ns = P$$

or
$$n\left(R_v - \frac{V}{2d_1}s\right) = P$$

or
$$n = \frac{P}{R_v - \dfrac{V}{2d_1}s} \qquad (9.36)$$

9.15 FLANGE PLATE SPLICE

Flange plates may have to be spliced if and when required. When the outermost cover plate is to be spliced, the splice plate area is kept equal to the area at the splice section. The length of the splice plate is governed by the number of rivets to be provided. This number of rivets required is computed from the force at the splice section. If some inner flange plate is to be spliced, it should be at the theoretical cut-off point of the upper plate because the extended portion of the curtailed plate may be assumed to serve as the splice plate. The inner flange plate can also be spliced by providing a splice plate on the top most flange cover plate. However, in this arrangement, the splice plates are not in direct contact with the flange plate to be spliced so the force for which rivets are to be designed is affected. But to simplify calculations it is assumed that the force is the same as that at the splice section and the rivets required are calculated.

9.16 DESIGN STEPS OF A PLATE GIRDER

1. The self weight of the plate girder is assumed. The live loads are estimated and the critical moment and shear force in the plate girder are computed.
2. The thickness of the web is assumed and the depth of the plate girder is determined from Eq. (9.1).
3. The flange area is computed using Eq. (9.3). Flange angles and flange cover plates furnishing the required area by flange area method are calculated.
4. The moment of inertia of the section and the bending stress in compression are computed. The gross area and net area of the section provided are computed. Also, the bending stress in tension is worked out which should be less than the permissible bending stress.
5. The flange cover plates are curtailed.
6. The connections of flange angles to web and flange angles to cover plates are designed in case of a riveted plate girder. For a welded plate girder the connection between flange plate and web plate are designed.
7. The bearing stiffeners and their connections are designed.
8. The intermediate stiffeners if required, and their connections are designed.
9. The web splice and its connection are designed.
10. The flange splice and its connection are designed.

9.17 COMPARISON OF WELDED AND RIVETED PLATE GIRDERS

1. Welded plate girders weigh 5%–15% less than riveted plate girders and are therefore economical.

2. Riveted plate girders involve making holes for the rivets, which affect the strength of the section and therefore more steel is required as compared to welded plate girders. Also the calculations are complicated for riveted plate girder (as the effect of holes on stresses are to be computed).
3. In riveted plate girders, flange angles are a must whereas in welded plate girders only flange plates are provided. Therefore, the effective depth for welded plate girder is increased. Due to increased effective depth, the moment of resistance of the section increases. In welded girders thick plates are provided which increase the lateral buckling strength of the compression flange.
4. Angle stiffeners require milling, crimping and filler plates, which further increase the cost of a riveted plate girder.
5. Splicing of riveted plate girder requires extra plates and extra holes for connectors reducing the plate girder net gross-section, whereas in the welded plate girder full gross-section is effective.
6. The clear depth of the web in a welded plate girder is more as compared to the riveted plate girder so more intermediate stiffeners are required.

Solved Examples

Example 9.1 Design a plate girder, 20 m span, to be provided in a hall of a restaurant. The superimposed load, exclusive of self weight is 100 kN/m. Design the web splice at one-third of span and flange angle splice at one-fourth of the span.

Solution

Design forces

$$\text{Total superimposed load} = 100 \times 20 = 2000 \text{ kN}$$

$$\text{Self-weight of plate girder} = \frac{Wl}{300} = \frac{2000 \times 20}{300}$$

$$= 133.33 \text{ kN} \approx 134 \text{ kN}$$

$$\text{Total uniformly distributed load, } W' = 2000 + 134 = 2134 \text{ kN}$$

$$\text{Maximum bending moment} = \frac{W'l}{8} = \frac{2134 \times 20}{8} = 5335 \text{ kNm}$$

$$\text{Maximum shear force} = \frac{W'}{2} = \frac{2134}{2} = 1067 \text{ kN}$$

Web plate Assume the plate girder to be laterally supported throughout. Let the structural steel be of yield stress 250 MPa.

$$\sigma_{bc} = 0.66 f_y = 165 \text{ N/mm}^2$$

$$\tau_{va} = 0.4 f_y = 100 \text{ N/mm}^2$$

The thickness of the web plate is assumed and the economical depth is worked out. The web is checked for shear and if found oversafe the thickness is reduced.

Assume the thickness of web plate $t = 12$ mm.

$$\text{Depth of the plate girder, } d = 1.1\sqrt{M/\sigma_{bc}\,t}$$

$$= 1.1\sqrt{5335 \times 10^6/(165 \times 12)}$$

$$= 1805.62 \text{ mm} \approx 1806 \text{ mm}$$

Let us assume that the flange angle leg connected to the web to be 100 mm.
Minimum thickness of web required from shear consideration,

$$t = \frac{\text{shear force}}{\text{shear stress} \times \text{effective depth}}$$

$$= \frac{1067 \times 10^3}{100 \times (1806 - 2 \times 100)} = 6.644 \text{ mm} < 12 \text{ mm}$$

This shows that the thickness of the web assumed is quite large. To achieve economy let us provide a web plate 8 mm thick.

$$d = 1.1\sqrt{5335 \times 10^6/(165 \times 8)} = 2211.428 \text{ mm} \approx 2212 \text{ mm}$$

$$\text{linimum required thickness of web} = \frac{1067 \times 10^3}{100\,(2212 - 2 \times 100)} = 5.303 \text{ mm} < 8 \text{ mm}$$

vhich is all right.

Hence, provide a web plate 2200 × 8 mm.

Design of flanges

$$\text{Approximate net flange area } A'_f = \frac{M}{\sigma_{bt}\,d_1} = \frac{5335 \times 10^6}{165 \times 2200}$$

$$= 14{,}696.96 \approx 14697 \text{ mm}^2$$

A minimum of one-third flange area should be provided by flange angles.

$$\text{Flange area required by flange angles} = \frac{14{,}697}{3} = 4899 \text{ mm}^2$$

$$\text{Flange area required by flange cover plates} = 14{,}697 - 4899 = 9798 \text{ mm}^2$$

Let us select 2 I.S.A. 125 × 95 × 12 mm with 95 mm leg parallel to the web and 2 plates 500 × 10 mm, from I.S. Handbook No. 1 and provide them as the flange section.

Provide 18 mm ϕ staggered, power driven rivets to connect flange angles with the flange plates and 18 mm ϕ, power driven rivets to connect flange angles with the web plate. A clearance of 5 mm may be provided between the flange cover plates and the top of the web plate.

The net area calculations have been done assuming that the tension flange section is weakened by two rivet holes (one in the flange angle and plates and the other in the flange angles and web).

Type of Section	Gross Flange Area, mm²	Rivet Hole Area, mm²	Net Flange Area, mm²
2 I.S.A. 125 × 95 × 12 mm	2 × 2498 = 4996	3 × 19.5 × 12 = 702	4996 – 702 = 4294
2 Plates 500 × 10 mm	2 × 500 × 10 = 10,000	2 × 19.5 × 10 = 390	10,000 – 390 = 9610
Equivalent web area	$A_{w_1} = (95 - 5) \times 8$ = 720	1 × 19.5 × 8 = 156	$A'_{w_1} = 720 - 156$ = 564
Total	15,716		14,468

$$\text{Gross flange area provided} = 15{,}716 \text{ mm}^2$$

$$\text{Net flange area provided} = 14{,}468 \text{ mm}^2$$

The flange area provided is checked by the moment of inertia method. The relevant properties of I.S.A. 125 × 95 × 12 mm, from I.S. Handbook No. 1, are

$$I_{xx} = 190.4 \times 10^4 \text{ mm}^4 \quad A = 2498 \text{ mm}^2 \quad C_{yy} = 24.7 \text{ mm}$$

$$I_{xx}(\text{gross}) = \left[\left(\frac{1}{12} \times 8 \times (2200)^3\right)\right] + 4 \times [190.4 \times 10^4 + 2498 \times (1105 - 24.7)^2]$$

$$+ 4 \times \left[\frac{500 \times 10^3}{12} + 500 \times 10 \times \left(1105 + \frac{20}{2}\right)^2\right]$$

$$= 43{,}632 \times 10^6 \text{ mm}^4$$

Note In the calculations, moment of inertia of plates about their own axis may be neglected (as the value will be insignificant) to save the calculation effort.

$$\sigma_{bc,\text{ cal}} = \frac{M}{I_{xx}} Y = \frac{5335 \times 10^6}{43{,}632 \times 10^6} \times (1105 + 20) = 137.56 \text{ N/mm}^2$$

$$\sigma_{bt,\text{ cal}} = 137.56 \times \frac{15{,}716}{14{,}468} = 149.42 \text{ N/mm}^2 < 165 \text{ N/mm}^2$$

which is all right.

The cross-section of the plate girder is shown in Fig. Ex. 9.1(a).

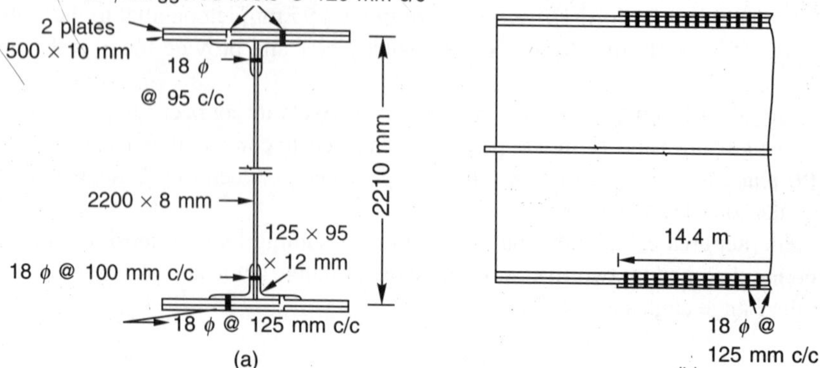

Fig. Ex. 9.1

Connections To achieve economy, the flange cover plates will be curtailed, leaving the section at support with one cover plate only on each flange. Since shear force is maximum at the support, design of the connection will be critical here rather than anywhere else along the span.

$$\text{Maximum shear force} = 1067 \times 10^3 \text{ N}$$

$$\text{Load per metre run} = 100 \text{ kN/m} = 100 \text{ N/mm}$$

Connection of flange angles to web in the compression flange Let us provide 18 mm ϕ, power driven rivets.

$$\text{Gross diameter} = 18 + 1.5 = 19.5 \text{ mm}$$

$$\text{Value or rivet in double shear} = 2 \times \frac{\pi}{4} \times 19.5^2 \times 100 \times 10^{-3} = 59.729 \text{ kN}$$

$$\text{Value of rivet in bearing} = 19.5 \times 8 \times 300 \times 10^{-3} = 46.8 \text{ kN}$$

$$\text{Rivet value, } R_v = 46.8 \text{ kN}$$

The distance of the C.G. of the compression flange from top,

$$\overline{Y} = \frac{500 \times 10 \times \dfrac{10}{2} + 2 \times 2498 \times (10 + 24.7)}{500 \times 10 + 2 \times 2498} = 19.84 \text{ mm}$$

Effective depth of the plate girder at the support,

$$d_1 = 2210 + 20 - 2 \times 19.84 = 2190.32 \text{ mm}$$

$$s = \frac{R_v}{\left\{\left(\dfrac{V}{d_1} \times \dfrac{A_f}{A_f + A_{w_1}}\right)^2 + w^2\right\}^{1/2}}$$

$$A_f = 4996 + 500 \times 10 = 9996 \text{ mm}^2$$

$$A_f + A_{w_1} = 9996 + 720 = 10716 \text{ mm}^2$$

$$s = \frac{46.8 \times 10^3}{\left\{\left(\dfrac{1067 \times 10^3}{2190.32} \times \dfrac{9,996}{10,716}\right)^2 + (100)^2\right\}^{1/2}}$$

$$= 100.58 \text{ mm} \approx 95 \text{ mm}$$

Provide 18 mm ϕ power driven rivets at a pitch of 95 mm c/c.

Connection of flange angles to flange plates in compression flange Let us provide 18 mm ϕ, power driven staggered rivets. The flange plates are stitched together, hence the rivets are in single shear.

Gross diameter or rivet = 18 + 1.5 = 19.5 mm

$$\text{Value of rivet in single shear } = \frac{\pi}{4} \times 19.5^2 \times 100/10^{-3} = 29.864 \text{ kN}$$

Value of rivet in bearing = 19.5 × 10 × 300 × 10^{-3} = 58.5 kN

Rivet value, R_v = 29.864 kN

$$s = \frac{R_v d_1}{V} \times \frac{A_f + A_{w_1}}{A_f} = \frac{29{,}864 \times 2190.32}{1067 \times 10^3} \times \frac{10{,}716}{9{,}996} = 65.72 \text{ mm}$$

As the rivets are staggered and are provided in two rows,

required pitch = 2 × 65.72 = 131.44 mm ≈ 125 mm

Also the pitch should not be greater than 12 t or 200 mm whichever is minimum, i.e. 12 × 10 = 120 mm.

Provide rivets at a pitch of 120 mm c/c.

Connection of flange angles to web in the tension flange

$$s = \frac{R_v (A'_f + A'_{w_1}) \times d_1}{V A'_f}$$

$$A'_f = 4294 + \frac{9610}{2} = 9099 \text{ mm}^2$$

$$A'_f + A'_{w_1} = 9099 + 564 = 9663 \text{ mm}^2$$

$$s = \frac{46{,}800 \times 9663 \times 2190.32}{1067 \times 10^3 \times 9099} = 102.02 \approx 100 \text{ mm}$$

Provide 18 mm ϕ, power driven rivets at a pitch of 95 mm.

Connection of flange angles to flange plates in the tension flange

Rivet value, R_v= 29.864 × 10^3 N

$$s = \frac{R_v (A'_f + A'_{w_1}) \times d_1}{V A'_f} = \frac{29.864 \times 10^3 \times 9663 \times 2190.32}{1067 \times 10^3 \times 9099} = 65.10 \text{ mm}$$

The rivets are provided in two rows and are staggered,

∴ pitch provided = 2 × 65.10 = 130.2 ≈ 125 mm

Also, it should not be greater than 16 t or 200 mm whichever is less, i.e. 16 × 10 = 160 mm.

Provide 18 mm ϕ, power driven rivets at a pitch of 125 mm.

The connection details are shown in Fig. Ex. 9.1(a).

Curtailment of flange cover plates The first cover plate will be carried throughout the span and the next plate is curtailed.

Net area of curtailed plate, $A'_1 = 500 \times 10 - 1 \times 19.5 \times 10 = 4805 \text{ mm}^2$

$$X_1 = \frac{l}{2}\sqrt{A_1'/A_{f_1}'} = \frac{20}{2}\sqrt{4805/14,468} = 5.76 \text{ m}$$

The variation of the bending moment diagram is parabolic.

$$Y = \frac{4a}{l^2}\, x(l - x)$$

where $a = 5335$ kNm $l = 20$ m

$$x = 10 - 5.76 = 4.24 \text{ m}$$

$$y = \frac{4 \times 5335}{20^2} \times 4.24 \times (20 - 4.24) = 3564.975 \text{ kNm}$$

Bending moment at the point of theoretical cut-off = 3564.975 kNm

At the supports there will be only two flange plates as the other two have been curtailed.

Hence,

$$I_{xx(\text{gross})} = \left[\frac{1}{12} \times 8 \times 2200^3\right] + 4 \times [190.4 \times 10^4 + 2498 \times (1105 - 24.7)^2]$$

$$+ 2 \times \left[\frac{500 \times 10^3}{12} + 500 \times 10 \times \left(1105 + \frac{10}{2}\right)^2\right]$$

$$= 31,088.5 \times 10^6 \text{ mm}^4$$

Distance of C.G. of the cover plate curtailed, from N.A.,

$$= 1105 + 10 + \frac{10}{2} = 1120 \text{ mm}$$

$$\sigma_{bc,\,\text{cal}} = \frac{3564.975 \times 10^6 \times 1120}{31,088.5 \times 10^6} = 128.43 \text{ N/mm}^2$$

$$\sigma_{bt,\,\text{cal}} = \sigma_{bc,\,\text{cal}} \times \frac{A_f}{A_f'}$$

As one plate is curtailed,

$$A_f = 4996 + 500 \times 10 + 720 = 10,716 \text{ mm}^2$$

and $$A'_f = 4294 + 4805 + 564 = 9663 \text{ mm}^2$$

∴ $$\sigma_{bt,\,\text{cal}} = 128.43 \times \frac{10,716}{9663} = 142.425 \text{ N/mm}^2$$

Tensile force on each cover plate at the edge of the cover plate

$= \sigma_{bt} \times$ area of cross-section of the cover plate

$= 142.425 \times 4805 = 684,352.1$ N

The curtailed plate is extended for some distance so as to accommodate the connectors to transfer this force.

Let us provide 18 mm ϕ, power driven rivets.

The rivets will be in single shear and bearing.

$$\text{Rivet value, } R_v = 29.864 \times 10^3 \text{ N}$$

$$\text{Number of rivets, } n = \frac{684.352.1}{29{,}864} = 22.915$$

Provide 24, 18 mm ϕ, staggered rivets, in two rows on both the curtailed ends at a pitch of $2.5\phi = 2.5 \times 18 = 45 \text{ mm} \approx 50 \text{ mm}$

Provide rivets at a pitch of 125 mm same as that for connecting the flange angle and cover plates (from fabrication view point) on tension flange and 120 mm on compression flange.

From I.S: 800–1984, Table 8.2,

Edge distance for 18 mm diameter rivets = 32 mm

$$\text{Actual length of curtailed plate} = 2 \times \left[5.76 + \frac{11 \times 125}{10^3} + \frac{32}{10^3}\right]$$

$$= 14.33 \text{ m} \approx 14.40 \text{ m}$$

The flange plate on the compression flange is curtailed at the same point as that on the tension flange. The details of the curtailed section is shown in Fig. Ex. 9.1(b).

Bearing stiffener

$$\text{Permissible bearing stress} = 187.5 \text{ N/mm}^2 \qquad (0.75\, f_y)$$

$$\text{Bearing area required} = \frac{1067 \times 10^3}{187.5} = 5690.66 \text{ mm}^2$$

Select four I.S.A. 100 × 75 × 12 mm with 12 mm packing plates.

For flange angle I.S.A. 125 × 95 × 12 mm, $r_1 = 9.0$ mm

$$\text{Bearing area provided} = 4 \times (100 - 9.0) \times 12 = 4368 \text{ mm}^2 < 5690 \text{ mm}^2$$

which is not safe.

From I.S. Handbook No. 1, select another section. It is found that no other section in the unequal angle section series fits in; therefore let us provide I.S.A. equal angle section 110 × 110 × 15 mm (Fig. Ex. 9.1(c)).

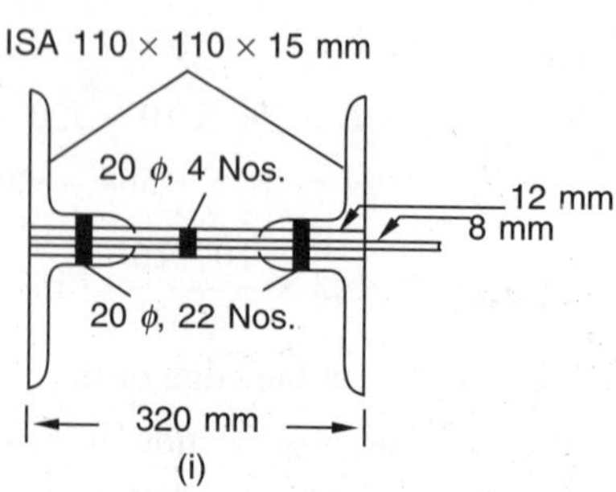

Fig. Ex. 9.1(c)(i)

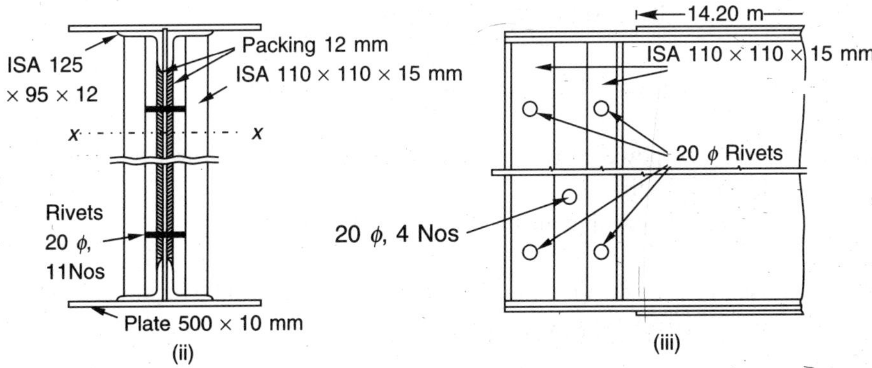

Fig. Ex. 9.1(c)(ii) and (iii)

Bearing area provided = $4 \times (110 - 9.0) \times 15 = 6060 \text{ mm}^2 > 5690 \text{ mm}^2$ which is all right.

Actual length of bearing stiffener,

$$d_1 = 2210 - 2 \times 12 = 2186 \text{ mm}$$

$$d_{1\text{ eff}} = 0.7 \times 2186 = 1530.2 \text{ mm}$$

Cross-sectional area of the bearing stiffener acting as column.

$$= 4 \times 3081 + 2 \times (20 \times 8) \times 8 = 14{,}884 \text{ mm}^2$$

$$I_{xx} = 4 \times 337.4 \times 10^4 + 4 \times 3081 \times (32.7 + 12 + 8/2)^2 + \frac{1}{12} \times (2 \times 20 \times 8) \times (8)^3$$

$$= 4273.83 \times 10^4 \text{ mm}^4$$

$$r_{xx} = \sqrt{I_{xx}/A} = \sqrt{4273.83 \times 10^4 / 14884} = 53.58 \text{ mm}$$

$$\lambda = \frac{d_{1,\text{eff}}}{r_{xx}} = \frac{1530.2}{53.58} = 28.55$$

From I.S: 800–1984, for $f_y = 250 \text{ N/mm}^2$ and $\lambda = 28.55$

$$\sigma_{ac} = 145.43 \text{ N/mm}^2$$

Load carrying capacity = $145.43 \times 14{,}884 = 2{,}164{,}580$ N $> 1067 \times 10^3$ N

which is sufficient.

The details of the bearing stiffener are shown in Fig. Ex. 9.1(c).

Connections Let us provide 20 mm ϕ power driven rivets.

Gross diameter = $20 + 1.5 = 21.5$ mm

Value of rivet in double shear = $2 \times \frac{\pi}{4} \times 21.5^2 \times 100 = 72{,}610.06$ N

Value of rivet in bearing = $21.5 \times 8 \times 300 = 51{,}600$ N

$$\text{Rivet value, } R_v = 51{,}600 \text{ N}$$

$$\text{Number of rivets } n = \frac{1{,}067 \times 10^3}{51{,}600} = 20.678$$

Provide 20 mm ϕ, power driven, 22 rivets in two vertical rows as shown in Fig. Ex. 9.1(c) (ii) & (iii).

Thickness of packing plate = thickness of flange angle
= 12 mm

As the packing is more than 6 mm thick, additional rivets will be required.

$$n' = \frac{2.5}{100} \times \frac{12}{2} \times 20.678 = 3.069$$

Provide 4 additional 20 mm ϕ rivets on packing as shown in Fig. Ex. 9.1(c)(iii).

Intermediate stiffeners For an unstiffened web, the minimum required thickness is as below,

$$\text{(i) } t_{w,\,\min} = \frac{d_1\sqrt{\tau_{va,\,\mathrm{cal}}}}{816}$$

where, $d_1 = (2200 + 10) - 2 \times 95 = 2020$ mm

and $$\tau_{va,\,\mathrm{cal}} = \frac{1067 \times 10^3}{2020 \times 8} = 66.0 \text{ N/mm}^2$$

$$t_{w,\,\min} = \frac{2020 \times \sqrt{66.0}}{816} = 20.11 \text{ mm}$$

$$\text{(ii) } t_{w,\,\min} = \frac{d_1\sqrt{f_y}}{1344} = \frac{2020\sqrt{250}}{1344} = 23.76 \text{ mm}$$

$$\text{(iii) } t_{w,\,\min} = \frac{d_1}{85} = \frac{2020}{85} = 23.76 \text{ mm}$$

The web thickness provided is 8 mm, which is lesser than the web thickness required if it is to be unstiffened. Therefore, intermediate web stiffeners will be required.

Vertical stiffeners are provided if the web thickness as calculated below is more than the thickness provided.

$$d_2 = 2 \times \frac{2020}{2} = 2020 \text{ mm}$$

$$\text{(i) } t_{w,\,\min} = \frac{d_2\sqrt{f_y}}{3200} = \frac{2020 \times \sqrt{250}}{3200} = 9.98 \text{ mm}$$

(ii) $t_{w,\,min} = \dfrac{d_2}{200} = \dfrac{2020}{200} = 10.1$ mm

The web thickness provided is 8 mm only which is less than 10.1 mm. Hence vertical stiffeners will be provided.

A horizontal stiffener will be required near the compression flange at a distance of 2/5 of the distance from the compression flange to the N.A. if the web thickness as calculated below is more than the provided one.

(i) $t_{w,\,min} = \dfrac{d_2\sqrt{f_y}}{4000} = \dfrac{2020\sqrt{250}}{4000} = 7.98$ mm

(ii) $t_{w,\,min} = \dfrac{d_2}{250} = \dfrac{2020}{250} = 8.08$ mm

The web thickness provided is approximately same as that required and the designer may not feel it necessary to provide a horizontal stiffener. Therefore, let us provide only vertical stiffeners.

Vertical stiffener

$$\frac{d}{t} = \frac{2020}{8} = 252.5$$

The smaller clear panel dimension $\not> 180\,t$

i.e. $180 \times 8 = 1440$ mm

Therefore, vertical stiffeners should be provided at a spacing less than 1440 mm. From I.S: 800–1984, Table 6.6 A, it is found that maximum clear spacing, for d/t ratio of 252.5, can be 0.6 d only. Therefore, provide vertical stiffeners at a spacing = 0.6 d = 1212 mm.

From Table 6.6 A of I.S: 800–1984, for d/t = 252.5, f_y = 250 MPa, and $c = 0.6\,d$

$$\tau_{va} = 81.75 \text{ N/mm}^2 > 66.0 \text{ N/mm}^2$$

which is all right.

$$\text{Maximum outstand of vertical stiffener} = \frac{256t}{\sqrt{f_y}} = \frac{256 \times 8}{\sqrt{250}} = 129.5 \text{ mm}$$

Let us choose I.S.A. 100 × 75 × 8 mm angle section as vertical stiffener.

$$I_{xx} = 131.6 \times 10^4 + 1336 \times 31.0^2 = 259.99 \times 10^4 \text{ mm}^4$$

$$\text{Minimum web thickness from shear consideration} = \frac{1067 \times 10^3}{2020 \times 81.75}$$

$$= 6.46 \text{ mm}$$

$$\text{Moment of inertia required by stiffener} = \frac{1.5\,d^3\,t^3}{c^2} = \frac{1.5 \times 2020^3 \times 6.46^3}{1212^2}$$

$$= 226.9 \times 10^4 \text{ mm}^4 < 259.99 \times 10^4 \text{ mm}^4$$

which is as it should be.

Connection

$$\text{Shear force} = \frac{125\,t^2}{h} = \frac{125 \times 8^2}{100} = 80 \text{ kN/m} = 80 \text{ N/mm}$$

Let us provide 18 mm ϕ, power driven rivets. The rivets will be in single shear.

$$\text{Rivet value} = 29.864 \text{ kN}$$

$$\text{Pitch of rivets} = \frac{29.864 \times 10^3}{80} = 373.3 \text{ mm}$$

This should not be greater than 200 mm or 12 t (12 × 8 = 96 mm) whichever is less. Therefore provide 18 mm ϕ, power driven rivets at a pitch of 95 mm.

(Refer to Fig. Ex. 9.1(d))

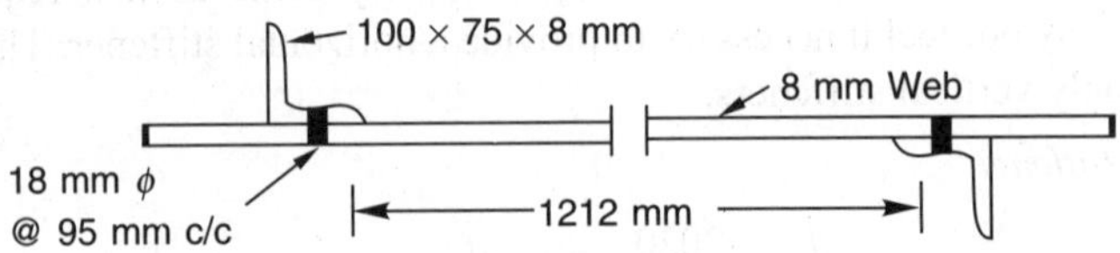

Fig. Ex. 9.1(d)

Web splice The web is to be spliced at 1/3rd of span $= \frac{L}{3} = \frac{20}{3} = 6.667$ m

$$\text{Maximum shear force} = \frac{1067 \times (10 - 6.667)}{10} = 355.63 \text{ kN} \approx 360 \text{ kN}$$

$$\text{Maximum bending moment, } y = \frac{4\,ax}{l^2}\,(l - x)$$

$$x = 6.667 \text{ m}, a = 5335 \text{ kNm}, l = 20 \text{ m}$$

$$y = \frac{4 \times 5335}{20^2} \times 6.667 \times (20 - 6.667)$$

$$= 4742.34 \text{ kNm} \approx 4743 \text{ kNm}$$

Bending moment and shear force at 1/3rd of span are 4743 kNm and 360 kN respectively.

Let us provide Moment Splice II.

Splice plate on web

Depth of splice plate = 2210 – 2 × 95 – 2 × 6 = 2008 mm

≈ 2000 mm (increase the clearance by 4 mm on each side)

$$\text{Thickness of splice plate} = \frac{1}{2} \times \text{thickness of web} = \frac{8}{2} = 4 \text{ mm} \nless 6 \text{ mm.}$$

Hence, provide two 6 mm thick splice plates one on each side of the web. Length of the splice plate can be found only after design of the connection.

Connection Let us provide 18 mm ϕ power driven rivets.

$$\text{Rivet value, } R_v = 46{,}800 \text{ N}$$

$$\text{Pitch of rivets, } s = \frac{R_v}{\sqrt{(\sigma_b \times t)^2 + (V/d_1)^2}}$$

where σ_b = the bending stress at the level of rivets connecting flange angle to the web. (It may be argued that the bending stress should be at the level of rivets under consideration, i.e. at the level of top most or lower most rivet line and if done so the rivets on the splice (on flange) will be overstressed. Therefore the value of bending stress is taken equal to the value at the level of rivets connecting flange angle to the web). Let the rivets be provided at 60 mm from the top of the flange angle.

$$\sigma_b = \sigma_{b2} = \frac{4743 \times 10^6 \times \left(\dfrac{2210}{2} - 60\right)}{43{,}630 \times 10^6} = 113.60 \text{ N/mm}^2$$

d_1 = effective depth of the plate girder

$= 2210 - 2 \times 24.7 = 2160.6$ mm

$$s = \frac{46{,}800}{\sqrt{(113.6 \times 8)^2 + (360 \times 10^3 / 2160.6)^2}} = 50.652 \text{ mm}$$

Provide rivets in two rows, pitch = 2 × 50.652 = 101.304 ≈ 100 mm

Provide 18 mm ϕ, power driven rivets at a pitch of 100 mm as shown in Fig. Ex. 9.1(e).

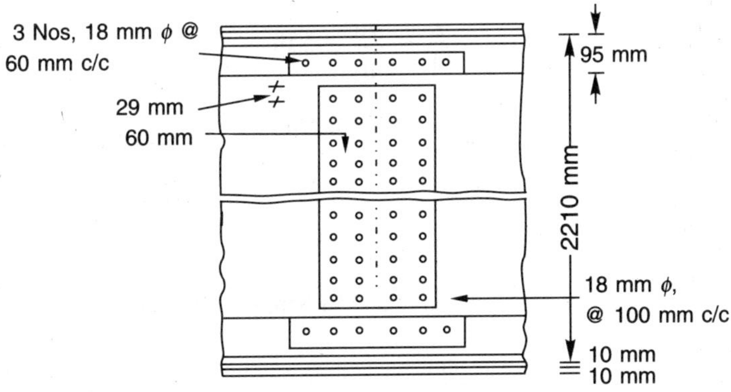

Fig. Ex. 9.1(e)

Let us provide a gauge distance of 60 mm and edge distance of 29 mm.

Width of the splice plate = 2 × (1 × 60 + 2 × 29) = 236 ≈ 240 mm

Provide two splice plates 2000 × 240 × 6 mm, one on each side of the web.

Splice over flange angle

Area of the web under flange angles = (95 – 5) × 8 = 720 mm^2

Area of the splice plate = area of the web under the flange angles

= 720 mm^2

Let us provide 70 mm wide splice plate.

$$\text{Thickness of the splice plate} = \frac{720}{70} = 10.29 \text{ mm}$$

∴ Thickness of each splice plate = 10.29/2 = 5.15 ≈ 6 mm

$$\text{Number of rivets, } n = \frac{\sigma_{b_2} A_{w_1}}{R_v - \dfrac{V}{d_1} \times \dfrac{A_f}{\left(A_f + A_{w_1}\right)} \times s}$$

Assuming pitch of the rivets to be 60 mm,

$$n = \frac{113.6 \times 720}{46{,}800 - \dfrac{360 \times 10^3}{2206.88} \times \dfrac{14{,}996 \times 60}{157{,}16}} = 2.18 \approx 3$$

Provide three rivets 18 mm ϕ, at a pitch of 60 mm and at an edge distance of 29 mm on each side of splice.

Length of the splice plate = 2 × (2 × 60 + 2 × 29)

= 356 mm ≈ 400 mm

Provide a splice plate 400 × 70 × 6 mm.

The details of the splice are shown in Fig. Ex. 9.1(e).

Flange angle splice The flange angle is to be spliced at 1/4th span.

$$\frac{L}{4} = \frac{20}{4} = 5 \text{ m}$$

$$\text{Maximum shear force} = \frac{1067 \times (10 - 5)}{10} = 533.5 \text{ kN} \approx 535 \text{ kN}$$

$$\text{Maximum bending moment, } y = \frac{4\,ax}{l^2} \times (l - x)$$

a = 5335 kNm, l = 20 m, x = 5 m

$$y = \frac{4 \times 5335 \times 5}{20^2} \times (20 - 5) = 4001.25 \approx 4005 \text{ kNm}$$

The bending moment and shear force, at 1/4th span, are 4005 kNm and 535 kN respectively,

Let us provide an angle section as a splice section.

Flange angle 125 × 95 × 12 mm to be spliced has an area = 2498 mm^2.

Let us provide 18 mm ϕ rivets for connection.

$$\text{Net area of I.S.A. } 125 \times 95 \times 12 \text{ mm} = 2498 - 2 \times 19.5 \times 12$$
$$= 2030 \text{ mm}^2$$

Net area of splice angle can be found by increasing the net area of flange angle at splice by 5%.

$$\text{Required net area of the splice angle} = 1.05 \times 2030 = 2131.5 \text{ mm}^2$$

The angle section to be used as a splice angle should furnish an area quite large as compared to the above calculated value. This is because the splice angle is cut to shape and the root of the angle is removed. Also the splice angle should not project beyond the flange angles.

Let us select I.S.A. 150 × 115 × 15 mm angle section.

It is cut to 113 × 83 × 15 mm size to fit in the flange angle.

$$\text{Gross area provided} = 113 \times 15 + (83 - 15) \times 15 = 2715 \text{ mm}^2$$

$$\text{Net area provided} = 2715 - 2 \times 19.5 \times 15 = 2130 \text{ mm}^2 > 2131.5 \text{ mm}^2$$

Since the angle has area marginally lesser than that required, the splice angle 150 × 115 × 15 mm may be provided. However, in case of a doubt angle section with 18 mm thick legs may be checked and provided.

$$\text{Rivet value, } R_v = 46{,}800 \text{ N} \qquad (19.5 \times 8 \times 300 = 46{,}800 \text{ N})$$

Bending stress at the C.G. of flange angles,

$$\sigma_b = \frac{4005 \times 10^6}{43{,}630 \times 10^6} \times \left(\frac{2250}{2} - 20 - 24.7\right) \times \frac{15{,}716}{14{,}468} = 107.71 \text{ N/mm}^2$$

$$\text{Force in flange angle} = 107.71 \times 2181.9 = 235017.88$$

$$\text{Required number of rivets, } n = \frac{235017.88}{46{,}800} = 5.02 \approx 6$$

Provide six rivets, 18 mm ϕ, on each side of the splice section. Three rivets are provided to connect the flange angle leg to the splice angle leg (perpendicular to web) and three rivets to connect the splice angle leg with the flange angle leg (on web). The details of flange splice are shown in Fig. Ex. 9.1(f).

Example 9.2 Design a welded plate girder 24 m in effective span and simply supported at the two ends. It carries a uniformly distributed load of 100 kN/m.

Solution

$$\text{Uniform load, } w = 100 \text{ kN/m}$$

$$\text{Self-weight of plate girder} = \frac{Wl}{400} = \frac{(100 \times 24) \times 24}{400} = 144 \text{ kN}$$

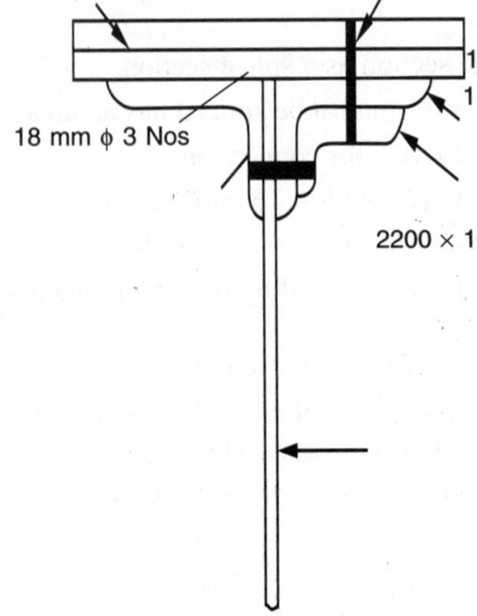

Fig. Ex. 9.1(f)

$$\text{Self-weight of plate girder per metre length} = \frac{144}{24} = 6.0 \text{ kN/m}$$

$$\text{Total uniform load, } w' = 100 + 6 = 106 \text{ kN/m}$$

$$\text{Maximum shear force} = \frac{w'l}{2} = \frac{106 \times 24}{2} = 1272 \text{ kN}$$

$$\text{Maximum bending moment} = \frac{w'l^2}{8} = \frac{106 \times 24^2}{8} = 7632 \text{ kNm}$$

Design of web Let the girder be laterally supported throughout the span. Let the thickness of the web plate be 8 mm and the structural steel be of yield stress 250 MPa.

$$\sigma_{bct} = 0.66 \times 250 = 165 \text{ N/mm}^2$$

$$\tau_{va} = 0.4 f_y = 100 \text{ N/mm}^2$$

$$d = 1.1\sqrt{M/\sigma_{bc}t} = 1.1\sqrt{7632 \times 10^6/(165 \times 8)} = 2644.99 \approx 2500 \text{ mm}$$

Let us select web plates 2500 × 8 mm size from I.S. Handbook No. 1.
Minimum web thickness from shear consideration.

$$t = \frac{1272 \times 10^3}{100 \times 2500} = 5.08 \text{ mm} < 8 \text{ mm}$$

which is as it should be.

Design of flanges Approximate flange area required,

$$A_f = \frac{M}{\sigma_{bct}\, d_1} = \frac{7632 \times 10^6}{165 \times 2500} = 18{,}501.8 \text{ mm}^2$$

The required flange area is provided by the plate sections. Let us provide a 40 mm thick flange plate.

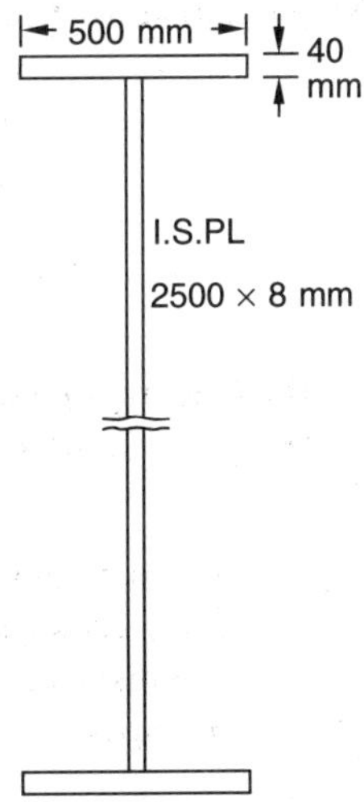

Fig. Ex. 9.2(a)

$$\text{Width of flange plate} = \frac{18,501.8}{40} = 462.54 \text{ mm}$$

$$\approx 500 \text{ mm}$$

$$\text{Flange outstand} = \frac{500-8}{2} = 246 \text{ mm}$$

$$\not> 640 \text{ mm } (16\,t = 16 \times 40)$$

which is all right.

Flange are provided by plate = 500 × 40
= 20,000 mm^2
> 18,501.8 mm^2

which is sufficient.

Hence provide a flange plate of dimensions 500 × 40 mm as shown in Fig. Ex. 9.2(a).

Check Moment of inertia of the plate girder,

$$I_{xx} = \frac{8 \times 2500^3}{12} + 2 \times \left[\frac{500 \times 40^3}{12} + 500 \times 40 \times \left(\frac{2500}{2} + 20\right)^2\right]$$

$$= 7,493,800 \times 10^4 \text{ mm}^4$$

Maximum bending stress,

$$\sigma_{bct} = \frac{7632 \times 10^6}{7,493,800 \times 10^4} \times \left(\frac{2500}{2} + 40\right)$$

$$= 131.37 \text{ N/mm}^2 < 165 \text{ N/mm}^2$$

which is safe.

Reduction of flange plate thickness The bending moment is maximum at the centre and reduces towards the support. Therefore, let us provide a 40 mm thick plate in the middle portion of the girder and reduce the thickness to 30 mm and 16 mm towards the support.

30 mm plate section

Maximum outstand = 246 mm

Permissible outstand = $16t = 16 \times 30 = 480$ mm

which is all right.

Moment of inertia of the plate girder,

$$I_{xx} = \frac{8 \times 2500^3}{12} + 2 \times \left[\frac{500 \times 30^3}{12} + 500 \times 30 \times \left(\frac{2500}{2} + 15\right)^2\right]$$

$$= 5,842,560 \times 10^4 \text{ mm}^4$$

Moment of resistance,

$$M_R = \frac{\sigma_{bct} I}{y}$$

$$= \frac{165 \times 5{,}842{,}560 \times 10^4}{\left(\frac{2500}{2} + 30\right)} = 7531.42 \times 10^6 \text{ Nmm}$$

The bending moment diagram is parabolic.

The equation of a parabola with supports as the origin is,

$$y = kx(l - x)$$

where y = maximum bending moment
k = a constant
x = distance of the section from support
l = span

$$7632 = k \times \frac{24}{2}\left(24 - \frac{24}{2}\right)$$

$$k = 53$$

x @ y = 7531.42 kNm:

$$7531.42 = 53\,x\,(24 - x)$$

or $$x^2 - 24x + 142.1 = 0$$

or $$x = 10.62 \text{ m} \approx 10.6 \text{ m}$$

Therefore, the plate thickness can be reduced to 30 mm from 40 mm for a length of 10.6 m from the supports. Provide 500 × 30 mm plates up to 10.6 m from the supports.

16 mm plate section

Maximum outstand = 246 mm

Permissible outstand = 16 × 16 = 256 mm

> 246 mm

which is sufficient.

Therefore, provide a 16 mm plate.

$$I_{xx} = \frac{8 \times 2500^3}{12} + 2\left[\frac{500 \times 16^3}{12} + 500 \times 16 \times \left(\frac{2500}{2} + 8\right)^2\right]$$

$$= 3{,}573{,}803 \times 10^4 \text{ mm}^4$$

Moment of resistance,

$$M_R = \sigma_{bct}\,\frac{I}{y} = \frac{165 \times 3{,}573{,}803 \times 10^4}{\left(\frac{2500}{2} + 16\right)}$$

$$= 4657.80 \times 10^6 \text{ Nmm} = 4657.80 \text{ kNm}$$

x @ $y = 4657.80$ kNm:

$$y = k\,x(l - x)$$
$$= 53 \times x \times (24 - x)$$

or $$x^2 - 24x + 87.88 = 0$$

or $$x = 4.508 \text{ m} \approx 4.5 \text{ m}$$

Therefore, the plate thickness can be reduced to 16 mm from 30 mm for a length of 4.5 m from the support. Provide flange plates as shown in Fig. Ex. 9.2(b).

500 × 16 mm	From supports up to 4.5 m
500 × 30 mm	From 4.5 m to 10.6 m
500 × 40 mm	From 10.6 m to midspan

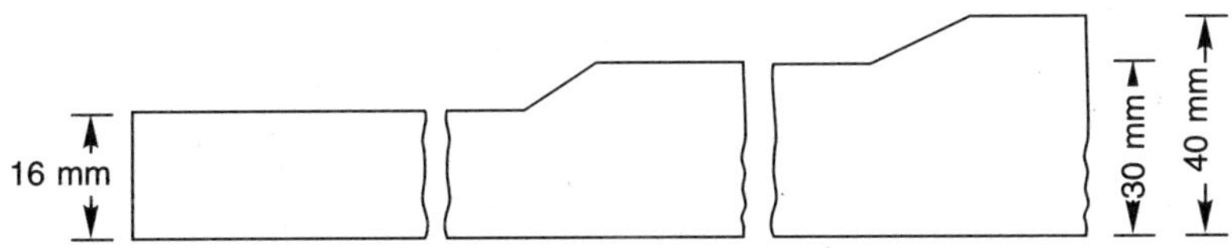

Fig. Ex. 9.2(b)

Connection

$$\text{Horizontal shear/mm} = \frac{VA\bar{Y}}{I_{xx}}$$

$$V = 1272 \text{ kN} = 1272 \times 10^3 \text{ N}$$

$$A\bar{Y} = 500 \times 16 \times \left(\frac{2500}{2} + \frac{16}{2}\right) = 1006.4 \times 10^4 \text{ mm}^3$$

$$I_{xx} = 3{,}573{,}803 \times 10^4 \text{ mm}^4$$

$$\text{Horizontal shear/mm} = \frac{1272 \times 10^3 \times 1006.4 \times 10^4}{3{,}573{,}803 \times 10^4} = 358.20 \text{ N/mm}$$

Size of weld Welding is done on both sides of the web

$$358.20 = 2 \times (0.7 \times s \times 1 \times 108)$$
$$s = 2.37 \text{ mm}$$

But the minimum size of a fillet weld for a 16 mm thick plate is 5 mm. Let us provide an intermittent fillet weld. The effective weld length is 4 s (4 × 5 = 20 mm) or 40 mm whichever is more. Therefore, provide 40 mm long intermittent fillet welds.

$$\text{Pitch of welds} = \frac{\text{strength of weld on both faces of web}}{\text{horizontal shear/mm length}}$$

$$= \frac{2 \times 40 \times 0.7 \times 5 \times 108}{358.20} = 84.42 \text{ mm} \approx 80 \text{ mm}$$

The clear spacing between the welds should not be more than 12 t or 200 mm whichever is less.

$$\text{Maximum clear spacing} = 12 \times 8 = 96 \text{ mm}$$

$$\text{Permissible pitch} = 96 + 40 = 136 \text{ mm}$$

Provide 40 mm long fillet welds at a pitch of 80 mm. The flange plates 40 mm, 30 mm and 16 mm thick are connected together by a double Vee butt weld.

Bearing stiffener

$$\text{Maximum shear force, } V = 1272 \text{ kN}$$

$$\text{Allowable bearing stress, } \sigma_p = 0.75\ f_y = 187.5 \text{ N/mm}^2$$

$$\text{Bearing area required } = \frac{1272 \times 10^3}{187.5} = 6784 \text{ mm}^2$$

Let us try two plates 200 mm wide as the stiffener.

$$\text{Thickness of plate } = \frac{6784}{2 \times 200} = 16.96 \text{ mm} \approx 20 \text{ mm}$$

$$\text{Outstand} = 200 \text{ mm}$$

Outstand should not be more than 12 t (= 240 mm).

$$\text{Bearing area provided} = 2 \times 200 \times 20 = 8000 \text{ mm}^2 > 6784 \text{ mm}^2$$

which is sufficient.

$$\text{Area of stiffener, } A = 2 \times 200 \times 20 + 2 \times 20 \times 8 \times 8 = 10{,}560 \text{ mm}^2$$

$$I_{xx} = 2 \times \left[\frac{20 \times 200^3}{12} + 20 \times 200 \times \left(100 + \frac{8}{2}\right)^2\right]$$

$$= 11{,}319.46 \times 10^4 \text{ mm}^4$$

$$\text{Radius of gyration, } r = \sqrt{I_{xx}/A} = \sqrt{11{,}319.46 \times 10^4 / 10{,}560} = 103.5 \text{ mm}$$

$$\text{Effective length, } l = 0.7 \times 2500 = 1750 \text{ mm}$$

$$\lambda = \frac{l}{r} = \frac{1750}{103.5} = 16.90$$

From I.S: 800–1984, for $\frac{l}{r} = 16.90$ and $f_y = 250$ N/mm^2,

$$\sigma_{ac} = 148.62 \text{ N/mm}^2$$

$$\text{Safe load} = 148.62 \times 10{,}560 = 1569.4 \text{ kN} > 1272 \text{ kN}$$

which is sufficient.

Therefore, provide 200 mm × 20 mm plates as stiffeners.

Connection Let us provide 6 mm size intermittent fillet welds.

$$\text{Strength of weld/mm} = 0.7 \times 6 \times 1 \times 108 = 453.6 \text{ N/mm}$$

$$\text{Required strength of weld/mm } = \frac{1272 \times 10^3}{4 \times 2500} = 127.2 \text{ N/mm}$$

The length of intermittent fillet weld is taken as $10\,t = 10 \times 20 = 200$ mm. Therefore, provide a 200 mm long fillet weld.

$$\text{c/c spacing of fillet weld} = \frac{200 \times 453.6}{127.2} = 713.20 \text{ mm} > 300$$

$(16 \times t = 16 \times 20$ mm or 300 mm, whichever is less)

Hence, provide 200 mm long fillet welds at a spacing of 300 mm. The welds are staggered as shown in Fig. Ex. 9.2(c).

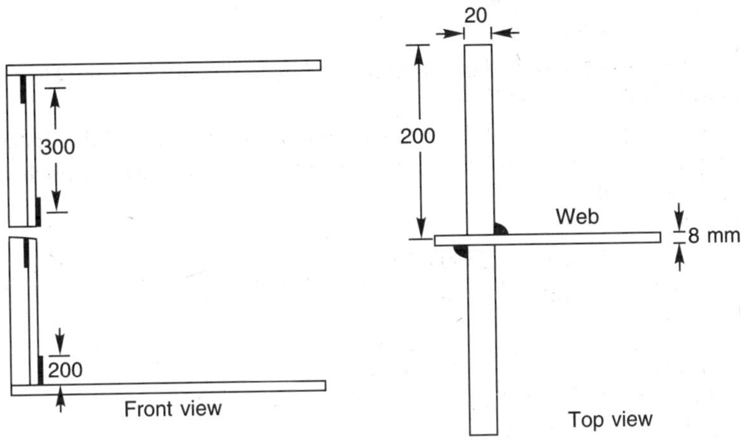

Fig. Ex. 9.2(c)

Intermediate stiffeners For an unstiffened web the minimum required thickness is as below:

(i) $$t_{w,\min} = \frac{d_1 \sqrt{\tau_{va,cal}}}{816}$$

$$\tau_{va,\,cal} = \frac{1272 \times 10^3}{2500 \times 8} = 63.6 \text{ N/mm}^2$$

$$t_{w,\min} = \frac{2500 \times \sqrt{63.6}}{816} = 24.43 \text{ mm}$$

(ii) $$t_{w,\min} = \frac{d_1 \sqrt{f_y}}{1344} = \frac{2500\sqrt{250}}{1344} = 29.41 \text{ mm}$$

(iii) $$t_{w,\min} = \frac{d_1}{85} = \frac{2500}{85} = 29.41 \text{ mm}$$

The web thickness provided is 8 mm, which is quite less than that required if it is to be unstiffened. Therefore, intermediate web stiffeners will be required.

Vertical stiffeners are provided if the web thickness as calculated below is more than the thickness provided.

$$d_2 = 2 \times \frac{2500}{2} = 2500 \text{ mm}$$

(i) $$t_{w,\,min} = \frac{d_2\sqrt{f_y}}{3200} = \frac{2500\sqrt{250}}{3200} = 12.35 \text{ mm}$$

(ii) $$t_{w,\,min} = \frac{d_2}{200} = \frac{2500}{200} = 12.5 \text{ mm}$$

The web thickness provided is 8 mm which is less than 12.5 mm. Hence vertical stiffeners will be provided.

A horizontal stiffener will be required at a distance of 2/5 of the distance from the compression flange to the N.A., if the web thickness as calculated below is more than that provided.

(i) $$t_{w,\,min} = \frac{d_2\sqrt{f_y}}{4000} = \frac{2500\sqrt{250}}{4000} = 9.88 \text{ mm}$$

(ii) $$t_{w,\,min} = \frac{d_2}{250} = \frac{2500}{250} = 10.0 \text{ mm}$$

Therefore, a horizontal stiffener will be provided at 500 mm $\left(\frac{2}{5} \times 1250\right)$ from the compression flange.

Another horizontal stiffener will be required at the N.A. if the web thickness as calculated below is more than the provided one.

(i) $$t_{w,\,min} = \frac{d_2\sqrt{f_y}}{6400} = \frac{2500\sqrt{250}}{6400} = 6.176 \text{ mm}$$

(ii) $$t_{w,\,min} = \frac{d_2}{400} = \frac{2500}{400} = 6.25 \text{ mm}$$

Web thickness provided is more than 6.25 mm. Hence the horizontal stiffener at N.A. is not required. Provide vertical stiffeners and a horizontal stiffener (at 500 mm from the compression flange).

Vertical stiffeners The smaller clear panel dimension $\not> 180\,t$

(i.e. 180 × 8 = 1440 mm)

Therefore, vertical stiffeners should be provided a spacing less than 1440 mm. Also, the spacing is chosen so as to provide a complete number of stiffeners between bearing stiffeners.

$$\text{Number of stiffeners} = \frac{24 \times 10^3}{1440} = 16.66 \approx 17$$

$$\text{Actual smaller clear panel dimension} = \frac{24 \times 10^3}{17} = 1411.76 \text{ mm}$$

$$\approx 1410 \text{ mm}$$

Provide vertical stiffeners at a spacing of 1410 mm. A horizontal stiffener will be provided at 500 mm from the compression flange.

$$\therefore \quad d = 2500 - 500 = 2000 \text{ mm}$$

$$\text{Spacing of stiffeners, } c = \frac{1410}{2000} = 0.705\,d \not> 0.70\,d \text{ (1400 mm)}$$

Therefore, provide vertical stiffeners at a spacing of 1400 mm.

$$\frac{d}{t} = \frac{2000}{8} = 250$$

From Table 6.6A of I.S: 800–1984, for $d/t = 250$ and $c = 0.70\,d$,

$$\tau_{va} = 74 \text{ N/mm}^2$$

$$> 63.6 \text{ N/mm}^2$$

which is all right.

Outstand of vertical stiffener $\not>$ 12 t, in case of flat sections.

$$\therefore \quad \text{Maximum outstand} = 12 \times 10 = 120 \text{ mm}$$

Minimum required web thickness from shear consideration,

$$t = \frac{1272 \times 10^3}{2500 \times 74.0} = 6.875 \text{ mm}$$

Let us try 110 I.S.F. 10 mm.

$$\text{Required moment of inertia} = \frac{1.5 d^3 t^3}{c^2} = \frac{1.5 \times 2000^3 \times 6.875^3}{1400^2}$$

$$= 198.90 \times 10^4 \text{ mm}^4$$

Moment of inertia of stiffener about the face of the web,

$$I_{xx} = \frac{10 \times 110^3}{12} + 10 \times 110.0 \times 55^2 = 443.66 \times 10^4 \text{ mm}^4$$

$$> 198.90 \times 10^4 \text{ mm}^4$$

which is as it should be.

Connections

$$\text{Shear force} = \frac{125 t^2}{h} = \frac{125 \times 8^2}{110} = 72.72 \text{ kN/m} = 72.72 \text{ N/mm}$$

Let us provide a 3 mm size the intermittent fillet weld.

$$\text{Strength of the weld/mm} = 0.7 \times 3 \times 1 \times 108 = 226.8 \text{ N/mm}$$

The length of the intermittent fillet weld is taken as $10t = 10 \times 10 = 100$ mm. Therefore, let us provide 100 mm long intermittent fillet welds.

$$\text{c/c spacing of welds} = \frac{2 \times 226.8 \times 100}{72.72} = 623.76 \text{ mm} > 160 \text{ mm}$$

$$(16t = 16 \times 10 = 160 \text{ mm or 300 mm, whichever is less})$$

Therefore, provide 3 mm size 100 mm long fillet welds at a centre-to-centre spacing of 160 mm on both sides of the web, (Fig. Ex. 9.2(d)).

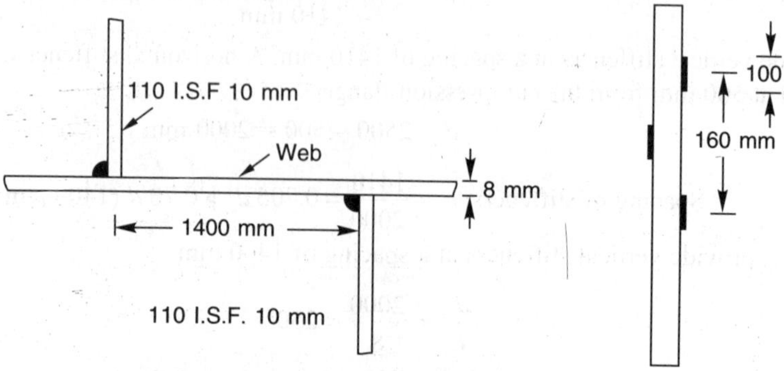

Fig. Ex. 9.2(d)

Horizontal stiffener A horizontal stiffener will be provided at 500 mm from the compression flange.

Let us try an 8 mm thick flat.

$$\text{Maximum outstand} = 12\,t = 12 \times 8 = 96 \text{ mm}$$

Try a flat section, 90 × 8 mm, as stiffener.

$$\text{Moment of inertia required} = 4\,c\,t^3 = 4 \times 1400 \times 6.875^3 = 181.90 \times 10^4 \text{ mm}^4$$

$$\text{Moment of inertia provided} = \frac{8 \times 90^3}{12} + 8 \times 90 \times 45^2 = 194.4 \times 10^4 \text{ mm}^4$$

$$> 181.90 \times 10^4 \text{ mm}^4$$

which is all right.

Connections Provide 3 mm size, 80 mm long fillet welds at a c/c spacing of 125 mm ($< 16 \times 8 = 128$ mm) (Fig. Ex. 9.2(e)).

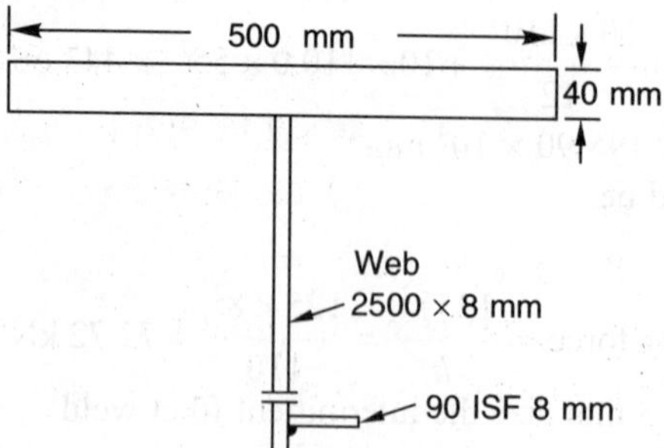

Fig. Ex. 9.2(e)

Exercises

Note Use the steel of yield stress 250 MPa if not specified in the problem.

9.1 A plate girder is made up of following elements.

Flange plates 400 × 20 mm one plate for each flange

Flange angles	$75 \times 75 \times 8$ mm	A pair of angles for each flange
Web	900×10 mm	

Flange plates are connected with flange angle legs by 18 mm ϕ, staggered power driven rivets in two rows. Flange angles legs are connected with web plates by 18 mm ϕ power driven rivets. Compute the sectional properties and the moment of resistance of the plate girder.

9.2 A welded plate girder has following elements.

Flange plates	400×16 mm	one plate for each flange
Web	$2000 \times 10 \times$ mm	

Compute the sectional properties and moment of resistance of the plate girder. Design also the bearing stiffeners, if the plate girder is to carry to uniformly distributed load of 120 kN/m.

9.3 Design horizontal and vertical stiffeners for the data given in Problem 9.2.

9.4 A welded plate girder consists of the following elements.

Flange plates	500×32 mm	one plate for each flange
Web	$2200 \times 8 \times$ mm	

The girder is subjected to a uniformly distributed load of 80 kN/m. It is also subjected to a lateral load of 2 kN at mid-height of the girder. Design the vertical stiffeners.

9.5 A plate girder consists of following elements.

Flange plates	450×12 mm	One plate for each flange
Flange angles	$100 \times 100 \times 10$ mm	A pair of angles for each flange
Web	$2000 \times 8 \times$ mm	

Design the following.

(a) Flange splice $M = 35000$ kNm
$V = 400$ kN

(b) Web splic M = 3000 kNm
$V = 300$ kN

9.6 Design a section of a plate girder to carry a uniformly distributed load of 1000 kN over a span of 10 m. A full lateral support is provided to the compression flange. Show the curtailment and also design the flange to web connections. Provide stiffeners, if required.

9.7 Design a plate girder spanning 14.0 m and loaded as shown in the Fig. Prob. 9.7. The self-weight of the plate girder may be assumed to be 60 kN distributed uniformly over the entire span. Use the steel of yield stress 250 MPa.

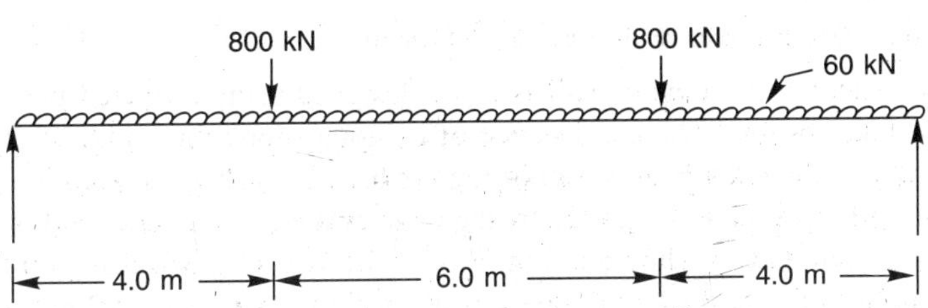

Fig. Prob. 9.7

9.8 Design a welded plate girder of span 30.0 m. It is subjected to a uniformly distributed load of 32 kN/m. Design also the stiffeners and their connections. Use the steel with yield stress 250 MPa.

9.9 It is proposed to construct a two storeyed structure as shown in the drawing (Fig. Prob. 9.9). The upper storey is designed for a live load of 4000 N/m^2 and is to be used for commercial purposes. The ground floor is to be used as a marriage hall. The arrangement of the structural framework is shown. Design the girders (f_y = 250 MPa) for the ground floor slab. The depth of the girder is restricted to 1.0 m. Ground floor height = 6.5 m, and first floor height = 5 m. Columns are shown by black dots.

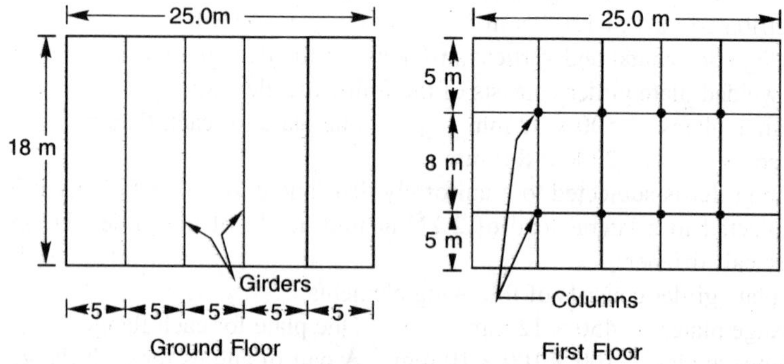

Fig. Prob. 9.9

9.10 Design a welded plate girder to carry a superimposed distributed load 50 kN/m. In addition to this the girder supports two concentrated loads of 600 kN each at the top flange at 1/3rd points from two secondary beams. The effective span of the girder is 15 m. The compression flange of the girder is laterally unsupported. Assume f_y =260 N/mm^2.

9.11 Design the most critical section of a welded plate girder of constant depth to carry a superimposed load of 100 kN/m in addition to its self-weight over a span of 20 m with two equal overhangs of 3 m on each side, thus making total length of the girder 26 m. Design the following:

1. Cross-section
2. Curtailment
3. Bearing stiffener
4. Intermediate stiffener

Note Use the steel of yield stress 260 N/mm^2.

9.12 The balcony of a cinema hall is to be designed using a riveted plated girder and raker beams. The c/c distance of columns along the width of the hall is 21 m and the raker beams supporting the balcony slab are spaced 3.5 m apart. one end of which is supported by the plate girder and the other end is attached to the column C as shown in Fig. Prob. 9.12. R.C.C. slab resting on the raker beam is 150 mm thick and the steps are 800 mm wide and 500 mm deep. The slab may be assumed to be continuous over the raker beams. Assume live load on the balcony as 4 kN/m^2 of the plan area. Design the following:

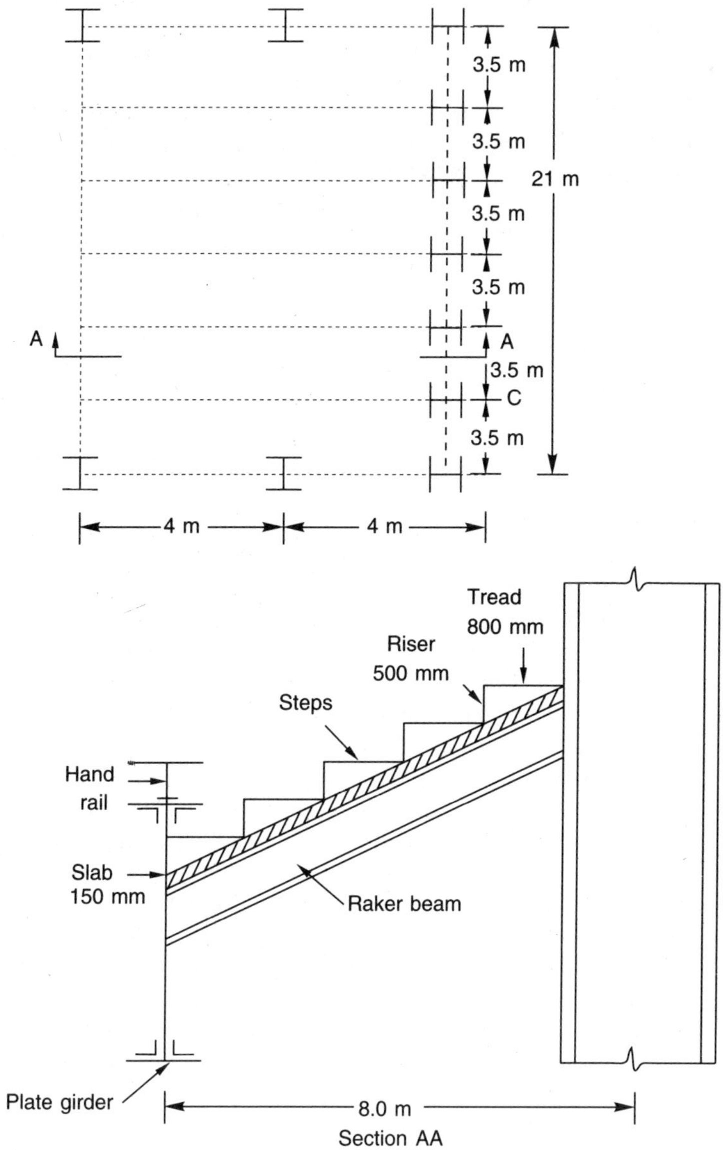

Fig. Prob. 9.12

(i) Raker beams
(ii) Mid-span section of the plate girder
(iii) Raker beam-plate girder connection.

Note Use the steel of yield stress 280 Mpa.

Eccentric Connections

10.1 INTRODUCTION

The design of fundamental structural elements viz. compression, tension and flexural members, and also the design of simple connections have been discussed in the previous chapters. In the frame of a steel building, a beam may be attached to another beam or to a column. In such cases, design of connections under a system of loads on the elements primarily depend on an understanding of the behaviour of the elements and their connections. The decision of the designer depends more upon this understanding rather than on sophisticated theoretical analysis. Therefore, a designer should be conversant with the different ways in which a connection can be made between beam to column and beam to beam, and the way it will behave under a given system of loads.

The beam column connections expected to resist and transfer end reactions only are termed as *shear connections or flexible connections*. These permit free rotation of the beam end and do not have any moment restraint. Other type of connections which do not permit any relative rotation between the beam and column and are expected to resist moments in addition to end reactions are termed as *moment connections* or *rigid connections*. An example of a 100 per cent rigid connection is a welded moment resistant connection. However, these connections do not behave as desired and are relatively flexible or rigid. A third type of connection which resists end moments and permits relative rotation between the beam and column is termed as a *semi-rigid connection*. Actually all beam to column riveted connections are of the semi-rigid type.

The moment-rotation relationship for different types of connections are shown in Fig. 10.1. This is a curve expressing the moment M transmitted by the connec-

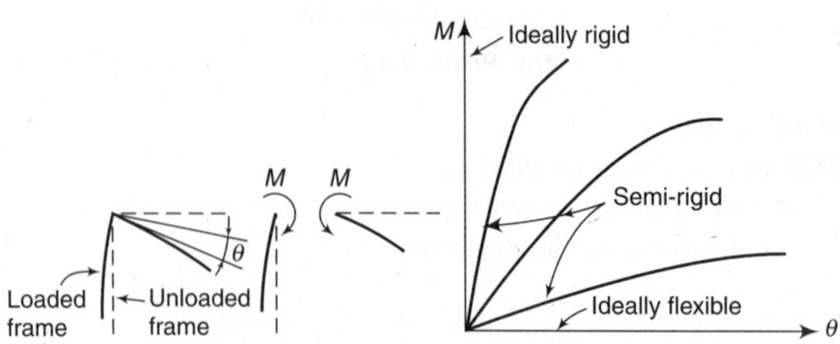

Fig. 10.1 *Theoretical Connection Characteristics*

tion as a function of the relative rotation θ of the elastic lines of the connected members at their point of intersection. An ideally rigid connection is one whose M-θ curve is a straight vertical line. Regardless of the moment acting on the joint, there will be no relative rotation between the two elastic lines. Likewise, an ideally simple connection is one having a horizontal M-θ curve. Any intermediate condition is semi-rigid. It will be appreciated that full rigidity are extreme conditions, never actually obtained. Generally, structures are designed as having rigid or simple joints only.

It is generally agreed that rigid connections are those which, when joined to absolutely fixed supports, would develop about 90 per cent or more of the fixed end moment. Similarly, simple connections are those which, when supported in the same fashion, would not develop more than about 25 per cent of the fixed end moment. Semi-rigid connections are in-between.

10.2 BEAM-COLUMN CONNECTIONS

Some of the ways to connect a beam with a column or with another beam are discussed as follows:

A beam may be supported over a bracket connected to a column. When the bracket is made up of two plates connected to the column flanges as shown in Fig. 10.2(a), it is called bracket connection-Type I. Figures 10.2(b) and (c) depict bracket connection-Type II wherein a bracket plate with a pair of angles one on each side of the bracket plate (Fig. 10.2(b)) or a bracket Tee is connected perpendicular to the coulmn flange (Fig. 10.2(c)). The joints in such cases transmit shear and moment due to eccentricity of the shear force. The moment may be in the plane of the connection, or in a plane perpendicular to the connection. In the bracket connection-Type I, the connectors are subjected to a shear force and additional shear due to torsion, whereas in bracket connection-Type II the connectors are subjected to a shear force and tension due to bending.

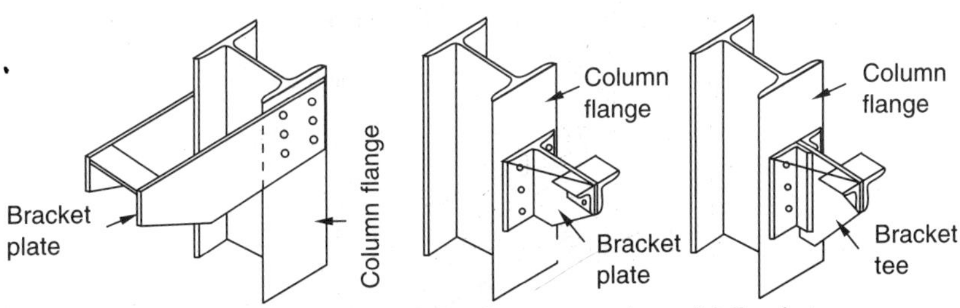

(a) Bracket Connection—Type I

(b) Bracket Connection—Type II (Using plate section)

(c) Bracket Connection—Type II (Using Tee-section)

Fig. 10.2 *Riveted Bracket Connections*

Another way could be to support the beam over an angle section connected to the column. In this case, one leg of the angle is used to make a seat for the beam and the other leg is connected to the column flange. Another angle, called cleat angle or clip angle, is provided on the top flange of the beam to hold it in position, as shown in Fig. 10.3(a). The cleat angle provides lateral stability to the compression flange at the ends by restraining the beam end against torsion. This connection is termed an *unstiffened seat connection* . When reactions are large, the seat angle may further be stiffened with angle section(s) and this connection is known as a *stiffened seat connection* (Fig. 10.3(b)).

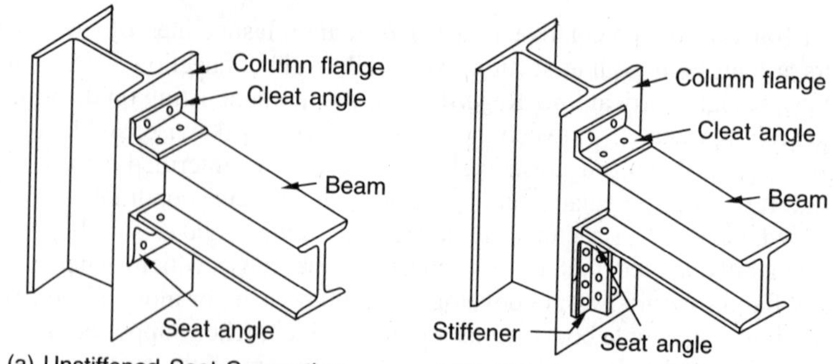

(a) Unstiffened Seat Connection (b) Stiffened Seat Connection

Fig. 10.3 *Riveted Seat Connections*

The web of the beam when connected to the column flange with angle sections (one on both side of the beam web) the connection is said to be a *framed connection* (Fig. 10.4). Seat angles may be provided to help in erection, but these do not transfer any load. When a beam frames at right angle to another beam, this connection is also called a *framed connection*. This is usually made with a pair of angles with their legs connected to the webs of the beams (Fig. 10.5). This connection is essentially designed for shear. Figure 10.5(a) shows a beam to beam connection when the compression flanges of the two beams are at different levels and Fig. 10.5(b) shows a beam to beam connection when the compression flanges of the two beams are at the same level. The essential condition of a frame connection is the connection of beam web with either of flange or web of column. All figures from 10.2 to 10.5 are made showing riveted or bolted connections only.

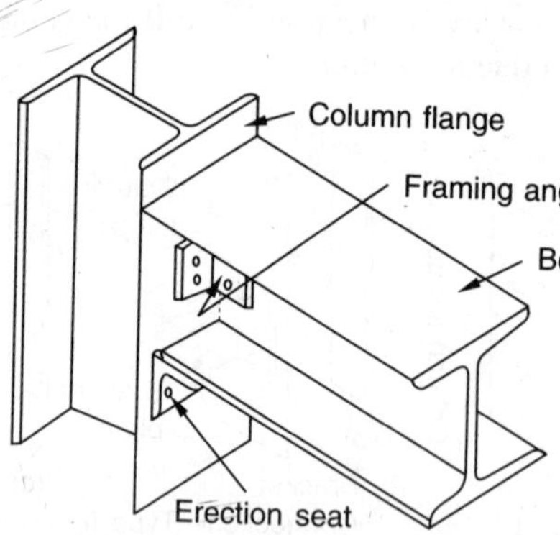

Fig. 10.4 *Riveted Framed Connection—Beam to Column*

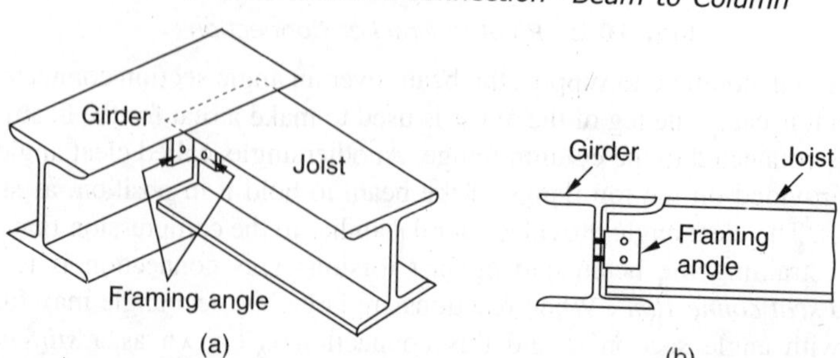

Fig. 10.5 *Riveted Framed Connection—Beam to Beam*

In the case of welded connections, the members may be welded directly to each other without any additional angles or plates. However, a direct connection requires butting of the sections together. This necessitates cutting of the members to an exact length. In the field it is difficult to achieve exact lengths and perfect fitting of the members simultaneously. Consequently, this generates undesirable and unpredictable stresses which may even change the behaviour of the connection, leading to its failure.

10.3 RIVETED SHEAR CONNECTIONS

Bracket Connections

Bracket type of connections are made whenever two members to be secured together do not intersect (Fig. 10.2). Structurally this is not a desirable type of connection because a pronounced eccentricity is introduced both in the connection and in one of the members to be joined.

Bracket Connection-Type I When the twisting moment is in the plane of connection the connection may be termed as a bracket connection Type I (Fig. 10.6(a)). The situation may arise when the line of action of load is in the plane of the riveted connection and the centre of gravity of the connection is the centre of rotation. The rivet group is subjected to shear and torsion.

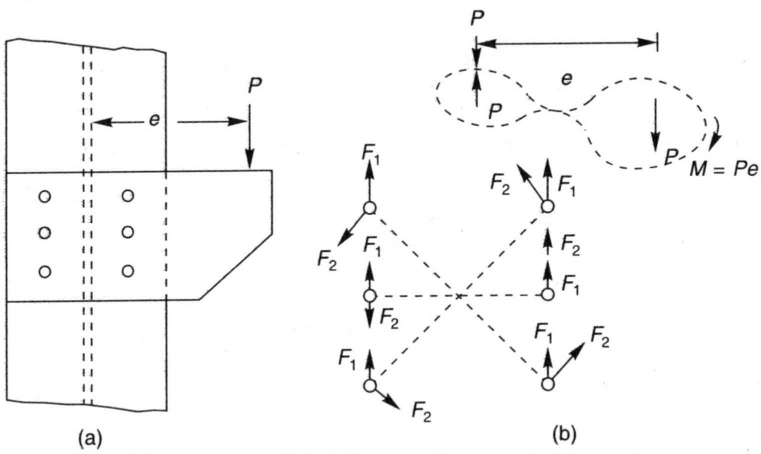

Fig. 10.6 *Forces on Rivets in Bracket Connection-Type 1*

Let M = torque in N mm, caused by the eccentric load
= load × eccentricity,

e = the perpendicular distance measured from the centre of gravity of the rivet group to the line of action of load, called eccentricity,

P = the eccentric load acting over the joint in N,

$r_1, r_2, r_3, \ldots, r_n$ = the distances of the rivets from the centre of gravity of the rivet group

n = number of rivets in the rivet group

F_1 = force in a rivet due to direct shear P

F_2 = force in a rivet due to torque M

K = constant.

The eccentric load P can be made concentric along with a torque M, as shown in Fig. 10.6(b). The forces in rivets due to direct shear and torque are F_1 and F_2 respectively as shown in Fig. 10.6(b). The total force in each rivet is the resultant of the direct shear and shear due to torque. This resultant force in each rivet should be less than the rivet value. From the assumption made in the concentric riveted joints, 'the load over the joint is shared equally by all the rivets', force F_1 in any rivet due to direct load is,

$$F_1 = \frac{P}{n} \tag{10.1}$$

Force F_2 in any rivet due to the torque is proportional to its distance from the centre of gravity of the connection.

$$F_2 \propto r \quad \text{or} \quad F_2 = Kr \tag{10.2}$$

or

$$K = \frac{F_2}{r} \tag{10.3}$$

Therefore, the torque about the C.G. of the rivet group

$$= F_2 r = Krr = Kr^2$$

$$\text{Total resisting torque} = \Sigma Kr^2$$

$$= K\Sigma r^2$$

$$= \frac{F_2}{r}\Sigma r^2 \tag{10.4}$$

This resisting torque should be equal to the torque over the connection. Hence,

$$M = \frac{F_2}{r}\Sigma r^2$$

or

$$Pe = \frac{F_2}{r}\Sigma r^2$$

or

$$F_2 = \frac{Per}{\Sigma r^2}$$

Force F_2 is maximum when distance r is maximum. Let the distance of the extreme rivet be r_n. Then,

$$F_2 = \frac{Per_n}{\Sigma r^2} \tag{10.5}$$

The two forces F_1 and F_2 act at some angle on various rivets in the connection. Let θ be the angle between these forces on the critical rivet. Then the resultant force F on the critical rivet can be computed by

$$F = \sqrt{F_1^2 + F_2^2 + 2F_1F_2 \cos\theta} \tag{10.6}$$

For the connection to be safe, this force F must be less than the rivet value.

From Eq. (10.6) it is evident that the resultant force F on a rivet depends upon F_1, F_2 and θ. For F to be maximum with F_1 being same for all the rivets and F_2 being maximum for the rivet farthest from the centre of gravity of the joint, the angle θ between the two forces has to be minimum.

Note A critical rivet is that which is subjected to the maximum resultant force. A close examination reveals that a rivet which is farthest from the centre of gravity of the rivet group and nearest to the applied load line is the most critical.

Design

1. To design a bracket connection Type I, the number and diameter of rivets for the connection are assumed and these are placed in two or more vertical rows at a suitable pitch and edge distance.
2. The resultant force on the critical rivet is worked out by Eq. (10.6).
3. The rivet value is computed. It should be more than the resultant force on the rivet.

Notes

1. For convenience the number of rivets required (in one vertical row) may be worked out from Eq. (10.7).

$$n = \sqrt{6M/n' s R_v} \tag{10.7}$$

where s is the pitch of the rivets and R_v is the rivet value. n' is the number of rows of rivets, e.g. a value of 2 is used for n' if rivets are provided in two vertical rows.
2. Equation (10.7) gives the number of the rivets for applied moment/torque. But the connection is subjected to shear along with the moment/torque. Therefore, the number of rivets may be revised, if required.

Bracket Connection-Type II When the moment is in a plane perpendicular to the plane of connection the bracket connection is termed as bracket connection-Type II. The line of action of load does not lie in the plane of the group of rivets and the line of rotation does not pass through the centre of gravity of the rivet group. The rivets are subjected to direct shear along with tension due to moment. Extra angle sections are connected to the bracket at the point where reactions are transferred (Fig. 10.7) to form a seat for the beam.

Let M = moment in N mm caused by the eccentric load

e = the eccentricity of the load P from the rivet plane to the line of action of load in mm

P = the load acting over the joint in N

M' = moment of resistance provided by rivets in tension

F_1 = force in the rivet due to direct shear P

F_2 = tensile force in the rivet due to bending moment

n = number of rivets in the rivet group

$Y_1, Y_2, Y_3, \ldots Y_n$ = distance of the rivets in tension from the axis of rotation

$\overline{Y}$ = the distance as shown in Fig. 10.7.

The eccentric load P can be made concentric along with a moment M (= Pe) as shown in Fig. 10.7. The rivets will be subjected to direct shear due to the load P and tension due to moment M. The critical (most highly stressed) rivet is investigated for combined stress due to direct shear and tension.

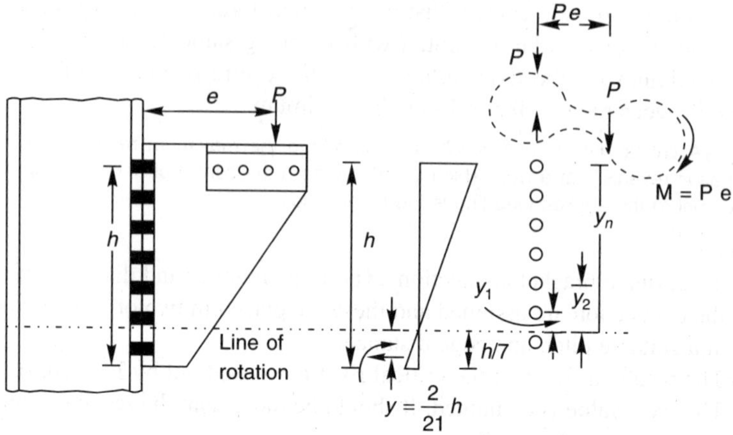

Fig. 10.7 *Bracket Type II connection*

From the assumption made in the concentric riveted joints, 'The load is shared equally by all the rivets', force F_1 in any rivet due to direct load, $F_1 = P/n$.

To estimate the amount of tension in the rivets, the position of N.A (line of rotation) has to be located first. The bracket section below the line of rotation presses against the compression flange of the column. Therefore, the bracket-section below the line of rotation will be in compression and the rivets above the line of rotation will be in tension (due to moment). Some designers assume the centre of rotation to be at the centre of the rivet group. In such a case the rivet above this line will be in tension. This assumption is not justified as the area of the bracket section below the centre of rotation will be in compression which will be more than the area of rivets in tension. Others assume the centre of rotation to be at the foot of the bracket. In such a case all the rivets will be in tension. This also is not justified as the bottom of the bracket and stanchion are not perfectly rigid. Therefore the correct centre of rotation lies somewhere between these two conditions. The exact calculation of the centre of rotation is a trial and error procedure, which involves lot of labour. Therefore, it is assumed that the bracket rotates about a line which lies at a distance of 1/7th of the depth of the bracket, from the base of the bracket. This depth is measured from the bottom of the bracket to the top most rivet in the connection. The rivets above the line of rotation will be in tension and in direct shear. The necessary compression to balance this tension will be provided by the bracket below the line. The tensile force in the rivet is proportional to the distance of the rivet from the line of rotation.

$$F_2 \propto Y \quad \text{or} \quad F_2 = KY \quad \text{or} \quad K = \frac{F_2}{Y} \tag{10.8}$$

Moment of resistance due to tensile force in a rivet,

$$= F_2 Y = KY^2 \tag{10.9}$$

Total moment of resistance due to tensile force,

$$M' = \Sigma KY^2 \quad \text{or} \quad M' = K\,\Sigma Y^2 \quad \text{or} \quad M' = \frac{F_2 \Sigma Y^2}{Y}$$

or
$$F_2 = \frac{M'Y}{\Sigma Y^2} \tag{10.10}$$

Tensile force in the extreme critical rivets is therefore,

$$F_2 = \frac{M'Y_n}{\Sigma Y^2} \tag{10.11}$$

Total tensile force in the riveted connection,

$$= \frac{M'\Sigma Y}{\Sigma Y^2} \tag{10.12}$$

For equilibrium,

Total tensile force = total compressive force

$$\text{Total compressive force, } C = \frac{M'\Sigma Y}{\Sigma Y^2}$$

External moment = moment resisted by rivets in tension
+ moment of the compressive force

or
$$M = M' + C\bar{Y}$$

$$= M' + \frac{M'\Sigma Y \times 2h}{\Sigma Y^2 \times 21}$$

$$= M'\left(1 + \frac{\Sigma Y \times 2h}{\Sigma Y^2 \times 21}\right)$$

or
$$M' = \frac{M}{1 + \frac{2h}{21} \times \frac{\Sigma Y}{\Sigma Y^2}} \tag{10.13}$$

Design

1. The diameter, pitch and edge distance of rivets is assumed and rivet value is computed.
2. The number of rivets required are determined from Eq. (10.7).
3. The shear force F_1 in the extreme critical rivet and the shear stress are calculated.
4. The tensile force F_2 in the extreme critical rivet is calculated from Eq. (10.11), and the tensile stress in the rivet is computed.
5. The connection is checked in combined tension and shear by using the following expression

$$\frac{\tau_{vf,\,\text{cal}}}{\tau_{vf}} + \frac{\sigma_{tf,\,\text{cal}}}{\sigma_{tf}} \le 1.4 \tag{10.14}$$

where $\tau_{vf,\,\text{cal}}$ = calculated shear stress in fastener
τ_{vf} = permissible shear stress in fastener
$\sigma_{tf,\,\text{cal}}$ = calculated tensile stress in fastener

σ_{tf} = permissible tensile stress in fastener.

Note Hot drivent rivets are not recommended for this type of connection. However, If hot driven rivets are used for making the connection, this type of a connection is avoided as the actual distribution of stresses in such rivets is not very certain. This is primarily due to the uncertain amount of initial tension existing in the rivets (because of shrinkage of hot driven rivets).

Seat Connections

Two types of seat connection are used, viz. unstiffened seat connection (Fig. 10.8) and stiffened seat connection (Fig. 10.9). In the unstiffened seat connection an angle is provided below the beam flange and is designed to transfer the end reactions of beam to the column through connections. The seat angles also provides a surface on which the beam flange may rest during erection, thus eliminating the need of erection bolts. Generally the outstanding (seating) leg is taken as 100 mm, which is sufficient to check the web crippling of the beam. A minimum of two rivets are provided for the connection of seat angle to beam flange. 10–15 mm clearance should be made between the beam and column. The seat is assumed to be flexible.

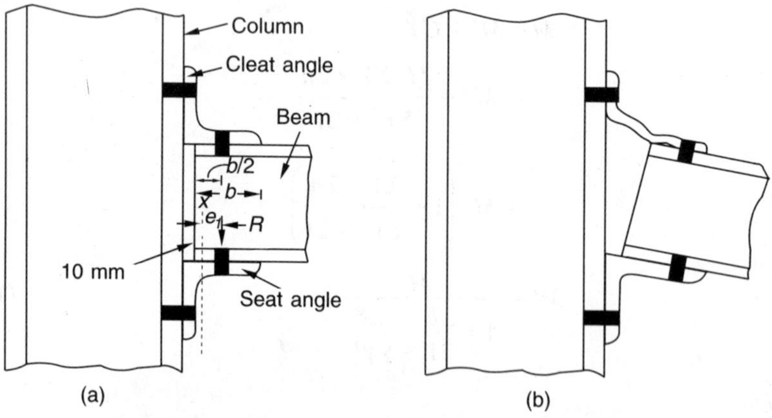

Fig. 10.8 *Riveted Unstiffened Seat Connection*

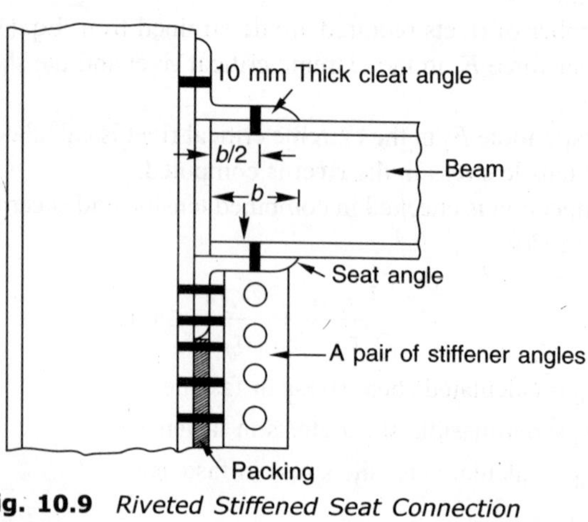

Fig. 10.9 *Riveted Stiffened Seat Connection*

An angle is provided at the upper flange of the beam and this is called *cleat* or *clip* angle. This provides lateral support to the compression flange of the beam at this point only and is assumed to carry no load. In the conventional design it is assumed that this type of connection does not restrain the rotation at the beam end in the vertical plane and the end reaction is the only force to be considered. But as shown in Fig. 10.8(b), the beam end rotates and exerts a moment on the connection, due to which the cleat angle is distorted and the rivets are elongated. In order to relieve the tension, the gauge should be large and the angle should be thin. This suggests that the cleat angle should be relatively flexible. When the beam tends to rotate, the cleat angle bends away from the column as well as the beam. Therefore, in the connectors of column to seat angle, no additional stresses are developed due to the eccentricity of the reaction. Flexible beam seats are the simplest and most desirable because the thickness of the seat angle provides the only resistance against bending in the outstanding leg.

In stiffened seat connection, the outstanding leg is stiffened by angle(s), thus the seat does not remain flexible. The outstanding leg of seat angle must be stiffened when the reaction to be transferred to the column is too large or when the seating leg of 100 mm cannot provide the required bearing area. Also, due to a large reaction the number of rivets required to join the connecting leg with the column may be large. In such cases more than two or three horizontal rows of rivets will be required. The rivets can be accommodated on an additional angle called a stiffener angle. Packings equal to the thickness of the seat angle are used between the column flange and the stiffener angle. Since the stiffener angle relieves the seat of the bending moments, it is desirable to have their outstanding legs extended as near as possible to the toe of the seat. However, it is difficult to cut the stiffener angle so as to make it fit tightly against the fillet of the seat; the result is that the effective width of the outstanding leg in bearing against the seat is understood to be its actual length less the fillet radius. In general, a pair of stiffener angles is provided and the outstanding legs are tack riveted. The rivets connecting the column flange to the stiffener angle are subjected to direct shear and bending moment (Re_{xx}). If the outstanding legs are not tack riveted the two stiffener angles act independently. In this case, the rivets connecting the stiffener angle with the column flange are designed for shear, bending moment (Re_{xx}) and twisting moment (Re_{yy}), where e_{xx}, e_{yy} are the eccentricities of the reaction parallel to the outstanding legs and connected legs respectively.

The procedure for the design of seated connections are outlined as follows.

Design of Unstiffened Seat Connections

1. A seat angle is chosen suitably on the following considerations:
   (a) The seat angle is assumed to have a length B, equal to the width of the beam flange.
   (b) Length of the outstanding leg of the seat angle may be calculated on the basis of the web crippling of the beam. The seat leg length is kept more than the calculated bearing length:

$$b = \frac{R}{\sigma_p \times t} - \sqrt{3}h_2 \qquad (10.15)$$

The equation may yield a negative value of b for a large beam with a small reaction. therefore a minimum bearing length is specified:

$$b \nless \frac{1}{2}\frac{R}{\sigma_p t}$$

where R = end reaction in N

σ_p = permissible bearing stress in MPa ($0.75f_y$)

t = thickness of the web of the beam in mm

h_2 = depth of the root of fillet from extreme fibre of flange for the beam in mm.

(c) The connected leg is so chosen that at least three horizontal rows of rivets can be accommodated. It can be assumed to be 150 mm or more as desirable.

(d) The thickness of the seat angle is chosen such that the outstanding leg does not fail in bending on a section at the toe of the fillet.

Let

e_1 = the distance from the critical section to the reaction (R) acting at the centre of the bearing length (Fig. 10.8(a))

$= 10 + \frac{b}{2} - t -$ radius of root of fillet

M = moment at the critical section (Re_1)

B = length of seat angle (equal to the width of beam flange)

σ_{bc} = bending stress in compression assumed to be equal to the bending stress in slab base, i.e. 185 N/mm^2.

$$\text{The bending stress} = \frac{M}{Z} = \frac{Re_1}{Bt^2/6} \tag{10.16}$$

or

$$\sigma_{bc} = \frac{6\,Re_1}{Bt^2}$$

or

$$t^2 = \frac{6\,Re_1}{B\sigma_{bc}}$$

or

$$t = \sqrt{6\,Re_1/(B\sigma_{bc})}$$

or

$$t = \sqrt{6\,Re_1/(B\sigma_{bs})} \tag{10.17}$$

(assuming $\sigma_{bc} = \sigma_{bs}$).

As the section of the angle is not known in the beginning the radius at the root of the fillet is also not known. Hence, eccentricity and thus the thickness cannot be worked out. The angle thickness is, therefore, assumed and the moment of resistance is computed, which should be more than the moment at the critical section.

2. Power driven rivets are used to connect the seat angle with the column flange. Two or three horizontal rows of rivets are provided. These are subjected to

direct shear forces. The rivet diameter is assumed and number of rivets required to connect the seat angle with the flange of column are determined. The outstanding leg of the seat angle is connected to the beam flange with two rivets of the same diameter as provided on the connected leg.

$$n = \frac{\text{end reaction}}{\text{rivet value}}$$

where n is number of rivets.

3. A cleat angle of nominal size is provided on the top flange of the beam and is connected by two rivets, on each of its legs, of the same diameter as provided on the seat angle.

Design of Stiffened Seat Connections

1. Size of the seat angle is assumed on the basis of the bearing length calculated from Eq. (10.15).
2. Suitable stiffener angle(s) is(are) selected. The bearing length is measured from the end of the stiffening leg. The outstanding leg of the stiffener angle must provide the bearing area required. Also, it should not exceed 16 times its thickness to avoid local buckling.

$$\text{Bearing area required} = \frac{R}{\sigma_p}$$

where R = end reaction,
σ_p = permissible bearing in MPa ($0.75 f_y$).

The thickness of the stiffener angle should not be less than the thickness of the web of the beam supported.

3. The seat is not flexible. Hence, the reaction is assumed to have a greater eccentricity. The rivets in the connecting legs are subjected to moments in addition to direct shear. The connection behaves as a bracket connection Type II. It is assumed that the reaction from the beam acts at the middle of the bearing length (the eccentricity in the case of a stiffened seat connection is greater as compared to an unstiffened seat connection, where the bearing length is measured from the supported end of the beam).
4. Bending moment about the face of the column flange is determined.
5. The number of rivets are computed and check similar to the bracket connection Type II is done.

$$\frac{\tau_{vf,\,\text{cal}}}{\tau_{vf}} + \frac{\sigma_{tf,\,\text{cal}}}{\sigma_{tf}} \le 1.4 \tag{10.18}$$

6. A cleat angle of nominal size is provided and is connected with rivets of the same diameter as on the seat angle. Two rivets are provided on each leg of the cleat angle.

Framed Connections

This type of connection is most frequently used to connect floor beams to other beams, girders or columns. The framing angles are used in pairs, i.e. one on each side of the web of the beam (Figs 10.4 and 10.5). One leg of the angle is connected to the web of the beam in the shop and is called the connected leg. The other leg of the angle is connected to the flange of the column or girder in the field and is called the outstanding leg. Standard sizes of framing angles are used, viz. 100 × 75 mm, 150 × 100 mm, 150 × 150 mm. Generally an 8 mm, 10 mm or 12 mm thick angle is used. The thickness is so selected that the rivet value is not governed by the bearing on framing angle legs. The thickness of the framing angles must also be limited to ensure necessary flexibility. As per I.S. specifications the thickness of the framing angle can be found by limiting the permissible shear stress to $0.4 f_y$ on the gross area of the vertical section through the connecting legs.

Design

1. The end reaction which the beam has to transfer is computed.
2. Connection of the framing angle with the beam web is designed. The connections are done in the shop and power driven rivets are used. The rivet diameter is assumed and the rivet value in double shear and bearing is computed. The number of rivets n, can be computed by

$$n = \frac{\text{end reaction}}{\text{rivet value}}$$

3. Power driven rivets are used to connect the outstanding leg of the angle with the web of the girder. The diameter of the rivets is assumed and the rivet value is computed in single shear and bearing.

   The number of rivets n' required to connect the outstanding leg with the flange of the beam are provided in one or more rows depending upon the space available:

$$n' = \frac{\text{end reaction}}{\text{rivet value}}$$

4. The size of the framing angle is governed by the number and rows of rivets on connected and outstanding legs.
5. The thickness of the angle can be computed as follows:

$$\text{Limiting shear stress, } 0.4f_y = \frac{\text{end reaction}}{2 \times h \times t}$$

or $$t = \frac{\text{end reaction}}{2h \times 0.4 f_y}$$

where h is the depth and t is the thickness of framing angle.

Note The depth h has been multiplied by 2 because a pair of angles is provided, one on each side of the beam web.

10.4 WELDED SHEAR CONNECTIONS

Welded joints are often subjected to loads which are eccentric. The welds in such cases are subjected to either shear and torsion or shear and bending. The bracket connection-Type I is an example of welds subjected to shear and torsion and the bracket connection-Type II is an example of welds subjected to shear and bending. Seated and framed connections may also be made as per the designer's choice.

Bracket Connections

Bracket Connection-Type I When the twisting moment is in the plane of the weld, i.e. the centre of gravity of the weld group lies in the plane of the line of action of the applied load, the weld is subjected to shear and torsion.

Let

M = twisting moment in N mm
P = the eccentric load acting over the joint
e = eccentricity of the load
t = effective throat thickness of the weld
l = length of the fillet weld $(2a + d)$ as shown in Fig. 10.10
d = depth of the bracket
r = the distance of the extreme weld from the centre of gravity of the weld group
τ_{vf} = shear stress in the weld
J = polar moment of inertia of the weld group
$= I_{xx} + I_{yy}$

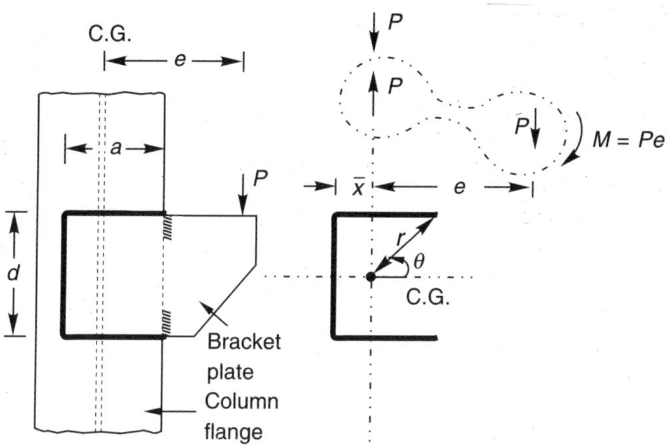

Fig. 10.10 *Welded Bracket Connection—Type I*

The eccentric load can be made concentric along with the twisting moment (Fig. 10.10). The weld group is subjected to direct shear stress due to the load P, and shear stress due to the twisting moment (Pe). The resulting stress should not be more than the permissible shear stress in the weld.

$$\text{Direct shear stress, } \tau_{vf_1} = \frac{\text{load}}{\text{effective area of weld}} = \frac{P}{(2a+d)t} \qquad (10.19)$$

The shear stress due to twisting moment can be computed from the torsion equation

$$\frac{T}{J} = \frac{\tau}{r}$$

$$\tau = \frac{Tr}{J}$$

$$\Rightarrow \qquad \tau_{vf_2} = \frac{Per}{J} \tag{10.20}$$

The resultant shear stress is obtained by adding the two stresses, as found above, vectorially.

$$\tau_{vf,\,\text{cal}} = \sqrt{(\tau_{vf_1})^2 + (\tau_{vf_2})^2 + 2\,\tau_{vf_1}\,\tau_{vf_2}\cos\theta} \tag{10.21}$$
$$< \tau_{vf}\,(108\ \text{N/mm}^2).$$

Design In the design of fillet welds for bracket connection Type I the overlap of the bracket plate is assumed (so the length of weld is fixed) and then the throat thickness or the size is computed. If required, the overlap can be readjusted to obtain the workable size of the weld. However, the size of weld can also be assumed and the overlap can be computed but difficulty in calculations will be there as the centre of gravity of the weld group, I_{xx}, I_{yy}, etc. will be in terms of the length, making the calculations cumbersome. The design of the connection is done as follows:

1. The overlap of the bracket plate is assumed and the length of the fillet weld is determined.
2. The distance of the centroid of the weld group from the edge of bracket plate is calculated. Let it be $\bar{x}$.
3. The polar moment of inertia of the weld group is computed.
4. The distance of the extreme weld from the centre of gravity of the weld group, r is calculated.
5. The shear stress τ_{vf_1} and τ_{vf_2} are computed from the Eqs (10.19) and (10.20).
6. The resultant shear stress is calculated from Eq. (10.21) and is equated to the allowable shear stress (108 N/mm^2). This gives the throat thickness of the weld.
7. The weld size can be found from the throat thickness by $t = KS$, where K is a constant 0.7 and S is the size of the weld.

Bracket Connection-Type II When the moment is in a plane perpendicular to the weld, i.e. the centre of gravity of the weld group lies in a plane perpendicular to the plane of line of action of the applied load, the weld is subjected to shear and bending. The eccentric load can be made concentric along with a bending moment (Fig. 10.11). The weld in such a case is subjected to direct shear stress due to a

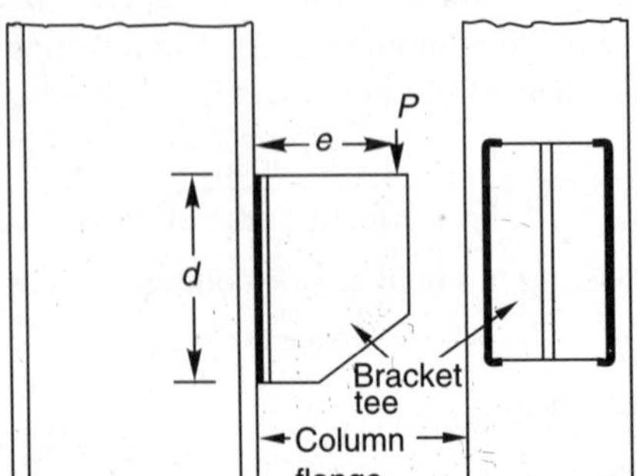

Fig. 10.11 *Welded Bracket Connection—Type II*

concentric load P and bending stress due to a bending moment (Pe). Either a butt weld or a fillet weld can be used in such situations.

$$\text{The direct shear stress in the weld} = \frac{\text{load}}{\text{effective area of weld}}$$

In the case of,

(i) Fillet weld, $$\tau_{vf,\,\text{cal}} = \frac{P}{2dt} \tag{10.22}$$

(ii) Butt weld, $$\tau_{vf,\,\text{cal}} = \frac{P}{dt} \tag{10.23}$$

$$\text{The bending stress in the weld} = \frac{\text{moment}}{\text{section modulus}}$$

In the case of,

(i) ***Fillet Weld*** When fillet welds are subjected to flexure the common relationship, $\sigma_b = My/I$ may be used to evaluate the stresses. The actual failure, and therefore the critical stresses exist on the throat of the weld (i.e. at 45° to the weld leg lengths). Thus,

$$\sigma_b = \frac{Pe \times \left(\frac{d}{2}\right)}{2 \times \left(\frac{td^3}{12}\right)} = \frac{Pe}{2 \times \left(\frac{td^2}{6}\right)}$$

The throat stress is then treated as a shear, since a 45° line of failure is indicative of a shear fracture. This horizontal shear stress is assumed to vary linearly from zero to mid-depth to maximum at the extreme fibres. Therefore,

$$\tau_{vf_1,\,\text{cal}} = \frac{Pe}{2 \times \frac{td^2}{6}} \tag{10.24}$$

(ii) ***Butt Weld***, $$\sigma_{b,\,\text{cal}} = \frac{Pe}{\frac{1}{6} \times td^2} \tag{10.25}$$

The combined stress in the fillet weld is given by the following equation,

$$\sqrt{\tau^2_{vf,\,\text{cal}} + \tau^2_{vf_1,\,\text{cal}}} < \tau_{vf}\ (108\ \text{N/mm}^2) \tag{10.26}$$

The combined bending and shear stress in the butt weld may be checked by the interaction formula,

$$\sqrt{\sigma^2_{b,\,\text{cal}} + 3\tau^2_{vf,\,\text{cal}}} < \sigma_e \tag{10.27}$$

where σ_e = equivalent stress (taken as equal to the yield stress of steel used).

Design In the design of such welded connections the size of the weld may be assumed and the length is computed. If the length of the weld is more than twice the depth of the bracket, the size of the weld is revised.

1. The size of the weld is assumed and the throat thickness is determined. The depth of the bracket is calculated from either of the following appropriate equations.
   In case of butt weld,
   $$d = \sqrt{6M/t\sigma_b} \tag{10.28}$$
   where $\sigma_b = 165$ N/mm^2.
   In case of fillet weld,
   $$d = \sqrt{6M/2t\tau_{vf}} \tag{10.29}$$
   where $\tau_{vf} = 108$ N/mm^2.
2. The direct shear stress is computed in terms of throat thickness from Eqs (10.22) or (10.23), as appropriate.
3. The stress due to bending moment is computed in terms of throat thickness from Eqs (10.24) or (10.25), as appropriate.
4. The equivalent stress is computed from Eqs (10.26) or (10.27), as appropriate.

Seat Connections

Unstiffened Seat Connection In the welded unstiffened seat connection two angles are used. One angle is welded to the column in the shop and forms the beam seat. A cleat angle of nominal size 100 × 100 × 6 mm is welded to the top of beam in the shop and to the column in the field. Here reference may be made to Fig. 10.12. The welded seat connection design is similar to the design of the riveted seat connection. Vertical welds are provided to connect the seat angle and these are turned at the ends. The resultant stress in the vertical weld is not uniform. Some codes recommend the neutral axis to be situated at the mid depth of the angle and others recommend the neutral axis to be at 1/3rd of the depth of seat angle from the bottom of the leg of the angle, connected to the column. The I.S. code recommends the former one. The resultant shear, comprising of horizontal shear per mm and vertical shear per mm, is computed and should be less than the allowable shear per mm of the weld.

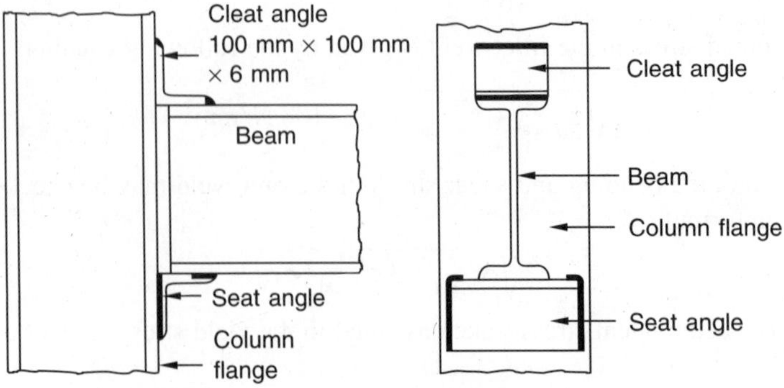

Fig. 10.12 *Welded Unstiffened Seat Connection*

Design The seat angle is designed in the same way as that discussed for unstiffened riveted seat connection. Welded connection is designed as follows.

1. The vertical shear per mm is calculated by $\tau_{vf_1} = \dfrac{R}{2d}$ $(t = \text{unity})$
2. The eccentricity of the reaction is determined and the bending moment is computed.
3. The horizontal shear per mm due to the bending moment is calculated,

$$\tau_{vf_2} = \frac{M}{2 \times \frac{1}{6} \times d^2 \times 1}, \ (t = \text{unity})$$

where d = length of the weld.

4. The resultant shear per mm will be,

$$\tau'_{vf} = \sqrt{\tau_{vf_1}^2 + \tau_{vf_2}^2}$$

5. This resultant shear per mm is equated to the strength of the weld per mm to find the size of the weld.

$$\tau'_{vf} = 0.7 \times S \times 1 \times \tau_{vf}$$

Stiffened Seat Connection In the stiffened seat connection, a T-Section built-up of two plates is used (Fig. 10.13). The bearing length of the seat plate is calculated as in the case of the unstiffened seat connection. The thickness of the seat plate is kept equal to the thickness of the flange of the beam and that of the stiffening plate is kept equal to the thickness of the web of the beam. The depth of the stiffening plate is decided depending upon the length of the vertical weld required. The seat plate is kept wider than the flange of the beam by at least twice the size of the weld on each side of the beam flange to facilitate welding.

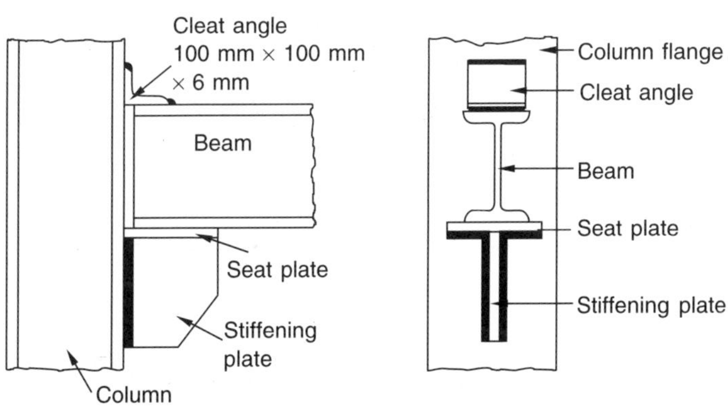

FIg. 10.13 *Welded Stiffened Seat Connection*

Design The design of the stiffened welded seat connection is almost similar to the design of the unstiffened welded seat connection except that the seat angle section is replaced by plate sections. The design is done as follows:

1. The width of seating plate is calculated from Eq. (10.15) in the same way as in the riveted unstiffened seat connection.

2. The thickness of the seat plate is assumed equal to the thickness of beam flange.
3. The thickness of the stiffening plate is assumed equal to the thickness of the web of the beam.
4. The eccentricity of load, and moment due to it are calculated.
5. Vertical and horizontal shears per mm length of weld τ_{vf_1} and τ_{vf_2} are computed.
6. The resultant shear per mm of weld length τ_{vf} is computed by the vector sum of τ_{vf_1} and τ_{vf_2}.
7. The size of the weld required to make the connection is calculated by equating the strength of the weld per mm length to the resultant shear per mm length.

Framed Connections

Beams are connected to the column or to the other beams by means of framing angles or plates. A pair of angles or plates are placed one on each side of the web of the beam to be connected. These angles or plates are used to transmit shear and moment. One of the legs of the angle is welded to the beam in the shop and the other leg is welded to the column in the field. Seats are only provided for erection and can be removed afterwards. The seat angle is set back from the beam web by 10–12 mm as shown in Fig. 10.14. The welds connecting the angles to the column are subjected to vertical shear ($V = R/2$) and to bending moment $(R/2)e_1$. It is assumed that the N.A. passes at a distance $d/6$ from the top of the angle where d, is the depth of the angle section. Hence, the maximum horizontal shear in the field weld occurs at the bottom of the angle section. The horizontal shear τ_{vh} will be concentrated at the centre of gravity of the triangle, i.e. at $(1/3)d$. Since the couple produced by these forces must be equal and opposite to the external moment, the value of τ_{vh} can be determined as follows:

$$\left(\frac{1}{2} \times \frac{5}{6}\, d\tau_{vh}\right)\frac{2d}{3} = \frac{R}{2} \times e_1$$

or

$$\tau_{vh} = 1.8 \times \frac{Re_1}{d^2} \tag{10.30}$$

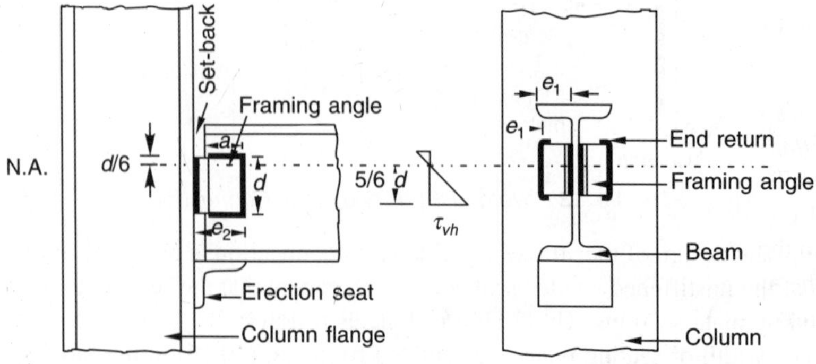

Fig. 10.14 *Welded Framed Connection*

$$\text{The vertical shear per mm, } \tau_{vs} = \frac{V}{d}$$

$$\text{Resultant shear, } \tau_v = \sqrt{(\tau_{vs})^2 + (\tau_{vh})^2}$$

The welds connecting the angle with the web of the beam are subjected to vertical shear $\left(V = \frac{R}{2}\right)$ and bending moment $\left(\frac{R}{2}e_2\right)$.

$$\text{Horizontal shear, } \tau_{vh} = \frac{Tr}{J} = \frac{Ve_2 r}{J}$$

and

$$\text{Vertical shear, } \tau_{vs} = \frac{V}{\text{effective area of weld}} = \frac{V}{dt + 2at}$$

where r is the extreme distance of the weld from the centre of gravity of the weld group and a and d are the weld lengths, and e_1 and e_2 are the eccentric distances as shown in Fig. 10.14.

$$\text{Resultant shear, } \tau_v = \sqrt{(\tau_{vs})^2 + (\tau_{vh})^2}$$

Design

1. The vertical reaction V, on each weld is calculated.
2. Vertical shear/mm length and horizontal shear/mm length for the weld connecting the framing angle leg to the column are computed and the resultant shear τ_v is determined.
3. The size of the weld is determined by equating the strength of weld/unit length to the computed value of τ_v.
4. The centre of gravity of the weld group to connect the angle section with the web of the beam is found.
5. The polar moment of inertia is calculated.
6. The twisting moment is computed.
7. The horizontal and vertical shear in the weld are calculated. The resultant shear is determined and equated to the allowable shear to obtain the size of weld.

10.5 MOMENT RESISTANT CONNECTIONS

Connections which are designed to transmit end moments in addition to the end shears are called *moment connections*. Such connections are also termed as *rigid connections*. These do not permit any relative rotation between the beam and the column. The moment resistant connections may be classified as riveted/bolted or welded connections. Following may be the cases where consideration has to be given to the end moments:

(a) the end reaction of the beams are eccentric, giving rise to end moments, and
(b) girders subjected to wind forces in multistorey buildings.

Riveted Moment Connections

Depending upon the amount of end moments to which the connection is subjected the riveted moment connections are divided into light moment connections and heavy moment connections. These are discussed in the articles to follow.

Light Moment Connections Connections subjected to small end moments and large end shears are called light moment connections. Four angles (two pairs) are used to make the connections. One pair of angles is used to connect the web of the beam with the column flange (one angle on each side of the web) and the other pair of angles called *clip angles*, is placed one on top of the beam flange and the second below the other flange of the beam (Fig. 10.15). If a clockwise end moment acts on the connection shown in Fig. 10.15, the rivets designated as 1 are subjected to tension and those designated as 2 and 4 to shear. Similarly, if an anticlockwise end moment acts on the connection, rivets 3 will be subjected to tension, and 2 and 4 to shear.

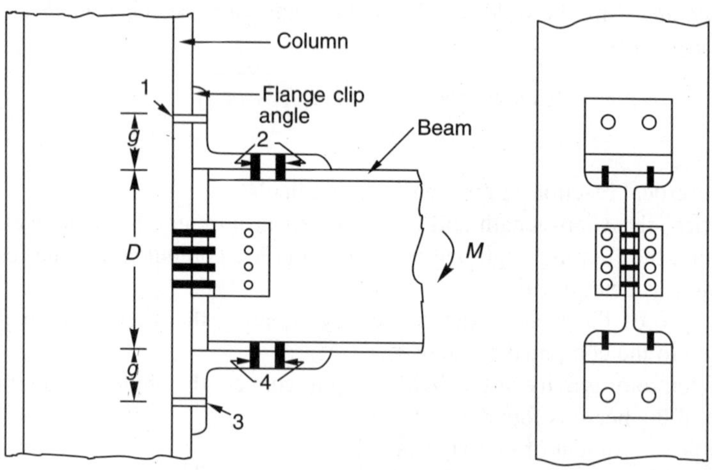

Fig. 10.15 *Clip Angle Connection*

Following assumptions are made in the analysis of small moment resistant connections:

1. The rivets due to end moment elongate under tension above the yield point, and the outstanding leg (leg connected to the flange of column) of the clip angle bends as a cantilever in single flexure. The rivets used in connection are assumed to have no initial tension (Fig. 10.16(a)).
2. The outstanding leg of the clip angle is held flat against the column flange due to initial tension in the rivets, and bends in double flexure (in a reverse curve) as shown in Fig. 10.16(b).

   Initial tension is developed in the rivets when the hot driven rivets cool down. Such rivets are not free to contract and large tensile forces are developed in them before they reach normal temperature. These rivets firmly clamp the sections jointed. Initial tension is, however, insignificant in cold driven rivets, unfinished bolts and turned bolts.

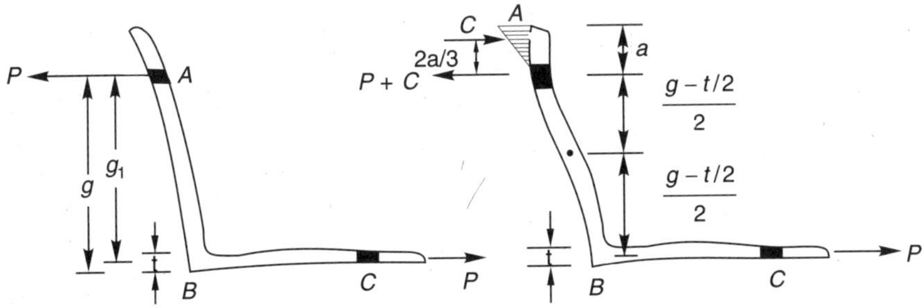

Fig. 10.16 *Clip Angle Flexure*

3. The angles provided on the beam web do not resist any bending moment and their connections transmit end shear only.
   In the case of single flexure (Fig. 10.16(a))

$$M = Pg_1$$

or
$$M = P(g - t/2) \tag{10.31}$$

In the case of double flexure, let us represent an angle section of Fig. 10.16(b) with a line diagram (the central line of an angle section) *ABC* (Fig. 10.17). *A* and *C* represent the centres of rivet holes and *B* represents the heel of the angle section.

Let g_1 = distance of rivet from centre line of angle leg *AB*, $\left(g - \frac{t}{2}\right)$

g_2 = distance of rivet from centre line of angle leg *BC*, $\left(g - \frac{t}{2}\right)$

where *g* is the gauge distance and *t* is the thickness of the angle section.

The leg *AB* (Fig. 10.17(b)) may be assumed as a cantilever beam with a moment M_B and a shear *P* at the end *B*.

$$\theta_B = \frac{Pg_1^2}{2\,EI} - \frac{M_B\, g_1}{EI} \tag{10.32}$$

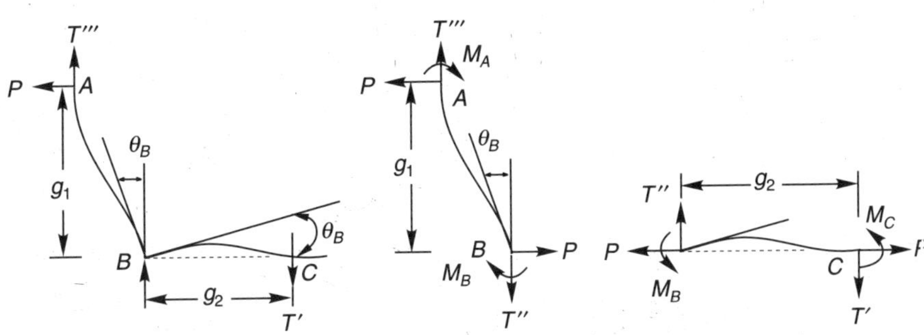

Fig. 10.17

The leg BC may be assumed to be a simple beam, subjected to moments M_B and M_C at the two ends B and C, respectively.

$$\theta_B = \frac{M_B g_2}{3EI} - \frac{M_C g_2}{6EI}$$

Since $\theta_C = 0$ and $M_C = \frac{M_B}{2}$,

$$\theta_B = \frac{M_B g_2}{3EI} - \frac{M_B}{2} \times \frac{g_2}{6EI}$$

or
$$\theta_B = \frac{M_B g_2}{4EI} \tag{10.33}$$

From Eqs (10.32) and (10.33),

$$\frac{Pg_1^2}{2EI} - \frac{M_B g_1}{EI} = \frac{M_B g_2}{4EI}$$

or
$$M_B = \frac{2Pg_1^2}{4g_1 + g_2}$$

Also,
$$M_A + M_B = Pg_1 \tag{10.34}$$

or
$$M_A + \frac{2Pg_1^2}{4g_1 + g_2} = Pg_1$$

or
$$M_A = \frac{Pg_1(2g_1 + g_2)}{4g_1 + g_2}$$

or
$$M_C = \frac{M_B}{2}$$

or
$$M_C = \frac{\frac{1}{2} \times 2 \times Pg_1^2}{4g_1 + g_2}$$

when
$$g_1 = g_2 = \left(g - \frac{t}{2}\right)$$

$$M_A = 0.6P\left(g - \frac{t}{2}\right) \tag{10.35}$$

$$M_B = 0.4P\left(g - \frac{t}{2}\right) \tag{10.36}$$

$$M_C = 0.2P\left(g - \frac{t}{2}\right) \tag{10.37}$$

The angle section should be of sufficient thickness to resist the maximum moment, which is found to be at point A.

Note

1. In the analysis it is assumed that the bent portion of the angle starts from the centre of the rivets, whereas a large portion of the angle is clamped by the rivet heads. Therefore, Eq. (10.35) overestimates the bending moment and the connection designed will be more rigid. Hence, for the design it may be assumed that the point of contraflexure lies at mid-height and the maximum bending moment will be,

$$M' = 0.5P\left(g - \frac{t}{2}\right) \tag{10.38}$$

2. In the case of double flexure, the design moment for the clip angle is reduced to 50% as compared to single flexure. But the tension T, on the rivets is increased from P to $P + C$ where P is the pull on the rivets, C is the compression under the toe of the outstanding leg of the flange clip angle to keep the far end of the leg in contact with the flange of the column, and T is the pull for which the rivets must be designed.

   For equilibrium,

$$C \times \frac{2}{3}a = 0.5 \times P\left(g - \frac{t}{2}\right)$$

or

$$C = 0.75\,P \times \frac{g - \frac{t}{2}}{a}$$

   Hence, the pull on the rivets,

$$T = P + C = P + 0.75P \times \frac{g - \frac{t}{2}}{a}$$

or

$$T = P\left[1 + 0.75 \times \frac{\left(g - \frac{t}{2}\right)}{a}\right]$$

3. If we assume single flexure the maximum bending moment for which the angle is designed will be twice the bending moment in double flexure. This results in thicker sections, difficult to punch and are therefore uneconomical.
4. From the free body diagrams of the two legs AB and BC of the angle, it can be observed that the tension T' at C is balanced by force T'' at B and T''' at A.

The rivets connecting the clip angle leg to the flange of the column are subjected to tension only. The shear on such rivets can be neglected as it is balanced. The rivets connecting the clip angle leg to the flange of the beam are subjected to shear only. The tension in such rivets can be neglected as it is balanced.

Limitations Clip angle connections have following limitations, due to which the number of rivets that can be provided to connect the clip angle with the column flange are limited to two only and thus the connection can transfer limited moments.

1. Only two rivets can be provided in one gauge line to connect the outstanding leg of clip angle with the column flange. If more rivets are provided in one gauge line, the rivets will be subjected to non-uniform forces (Fig. 10.18(a)).
2. Rivets can be provided in only one gauge line. If more than one gauge line is provided, rivets will be subjected to non-uniform forces (Fig. 10.18(b)).

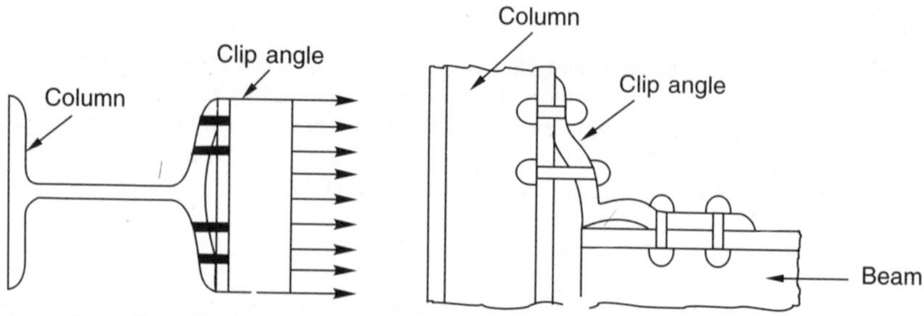

(a) More Than Two Rivets in One Row (b) More Than One Row of Rivets

Fig. 10.18

Design The clip angle connection is designed as follows:

1. The diameter of rivets is assumed and the value of rivet in tension, bearing and shear is determined.
2. As only two rivets can be provided over the clip angle outstanding leg, the maximum tension resisted by the connection will be twice the value of the rivet in tension.
3. The size of the angle and the gauge distance of the clip angle outstanding leg are assumed. The pull to be transmitted by the rivets on the outstanding leg,

$$P = \frac{M}{\text{lever arm}}$$

$$< 2 \times \text{value of rivet in tension}$$

where P is the pull to be transmitted by rivets in tension, M is the end moment, and lever arm is the depth of beam + 2 times gauge distance. If the effect of initial tension in rivets is also considered, the rivets are designed for maximum tension,

$$T = P\left[1 + 0.75 \times \frac{g - \frac{t}{2}}{a}\right]$$

4. The maximum moment induced in the clip angle is calculated with the assumed thickness of clip angle by $M' = 0.5P(g - t/2)$
5. The moment of resistance of the clip angle is given by,

$$M_R = \sigma_{bct} Z = \sigma_{bct} \times \frac{1}{6} lt^2$$

where

σ_{bct} = bending stress (it is taken equal to bending stress in slab base = 185 N/mm^2)

l = length of the clip angle
≈ c/c distance between the two rivets
= rivet spacing on column flange from I.S. Handbook No. 1

t = thickness of angle.

6. The thickness of the angle section can be computed by equating the moment M' to moment of resistance M_R,

$$M' = M_R$$

or $$0.5\,P\left(g - \frac{t}{2}\right) = \sigma_{bct}\,\frac{1}{6}\,lt^2$$

$$= \sigma_{bs}\,\frac{1}{6}\,lt^2 \text{ (assuming } \sigma_{bct} = \sigma_{bs}\text{)}$$

or $$t = \sqrt{6 \times 0.5\,P(g - t/2)/(\sigma_{bs}\,l)}$$

or $$t = \sqrt{6M'/(l\sigma_{bs})}$$

7. The horizontal shear in rivets connecting clip angles to the flange of beam will be, $V = \dfrac{\text{moment}}{\text{depth of beam}}$.
8. The number of rivets required are determined by dividing the horizontal shear by the rivet value. A minimum of 4 rivets (in two rows) are provided to make a rigid connection.
9. The angles connecting the web of the beam to the column flange are designed to transmit end shear. The size of the angle section will depend upon the number of rivets required to connect the two legs of the angle with the web of the beam and flange of the column.
10. The rivets connecting angle with the beam web will be in double shear. The rivet value is found and the number of rivets required to transfer the end shear are calculated by,

$$n = \frac{\text{end shear}}{\text{rivet value}}$$

11. The rivets connecting the angle with the flange of the column will be in single shear. The rivet value is found and the number of rivets required to transfer the end shear are calculated by

$$n' = \frac{\text{end shear}}{\text{rivet value}}$$

Heavy Moment Connections The moment carrying capacity of clip angle connections is limited. When the connections are subjected to heavy moments any of the following types of connections may be provided:

1. The split beam connection, and
2. The bracket connection.

Split Beam Connections Split beams are made by cutting I-Sections at the mid of the web (T-Sections may also be provided to make the connection) as shown in Fig. 10.19(a). The bending moment in split beam connection is resisted by the rivets on split beams and the end shear is resisted by the rivets on the web clip angles. A limitation of the split beam connection is that only four rivets can be used to resist the tension, the reason being the same as discussed in the case of a clip angle

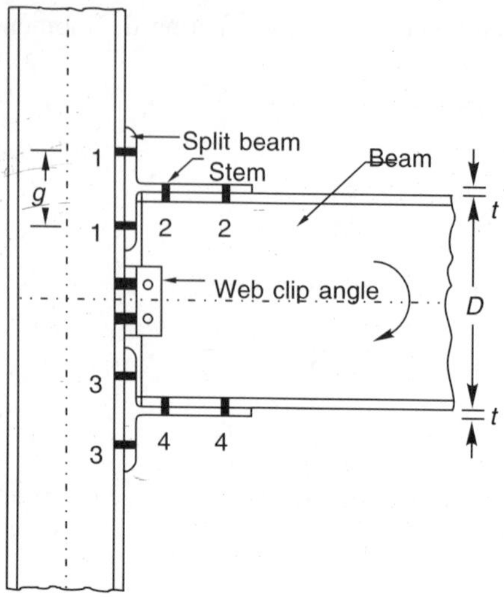

(a) Split Beam Connection

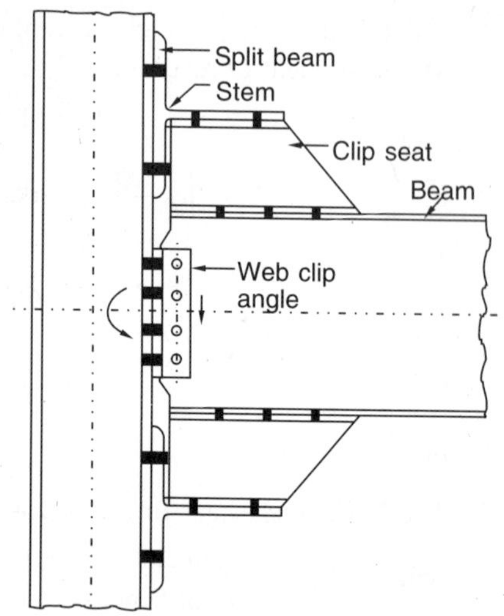

(b) Split Beam with Clip Seat

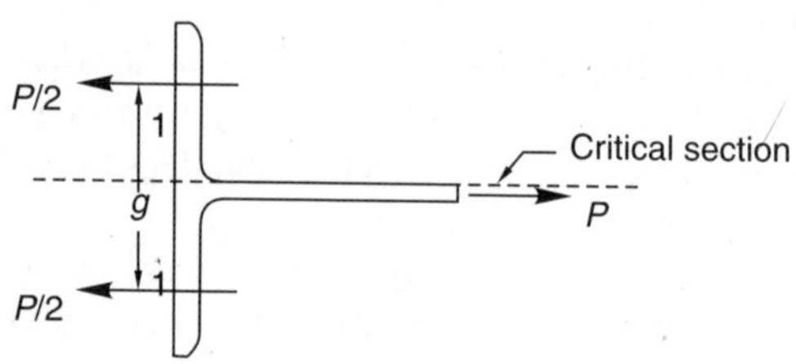

Fig. 10.19

connection. When the end moment produces a pull greater than the value of the four rivets in tension, the split beam connection can be modified by introducing flange clip seats as shown in Fig. 10.19(b). The flange clip seat is usually an I-Section. As the lever arm increases, the value of the pull to be transmitted by the split beam is reduced. Therefore, less rivets are required to transmit the same moment.

When a clockwise moment acts on the connection, rivets, 1, 1 are subjected to tension and rivets 2, 2 to shear. When an anticlockwise moment acts on the connection rivets 3, 3 are subjected to tension and rivets 4, 4 to shear.

Let us investigate the connection under a clockwise end moment M. It is assumed that because of symmetry the stem of the split beam does not bend. Rivets 1, 1 will be subjected to a tensile force P, and rivets, 3, 3 will be subjected to a compressive force P. These two forces from a couple.

Hence, $$M = P(D + t)$$

or $$P = \frac{M}{D+t} \tag{10.39}$$

where D is the depth of the beam section, and t is the thickness of the split beam stem. The rivets, 2, 2 are subjected to shear. The lever arm for calculating the force in the rivets is equal to the depth of the beam,

Hence,
$$M = VD$$

or
$$V = \frac{M}{D} \tag{10.40}$$

The moment at the critical section,
$$M' = \left[0.5P \times \left(\frac{g-t}{2}\right)\right]$$

The moment of resistance of the flange is
$$M_R = \frac{1}{6} BT^2 \sigma_{bt}$$

where, V = design shear force for rivets 2, 2
g = gauge distance
t = thickness of stem of split beam
T = thickness of flange of split beam
B = flange width of split beam
σ_{bt} = bending stress.

Note The bending stress is assumed to be equal to the bending stress in the slab base = 185 N/mm^2.

Equating the moment to moment of resistance,
$$M' = 0.5P \times \left(\frac{g-t}{2}\right) = \frac{1}{6} BT^2 \sigma_{bs}$$

or
$$T = \sqrt{6M'/(B\sigma_{bs})} \tag{10.41}$$

The thickness of the stem of the split beam is kept sufficient to provide the value of the rivet in bearing. It should also be able to transmit the pull. The thickness of the stem can be worked out from the following equation:
$$b \times t \times \sigma_{at} = P$$

or
$$t = \frac{P}{b\sigma_{at}} \tag{10.42}$$

where b is the net width of the stem, and σ_{at} is the permissible axial stress in tension. The web clip angles are designed to resist end shear.

Design A split beam connection is designed as follows:

1. The diameter of the rivets is assumed and the value of rivet in tension is determined.
2. As four rivets in two gauge lines are provided over the split beam flange, the maximum pull resisted by the connection will be four times the value of a rivet in tension.
3. The thickness of the stem of the split beam section is assumed and the pull to be transmitted by the split beam flange rivets is determined by Eq. (10.39).
4. The maximum value of the moment in split beam is calculated and the thickness of the split beam flange is found from Eq. (10.41).
5. The thickness of the stem of the split beam section is found from by Eq. (10.42).

6. A suitable split beam/T-Section is selected from I.S. Handbook No. 1 and the strength of the split beam section in axial tension is computed, which should be more than the pull over it.
7. The shear is calculated by Eq. (10.40) and the number of rivets required to connect the split with the beam flange are determined by $n = V/R_v$.
8. The design of angles connecting the web of the beam to the flange of the column and their connections are done in a way similar to as in the case of a clip angle connection.

Bracket Connections Bracket connections reduce the girder moments considerably and are more rigid as compared to split beam connections. But fabrication cost is quite high and more than that for split beam connections. The connection is made with a bracket plate and three pairs of angles. Two pairs (one pair on each flange) are connected to the beam flanges and with the bracket plate. The third pair is kept continuous across the beam and is connected to the flange of the column and to the bracket plate (Fig. 10.20). The pitch of the rivets connecting the column flange with the bracket plate cannot be kept uniform as rivets are not provided opposite the beam flange, but for analysis purpose the pitch is assumed to be uniform. For compatibility and to faciliate fabrication, the thickness of the bracket plate is kept equal to the thickness of the beam web. In the case of rigid connections, the column flange may deflect outwards under the pull from the flange clip. The effect is more pronounced in split beams than in bracket connections. To reduce the effect, the diaphragms are connected between the column flanges opposite the beam flanges.

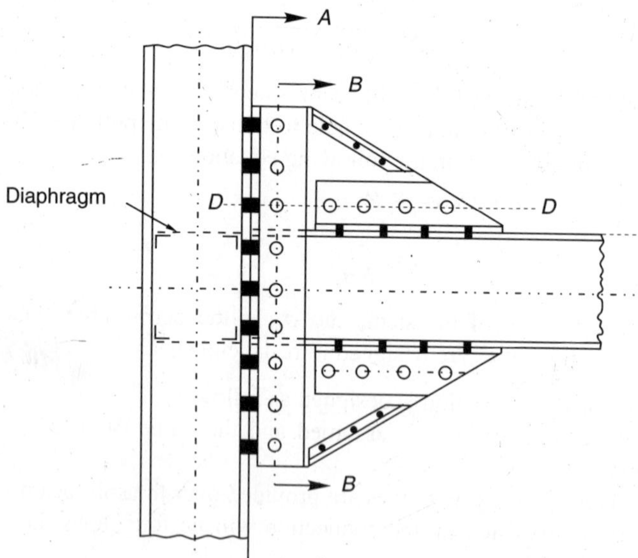

Fig. 10.20 *Riveted Bracket Connection*

Design of Connection

Rivets Along Line AA The rivets are subjected to shear and tension. The length of the angle section is computed from the maximum number of rivets to be provided as worked out from either bearing or tension consideration.

The rivet value in tension is calculated assuming rivets to be in two rows by

$$n' = \sqrt{6M/(2sR_{v_1})}$$

where R_{v1} is the rivet value in tension.

The rivets provided are checked in combined shear and tension by the following equation

$$\frac{\tau_{vf,\,cal}}{\tau_{vf}} + \frac{\sigma_{tf,\,cal}}{\sigma_{tf}} \le 1.4$$

The thickness of the angle is found for the maximum bending moment $= 0.5\,P\left(g - \frac{t}{2}\right)$, where P is the pull to be transmitted by the rivets in tension.

Rivets Along Line BB The rivets on line *BB* are subjected to combined shear due to direct shear force and shear due to bending moment. Therefore the check for combined shear is applied.

The rivet value in shear is calculated and the number of rivets are determined by $n = \sqrt{6M/(sR_v)}$ and the rivets are adjusted in one row.

where R_v is the rivet value in bearing

Rivets Along Line CC The rivets along line *CC* are subjected to shear and tension (because of bending moment about *CC* due to the horizontal force in rivets along *BB*). Therefore the check for combined shear and tension is applied.

Rivets Along Line DD The rivets along line *DD* are subjected to direct shear and shear because of bending moment about *DD* due to the horizontal force in rivets along *BB*. Therefore, the check for combined shear is applied.

Design of the Bracket Plate The bracket plate along with the beam web acts as a rectangular beam. The critical section is along line *BB*. At this section the plate should be able to develop sufficient moment of resistance to resist the moment at the section. The bending stress at the extreme fibre,

$$\sigma_{bt} = \frac{M}{I} Y \times \frac{A_{\text{gross}}}{A_{\text{net}}}$$
$$< 0.66\, f_y$$

When a clockwise moment acts, the inclined edge of the bracket plate on top is in tension and that of the bottom of bracket plate is in compression, whereas in the case of an anticlockwise moment it is vice-versa. When the inclined edge of the bracket plate is in compression it behaves as a compression member and lateral buckling of the plate may take place.

To check lateral buckling stiffening angles are provided on both sides of the bracket plate.

$$\frac{l}{r} = \frac{l}{t}\sqrt{12} \text{ (for bracket plates with ends free to rotate)}$$
$$< 60$$

$$\frac{l}{r} = \frac{l/2}{t}\sqrt{12} \text{ (for bracket plates with ends fixed)}$$
$$< 60$$

Welded Moment Connections

Moment resistant connections with weld are made using plates only. A plate is provided over the top flange of the beam. The length of the plate is kept about 1.5 times the width of the beam flange. The length should be such as to ensure that the plate may yield before the connection. The plate is connected to the beam flange by a fillet weld and to the column flange by a butt weld (Fig. 10.21). The plate is tapered to facilitate fillet welding. When the beam is connected to the web of the column the connected top plate is flared to connect the plate with the column flange by side welds.

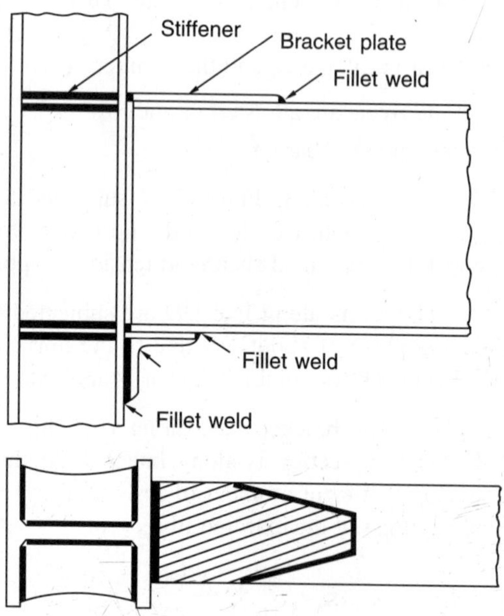

Fig. 10.21 *Welded Moment Connection*

The connecting plate has little resistance to shear. Therefore, a seat plate is provided to resist it. If the end reactions are large, stiffening plates or angles may also be provided. Under heavy moments the column flange may deflect away due to the pull and its moment carrying capacity is reduced. To check this, additional plates called stiffener plates are welded between column flanges opposite the beam flanges as shown in Fig. 10.21. Some codes specify the length of stiffener plates to be equal to half the depth of the column. Generally the plates are provided for the full depth. The design of a welded moment connection is illustrated by the Solved Example 10.19.

10.6 SEMI-RIGID CONNECTIONS

Semi-rigid connections are very intriguing, but unfortunately they can be misleading. A semi-rigid connection resists end moments but gives a relative rotation between the beam and column as shown in Fig. 10.22. This means that the member ends will rotate slightly more than the rest of the joint. In fact, the joint as a whole will rotate through θ radians whereas the end of the beam will rotate through ϕ radians. $(\phi - \theta)$ is thus the angle of strain for the semi-rigid connection. However, in case of rigid connection both the end of beam and joint rotate by same angle θ. A beam having a flexible connection (simply supported) with a column and subjected to uniformly distributed load, w per unit length, has the greatest moment at the centre section: $M = wl^2/8$ (Fig. 10.23(a)). For the same beam having fully rigid connections (ends fixed) the central moment reduces to $wl^2/24$ and the end moments increase to $wl^2/12$ (Fig. 10.22(b)). Thus, the beam needs a section modulus just 2/3 $\left(\frac{wl^2/12}{wl^2/8} = \frac{2}{3}\right)$ of that required for a simple beam.

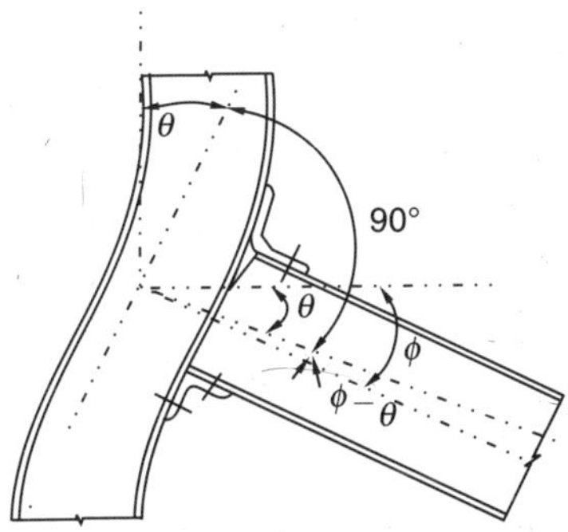

Fig. 10.22 *Semi-rigid Connection*

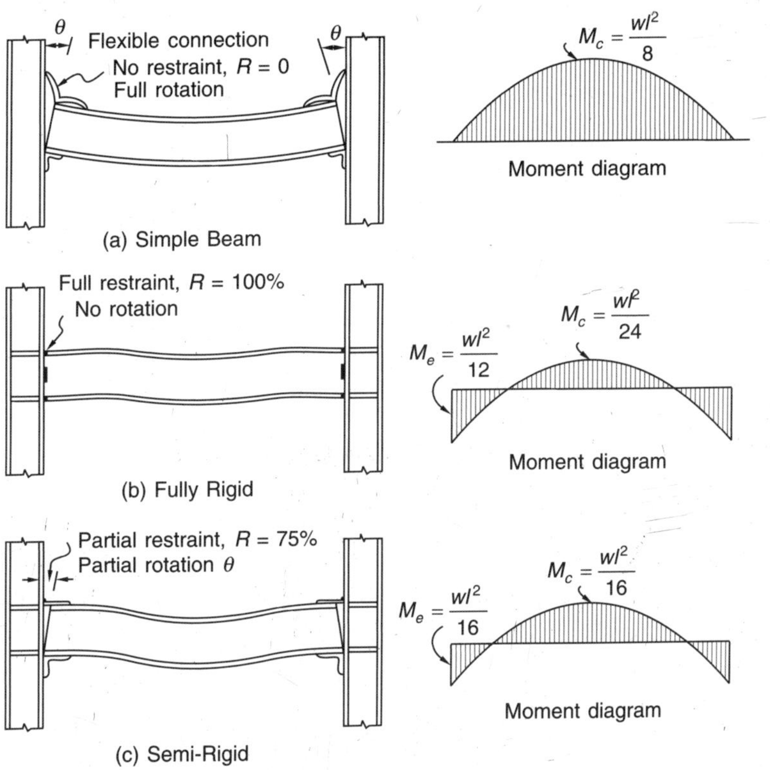

Fig. 10.23 *Moment Diagrams for Beams with Different End Conditions*

If a semi-rigid connection is used as shown in Fig. 10.22(c) the centre moment and end moments may be made equal to $wl^2/16$ each, provided, the connections have an end restraint $R = 75\%$. The section modulus requirement for the section reduces to 50% of that needed for a simple beam. Also, this is true provided the supports to which the connections join the beam are unyielding, and the beam is not influenced by adjacent spans from which additional moments might be carried over through the connection. In addition to the economy to be achieved in beam design through the semi-rigid type of connection, it should be evident that there is also less moment to be resisted by the column. In short, the semi-rigid connection, if properly analyzed and designed, will result in an overall economy. Since the condition of $R = 75\%$ does not offer any range of rigidity, a little error may lead to hazard. To take care of this, it is suggested that the beam is designed for an end restraint of $R =$ 50% (centre moment of $wl^2/12$) and the connection for a restraint of $R = 75\%$ (end moment of $wl^2/16$).

This appears to be good until the resulting design moment of $wl^2/12$ is not lower than if fully rigid welded connections were used. So there is no saving in beam requirements by using semi-rigid connections. However, there is a slight reduction in the amount of connecting weld. This might be true for a fully rigid connection using a top connecting plate (groove welded to the column and fillet welded to the beam) but would not be true if the beam flanges are groove welded directly to the column without additional fillet welds. Since for semirigid connections each span has to be dealt with separately, it is difficult to apply the principles to actual frames of different spans and storeys and different span loadings. Further, for a frame using semi-rigid connections, since the end moments, distribution factors, and carry over factors, will all be affected by the degree of rigidity of the connections, the analysis becomes too much complicated. Therefore, certain cleat angles have been standardized after extensive experimentation (by determining their $M - \theta$ curves) for calculating the end moments developed for a given span and depth of the beams.

An analytically troublesome complication in semi-rigid connection performance is the nonlinearity of the M–θ diagram. Almost all connections behave inelastically—flexible ones practically from the start of loading, rigid one sooner or later. Methods have been proposed for calculating the M–θ relationship for some types of semi-rigid connections, but most M–θ curves must be determined experimentally. However, the scope of the book limits the analysis and design of semi-rigid connections.

Solved Examples

Example 10.1 Find out the safe load P carried by the joint shown in Fig. Ex. 10.1. The rivets are power driven 20 mm ϕ at a pitch of 80 mm. The thickness of the flange is 9.1 mm and that of the bracket plate 10 mm.

Also find the safety of the joint if instead of rivets, same numbers of M20–8.8G HSFG bolts are used for making the connection.

Solution

$$\text{Nominal diameter of rivet, } \phi = 20 \text{ mm}$$

$$\text{Gross diameter} = 20 + 1.5 = 21.5 \text{ mm}$$

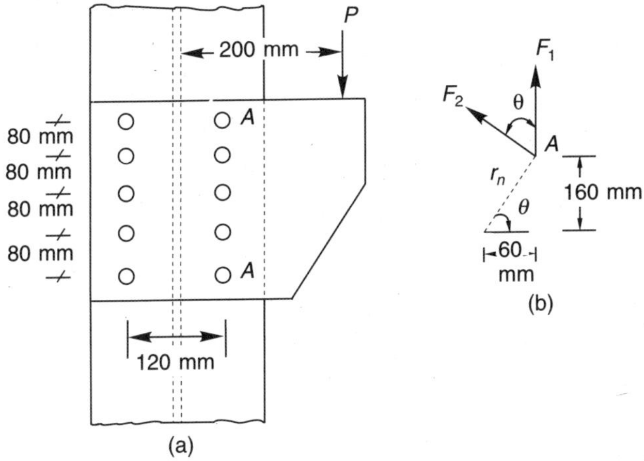

Fig. Ex. 10.1

$$\text{Rivet value in single shear} = \frac{\pi}{4} \times 21.5^2 \times 100 = 36{,}305.03 \text{ N}$$

$$\text{Rivet value in bearing} = 21.5 \times 9.1 \times 300 = 58{,}695 \text{ N}$$

$$\text{Hence, the value of rivet, } R_1 = 36{,}305 \text{ N}$$

$$\text{Number of rivets, } n = 10$$

Rivet which is stressed maximum is A.

$$\therefore \quad F_1 = \frac{P}{10}$$

$$F_2 = \frac{Per_n}{\Sigma r^2}$$

$$r_n = \sqrt{160^2 + 60^2} = 170.88 \text{ mm}$$

$$\Sigma r^2 = 4 \times [(160^2 + 60^2) + (80^2 + 60^2)] + 2 \times 60^2 = 164{,}000 \text{ mm}^2$$

$$e = 200 \text{ mm}$$

$$\therefore \quad F_2 = \frac{P \times 200 \times 170.88}{164{,}000} = 0.20839P$$

$$\cos\theta = \frac{60}{\sqrt{60^2 + 160^2}} = 0.3511$$

The resultant force on the rivet should be equal to the rivet value.

or $\quad 36{,}305.03 = \sqrt{(P/10)^2 + (0.20839P)^2 + 2 \times (P/10) \times 0.20839P \times 0.3511}$

or $\quad 36{,}305.03 = 0.26088223P$

or $\quad P = 139{,}162.525 \text{ N} = 139.162 \text{ kN}$

(b) There is only one interface.

Proof load for 1, M20–8.8G bolt = 147 kN.

$$F = \frac{n\mu T}{F_s}$$

$$\text{Maximum shear } F \text{ taken by one bolt} = \frac{1 \times 0.50 \times 147}{1.4} = 52.5 \text{ kN}$$

Equating this value to the load in the most critical bolt which is 0.26088223 P (already calculated for riveted connection),

$$52.5 = 0.26088223\,P$$

or, $$P = 201.24 \text{ kN} > 139.162$$

Hence, the joint to safe.

Example 10.2 Design a bracket connection to support an end reaction of 200 kN. The eccentricity of the load is 250 mm as shown in Fig. Ex. 10.2.

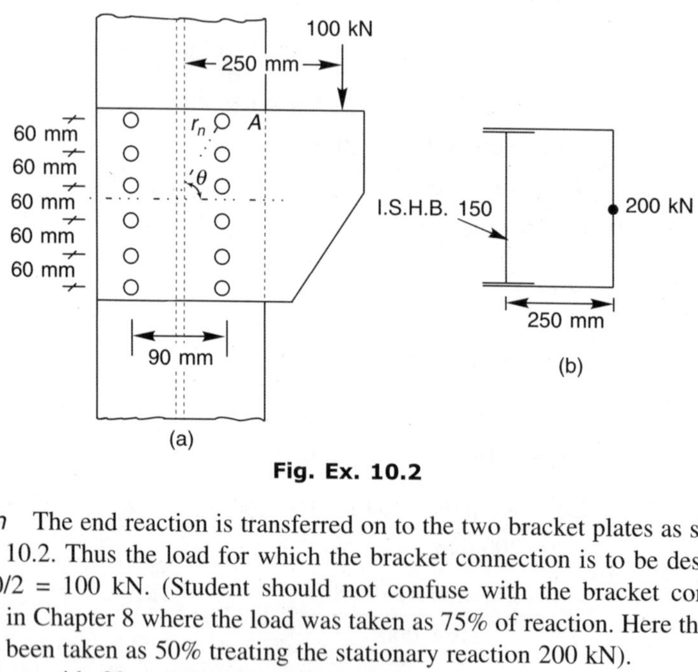

Fig. Ex. 10.2

Solution The end reaction is transferred on to the two bracket plates as shown in Fig. Ex. 10.2. Thus the load for which the bracket connection is to be designed is P = 200/2 = 100 kN. (Student should not confuse with the bracket connection example in Chapter 8 where the load was taken as 75% of reaction. Here the design load has been taken as 50% treating the stationary reaction 200 kN).

Let us provide 20 mm ϕ power driven rivets.

Assuming the rivet value of the rivet in shear to be less than the rivet value of rivet in bearing.

$$\text{Rivet value, } R_v = \frac{\pi}{4} \times 21.5^2 \times 100 = 36{,}305.03 \text{ N}$$

The thickness of the bracket plate required,

$$36{,}305.03 = 21.5 \times t \times 300$$

or $$t = 5.628 \text{ mm}$$

Let us provide an 8 mm thick bracket plate. The minimum pitch of the rivets

$$s = 2.5\,\phi = 2.5 \times 20 = 50 \text{ mm}$$

Provide a pitch of 60 mm.

Let us provide rivets in two vertical rows.

$$\text{Number of rivets, } n = \sqrt{6M/(sn'R_v)}$$

$$= \sqrt{(6 \times 100 \times 10^3 \times 250)/(60 \times 2 \times 36{,}305.03)} = 5.867 \approx 6$$

Provide a total of 12 rivets on each bracket plate. Number of rivets in each row is 6. For I.S.H.B. 150, the gauge distance = 90 mm.

Force on critical rivet A

$$F_1 = \frac{P}{n} = \frac{100 \times 10^3}{12} = 8333.33 \text{ N}$$

$$F_2 = \frac{Per_n}{\Sigma r^2}$$

$$e = 250 \text{ mm}$$

$$r_n = \sqrt{150^2 + 45^2} = 156.60 \text{ mm}$$

$$\Sigma r^2 = 4 \times [(150^2 + 45^2) + (90^2 + 45^2) + (30^2 + 45^2)]$$

$$= 4 \times [24{,}525 + 10{,}125 + 2925] = 150{,}300$$

$$F_2 = \frac{100 \times 10^3 \times 250 \times 156.60}{150{,}300} = 26{,}047.9 \text{ N}$$

$$\cos\theta = \frac{45}{\sqrt{150^2 + 45^2}} = 0.2873$$

Resultant force F on the critical rivet,

$$F = \sqrt{(8333.33)^2 + (26{,}047.9)^2 + 2 \times 8333.33 \times 26{,}047.9 \times 0.2873}$$

$$= 2.95408 \times 10^4 \text{ N} = 29540.8 \text{ N} < 36{,}305.03 \text{ N}$$

which is as it should be.

Example 10.3 Estimate the safe inclined load as shown in Fig. Ex. 10.3. The rivets are 22 mm ϕ, eight in number and are hand driven.

Solution The critical rivet will be rivet A as shown in Fig. Ex. 10.3 being nearest to the load line.

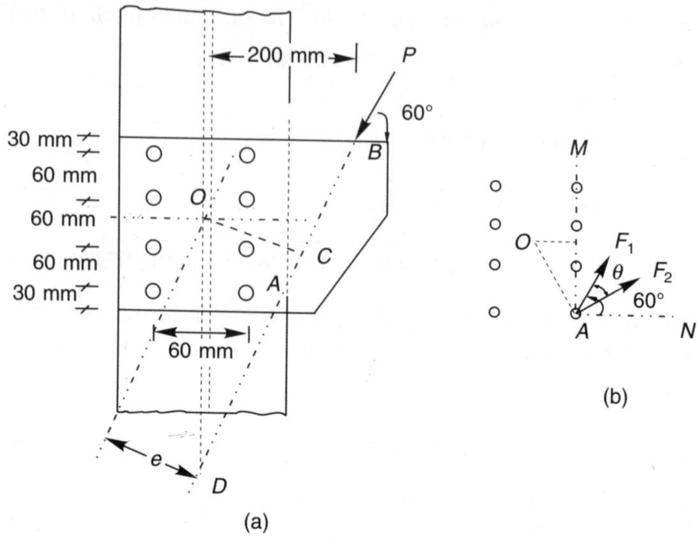

Fig. Ex. 10.3

Calculation of eccentricity Fig. Ex. 10.3(a)

$$OD = FD - OF$$

$$FD = FB \tan 60° = 200 \tan 60° = 346.410 \text{ mm}$$

$$OD = 346.410 - 120 = 226.410 \text{ mm}$$

∴ $$e = OC = OD \sin 30° = 226.410 \sin 30° = 113.205 \text{ mm}$$

$$F_1 = \frac{P}{8} = 0.125P \text{ N}$$

$$F_2 = \frac{Per_n}{\Sigma r^2}$$

$$= \frac{P \times 113.205 \times \sqrt{30^2 + 90^2}}{4 \times [(30^2 + 90^2) + (30^2 + 30^2)]}$$

$$= \frac{10,739.569\,P}{43,200} = 0.2486P$$

From Fig. Ex. 10.3(b),

$$\theta = \angle F_1AN - \angle F_2AN$$

$$\angle F_2AN + \angle F_2AM = \angle F_2AM + \angle MAO = 90°$$

Hence

$$\angle F_2AN = \angle MAO = \tan^{-1}\frac{30}{90} = 18.4349°$$

∴ $$\theta = 60° - 18.4349° = 41.5650°$$

$$\text{Resultant force} = \sqrt{(0.125\,P)^2 + (0.2486\,P)^2 + 2 \times 0.125P \times 0.2486P \times 0.748}$$
$$= 0.35201P$$

Gross diameter of the rivet = 22 + 1.5 = 23.5 mm

$$\text{Value of the rivet in shear} = \frac{\pi}{4} \times 23.5^2 \times 80 = 34{,}698.89 \text{ N}$$

$$34{,}698.89 = 0.35201P$$

or $$P = 98{,}573.59 \text{ N} = 98.573 \text{ kN}$$

Example 10.4 Design a bracket connection to transfer a load of 150 kN as shown in Fig. Ex. 10.4(a). The load is acting at an eccentricity of 300 mm from the face of the column.

(a) Design riveted connections at sections *A–A*, *B–B* and *C–C*.
(b) Design HSFG bolted connection at section *A–A*.

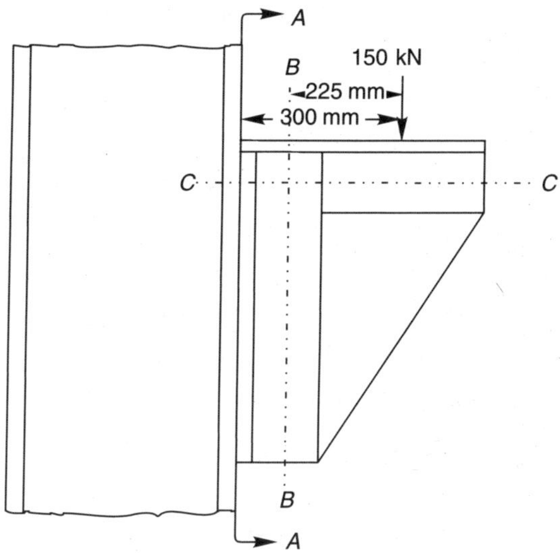

Fig. Ex. 10.4(a)

Solution

Rivets along line A–A The rivets along line *AA* are subjected to shear due to the load of 150 kN and tension due to the bending moment

$$= 150 \times 300 = 45{,}000 \text{ kN mm}$$

Let us provide rivets in two rows.

Assume 22 mm ϕ power driven cold rivets.

Gross diameter of rivet = 22 + 1.5 = 23.5 mm

Let the rivet value be governed by the tension of rivets.

$$\text{Rivet value, } R_v = \frac{\pi}{4} \times 23.5^2 \times 100 = 43{,}373.614 \text{ N}$$

$$s = 2.5\phi = 2.5 \times 22 = 55 \approx 65 \text{ mm}$$

$$n = \sqrt{6M/(n'sR_v)}$$

$$= \sqrt{(6 \times 45{,}000 \times 10^3)/(2 \times 65 \times 43373.624)} = 6.919 \approx 7$$

Provide seven rivets in each row.

Depth of bracket plate For 22 mm ϕ rivets, minimum edge distance is 32 mm. Let us provide 40 mm edge distance.

$$\text{Depth of bracket plate} = 6 \times 65 + 2 \times 40 = 470 \text{ mm}$$

$$h = 470 - 40 = 430 \text{ mm}$$

The neutral axis is assumed to lie at $h/7$ from the bottom, i.e. at

$$\frac{430}{7} = 61.42 \text{ mm}$$

$$\Sigma y = 2 \times [(65 + 40 - 61.42) + (130 + 40 - 61.42) + (195 + 40 - 61.42) + (260 + 40 - 61.42) + (325 + 40 - 61.42) + (390 + 40 - 61.42)]$$
$$= 2472.96 \text{ mm}$$

$$\Sigma y^2 = 2 \times [43.58^2 + 108.58^2 + 173.58^2 + 238.58^2 + 303.58^2 + 368.58^2]$$
$$= 657502.6 \text{ mm}^2$$

$$M' = \frac{M}{1 + \dfrac{2h}{21} \times \dfrac{\Sigma y}{\Sigma y^2}}$$

$$M' = \frac{45{,}000 \times 10^3}{1 + \dfrac{2 \times 430}{21} \times \dfrac{2472.96}{657{,}502.6}} = 38993.865 \times 10^3 \text{ Nmm}$$

Tensile stress in the critical rivet,

$$\sigma_{tf,\,\text{cal}} = \frac{M' y_n}{A \times \Sigma y^2} = \frac{38993.865 \times 10^3 \times 368.58}{\dfrac{\pi}{4} \times 23.5^2 \times 657{,}502.6} = 50.397 \text{ N/mm}^2$$

Shear stress in the critical rivet,

$$\tau_{vf,\,\text{cal}} = \frac{150 \times 10^3}{\dfrac{\pi}{4} \times 23.5^2 \times 7 \times 2} = 24.70 \text{ N/mm}^2$$

Check

$$\frac{\tau_{vf,\,\text{cal}}}{\tau_{vf}} + \frac{\sigma_{tf,\,\text{cal}}}{\sigma_{tf}} \le 1.4$$

or

$$\frac{24.70}{100} + \frac{50.397}{100} = 0.7509 \le 1.4$$

which is safe.

Rivets along line B–B The rivets are subjected to shear due to the eccentric load of 150 kN.

$$e = 225.0 \text{ mm}$$
$$M = 150 \times 225 = 33{,}750 \text{ kNmm}$$
$$n = \sqrt{6M/(sn'R_v)}$$

The rivets will be in double shear. Assume the thickness of angles and bracket plate to be 10 mm.

$$\text{Assume diameter of rivet} = 22 \text{ mm}$$
$$\text{Gross diameter} = 22 + 1.5 = 23.5 \text{ mm}$$
$$\text{Value of rivet in shear} = 2 \times \frac{\pi}{4} \times 23.5^2 \times 100 = 86{,}747.227 \text{ N}$$
$$\text{Value of rivet in bearing} = 23.5 \times 10 \times 300 = 70{,}500 \text{ N}$$
$$\text{Value of rivet, } R_v = 70500 \text{ N}$$

Assume pitch, $s = 65$ mm.

$$n = \sqrt{6 \times 33{,}750 \times 10^3/(65 \times 1 \times 70{,}500)} = 6.64 \approx 7$$

Provide seven rivets at a pitch of 65 mm and edge distance of 40 mm.

$$F_1 = \frac{150}{7} = 21.42 \text{ kN}$$

$$F_2 = \frac{Mr_n}{\Sigma r^2} = \frac{33{,}750 \times 195}{2 \times [(65.0)^2 + (2 \times 65)^2 + (3 \times 65)^2]} = 55.63 \text{ kN}$$

$$\text{Resultant force} = \sqrt{F_1^2 + F_2^2}$$
$$= \sqrt{(21.42)^2 + (55.63)^2} = 59.61 \text{ kN} < 70.5 \text{ kN}$$

which is safe.

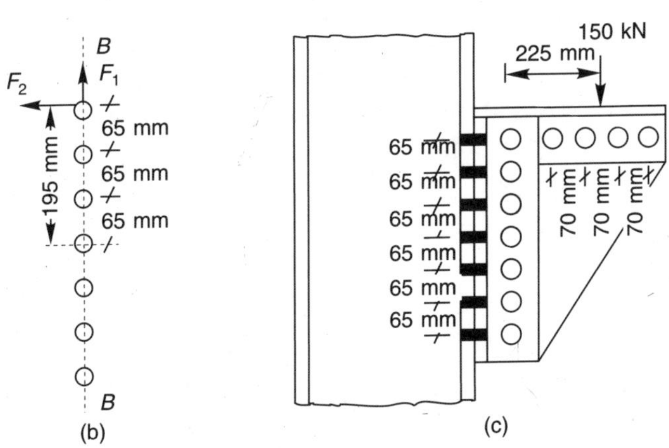

Fig. Ex. 10.4 (b) and (c)

Rivets along line C–C The rivets are subjected to an eccentric load of 150 kN. The rivet group may be assumed to be subjected to a concentrated load of 150 kN and a bending moment of 150 × *e*.

Let us provide 22 mm ϕ rivets, five in number at a pitch of 70 mm.

$$e = 225 - 2 \times 70 = 85 \text{ mm}$$

$$M = 150 \times 85 = 12{,}750 \text{ kNmm}$$

$$F_1 = \frac{150}{5} = 30 \text{ kN}$$

$$F_2 = \frac{Mr_n}{\Sigma r^2} = \frac{12{,}750 \times 140}{2 \times (70^2 + 140^2)} = 36.42$$

$$\text{Resultant force} = F_1 + F_2$$

$$= 30 + 36.42 = 66.42 \text{ kN} < 70.5 \text{ kN}$$

which is safe.

(b) Design connection at section *A–A* of the bracket connection using HSFG bolts.

$$M = 45 \text{ kNm}$$

Let us try M20–8.8G HSFG bolts (Proof load = 147 kN)

There is only one interface.

$$\text{Shear per bolt} = \frac{n\mu T}{F_s} = \frac{1 \times 0.50 \times 147}{1.4} = 52.5 \text{ kN}$$

Number of bolts from moment consideration, assuming 2 rows of bolts at pitch of 50 mm and edge distance of 40 mm.

$$= \sqrt{\frac{6 \times 45 \times 10^3}{2 \times 50 \times 52.5}} = 7.17 \approx 7$$

Let us try 7, M20–8.8G HSFG bolts as shown in Fig. Ex. 10.4(d).

$$h = 6 \times 50 + 2 \times 40 = 380 \text{ mm}$$

$$\text{Distance of N.A. from bottom of bracket} = \frac{h}{7} = \frac{380}{7} = 54.28 \text{ mm}$$

$$y_n = 40 + 50 - 54.28 + 5 \times 50 = 285.72 \text{ mm}$$

$$\Sigma y = 2 \times [(50 + 40 - 54.28) + (100 + 40 - 54.28) + (150 + 40 - 54.28) + (200 + 40 - 54.28) + (250 + 40 - 54.28) + (300 + 40 - 54.28)$$

$$= 2[35.72 + 85.72 + 135.72 + 185.72 + 235.72 + 285.72]$$

$$= 1928.64 \text{ mm}$$

$$\Sigma y^2 = 2 \times [35.72^2 + 85.72^2 + 135.72^2 + 185.72^2 + 235.72^2 + 285.72^2]$$

$$= 397{,}471.02$$

$$M' = \frac{M}{1 + \dfrac{2h}{21} \times \dfrac{\Sigma y}{\Sigma y^2}} = \frac{45}{1 + \dfrac{2 \times 380}{21} \times \dfrac{1928.64}{397{,}471.02}} = 38.278 \text{ kNm}$$

$$T = \frac{M'y_n}{\sum y^2} = \frac{38.278 \times 10^3 \times 285.72}{397{,}471.02} = 27.515 \text{ kN}$$

$$\text{Direct shear} = \frac{150}{2 \times 7} = 10.71 \text{ kN}$$

$$\frac{\text{Calculated shear}}{\text{Slip factor} \times \text{Number of interfaces}} \le \frac{\text{Proof load} - \text{Calculated tension} \times f}{\text{Factor of safety}}$$

$$\frac{10.71}{0.50 \times 1} \le \frac{\text{Proof load} - 27.515 \times 1.7}{1.4}$$

$$\Rightarrow \quad \text{Proof load} = 76.76 \text{ kN} < 147 \text{ kN}$$

Hence safe.

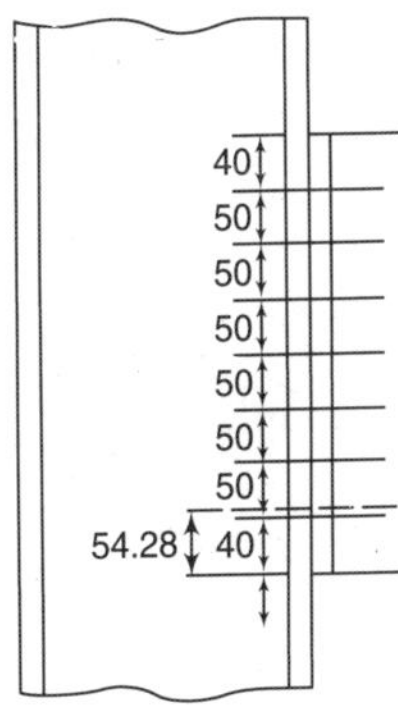

Fig. Ex. 10.4(d)

Example 10.5 A flat steel bar is attached to a gusset plate by eight rivets as shown in Fig. Ex. 10.5. At the section *AB*, the gusset plate exerts on the flat a vertical shear force *V* and an anticlockwise bending moment *M*.

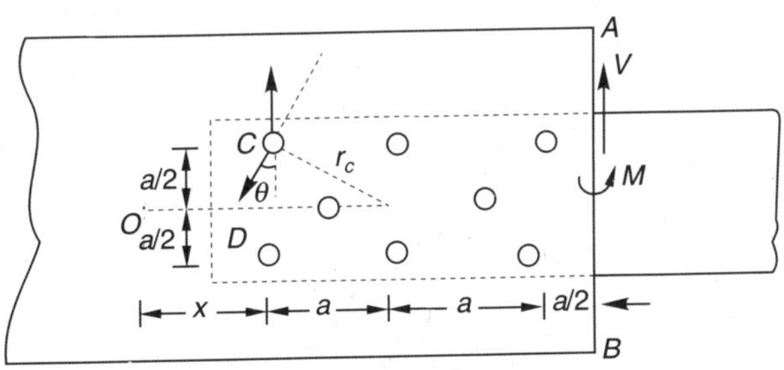

Fig. Ex. 10.5

Assuming that the gusset plate relative to the flat undergoes a minute rotation about a point *O* on the line of the two middle rivets, and that the load on the rivets are due to and proportional to the relative movement of the plates at the rivet holes, prove that

$$x = -a\,\frac{3Va + 4M}{6Va + 4M}$$

and that the horizontal and vertical components of the load on the top left-hand rivet C are,

$$\frac{3Va+2M}{24a} \text{ and } \frac{3Va+4M}{24a} \text{ respectively}.$$

Solution The rivet C carries a shear load due to shear V and a shear load due to couple $\left(V\frac{3a}{2}+M\right)$.

Shear load due to the direct shear V, $F_1 = \frac{V}{8}$

$$\text{Shear load due to the couple} = \left(V\frac{3a}{2}+M\right)\frac{r_c}{\Sigma r^2}$$

$$r_c = \sqrt{\left(\frac{a}{2}\right)^2 + a^2} = \frac{\sqrt{5}}{2}a$$

$$\Sigma r^2 = 4\times\frac{a^2\times 5}{4} + 4\times\frac{a^2}{4} = 6a^2$$

$$\text{Hence shear load due to couple} = \left(\frac{3Va}{2}+M\right)\times\frac{\frac{\sqrt{5}}{2}a}{6a^2} = \frac{3Va+2M}{24a}\sqrt{5}$$

From Fig. Ex. 10.5; $\tan\theta = 1/2$

The vertical component of the load on rivet C

$$= \frac{V}{8} - \frac{3Va+2M}{24a}\sqrt{5}\cos\theta$$

$$= \frac{V}{8} - \frac{3Va+2M}{24a}\sqrt{5}\times\frac{2}{\sqrt{5}}$$

$$= \frac{V}{8} - \frac{6Va+4M}{24a} = \frac{3Va-6Va-4M}{24a} = -\frac{3Va+4M}{24a}$$

The horizontal component of the load on rivet C

$$= \frac{3Va+2M}{24a}\sqrt{5}\sin\theta$$

$$= \frac{3Va+2M}{24a}\sqrt{5}\times\frac{1}{\sqrt{5}} = \frac{3Va+2M}{24a}$$

If α is the angle of inclination of the resultant with the horizontal,

$$\tan\alpha = -\frac{3Va+4M}{3Va+2M}$$

The resultant when produced should pass through O.

Hence
$$x = \frac{a}{2}\left(-\frac{3Va+4M}{3Va+2M}\right) = -a\left(\frac{3Va+4M}{6Va+4M}\right)$$

Example 10.6 Design a seat connection for a beam end reaction of 84 kN. The beam section is I.S.M.B. 250 @ 365.9 N/m connected to the flange of the column section I.S.H.B. 200 @ 365.9 N/m.

(a) Provide riveted connections
(b) Provide HSFG botls

Solution The relevant properties of the section from I.S. Handbook No. 1 are:

	I.S.M.B. 250	*I.S.H.B. 200*
B	125 mm	200 mm
t	6.9 mm	6.1 mm
h_2	27.95 mm	
T	12.5 mm	9.00 mm
g	65 mm	55 mm

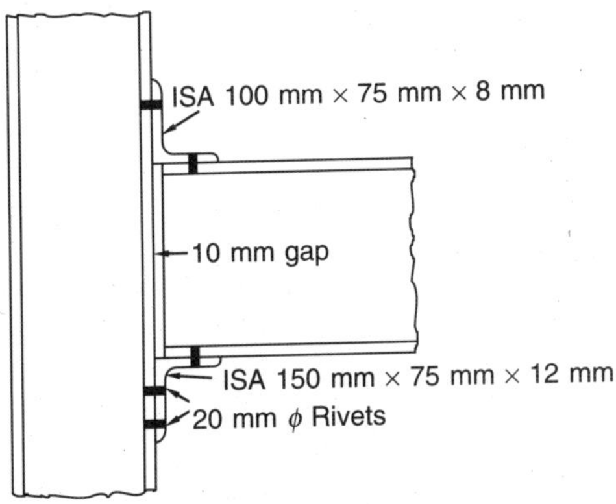

Fig. Ex. 10.6

Seat angle

Length of the seat angle = width of beam flange
= 125 mm

$$\text{Bearing length, } b = \frac{R}{\sigma_p t} - \sqrt{3}h_2 \not< \frac{1}{2} \times \frac{R}{\sigma_p t}$$

$$\sigma_p = 0.75 f_y = 0.75 \times 250 = 187.5 \text{ N/mm}^2$$

$$\frac{1}{2} \times \frac{R}{\sigma_p t} = \frac{1}{2} \times \frac{84 \times 10^3}{187.5 \times 6.9} = 32.46 \text{ mm}$$

$$b = \frac{84 \times 10^3}{187.5 \times 6.9} - \sqrt{3} \times 27.95 = 16.516 \text{ mm} \not< 32.46 \text{ mm}$$

Therefore, the bearing length required is 32.46 mm.

Provide a seat leg length of 75 mm. Provide a clearance of 10 mm between the beam and the column flange.

Let us choose an angle section 125 × 75 × 10 mm. The relevant properties of the angle section from I.S. Handbook No. 1.

$$r_1 = 9 \text{ mm}, A = 1902 \text{ mm}^2$$

$$e_1 = \text{eccentricity} = 10 + \frac{b}{2} - t - r_1$$

Assume a 10 mm thick seat angle

$$e_1 = 10 + \frac{32.46}{2} - 10 - 9 = 7.23 \text{ mm}$$

$$t = \sqrt{6Re_1/(B\sigma_{bs})} = \sqrt{(6 \times 84 \times 10^3 \times 7.23)/(125 \times 185)} = 12.55 \text{ mm}$$

$$> 10 \text{ mm}$$

which is not safe.

Note

1. The leg connected to the column was assumed to be 125 mm to accomodate the connectors.
2. Let us try some other angle. Try an angle section having thickness more than 12 mm.

From I.S. Handbook No. 1, let us provide I.S.A. 150 × 75 × 12 mm.
The relevant properties of the angle section are

$$r_1 = 10 \text{ mm}, A = 2562 \text{ mm}^2$$

$$e_1 = 10 + \frac{32.46}{2} - 12 - 10 = 4.23 \text{ mm}$$

$$t = \sqrt{(6 \times 84 \times 10^3 \times 4.23)/(125 \times 185)} = 9.60 \text{ mm}$$

$$< 12 \text{ mm}$$

which is safe.
Hence, provide a seat angle of 150 mm × 75 mm × 12 mm.
Connections
(a) Let us provide 20 mm ϕ power driven rivets.

Gross diameter = 20 + 1.5 = 21.5 mm

$$\text{Strength of the rivet in single shear} = \frac{\pi}{4} \times 21.5^2 \times 100 = 36{,}305.03 \text{ N}$$

Strength of the rivet in bearing = 21.5 × 9.0 × 300 = 58,050 N

Value of the rivet = 36,305.03 N

The number of rivets required to connect the flange of the column with the seat angle,

$$n = \frac{84 \times 1000}{36{,}305.03} = 2.313 \approx 4 \text{ rivets}$$

The minimum pitch = 2.5 ϕ = 2.5 × 20 = 50 mm

Minimum edge distance for 20 mm ϕ rivets is 29 mm.

Provide the rivets in two rows (two rivets in each row). Total number of rivets provided is four.

Provide two rivets, 20 mm ϕ, to connect the seat angle leg with the beam flange. Provide a nominal size cleat angle 100 × 75 × 8 mm. To connect the cleat angle with the beam flange and column flange provide two 20 mm ϕ rivets on each leg.

(b) The reaction from beam = 84 kN

Let us provide HSFG bolts.

The number of interface = 1

$$F = \frac{n\mu T}{F_s}$$

$$84 = \frac{1 \times 0.50 \times T}{1.4}$$

Proof load, T = 235.2 kN

Let us try M20–8.8G HSFG bolts.

Proof load of 2, M20–8.8G HSFG bolts = 2 × 147 = 294 kN > 235.2 kN.

Hence, provide 2, M20–8.8G HSFG bolts to connect the seat angle leg with the column flange. Also, the seat angle leg for making connection with flange can be reduced from 125 mm to 100 mm to accommodate one row of bolts. Hence, provide seat angle 100 × 75 × 10 mm.

Example 10.7 An I.S.M.B. 350 @ 514 N/m is to join a column I.S.H.B. 300 @ 576.8 N/m. The beam has to transmit an end reaction of 230 kN. Design a stiffened seat connection.

Solution The relevant properties of the sections from I.S. Handbook No. 1 are:

	I.S.M.B. 350	*I.S.H.B. 300*
B	140	250 mm
T	14.2	10.6 mm
t	8.1	7.6 mm
h_2	31	
g	80	

Length of seat angle = width of beam flange = 140 mm

$$b \nless \frac{1}{2} \times \frac{R}{\sigma_p t}$$

or $$\frac{1}{2} \times \frac{R}{\sigma_p t} = \frac{1}{2} \times \frac{230 \times 10^3}{187.5 \times 8.1} = 75.72 \text{ mm}$$

$$b = \frac{R}{\sigma_p t} - \sqrt{3}h_2 = \frac{230 \times 10^3}{187.5 \times 8.1} - \sqrt{3} \times 31 = 97.746 \text{ mm}$$

$$\nless 75.72 \text{ mm}$$

Provide a clearance of 10 mm.

Length of the outstanding leg required = 97.746 + 10 = 107.746 mm

Provide a leg length of 115 mm.

Provide a seat angle 150 × 115 × 12 mm.

Stiffener angle

$$\text{Bearing area required for stiffener angle} = \frac{230 \times 1000}{187.5} = 1226.66 \text{ mm}^2$$

Let us select two I.S.A. 90 × 60 × 8 mm from I.S. Handbook No. 1. Area provided by stiffening legs of the angles

$$= 2 \times (90 \times 8) = 1440 \text{ mm}^2$$

$$\text{Length of the outstanding leg} = 90 - 8 = 82 \text{ mm}$$

$$\text{Thickness of the angle, } t = 8 \text{ mm}$$

$$\text{Outstanding ratio} = \frac{82}{8} = 10.2$$

$$< 16$$

which is safe.

$$\text{Distance of end reaction from column flange} = \frac{115}{2} = 57.5 \text{ mm}$$

$$e_{xx} = 57.5 \text{ mm}$$

Stiffener angles provide some rigidity to the seat angle and the reaction is assumed to act at the middle of the seat leg. Thus the eccentricity is increased.

Let us provide 22 mm ϕ rivets at 70 mm pitch.

$$\text{Value of rivet in shear} = \frac{\pi}{4} \times 23.5^2 \times 100 = 43{,}373.61 \text{ N}$$

$$\text{Value of rivet in bearing} = 23.5 \times 8 \times 300 = 56{,}400 \text{ N}$$

Hence,

$$\text{The rivet value, } R_v = 43373.61 \text{ N}$$

Let us provide rivets in two rows.

$$n = \sqrt{6M/(s\, n' R_v)}$$

$$= \sqrt{(6 \times 230 \times 10^3 \times 57.5)/(70 \times 2 \times 43373.61)} = 3.614 \approx 4$$

Provide four rivets in each row. For 22 mm ϕ rivets, minimum edge distance is 32 mm. Let us provide edge distance of 40 mm.

$$\text{The depth of stiffener angle} = 3 \times 70 + 2 \times 40 = 210 + 80 = 290 \text{ mm}$$

$$h = 290 - 40 = 250 \text{ mm}$$

$$\text{The neutral axis lies at } \frac{h}{7} = \frac{250}{7} = 35.714 \text{ mm}$$

$$\text{Shear force in each rivet, } F_1 = \frac{230 \times 10^4}{8} = 28.750 \times 10^3 \text{ N}$$

$$\tau_{vf,\,cal} = \frac{28.750 \times 10^3}{\frac{\pi}{4} \times 23.5^2} = 66.28 \text{ N/mm}^2$$

The critical rivets are *A*, as shown in Fig. Ex. 10.7.

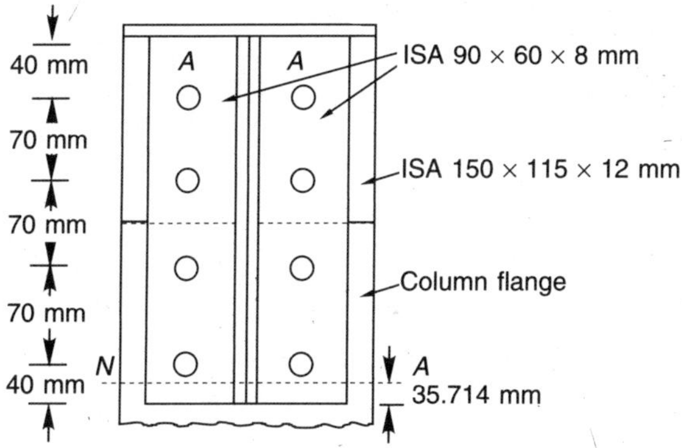

Fig. Ex. 10.7

$$\Sigma y = 2 \times [(40 - 35.714) + (110 - 35.714) + (180 - 35.714) + (250 - 35.714)]$$
$$= 874.288 \text{ mm}$$

$$\Sigma y^2 = 2 \times [4.286^2 + 74.286^2 + 144.286^2 + 214.286^2] = 144{,}547.44 \text{ mm}^2$$

$$M' = \frac{M}{1 + \frac{2h}{21} \times \frac{\Sigma y}{\Sigma y^2}} = \frac{230 \times 10^3 \times 57.5}{1 + \frac{2 \times 250}{21} \times \frac{874.288}{144{,}547.44}} = 11{,}560.206 \times 10^3 \text{ Nmm}$$

$$F_2 = \frac{M' y_n}{\Sigma y^2} = \frac{11{,}560.206 \times 10^3 \times 214.286}{144{,}547.44} = 17{,}137.56 \text{ N}$$

$$\sigma_{tf,\,\text{cal}} = \frac{17{,}137.56}{\frac{\pi}{4} \times 23.5^2} = 39.51 \text{ N/mm}^2$$

Check
$$\frac{\sigma_{tf,\,\text{cal}}}{\sigma_{tf}} + \frac{\tau_{vf,\,\text{cal}}}{\tau_{vf}} \le 1.4$$

or
$$\frac{39.51}{100} + \frac{66.28}{100} = 1.0579 < 1.4$$

which is safe.

Example 10.8 An I.S.L.B. 300 @ 369.8 N/m transmits an end reaction of 260 kN to the web of I.S.M.B. 450 @ 710.2 N/m. Design a framed connection. See Fig. Ex. 10.8.

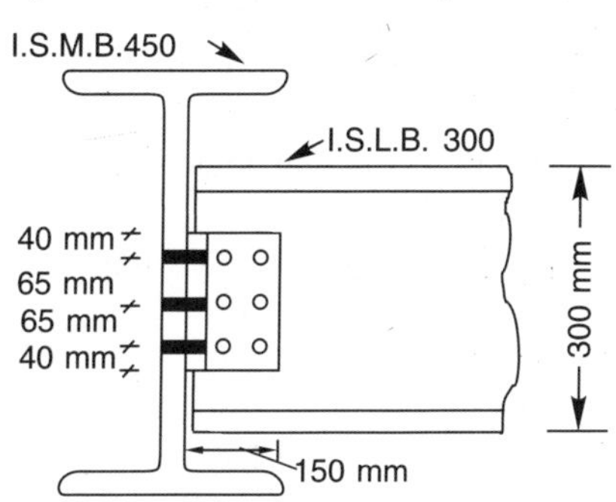

Fig. Ex. 10.8

Solution The relevant properties of the sections from I.S. Handbook No. 1 are:

I.S.L.B. 300	*I.S.M.B. 450*
$B = 150$ mm	150 mm
$T = 9.4$ mm	17.4 mm
$t = 6.7$ mm	9.4 mm
$g = 90$ mm	90 mm
$h = 300$ mm	450 mm

Connection of web of I.S.L.B 300 with angle

Rivets are in double shear. Provide 22 mm ϕ rivets.

$$d = 22 + 1.5 = 23.5 \text{ mm}$$

$$\text{Value of rivet in double shear} = 2 \times \frac{\pi}{4} \times 23.5^2 \times 100 = 86{,}747.22 \text{ N}$$

$$\text{Value of rivet in bearing} = 23.5 \times 6.7 \times 300 = 47{,}235 \text{ N}$$

Hence Rivet value, $R_v = 47{,}235$ N

$$n = \frac{260 \times 10^3}{47{,}235} = 5.504 \approx 6$$

$$\text{Minimum pitch} = 2.5\phi = 2.5 \times 22 = 55 \text{ mm}$$

Provide a pitch of 65 mm.

Minimum edge distance from I.S: 800–1984 for 22 mm ϕ rivets is 32 mm.

Two framing angles are provided, one on each side of the web.

Provide the rivets in two vertical rows on the web cleat angles.

$$\text{Minimum size of angle leg} = 2 \times 32 + 65 = 129 \text{ mm}$$

Provide two angles, 150 mm × 115 mm × 10 mm.

Connection of framing angle to web of I.S.M.B. 450

The rivets are in single shear. Provide 22 mm ϕ rivets.

$$\text{Gross diameter} = 22 + 1.5 = 23.5 \text{ mm}$$

$$\text{Value of rivet in single shear} = \frac{\pi}{4} \times 23.5^2 \times 100 = 43{,}373.61 \text{ N}$$

$$\text{Value of rivet in bearing} = 47{,}235 \text{ N}$$

Hence, the rivet value, $R_v = 43{,}373.61$ N

$$n' = \frac{260 \times 10^3}{43{,}373.61} = 5.994 \approx 6$$

Provide 3 rivets on each framing angle at a pitch of 65 mm and edge distance of 40 mm.

Check h = depth of framing angle = $2 \times 40 + 2 \times 65 = 210$ mm

$$\tau_{va} = \frac{260 \times 10^3}{2 \times 210 \times 10} = 61.90 \text{ N/mm}^2$$

$$< 100 \text{ N/mm}^2 \qquad (0.4 \times 250 = 100 \text{ N/mm}^2)$$

which is safe.

Example 10.9 A bracket plate is used to transfer the reaction of a beam to a column flange as shown in Fig. Ex. 10.9. The bracket plate is connected to the column flange by a 6 mm fillet weld. Compute the maximum load that can be placed over the bracket plates at a distance of 75 mm from the flange of the column section.

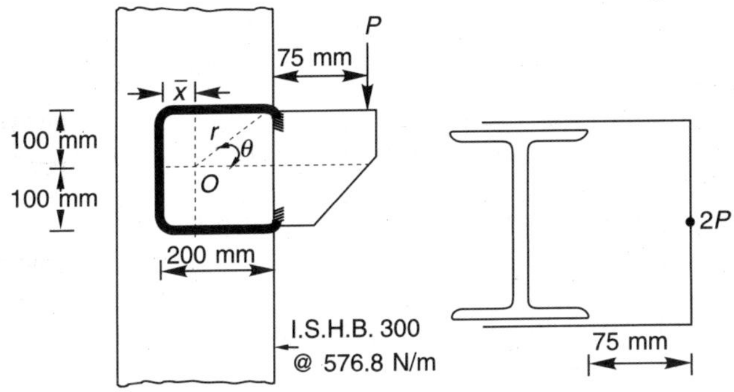

Fig. Ex. 10.9

Solution Let P be the load which the joint can transfer through one flange. The total load transferred by the bracket connection will be $2P$.

$$\text{Size of weld} = 6 \text{ mm}$$

$$\text{Effective throat thickness, } t = 0.7 \times 6 = 4.2 \text{ mm}$$

Let $\bar{x}$ = the distance of the centroid of the weld group from the left edge of the bracket plate.

$$\bar{x} = \frac{(2 \times 200 \times 4.2) \times 100 + (200 \times 4.2) \times 0}{2 \times 200 \times 4.2 + 200 \times 4.2} = 66.66 \text{ mm}$$

Polar moment of inertia of weld group

$$I_{xx} = \frac{4.2 \times 200^3}{12} + 2 \times \left[\frac{200 \times 4.2^3}{12} + 200 \times 4.2 \times 100^2\right]$$

$$= 1960.2 \times 10^4 \text{ mm}^4$$

$$I_{yy} = \frac{200 \times 4.2^3}{12} + 2 \times \left[\frac{4.2 \times 200^3}{12} + 4.2 \times 200(100 - 66.66)^2\right] + 200 \times 4.2 \times 66.66^2$$

$$= 1120.12 \times 10^4 \text{ mm}^4$$

$$J = I_{xx} + I_{yy}$$

$$= 1960.2 \times 10^4 + 1120.12 \times 10^4 = 3080.32 \times 10^4 \text{ mm}^4$$

r = distance of extreme point of weld from the C.G. of the weld group

$$= \sqrt{100^2 + (200 - 66.66)^2} = 166.672 \text{ mm}$$

$$\text{Direct shear stress } \tau_{vf_1} = \frac{P}{(2 \times 200 + 200) \times 4.2} = 0.0003968P \text{ N/mm}^2$$

Shear stress due to twisting moment

$$= \frac{P \times (200 - 66.66 + 75)}{3080.32 \times 10^4} \times 166.672 = 0.0011313P \text{ N/mm}^2$$

$$\cos \theta = \frac{200 - 66.66}{166.672} = 0.80$$

Resultant shear stress

$$= \sqrt{(0.0003968\,P)^2 + (.0011313\,P)^2 + 2 \times .0003968\,P \times .0011313\,P \times 0.8}$$

$$= P\sqrt{0.15 \times 10^{-6} + 1.2789 \times 10^{-6} + 0.7182 \times 10^{-6}}$$

$$= 1.4656 \times 10^{-3}P$$

Resultant shear stress should not exceed 108 N/mm^2.

or $\quad 1.4656 \times 10^{-3}P = 108$

or $$P = \frac{108}{1.4656 \times 10^{-3}} = 73689.95 \text{ N} = 73.689 \text{ kN}$$

∴ The total load that the bracket can support = 2 × 73.689 = 147.379 kN

Example 10.10 A bracket plate is welded to the flange of a column section I.S.H.B. 300 @ 618 N/m, as shown in Fig. Ex. 10.10. Calculate the size of the weld required to support a load of 110 kN.

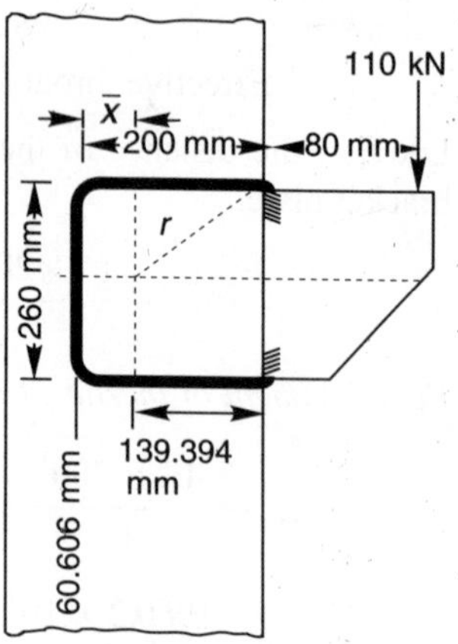

Fig. Ex. 10.10

Solution

Let

$\bar{x}$ = the distance of the centroid of the weld group from the left edge of the bracket plate.

t = throat thickness

$$\bar{x} = \frac{(2 \times 200 \times t) \times 100 + 260 \times t \times 0}{2 \times 200 \times t + 260 \times t}$$

$$= 60.606 \text{ mm}$$

Hence, eccentricity of the load = (200 – 60.606) + 80 = 139.394 + 80 = 219.4 mm.

Polar moment of inertia

$$I_{xx} = 2 \times \left[\frac{200 \times t^3}{12} + (200 \times t) \times 130^2\right] + \frac{t \times 260^3}{12}$$

$$= 822.466 \times 10^4 t \text{ mm}^4$$

$$I_{yy} = 2 \times \left[\frac{t \times 200^3}{12} + t \times 200\,(100 - 60.606)^2\right]$$

$$+ \frac{260 \times t^3}{12} + 260 \times t \times 60.606^2 = 290.9 \times 10^4\, t \text{ mm}^4$$

In the calculation of I_{xx} and I_{yy}, the term $\frac{260 \times t^3}{12}$ has been neglected. This is because its numerical value will be insignificant, and will also lead to unnecessary complications in calculation if considered.

$$J = I_{xx} + I_{yy}$$
$$= (822.466\,t + 290.9\,t) \times 10^4 = 1113.366 \times 10^4\,t \text{ mm}^4$$
$$r = \sqrt{130^2 + 139.394^2} = 190.606 \text{ mm}$$
$$\cos\theta = \frac{139.394}{190.606} = 0.73132$$

Direct shear stress, $\tau_{vf_1} = \dfrac{110 \times 10^3}{(2 \times 200 + 260)\,t} = \dfrac{166.66}{t}$ N/mm^2

Shear stress due to twisting moment, $\tau_{vf_2} = \dfrac{110 \times 10^3 \times 219.4}{1113.366 \times 10^4\,t} \times 190.606$

$$= \frac{413.15}{t} \text{ N/mm}^2$$

Combined stress

$$= \left[\left(\frac{166.66}{t}\right)^2 + \left(\frac{413.15}{t}\right)^2 + 2 \times \frac{166.66}{t} \times \frac{413.15}{t} \times 0.73132\right]^{1/2}$$

$$= \frac{546.97}{t}$$

The combined stress should not exceed 108 N/mm^2.

$\therefore$ $$\frac{546.97}{t} = 108$$

or $$t = 5.06 \text{ mm}$$

Hence, size of weld $= \dfrac{5.06}{0.7} = 7.23$ mm ≈ 8 mm

Provide an 8 mm fillet weld.

Example 10.11 A bracket plate 10 mm thick is used to transmit a reaction of 100 kN at an eccentricity of 100 mm from the column flange as shown in Fig. Ex. 10.11. Design the weld.

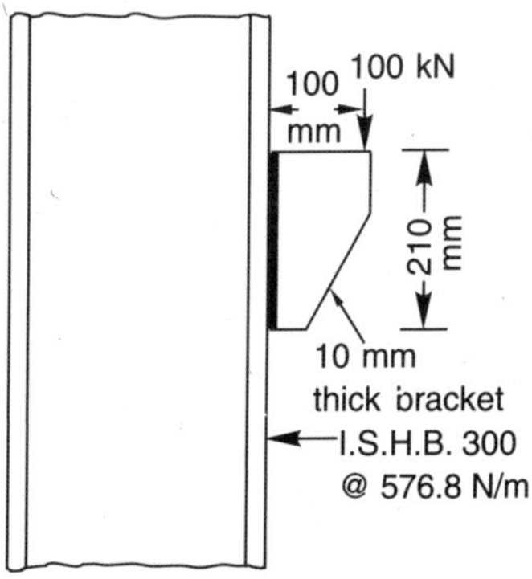

Fig. Ex. 10.11

Solution The joint can be designed as a butt welded or a fillet welded joint.

(a) *Butt weld*

Let us provide a double *J*-butt weld. Effective throat thickness,

$$t = \text{thickness of bracket plate} = 10 \text{ mm}$$

Depth of the bracket plate from Eq. 10.28,

$$d = \sqrt{6M/(t\sigma_b)} = \sqrt{6 \times 100 \times 10^3 \times 100/(10 \times 165)} = 190.06 \text{ mm}$$

Provide a depth of 210 mm.

Direct shear stress,

$$\tau_{vf} = \frac{100 \times 10^3}{210 \times 10} = 47.619 \text{ N/mm}^2$$

Actual bending stress,

$$\sigma_b = \frac{6M}{td^2} = \frac{6 \times 100 \times 10^3 \times 100}{10 \times 210^2} = 136.05 \text{ N/mm}^2$$

Check

$$\sqrt{\sigma_{bc,\,t}^2 + 3\tau_{vf}^2} < \sigma_e$$

$$= \sqrt{136.05^2 + 3 \times 47.619^2} = 159.098 \text{ N/mm}^2 < 250 \text{ N/mm}^2$$

which is safe.

Fillet weld

Maximum size of the fillet weld = 10 – 1.5 = 8.5 mm

Minimum size of the fillet weld = 5 mm

Let us provide a 6 mm fillet weld.

Effective throat thickness = 0.7 × 6 = 4.2 mm

$$d = \sqrt{6M/(2 \times t \times \tau_{vf})}$$

$$= \sqrt{6 \times 100 \times 10^3 \times 100/(2 \times 4.2 \times 108)} = 257.17 \text{ mm} \approx 280 \text{ mm}$$

Let us provide depth of the bracket plate = 280 mm.

$$\tau_{vf,\,\text{cal}} = \frac{100 \times 10^3}{2 \times 4.2 \times 280} = 42.517 \text{ N/mm}^2$$

$$\tau_{vf_1,\,\text{cal}} = \frac{6 \times 100 \times 10^3 \times 100}{2 \times 4.2 \times 280^2} = 91.10 \text{ N/mm}^2$$

$$\text{combined stress} = \sqrt{\tau_{vf,\,\text{cal}}^2 + \tau_{vf_1,\,\text{cal}}^2}$$

$$= \sqrt{91.10^2 + 42.517^2} = 100.53 \text{ N/mm}^2$$

$$< 108 \text{ N/mm}^2$$

which is safe.

Example 10.12 A joist cutting is used as a bracket to support a load of 135 kN. It is welded to the column flange as shown in Fig. Ex. 10.12. Compute the size of the fillet weld.

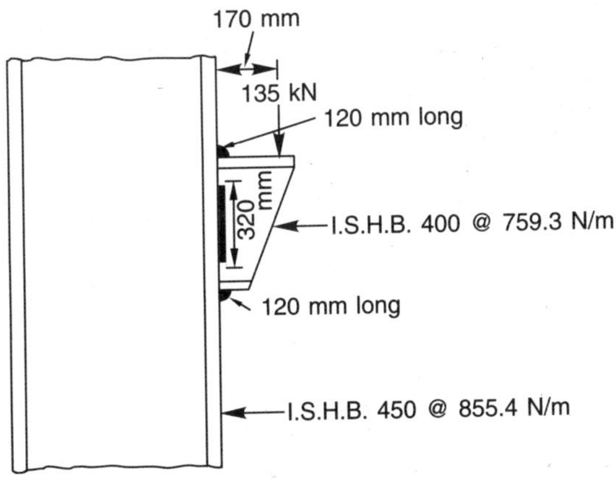

Fig. Ex. 10.12

Solution

Let t = effective throat thickness of the weld on the web

$2t$ = effective throat thickness of weld on the flange

$$I_{xx} \text{ of the weld} = \frac{2 \times 320^3 \times t}{12} + (2 \times t \times 120) \times 2 \times 200^2$$

$$= 546.133 \times 10^4 t + 1920 \times 10^4 t \text{ mm}^4$$

$$= 2466.133 \times 10^4 t \text{ mm}^4$$

Note In the calculation of the moment of inertia I_{xx} of the weld, the moment of inertia of the flange weld about its own axis is insignificant and has been therefore, neglected.

Direct shear stress in weld

Total area of the welds = area of weld on the flanges + areas of weld on the web

$$= 2 \times 120 \times 2t + 2 \times 320 \times t$$

$$= 1120\, t \text{ mm}^2$$

Load is shared in the proportion of areas of welds.

$$\text{Load shared by the flange welds} = \frac{2 \times 120 \times 2t}{1120t} \times 135 = 57.857 \text{ kN}$$

$$\text{Load shared by the web welds} = \frac{2 \times 320 \times t}{1120t} \times 135 = 77.143 \text{ kN}$$

$$\text{Shear stress in the flange weld} = \frac{57.857 \times 10^3}{2 \times 120 \times 2t} = \frac{120.53}{t} \text{ N/mm}^2$$

$$\text{Shear stress in the web welds} = \frac{77.143 \times 10^3}{2 \times 320 \times t} = \frac{120.53}{t} \text{ N/mm}^2$$

The moment is shared in proportion of the extreme weld distance from the neutral axis. Horizontal shear stress in the flange weld,

$$\tau_{fv1,cal} = \frac{135 \times 10^3 \times 170}{2466.133 \times 10^4 t} \times 200 = \frac{186.12}{t} \text{ N/mm}^2$$

Taking the worst combination,

$$108 = \left[\left(\frac{186.12}{t}\right)^2 + \left(\frac{120.53}{t}\right)^2\right]^{1/2}$$

or

$$108 = \frac{221.738}{t}$$

or

$$t = 2.053 \text{ mm}$$

$$\text{Size of the web weld} = \frac{2.053}{0.7} = 2.93 \text{ mm} \approx 5 \text{ mm}$$

Hence, provide the size of weld on web as equal to 5 mm.

$$\text{Size of weld on flange} = 2 \times 2.93 = 5.86 \text{ mm} \approx 6 \text{ mm}$$

Note The maximum size of the weld which can be provided on flange is

$$= 12.7 - 1.5 = 11.2 \text{ mm} > 6 \text{ mm}$$

which is all right.

Example 10.13 An I.S.A., 150 × 115 × 10 mm is welded to the column flange as shown in Fig. Ex. 10.13. It is to support and transfer an end reaction of 110 kN to the column section I.S.H.B. 250 @ 536.6 N/m. Design the weld.

Solution

For I.S.H.B. 250, $B = 250$ mm.

Therefore the maximum length of weld = 250 mm on each side.

Let t = effective throat thickness.

$$\tau_{vf,\,cal} = \frac{110 \times 10^3}{2 \times (250 \times t)} = \frac{220}{t} \text{ N/mm}^2$$

$$\tau_{vf1,\,cal} = \frac{M}{\text{lever arm} \times \text{area}}$$

$$= \frac{110 \times 10^3 \times 55}{150 \times (250 \times t)} = \frac{161.33}{t} \text{ N/mm}^2$$

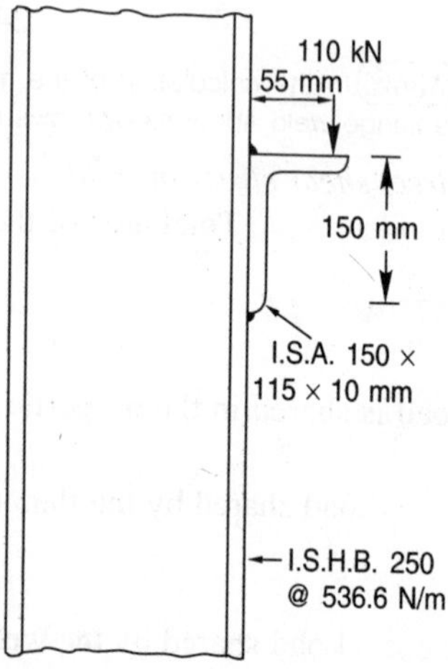

Fig. Ex. 10.13

The combined stress in fillet weld should be less than 108 N/mm².

$$108 = \left[\left(\frac{161.33}{t}\right)^2 + \left(\frac{220}{t}\right)^2\right]^{1/2}$$

or

$$t = 2.526 \text{ mm}$$

$$\text{Size of the weld} = \frac{2.526}{0.7} = 3.60 \text{ mm} \approx 5 \text{ mm}$$

Provide a 5 mm weld.

Example 10.14 Design an unstiffened welded seat connection for a beam I.S.M.B. 250 @ 365.9 N/m, transmitting 84 kN to the I.S.H.B. 200 @ 365.9 N/m.

Solution See Example 10.6.

$$\text{Bearing length, } b = 32.46 \text{ mm}$$

The seat angle required will be same, i.e., 150 × 75 × 12 mm.

$$e_1 = 4.23 \text{ mm}$$

Connection

$$\text{Length of vertical legs} = 150 \text{ mm}$$

$$\text{Vertical shear/mm, } \tau_{vf_1} = \frac{84 \times 10^3}{2 \times 150 \times 1} = 280 \text{ N/mm}$$

$$\text{Horizontal shear/mm, } \tau_{vf_2} = \frac{84 \times 10^3 \times 4.23}{2 \times \frac{1}{6} \times 150^2 \times 1} = 47.376 \text{ N/mm}$$

$$\text{Resultant shear stress/mm} = \sqrt{280^2 + 47.376^2} = 283.98 \text{ N/mm}$$

The resultant shear/mm must be equal to the strength of weld/mm

$$= 0.7 \times S \times 108 = 75.6 \times S \text{ N/mm}$$

or $$75.6\,S = 283.98$$

or $$S = 3.756 \text{ mm}$$

$$\nless 5 \text{ mm}$$

Provide a 5 mm fillet weld.

Example 10.15 Design a stiffened seat connection to join I.S.M.B. 350 @ 514 N/m with a column section I.S.H.B. 300 @ 576.8 N/m. The beam has to transmit an end reaction of 230 kN.

Solution See Example 10.7.

$$\text{Bearing length } b = 97.746$$

Provide a clearance of 10 mm.

$$\text{Width of seat plate} = 97.746 + 10 = 107.746 \text{ mm}$$

Adopt width of seat as 120 mm.

$$\text{Thickness of seat plate} = \text{thickness of beam flange} = 14.2 \text{ mm} \approx 16 \text{ mm}$$

Provide a seat plate 120 × 16 mm.

$$\text{Thickness of the stiffening plate} = \text{thickness of beam web} = 8.1 \text{ mm} \approx 10 \text{ mm}$$

Provide a stiffener plate 200 mm × 10 mm. Assume length of weld as shown in Fig. Ex. 10.15(a). Distance of end reaction from outer end of seat plate,

$$= 120 - \frac{1}{2} \times 97.746 = 71.127 \text{ mm}$$

Bending moment, $M = 230 \times 71.127$
$= 16{,}359.21$ kNmm

Width of the flange of beam = 140 mm

C.G. of the weld group,

$$\bar{y}_1 = \frac{2 \times 200 \times 100 + 0}{2 \times 200 + (120 - 10)} = 78.43 \text{ mm}$$

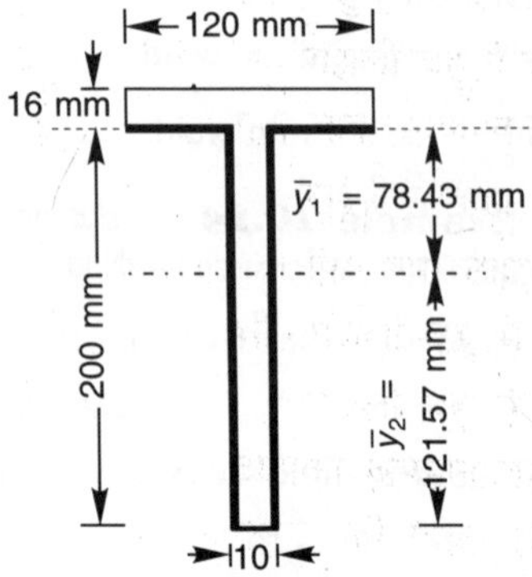

Fig. Ex. 10.15(a)

$$\bar{y}_2 = 200 - 78.43 = 121.57 \text{ mm}$$

$$I_{xx} = \left[1 \times (120 - 10) \times 78.43^2 + 2 \times \frac{200^2}{12} \times 1 + 2 \times 200 \times (121.57 - 100)^2\right]$$

$$= 219.62 \times 10^4 \text{ mm}^4$$

$$\text{Vertical shear, } \tau_{vf_1} = \frac{230 \times 10^3}{2 \times 200 + 110} = 450.98 \text{ N/mm}$$

$$\text{Horizontal shear, } \tau_{vf_2} = \frac{16{,}359.21 \times 10^3}{219.607 \times 10^4} \times 78.43 = 584.249 \text{ N/mm}$$

Note The horizontal shear is calculated assuming that the weld will not fail in thrust.

Resultant shear stress/mm = length of weld/mm

$$\sqrt{450.98^2 + 584.249^2} = 0.7 \times S \times 108 \times 1$$

or $S = 9.76 \text{ mm} \approx 10 \text{ mm}$

Provide a 10 mm fillet weld to make the connection.

Provide a nominal size flange cleat 100 × 100 × 6 mm and connect it by a 5 mm fillet weld as shown in Fig. Ex. 10.15(b).

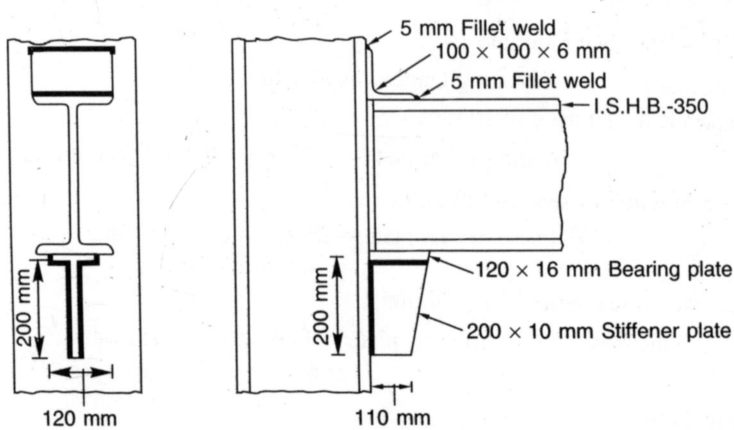

Fig. Ex. 10.15(b)

Example 10.16 In a framed connection an I.S.L.B. 350 @ 485.6 N/m transmits an end reaction of 150 kN and a moment of 15 kNm to a column I.S.H.B. 300 @ 576.8 N/m. Design the connections.

Solution As the moment is small, a clip angle connection may be provided. The relevant properties of I.S.L.B. 350 @ 485.6 N/m and I.S.H.B. 300 @ 576.8 N/m from I.S. Handbook No. 1 are:

	I.S.L.B. 350	*I.S.H.B. 300*
g	—	140 mm
B	165 mm	—
T	11.4 mm	10.6 mm
t	7.4 mm	7.6 mm
D	350 mm	—

Let us provide 22 mm ϕ power driven rivets.

$$\text{Gross diameter} = 22 + 1.5 = 23.5 \text{ mm}$$

$$\text{Value of the rivet in tension} = \frac{\pi}{4} \times 23.5^2 \times 100 = 43{,}373.6 \text{ N}$$

$$\text{Value of rivet in single shear} = 43{,}373.6 \text{ N}$$

$$\text{Value of the rivet in double shear} = 2 \times \frac{\pi}{4} \times 23.5^2 \times 100 = 86{,}747.2 \text{ N}$$

Value of the rivet in bearing (for web clip angle connecting beam web)

$$= 23.5 \times 7.4 \times 300 = 52{,}170 \text{ N}$$

Value of the rivet in bearing (for web clip angle connecting column flange)

$$= 23.5 \times 10.6 \times 300 = 74{,}730 \text{ N}$$

Flange clip angles Provide two flange clip angles, I.S.A. 200 × 100 mm, with 100 mm leg connected to the column flange.

Assume the gauge distance for 100 mm leg, $g = 60$ mm

$$\therefore \quad \text{Lever arm} = 60 + 350 + 60 = 470 \text{ mm}$$

$$\text{Pull, } P = \frac{15 \times 10^3}{470} = 31.914 \text{ kN}$$

Consider the effects of initial tension (asume thickness of angle = 15 mm).

Then,

$$\text{Maximum pull on rivet, } T = P\left[1 + 0.75\left(\frac{g - t/2}{a}\right)\right]$$

$$= 31.914 \times \left[1 + 0.75\left(\frac{60 - 15/2}{40}\right)\right]$$

$$= 63.329 \text{ kN} = 63329 \text{ N}$$

Provide two rivets to connect the flange clip leg with the column flange.

Safe pull transmitted by the rivets = 2 × 43,373.614 = 86,747.228 N

> 63329 N

which is safe.

Maximum moment in clip angle,

$$M_1 = 0.5P\left(g - \frac{t}{2}\right) = 0.5 \times 31.914 \times (60 - 15/2) = 837.7425 \text{ kN mm}$$

Here, $l = g = 140$ mm

$$t = \sqrt{6M/(l\sigma_{bs})} = \sqrt{6 \times 837.7425 \times 10^3/(140 \times 185)} = 13.93 \text{ mm}$$

< 15 mm

Provide I.S.A. 200 × 100 × 15 mm clip angles.

Number of rivets to connect the flange clip with beam flange

$$\text{Horizontal shear, } V = \frac{15 \times 10^3 \times 10^3}{350} = 42,857 \text{ N}$$

$$n = \frac{42,857}{43,373.614} = 0.9880 \simeq 4$$

Provide four rivets to make the connection. The rivets are provided in two rows.

Web clip angles Number of rivets required to connect the web angle to column flange (R_v = 43,373.6 N)

$$= \frac{150 \times 10^3}{43,373.6} = 3.458 \approx 4$$

Provide four rivets of 22 mm ϕ in two rows, at a pitch of 70 mm and an edge distance of 40 mm.

Number of rivets required to connect web clip angle to web of beam (R_v = 52,170 N)

$$= \frac{150 \times 10^3}{52,170} = 2.8752 \approx 4$$

Provide four rivets of 22 mm ϕ in two rows, at a pitch of 70 mm and edge distance of 32 mm.

Size of web clip angle

Size of each leg = s + 2 × edge distance

= 70 + 2 × 32 = 134 mm ≈ 150 mm

Provide two web clip angles of 150 × 150 × 10 mm. (See Fig. Ex. 10.16)

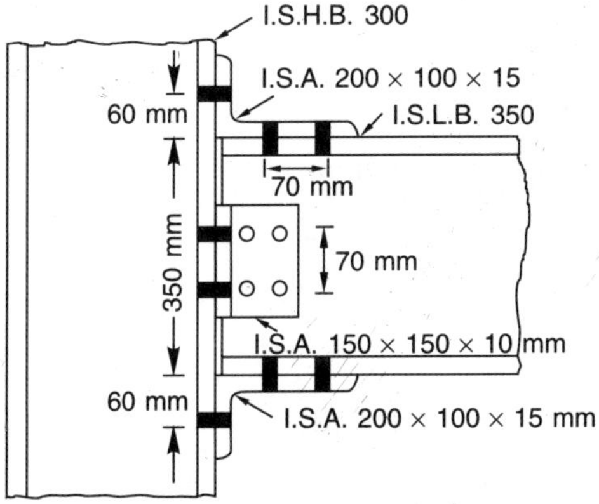

Fig. Ex. 10.16

Example 10.17 Design a split beam connection to connect an I.S.L.B. 600 @ 976.1 N/m to a column I.S.H.B. 350 @ 661.2 N/m. The end reaction to be resisted is 150 kN and the bending moment is of 80 kNm.

Solution The relevant properties of the sections from I.S. Handbook No. 1 are:

	I.S.L.B. 600	*I.S.H.B. 350*
B	210 mm	250 mm
T	15.5 mm	11.6 mm
t	10.5 mm	8.3 mm

Let us provide 22 mm ϕ power driven rivets.

$$\text{Gross diameter} = 22 + 1.5 = 23.5 \text{ mm}$$

$$\text{Value of the rivet in single shear} = \frac{\pi}{4} \times 23.5^2 \times 100 = 43{,}373.6 \text{ N}$$

$$\text{Value of the rivet in double shear} = 2 \times \frac{\pi}{4} \times 23.5^2 \times 100 = 86{,}747.2 \text{ N}$$

$$\text{Value of the rivet in tension} = \frac{\pi}{4} \times 23.5^2 \times 100 = 43{,}373.6 \text{ N}$$

Value of the rivet in bearing (t = 10 mm for web angle)

$$= 23.5 \times 10 \times 300 = 70{,}500 \text{ N}$$

Value of the rivet in bearing (t = 10.5 mm for beam web)

$$= 23.5 \times 10.5 \times 300 = 74{,}025 \text{ N}$$

The rivets are subjected to a pull acting through the centroid of the rivets.

Let thickness of split beam = 10 mm.

$$P = \frac{M}{D+t} = \frac{80 \times 10^3 \times 10^3}{600 \times 10} = 131{,}147.54 \text{ N}$$

$$\text{Number of rivets required to resist the pull} = \frac{131{,}147.54}{43{,}373.6} = 3.0236 \approx 4$$

Provide 4 rivets of 22 mm ϕ. Let us provide two rivets in each row.

Thickness of T-flange Assume a gauge distance for split beam = 140 mm.

$$M' = 0.5P \times \frac{(g-t)}{2}$$

$$= 0.5 \times 131{,}147.54 \times \left(\frac{140-10}{2}\right) = 4{,}262{,}295.1 \text{ Nmm}$$

$$T = \sqrt{6M'/B\sigma_{bs}} = \sqrt{6 \times 4{,}262{,}295.1/(250 \times 185)} = 23.51 \text{ mm}$$

Thickness of T-web

$$t = \frac{P}{b \times \sigma_{at}} = \frac{131{,}147.54}{(210 - 2 \times 23.5) \times 150} = 5.36 \text{ mm}$$

From I.S. Handbook No. 1, select I.S.W.B. 600 @ 1423.4 N/m. Cut this section from the centre of the web to get split beams. The relevant properties of this section are:

$$T = 23.6 \text{ mm}, t = 11.8 \text{ mm}$$

$$\text{Length of the web of split beam} = \frac{600}{2} = 300 \text{ mm}$$

The rivets connecting the split beam web to the flange of the beam section will be in single shear and bearing.

$$\therefore \qquad R_v = 43{,}373.6 \text{ N}$$

$$V = \frac{M}{D} = \frac{80 \times 10^3 \times 10^3}{600} = 133{,}333.33 \text{ N}$$

$$\text{Number of rivets, } n' = \frac{133{,}333.33}{43{,}373.6} = 3.07 \approx 4$$

Provide four rivets, 22 mm ϕ in two rows. The length available for connection is 300 mm.

Web cleats Two angles, one on each side of the web are provided. The rivets are in double shear and bearing.

$$R_v = 70{,}500 \text{ N}$$

$$\text{End shear} = 150 \text{ kN}$$

$$\text{Number of rivets required} = \frac{150 \times 10^3}{70{,}500} = 2.1276 \approx 3$$

Provide three rivets to connect the web cleat angle to the web of the beam. The same number of rivets are provided to connect the web cleat with the flange of the column. As the rivets can be accommodated in a single row, provide a 90 × 90 × 10 mm angle section. See Fig. Ex. 10.17.

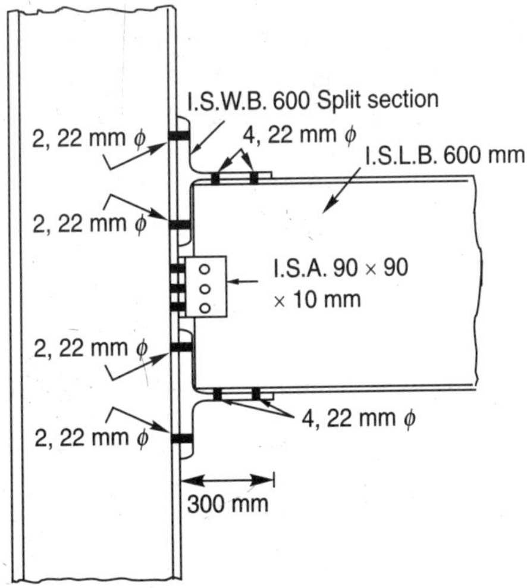

Fig. Ex. 10.17

Example 10.18 Design a split beam connection to connect an I.S.L.B. 550 @ 846.6 N/m to a column I.S.H.B. 300 @ 618 N/m. The end reaction and the end moment are 165 kN and 200 kNm respectively.

Solution The relevant properties of the sections from I.S. Handbook No. 1 are:

	I.S.L.B. 550	*I.S.H.B. 300*
B	190 mm	250 mm
T	150 mm	10.6 mm
t	9.9 mm	9.4 mm
D	550 mm	300 mm

Let us provide 22 mm ϕ power driven rivets.

$$\text{Gross diameter} = 22 + 1.5 = 23.5 \text{ mm}$$

$$\text{Value of the rivet in single shear} = \frac{\pi}{4} \times 23.5^2 \times 100 = 43{,}373.6 \text{ N}$$

$$\text{Value of the rivet in double shear} = 2 \times \frac{\pi}{4} \times 23.5^2 \times 100 = 86{,}747.22 \text{ N}$$

$$\text{Value of the rivet in tension} = \frac{\pi}{4} \times 23.5^2 \times 100 = 43{,}373.6 \text{ N}$$

Value of the rivet in bearing (assuming web angle, 10 mm thick)

$$= 23.5 \times 9.9 \times 300 = 69{,}795 \text{ N}$$

Let thickness of the split beam = 10 mm

$$\text{Pull on the rivets, } P = \frac{M}{D+t} = \frac{200 \times 10^3 \times 10^3}{550 + 10} = 357{,}142 \text{ N}$$

$$\text{Number of rivets required, } n = \frac{357{,}142}{43{,}373.6} = 8.234 \approx 9$$

As not more than four rivets can be provided on a split beam, let us provide flange clip seats to increase the lever arm.

Provide two flange clip seats of I.S.L.B. 350 @ 485.6 N/m.

$$\text{Pull on the rivets, } P = \frac{M}{D+t} = \frac{200 \times 10^3 \times 10^3}{(550 + 2 \times 350) + 10} = 158{,}730 \text{ N}$$

$$\text{Number of rivets} = \frac{158{,}730}{43{,}373.6} = 3.659$$

Provide four rivets, 22 mm ϕ, in two rows to connect it with the column flange.

Thickness of T-flange

$$M' = 0.5\,P \times \frac{(g-t)}{2} = 0.5 \times 158730 \times \left(\frac{100-10}{2}\right) = 3{,}571{,}425 \text{ Nmm}$$

$$T = \sqrt{6\,M'/B\sigma_{bs}} = \sqrt{(6 \times 3{,}71{,}425)/(250 \times 185)} = 21.52 \text{ mm}$$

Thickness of T-web

$$t = \frac{P}{b \times \sigma_{at}} = \frac{158{,}730}{(190 - 2 \times 23.5) \times 150} = 7.4 \text{ mm}$$

From I.S. Handbook No. 1, select an I-section providing T = 21.52 mm and t = 7.4 mm. Provide I.S.W.B. 600 @ 1423.4 N/m.

$$T = 23.6 \text{ mm}, t = 11.8 \text{ mm}$$

Length of the web of split beam = 600/2 = 300 mm

Rivets connecting split beam to flange clip seat Rivets will be in single shear.

$$R_v = 43{,}373.6 \text{ N}$$

$$V = \frac{M}{D}$$

where $D = 550 + 2 \times 350 = 1250$ mm

$$\therefore \quad V = \frac{200 \times 10^3 \times 10^3}{1250} = 160{,}000 \text{ N}$$

$$\text{Number of rivets} = \frac{160{,}000}{43{,}373.6} = 3.688$$

Provide four rivets in two rows. The length available for connection is 300 mm and is more than sufficient for the connection.

Rivets connecting flange clip seat to beam flange

$$V = \frac{M}{D}$$

where $D = 550$ mm

$$\therefore \quad V = \frac{200 \times 10^3 \times 10^3}{550} = 363{,}636.36 \text{ N}$$

$$\text{Number of rivets} = \frac{363{,}636.36}{43{,}373.6} = 8.38 \approx 10$$

Provide ten rivets in two rows, i.e., five rivets in each row (Fig. Ex. 10.18).

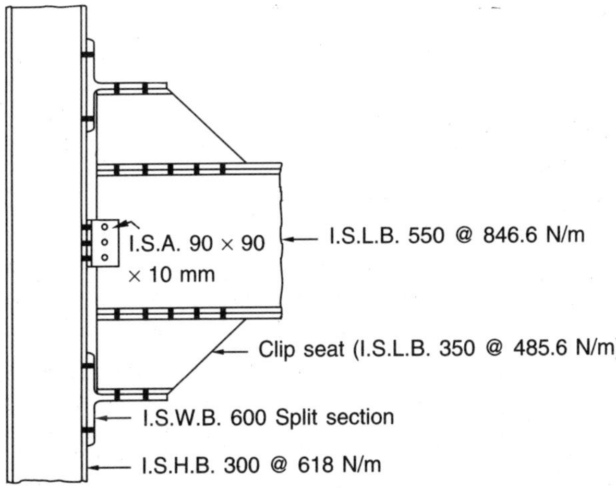

Fig. Ex. 10.18

Web cleat

The rivet value = 69,795 N

End reaction = 165×10^3 N

$$\text{Number of rivets} = \frac{165 \times 10^3}{69{,}795} = 2.36$$

Provide three rivets to connect the web cleat with the beam web. The same number of rivets are required to connect the column flange with the web cleat. As the rivets can be accommodated in one row provide 90 × 90 × 10 mm angle section on each side of the web.

Example 10.19 An I.S.M.B. 500 @ 852.5 N/m transmits an end reaction of 200 kN and a bending moment of 110 kNm to the flange of a column I.S.H.B. 300 @ 576.8 N/m. Design a welded connection.

Solution As the moment is quite high, a moment connection is provided. The relevant properties of the sections from I.S. Handbook No. 1 are:

	I.S.M.B. 500	*I.S.H.B. 300*
D	500 mm	300 mm
B	180 mm	250 mm
T	17.2 mm	10.6 mm
t	10.2 mm	7.6 mm
h_2	37.95 mm	

Provide a plate at the top of the beam flange.

$$\text{Force in top plate} = \frac{110 \times 10^3}{\text{lever arm}} = \frac{110 \times 10^3}{500} = 220 \text{ kN}$$

Allowable axial tensile stress, $0.60 f_y = 150 \text{ N/mm}^2$.

$$\text{Cross-sectional area of the plate} = \frac{220 \times 10^3}{150} = 1466.66 \text{ mm}^2$$

Now $B = 180$ mm. Let us provide a 180 mm wide plate. The plate over the flange is tapered to 160 mm to facilitate connection.

$$\text{Thickness of the top plate} = \frac{1466.66}{160} = 9.16 \text{ mm} \approx 10 \text{ mm}$$

Provide a plate 180 mm wide, tapered to 160 mm at the end and let it be 10 mm thick. Let us provide a 6 mm weld.

$$\text{Strength of the weld/mm} = 0.7 \times 6 \times 108 = 453.6 \text{ N}$$

$$\text{Length of the weld required} = \frac{220 \times 10^3}{453.6} = 485 \text{ mm}$$

$$\text{Provide side filled weld} = \frac{485 - 160}{2} = 162.5 \text{ mm} \approx 170 \text{ mm}$$

$$\text{Total length of the weld} = 2 \times 170 + 160 = 340 + 160 = 500 \text{ mm}$$

The length of the unwelded portion of the plate should be equal to the width of plate = 180 mm

$$\text{Length of the top plate} = 180 + 170 = 350 \text{ mm}$$

Connect this plate to the column flange by complete penetration butt weld.

The end reaction is 200 kN. A seat angle may be designed for resisting this end reaction.

$$\text{Bearing length, } b \nless \frac{1}{2} \times \frac{R}{\sigma_p t}$$

$$\nless \frac{1}{2} \times \frac{200 \times 10^3}{187.5 \times 10.2} \nless 52.28 \text{ mm}$$

$$b = \frac{R}{\sigma_p t} - \sqrt{3} h_2$$

$$= \frac{200 \times 10^3}{187.5 \times 10.2} - \sqrt{3} \times 37.95 = 38.82 \text{ mm}$$

Provide a bearing length of 52.28 mm.

Provide a clearance of 10 mm.

$$\text{Minimum angle leg} = 52.28 + 10 = 62.28 \text{ mm}$$

Select I.S.A. 200 × 150 × 18 mm.

$$e_1 = 10 + \frac{52.28}{2} - 18 - 13.5 = 4.64 \text{ mm}$$

$$t = \sqrt{6 \times 200 \times 10^3 \times 4.64/180 \times 185} = 12.93 \text{ mm} < 18 \text{ mm}$$

So provide 200 × 150 × 18 mm seat angle with 200 mm leg as the vertical leg.

$$\text{Vertical shear/mm, } \tau_{vf_1} = \frac{200 \times 10^3}{2 \times 200} = 500 \text{ N/mm}$$

$$\text{Horizontal shear/mm, } \tau_{vf_2} = \frac{200 \times 10^3 \times 4.64}{2 \times 1/6 \times 200^2 \times 1} = 69.6 \text{ N/mm}$$

The resultant shear/mm = strength of the weld/mm

$$\sqrt{500^2 + 69.6^2} = 0.7 \times S \times 108 \times 1$$

or

$$S = 6.67 \text{ mm} \simeq 7 \text{ mm}$$

Hence, provide a 7 mm fillet weld.

$$\text{Force in the two stiffeners} = \frac{M}{D} = \frac{110 \times 10^3 \times 10^3}{500} = 220 \text{ kN}$$

Pull in stiffeners should be equal to the strength of the weld.

Let us assume the size of weld as 6 mm.

$$220 \times 10^3 = l \times 0.7 \times 6 \times 108$$

or

$$l = \frac{220 \times 10^3}{0.7 \times 6 \times 108} = 485.0 \text{ mm}$$

Let us provide stiffener plates for half the depth of the column section. The available places to weld the stiffener plate with the column flange will be four in number, Fig. Ex. 10.19.

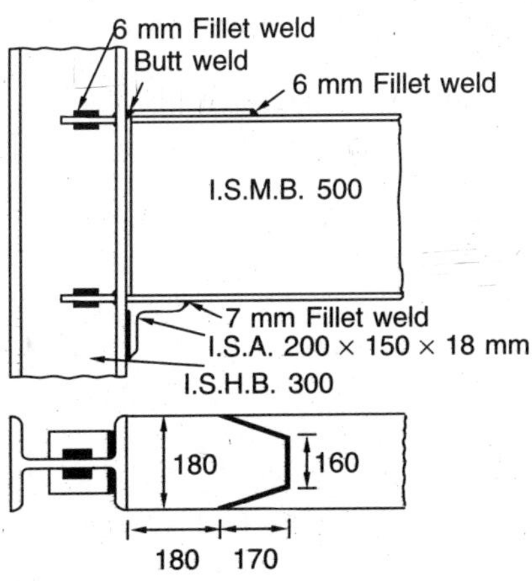

Fig. Ex 10.19

Maximum column flange length available for connection

$$= \frac{250-(7.6+2\times5.3)}{2} = 115.9 \text{ mm}$$

Let us provide a stiffener plate 100 mm wide and 10 mm thick.

Total length of the weld that can be accommodated to connect the stiffener with the column flange

$$= 4 \times (100 - 5.3) = 378.8 \text{ mm} < 476.19 \text{ mm}$$

The remaining length of the weld can be adjusted on the stiffener plate connecting the web of column.

Let us provide 40 mm of weld length on each side of the stiffener plate joining it with column web.

$$\text{Total length of weld provided} = 378.8 + 4 \times 40 = 538.8 \text{ mm}$$
$$> 485.0$$

which is safe.

Exercises

10.1 Design a bracket connection as shown in Figs Prob. 10.1(a) and (b). Provide mm 22 ϕ power driven rivets.

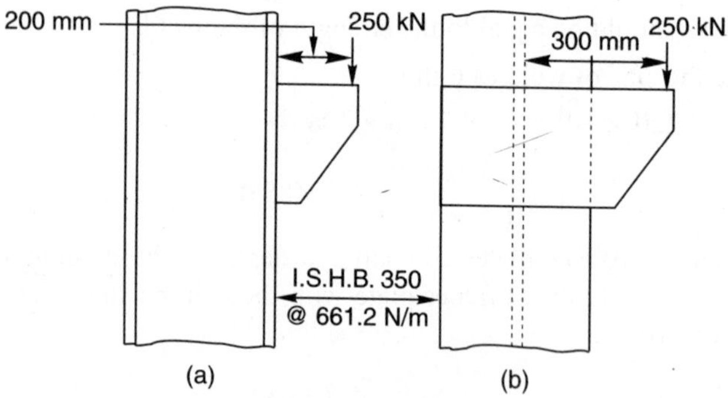

Fig. Prob. 10.1

10.2 Design the column to bracket connection with rivets. The bracket type is shown in Fig. Prob. 10.2.

10.3 A load of 150 kN is to be transferred through a bracket plate 12 mm thick welded (normal) to the flange of a column section I.S.H.B. 300 @ 618 N/m. The load is acting at an eccentricity of 100 mm from the column face. Design the connection using HSFG bolts.

10.4 A bracket plate is welded to the flange of a column as shown in Fig. Prob. 10.4. Design the connection.

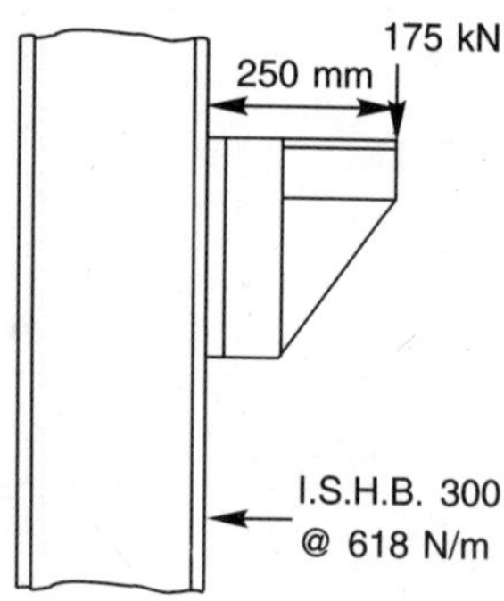

Fig. Prob. 10.2

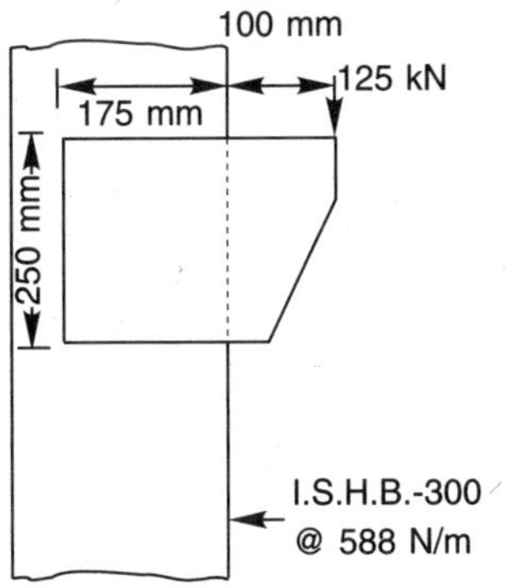

Fig. Prob. 10.4

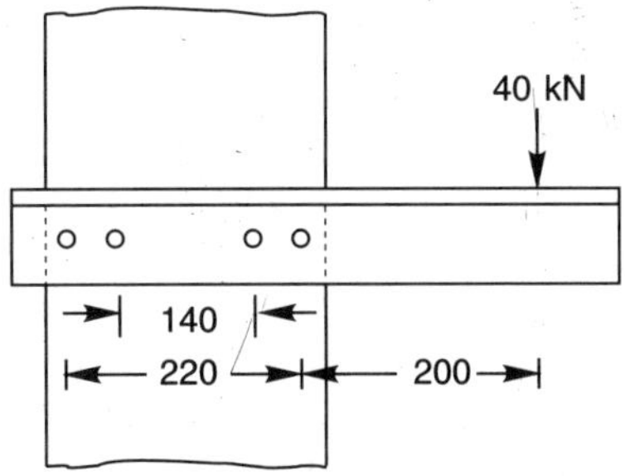

Fig. Prob. 10.5

10.5 A 8 mm thick angle is riveted to the flange of an I.S.H.B. 350 @ 661.2 N/m column as shown in Fig. Prob. 10.5. Both members have $f_y = 250$ N/mm^2. Determine which rivet is critical. What the maximum total stress will be when the 40 kN load is applied ? Use 25 mm ϕ power driven shop rivets. Is the connection safe?

10.6 Determine the maximum permissible load P which can be supported by the 10 mm plate connected to the I.S.H.B. 250 @ 536.6 N/m column flange shown in Fig. Prob. 10.6. Use 16 mm ϕ power driven field rivets.

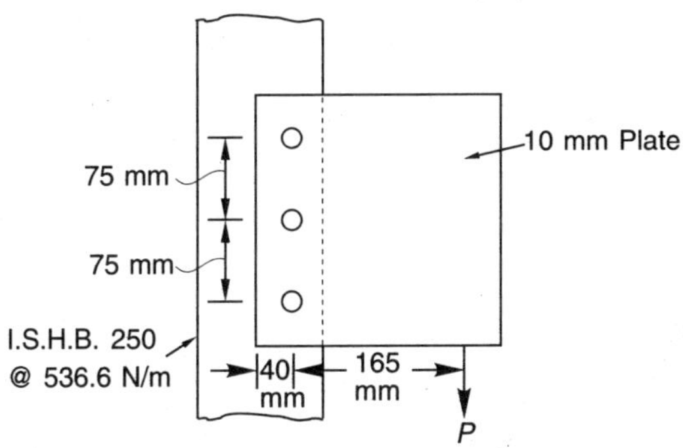

Fig. Prob. 10.6

10.7 Design a framed connection to connect an I.S.L.B. 350 @ 485.6 N/m transmitting an end reaction of 300 kN to the web of I.S.M.B. 550 @ 1017.3 N/m.
(i) Design the riveted connection.
(ii) Design the welded connection.

10.8 An I.S.M.B. 400 @ 604.3 N/m is connected to the web of a girder I.S.M.B. 550 @ 1017.3 N/m. The maximum end shear is 120 kN. Design the framed connection. If another I.S.M.B. 400 @ 604.3 N/m with the same end shear joins the web of I.S.M.B. 550 @ 1017.3 N/m from the other side of the web, redesign the connection.

10.9 An I.S.L.B. 225 @ 230.5 N/m and 1 m long is connected at one end to the column section I.S.H.B. 200 @ 365.9 N/m. It supports a load of 300 kN at its free end. Design the riveted connection.

10.10 An I.S.L.B. 450 @ 640.6 N/m transmits an end reaction of 125 kN to the flange of stanchion I.S.H.B. 250 @ 500.3 N/m. Design an unstiffened seat connection.

10.11 An I.S.M.B. 400 @ 604.3 N/m transmits an end reaction of 270 kN to the flange of I.S.H.B. 300 @ 618 N/m. Design a suitable stiffened seat connection.

10.12 Design a clip angle connection to transmit a moment of 15 kNm and a reaction of 150 kN from an I.S.L.B. 400 @ 558.2 N/m to a column section I.S.H.B. 300 @ 576.8 N/m. In the same case if the moment is increased from 15 kN m to 125 kN m redesign the connection as a bracket connection.

10.13 In a framed connection an I.S.L.B. 500 @ 735.8 N/m transmits an end reaction of 150 kN and an end moment of 45 kNm to an I.S.H.B. 400 @ 806.4 N/m column section. Design a split beam connection.

Check the safety of the connection if the moment increases to 80 kNm. Suggest and design the connection to transmit the new forces.

10.14 An I.S.L.B. 325 @ 422.8 N/m transmits an end reaction of 10 kN and an end moment of 70 kNm to the flange of a column I.S.H.B. 300 @ 576.8 N/m. Design the welded connection.

10.15 An I.S.M.B. 550 @ 1017.3 N/m transmits an end reaction of 150 kN and an end moment of 125 kNm to the flange of a column I.S.H.B. 250 @ 500.3 N/m. Design the welded connection.

Industrial Buildings

11.1 INTRODUCTION

Steel building structures can be classified into shell structures and framed structures. A framed structure can further be classified as a single storey structure (e.g. low-rise steel structure) and a multi-storey (high-rise) steel structure. Industrial buildings are low rise steel structures characterised by their comparatively low height and lack of interior floors, walls and partitions. Invariably, the roofing system for such buildings is truss with roof coverings. A single storey or a single storey with mezzanine type of steel structure is most suitable. To provide a free working space, the lockers, cabins, tool counter and rest rooms for the staff may be located on the mezzanine floor as they will not interfere with the production layout. Mezzanine floors are usually cast in situ slabs which rest on steel joists supported by girders of the building frame. Sometimes rolled floor plates 5–20 mm thick are also provided. The enclosure of an industrial building may be brick masonry, concrete walls or G.I. sheet coverings. The walls are usually non-bearing but these must be adequately strong to resist lateral forces due to the wind or an earthquake. Some examples of such buildings are workshops, steel mills, machine plants, warehouses, hangars, etc.

The roof truss together with the supporting columns constitutes a bent (Fig. 11.1). The width of an industrial building may consist of one or more

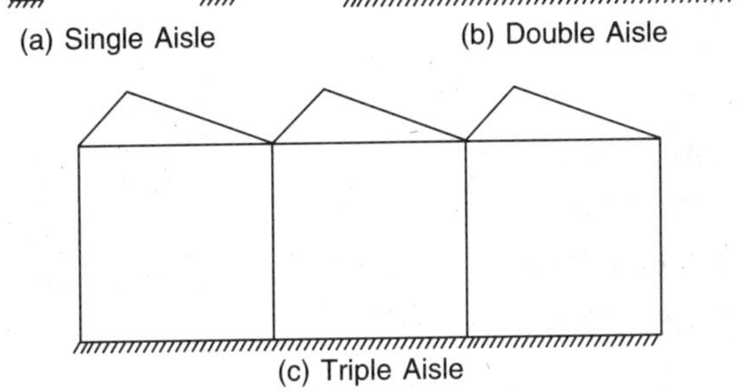

Fig. 11.1 *Industrial Building Bents*

enclosures called aisles. Single aisle, double aisle and triple aisle bents have been shown in Figs 11.1(a), (b) and (c) respectively. The length of an industrial building is divided into bays. A bay is defined as the space between two adjacent bents. Overall economy in the design of an industrial building is achieved when bay lengths are much smaller than the span of the truss. Usually bay lengths are kept between 4–8 m, whereas a truss span may range from 10–25 m or even more if required.

11.2 PLANNING

The planning of an industrial building is based on functional requirements, i.e. on the operations to be performed inside the building. In the planning of a particular type of industrial building, due consideration should be given to factors such as wide area free of columns, large height, large doors and openings, large span of trusses, consistent with the operations to be performed, a compromise with all the listed factors to give minimum weight of trusses, purlins, columns, etc. and lighting and sanitary arrangements.

The site for a proposed plant is in general, pre-selected before it comes for design. But it is better to discuss with the designer the preliminary plans in advance. This gives the designer an opportunity to choose a suitable site giving due consideration to future developments. Some of the factors governing the site selection are as listed below:

1. The site should be located on an arterial road
2. Local availability of raw materials
3. Facilities like water, electricity, telephone, etc.
4. Topography and water drainage
5. Soil condition with reference to foundation design
6. Sufficient space should be available for storage of raw materials and finished products
7. Sufficient space should be available for transportation facilities to deliver raw materials and collect the finished products
8. Waste disposal facilities

A schematic layout of an industrial building is shown in Fig. 11.2.

11.3 STRUCTURAL FRAMING

After making the schematic layout of an industrial building the structural framing scheme is developed. A structural framework consists of steel trusses supported over columns making a transverse bent, whereas a longitudinal bent is formed by joining transverse bents (Fig. 11.2). Structural framing consists of the following steps:

1. Once the production layout is decided the column free area requirement is known. The column rows can be located according to clearance requirements. Column rows at wide distances and with large trusses are economical as compared to closely spaced more number of rows of columns with short span trusses. Also, if the aisle is wide, alterations or modifications can be easily accommodated in future.

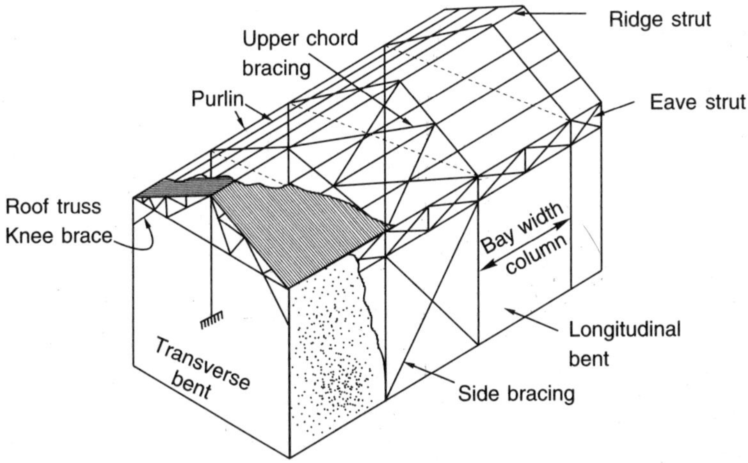

Fig. 11.2 *Structural Frame of an Industrial Building*

2. The columns in each row are fixed so as not to interfere with the mechanical layout. Generally, a column spacing of 4 m to 8 m is provided. If larger column spacing is required in some portion of a multi-bay system industrial building, the trusses may be supported over girders or supporting trusses and the columns at *a* and *b* can be omitted (Fig. 11.3).
3. A sketch of the members, openings, lintels, doors, windows, girts (required to support the wall surface) and end wall columns is made.
4. As the column spacing is fixed by this time, the pitch of the truss is selected and the type of truss to be provided is decided. The panel lengths are worked out and purlin locations are marked.
5. A plan is prepared to show the lateral bracing system. It may be done either in the plane of the bottom chord or in the plane of the top chord and some times in both the planes.
6. Various materials to be used for the floor, roof walls and partition walls are decided.

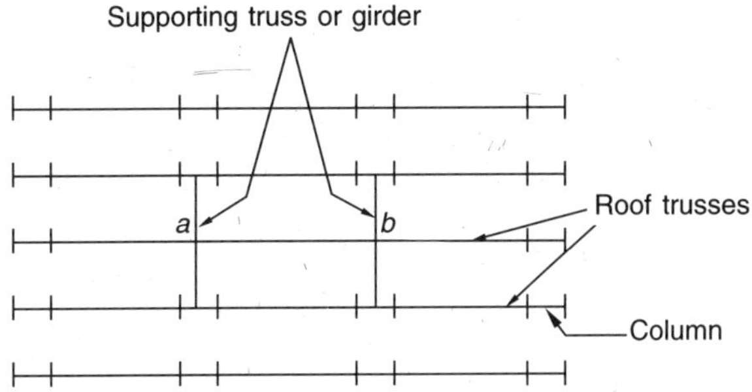

Fig. 11.3 *Trusses Supported Over Girders*

11.4 TYPES

A single storey single aisle industrial building can be of any one of the types shown in Fig. 11.4.

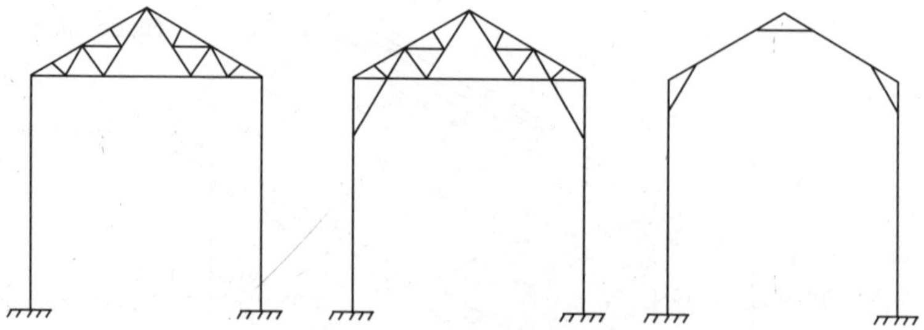

Fig. 11.4 *Shape of Single Aisle Industrial Buildings*

In the truss column type of industrial building (Fig. 11.4(a)), the roof truss is directly placed over the columns. The column may be an I-section with a bracket or a built-up section. A built-up column consists of two separate columns termed as legs. One leg of the column may be used to support the roof truss and the other is used to support the gantry girder. The two legs of the column are suitably connected. Normally the column is assumed to be fixed at the base and is hinged at the top with the shoe of the truss (Fig. 11.5). In the knee braced truss and column type of industrial building, the roof truss is connected with the columns by knee braces (Fig. 11.4(b)).

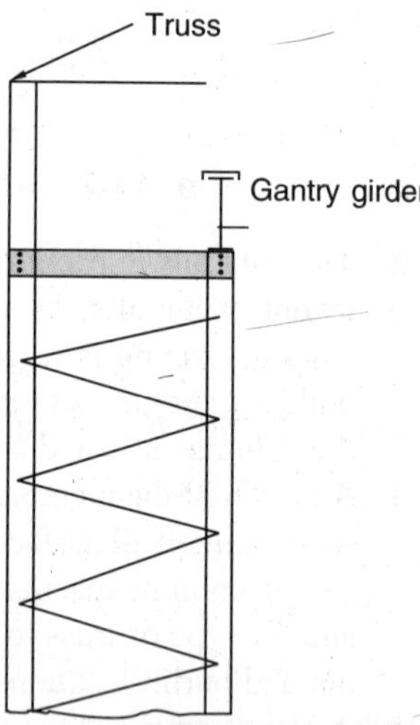

Fig. 11.5 *An Industrial Building Column Supporting Truss and Gantry Girder*

In the rigid frame portal type of industrial building the bent is made up of a continuous member, bent to shape. The rigid frame can be fixed or hinged at the base depending upon the conditions provided at the base (Fig. 11.4(c)). It can also be hinged at the crown. A rigid frame portal is essentially a welded frame structure.

Suitability The behaviour of an industrial bent under lateral loads depends upon two factors. Firstly, on the degree of fixity of the column at the base and, secondly on the provision of a rigid connection between the column and truss. When the column bending moments due to lateral loads are small along with large axial loads and if, large foundations can be provided, columns hinged at the top and fixed at the base can be provided (Fig. 11.6(a)). The moments are small when the height of the building is small. Therefore, for small heights and wide industrial buildings, such end conditions are most suitable. When the column bending moments due to lateral loads are large along with large axial loads, knee braces as shown in Figs 11.6(b), (c) are provided. This arrangement reduces the bending moments at the base when the column is fixed at the base and consequently reduces the size of the foundation. When the column is hinged at the base (Fig. 11.6(c)), no bending moment is transferred at the base and the size of foundation can be reduced further. Moments in

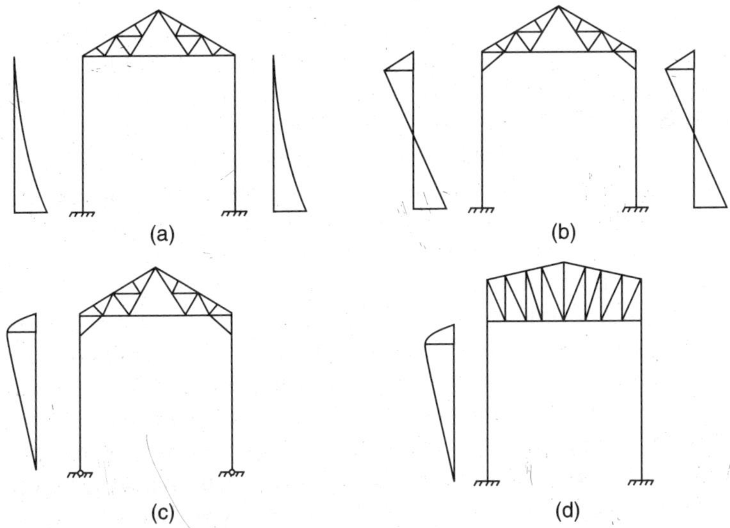

Fig. 11.6 *End Conditions of Industrial Building Columns and Bending Moment Diagrams*

such an arrangement are transferred only at the junction of the knee brace and column. When it is not possible to brace the column with knee braces due to clear head room requirements either rigid frame portals may be provided (Fig. 11.4(c)) or the columns may be extended up to the top of the truss and suitably jointed to the truss chords (Fig. 11.6(d)).

11.5 ROOF AND SIDE COVERINGS

Corrugated galvanised iron (G.I.) sheets are usually adopted as coverings for roofs and sides of industrial buildings. Sometimes, asbestos cement (A.C.) sheets are also provided as roof coverings. A.C. sheets are better insulators of heat for roofing. Two corrugation laps are usually provided for roofing and one corrugation lap is provided for side coverings. At the ends a minimum of 15 cm overlap is given for roof coverings for slopes of 20° or more and for flatter slopes this overlap is increased. For side coverings, an end overlap of 10 cm is sufficient. The sheets are fastened to purlins or girts by 8 mm diameter hook bolts at a maximum pitch of 350 mm.

11.6 ELEMENTS OF AN INDUSTRIAL BUILDING

The elements of an industrial building are as listed below. These are discussed one by one in the sections to follow.

1. Purlins
2. Sag rods
3. Principal rafter
4. Roof truss
5. Gantry girder
6. Bracket
7. Column and column base
8. Girt
9. Bracing

11.7 PURLINS

Purlins are beams which are provided over roof tusses to support the roof coverings. Purlins span between the top chords of two adjacent roof trusses and transmit the loads from roof coverings to the trussses. Purlins are usually placed over the panel points of roof trusses. However, structurally the purlins should be placed slightly away from the panel points so as to keep the centroids of the purlin sections vertically over the joint centre (panel point). When sheeting is directly placed over the purlins, these can be placed at intermediate points along the top chord of the trusses. The top chord of the truss in such a case has a disadvantage as it has to support an axial compressive load alongwith the reactions from purlins and behaves like a beam-column. A channel section is best suited for a purlin. The other types of sections used are an I-section and an angle section.

Purlins should be designed carefully as these constitute a large proportion of the steel dead weight in a structure. Therefore, these affect the overall economy of the structure. Purlins can be designed as simple, continuous or cantilever beams. As per Indian Standard 800 : 84 purlins should be designed as continuous beams. But the drawback in designing them like this is that purlins may have to be spliced and continuity cannot be assured through splices. Therefore, the purlin section provided should be as long in length as possible. The exterior bay width is kept at about 80% of the interior bay width to reduce the maximum moment in the end span of the purlins. This makes the moments in the end span equal to those in the interior spans. Purlins are often designed for only the normal component of forces. The justification given for this is that purlins with roof coverings act like a deep plate girder and due to this the stress from parallel components can be neglected.

11.8 SAG RODS

A purlin section may result in a considerable saving in steel if an efficient and economical method of absorbing a part of the tangential component of the roof load outside the section itself can be devised. Some designers rely on closely placed sheeting to take the tangential component of roof loads. This could be justified only for relatively flat roofs. When the slope of the roof is appreciable, the component of the vertical load acting in the weak plane of the purlin may demand consideration. In such situations, sag rods are provided between adjacent purlins to extend lateral support for the purlins in their weaker directions and enhance the rigidity of purlin sections about their weak axis.

These are round section rods and are fastened to the web of the purlins. A sag rod is designed as a tension member to resist the tangential component of the resultant of the roof load and purlin dead load. The tangential component of the roof load is considered to be acting at the top flange of purlins, whereas the normal component and purlin's dead load is assumed to act at its centroid. Therefore, the sag rods should be theoretically placed at the point where the resultant of these forces act. But this is not practicable and sag rods are placed at the minimum gauge distance below the top.

The sag rod provided at the crown is termed as *tie rod*. This resists the tangential components from the two sides of the roof truss (Fig. 11.7).

The number of sag rods to support each purlin depends upon the length of the purlin and load to be supported. In general, a single line of sag rods at the centre of the purlins is sufficient. Sag rods also help to hold the purlins in line until the roof covering is applied. The sag rods should not be terminated at the ridge as the ridge

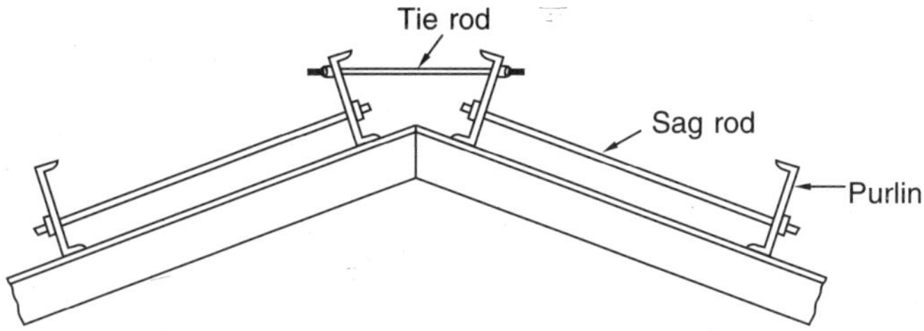

Fig. 11.7 *Sag Rod Connection*

purlin in such a case is subjected to excessive pull. Therefore, the sag rods are extended over the ridge forming a continuous line between the eaves. When the purlins are supported by sag rods, the span of the purlins for the load component parallel to the roof truss is taken as the distance between the sag rods. The span of the purlins for the load component perpendicular to the roof truss is taken to be the distance between the adjacent trusses.

11.9 PRINCIPAL RAFTER

The principal rafter of a roof truss is designed as a continuous strut. Generally, a double unequal angle section with long legs back-to-back is most suited for an industrial building structure. This is due to two reasons. A rafter section should have the same radius of gyration about the main axes ($r_{xx} = r_{yy}$), so as to achieve the same bending strength about the two axes and to have the value of the least radius of gyration as large as possible. Also, the double angle section provides extra strength which is useful during erection when the trusses are unbraced.

The heel of the rafter angle is placed on top due to the following two reasons:

1. The rafter angle acts as a beam in carrying its self weight. Therefore, the compression edge should be wide.
2. If the heel is kept down the rafter angle will act as a gutter and in exposed conditions will lead rain water down to the connections.

Although the design forces in various panels of the rafter vary considerably, they are still made of uniform cross-section. There are two reasons for this:

1. The saving achieved by changing the sizes in various panels is offset by the cost of splice and connections.
2. There is a loss of lateral stiffness of the structure otherwise.

11.10 ROOF TRUSSES

Roof trusses are elements of the structure composed of members subjected to direct stresses. Sometimes the truss is also called an *open web beam*. It consists of a triangular network of compression and tension members. On the basis of structural behaviour, roof trusses can be classified as simple roof trusses supported over masonry walls, and roof trusses supported over columns and connected to it with knee braces. Theoretically, truss members are subjected only to direct tension and direct compression. But in reality some shear and moment is always there at the joints which is neglected except for large bridge trusses. Various types of roof trusses are shown in Fig. 11.8.

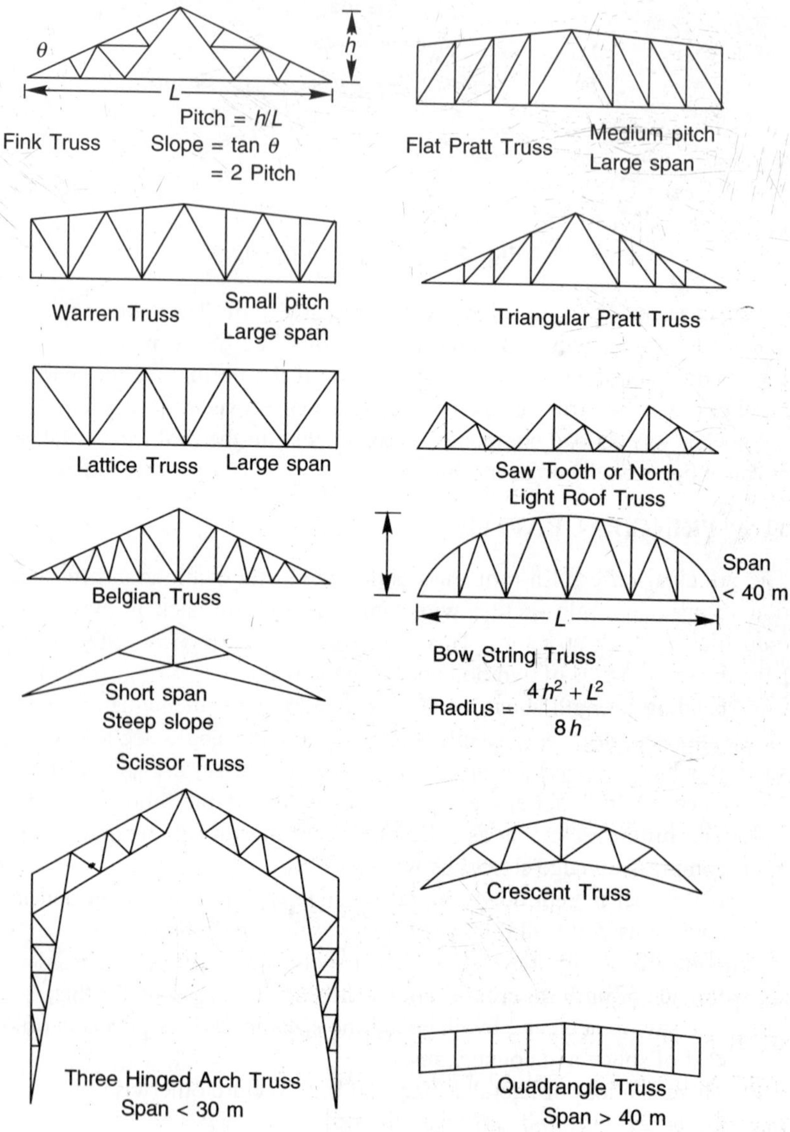

Fig. 11.8 *Typical Roof Trusses*

Selection of the Type of Truss The type of roof truss to be provided depends primarily upon the pitch of the truss. Fink truss, Pratt and Howe truss, Warren truss are provided for large, medium and small pitch, respectively. A skylight can be fitted on them for daylight. When the layout of an industrial building is such that more daylight is required, a North light truss is most suitable, as natural light can be obtained from its geometry. The pitch of the roof truss is the height of the truss divided by the span. One must not confuse it with the slope of the truss (the slope is numerically twice the pitch). Some of the other factors which may affect the selection of a particular type of roof truss are:

(a) Roof coverings The pitch of the truss depends upon the roofing material. The minimum recommended rise of trusses with G.I. sheets is 1 in 6 and with A.C. sheets is 1 in 12.

(b) Fabrication and Transportation This often guides the type of truss to be selected. Normally trusses are fabricated in the workshop and are transported to the site for erection. From the transportation consideration, depth of the truss becomes a controlling factor as it will not be feasible to transport a very deep truss.

(c) Aesthetic From the aesthetic point of view the architect may give a very flat or deep truss, thereby limiting the choice.

(d) Climate The climate of a particular area plays an important role in the selection of truss. Drainage of water, ice and snow retention, etc. will have to be given due consideration.

Spacing of Truss The economic spacing of the truss is the spacing that makes the overall cost of trusses, purlins, roof coverings columns, etc. the minimum. It depends upon the relative cost of the trusses, purlins, roof coverings, spacing of columns, etc. If the spacing of trusses is large, the cost of these trusses per unit area decreases but the cost of purlins increases. If the spacing of trusses is small, the cost of trusses per unit area increases. Roof coverings cost more if the spacing of trusses is large.

Let S = spacing of the trusses
t = cost of the trusses/unit area
p = cost of the purlins/unit area
r = cost of the roof coverings/unit area
x = overall cost of the roof system/unit area.

The cost of the truss is inversely proportional to the spacing of trusses,

$$t \propto \frac{1}{S}$$

$$t = \frac{c_1}{S}$$

The cost of the purlins is directly proportional to the square of spacing of trusses,

$$p \propto S^2$$

$$p = c_2 S^2$$

The cost of roof coverings is directly proportional to the spacing of trusses,

$$r \propto S$$

$$r = c_3 S$$

$$x = t + p + r$$

$$x = \frac{c_1}{S} + c_2 S^2 + c_3 S$$

For the overall cost to be minimum,

$$\frac{dx}{dS} = 0$$

or

$$\frac{dx}{dS} = 0 = -\frac{c_1}{S^2} + 2c_2 S + c_3$$

or

$$-\frac{c_1}{S^2} + 2c_2 S + c_3 = 0$$

or

$$-\frac{c_1}{S} + 2c_2 S^2 + c_3 S = 0$$

or

$$-t + 2p + r = 0$$

or

$$t = 2p + r \tag{11.1}$$

Therefore, for economic spacing of roof trusses, the cost of trusses should be equal to twice the cost of purlins plus the cost of roof coverings. The above expression is used to check the spacing of the roof trusses. It cannot be used to design the spacing as the spacing occurs nowhere in the final equation. As a guide, the spacing of the roof trusses can be kept 1/4 of the span for up to 15 m, and 1/5 of span from 15–30 m spans of roof trusses.

Loads on the Roof Truss Roof trusses are subjected to dead load, live load, snow load and wind load. These are discussed in detail in the paragraphs to follow:

(a) Dead loads The dead loads of the truss include the dead load of roofing materials, purlins, trusses and bracing systems. The unit weight of roofing materials are prescribed by the I.S: 1911. Some of the roofing materials and their weights are listed in Appendix II. The weight of purlins is known in advance as these are designed prior to the trusses. The dead weight of the trusses may be assumed to be equal to 10% of the loads on the truss. Some formulae have also been developed and the designer may choose any one of them. In any case, the dead load of the truss after the design should not vary by more than 10%. A rough estimate of the dead weight of the roof truss can be 60% of the dead weight of a girder subjected to the same design loads, but distributed uniformly. The weight of the bracings may be assumed to be 12–15 N/m^2 of the plan area. A simple formula for estimation of the approximate dead weight of roof truss in N/m^2 is $\left(\frac{\text{span}}{3} + 5\right) \times 10$

(b) Live loads I.S: 875 specifies the following live loads to be assumed in the analysis of an industrial building.

Roof slope	*Access*	*Live load*
$\le 10°$	Provided	1.5 kN/m^2 of plan area
	Not provided	0.75 kN/m^2 of plan area
$\ge 10°$		For roof membrane sheets or purlins 0.75 kN/m^2 less 0.02 kN/m^2 for every degree increased in slope over 10 degrees $\nless$ 400 N/m^2

(c) Snow load If the structure is situated in an area where the roof is subjected to snow, the load considered for design should be maximum of the live load or snow load. The load due to snow depends upon the pitch of the roof, shape of the roof and roofing material. Snow load may be assumed to be 2.5 N/m^2 per mm depth of snow. When the roof slope is greater than 50°, the snow load may be neglected.

Wind load The most critical load on an industrial building is the wind load. The design wind pressure is calculated in a way similar to that discussed earlier (Section 1.10). However, for the roofs and walls of an industrial building, consideration must be made for the pressure difference between the opposite faces of such elements to account for external and internal air pressures exerted by the wind blowing against the building. The internal air pressure may be positive or negative depending upon the direction of flow of air with respect to the openings in the building. When the negative air pressure is less than the atmospheric pressure it is known as *suction.* The openings in relation to the wall area define the permeability of the building. Table 11.1 gives the degree of permeability and corresponding internal air pressure coefficients. For the external air pressure coefficient and coefficients for local effects, reference may be made to Appendices IX and X. Local effects are because of the high local suction in areas such as ridges, cornices, corners of roofs. etc. For pitched and saw tooth-roofs of multispan buildings reference may be made to IS 875 (Part 3)-1987.

Table 11.1 *Internal Air Pressure Coefficients for Roofs and Walls of an Industrial Building*

Permeability	*Openings in Relation to Wall Area (%)*	*Internal Air Pressure Coefficient (c_{pi})*
Zero	0	± 0
Normal	5	± 0.2
Medium	5 – 20	± 0.5
Large	> 20	± 0.7

Notes

(i) Positive wind load indicates that the force is acting towards the structural element and negative away from it.
(ii) Buildings with normal, medium or large permeability should be examined for their respective internal air pressure coefficients with both positive and negative signs and the one producing greater distress on the member should be adopted.

The wind force F is obtained by

$$F = (c_{pe} - c_{pi})\, A\, P_d$$

where c_{pe} = external pressure coefficient
c_{pi} = internal pressure coefficient
A = surface area of the element under consideration
P_d = design wind pressure

Load Combinations Various combinations of the loads on roof trusses are considered, and the critical condition is arrived at for the design. The following load combinations may be worked out:

1. Dead load + snow load
2. Dead load + live load
3. Dead load + partial or full live load (whichever causes the maximum stress in the member)
4. Dead load + wind load + internal positive air pressure
5. Dead load + wind load + internal suction air pressure

Analysis of Roof Truss The analysis of roof truss consists in determining the loads at the panel ponts and then analysing the internal forces in the various members of the roof truss. The critical stresses generated out of the various possible load combinations are computed. Any of the following methods can be used for computing the stresses:

(i) The algebraic method of joints
(ii) The graphic method of joints
(iii) The method of sections

All the above three methods are equally good for computing the stresses due to gravity loads. For complicated trusses or for the trusses subjected to gravity and lateral loads, the graphic method of joints is particularly advantageous.

Selection of Section Angles or tubular sections may be used for compression members and flats, angles or tubular sections for tension members. However, flats for tension members are not recommended for the following reasons. Suction on the leeward side of the roof truss causes reversal of stress in some members of the roof truss and flats are not suitable for compressive forces. Also great care is required to make the flat perfectly straight to maintain the correct distance between the holes for rivets. Otherwise, the flat will be distorted when the truss is riveted up and the truss must deform before tension can be developed in the member. Further, flats get easily damaged during transit and erection restricting their use in trusses. Double angles are used for members forming the skeleton of the truss (e.g., rafter, principal tie, etc.) and single angles for filler members.

Assembling of Filler Members The usual practice is to place the outstanding legs of the strut members facing the apex of the truss. This is to reduce the bending moments generated because of the centre of gravity of the rivets not coinciding with the centre of gravity of the members. Consider a rafter of two panels as shown in Fig. 11.9. The eccentricity in the rafter will be uniform and the direct compressive forces in the panels usually decrease as they approach the apex. The rafter tends to deform in a direction away from the apex and if the outstanding legs of the struts are placed facing the apex of the truss, on their bending as shown by dotted lines, they tend to move the centre of gravity of the strut towards the centre of gravity of the rivets. This helps to reduce eccentricity and secondary stresses.

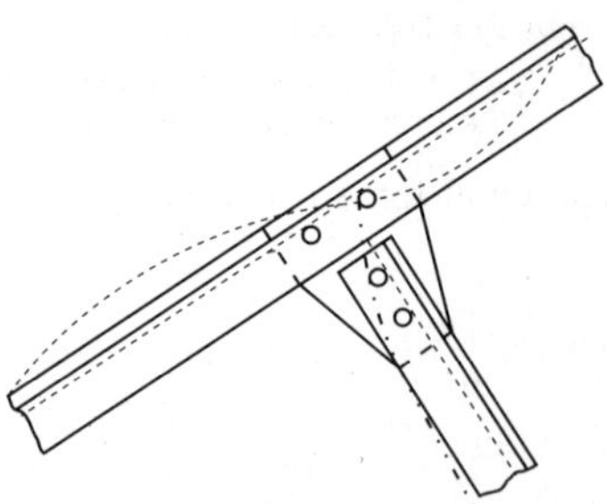

Fig. 11.9 *Correct Placement of Outstanding Leg of Strut*

Design of a Roof Truss The design of a roof truss consists in selecting a suitable type of truss, estimation of loads and design of purlins, members of roof truss and their connections. The design is done in the following steps.

1. Depending upon the span, lighting, roofing material, etc. available, the type of truss is decided and a line diagram of the truss is prepared.
2. Various loads acting over the roof truss are estimated.
3. The purlins are designed and the loads acting at the panel points of the truss are computed.
4. The roof truss is analysed by any suitable method.
5. The compression members are designed as discussed in Chapter 4. The principal rafter is designed as a continuous strut and the other compression members are designed as discontinuous struts.
6. The tension members are designed as discussed in Chapter 6. Single angle tension members have twisting tendency and produce eccentric forces at the joints. Therefore, double angle sections are preferred. A minimum of $50 \times 50 \times 6$ mm angle section is provided.
7. The width of the truss members should be kept minimum as far as possible, because wide members have greater secondary stresses.
8. If purlins are placed at intermediate points, i.e. between the joints of the top chord, the top chord will be subjected to moments. The member in such a case, is designed for the direct stresses and moments.
9. The members meeting at a joint are so proportioned that their centroidal axes intersect at one point (to avoid eccentric conditions).
10. The joints of the truss are designed.

11.11 GANTRY GIRDERS

These are laterally unsupported beam and are provided in industrial buildings to carry cranes. The design of a gantry girder has already been discussed in Chapter 8.

11.12 BRACKETS

Brackets are used to support the gantry girders. The design of bracket has been discussed in Chapter 10.

11.13 CRANE COLUMNS

The columns which support the gantry girders are in general, called gantry columns or crane columns. Depending upon the arrangement of the connection and crane load to be transmitted the columns are divided into three categories:

1. Column with bracket,
2. Stepped column or notched column,
3. Double column.

A column with bracket is the simplest arrangement which is provided to support a crane girder (Fig. 11.10(a)). Laced angle struts at the gantry girder level are provided to resist the longitudinal forces due to crane movement. This also braces the crane columns and acts as a part of the bracing system in a braced bay of an industrial building. A stepped column is made as shown in Fig. 11.10(b), by step-

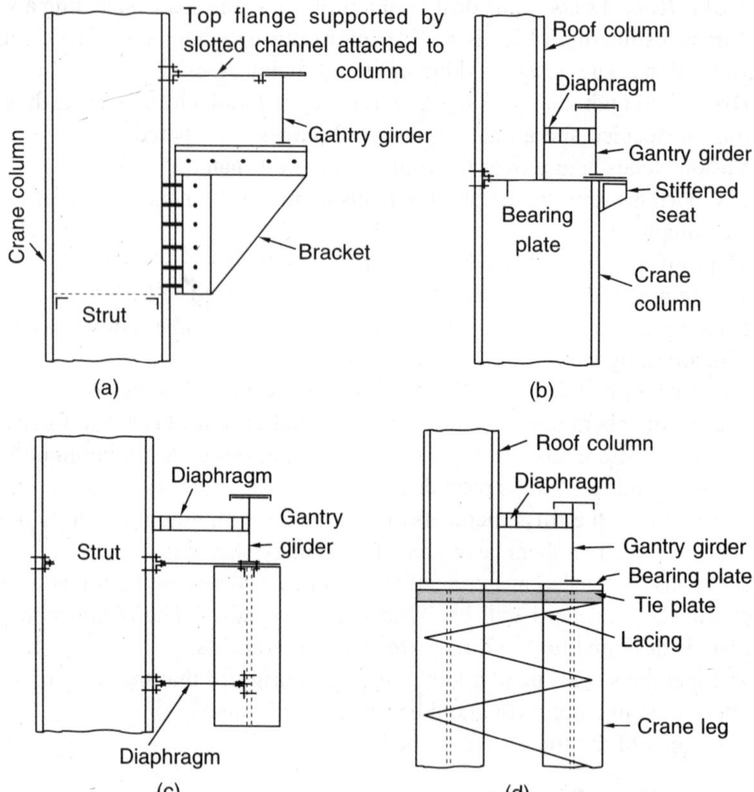

Fig. 11.10 *Shapes of Crane Columns*

ping the column at the crane level. This step, on the inner flange of the I-section column, is then enlarged to form a seat for the gantry girder. The seat may be stiffened by means of an angle section. The outer flange of the I-section supports the truss. The two-column section is provided to support heavy cranes (Fig. 11.10(c)). The two columns are designed separately and are connected by diaphragms. The outer column section is designed for gravity and wind loads from industrial buildings and lateral thrust from the cranes. The inner column section is designed for loads transmitted from gantry girders. In the case of stepped columns or double column section, double angle struts are provided on columns (Figs 11.10(b), (c)). These struts resist longitudinal forces at the gantry level and also act as bracing. Diagonal bracing of outer flange is also done by a single angle section above the gantry level.

The roof leg of the column (Fig. 11.10(c)) is strong in the direction across the shop but has less resistance against deflection in the direction parallel to the length of the roof leg. The roof leg of the column (Fig. 11.10(d)) is, unlike the other three examples (Figs 11.10(a)–(c)), not continuous from roof to base but sheds the outstanding joist and plate at the crane cap level. This roof leg of the column has a large radius of gyration about both the *xx*- and *yy*-axes and is fully capable of carrying the bending moment due to the portal action, in addition to the direct

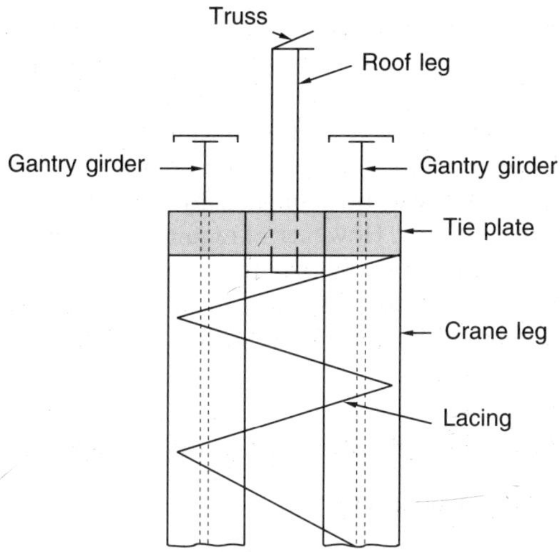

Fig. 11.11 *Crane Column with Cranes in Adjacent Aisles*

vertical load from the roof. The connectors at the crane cap level should be sufficient to resist the moment and also to transfer the direct load into side plates. Below the crane cap level in the crane leg of the column, any bending moment due to a portal action or other causes is resisted by the widely spaced joists suitably connected by lacing.

Unless the cranes are very light it is more economical to have a separate column leg to support the crane girders than to employ a crane bracket from the roof leg.

The connections of columns with gantry girders are made flexible by providing slotted holes, parallel to the girder. This allows free rotation in the vertical plane while the girder is restrained against twisting at the support. In the stepped column the connections are less flexible since the rotation in the vertical plane will depend on the warping of the diaphragm. When the cranes are to be provided at the same level in the adjacent aisles (Fig. 11.11), girders are made entirely self-supporting both vertically and laterally and are not tied with diaphragms or lacing.

Column Loads The problem of determining the column load distribution in an industrial building column is statically indeterminate. To simplify the analysis, the column is isolated from the space frame and is analysed as a column subjected to axial load and bending moment.

An industrial building column is subjected to the following loads in addition to its self-weight.

1. Dead loads from roof truss
2. Live loads on roof truss
3. Crane load which consists of dead weight of gantry girder, rail, bracket, wheel load and impact load
4. Loads due to wind (i.e. the horizontal and vertical component of wind reaction) and knee brace loads. This can be found from the wind stress analysis of bents. The wind load is assumed to be uniformly distributed over the columns.

The horizontal components of the reactions at the lower ends of the windward and leeward columns are found and the shear force and moment diagrams are constructed. In case of knee braced bents, the moment diagram is constructed by assuming the moment to be zero at 1/3 of the height of the knee brace above the foot of the column. Depending upon the end conditions the forces and moments in an industrial building column can be estimated by approximate methods as discussed in the paragraphs to follow. However, if required exact analysis can be done. In the analysis, it is assumed that the horizontal component of the lateral loads are equally divided between the points of support of both the bent columns. The assumption is true if the columns have the same stiffness and the bent is symmetrical.

Analysis

(i) Column hinged at base (Fig. 11.12(a)) From the assumption,

$$\text{Horizontal reactions, } H_A = H_B = \frac{P}{2}$$

and

$$\text{Vertical reactions, } R_A = R_B = \frac{P(h_1 + h_2)}{L}$$

where H_A, H_B = horizontal reactions at the supports, A and B

P = lateral load

h_1 = height of knee brace from the foot of the column

h_2 = height of the top of the column from the knee brace

L = span of the bent.

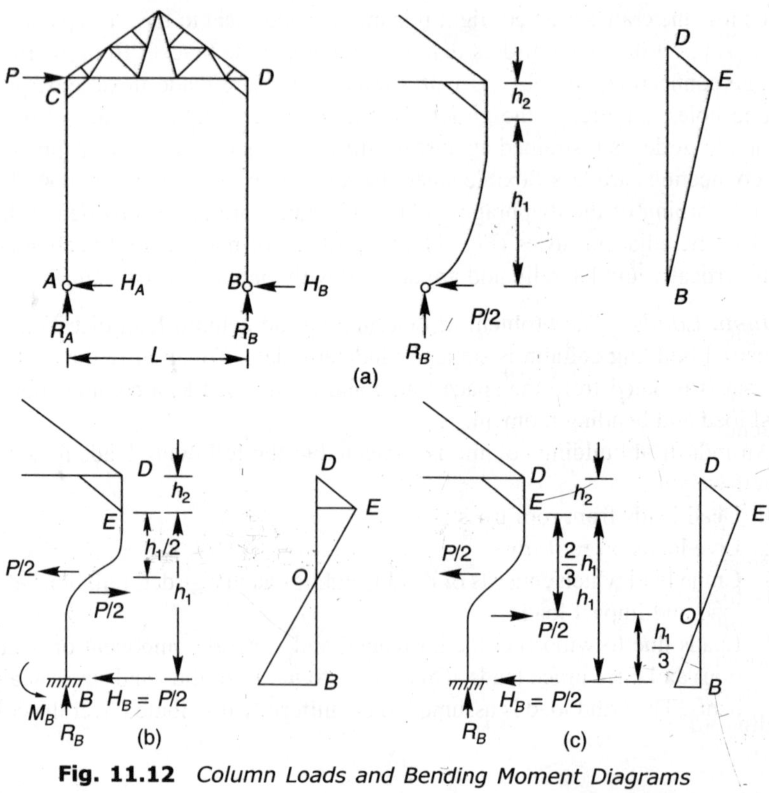

Fig. 11.12 *Column Loads and Bending Moment Diagrams*

Let the maximum bending moment be M, at the knee brace; then

$$M = \frac{P}{2} h_1$$

(ii) Column fixed at base (Fig. 11.12(b)) From the assumption,

$$\text{Horizontal reactions, } H_A = H_B = \frac{P}{2}$$

It is assumed that the point of inflexion lies at half the height between column foot and knee brace foot,

$$\text{Vertical reactions, } R_A = R_B = \frac{P \times \left(\frac{h_1}{2} + h_2 \right)}{L}$$

$$\text{Moments, } M_B = M_E = \frac{P}{2} \times \frac{h_1}{2} = \frac{Ph_1}{4}$$

(iii) Columns partially fixed at base (Fig. 10.12(c)) From the assumption,

$$\text{Horizontal reactions, } H_A = H_B = \frac{P}{2}$$

As the end fixity can never be assured cent per cent, the point of inflexion lies somewhere below the mid-point of the height from the foot of the column to the foot of the knee brace. It is assumed that the point of inflexion lies at 1/3 of height from the foot of the column to the foot of the knee brace above the foot of column.

$$\text{Vertical reaction, } R_A = R_B = \frac{P \times \left(\frac{2}{3} h_1 + h_2 \right)}{L}$$

$$\text{Bending moment at the foot of the knee, } M_E = \frac{Ph_1}{3}$$

$$\text{Bending moment at the foot of the column, } M_B = \frac{P}{2} \times \frac{h_1}{3} = \frac{Ph_1}{6}$$

11.14 GIRTS

These are beams subjected to unsymmetrical bending. These support vertical dead loads from the siding (sheeting or covering) and horizontal wind loads. Usually these are unequal angle sections with the longer leg as the outstanding (horizontal) leg to withstand the effect of wind. Sometimes a light channel or a zee section is also provided. Girts are assumed to be continuous (In such a case, the maximum bending moment for the design may be assumed to be $WL/10$). Sometimes arrangements are done to support the weight of the siding by sag rods suspended from the eaves strut. The girts in such cases are subjected to wind loads only.

The girts are bolted to the bent columns and the verticals. The sheeting is fixed to them in an identical manner as in the case of roofing.

11..15 BRACING

A simple industrial building bent can effectively resist gravity loads but deforms under lateral loads. The bracings are so planned and provided that lateral forces due to wind, earthquake and crane surge, etc. are transmitted efficiently to the foundation of the building. An individual bent may be stable against lateral loads in its own plane. However, it has little resistance to the forces acting normal to its plane. Also, a series of bents constituting the industrial building should be stable. Therefore, a system of lateral or diagonal bracing is provided to prevent the building from twisting under the action of wind. Various types of bracings, their locations and design are discussed in the following sections.

Transverse Bracing A bent should be braced transversely (Fig. 11.13(a)) to preclude collapse even during erection. Each bent after erection should be stable in the transverse direction. The end conditions of the columns have large bearing over the stability of the bent. Various types of transverse bracings of a bent have been shown in Figs 11.4(b), (c). The ultimate aim of the bracing in the transverse direction is to reduce the end moments of the columns. This is achieved either by providing knee braces or by rigid frame portals. The knee braced bents, though popular, have a disadvantage where travelling cranes are to be used.

Longitudinal Bracing The wind may be normal to the transverse bent. Therefore it becomes essential to brace the industrial building in the longitudinal direction. A bracing system (Fig. 11.13(b)) shows bracing in the plane of the top chord of the roof truss and in the vertical plane of the longitudinal bent. The diagonals in the plane of the top chord of the truss are usually bolted to purlins wherever possible. This arrangement reduces the unsupported length of the diagonals and makes them more effective in compression. Sagging due to self-weight is also reduced and makes them less prone to vibrations. Bracing may also be done in the plane of the lower chord of the roof truss. The truss bracing system transfers the longitudinal loads to the columns and the vertical column bracing transfers them to the foundation.

Every fourth or fifth bay may be braced (Fig. 11.13(c)). But no industrial building should have less than two braced bays. Some designers feel that the longitudinal vertical bracing shown in Fig. 11.10(e) is better than that shown in Fig. 11.13(c), because it does not interfere with the openings. But the columns with a bracing system shown in Fig. 11.13 (e) are subjected to biaxial bending. Therefore, side bracing as shown in Fig. 11.13(c) is preferred. However, in long industrial buildings the end panels should not be braced. One reason may be that due to increase in temperature the building expands but the column bases are fixed in position and consequently the column tops are forced outwards. This effect is more pronounced in the end columns and if the end panels are braced diagonally, it is overstressed.

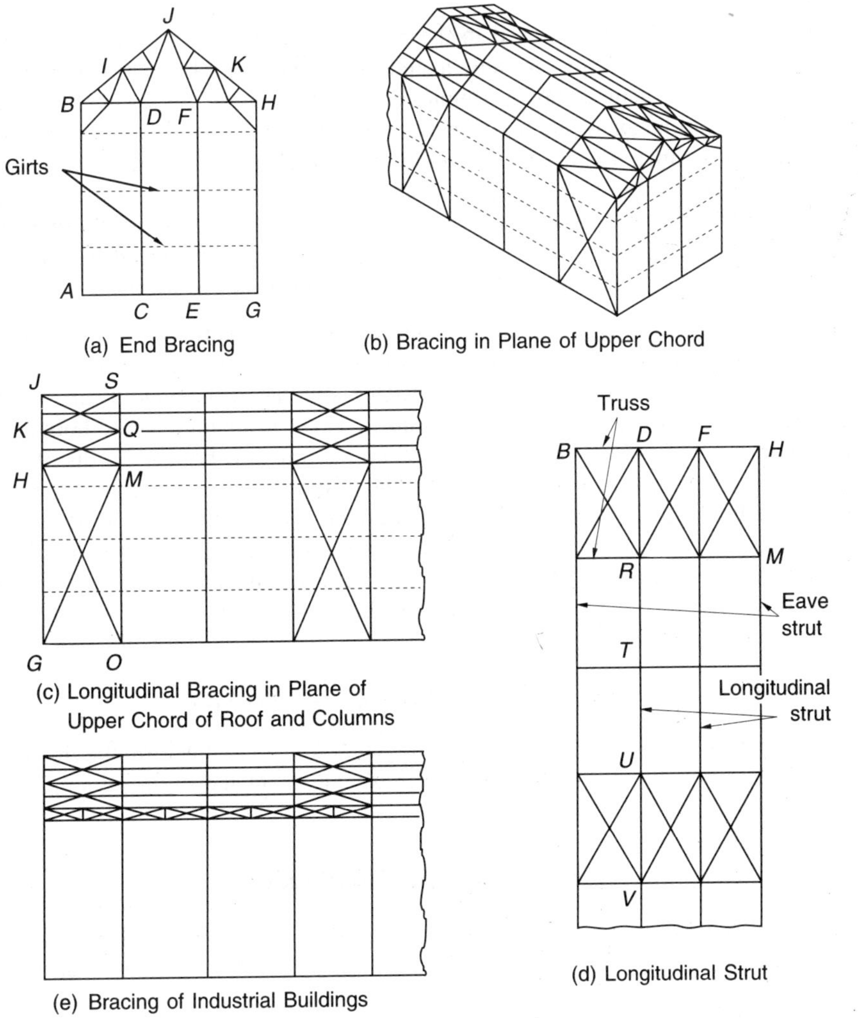

Fig. 11.13 *Types of Bracings*

Hence, it is not advisable to brace the end panels. The diagonals are kept short of fit for about 0.225 mm/m length. While making the connections the diagonals are drifted and thus, an initial tension is induced in them. These prestressed diagonals hold the structure rigidly. Also this initial tension decreases the compression in the diagonal and thus reduces its buckling.

Diagonal cross-bracing is used in the planes of the upper chord member. These braces together with the purlins (acting as a strut to the wind) make a complete upper chord bracing system. This bracing is provided in every fourth or fifth bay. In addition to the top chord bracing the bottom chords of the trusses may be tied together at the columns by longitudinal struts (e.g. *DRTUV*) (Fig. 11.13(d)). These struts can be provided at the panel points. The ridge strut and eaves strut also form part of the bracing system.

Diagonal cross bracing is used in the planes of the lower chord member only when heavy vibrating loads are anticipated in the industrial buildings. For small span trusses and structures having simple loads the bracing of the lower chord may be omitted. The diagonal bracing in either the upper or lower chord must form a horizontal or inclined truss, whose reactions are furnished by the end frames.

The complete longitudinal bracing system is seldom necessary. It can be provided either in the plane of the top chord or in the plane of the bottom chord of the roof truss. Purlins, eave struts and wall struts together with a braced bay form a structure capable of resisting forces acting along the axis of the building.

Design The stresses in the members of the bracing system are small in magnitude and even a nominal section is sufficient to resist the stresses induced. Therefore, the slenderness ratio is the only controlling design factor. The wind panel loads at *B, I, J, K, H* and *D, F* (Fig. 11.13(a)) are calculated. These loads give rise to horizontal shear across the length of the building and are assumed to be equally distributed among those bays that have cross bracing. The design force in each diagonal member of the bracing is assumed to be the horizontal shear in the panel times sec θ, where θ is the inclination of the horizontal shear with the diagonal member. Both diagonal members in the panel are designed as tension members with a maximum slenderness ratio, $\lambda = 350$. When one is functioning the other is assumed to have buckled out of action.

11.16 DESIGN STEPS OF INDUSTRIAL BUILDING

An economical design is obtained when bay lengths are kept much smaller than the span of trusses. The steps involved in the design of an industrial buildings are illustrated with the help of Examples 11.4 to 11.7.

Solved Examples

Example 11.1 Design a roof truss for a railway platform 30 × 12 m wide. The type of roof truss to be provided is shown in the Fig. Ex. 11.1(a). The roofing is to be done with asbestos cement sheets. Assume the necessary relevant data.

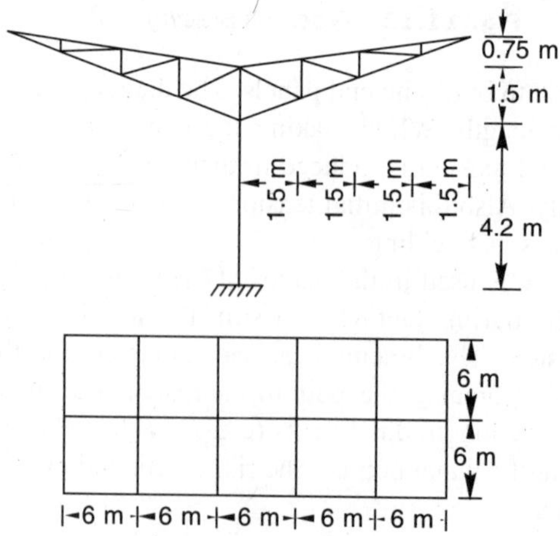

Fig. Ex. 11.1(a)

Solution As the truss is symmetrical about the column, only half the truss is analysed.

(i) *Dead load*

Let Weight of A.C. sheets = 171 N/m^2

Weight of purlins = 200 N/m^2

Weight of bracing = 12 N/m^2

$$\text{Self-weight of the roof truss} = \left(\frac{\text{span}}{3} + 5\right) \times 10 = \left(\frac{12}{3} + 5\right) \times 10 = 90 \text{ N/m}^2$$

Total dead load = 171 + 200 + 12 + 90 = 473 N/m^2 = 500 N/m^2

Note The sheeting rest on the purlins which in turn are supported on trusses.

Spacing of truss = 6.0 m

Length of each panel = 1.5 m

∴ Load on each intermediate panel = 500 × 6 × 1.5 = 4500 N = 4.5 kN

$$\text{Load on end panels} = \frac{4.5}{2} = 2.25 \text{ kN}$$

(ii) *Live load* It is assumed that the platform trusses are not subjected to live loads. Therefore, live load = 0.0.

(iii) *Wind load*

Let the design wind pressure be 1.05 kN/m^2

Refer to IS 875 (Part 3) 1987.

Angle of inclination, $\alpha = 7.125°$

For $\alpha = 7.125°$, $c_{pe} = +0.3425$ (Table 8, IS 875 (Part 3))

$$\text{Solidity factor, } \phi = \frac{\text{area of obstruction under the canopy}}{\text{gross area under the canopy}}$$

$$\alpha = 7.125 \left| \begin{matrix} \phi = 0 \\ \phi = 1 \end{matrix} \right| \begin{matrix} C_{pe} = -0.5425 \\ C_{pe} = -0.8 \end{matrix}$$

The butterfly truss has a large opening (> 20%). Hence, $C_{pi} = \pm 0.7$

Maximum positive wind load = (+0.3425 ± 0.7) × (6 × 6) × 1.05
= ±39.4 kN

Maximum negative wind load = (–0.8 ± 0.7) × (6 × 6) × 1.05
= –56.7 kN

$$\text{Wind load on half the truss} = \frac{56.7}{2} = 28.35 \text{ kN} \simeq 30 \text{ kN}$$

Let wind load at intermediate panel point be W_1.

$$\text{Wind load at the end panel points of the truss} = \frac{W_1}{2}$$

$$\therefore \qquad \frac{W_1}{2} + W_1 + W_1 + W_1 + \frac{W_1}{2} = 30 \text{ kN}$$

$$\Rightarrow \qquad W_1 = 7.5 \text{ kN}$$

Member stresses due to dead load and wind load from Fig. Ex. 11.1(b)(i) & (ii) respectively are listed in Table 11.2.

Table 11.2

Stress on Force Polygon	*Stress due to Dead Load kN*	*Stress due to Wind Load kN*	*Combined Stress kN*	*Design Stress kN*	*Member*
e–1	+9.00	+15.00	+24.00		L_4U_4
d–2	+9.00	+16.1	+25.10	73.5	U_4U_3
c–4	+17.60	+31.9	+49.5	Tension	U_3U_2
b–6	+26.60	+46.9	+73.50	U_2U_1	U_2U_1
f–1	–9.60	–15.8	–25.40		L_4L_3
f–3	–18.70	–31.5	–50.20	99.1	L_3L_2
				Compression	
f–5	–28.00	–46.6	–74.60	L_1L_0	L_2L_1
f–7	–37.20	–61.9	–99.10		L_1L_0
1–2	–4.30	–7.5	–11.80	23.80	U_4L_3
3–4	–6.50	–11.25	–17.75	Compression	U_3L_2
5–6	–8.80	–15.0	–23.80	U_2L_1	U_2L_1
2–3	+8.6	+14.3	+22.9	27.40	U_3L_3
				Tension	
4–5	+9.20	+15.0	+24.20		U_2L_2
6–7	+10.10	+17.3	+27.40	U_1L_1	U_1L_1

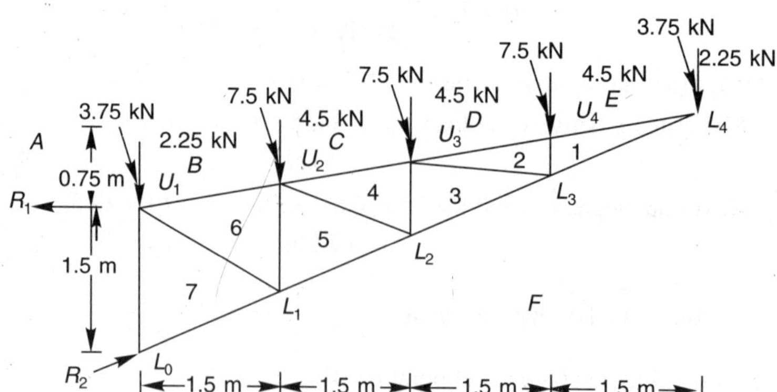

Truss Diagram

Fig. Ex. 11.1(b)

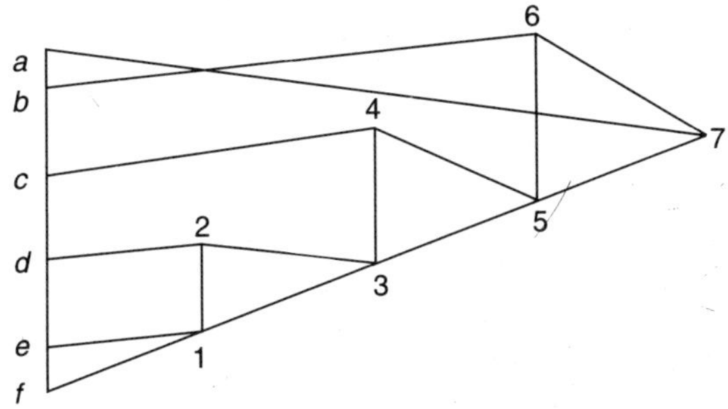

(i) Dead Load Stress Diagram

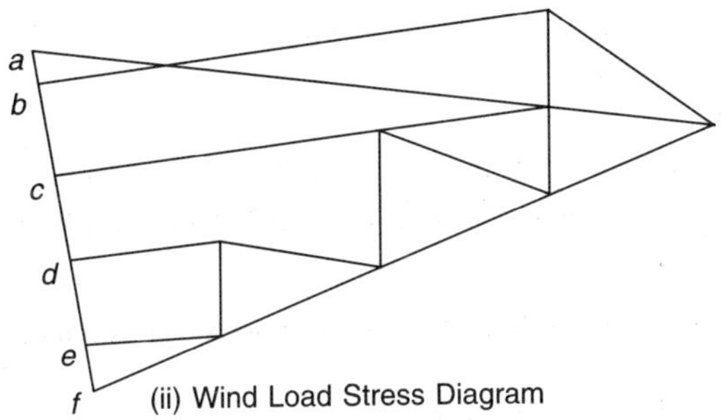

(ii) Wind Load Stress Diagram

Fig. Ex. 11.1(b) *Graphical Analysis of Truss*

The wind load acts normal to the principal rafter $U_1 L_4$

Wind load at intermediate panel points = 7.5 kN

Wind load at end panel points = 3.75 kN

The stresses in the various members of the truss due to the above loads, found graphically, are tabulated in Table 11.2.

Note As per I.S. specifications, the wind load on the roof acts normal to the roof surface. When the inclination of the roof is < 10° and the roofing system is of the shape shown, the intensity of wind pressure on both the leeward and windward side is +0.55*P* ↓.

Design of truss members

Member L_1L_0

Design load = 99.1 kN

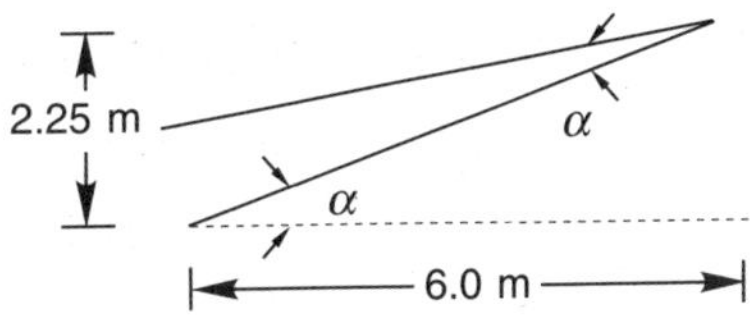

Fig. Ex. 11.1 (c)

$$\tan\alpha = \frac{2.25}{6.00}$$

$\alpha = 20.556°$ (Refer Fig. Ex. 11.1(c))

Length of the member, $L_1L_0 = \dfrac{1.50}{\cos 20.556} = 1.60$ m

The effective length of the continuous member = 0.75 L to L

Let us assume the effective length, l = 1.60 m ($l = L$).

Let us assume yield stress of steel, f_y = 250 MPa.

$$\text{Cross-sectional area required} = \frac{99.1 \times 10^3}{60} = 1651.6 \text{ mm}^2$$

Let us provide two I.S.A. 55 × 55 × 6 mm from I.S. Handbook No. 1. Provide the angles on the opposite side of the gusset plate.

$$\text{Area provided} = 1252 \text{ mm}^2$$

$$r_{min} = 16.6 \text{ mm}$$

$$\frac{l}{r_{min}} = \frac{1600}{16.6} = 96.38 < 250$$

which is all right.

From I.S: 800 – 1984, for $l/r = 96.38$ and $f_y = 250$ MPa,

$$\sigma_{ac} = 83.62 \text{ N/mm}^2$$

Increase the stress by $33\frac{1}{3}\%$ to account for wind.

$$\text{Safe compressive load} = \sigma_{ac} \times \text{area provided}$$

$$= 1.33 \times 83.62 \times 1252 \times 10^{-3} = 139.23 \text{ kN} > 99.1 \text{ kN}$$

which is safe.

Member U_2L_1 Provide a single angle discontinuous strut connected by more than one rivet.

$$\text{Load} = 23.8 \text{ kN}$$

Let L = the length of the strut U_2L_1.

$$= 1.5 - \frac{1.5 \times 1.5}{6} = 1.125 \text{ m}$$

$$\text{Effective length, } l = 0.85 \times 1.125 \times 10^3 = 956.23 \text{ mm} \simeq 956.3 \text{ mm}$$

Let us assume an allowable average compressive stress = 60 MPa

$$\text{Cross-sectional area required} = \frac{23.8 \times 10^3}{60} = 396.66 \text{ mm}^2$$

Let us provide an I.S.A. 50 × 50 × 5 mm section from I.S. Handbook No. 1.

$$\text{Area provided} = 479 \text{ mm}^2$$

$$r_{min} = 9.7 \text{ mm}$$

$$\lambda = \frac{l}{r} = \frac{956.3}{9.7} = 98.58 < 250$$

which is all right.

For $\frac{l}{r} = 98.58$ and $f_y = 250$ MPa, from I.S: 800 – 1984,

$$\sigma_{ac} = 81.42 \text{ N/mm}^2$$

$$\text{Load carrying capacity} = \text{allowable stress} \times \text{area provided}$$

$$= 1.33 \times 81.42 \times 479 \times 10^{-3} = 51.86 \text{ kN} > 23.8 \text{ kN}$$

which is safe.

Member U_1L_1

Design load = 27.4 kN

Let us provide a single angle section for the tie member.

Allowable tensile stress $\sigma_{at} = 0.6 \times 250 = 150$ N/mm^2.

$$\text{Net cross-sectional area required} = \frac{27.4 \times 10^3}{150} = 182.66 \text{ mm}^2$$

Increase the net area by about 40% (to account for rivet holes).

Gross cross-sectional area required = $1.4 \times 182.66 = 255.73$ mm^2

Let us provide an I.S.A. 50 × 30 × 5 mm angle section from I.S. Handbook No. 1.

Area provided = 378 mm^2

Let us provide 14 mm diameter rivets for the connection.

$$\text{Area of connected leg, } A_1 = \left(50 - 15.5 - \frac{5}{2}\right) \times 5 = 160 \text{ mm}^2$$

$$\text{Area of outstanding leg, } A_2 = \left(30 - \frac{5}{2}\right) \times 5 = 137.5 \text{ mm}^2$$

$$k = \frac{3 \times 160}{3 \times 160 + 137.5} = 0.7773$$

Net area provided, $A = A_1 + kA_2$

$= 160 + 0.7773 \times 137.5 = 266.88$ mm^2

Load carrying capacity = allowable stress × area provided

$= 1.33 \times 150 \times 266.88 \times 10^{-3} = 53.24$ kN

> 27.4 kN

Check The length of the member as scaled off from the space diagram,

$l = 1.8$ m

$$\therefore \quad \lambda = \frac{l}{r} = \frac{1.8 \times 10^3}{6.3} = 285.71 < 400$$

which is all right.

Member U_2U_1

Design load = 73.5 kN

Allowable tensile stress, $\sigma_{at} = 150$ N/mm^2.

$$\text{Net area required} = \frac{73.5 \times 10^3}{150} = 490 \text{ mm}^2$$

Increase the net area by about 40%.

Gross cross-sectional area required = $1.4 \times 490 = 686$ mm^2

From I.S. Handbook No. 1, select two I.S.A. 50 × 30 × 6 mm with shorter legs as outstanding legs (tack riveted). The angles are placed back-to-back on the opposite side of the gusset plates and are connected by 14 mm ϕ rivets.

Net area of the angles = 8°4 – 2 × 15.5 × 6 = 708 mm^2

Load carrying capacity = Allowable stress × area provided

$$= 1.33 \times 150 \times 708 \times 10^{-3} = 141.24 \text{ kN} > 73.5 \text{ kN}$$

which is safe.

Check $\lambda = \dfrac{l}{r}$

The length of the member $U_2U_1 \approx 1.5$ m

$$\therefore \quad \lambda = \frac{1.5 \times 10^3}{8.1} = 185.185 < 400$$

which is safe.

Design of joints Let us provide 8 mm thick gusset plate to connect the members. Provide 14 mm ϕ power driven rivets to connect the members with the gusset plate.

Number of rivets

Member L_1L_0

Load = 99.1 kN

Diameter of rivet = 14 mm

Gross diameter = 14 + 1.5 = 15.5 mm

$$\text{Strength of the rivet in double shear} = 2 \times \frac{\pi}{4} \times d^2 \times \tau_{vf}$$

$$= 2 \times \frac{\pi}{4} \times 15.5^2 \times 1.25 \times 100 = 434672.97 \text{ N}$$

$$\text{Strength of the rivet in bearing} = dt\sigma_{pf}$$

$$= 15.5 \times 6 \times 1.25 \times 300 = 34875 \text{ N}$$

$$\therefore \quad \text{Rivet value} = 34875 \text{ N} = 34.8 \text{ kN}$$

$$\therefore \quad \text{Number of rivets} = \frac{99.1}{34.8} = 2.847 \approx 4$$

Provide four rivets 14 mm ϕ for connection.

Member U_2L_1

Load = 23.8 kN

$$\text{Strength of the rivet in single shear} = \frac{\pi}{4} \times d^2 \times \tau_{vf}$$

$$= \frac{\pi}{4} \times 15.5^2 \times 1.25 \times 100 = 23586.4 \text{ N}$$

$$\text{Strength of the rivet in bearing} = dt\sigma_{pf} = 15.5 \times 5 \times 1.25 \times 300 = 29062.5 \text{ N}$$

$$\text{Rivet value} = 23586.4 \text{ N} = 23.58 \text{ kN}$$

$$\text{Number of rivets required} = \frac{23.8}{23.58} = 1.009 \approx 2$$

Provide two rivets 14 mm ϕ for the connection.

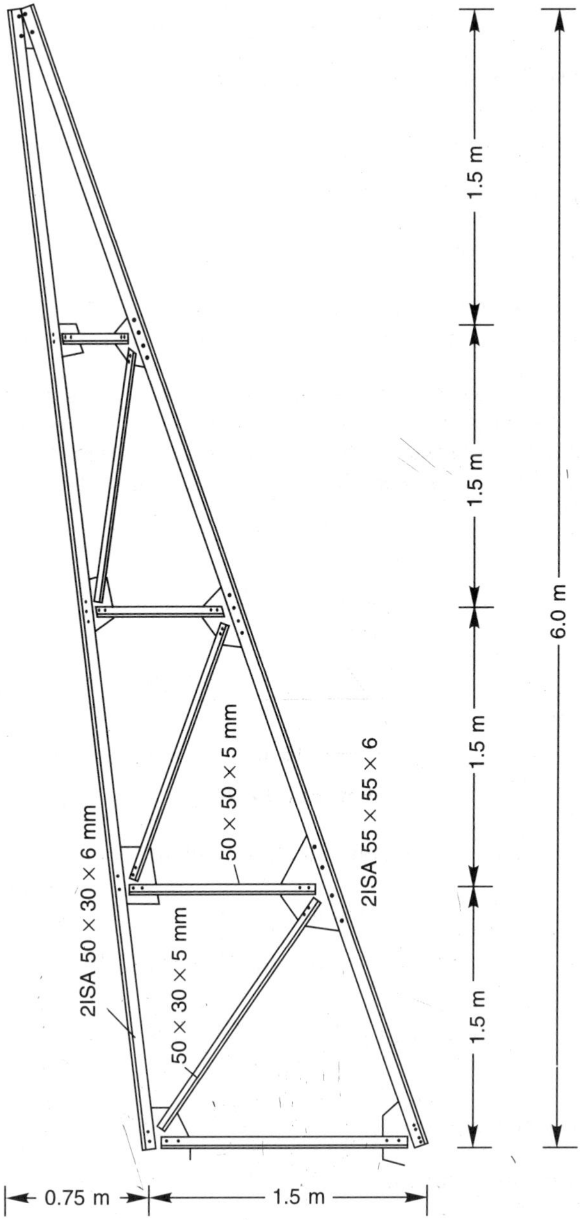
2ISA 50 × 30 × 6 mm
50 × 30 × 5 mm
50 × 50 × 5 mm
2ISA 55 × 55 × 6
1.5 m
1.5 m
1.5 m
1.5 m
6.0 m
0.75 m
1.5 m

Fig. Ex. 11.1(d)

Member U_1L_1

$$\text{Load} = 27.4 \text{ kN}$$

$$\text{Rivet value} = 18.869 \text{ kN}$$

$$\therefore \quad \text{Number of rivets} = \frac{27.4}{23.58} = 1.16 \approx 2$$

Provide two rivets 14 mm ϕ for connection.

Member U_2U_1

$$\text{Load} = 73.5 \text{ kN}$$

$$\text{Rivet value} = 34.5 \text{ kN}$$

$$\text{Number of rivets} = \frac{73.5}{34.5} = 2.13 \approx 3$$

Provide three rivets 14 mm ϕ for connection.

Member sections and their connections are shown in Fig. Ex. 11.1(d).

Example 11.2 Determine the design forces in the members of the steel roof truss (Fig. Ex. 11.2(a)) 16.0 m in span and resting on brick masonry walls. The trusses are placed 8 m c/c. The rise of the truss is 1/4 of span. Roofing is of asbestos cement sheets of dead weight 171 N/m^2. The wind load normal to roof truss is 940 N/m^2. One end of the truss is hinged and the other end is supported on rollers.

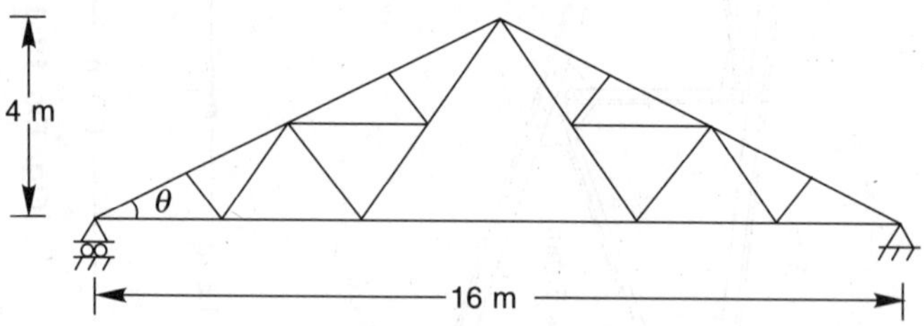

Fig. Ex. 11.2(a)

Solution Let θ be the inclination of the roof with the horizontal.

$$\tan\theta = \frac{4}{8} = \frac{1}{2}$$

or

$$\theta = 26° \, 34' = 26.566°$$

$$\text{Length of rafter} = \sqrt{\left(\frac{16}{2}\right)^2 + 4^2} = 8.94 \text{ m}$$

$$\text{Distance between panel points} = \frac{8.94}{4} = 2.235 \text{ m}$$

Load on panel points (Refer Fig. Ex. 11.2(b))

(i) *Dead load*

$$\text{Weight of A.C. sheets} = 171 \text{ N/m}^2$$

Assume

$$\text{Weight of purlins} = 318 \text{ N/m}$$

$$\text{Weight of bracing} = 12 \text{ N/m}^2$$

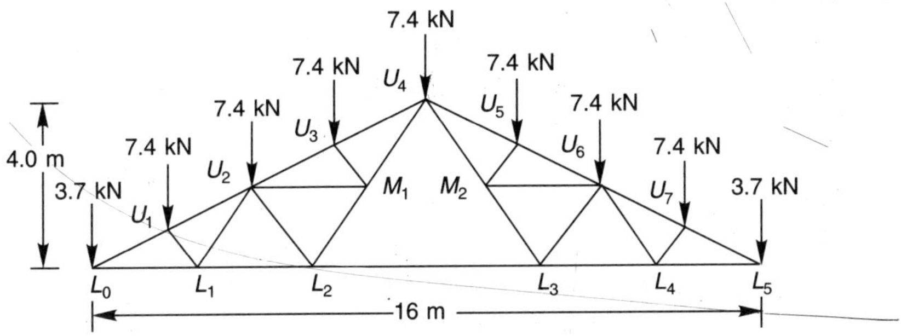

(i) Dead Loads at Panel Points

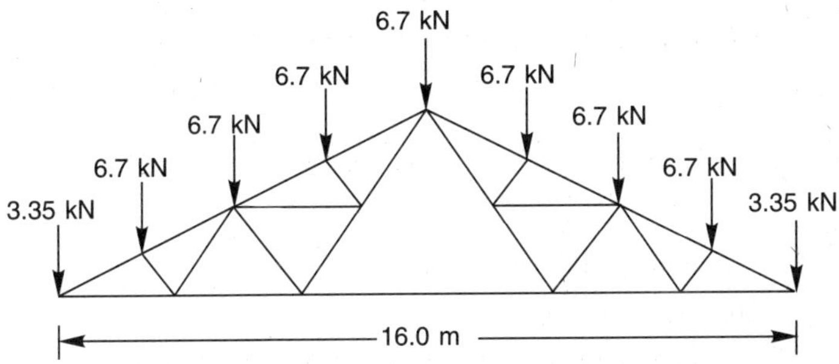

(ii) Live Loads at Panel Points

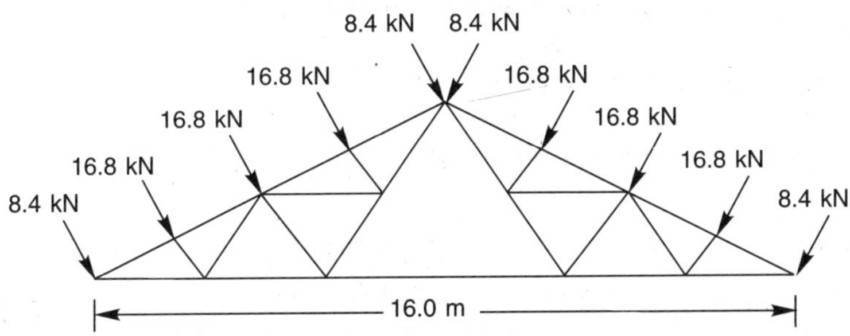

(iii) Wind Loads at Panel Points

Fig. Ex. 11.2(b)

$$\text{Self-weight of the roof truss} = \left(\frac{\text{span}}{3} + 5\right) \times 10$$

$$= \left(\frac{16}{3} + 5\right) \times 10 = 103.33 \text{ N/m}^2 \approx 110 \text{ N/m}^2$$

$$\text{Total dead load} = 171 + 12 + 110 = 293 \text{ N/m}^2$$

$$\text{Dead load of purlin} = 318 \times 8 = 2544 \text{ N}$$

$$\text{The panel length } L_0U_1 = U_1U_2 = U_2U_3 = U_3U_4 = 2.235 \text{ m}$$

$$\text{The panel length in plan} = 2.235 \cos 26° \, 24' = 2.000 \text{ m}$$

$$\text{Spacing of truss} = 8.0 \text{ m}$$

Load on each intermediate panel due to dead load

$$= 293 \times 8.0 \times 2.0 + 2544 = 7232 \text{ N} \approx 7.4 \text{ kN}$$

Load on end panel points is half of this load

$$= \frac{7.4}{2} = 3.7 \text{ kN}$$

(ii) *Live load*

$$\theta = 26° \, 34' = 26.566°$$

Let us assume that no access is provided to the roof. The live load is reduced by 20 N/m^2 for each one degree above 10° slope.

$$\therefore \quad \text{Live load} = [750 - 20 \times (26.566 - 10)] = 418.68 \text{ N/m}^2$$

The load on each intermediate panel $= 418.68 \times 8 \times 2$

$$= 6698.88 \text{ N} \approx 6700 \text{ N} = 6.7 \text{ kN}$$

The load on each end panel point $= \dfrac{6700}{2} = 3350 \text{ N} = 3.35 \text{ kN}$

(iii) *Wind load*

Length of rafter = 8.94 m

Spacing of trusses = 8.00 m

$$\therefore \quad \text{Total wind load} = 8.94 \times 8 \times 940 = 67{,}228.8 \text{ N}$$

Let the wind load on intermediate panel points = W_1

Wind load on end panel points = $\dfrac{W_1}{2}$

$$\therefore \quad \frac{W_1}{2} + W_1 + W_1 + W_1 + \frac{W_1}{2} = 67{,}228.8 \text{ N}$$

or

$$W_1 = 16{,}807.2 \text{ N} \approx 16.8 \text{ kN}$$

The resultant panel point loads due to dead loads and live loads are shown in Figs Ex. 11.2(b) (i) & (ii) respectively, and due to wind load are shown in Fig. Ex. 11.2(b) (iii). The member forces are obtained graphically.

Note The load line can be drawn and points 1, 2, 3 can be obtained in usual way. The problem arises in locating point 4. To obtain it, remove member 4–5, 5–6. Insert a substitute member *m–n*. A line on force polygon can be drawn parallel to member 3–m intersecting in point *m*, on a line from *d*, parallel to member *m–D*. From *m*, draw a line parallel to member *m–n*. *n* is obtained by the intersection of this line with a line parallel to member *E*–6. From n a line parallel to member *m*–7 is drawn, intersecting the line *k*–1 in 7. The point 4 can be obtained by extending the line 3–*m*. From point 4, draw a line 4–5 parallel to the member 4–5 to intersect line from *d*, parallel to the member *D*–5. From point 5, draw a line 5–6 parallel to the member 5–6. Reference may be made to Fig. Ex. 11.2(c).

To analyse the truss for wind loads, the load line is plotted parallel to the wind loads. The process for the graphical plot is same except that all the lines in the force diagram are drawn perpendicular to the members (Fig. Ex. 11.2(d)).

Reactions The truss is symmetrical and therefore, the dead load and live load reactions will be same on both the supports but the reactions due to wind load will be different on the two supports.

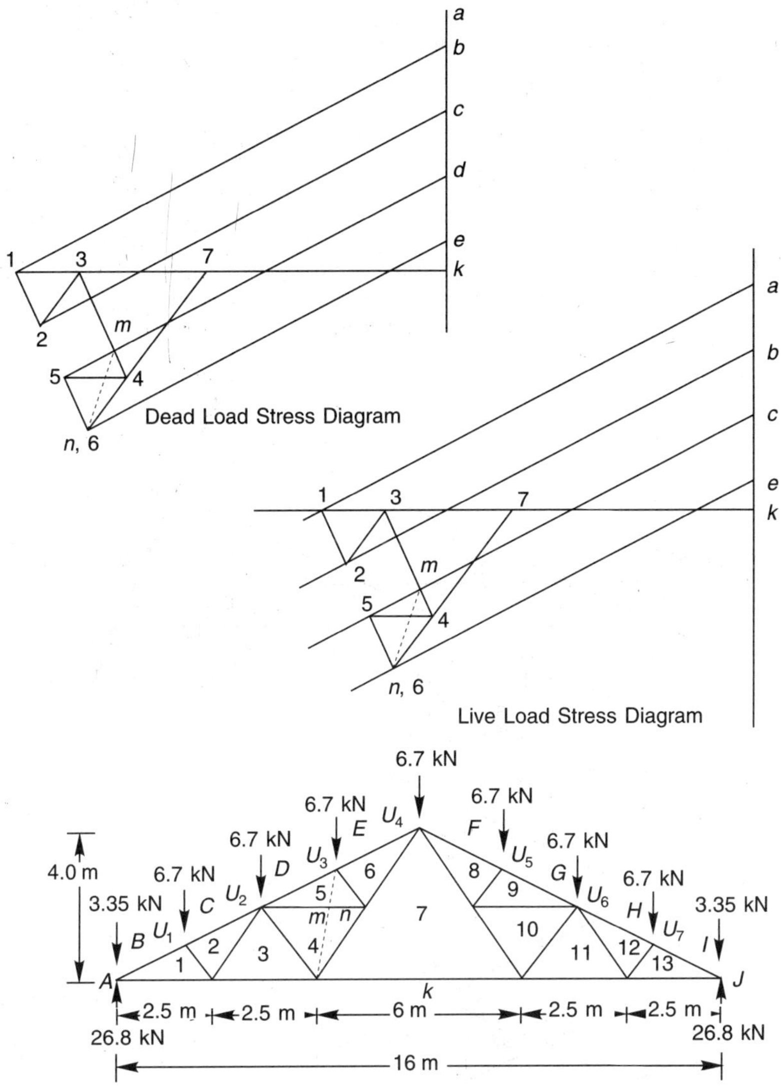

Fig. Ex. 11.2(c)

Live load reactions Taking moments about L_0,

$$6.7 \times 2 + 6.7 \times 4 + 6.7 \times 6 + 6.7 \times 8 + 6.7 \times 10$$
$$+ 6.7 \times 12 + 6.7 \times 14 + 3.35 \times 16 = R_{L_5} \times 16$$

$\therefore$ $$R_{L_5} = 26.8 \text{ kN}$$

By symmetry,

$$\Rightarrow R_{L_0} = 26.8 \text{ kN}$$

Dead load reactions

$$7.4 \times 2 + 7.4 \times 4 + 7.4 \times 6 + 7.4 \times 8 + 7.4 \times 10 + 7.4 \times 12$$
$$+ 7.4 \times 14 + 3.7 \times 16 = R_{L_5} \times 16$$

$\Rightarrow$ $$R_{L_5} = 29.6 \text{ kN}$$

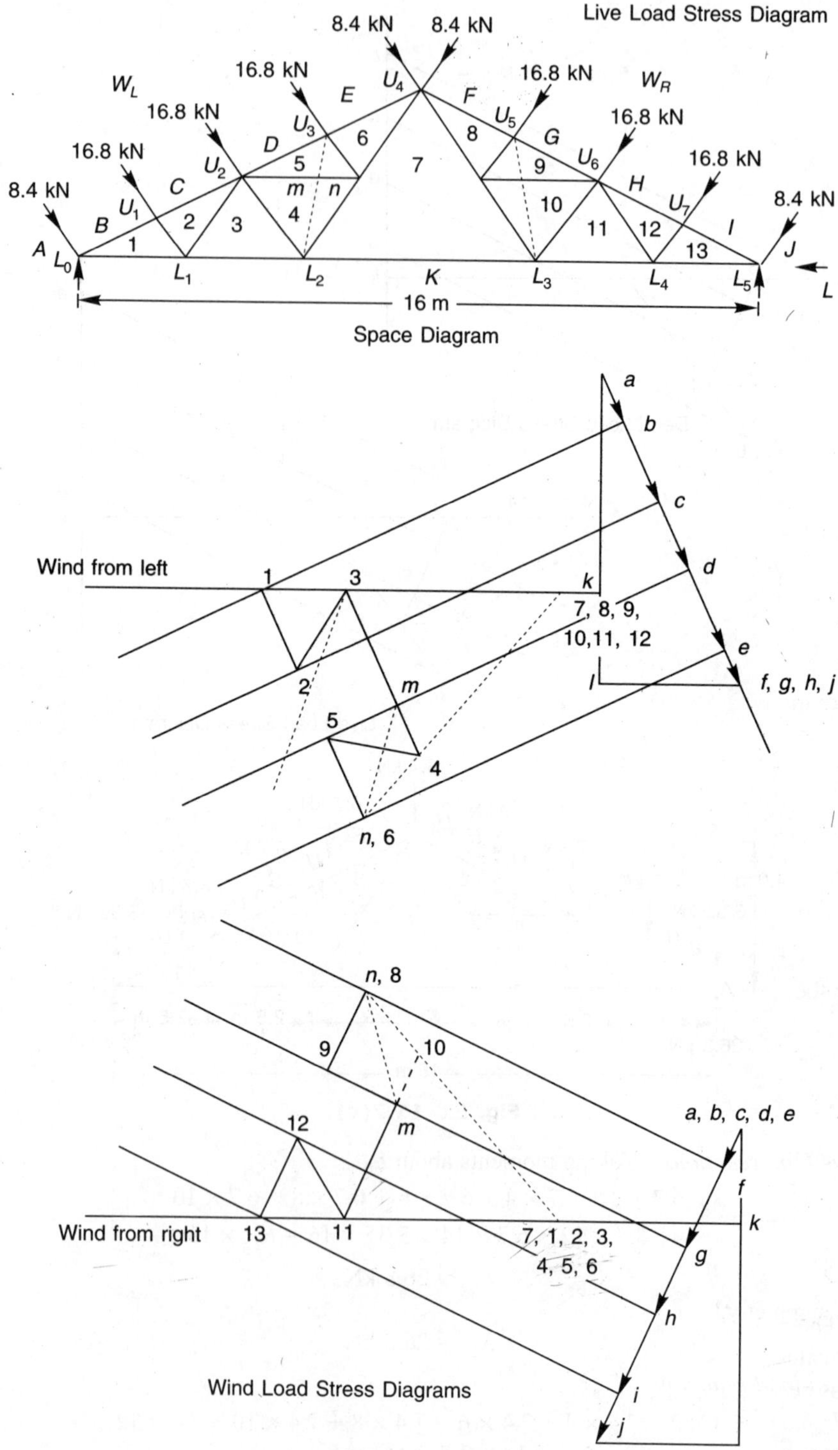

Live Load Stress Diagram
8.4 kN
8.4 kN
16.8 kN
16.8 kN
16.8 kN
16.8 kN
16.8 kN
16.8 kN
8.4 kN
8.4 kN
W_L
W_R
16 m
Space Diagram
Wind from left
7, 8, 9,
10,11, 12
f, g, h, j
n, 6
n, 8
a, b, c, d, e
Wind from right
7, 1, 2, 3,
4, 5, 6
Wind Load Stress Diagrams

Fig. Ex. 11.2(d)

By symmetry,

$$\Rightarrow \qquad R_{L_0} = 29.6 \text{ kN}$$

Wind load reactions (Fig. Ex. 11.2(e)) The wind is blowing from the left.

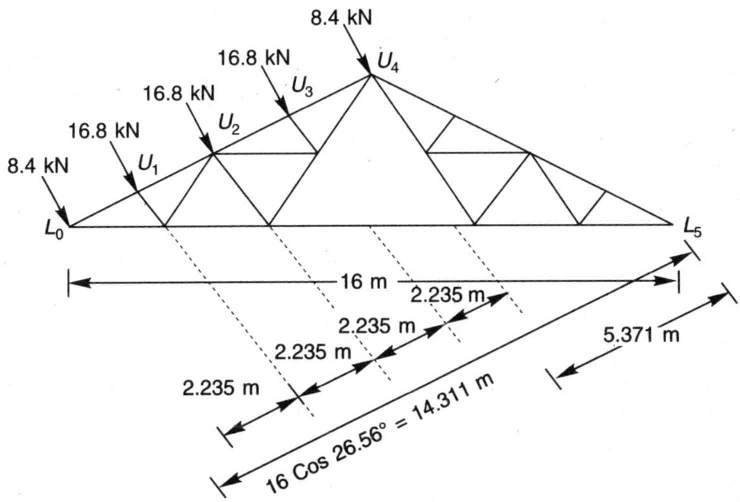

Fig. Ex. 11.2(e)

Take moment about L_0

$$R_{L_5} \times 16 = 16.8 \times 2.235 + 16.8 \times 4.47 + 16.8 \times 6.705 + 8.4 \times 8.94$$

$$\Rightarrow \qquad R_{L_5} = 18.774 \text{ kN}$$

Take moment about L_5

$$R_{L_0} \times 16 = 8.4 \times 14.311 + 16.8 \times 12.076 + 16.8 \times 9.84 + 16.8 \times 7.606 + 8.4 \times 5.371$$

$$\Rightarrow \qquad R_{L_0} = 41.33 \text{ kN}$$

For the horizontal reaction at L_5, take moment about ridge U_4

$$H_{L_5} \times 4 = 8.4 \times 8.94 + 16.8 \times 6.705 + 16.8 \times 4.47 + 16.8 \times 2.235 - 41.33 \times 8 + 18.774 \times 8$$

or $\qquad H_{L_5} = 29.984$ kN

When the wind blows from right, the reactions will be

$$R_{L_0} = 18.774 \text{ kN}$$

$$R_{L_5} = 41.33 \text{ kN}$$

$$H_{L_5} = -29.984 \text{ kN}$$

The results of the graphical analysis are tabulated in Table 11.3. The plus sign indicates tension and the minus sign indicates compression. All the members of the truss need not be designed, e.g., the main rafter L_0U_4 consists of L_0U_1, U_1U_2, U_2U_3, U_3U_4 and the stresses as calculated are found to be different. So the member L_0U_4 is designed with a uniform cross-section for the above members with the maximum force occurring in any of its element as the cost of splicing, arrangements for roofing, etc. will nullify the economy achieved by providing different cross-sections for L_0U_1, U_1U_2,...., etc. In the last column of the table maximum design forces are listed for each group of members.

Table 11.3

Stresses	*Member*	*Dead Load*	*Live Load*	*Wind Load From* *Left*	*Wind Load From* *Right*	*Design Stresses*
	Rafter					
b–1	L_0U_1	–58.0	–52.5	–76.0	–42.8	
c–2	U_1U_2	–55.6	–50.3	–76.4	–42.8	–187.4 kN
d–5	U_2U_3	–51.4	–46.5	–37.0	–42.8	
e–6	U_3U_4	–48.0	–43.5	–75.0	–42.8	
f–8	U_4U_5	–48.0	–43.5	–37.0	–77.3	
g–9	U_5U_6	–51.4	–46.5	–37.0	–77.3	
h–12	U_6U_7	–55.6	–50.3	–37.0	–77.3	
i–13	U_7L_5	–58.0	–52.5	–37.0	–77.3	
	Main tie					
1–k	L_0L_1	+52.0	+47.0	+64.7	+38.6	
3–k	L_1L_2	+45.0	+40.7	+45.4	+38.6	+163.7 kN
11–k	L_3L_4	+45.0	+40.7	+8.4	+76.4	
13–k	L_4L_5	+52.0	+47.0	+8.4	+16.8	
7–k	L_2L_3	+31.1	+28.2	+8.0	+38.6	
	Inclined members					
1–2	U_1L_1	–6.2	–5.6	–16.8	0	
3–4	L_2U_2	–12.4	–11.2	–32.8	0	
5–6	U_3M_1	–6.6	–6.0	–16.8	0	
13–12	L_4U_7	–6.2	–5.6	0	–16.8	57.2 kN
11–10	L_3U_6	–12.4	–11.2	0	–33.6	
9–8	U_5M_2	–6.6	–6.0	0	–16.8	
2–3	L_1U_2	+7.0	+6.3	+18.9	0	
4–5	U_2M_1	+6.0	+5.4	+17.7	0	30.7 kN
12–11	L_4U_6	+7.0	+6.3	0	+18.5	
10–9	U_6M_2	+6.0	+5.4	0	+19.3	
6–7	U_4M_1	+20.3	+18.4	+55.4	0	
4–7	L_2M_1	+13.8	+12.5	+37.0	0	95.5 kN
8–7	U_4M_2	+20.3	+18.4	0	+56.3	
10–7	L_3M_2	+13.8	+12.5	0	+37.8	

Example 11.3 Design a fink type roof truss for an industrial building given the following data:

Overall length	48 m
Overall width	16.5 m
Width c/c of roof columns	16.0 m

Height of columns — 11.0 m
Roofing material — Asbestos cement sheets
Side covering — Asbestos cement sheets

The industrial building is situated in Naini, Allahabad.
Assume suitably any data that is missing.

Solution Let us assume the life of the industrial building to be 50 years and the land to be plain and surrounded by small buildings.

$k_1 = 1.0$ (for 50 years)

$k_2 = 0.89$ (for terrain category 3, building height = 11 m)

$k_3 = 1.0$ (for plain land)

$V_b = 47$ m/s (for zone 3)

$$\text{Design wind speed, } V_z = k_1\, k_2\, k_3\, V_b = 1.0 \times 0.89 \times 1.0 \times 47 = 41.83 \text{ m/s}$$

$$\text{Design wind pressure, } p_d = 0.6\, V_z^2 = 0.6 \times 41.83^2 = 1049.8 \text{ N/m}^2 \simeq 1.05 \text{ kN/m}^2$$

Purlins Prior to the design of purlins it is necessary to decide the spacing of the columns so that the spacing of trusses is fixed. Let us provide columns 8 m c/c and fix the geometry of the roof truss so that the spacing of the purlins can be fixed.

Let us provide a fink roof truss shown in Fig. Ex. 11.3(a)

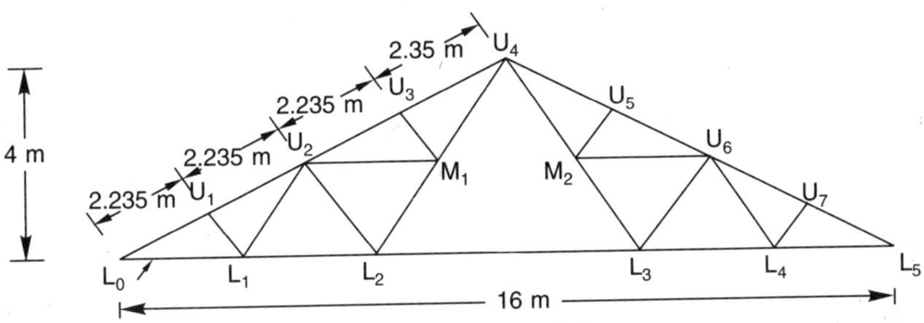

Fig. Ex. 11.3(a)

Let us assume the rise of the truss = 4.0 m.

Let the inclination of the roof with the horizontal = θ

$$\theta = \tan^{-1}\frac{4}{8} = 26° \, 34'$$

$$\text{Length of the rafter, } L_0U_4 = \sqrt{\left(\frac{16}{2}\right)^2 + 4^2} = 8.94 \text{ m}$$

Distance between panel points of the top chord of roof truss,

$$L_0U_1 = U_1U_2 = U_2U_3 = U_3U_4 = \frac{8.94}{4} = 2.235 \text{ m}$$

Let us place the purlins at the panel points.

Design data

Span of the purlin = c/c spacing of roof truss = 8.0 m

c/c spacing of purlins = 2.235 m

Slope of the roof truss = 26° 34′

Wind load on the roof normal to the roof surface = 1050 N/m^2

Assume Dead load of the roof covering asbestos cement sheets = 171 N/m^2

Dead load of purlin = 300 N/m

Dead load of A.C. sheets/m = 171 × 2.235 = 382.185 N/m

Total dead load = 382.185 + 300 = 682.185 N/m

Live load For $\theta = 26.566°$,

Live load = 418.68 N/m^2 (From Example 11.2)

Live load = 418.68 × 2.235 cos 26.566° = 836.95 N/m ≃ 880 N/m

Wind load Height of building column above ground level, h = 11 m

Width of building, w = 16 m

$$\frac{h}{w} = \frac{11}{16} = 0.6875$$

The external air pressure coefficients C_{pe} for the condition $\frac{1}{2} < \frac{h}{w} < \frac{3}{2}$ and for $\theta = 26.566$, from Appendix IX are –0.37 for windward side and –0.8 for leeward side.

Let us assume the building to have normal permeability. The internal air pressure coefficients C_{pi} are ±0.02 for both the windward and leeward sides.

$$\begin{aligned}\text{Wind load} &= (C_{pe} - C_{pi})\, p_d\, A \\ &= (-0.8 - (\pm\, 0.2)) \times 1.05 \times 8 \times 2.235 \simeq -18.774 \text{ kN} \\ &= 18.8 \text{ kN} \quad \text{(Uplift)}\end{aligned}$$

$$\text{Wind load per unit length of purlins} = \frac{18.8}{8} = 2.35 \text{ kN/m} = 2350 \text{ N/m}$$

Let us provide angle section purlins.

(i) As the roof slope is less than 30° the dead load, live load and the wind load are assumed to act normal to the roof surface.

(ii) The wind load will act in the opposite direction to that of the dead and live loads on the roof surface.

Therefore, the effect of dead loads will be to reduce the wind load effect on the purlins. The critical design load for the purlins will therefore, be the maximum of either

(i) the live load and dead load or

(ii) the difference of wind load and dead load, as the two loads are acting in the opposite direction.

Truss Let us provide a fink roof truss, hinged at the top of the columns. The dead load and live load computations are same as that of Example 11.2.

Dead load

Load on the intermediate panel points = 7.4 kN

Load on the end panel points = 3.7 kN

Live load

Load on the intermediate panel points = 6.7 kN

Load on the end panel points = 3.35 kN

Wind load Let us assume the industrial building to have normal permeability.

Wind load on panel points

Windward side

$$F = (C_{pe} - C_{pi})\, p_d\, A$$
$$= (-0.37 \pm 0.2) \times 1.05 \times (8 \times 2.235) = -10.71 \text{ kN} \simeq -10.8 \text{ kN}$$

Wind load on each intermediate panel point = –10.8 kN

$$\text{Wind load on each end panel point} = \frac{-10.8}{2} = -5.4 \text{ kN.}$$

Leeward side

$$F = (C_{pe} - C_{pi})\, p_d\, A$$
$$= (-0.8 \pm 0.2) \times 1.05 \times (8 \times 2.235) = -18.77 \text{ kN} \simeq -18.8 \text{ kN}$$

Wind load on each intermediate panel point = –18.8 kN

$$\text{Wind load on each end panel point} = \frac{-18.8}{2} = -9.4 \text{ kN}$$

$\therefore$ Design load normal to the roof surface = net uplift force

$$= 2350 - 682.185 = 1667.82 \text{ N/m}$$

$$\text{Maximum moment, } M = \frac{WL}{10} = \frac{(1667.82 \times 8) \times 8 \times 10^3}{10}$$
$$= 1067 \times 10^4 \text{ Nmm}$$

$$\text{Required section modulus} = \frac{1067 \times 10^4}{1.33 \times 165} = 48640 \text{ mm}^3$$

Approximate depth of the angle section purlin

$$= \frac{L}{45} = \frac{8000}{45} = 177.77 \text{ mm} \approx 200.0 \text{ mm}$$

Approximate width of the angle section purlin

$$= \frac{L}{60} = \frac{8000}{60} = 133.33 \text{ mm} \approx 150 \text{ mm}$$

Let us provide I.S.A. 200 × 150 × 12 mm angle section as purlin and check its safety.

The self-weight of the section = 312 N/m.

Section modulus provided = 117,400 mm^3

> 48640 mm^3

which is sufficient.

$$\text{Bending stress, } \sigma_{bc,\text{cal}} = \frac{1067 \times 10^4}{117{,}400} = 90.88 \text{ N/mm}^2$$

$$< 1.33 \times 165 \; (219.45 \text{ N/mm}^2)$$

which is all right.

Note The dead weight was assumed to be 300 N/m and the section provided is of dead weight 312 N/m. As the design load is net-uplift, the design moment will be reduced slightly. The section modulus will not be affected much if the revised dead weight is used in the computations. Moreover, the design is conservative; therefore, provide I.S.A. 200 × 150 × 12 mm as the purlin section.

The design loads for the truss are shown in the space diagram, Fig. Ex. 11.3(b). The dead loads and live loads are same as in Example 11.2 and therefore, the same results are used here. For wind load analysis the graphical method is adopted. The stress diagram is plotted and scaled off stresses are presented in Table 11.4.

Reactions at truss shoe due to wind load For analysis of the truss in case of industrial buildings the truss is assumed to be simply supported over its supports. In such a case the truss becomes indeterminate to the first degree. A very close approximation is made further to obtain the reactions. "The horizontal components of both the shoe reactions are equal to each other." This assumption in fact, is consistent with the usual assumption that the supporting columns share the horizontal side thrust equally between them.

From the truss geometry, the line of action of 43.2 kN resultant (of the forces on the left hand roof slope) cuts the horizontal member L_0L_5 in L_2, and the line of action of 75.2 kN resultant (of the forces on the right hand roof slope) cuts the horizontal member L_0L_5 in L_3. The unbalanced horizontal force is worked out and is distributed equally at each shoe.

Components of resultants

Force 43.2 kN:

$$\text{Vertical component} = 43.2 \cos 26.566° = 38.639 \text{ kN} \approx 38.64 \text{ kN} \uparrow$$

$$\text{Horizontal component} = 43.2 \sin 26.566° = 19.32 \text{ kN} \leftarrow$$

Force 75.2 kN:

$$\text{Vertical component} = 75.2 \cos 26.566° = 67.26 \text{ kN} \uparrow$$

$$\text{Horizontal component} = 75.2 \sin 26.566° = 33.63 \text{ kN} \rightarrow$$

$$\text{Net horizontal component} = 33.63 - 19.32 = 14.31 \text{ kN} \rightarrow$$

$$\text{Horizontal force at each shoe} = \frac{14.31}{2} = 7.155 \approx 7.16 \text{ kN} \rightarrow$$

Note The unbalanced horizontal force is 14.31 kN. Therefore, the equal and opposite horizontal force to be applied at each shoe is 7.16 kN. The horizontal component of reactions at supports is therefore = 7.16 kN ←.

To find the vertical component of reactions at support, take moment about either shoe.

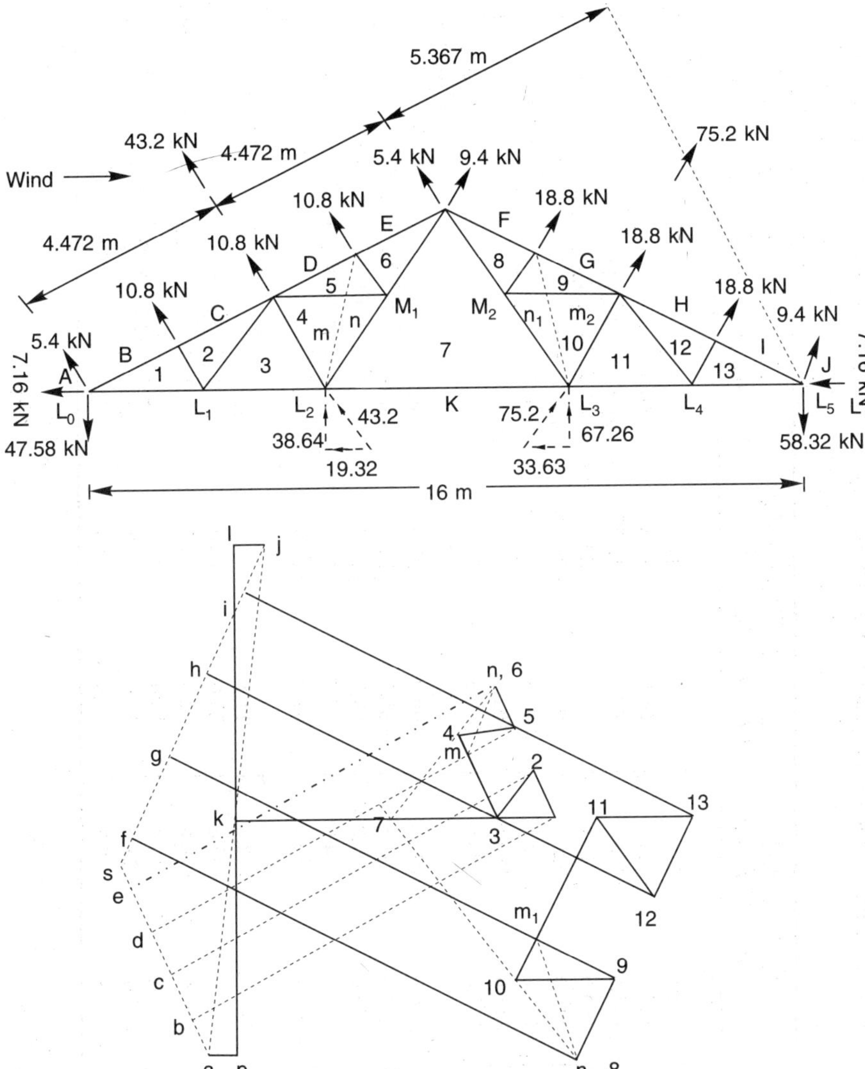

Fig. Ex. 11.3(b) *Wind Load Stress Diagram (Wind from Left)*

Take moment about L_5 Fig. Ex. 11.3(b)

$$R_{L_0} \times 16 = 43.2 \times (4.472 + 5.367) + 75.2 \times 4.472$$

$$\Rightarrow \quad R_{L_0} = 47.584 \text{ kN} \approx 47.58 \text{ kN}$$

Take moment about L_0

$$R_{L_5} \times 16 = 43.2 \times 4.472 + 75.2 \times (4.472 + 5.367)$$

$$\Rightarrow \quad R_{L_5} = 58.317 \text{ kN} \approx 58.32 \text{ kN}$$

Design

Main tie The main tie is a tension member under constant dead load. When wind blows, a reversal of stress occurs and the tie becomes a strut for a part of its length.

The results of the load analysis of Case I, dead load and live load analysis (from Example 2), and the design stresses are tabulated below:

Table 11.4

Stress Member	Dead Load Stress (kN)	Live Load Stress (kN)	Wind Load Stress		Maximum Stress			Design Force	Member
			Wind Left	Wind Right	DL + LL	DL + LL + WL	DL + WL		
									Rafter
b–1	–58.0	–52.5	+95.6	+111.6	–110.5	–14.9	+53.6		L_0U_1
c–2	–55.6	–50.3	+95.6	+111.6	–105.9	–10.3	+56.0		U_1U_2
d–5	–51.4	–46.5	+95.6	+111.6	–97.9	+13.7	+60.2	110.5 kN	U_2U_3
e–6	–48.0	–43.5	+95.6	+111.6	–91.5	+20.1	+63.6	Compression	U_3U_4
f–8	–48.0	–43.5	+111.6	+95.6	–91.5	+20.1	+63.6	63.6 kN Tension	U_4U_5
g–9	–51.4	–46.5	+111.6	+95.6	–97.9	+13.7	+60.2		U_5U_6
h–12	–55.6	–50.3	+111.6	+95.6	–105.9	–10.3	+56.0		U_6U_7
i–13	–58.0	–52.5	+111.6	+95.6	–110.5	–14.9	+53.6		U_7L_5
									Main Tie
1–k	+52.0	+47.0	–76.0	–102.4	+99.0	+23.0	–50.4		L_0L_1
3–k	+45.0	+40.7	–64.0	–81.8	+85.7	+21.7	–36.8	99.0 kN Tension	L_1L_2
7–k	+31.1	+28.2	–39.8	–39.8	+59.3	+19.5	–8.7	50.4 kN	L_2L_3
11–k	+45.0	+40.7	–81.8	–64.0	+85.7	+21.7	–36.8	Compression	L_3L_4
13–k	+52.0	+47.0	–102.4	–76.0	+99.0	+23.0	–50.4		L_4L_5
									Struts
1–2	–6.2	–5.6	+10.8	+18.8	–11.8	+7.0	+12.6		U_1L_1
3–4	–12.4	–11.2	+21.6	+37.6	–23.6	+14.0	+25.2	23.6 kN	U_2L_2
5–6	–6.6	–6.0	+10.8	+18.8	–12.6	+6.2	+12.2	Compression	U_3M_1
8–9	–6.6	–6.0	+18.8	+10.8	–12.6	+6.2	+12.2	25.2 kN Tension	U_5M_2
10–11	–12.4	–11.2	+37.6	+21.6	–23.6	+14.0	+25.2		U_6L_3
12–13	–6.2	–5.6	+18.8	+10.8	–11.8	+7.0	+12.6		U_7L_4

(*Contd.*)

Table 11.4 *(Contd.)*

Stress Mmember	Dead Load Stress (kN)	Live Load Stress (kN)	Wind Load Stress		Maximum Stress			Design Force	Member
			Wind Left	Wind Right	DL + LL	DL + LL + WL	DL + WL		
									Minor Sling
2–3	+7.0	+6.3	−12.1	−21.0	+13.3	−7.7	−14.0	13.3 kN Tension	U_2L_1
4–5	+6.0	+5.4	−12.1	−21.0	+11.4	−9.6	−15.0	15.0 kN	U_2M_1
9–10	+6.0	+5.4	−21.0	−12.1	+11.4	−9.6	−15.0	Compression	U_6M_2
11–12	+7.0	+6.3	−21.0	−12.1	+13.3	−7.7	−14.0		U_6L_4
									Main Sling
6–7	+20.3	+18.4	−36.3	−63.0	+38.7	−24.3	−42.7	38.7 kN Tension	U_4M_1
4–7	+13.8	+12.5	−24.2	−42.0	+26.3	−15.7	−28.2	42.7 kN	L_2M_1
7–8	+20.3	+18.4	−63.0	−36.3	+38.7	−24.3	−42.7	Compression	U_4M_2
7–10	+13.8	+12.5	−42.0	−24.2	+26.3	−15.7	−28.2		L_3M_2

Therefore, primarily it is designed as a tension member and then the section is checked for the maximum compression in it. The tie also finds support from the horizontal bracing angles between the trusses (when braced in the plane of lower chord). While checking it as a compression member it is treated as a strut.

$$\text{Design tensile force} = 99.0 \text{ kN}$$

$$\text{Design compressive force} = 50.4 \text{ kN}$$

$$\text{Net area required} = \frac{99.0 \times 10^3}{150} = 660 \text{ mm}^2$$

Increase it by 40%.

$$\text{Gross area required} = 1.4 \times 660 = 924 \text{ mm}^2$$

From I.S. Handbook No. 1, let us provide two unequal angles 65 × 45 × 8 mm, back-to-back, on the opposite side of the gusset plate.

$$\text{Area provided} = 1634 \text{ mm}^2$$

Provide 14 mm rivets for connection.

$$\text{Net area provided} = 1634 - 2 \times 8 \times 15.5 = 1386 \text{ mm}^2$$

$$\therefore \quad \text{Safe tensile load} = \sigma_{at} \times \text{net area provided}$$

$$= 1.33 \times 150 \times 1386 = 276{,}507 \text{ N} = 276.5 \text{ kN}$$

$$> 99.0 \text{ kN}$$

which is as it should be.

Due to the reversal of stresses the main tie is checked as a compression member.

Length of the tie between node points, $L \simeq 6.0$ m

$$\text{Effective length, } l_{xx} = 0.85 \times 6.0 = 5.10 \text{ m} = 5100 \text{ mm}$$

Two angle sections are provided for bracing, one at L_2 and the other at L_3 node in the trusses. Assuming longitudinal tie at main node points,

$$\text{Effective length } l_{yy} = 1.0 \times 6.0 = 6.0 \text{ m} = 6000 \text{ mm}$$

The relevant properties of the built-up section are:

$$A = 1634 \text{ mm}^2,\ r_{xx} = 20.2 \text{ mm};\ r_{yy} = 21.1 \text{ mm}$$

$$\frac{l_{xx}}{r_{xx}} = \frac{5100}{20.2} = 252.47$$

$$\frac{l_{yy}}{r_{yy}} = \frac{6000}{21.1} = 284.36$$

From IS: 800–1984, for $f_y = 250$ N/mm^2 $\frac{l}{r} = 284.36$

$$\therefore \quad \sigma_{ac} = 14.26 \text{ N/mm}^2$$

Safe compressive load carrying capacity

$$= \sigma_{ac} \times \text{area provided}$$

$$= 1.33 \times 14.26 \times 1634 \times 10^{-3} = 30.99 \text{ kN}$$

$$< 50.4 \text{ kN}$$

Hence, the section is not safe.

Revise the section and provide two I.S.A. 100 × 65 × 8 mm.

The relevant properties are

$$A = 2514 \text{ mm}^2, \; r_{xx} = 31.6 \text{ mm}, \; r_{yy} = 27.5 \text{ mm}$$

As we have chosen a higher section than 65 × 45 × 8 mm there is no need to check it as a tension member (as the section smaller than this was found safe).

Let us check it for compression under reversal due to wind.

$$\frac{l_{xx}}{r_{xx}} = \frac{5100}{31.6} = 161.39$$

$$\frac{l_{yy}}{r_{yy}} = \frac{6000}{27.5} = 218.18$$

Therefore, the critical slenderness ratio is 250.

From I.S: 800–1984, for $\frac{l}{r} = 218.18, f_y = 250$ MPa,

$$\sigma_{ac} = 23.36 \text{ N/mm}^2$$

$$\text{Safe compressive load} = 1.33 \times 18 \times 2514 \times 10^{-3} = 78.10 \text{ kN}$$

$$> 50.4 \text{ kN}$$

which is sufficient.

Rafter

$$\text{Panel length, } L = 2.235 \text{ m}$$

$$\text{Effective length about } xx\text{-axis} = 0.85L$$

$$= 0.85 \times 2.235 = 1.8997 \text{ m} = 1899.7 \text{ mm}$$

The purlins are placed at the panel points. Hence, the length of the rafter between purlins is equal to the panel length.

$$\text{Effective length about } yy\text{-axis} = 1.0 \times L = 2.235 \text{ m} = 2235 \text{ mm}$$

$$\text{Design compressive load} = 110.5 \text{ kN}$$

Let, us try some section and check its safety. From I.S. Handbook No. 1 select two unequal angle sections I.S.A. 65 × 45 × 8 mm. The relevant properties of the section provided are,

$$A = 1634 \text{ mm}^2, \; r_{xx} = 20.2 \text{ mm}, \; r_{yy} = 21.1 \text{ mm}$$

$$\frac{l_{xx}}{r_{xx}} = \frac{1899.7}{20.2} = 94.04$$

and

$$\frac{l_{yy}}{r_{yy}} = \frac{2235}{21.1} = 105.924$$

The slenderness ratio l_{yy}/r_{yy} is critical.

From I.S: 800–1984, for $\lambda = 105.924$ and $f_y = 250$ MPa,

$$\sigma_{ac} = 75.26 \text{ N/mm}^2$$

$$\text{Safe load carrying capacity} = \sigma_{ac} \times \text{area provided}$$

$$= 1.33 \times 75.26 \times 1634 \times 10^{-3} = 163.55 \text{ kN}$$

$$> 110.5 \text{ kN}$$

which is sufficient.

Strut

$$\text{Design compressive force} = 23.6 \text{ kN}$$

$$\text{Design tensile force} = 25.2 \text{ kN}$$

$$\text{Length of member (scaled off from diagram)} = 2.235 \text{ m}$$

$$\text{Let us assume allowable compressive stress} = 60 \text{ MPa}$$

$$\text{Area required} = \frac{23.6 \times 10^3}{60} = 393.33 \text{ mm}^2$$

Let us try an I.S.A. 65 × 65 × 5 mm section from I.S. Handbook No. 1. The relevant properties of the section are:

$$A = 625 \text{ mm}^2,\ r_{xx} = r_{yy} = 16.6 \text{ mm},\ r_{vv} = 12.6$$

$$\frac{l}{r} = \frac{2235}{12.6} = 177.38 < 180$$

From I.S: 800–1984, for $l/r = 177.38$, and $f_y = 250$ MPa,

$$\sigma_{ac} = 34.05 \text{ N/mm}^2$$

$$\begin{aligned}\text{Safe compressive load} &= \text{permissible compressive stress} \times \text{area provided}\\ &= \sigma_{ac} \times \text{area provided}\\ &= 1.33 \times 34.05 \times 625 \times 10^{-3} = 28.20 \text{ kN} > 23.6 \text{ kN}\end{aligned}$$

Check as a tension member Let us provide 14 mm ϕ power driven rivets for connection.

$$\begin{aligned}A_1 &= \text{area of connected leg}\\ &= \left(65 - 15.5 - \frac{5}{2}\right) \times 5 = 260 \text{ mm}^2\end{aligned}$$

$$\begin{aligned}A_2 &= \text{area of outstanding leg}\\ &= \left(65 - \frac{5}{2}\right) \times 5 = 312.5 \text{ mm}^2\end{aligned}$$

$$k = \frac{3A_1}{3A_1 + A_2} = \frac{3 \times 260}{3 \times 260 + 312.5} = 0.714$$

$$\therefore \quad \text{Neat area provided} = A_1 + kA_2 = 260 + 0.714 \times 312.5 = 483.125$$

$$\begin{aligned}\text{Safe tensile load} &= \sigma_{at} \times \text{net area provided}\\ &= 1.33 \times 150 \times 483.125 \times 10^{-3} = 96.38 \text{ kN} > 25.2 \text{ kN}\end{aligned}$$

which is sufficient.

Minor sling

$$\text{Design tensile force} = 13.3 \text{ kN}$$

$$\text{Design compressive force} = 15.0 \text{ kN}$$

$$\text{Net cross-sectional area required} = \frac{13.3 \times 10^3}{150} = 88.67 \text{ mm}^2$$

$$\text{Gross cross-sectional area required} = 1.4 \times 88.67 = 124 \text{ mm}^2$$

$$\text{Length (scaled off from the space diagram)} = 2.48 \text{ m}$$

Let us try an I.S.A. 50 × 50 × 5 mm section from I.S: Handbook No. 1. The relevant properties of the section are:

$$A = 479 \text{ mm}^2, r_{xx} = r_{yy} = 15.2 \text{ mm}, r_{vv} = 9.7 \text{ mm}$$

Let us provide 14 mm diameter rivets for connection.

$$A_1 = \text{area of connected leg} = \left(50 - 15.5 - \frac{5}{2}\right) \times 5 = 160 \text{ mm}^2$$

$$A_2 = \text{area of outstanding leg} = \left(50 - \frac{5}{2}\right) \times 5 = 237.5 \text{ mm}^2$$

$$k = \frac{3A_1}{3A_1 + A_2} = \frac{3 \times 160}{3 \times 160 + 237.5} = 0.669$$

$$\text{Net area provided} = A_1 + kA_2 = 160 + 0.669 \times 237.5 = 318.88 \text{ mm}^2$$

$$\text{Safe tensile load} = \sigma_{at} \times \text{net area provided}$$

$$= 1.33 \times 150 \times 318.88 \times 10^{-3} = 63.62 \text{ kN} > 13.3 \text{ kN}$$

which is sufficient.

Check for compression

Length of the member (as scaled from space diagram)

$$L = 2.48 \text{ m} = 2480 \text{ mm}$$

$$\therefore \quad l = 0.85 \times 2480 = 2108 \text{ mm}$$

$$\frac{l}{r} = \frac{2108}{9.7} = 217.3$$

From I.S.: 800–1984, for $\frac{l}{r} = 217.3$ and $f_y = 250$ MPa,

$$\sigma_{ac} = 23.54 \text{ N/mm}^2$$

$$\text{Safe compressive load} = \sigma_{ac} \times \text{area provided}$$

$$= 1.33 \times 23.54 \times 479 \times 10^{-3} = 15 \text{ kN}$$

$$\geq 15.0 \text{ kN}$$

which is sufficient.

Main sling

$$\text{Design tensile force} = 38.7 \text{ kN}$$

$$\text{Design compressive force} = 42.7 \text{ kN}$$

The lengths of the member about *xx* and *yy*-axes are as below.

$$U_4M_1 = 2.48 \text{ m}$$

$$U_4L_2 = 4.96 \text{ m}$$

$$l_{xx} = 0.85 \times 2.48 \text{ mm} = 2108 \text{ mm}$$

$$l_{yy} = 4960 \text{ mm}$$

There is no need to design the member as a close observation shows that the same section will have to be provided for the main sling as that of the main tie. (The safe compression for the above conditions, with 2 I.S.A. 90 × 60 × 8 mm back-to-back is 75.0 kN.)

Hence, provide two I.S.A. 90 × 60 × 8 mm sections for the main sling.

Joints Let us provide 14 mm diameter power driven rivets.

Gross diameter of rivet = 15.5 mm

$$\text{Strength of the rivet in single shear} = \frac{\pi}{4} d^2 \tau_{vf}$$

$$= \frac{\pi}{4} \times 15.5^2 \times 1.25 \times 100 \times 10^{-3} = 23.58 \text{ kN}$$

Strength of the rivet in double shear = 2 × 23.58 = 47.16 kN

Strength of the rivet in bearing over 6 mm = $dt\sigma_{pf}$

$$= 15.5 \times 6 \times 1.25 \times 300 \times 10^{-3} = 34.8 \text{ kN}$$

Strength of the rivet in bearing over 8 mm = $15.5 \times 8 \times 1.25 \times 300 \times 10^{-3} = 46.5$ kN

Number of rivets (Refer Fig. Ex. 11.3(a))

Rafter

Joint U_4

Maximum force = 91.5 kN

Rivet value = 46.5 kN

(minimum of 47.16, 46.5)

$$\text{Number of rivets} = \frac{91.5}{46.5} = 1.96 \simeq 3$$

Joint L_0, L_5

Maximum force = 110.5 kN

$$\text{Number of rivets} = \frac{110.5}{46.5} = 2.37 \simeq 3$$

Joint U_1, U_7

Force in the gusset plate parallel to rafter = 110.5 – 105.9 = 4.6 kN

$$\text{Number of rivets} = \frac{4.6}{23.58} = 0.19 \simeq 2$$

Provide two rivets to connect the angle with the rafter.

Maximum force in the gusset plate normal to rafter

= maximum force in member U_1L_1

= 18.8 kN

$$\text{Therefore, number of rivets required} = \frac{18.8}{23.58} = 0.79 \simeq 3$$

Provide three rivets to connect the rafter with the gusset plate.

Joint U_2, U_6

Force in the gusset plate parallel to rafter = 105.9 – 97.9 = 8 kN

$$\text{Number of rivets} = \frac{8}{23.58} = 0.34 \simeq 2$$

Provide two rivets to connect the angle with the rafter.

Joint U_3, U_5

Force in the gusset plate parallel to the rafter = difference in member forces

$$= 97.9 - 91.5 = 6.4 \text{ kN}$$

$$\text{Number of rivets} = \frac{6.4}{23.58} = 0.27 \simeq 2$$

Provide two rivets to connect angle with the rafter.

Maximum force in the gusset plate normal to the rafter

$$= 18.8 \text{ kN}$$

$$\text{Number of rivets} = \frac{18.8}{23.58} = 0.79 \simeq 2$$

Provide minimum two rivets.

Main tie

Joint L_0, L_5

$$\text{Maximum force} = 99.0 \text{ kN}$$

$$\text{Rivet value} = 37.2 \text{ kN}$$

$$\text{Number of rivets} = \frac{99.0}{46.5} = 2.13 \simeq 3$$

Joint L_1, L_4

Force in the gusset plate parallel to main tie = difference in member forces

$$= 99 - 85.7 = 13.3 \text{ kN}$$

$$\text{Number of rivets} = \frac{13.3}{46.5} = 0.286$$

Provide minimum two rivets.

Joint L_2, L_3

Force in the gusset plate parallel to main tie = difference in member forces

$$= 85.7 - 59.3 = 26.4 \text{ kN}$$

$$\text{Number of rivets} = \frac{26.4}{46.5} = 0.56$$

Provide minimum two rivets.

Strut

$$\text{Maximum force} = 25.2 \text{ kN}$$

Rivet value = 18.869 kN (minimum of 18.869, 27.9)

$$n = \frac{25.2}{23.58} = 1.06 \simeq 2$$

Provide minimum two rivets.

Minor sling

$$\text{Maximum force} = 15.0 \text{ kN}$$

Rivet value = 18.869 kN (minimum of 18.869, 27.9)

$$\text{Number of rivets} = \frac{15.0}{23.58} = 0.63 \simeq 2$$

Provide minimum two rivets.

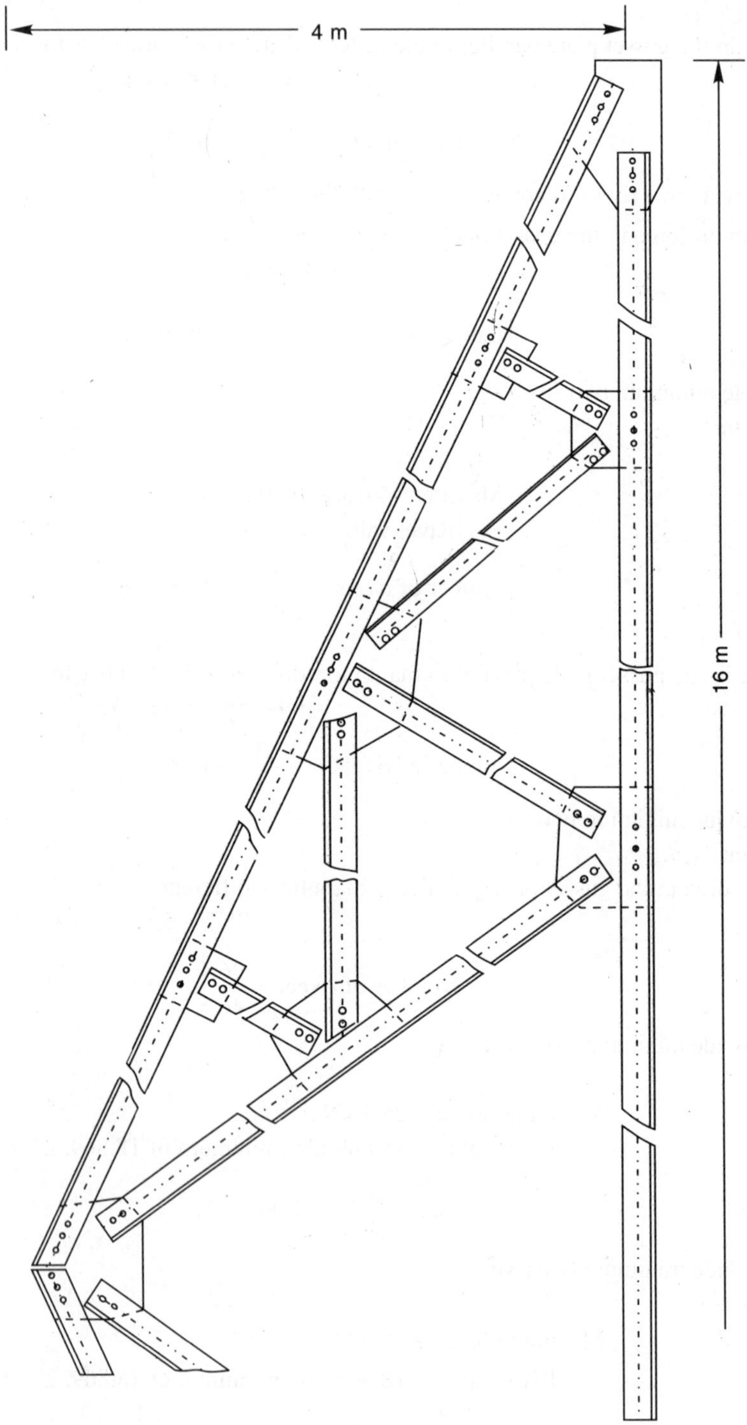

Fig. Ex. 11.3(c)

Main sling

$$\text{Maximum force} = 42.7 \text{ kN}$$

$$\text{Rivet value} = 46.5 \text{ kN (minimum of 46.5, 47.16)}$$

$$n = \frac{42.7}{46.5} = 0.98 \simeq 2$$

Provid minimum two rivets.

The detailing of the roof truss is shown in Fig. Ex. 11.3(c).

Example 11.4 Design an industrial building column 11 m high for the following data.

Height of the column from ground to top of the gantry girder	9 m
Height of the column from the bottom of crane rail to the column top	2 m
c/c spacing of the columns	8 m
c/c spacing of truss columns	16 m
Overall depth of gantry girder	0.675 m
Reaction from the roof truss	29.6 kN
Reaction from the gantry girder	15.96 kN
Maximum reaction on the left column bracket	251.56 kN
Minimum reaction on the right column bracket	80.44 kN
Self-weight of the roof coverings	171 N/m^2
Self-weight of girts	200 N/m
Spacing of girts	1.6 m c/c

Solution It is assumed that the column is of a uniform cross-section. The condition for maximum B.M. and thrust is derived and then the column section is designed.

Dead loads

$$\text{Reaction from truss} = 29.6 \text{ kN}$$

$$\text{Reaction from gantry girder} = 15.96 \text{ kN}$$

Reaction from side rails (girts) and sheeting

$$= \frac{200 \times 8 \times 11}{1.6} + 171 \times 8 \times 11$$

$$= 26{,}048 \text{ N} = 26.048 \text{ kN}$$

$$\text{Assume self-weight of the column} = 20 \text{ kN}$$

$$\text{Total dead load} = 29.6 + 15.96 + 20 + 26.048 = 91.608 \text{ kN}$$

The reaction due to the dead load from the gantry girder is eccentric with the column axis and it develops a moment at the gantry level. Assuming the bracket plate to be 500 mm deep, the distance of the reaction from the inner column face to be 250 mm and the approximate eccentricity of reaction from column axis to be 500 mm,

$$\text{Moment at the crane level, } M = 15.96 \times 0.5 = 7.98 \text{ kNm}$$

$$\text{Horizontal reaction at the top of the column} = \frac{3M\left[1-\left(\frac{a}{h}\right)^2\right]}{2h(1+m)}$$

where, a = 11 – (9.00 – 0.675 – 0.5) = 3.175 m

h = 11 m

m = 0 $\quad (I_1 = I_2)$

∴ Horizontal reaction at the top of the column

$$= \frac{3 \times 7.98 \times \left[1 - \left(\frac{3.175}{11}\right)^2\right]}{2 \times 11} = 0.9975 \text{ kN}$$

Moment at the base = 0.9975 × 11 – 7.98

= 2.9925 kNm

Moment at crane level due to horizontal reaction at top

= 0.9975 × 3.175 = 3.167 kNm

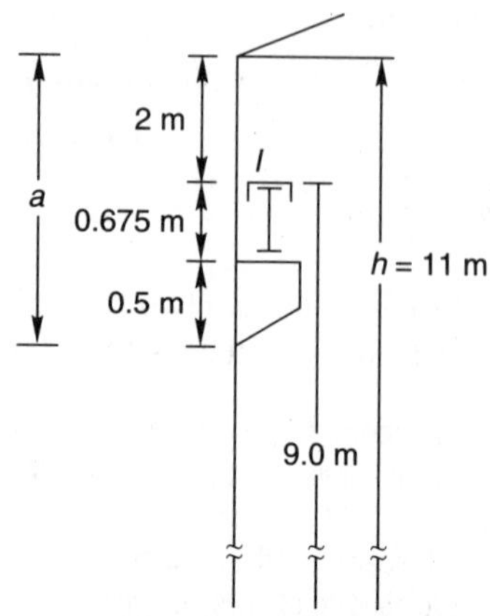

Fig. Ex. 11.4(a)

Due to symmetry there will be no side sway and therefore, side rails (girts) and sheeting do not produce any bending moment.

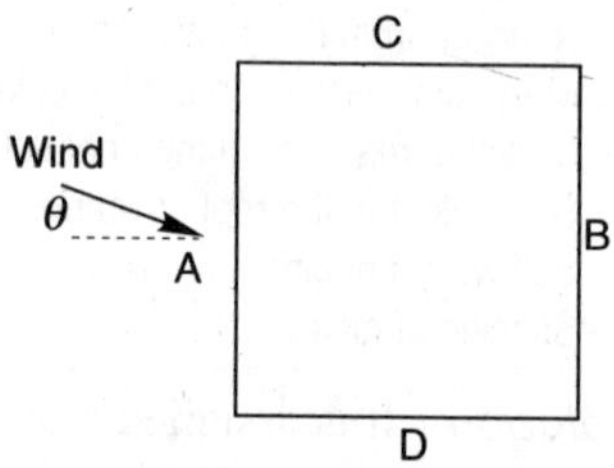

Fig. Ex. 11.4(b)

Live load

The live load reaction from Example 2

= 4 × 6.7 = 26.8 kN

Wind load

Refer to Fig. Ex. 11.4(b), Table 4 (IS: 875 (Port 3))

$$\frac{h}{w} = \frac{11}{8} = 1.375 \qquad \left(1 \le \frac{h}{w} \le \frac{3}{2}\right)$$

$$\frac{l}{w} = \frac{8.94}{8} = 1.12 \qquad \left(1 \le \frac{l}{w} \le \frac{3}{2}\right)$$

In Fig. Ex. 11.4(b), A, B, C and D represent the four surfaces of the rectangular clad building.

For the windward surface C_{pe} = + 0.7 and for the leeward surface C_{pe} = –0.25

$C_{pi} = \pm 0.2$

$P_1 = (0.7 \pm 0.2) \times 8 \times 1.05 = 7.56$ kN/m

$P_2 = (-0.25 \pm 0.2) \times 8 \times 1.05 = -3.78$ kN/m

$$H_A = \frac{3}{8} P_1 \times h = \frac{3}{8} \times 7.56 \times 11 = 31.185 \text{ kN} \simeq 31.19 \text{ kN} \leftarrow$$

$$H_B = \frac{3}{8} P_2\, h = \frac{3}{8} \times 3.78 \times 11 = 15.59 \text{ kN} \rightarrow$$

Total horizontal shear on the top of column = $H_A - H_B$ = 31.19 – 15.59

= 15.6 kN ←

Net horizontal reaction at the top of column A = $31.19 - \frac{1}{2} \times 15.6$

= 23.39 kN ←

$$\text{Net horizontal reaction at top of column } B = 15.59 - \frac{1}{2} \times 15.6$$
$$= 7.8 \text{ kN} \rightarrow$$

Bending moment

Wind ward column A

$$\text{Bending moment at the crane level} = 23.39 \times 3.175 - 7.56 \times \frac{3.175^2}{2}$$
$$= 36.158 \text{ kNm}$$

$$\text{Bending moment at the base} = 23.39 \times 11 - \frac{7.56 \times 11^2}{2} = 200.10 \text{ kNm}$$

Leeward column B

$$\text{Bending moment at the crane level} = 7.8 \times 3.175 - \frac{3.78 \times 3.157^2}{2}$$
$$= 5.7 \text{ kNm}$$

$$\text{Bending moment at the base} = 7.8 \times 11 - \frac{3.78 \times 11^2}{2}$$
$$= 142.89 \text{ kNm}$$

Maximum shear force on the left side column bracket = 251.56 kN

∴ Maximum reaction on the left side column bracket = 1.25 × 251.56
= 314.45 kN

Maximum shear force on the right side column bracket = 80.44 kN

∴ Maximum shear force on the right side column bracket = 1.25 × 80.44
= 100.55 kN

The forces are shown in Fig. Ex. 11.4(c)(i).

Distance between c/c of truss columns = 16 m

The bracket plate is assumed to have a projection of 500 mm from the column face.

∴ Distance between c/c of gantry girders = 16 – 0.5 = 15.5 m

Maximum moment on the left side column = 314.45 × 0.5 = 157.225 kNm

Maximum moment on the right side column = 100.55 × 0.5 = 50.275 kNm

Assuming no sway condition,

Horizontal reaction on top of the left side column

$$= \frac{3 \times 157.225 \times \left[1 - \left(\frac{3.175}{11}\right)^2\right]}{2 \times 11} = 19.655 \text{ kN} \leftarrow$$

Horizontal reaction at the top of the right side column

$$= \frac{3 \times 50.275 \times \left[1 - \left(\frac{3.175}{11}\right)^2\right]}{2 \times 11} = 6.284 \text{ kN} \rightarrow$$

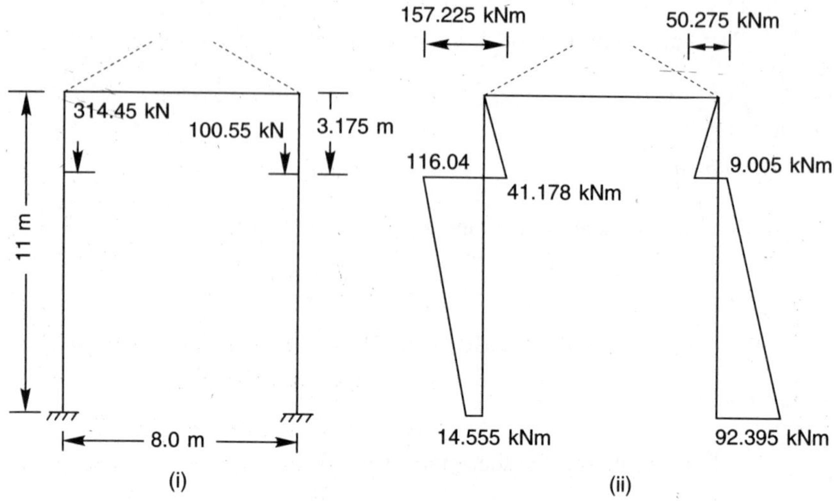

Fig. Ex. 11.4(c)

Unbalanced shear at the top of the columns = 19.655 – 6.284 = 13.371 kN ←

Apply a shear of $\frac{13.371}{2}$ = 6.6855 kN towards the right on each column at the top.

Force on top of the left side column = 19.665 – 6.6855 = 12.97 kN ←

Force on top of the right side column = 6.284 + 6.6855 = 12.97 kN →

Moment at the base of the left side column = 12.97 × 11 – 157.225

= –14.555 kNm

Moment at the base of the right side column = 12.97 × 11 – 50.275

= 92.395 kNm

Moment at crane level due to horizontal reaction at column top

= (19.655 – 6.6855) × 3.175

= 41.178 kNm

All the above calculated moments are shown in Fig. Ex. 11.4(c)(ii).

Lateral loads

Lateral load on each wheel = 12.0 kN

Maximum S.F. due to lateral load = 18.75 kN

Bending moment at the base = 18.75 × (11 – 3.175) = 146.718 kNm

The bending moment and thrust are tabulated for left side and right side columns in Table 11.4. The column section designed is checked for the critical combinations.

Design of the column From Table 11.5, the critical combinations of thrust and B.M. are as follows:

Thrust (kN)	*Bending moment (kN m)*	*Column*
(i) 409.468	334.362	A—Left side
(ii) 226.758	384.995	B—Right side

Table 11.5

Loads	Column A (left side)				Column B (right side)			
	Crane Level		Bottom		Crane Level		Bottom	
	Thrust (kN)	Moment (kNm)	Thrust (kN)	Moment (kNm)	Thrust (kN)	Moment (kNm)	Thrust (kN)	Moment (kNm)
Dead load	+91.608	+4.813	+91.608	+2.9925	+91.608	–4.813	+91.608	–2.9925
Live load	+26.8		+26.8		+26.8		+26.8	
Wind Load	–23.39	±36.158	–23.39	±200.10	±7.8	±5.7	+7.8	±142.89
Crane Load								
Vertical	+314.45	+116.04	+314.45	–14.555	+100.55	+9.095	+100.55	–92.395
Lateral				±146.718				±146.718
Dead Load + Live Load + Crane Load								
	+432.858	+120.853	+432.858	–158.28	+218.958	+4.282	+218.958	–242.105
Dead Load + Live Load + Crane Load + Wind Load								
	+409.468	+157.011	+409.468	–334.362	+226.758	+9.982	+226.758	–384.995

The column section is assumed and is checked for the above critical conditions. It is found that even the largest I.S.H.B. section does not satisfy the above conditions. Therefore, a built-up I-section with cover plates is selected and is checked for safety.

From I.S. Handbook No. 1 (Table XXXIII), let us select a section I.S.H.B. 450 (Table XXXIII) with cover plates 400 × 20 mm on each flange @ 2087.6 N/m. The relevant properties of the section are,

$$A = 27114 \text{ mm}^2,\ Z_{xx} = 5209.1 \times 10^3 \text{ mm}^3,\ r_{xx} = 217 \text{ mm},\ r_{yy} = 94.7 \text{ mm}$$

Case I

$$\text{Thrust} = 409.468 \text{ kN}$$

$$\text{B.M.} = 334.362 \text{ kNm}$$

$$\sigma_{ac,\,cal} = \frac{409.468 \times 10^3}{27{,}114.0} = 15.10 \text{ N/mm}^2$$

$$\sigma_{bcx,\,cal} = \frac{334.362 \times 10^6}{5209.1 \times 10^3} = 64.18 \text{ N/mm}^2$$

Let l_1 = effective length of the column about xx-axis,

l_2 = effective length of the column about yy-axis.

Let us provide a column section with a bracket plate to support the gantry girder. As an alternative, a stepped column may also be provided.

Assuming l_1 = 1.5 × length of column from its base up to the truss

= 1.5 × 11 m

l_2 = 0.75 × length of column from its base up to the gantry girder

= 0.75 × 9 m

$$\frac{l_1}{r_{xx}} = \frac{1.5 \times 11 \times 10^3}{217} = 76.03$$

$$\frac{l_2}{r_{yy}} = \frac{0.75 \times 9.0 \times 10^3}{94.7} = 71.27$$

From I.S: 800–1984, for $\lambda = 76.03$ and $f_y = 250$ MPa,

$$\sigma_{ac} = 105.367 \text{ N/mm}^2$$

$$\frac{\sigma_{ac,\,cal}}{\sigma_{ac}} = \frac{15.10}{105.367} = 0.143 < 0.15$$

Check:

$$\sigma_{bcx} = 0.66 \times \frac{(\pi^2 \times 2 \times 10^5 / 76.03^2) \times 250}{\{(\pi^2 \times 2 \times 10^5 / 76.03^2)^{1.4} + (250)^{1.4}\}^{1/1.4}} = 115.568$$

$$\frac{\sigma_{ac,\,cal}}{\sigma_{ac}} + \frac{\sigma_{bc\,x,\,cal}}{\sigma_{bc\,x}} \le 1$$

or

$$\frac{15.10}{105.367} + \frac{64.18}{115.568} = 0.698 < 1$$

which is safe.

Hence, provide column section I.S.H.B. 450 with cover plates 400 mm × 20 mm on each flange.

Example 11.5 Design gable and components for an industrial building for the data given in Example 11.3.

Solution At the extreme end of an industrial building, a truss is not provided. At the eaves level, a gable wind girder is provided. This is made in the form of a truss by joining the gable columns with the panel points of the lower chord of the adjacent roof trusses. Struts are provided at the eaves level and at the ridge. The gable columns carry the wind loads from the respective hatched area as shown in Fig. Ex. 11.5(a). The wind load up to half the height of the eaves from the ground is assumed to be transferred to the foundation directly by the stanchion. The wind load of area IV is assumed to act at the ridge F which is resisted by the strut at the ridge.

$$\frac{h}{w} = \frac{(5.5+2)}{16} = 0.468 \qquad \left(\frac{h}{w} \le \frac{1}{2}\right)$$

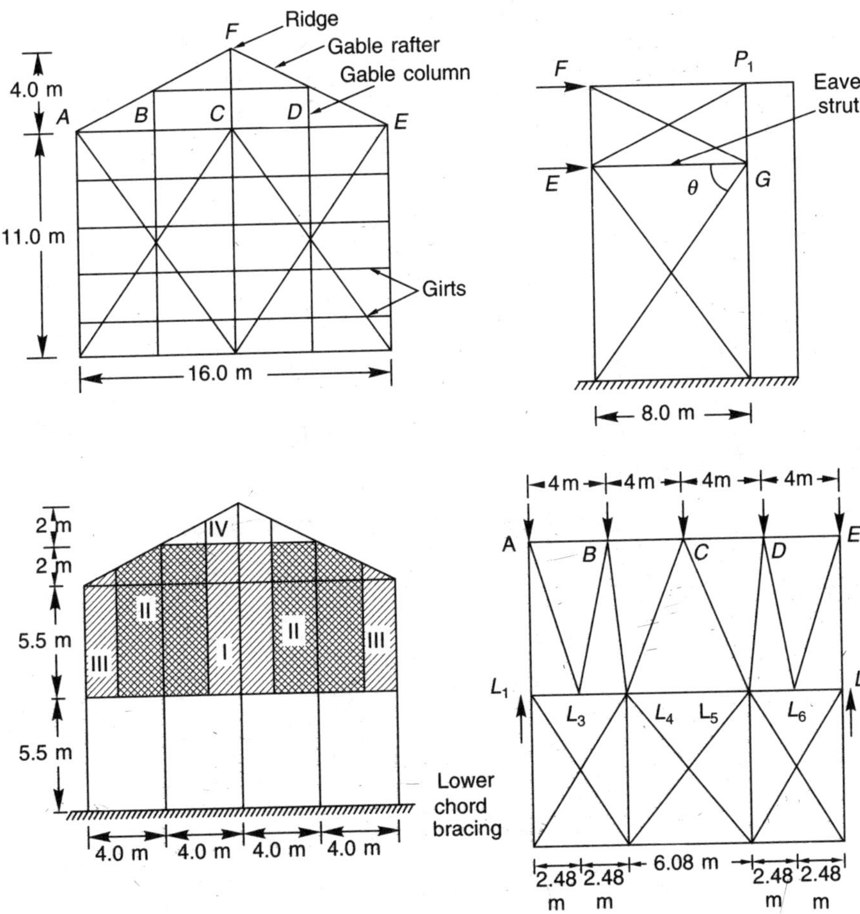

Fig. Ex. 11.5(a)

$$\frac{l}{w} = \frac{48}{16} = 3 \qquad \left(\frac{3}{2} < \frac{l}{w} < 4\right)$$

For gable end, $\theta = 90°$ (Refer Fig. Ex. 11.5(b)).

Note A, B, C and D in Fig. Ex. 11.5(b) are the surfaces of rectangular clad building (shown in plan). One must not get confused with A, B, C and so on of Fig. Ex. 11.5(a).

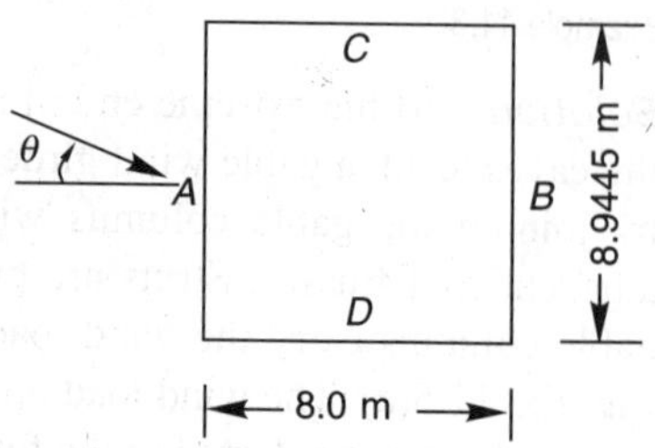

Fig. Ex. 11.5(b)

$c_{pe} = -0.5$ for surface A as well as B

and $c_{pi} = \pm 0.2$

Wind force at point C

$$= (c_{pe} - c_{pi})\, P_d \times A$$

$$= (-0.5 \pm 0.2) \times 1.05 \times (5.5 + 2) \times 4$$

$$= 22.05 \text{ kN} \simeq 22.5 \text{ kN}$$

Wind force at point D

$$= (-0.5 \pm 0.2) \times 1.05 \times \left[(5.5+2)\times 2 + \frac{(5.5+1)+(5.5+2)}{2} \times 2\right]$$

$$= 21.31 \text{ kN} \simeq 21.75 \text{ kN}$$

$$\text{Wind force at points } E \text{ and } A = (-0.5 \pm 0.2) \times \left[\frac{(5.5+1)+5.5}{2} \times 2\right]$$

$$= 8.82 \text{ kN} \simeq 9 \text{ kN}$$

$$\text{Wind force at point } F = (-0.5 \pm 0.2) \times 1.05 \times \frac{8 \times 2}{2} = 5.88 \text{ kN} \simeq 6 \text{ kN}$$

Reactions at L_1, L_2 will be same as the geometry of the girder and the loading is symmetrical (Fig. Ex. 11.5(c)).

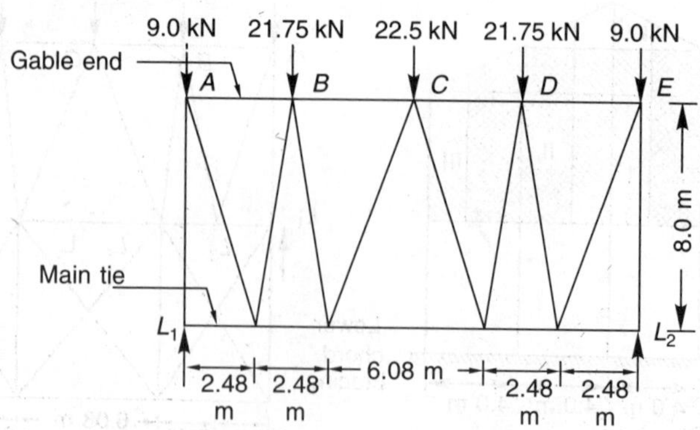

Fig. Ex. 11.5(c) *Gable End Girder (Plan)*

$$R_{L_1} = R_{L_2} = \frac{9.0 \times 2 + 21.75 \times 2 + 22.5}{2} = 42 \text{ kN}$$

The forces in the members of the truss are calculated graphically (Fig. Ex. 11.5(d)) and are tabulated in Table 11.6. Due to symmetry only half the girder is solved. The girder on the windward side is analysed. As the wind can blow in a reverse direction, the stresses can be reversed.

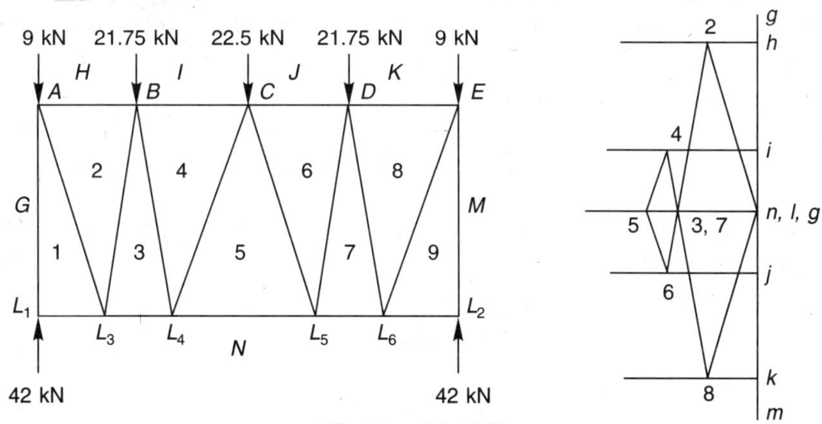

Fig. Ex. 11.5(d)

Table 11.6

Stress	*Wind Load (kN) from Left* — *Windward Girder*	*Leeward Girder*	*Design Load kN*	*Member*	*Remarks*
h–2	–9.9	+9.9		AB tension	+ive sign indicates
i–4	–18.0	+18.0	±18.0	BC	
j–6	–18.0	+18.0		CD	–ive sign indicates Compression
k–8	–9.9	+9.9		DE	
1–n	0.0	0.0		L_1L_3	
3–n	+16.0	–16.0		L_3L_4	
5–n	+22.5	–22.5	±22.5	L_4L_5	
7–n	+16.0	–16.0		L_5L_6	
9–n	0.0	0.0		L_6L_2	
g–1	–42.0	+42.0	±42.0	L_1A	Eave strut
1–2	+34.4	–34.4	±34.4	AL_3	
3–4	+10.8	–10.8		BL_4	
2–3	–33.3	+33.3	±33.3	BL_3	
4–5	–11.8	+11.8		CL_4	
5–6	–11.8	+11.8		CL_5	
7–8	–33.3	+33.3		DL_6	
6–7	+10.8	–10.8		DL_5	
8–9	+34.4	–34.4		EL_6	
9–m	–42.0	42.0		EL_2	Eave strut

Design The members of a girder are subjected to a reversal of loading. Therefore, these are designed as compression members and are checked for tension. As per I.S. specifications the stresses are increased by $33\frac{1}{3}\%$ due to the wind force.

Diagonal members

$$\text{Length of the member} = \sqrt{8^2 + 2.48^2} = 8.375 \text{ m}$$

$$\text{Design force} = \pm 34.4 \text{ kN}$$

Let us try a single angle section connected by more than one rivet. Limiting the slenderness ratio to 250,

$$\text{Effective length, } l = 0.85 \times 8.375 = 7.12 \text{ m}$$

$$r_{vv} = r_{\min} = \frac{7.12 \times 10^3}{250} = 28.48 \text{ mm}$$

From I.S. Handbook No. 1, let us select an I.S.A. 150 × 150 × 10 mm section. The relevant properties are,

$$A = 2903 \text{ mm}^2$$

$$r_{\min} = r_{vv} = 29.3 \text{ mm}$$

$$\lambda = \frac{l}{r_{\min}} = \frac{7.12 \times 10^3}{29.3} = 243.0$$

From I.S: 800–1984, for $\lambda = 243.0, f_y = 250$ MPa,

$$\sigma_{ac} = 19.4 \text{ N/mm}^2$$

$$\sigma_{ac,\text{ all}} = 1.33 \times 19.4 = 25.802 \text{ N/mm}^2$$

$$\text{Safe compressive load carrying capacity} = \sigma_{ac,\text{ all}} \times \text{area provided}$$

$$= 25.802 \times 2903 \times 10^{-3}$$

$$= 74.9 \text{ kN} > 34.4 \text{ kN}$$

which is sufficient.

Check as a tension member Let us connect the angle by 14 mm ϕ power driven rivets.

$$A_1 = \text{area of connected leg}$$

$$= \left(150 - 15.5 - \frac{10}{2}\right) \times 10 = 1295 \text{ mm}^2$$

$$A_2 = \text{area the outstanding leg}$$

$$= \left(150 - \frac{10}{2}\right) \times 10 = 1450 \text{ mm}^2$$

$$k = \frac{3A_1}{3A_1 + A_2} = \frac{3 \times 1295}{3 \times 1295 + 1450} = 0.7282$$

$$\text{Net area of the section} = A_1 + kA_2$$

$$= 1295 + 0.7282 \times 1450 = 2350.89 \text{ mm}^2$$

Tensile load carrying capacity = σ_{at} × net area provided

$$= 1.33 \times 150 \times 2350.89 \times 10^{-3} = 469.00 \text{ kN}$$

$$> 34.4 \text{ kN}$$

which is safe.

Design of members AB, BC

Maximum design force = ±18 kN

Length of the member L = 4.00 m

Let us provide a single angle continuous strut.

$$\text{Effective length, } l = 0.85 \times 4.00 \times 10^3 = 3400 \text{ mm}$$

Limiting the slenderness ratio to 250,

$$r_{\text{min, req}} = \frac{3400}{250} = 13.6 \text{ mm}$$

From I.S. Handbook No. 1, let us try an I.S.A. 70 × 70 × 6 mm section.

Its relevant properties are:

$$A = 806 \text{ mm}^2, r_{xx} = r_{yy} = 21.4 \text{ mm}, r_{vv} = 13.6 \text{ mm}$$

$$r_{\min} = 13.6 \text{ mm}$$

$$\lambda = \frac{l}{r_{\min}} = \frac{3400}{13.6} = 250$$

From I.S: 800–1984, for $\lambda = 250, f_y = 250$ N/mm²,

$$\sigma_{ac} = 18 \text{ N/mm}^2$$

The allowable stress is increased by $33\frac{1}{3}\%$ to account for wind.

$$\sigma_{ac} = 1.33 \times 18 = 23.94 \text{ N/mm}^2$$

Safe compressive load carrying capacity

$$= \sigma_{ac} \times \text{area provided} = 23.94 \times 806 \times 10^{-3} = 19.295 \text{ kN}$$

$$> 18 \text{ kN}$$

which is sufficient.

Check as a tension member Let us connect the angle by 14 mm ϕ power driven rivets.

A_1 = area of connected leg

$$= \left(70 - 15.5 - \frac{6}{2}\right) \times 6 = 309 \text{ mm}^2$$

A_2 = area of outstanding leg

$$= \left(70 - \frac{6}{2}\right) \times 6 = 402 \text{ mm}^2$$

$$k = \frac{3A_1}{3A_1 + A_2} = \frac{3 \times 309}{3 \times 309 + 402} = 0.6975$$

$$\text{Net area provided} = A_1 + kA_2 = 309 + 0.6975 \times 402 = 589.40 \text{ mm}^2$$

$$\text{Safe tensile load carrying capacity} = \sigma_{at} \times \text{net area provided}$$

$$= 1.33 \times 150 \times 589.40 \times 10^{-3} = 117.5 \text{ kN}$$

$$> 18.0 \text{ kN}$$

which is as it should be.

Connections The connection is made with 14 mm ϕ rivets. The rivet value is 18.86 kN (refer connection of truss members Fig. Ex. 11.5(e)).

$$\text{Number of rivets to connect diagonals} = \frac{34.4}{18.86} = 1.82 \approx 2$$

$$\text{Number of rivets to connect } AB, BC = \frac{18.00}{18.86} = 0.954 \approx 2$$

The members and their connections are shown in Fig. Ex. 11.5(e).

Note Members L_3L_4, L_4L_5, L_5L_6, ... are not designed as these are the lower chord members of the roof truss and have already been designed.

The maximum force due to the gable end effect is ±22.5 kN, whereas while designing as a tie member in the truss, the design force is +109.9, –40.7 kN. Therefore, the members will be safe under gable wind loads.

Eaves strut

$$\text{Design compressive load} = 42 \text{ kN}$$

$$\text{Length of the strut } AL_1 = 8.0 \text{ m}$$

$$\text{Effective length } l = 0.85 \times 8 = 6.8 \text{ m}$$

Limiting the slenderness ratio to 250,

$$r_{min} = \frac{6.8 \times 10^3}{250} = 27.2 \text{ mm}$$

From I.S. Handbook No. 1, it is found that the angle section with the radius of gyration next to 27.2 mm is the section I.S.A 150 × 150 × 10 mm.

Let us select this section and provide it as an eave strut.

$$A = 2903 \text{ mm}^2, r_{min} = 29.3 \text{ mm}$$

$$\lambda = \frac{l}{r_{min}} = \frac{6.8 \times 10^3}{29.3} = 232.08$$

From I.S: 800–1984, for $\lambda = 232.08$ and $f_y = 250$ MPa,

$$\sigma_{ac} = 20.792 \text{ N/mm}^2$$

$$\sigma_{ac,\text{ all}} = 1.33 \times 20.792 = 27.65 \text{ N/mm}^2$$

$$\text{Safe compressive load} = \sigma_{ac,\text{ all}} \times \text{area provided}$$

$$= 27.65 \times 2903 \times 10^{-3} = 80.267 \text{ kN} > 42 \text{ kN}$$

which is sufficient.

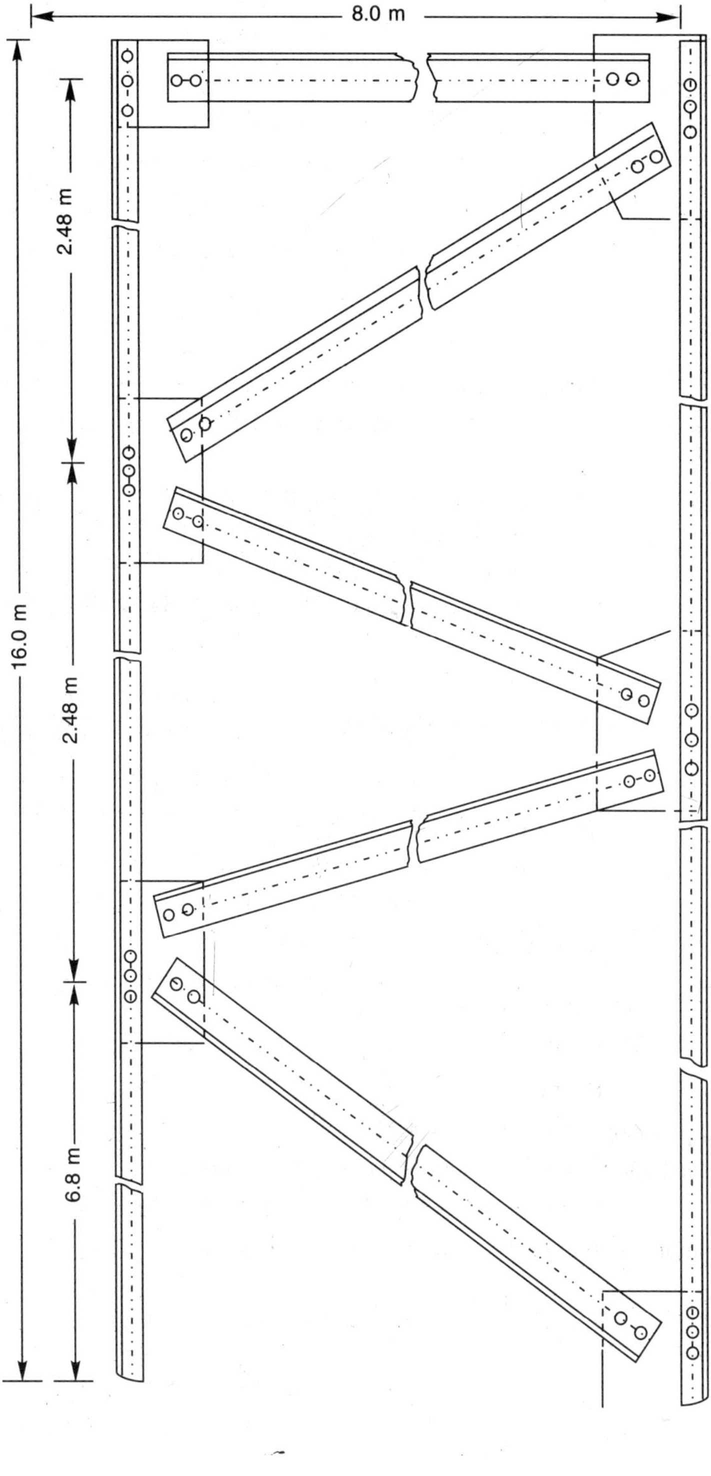

Fig. Ex. 11.5(e)

Example 11.6 Design girts in an industrial building for the following data

Height of columns	11 m
c/c Spacing of columns	8 m
Span of truss	16 m
Side coverings	AC sheets
Intensity of wind pressure	1.05 kN/m^2

Solution Girts are the side rails to support the vertical sheeting (asbestos cement sheets or G.I. sheets), used to cover the sides of the industrial building. These are designed as beams.

$$\text{Span of girt} = \text{spacing of roof trusses}$$
$$= \text{spacing of columns}$$
$$= 8.0 \text{ m} = 8 \times 10^3 \text{ mm}$$

Let the vertical spacing of girts = 1.60 m

Let the weight of the asbestos cement sheets be 171 N/m^2.

Vertical loads

Dead weight of asbestos cement sheets = 8.0 × 1.60 × 171 = 2188.8 N

Let Dead weight of girts = 200 N/m

∴ Self-weight of girt = 200 × 8 = 1600 N

Total vertical load = 2188.8 + 1600 = 3788.8 N

$$W \simeq 3800 \text{ N}$$

$$\textit{Bending moment in vertical plane} = \frac{WL}{10} = \frac{3800 \times 8}{10} = 3040 \text{ Nm}$$

Horizontal loads

$$\frac{h}{w} = \frac{1.6}{8.0} = 0.2 \qquad \left(\frac{h}{w} \le \frac{1}{2}\right)$$

$$\frac{l}{w} = \frac{16}{8} = 2 \qquad \left(\frac{3}{2} < \frac{l}{w} \le 4\right)$$

For surface C (refer Fig. Ex. 11.5(b))

when $\theta = 0°$; $C_{pe} = -0.5$

$\theta = 90°$; $C_{pe} = +0.7$

and internal pressure coefficient, $c_{pi} = \pm 0.2$

Horizontal load due to wind considering the worst condition

$$P = (0.7 \pm 0.2) \times (8 \times 1.6) \times 1.05 = 12.1 \text{ kN}$$

Bending moment in the horizontal plane (due to wind)

$$= \frac{PL}{10} = \frac{12.1 \times 10^3 \times 8}{10} = 9680 \text{ Nm}$$

From IS Handbook No. 1, try an ISA 150 × 115 × 10 mm section. The relevant properties are:

$C_{xx} = 45.5$ mm, $c_{yy} = 28.2$ mm, $I_{xx} = 573.3 \times 10^4$ mm^4

$I_{yy} = 293.4 \times 10^4$ mm^4, $w = 200$ N/m

Stresses due to moment in the vertical plane

Refer to Fig. Ex. 11.6(a)

$$\text{Maximum bending compressive stress} = \frac{3040 \times 10^3 \times 28.2}{293.4 \times 10^4} = 29.22 \text{ N/mm}^2$$

$$\text{Maximum bending tensile stress} = \frac{3040 \times 10^3 \times (115.0 - 28.2)}{293.4 \times 10^4} = 89.94 \text{ N/mm}^2$$

Stresses due to moment in the horizontal plane

Refer to Figs Ex. 11.6(b) and (c).

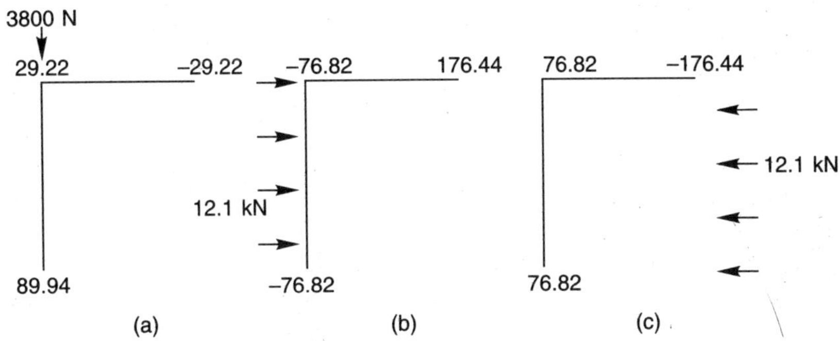

Fig. Ex. 11.6

The wind can blow in any direction so the stresses can reverse.

Maximum bending stress (compressive/tensile)

$$= \frac{9680 \times 10^3 \times 45.5}{573.3 \times 10^4} = 76.82 \text{ N/mm}^2$$

Maximum bending stress (compressive/tensile)

$$= \frac{9680 \times 10^3 \times (150 - 45.5)}{573.3 \times 10^4} = 176.44 \text{ N/mm}^2$$

The combined stresses should be less than $0.9 f_y$ ($0.9 \times 250 = 225$ N/mm^2).

The various combinations to be checked are:

(i) -29.22 ± 76.82; i.e. -106.04 or 47.6 N/mm^2

(ii) -29.22 ± 176.44; i.e. -205.66 or 147.22 N/mm^2

(iii) 89.4 ± 76.82; i.e. -13.12 or 166.76 N/mm^2

which is safe

Example 11.7 Design the following elements of an industrial building for the data given below.

(i) Ridge strut
(ii) Rafter bracing
(iii) Gable rafter
(iv) Vertical bracing of side frames.

Height of columns	11 m
c/c spacing of columns	8.0 m
Length of roof truss rafter	8.94 m
Gable columns spacing	4.0 m
Number of purlins over roof trusses	10
Self-weight of purlins	318 N/m
Design load for ridge strut	6.0 kN
Live load on gable rafter	39.595 kN
Reaction from the gable wind girder	42 kN
Longitudinal surge from the crane	8 kN
Intensity of wind pressure	1.05 kN/m^2

Solution

(i) *Ridge strut*

Design load = 6 kN

As the load is very small no additional section is required. The force can be resisted by the purlin section at the ridge.

(ii) *Rafter bracing*

The upper chord of the truss is braced as shown in Fig. Ex. 11.7(a).

The force due to wind at point A = 6.0 kN. The bracing system is provided on both the slopes of the trusses.

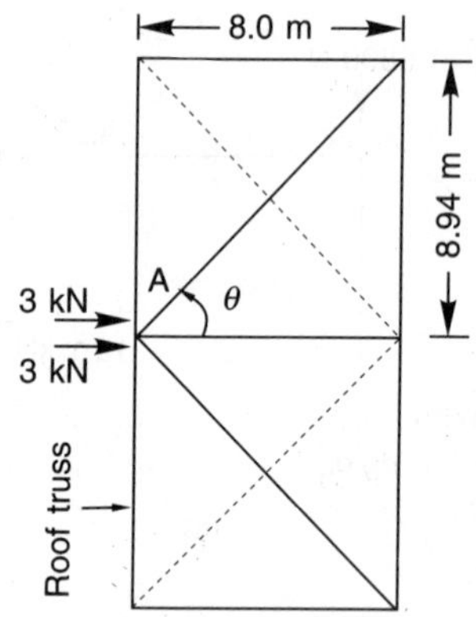

Fig. Ex. 11.7(a)

$$\therefore \quad \text{Wind force to be resisted} = \frac{6.0}{2} = 3.0 \text{ kN}$$

$$\text{Length of the bracing} = \sqrt{8^2 + 8.94^2} = 12.00 \text{ m}$$

$$\text{Force in bracing member} = 3 \sec \theta = 3 \times \frac{12}{8} = 4.5 \text{ kN}$$

As the force is very small let us provide I.S.A. 65 × 45 × 6 mm angle for braces.

Gable rafter The purlins transfer concentrated loads to the rafter but it is assumed that the load is uniformly distributed.

Gable column spacing = 4.0 m

$\therefore$ Horizontal span of gable rafter = 4.0 m

Length of rafter = 8.94 m

Loads

$$\text{Weight of sheetings} = 171 \times \left(\frac{8}{2}\right) \times (8.94 \times 2) = 12{,}229.9 \text{ N} = 12.23 \text{ kN}$$

Total number of purlins supported by rafters = 10

$$\text{Weight of purlins} = 318 \times \left(\frac{8}{2}\right) \times 10 = 12{,}720 \text{ N} = 12.720 \text{ kN}$$

$$\text{Live load} = 39.595 \text{ kN}$$

$$\text{Let self-weight} = 3.0 \text{ kN}$$

$$\text{Total load} = 12.23 + 12.72 + 39.595 + 3 = 67.545 \text{ kN}$$

Total number of spans formed by gable columns = 4

$$\frac{\text{Load}}{\text{Span}} = \frac{67.545}{4} = 16.886 \text{ kN}$$

$$M_{\max} = \frac{16.886 \times 4.00}{8} = 8.443 \text{ kNm}$$

Let the bending stress = 0.66 × 250 = 165 MPa

$$Z_{\text{req}} = \frac{8.443 \times 10^6}{165} = 51169.69 \text{ mm}^3$$

Let us select two I.S.A. 100 × 75 × 12 mm from I.S. Handbook No. 1.

$$Z_{\text{provided}} = 27.9 \times 10^3 \times 2 = 55{,}800 \text{ mm}^2$$

$$\text{Bending stress} = \frac{8.443 \times 10^6}{55{,}800} = 151.308 \text{ N/mm}^2$$

$$< 1.33 \times 165 \text{ N/mm}^2$$

which is safe.

Gable columns

Loads

(i) *Vertical load*

Reaction from gable rafter = 16.886 kN

Side sheeting = 171 × 4 × (11 + 4) = 10260 N = 10.26 kN

(Number of girts in 11 m = 7)

Girts or side rails @ 1.60 m c/c = 200 × 4 × 7 = 5600 N = 5.6 kN

Self weight of column (say) = 5.0 kN

Total load = 16.886 + 10.26 + 5.60 + 5.0 = 37.746 kN

(ii) *Horizontal load*

$$\text{Wind pressure} = 1.05 \text{ kN/m}^2$$

$$\text{Total wind load} = 11 \times 4 \times 1.05 = 46.2 \text{ kN}$$

$$\text{Bending moment due to wind load} = \frac{WL}{8} = \frac{46.2 \times 11}{8} = 63.525 \text{ kNm}$$

The gable columns are braced at mid-height, i.e. 5.5 m above the ground. Therefore, these are assumed to be laterally supported.

∴ Effective length of gable column l = 5.5 m

Let us try I.S.H.B. 300 @ 576.8 N/m.

The relevant properties are,

$$A = 7485\ \text{mm}^2,\ Z_{xx} = 836.3 \times 10^3\ \text{mm}^3,\ r_{\min} = 54.1\ \text{mm}$$

$$\sigma_{ac,\ \text{cal}} = \frac{37.746 \times 10^3}{7485} = 5.04\ \text{N/mm}^2$$

$$\sigma_{bc,\ \text{cal}} = \frac{63.525 \times 10^3}{836.3 \times 10^3} = 75.95\ \text{N/mm}^2$$

$$\lambda = \frac{l}{r_{\min}} = \frac{5.5 \times 10^3}{54.1} = 101.66$$

$$\Rightarrow \quad \sigma_{ac} = 1.33 \times 78.66 = 104.61\ \text{N/mm}^2$$

$$\frac{\sigma_{ac,\text{cal}}}{\sigma_{ac}} = \frac{5.04}{104.61} = 0.48 < 0.15$$

$$\frac{\sigma_{ac\ \text{cal}}}{\sigma_{ac}} + \frac{\sigma_{bc,\ \text{cal}}}{\sigma_{bc}} = 0.064 + \frac{75.95}{1.33 \times 165}$$

$$= 0.048 + 0.346$$

$$= 0.394 < 1$$

which is safe.

Vertical bracing (Fig. Ex. 11.7(b))

Reaction from gable wind girder = 42 kN

Longitudinal surge from crane = 16 kN

Half of this longitudinal surge is taken at each end

$$= \frac{16}{2} = 8\ \text{kN}$$

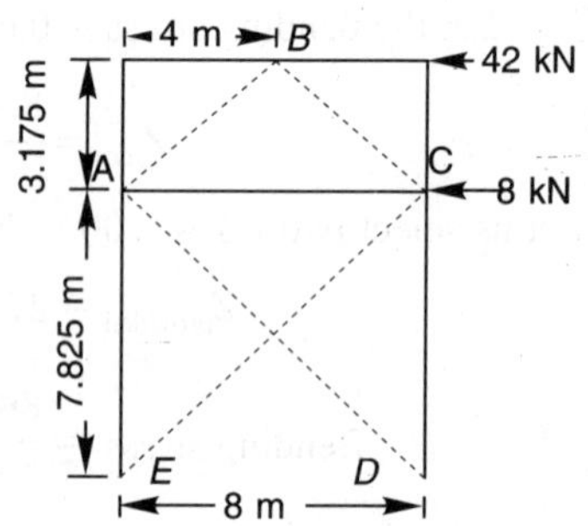

Fig. Ex. 11.7(b) *Bracing in Vertical Plane*

Above crane level

$$\text{Length of bar } AB = \sqrt{4^2 + 3.175^2} = 5.106\ \text{m}$$

$$\text{Force in the bar} = \frac{42 \times 5.106}{3.175} = 67.55\ \text{kN}$$

The section is designed as a compression member.

Effective length of bar = 0.85 × 5.106 = 4.34 m

From I.S. Handbook No. 1, let us select two I.S.A. 100 × 65 × 6 mm placed back-to-back.

$$A = 3820\ \text{mm}^2,\ r_{\min} = 18.4\ \text{mm}$$

$$\lambda = \frac{l}{r_{\min}} = \frac{4.34 \times 10^3}{18.4} = 235.86$$

From I.S: 800–1984, for $\lambda = 235.86$ and $f_y = 250$ MPa

$$\sigma_{ac} = 20.413\ \text{N/mm}^2$$

The allowable stresses are increased by $33\frac{1}{3}\%$

$$\sigma_{ac} = 1.33 \times 20.413 = 27.149\ \text{N/mm}^2$$

Safe load carrying capacity = 27.149 × 3820 × 10^{-3} = 103.71 kN > 67.55 kN which is safe.

Below crane level (The member is subjected to tension.)

$$CE = \sqrt{8^2 + 7.825^2} = 11.19 \text{ m}$$

$$\text{Design load} = 8 + 42 = 50 \text{ kN}$$

$$\text{Tension in CE} = 50 \times \left(\frac{11.19}{8}\right) = 69.93 \text{ kN}$$

From I.S. Handbook No. 1, let us try I.S.A. 100 × 65 × 6 mm. The relevant properties are

$$A = 1910 \text{ mm}^2$$

$$A_1 = \text{area of connected leg} = \left(100 - 15.5 - \frac{6}{2}\right) \times 6 = 489 \text{ mm}^2$$

$$A_2 = \text{area of outstanding leg} = \left(65 - \frac{6}{2}\right) \times 6 = 372 \text{ mm}^2$$

$$k = \frac{3A_1}{3A_1 + A_2} = \frac{3 \times 489}{3 \times 489 + 372} = 0.7977$$

$$\text{Net area provided} = 489 + 0.7977 \times 372 = 785.75 \text{ mm}^2$$

$$\text{Safe load carrying capacity} = \sigma_{at} \times \text{net area provided}$$

$$= 1.33 \times 150 \times 785.75 \times 10^{-3} = 156.75 \text{ kN}$$

$$> 69.93 \text{ kN}$$

Exercises

Note Use the steel of yield stress 250 MPa if not specified in the problem. Assume suitable data if missing or not supplied in the problems.

11.1 Analyze a steel roof truss shown in Fig. Prob. 11.1 for a clear span of 12.0 m. The truss is supported over masonry columns 45 cm × 45 cm. The trusses are placed 3 m c/c and support galvanised iron sheet on rafters and steel purlins. The rise of the truss is 1/3 of span. The design wind pressure may be assumed to be 1000 N/m^2.

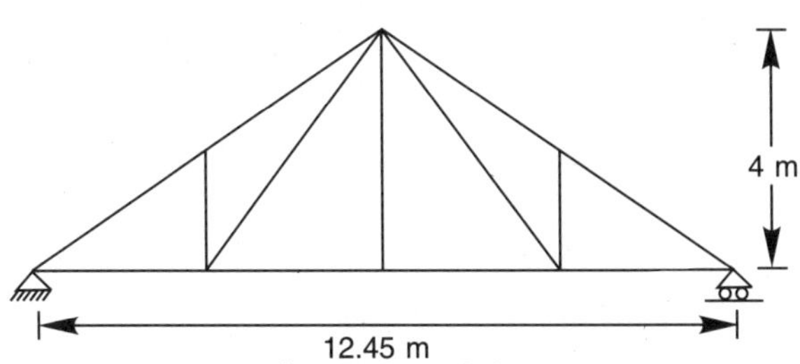

Fig. Prob. 11.1

11.2 Design a Pratt truss as shown in Fig. Prob. 11.2. The design wind pressure is 1200 N/m^2. The trusses are covered with A.C. sheets and the centre-to-centre spacing of trusses is 6 m.

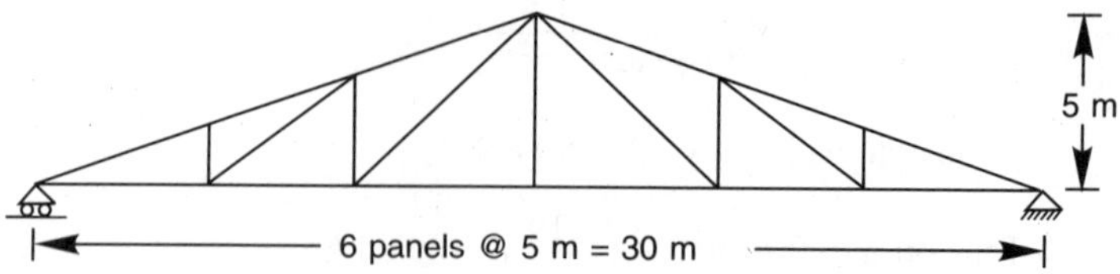

Fig. Prob. 11.2

11.3 Design a roof truss for an island platform roof as shown in Fig. Prob. 11.3.

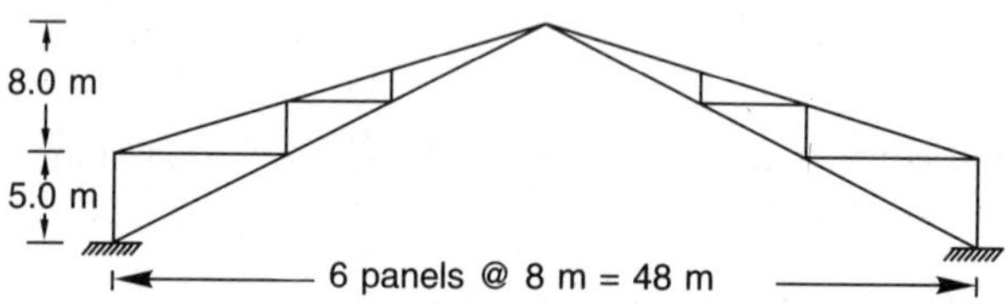

Fig. Prob. 11.3

11.4 A member of a riveted truss consists of two angles connected back-to-back with a gap of 12 mm. The axial force in the member is 22 kN (comp.) due to dead load, 25 kN (comp.) due to live load and 50 kN (Tens.) due to wind load. The distance between the centres of end connections is 2.5 m. Design the member and the end connection.

11.5 Find out the design stresses in the members of a three-hinged arch as shown in Fig. Prob. 11.5. The trusses are placed 8 m c/c. The design wind pressure is 1500 N/m^2 and the roofing material is asbestos cement sheets.

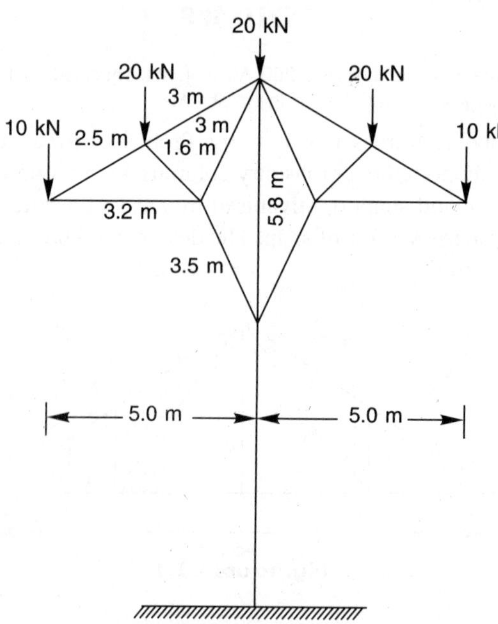

Fig. Prob. 11.5

11.6 The joint A of a roof truss where the main rafter and the main tie intersect is hinged as shown in Fig. Prob. 11.6. It is subjected to end reactions as under (a) Dead load + Live load and (b) Dead load + Wind load as shown in the figure. Design the members meeting at the joint A and the joint A itself using riveted connection.

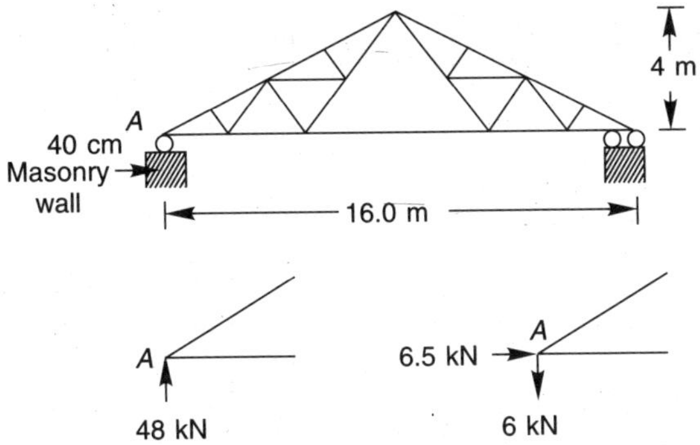

Fig. Prob. 11.6

Chapter 12

Water Tanks

12.1 INTRODUCTION

Steel tanks are used for the storage of water and other liquids, e.g. acids, alcohols, gasolin etc. The tank may either be supported on the ground in which case it is known as a stand-pipe, or it may be elevated and supported on a staging, the examples being rectangular and circular tanks. Tanks on the ground are generally supported on mass concrete and in the case of low bearing capacity soils they are supported on grillage foundation. These are economically used where capacity is more important than pressure, or where local conditions do not require an elevated tank. The principal reasons for the use of elevated tanks are as follows.

1. Their capacity can be stored economically at any desired pressure.
2. The entire capacity is available for use.
3. The maximum pressure will be slightly above the minimum required pressure, thus giving efficient operations of the pumps.
4. The tank when filled depends only upon the unfailing law of gravitation for its pressure.

Steel plates are used to form the container. The plates are designed and detailed so as to be readily made liquid tight by ordinary shop and erection methods. The design should be such that the unit stresses are within permissible limits and a continuous caulking is also achieved as the work progresses. Precautions should be taken to see that all rivets in seams are placed before it is caulked, as the driving of rivets is liable to spring the caulking and may cause leaks that are hard to correct. It should also be ensured that the newly formed head of the rivet is on the caulked side.

Following codes should be referred to for the design of steel water storage tanks.

1. IS 800–1984 Code of practice for general construction in steel.
2. IS 804–1998 Code of practice for rectangular pressed steel tanks.
3. IS 805–1995 Code of practice for use of steel in gravity water tanks.

12.2 PERMISSIBLE STRESSES

Since the water tanks remain in contact with water, the working stress method of design is followed. However, the allowable stresses are reduced to 0.8 times of those specified by I.S. 800–1984. The permissible stresses for members are increased by $33\frac{1}{3}\%$ when wind or seismic load are considered and for connections this increase is 25%. Following stress values are recommended for mild steel with yield stress, $f_y = 250$ N/mm^2.

$$\text{Direct tension } \sigma_{at} = 0.8 \times 0.6 f_y = 120 \text{ N/mm}^2$$

$$\text{Bending tension/compression } \sigma_{bc,t} = 0.8 \times 0.66 f_y = 132 \text{ N/mm}^2$$

$$\text{Connections } 0.8\ \tau_{vf},\ 0.8\ \sigma_{tf},\ 0.8\ \sigma_{pf}$$

Note For the design of the supporting structure, permissible stresses given in IS 800–1984 are not reduced to 80%.

12.3 THICKNESS SPECIFICATIONS

To account for corrosion, a minimum thickness of the steel section is specified (I.S. 805–1995) as below.

Section	Thickness in mm
1. Plates, angles, etc.	(Designed thickness + corrosion allowance) $\nless$ 6
2. Plates of suspended bottoms	
(i) 5×10^5 litres to 7×10^5 litres	8
(ii) 7×10^5 litres to 10×10^5 litres	10
3. Roof Plates	
(i) Tanks up to 9.0 m diameter	3
(ii) Tanks over 9.0 m diameter	5
4. Pressed steel tank plates	specified in Tables 2 to 5

12.4 STIFFENING ANGLE

A Stiffening angle is provided around the top of tanks without roofs. It should have a minimum section modulus, $Z = 0.0578\ D^2H$

where D = diameter of the tank

H = height of the cylindrical portion of the tank shell.

12.5 STAND-PIPE

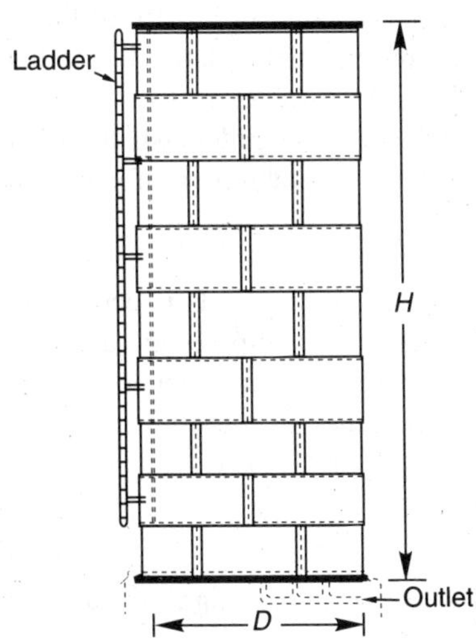

Fig. 12.1 Stand Pipe

A stand-pipe (Fig. 12.1) may be defined as a tank resting on ground, generally cylindrical, and used for the storage of water, oil etc. Its height, in most cases, is considerably greater than its diameter; it has a flat bottom, and it rests directly on its foundation. Stand-pipes for storage of oil are an exception since these are of large diameter and small height (not exceeding 10 m) and are usually referred to as tanks.

Stand-pipes may or may not have roofs. In places with cold climate the roof may be avoided, since the water stored may freeze. However, at places with hot and humid climate the stored water may become an ideal place for mosquitoes; oil may get spoilt by rain water and a stand-pipe with a roof is the choice.

The plates forming the sides of the stand-pipe are made of different diameters, so that the courses overlap each other, inside and outside alternatively. The joints for the horizontal seams in the sides, and for the bottom plates, should preferably be lap joints. The thickness of side plates should not be less than 6 mm and that of the bottom plates 8 mm. The bottom and the side plates are jointed together with the help of an angle section, riveted from inside the tank, bevel sheared for caulking along both legs. Where this is done, it is necessary to splice the bottom angle by means of a wrapper plate made in the shape of an angle curved to the diameter of the tank and having its ends tapered gradually to a thickness of 3 mm or even less. The wrapper plates ar usually 10 mm thick (Fig. 12.2). A manhole 30 cm × 45 cm of elliptical shape is provided on the side near the bottom. Inlet and outlet pipes are provided near the top and bottom of the tank respectively. All openings in the stand-pipe are properly reinforced with plates.

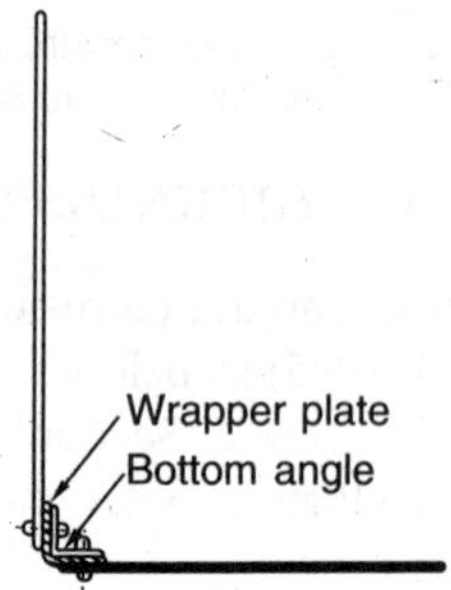

Fig. 12.2 *Bottom Angle and Wrapper Plate*

Suitable circular stiffening angles are provided to prevent the plates from buckling during wind storms. An outside ladder 40 cm wide, extending from a point 2.4 m above the foundation to the top of tank is provided. Stand-pipes storing water need a proper anchorage and for this reason a minimum of 32 mm ϕ bolts with anchor plates are provided. All the bolts are connected directly to the side plates with bent plates. The roof of the tank should be constructed to last long. The tendency now is to use however plates since it adds merely to the cost. A minimum of 5 mm thick plates are used. Usually a conical roof is provided. For small diameter tanks a pitch of 3 vertical to 4 horizontal is recommended whereas for large diameter tanks (10 to 15 m) flat roofs with a pitch of 1 vertical to 3 horizontal may be provided. The latter has radial rafters and roof trusses carrying the entire roof load. The trusses rest on a central column, while in quite larger tanks the radial rafters rest on girders which in turn are connected to a series of columns. Sometimes a dome or umbrella roof is provided for tanks up to 12 m diameter.

Design

Size of the tank It is economical to build large diameter and low height tanks when the stand-pipes are at such an elevation that these can supply water at sufficient pressure to meet local requirements. Usual height to diameter ratio of the stand-pipes are as follows:

$H = 0.4\,D$	stand-pipes with inclined roof
$H \le D$	stand-pipes without roof
$H > D$	stand-pipes with a particular water head

where, H is the height of stand-pipe, and D is its diameter.

Size of plates For a pleasent look, the net width of the various rings of the plates is kept same (about 2 m). The length of the plates is kept 5–6 m.

Having selected the diameter of the stand-pipe, its height, the number of rings of plates, and the number of plates per ring, the thickness of plates can be found by the hoop stress theory as applicable to thin cylinders.

$$\sigma_{at} = \frac{pD}{2t}$$

or

$$t = \frac{pD}{2\sigma_{at}}$$

or

$$t = \frac{\gamma HD}{2\sigma_{at}} \quad (12.1)$$

where γ = unit weight of the liquid (9.81×10^{-6} N/mm^3 for water)

To account for the efficiency of the joint (η) Eq. (12.1) is modified as below

$$t = \frac{\gamma HD}{2\eta\sigma_{at}} \quad (12.2)$$

Since the steel plates will be in contact with water, which may lead to corrosion, the thickness so found is increased by 1.5 mm. But in no case the thickness of the plates should be less than 6 mm.

Anchor bolts Anchor bolts are designed for the uplift caused by wind when the stand-pipe is empty.

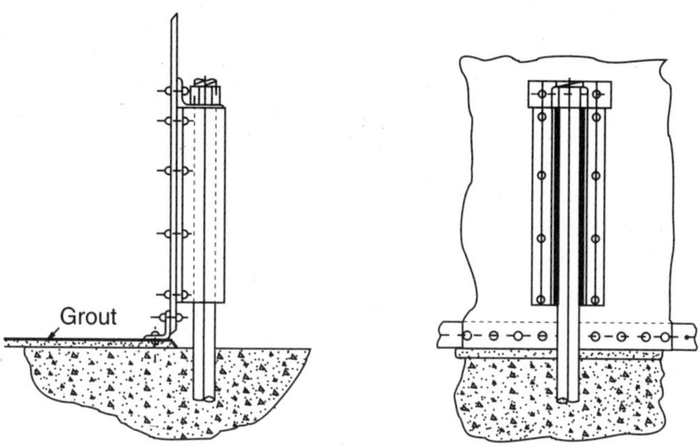

Fig. 12.3 *Anchor Bolt Connection*

Wind moment, $M_w = P \times \frac{H}{2}$

where P = wind force

$= K\, p_z\, A_P$

K = shape factor

p_z = design wind pressure = $0.6\, V_Z^2$

A_P = projected area

V_z = design wind velocity.

We know, $$\sigma_{bc,t} = \frac{M_w}{I} \times \frac{D}{2}$$

$$= \frac{PH/2}{\pi D^3 t/8} \times \frac{D}{2}$$

$$= \frac{2PH}{\pi D^2 t}$$

The uplift force/unit circumferential length, $F_1 = \sigma_{bc,t} \times t \times 1$

$$= \frac{2PH}{\pi D^2 t} \times t = \frac{2PH}{\pi D^2}$$

The maximum uplift force in each anchor bolt, $F_2 = F_1 g$

where g = c/c distance of the anchor bolts.

Area at the roof of the anchor bolt $= \dfrac{F_2}{\sigma_{tf}}$ (12.3)

12.6 ELEVATED TANKS

Tanks elevated on staging or towers are generally provided for water storage and supply. The capacity and height to the bottom are determined from a consideration of the service of the tank. Normally the capacity is based on the hourly consumption to meet out the peak loads. The elevation of the bottom depends upon local conditions, about 10 to 15 m being a common height for a city like Allahabad. In any case the tank height should be such as to furnish proper service at each and every point.

Once the capacity and the height to the bottom of the tank is fixed, the general shape can be decided. Basically there are two types of elevated tanks, the flat bottom tanks resting on a set of beams, e.g. rectangular tanks (Plate 1), pressed steel tanks (Plate 2), and those with suspended bottoms, e.g. circular tanks. The suspended bottoms may be classed as segmental, hemispherical, elliptical and conical. The former are much more commonly used, due to a saving in metal and cost for a given capacity, and because the bottom can be inspected and painted more readily and thoroughly.

The hemispherical bottom is the most common. Conical bottoms are rarely used on account of the difficulty in making a satisfactory connection to the tank cylinder. However, these find their place in railways where water carrying considerable sediment is frequently found. The sediment deposited at the apex of the bottom can be removed from time to time.

12.7 CIRCULAR TANKS

In general all circular tanks have a cylindrical shell with a suspended bottom. Occasionally these may be built with a special shell or bottom to represent some object like a beer bottle, a tub, or an electric bulb etc.

The hemispherical and elliptical or segmental bottoms (Fig. 12.4(a)) are more commonly used, the former being the most common. The hemispherical bottom consists of a dished circular plate in the bottom called a *saucer plate* and plates with radial seams make up the rest of the bottom. In hemispherical bottom tanks, the ratio of the height of the cylindrical shell to the diameter is approximately 1:1 for small capacities and 1.25:1 for capacities over 5×10^5 litres. In elliptical bottom tanks of capacity less than 7.5×10^5 litres, the ratio of the shell height to the diameter is 0.6:1, and 0.5:1 for capacities over 7.5×10^5 litres. Conical bottom tanks (Fig. 12.4(b)) are rarely in excess of 9 m diameter. The slope of the cone is usually 45° from the vertical.

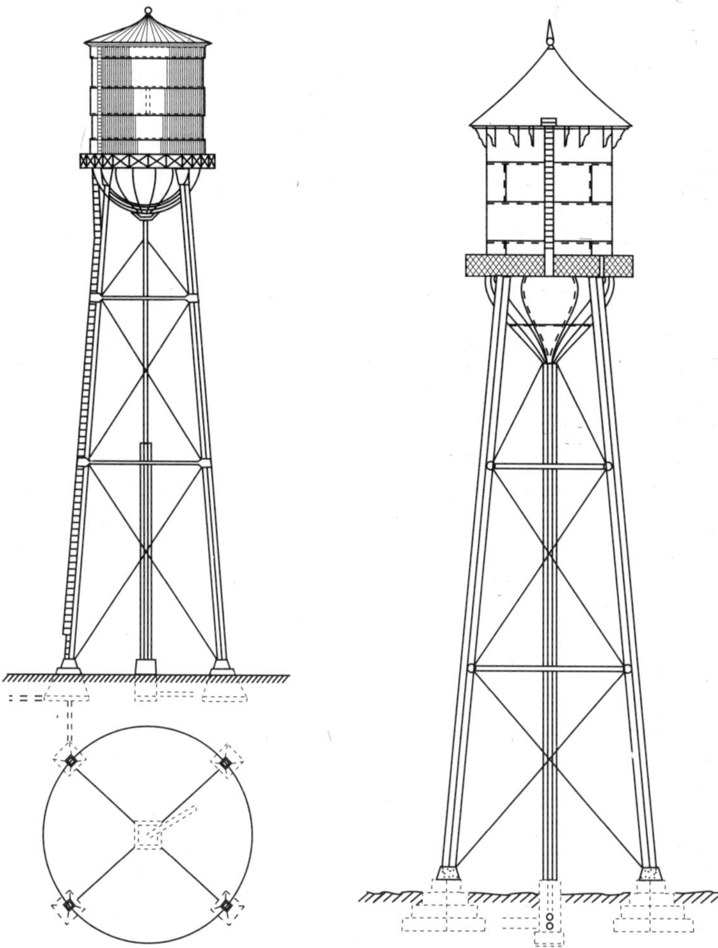

(a) (b)

Fig. 12.4 *Circular Water Tank with Segmental Bottom and (b) Circular Water Tank with Conical Bottom*

Tanks up to 9 m in height commonly have vertical columns. For higher tanks, the columns in the tower supporting the tank, are battered. The batter is about 1.25 to 1.5:12 for hemispherical bottom tanks upto 25×10^5 litre capacity and 1:12 for larger capacities. This batter decreases wind stresses in the tower.

The plates forming the sides of cylindrical shell are of different diameters, so that the courses can lap over each other, inside and outside, alternately. The joints for the horizontal seams, and for the radial seams in spherical bottom, are preferably lap joints. For vertical seams lap or butt joint may be used the latter being for large thickness plates. The minimum thickness of plates in cylindrical portion is 6 mm. The thickness of the plates in spherical bottom in no case should be less than that of the lower course in the cylindrical portion of the tank. The radial sections of spherical bottoms are made in multiples of the number of columns supporting the tank, and should be reinforced at the lower parts where holes are made for piping. When the centre of the spherical bottom is above the point of connection with the cylindrical portion of the tank, a girder is provided at the said point of connection to take the horizontal thrust. This also applies when the tank is supported by inclined (battered) columns. A balcony is provided around the tank. A railing, 1 m high, should also be provided. In order to avoid eccentric loading over tower columns, and local stresses in spherical bottoms, the connections between the column and sides of the tank are made in such a manner that the centre of gravity of the column section intersects the centre of connection between the spherical bottom and sides of the tank. If the tank is supported on the columns riveted directly to the sides, additional material is provided in the tank plates to take the shear and for this reason thicker tank plates or otherwise reinforcement plates at the column connections are desirable. The bending moments are taken by the upper and lower flange angles. On large tanks circular stiffening angles are provided in order to prevent the plates from buckling during wind storms. The top of the tank is generally covered with thin plates with a pitch of 1 to 6. For tanks up to 7.0 m diameter the roof plates are assumed to be self-supporting and for larger diameters angle rafters are used to support the plates. A free board of 15 cm is provided.

Circular Girder A circular girder of an angle or channel section is provided at the junction of the cylindrical shell and suspended bottom. It has to support the weight of the tank, the weight of the water stored and its own weight. The total load acts as a uniformly distributed load over the girder. The girder behaves as a beam curved in plan subjected to shear, bending moment and torsion. In turn it transmits loads to the vertical column over which it is simply supported. Even number of columns, spaced at an equal distance along the circumference (Fig. 12.5) are provided. Consequently all the columns are subjected to same vertical reaction, and the shear force. Torsion at the centre of each span will be zero and that at the supports also becomes zero. At the points of zero bending moment, torsion will be maximum. The following expressions may be used to obtain the forces in the circular girder.

$$\text{Vertical reaction in each column} = \frac{W}{n}$$

$$\text{Bending moment at supports, } M_0 = -\frac{WR}{n}\left(\frac{1}{\phi} - \frac{1}{2}\cot\frac{\phi}{2}\right) \qquad (12.4)$$

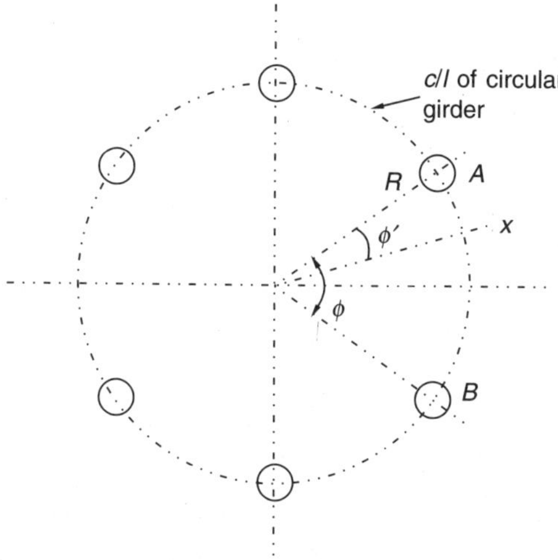

Fig. 12.5 *Circular Girder over Even Number of Columns*

Maximum bending moment, $$M_1 = M_0 \cos\frac{\phi}{2} + \frac{WR}{2n}\left[\sin\phi - \frac{2\sin^2\frac{\phi}{4}}{\phi/2}\right] \quad (12.5)$$

Maximum torsion (at a point x at an angle ϕ' from the column (Fig. 12.5)

$$T = M_0 \sin\phi' - \frac{WR}{2n}(1 - \cos\phi') + \frac{W\phi' R}{4}\left(1 - \frac{\sin\phi'}{\phi}\right) \quad (12.6)$$

where W = total vertical load,
n = number of columns (4, 6, 8, 12)
R = radius of circular girder.

The values of maximum shear, maximum positive and negative bending moments and maximum torsion are given in Table 12.1.

Table 12.1 *Forces in Circular Girder*

Number of Columns	*Load on Columns*	*Maximum Shear (N)*	*Bending Moment* at Column (Nm)	*Bending Moment* at Centre (Nm)	ϕ	*Maximum Torsion (Nm)*
4	$W/4$	$W/8$	–0.03415 WR	+0.01762 WR	19°20′	0.0053 WR
6	$W/6$	$W/12$	–0.01482 WR	+0.00751 WR	12°44′	0.00151 WR
8	$W/8$	$W/16$	–0.00827 WR	+0.00416 WR	9°33′	0.00063 WR
12	$W/12$	$W/24$	–0.00365 WR	+0.00190 WR	6°21′	0.000185 WR

When battered columns are used to support the tank, the columns exert an inward thrust which is resisted by the circular girder. The force due to the horizontal thrust (P) may be obtained from Table 12.2.

Table 12.2 *Forces on the Circular Girder Over Battered Columns*

Number of Columns	Bending Moment (Nm)		Shear (N)		Compression (N)	
	Under the Load	Midway between the Loads	Under the Load	Midway between the Loads	Under the Load	Midway between the Loads
4	$+0.68\ PD_1$	$-0.352\ PD_1$	5.0 P	0	5.0 P	7.07 P
6	$+0.45\ PD_1$	$-0.225\ PD_1$	5.0 P	0	5.0 P	10.0 P
8	$+0.34\ PD_1$	$-0.165\ PD_1$	5.0 P	0	5.0 P	13.1 P
12	$+0.22\ PD_1$	$-0.11\ PD_1$	5.0 P	0	5.0 P	19.3 P

where P = horizontal thrust at the top of each column
D_1 = diameter of the circular girder.

Stresses The following formulae based on the hoop stress theory of thin cylinders may be used.

Stress in the segmental bottom

$$\text{Stress in the radial joints} = \frac{wh\,D_1}{2t} \tag{12.7}$$

$$\text{Stress in the radial joints of the hemispherical bottom} = \frac{wh\,D}{4t} \tag{12.8}$$

Note Stress in the circumferential joints is same as that in the radial joints.

Stress in the conical bottom

$$\text{Stress in the radial joints} = \frac{wh\,D_1}{t}\ \text{cosec}\ \theta \tag{12.9}$$

$$\text{Stress in the circumferential joints} = 2\ \frac{wh\,D_1}{t}\ \text{cosec}\ \theta \tag{12.10}$$

where w = unit weight of water (9.81×10^{-6} N/mm^3)
D_1 = diameter of the segmental bottom (mm)
D = diameter of the cylindrical shell (mm)
θ = inclination of the cone to the horizontal
h = height of the top of the cylindrical shell from the lowest point of the suspended bottom.

Note The maximum stress in the suspended bottom occurs at the junction of the cylindrical shell and conical bottom.

Design

Size Depending upon the capacity and the shape of the suspended bottom, the height to diameter ratio (usually 0.8) of the cylindrical shell is assumed, and the diameter of the shell is worked out.

Thickness of the plates in the cylindrical shell is obtained from the hoop stress theory of thin shells.

$$t = \frac{wHD}{2\eta\sigma_{at}} \tag{12.11}$$

where H is the height of the cylindrical portion and D the diameter.

The thickness obtained so is increased by 1.5 mm to account for possible corrosion and in no case the thickness provided should be less than 6 mm.

Thickness of the plates in the suspended bottom is governed by stress in the radial joints and is given by,

$$t = \frac{whD}{4\eta\,\sigma_{at}} \tag{12.12}$$

the thickness obtained so should not be less than 6 mm after making allowance for corrosion.

Connections A suitable rivet diameter is assumed and the rivet value is calculated.

Hoop stress per linear vertical height in the vertical joints of cylindrical shell is given by

$$\sigma_{at_1,\text{cal}} = \frac{whD}{2}$$

$$\text{pitch of the rivets} = \frac{\sigma_{at_1,\text{cal}}}{R_V} \tag{12.13}$$

$$\not> 10\,t$$

where t = thickness of the plates.

Hoop stress per unit length in the radial joint is given by,

$$\sigma_{at_2,\text{cal}} = \frac{\omega h D}{4}$$

$$\text{Pitch of the rivets} = \frac{\sigma_{at_2,\text{cal}}}{R_V} \tag{12.14}$$

Circular girder The number of columns to be provided are fixed. The total uniform load is estimated and the values of maximum negative and positive bending moments, torsion and maximum shear force are obtained from Table 12.1. A tentative section of the girder is assumed and the total shear stress because of the direct shear and torsion is obtained. The maximum bending stress is also calculated and the combined stress is then determined using the principal stress formula and the resultant stress is checked to be less than $0.66\,f_y$.

Complete design of the circular tank and supporting structure are illustrated with the help of Examples 12.1 and 12.2 respectively.

12.8 RECTANGULAR TANKS

A rectangular tank as shown in Fig. 12.6 is built with plates 1.20 m, 1.25 m, 1.40 m, 1.50 m, ..., 2.50 m in width and has a flat bottom. The length of the bottom plates is kept parallel to the shorter dimension of the tank. These plates at the ends are turned up to form a butt joint with the vertical side plates. The joints of bottom plates are covered with Tee-sections from inside the tank. The Tee-sections are turned and carried vertical along the vertical joints of the side plates of the tank. All the vertical and horizontal side plate and bottom plate joints are covered with flats from outside

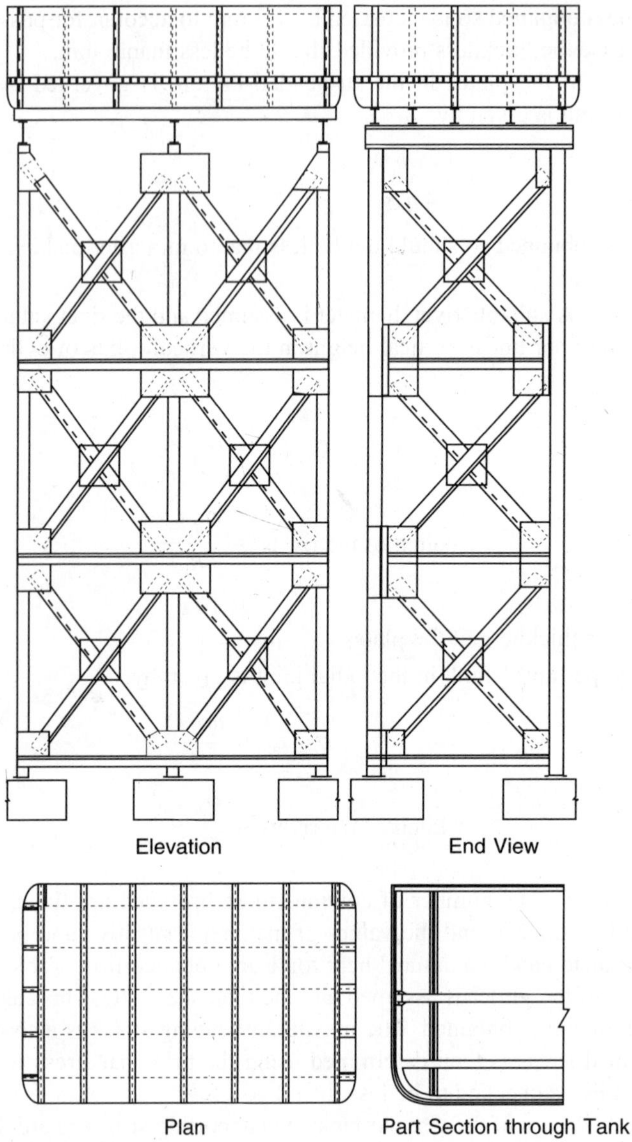

Fig. 12.6 *Elevated Rectangular Tank*

the tank. The Tee-section together with the flat section (Fig. 12.7) stiffens the side plates. The end and bottom corners of the tank are curved.

A horizontal stay arrangement at mid-height of the tank is provided inside the tank, in both the longitudinal and transverse directions. The section for the stay may be round or flat. These are connected to the vertical stiffeners (Tee-sections). To relieve the longitudinal stays partially from bending stress originating from their self-weight, they are made to rest over the transverse stays. Stays are subjected to tension and are designed to resist the overturning effect due to water pressure and weight of the overhang portion of the tank. A similar stay arrangement may also be

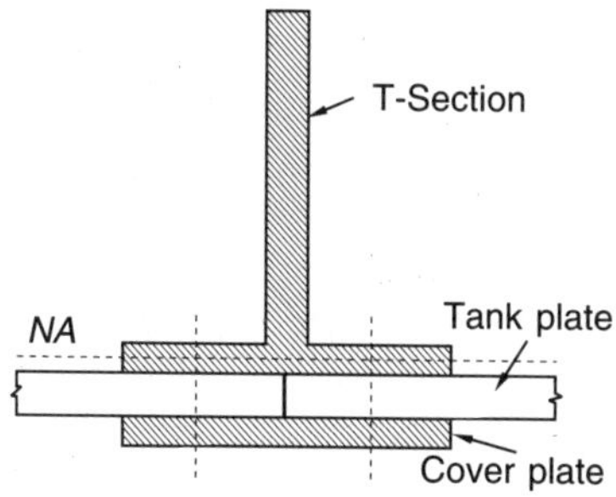

Fig. 12.7 *Stiffener Section for Rectangular Tank*

provided at the the top of the tank. In case it is not provided, the upper half of side plate (above the stays) resists the water pressure by cantilever action.

Two-tier arrangement of steel beams supporting the steel tank is made. These beams are supported over suitable arrangement of braced columns.

Design

Roof 5 mm thick roof plates supported over Tee-sections are provided. These Tee-sections rest on a beam or a truss.

Size The size of a rectangular tank is a function of capacity for which it is to be designed. The width of the tank depends upon the length of the plates available. A free board of 150 mm is assumed while ascertaining the tank dimensions.

Thickness of plates The thickness of the bottom plates is found from the bending theory. The weight of the water column on unit width of plate is obtained and self-weight of the plates (about 1/20 of weight of water) is added to it. Let this be W. Then the maximum bending moment, $M = Wb^2/8$.
where, b = the width of plates used (= spacing of longitudinal beams).

By equating the maximum bending moment to the moment of resistance, the thickness of bottom plates can be obtained.

$$M = M_R$$

or
$$M = \frac{1}{6} \times 1 \times t^2 \times \sigma_{bct}$$

Thickness of the side plates is kept same as that of the bottom plates.

Stiffener A stiffener section for rectangular tanks is shown in Fig. 12.7. The stiffener is assumed to be continuous and is subjected to self-weight and weight of the water column. A tentative stiffener section is assumed in the beginning. The section modulus required to resist the bending moment is worked out. This should be less than the section modulus provided. The rivets connecting the stiffener to the tank plates will be in double shear and are designed for longitudinal shear.

Stays Stays resist the water pressure and the weight of the overhanging potion of the tank (Fig. 12.8).

Water pressure, $P = \frac{1}{2} wh^2 b$, acts at $\frac{h}{3}$ from the bottom.

Weight of water column,

$W_1 = wabh$, acts at x_1 from L_0

Weight of bottom plates,

$W_2 = \rho abt$, acts at x_2 from L_0

Weight of side plates,

$W_3 = \rho bht$, acts at a from L_0

where, ρ = unit weight of steel,

a = overhang of the tank,

b = width of the tank plate (spacing between longitudinal beams) and

h = height of the tank

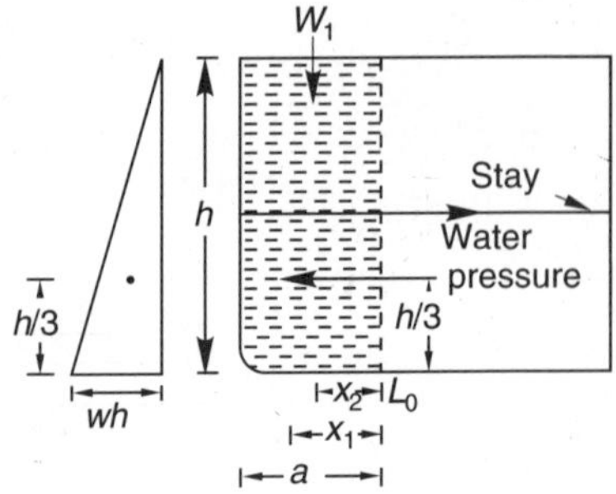

Fig. 12.8 *Water Pressure Diagram on Tank Walls*

The tension T in the stay can be obtained by taking the moment of all the above loads about L_0.

$$\text{Area of the stay} = \frac{T}{\sigma_{at}} \tag{12.15}$$

Beams Two tier of beams are provided below the tank as shown in Fig. 12.6. Stiffeners transfer the loads to the longitudinal beams. A longitudinal beam is thus subjected to the weight of the water column, weight of the vertical plates, weight of the bottom plates and its own weight. Transverse beams are subjected to the reactions from the longitudinal beams and its own weight. Both the longitudinal and transverse beams are not braced and are therefore designed as laterally unsupported beams. The design of a rectangular tank is illustrated with the help of Example 12.3.

12.9 PRESSED STEEL TANK

Pressed steel tanks have come into use to a large extent because of their ease in erection, facility in transport, standard and fast construction, ease in dismantling and re-erection. These are widely used for storage of cold and hot (40°–100°C) water and certain other liquids under pressure not greater than the static head corresponding to the depth of the tank.

A typical pressed steel tank is shown in Fig. 12.9. These are designed in accordance with the specifications laid down in IS 804–1998. Pressed steel tanks are formed by joining pressed mild steel plates (Fig. 12.10(a)) of a nominal size 1.25 m × 1.25 m. These plates are either heated uniformly in a furnace and formed in a press, or may be pressed cold. The plates are jointed together by means of mild steel hexagonal, black finished bolts conforming to IS 1363. 12 mm ϕ bolts are used for tanks up to 1.25 cu m and for higher capacities 14 mm ϕ bolts are used. Spacing of the bolt holes should not exceed 80 mm. The plates are jointed together at the flanges, 40 to 60 mm wide, and are pressed square (Fig. 12.10(b)) or partly square and partly at an angle of 45° as shown in Fig. 12.10(c), with the help of a vee-die and the corners are welded. It should be ensured than cracks do not develop in the welds on cooling.

A typical stay arrangement for a pressed steel tank is shown in Fig. 12.11. All the four sides of the tank should be supported at the junction of two or more plates. Stays connecting the side plates to the bottom plates are called *diagonal stays* and

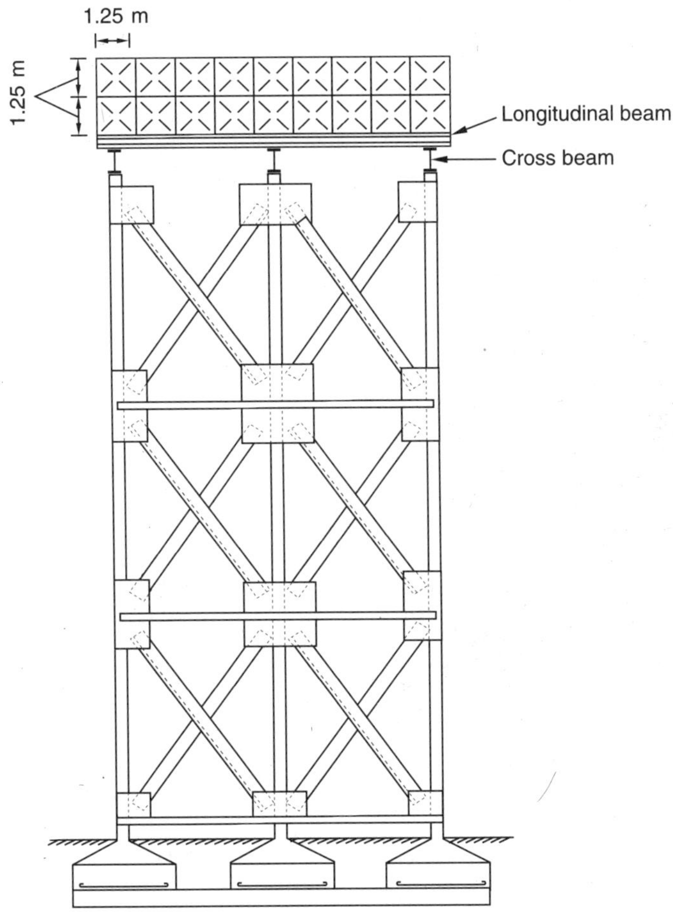

Fig. 12.9 *Pressed Steel Tank*

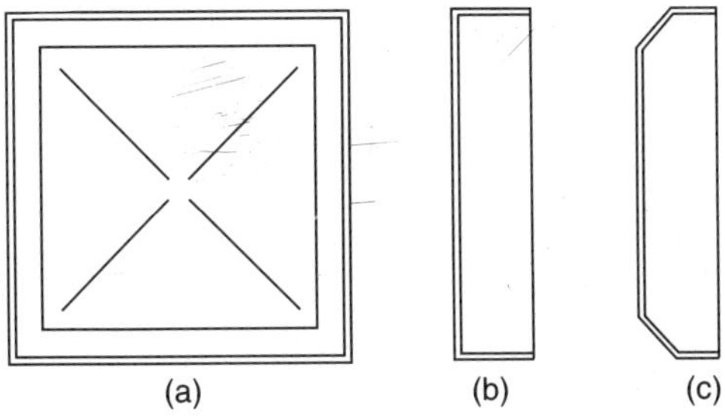

Fig. 12.10 *Pressed Steel Plate*

those connecting the two square sides are called *horizontal stays*. The diagonal stays are arranged so as not to intersect each other. The top horizontal stays may be omitted if the top cover acts as an effective tie. Stays are made of mild steel rolled sections—usually rods and are either bolted to the flanges or to the mild steel cleats attached to the tank plates.

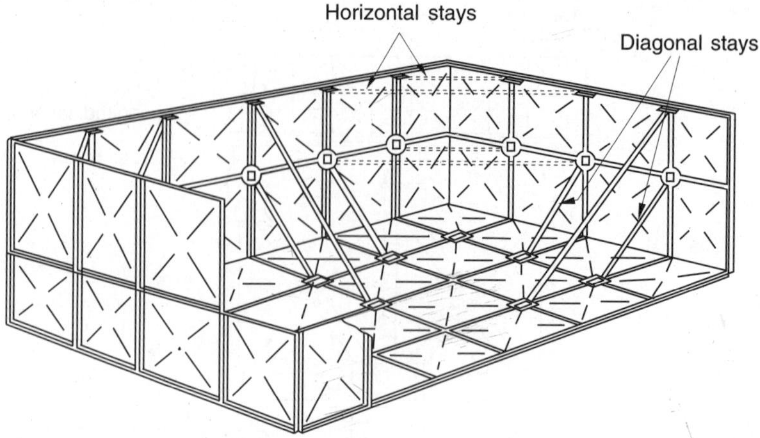

Fig. 12.11 *Pressed Steel Tank Container Showing Stay Arrangement*

The edges of the flanges should be finished flush with each other and all the joints are made leak proof. Typical flange joints and corner joint are shown in Fig. 12.12. The material used in jointing should be insoluble in the liquid to be stored and capable of withstanding any temperature variation in the stored liquid.

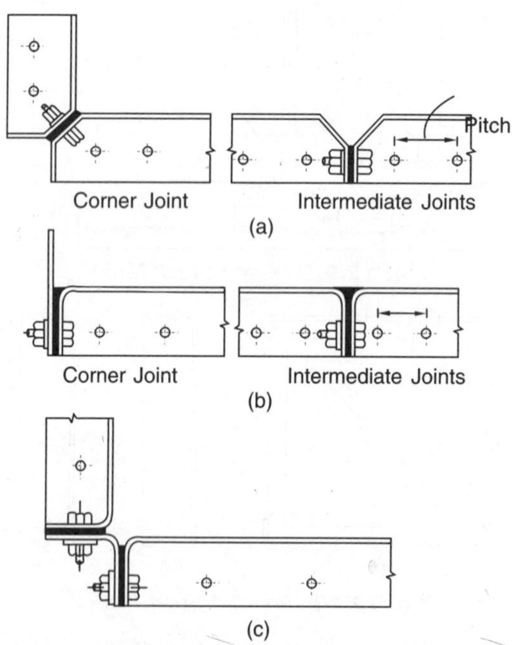

Fig. 12.12 *Corner and Intermediate Joints Details of Pressed Steel Tanks*

The tank is supported effectively under each transverse joint and at the two corresponding ends of the tank by any of the following arrangements.

1. By supporting the tank over rolled steel joists forming beams and architraves mounted on dwarf pillars as shown in Fig. 12.13.

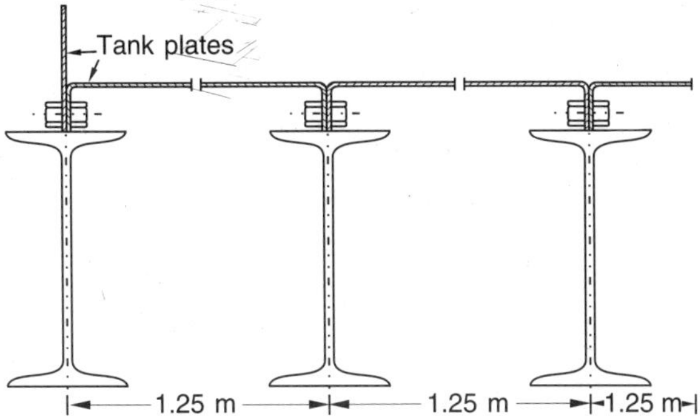

Fig. 12.13 *Rolled Steel Joist Bearers for Pressed Steel Tanks*

2. By providing dwarf walls of brick or concrete with footings to suit the depth of tank and nature of the ground. Dwarf walls for 3.75 m to 5.00 m deep tanks should have a mild steel capping plate to provide bearing surface on the wall (Fig. 12.14).

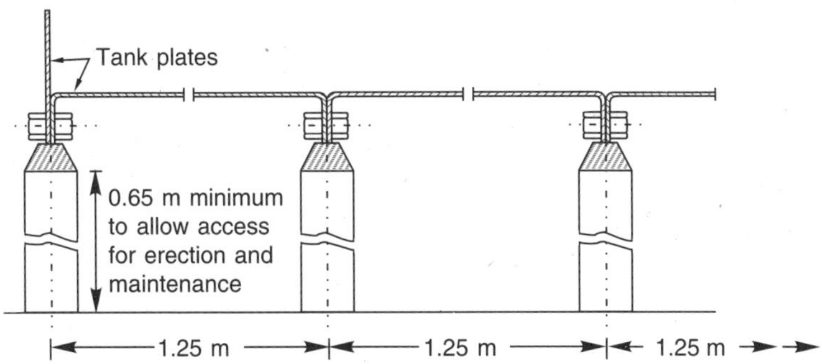

Fig. 12.14 *Dwarf Wall Supports for Pressed Steel Tanks*

3. By providing two tiers of beams, below the tank, and placed over suitably spaced columns.

Pressed steel tanks over 5 m deep are not recommended. The tanks may either be with open top or covered top. In the latter case covering may be mild steel plates (3.15 mm thick) conforming to IS 226, galvanized sheets conforming to IS 1079, or asbestos sheets conforming to IS 459.

A manhole of 450 mm diameter or 450 × 450 mm in size is provided in the roof. Rolled steel bearers or trusses are used to support the roof coverings. A 40 cm wide mild steel ladder is provided inside the tanks which are 2.5 m deep or more.

Classification Depending upon the position of the flanges pressed steel tanks are classified as Type-I, Type-II or Type-III. Type-I tank has all the flanges external whereas Type II has all the flanges internal. Type-III tank has all the bottom flanges internal and side flanges external.

Type-I tanks are used where external access is not restricted, or where the exterior of the tank is to be lagged. Type-II tank is used where access to the exterior for erection is precluded due to insufficient space inside a building. Both Type-II and Type-III are suitable for use when these are to be erected on a level floor.

Typical sizes, approximate weights and nominal capacities of Type-I tanks with open tops for depths 1.25 m to 5.0 m are given in Tables 12.3 to 12.6. Suitable approximations may be made for tanks with covered tops. The minimum nominal thickness of plates used for different depths of tanks, used for storage of cold liquids with sp. gr. not exceeding 1.0 is given in Table 12.7. In the case of hot liquids (sp. gr. not exceeding 1.0), the plate thickness should conform to that given in Table 12.7.

Table 12.3 *Typical Sizes, Approximate Weights and Nominal Capacities of Tanks with External Flanges*

Tanks 1.25 Metres Deep, Plate 5.0 mm Thick Throughout

Nominal Size	*Plates in Tank (Number)*	*Nominal Capacity (Litres)*	*Approximate Total Weight Empty (kg)*	*Approximate Outside Dimensions (m)*		
Length × Breadth m × m				*Length*	*Breadth*	*Depth*
1.25 × 1.25*	5	1 950	400	1.40	1.40	1.33
2.50 × 1.25	8	3 900	700	2.65	1.40	1.33
3.75 × 1.25	11	5 850	900	3.90	1.40	1.33
5.00 × 1.25	14	7 800	1 200	5.15	1.40	1.33
2.50 × 2.50	12	7 800	1 000	2.65	2.65	1.33
3.75 × 2.50	16	11 700	1 350	3.90	2.65	1.33
5.00 × 2.50	20	15 600	1 650	5.15	2.65	1.33
6.25 × 2.50	24	19 500	2 050	6.40	2.65	1.33
3.75 × 3.75	21	17 550	1 750	3.90	3.90	1.33
5.00 × 3.75	26	23 400	2 200	5.15	3.90	1.33
6.25 × 3.75	31	29 250	2 600	6.40	3.90	1.33
7.50 × 3.75	36	35 100	3 050	7.65	3.90	1.33
8.75 × 3.75	41	40 950	3 500	8.90	3.90	1.33
5.00 × 5.00	32	31 200	2 700	5.15	5.15	1.33
6.25 × 5.00	38	39 000	3 200	6.40	5.15	1.33
7.50 × 5.00	44	46 800	3 700	7.65	5.15	1.33
8.75 × 5.00	50	54 600	4 250	8.90	5.15	1.33
10.0 × 5.00	56	62 400	4 800	10.15	5.15	1.33
6.25 × 6.25	45	48 750	3 800	6.40	6.40	1.33
7.50 × 6.25	52	58 500	4 400	7.65	6.40	1.33
7.50 × 7.50	60	70 200	4 950	7.65	7.65	1.33
8.75 × 8.75	77	95 550	6 450	8.90	8.90	1.33

*Thickness of sheets of this tank only is 3.15 mm throughout.

Note—The capacity of tank with joints as in Fig. 12.12(a) will have slight reduction in capacity.

Table 12.4 *Typical Sizes, Approximate Weights and Nominal Capacities of Tanks with External Flanges*

Tanks 2.50 Metres Deep, Plates 5.0 mm and 6.0 mm Thick

Nominal Size *Length × Breadth m × m*	*Plates in Tanks (Number)*	*Nominal Capacity (Litres)*	*Approximate Total Weight Empty (kg)*	*Approximate Outside Dimensions (m)*		
				Length	*Breadth*	*Depth*
2.50 × 1.25	14	7 800	1 400	2.65	1.40	2.58
3.75 × 1.25	19	11 700	1 900	3.90	1.40	2.58
5.00 × 1.25	24	15 600	2 450	5.15	1.40	2.58
2.50 × 2.50	20	15 600	2 000	2.65	2.65	2.58
3.75 × 2.50	26	23 400	2 650	3.90	2.65	2.58
5.00 × 2.50	32	31 200	3 300	5.15	2.65	2.58
6.25 × 2.50	38	39 000	3 900	6.40	2.65	2.58
3.75 × 3.75	33	35 100	3 350	3.90	3.90	2.58
5.00 × 3.75	40	46 800	4 100	5.15	3.90	2.58
6.25 × 3.75	47	58 500	4 850	6.40	3.90	2.58
7.50 × 3.75	54	70 200	5 550	7.65	3.90	2.58
8.75 × 3.75	61	81 900	6 300	8.90	3.90	2.58
5.00 × 5.00	48	62 400	4 950	5.15	5.15	2.58
6.25 × 5.00	56	78 000	5 800	6.40	5.15	2.58
7.50 × 5.00	64	93 600	6 700	7.65	7.65	2.58
8.75 × 5.00	72	109 200	7 550	8.90	5.15	2.58
10.00 × 5.00	80	124 800	8 450	10.15	5.15	2.58
6.25 × 6.25	65	97 500	6 850	6.40	6.40	2.58
7.50 × 6.25	74	117 000	7 700	7.65	6.40	2.58
7.50 × 7.50	84	140 400	8 850	7.65	7.65	2.58
8.75 × 8.75	105	191 100	11 000	8.90	8.90	2.58

Note—The capacity of tank with joints as in Fig. 12.12(a) will have slight reduction in capacity.

Table 12.5 *Typical Sizes, Approximate Weight and Nominal Capacities of Tanks with External Flanges*

Tanks 3.75 Metres Deep, Plates 5.0 mm and 6.0 mm Thick

Nominal Size *Length × Breadth m × m*	*Plates in Tank (Number)*	*Nominal Capacity (Litres)*	*Approximate Total Weight Empty (kg)*	*Approximate Outside Dimensions (m)*		
				Length	*Breadth*	*Depth*
3.75 × 3.75	45	52 650	5 050	3.90	3.90	3.83
5.00 × 3.75	54	70 200	6 100	5.15	3.90	3.83
5.00 × 5.00	64	93 600	7 200	5.15	5.15	3.83
6.25 × 5.00	74	117 000	8 400	6.40	5.15	3.83
6.25 × 6.25	85	146 250	9 550	6.40	6.40	3.83
7.50 × 6.25	96	175 500	10 900	7.65	6.40	3.83
7.50 × 7.50	108	210 600	12 350	7.65	7.65	3.83

(Contd.)

Table 12.5 (*Contd.*)

Nominal Size *Length × Breadth* *m × m*	*Plates in Tank (Number)*	*Nominal Capacity (Litres)*	*Approximate Total Weight Empty (kg)*	*Approximate Outside Dimensions (m)*		
				Length	*Breadth*	*Depth*
8.75 × 7.50	120	245 700	13 650	8.90	7.65	3.83
8.75 × 8.75	133	286 650	15 100	8.90	8.90	3.83
10.00 × 8.75	146	327 600	16 600	10.15	8.90	3.83
11.25 × 8.75	159	368 550	18 150	11.40	8.90	3.83
10.00 × 10.00	160	374 400	18 200	10.15	10.15	3.83
11.25 × 10.00	174	421 200	20 000	11.40	10.15	3.83
12.50 × 10.00	188	468 000	21 500	12.65	10.15	3.83
11.25 × 11.25	189	473 850	21 600	11.40	11.40	3.83
12.50 × 11.25	204	526 500	23 300	12.65	11.40	3.83
12.50 × 12.50	220	585 000	25 200	12.65	12.65	3.83
13.75 × 12.50	236	643 500	27 000	13.90	12.65	3.83
13.75 × 13.75	253	707 850	29 900	13.90	13.90	3.83
15.00 × 13.75	270	772 200	30 950	15.15	13.90	3.83
15.00 × 15.00	288	842 400	33 000	15.15	15.15	3.83
16.25 × 15.00	306	912 600	35 000	16.40	15.15	3.83
16.25 × 16.25	325	988 650	37 300	16.40	16.40	3.83

Note—The capacity of tank with joints as in Fig. 12.12(a) will have slight reduction in capacity.

Table 12.6 *Typical Sizes, Approximate Weight and Nominal Capacities of Tanks with External Flanges*

Tanks 5 Metres Deep, Plates 5.0 mm, 6.0 mm and 8.0 mm Thick

Nominal Size *Length × Breadth* *m × m*	*Plates in Tank (Number)*	*Nominal Capacity (Litres)*	*Approximate Total Weight Empty (kg)*	*Approximate Outside Dimensions (m)*		
				Length	*Breadth*	*Depth*
3.75 × 3.75	57	70 200	7 100	3.90	3.90	5.08
5.00 × 3.75	68	93 600	8 850	5.15	3.90	5.08
5.00 × 5.00	80	124 800	10 250	5.15	5.15	5.08
6.25 × 5.00	92	156 000	11 800	6.40	5.15	5.08
6.25 × 6.25	105	195 000	13 550	6.40	6.40	5.08
7.50 × 6.25	118	234 000	15 300	7.65	6.40	5.08
7.50 × 7.50	132	280 800	17 200	7.65	7.65	5.08
8.75 × 7.50	146	327 600	19 050	8.90	7.65	5.08
8.75 × 8.75	161	382 200	21 000	8.90	8.90	5.08
10.00 × 8.75	176	436 800	23 150	10.15	8.90	5.08
11.25 × 8.75	191	491 400	25 100	11.40	8.90	5.08
10.00 × 10.00	192	499 200	25 300	10.15	10.15	5.08
11.25 × 10.00	208	561 600	27 400	11.40	10.15	5.08
12.50 × 10.00	224	624 000	29 600	12.65	10.15	5.08
11.25 × 11.25	225	631 800	30 000	11.40	11.40	5.08

(*Contd.*)

Table 12.6 (*Contd.*)

Nominal Size Length × Breadth m × m	*Plates in Tank (Number)*	*Nominal Capacity (Litres)*	*Approximate Total Weight Empty (kg)*	*Approximate Outside Dimensions (m)*		
				Length	*Breadth*	*Depth*
12.50 × 11.25	242	702 000	32 100	12.65	11.40	5.08
12.50 × 12.50	260	780 000	34 500	12.65	12.65	5.08
13.75 × 12.50	278	858 000	37 000	13.90	12.65	5.08
13.75 × 13.75	297	943 800	39 600	13.90	13.90	5.08
15.00 × 13.75	316	1 029 600	42 250	15.15	13.90	5.08
15.00 × 15.00	336	1 123 200	45 000	15.15	15.15	5.08
16.25 × 15.00	356	1 216 800	47 700	16.40	15.15	5.08
16.25 × 16.25	377	1 318 200	50 700	16.40	16.40	5.08

Note—The capacity of tank with joints as in Fig. 12.12(a) will have slight reduction in capacity.

Table 12.7 *Minimum Nominal Thickness of Plates*

Depth of Tank (m)	*Description of Plates with Regard to their Location*	*Thickness of Plates Sheets (mm)*
1.25	Bottom, sides (cubic tank only)	3.15
2.50	Bottom and first tier of sides	6.0
	Top tier of sides	5.0
3.75	Bottom and first tier of sides	6.0
	Second tier of sides	6.0
	Top tier of sides	5.0
5.00	Bottom and first tier of sides	8.0
	Second and third tier of sides	6.0
	Top tier of sides	5.0

Design

Size From the given capacity, a tank of suitable size with all dimensions being multiples of 1.25 m is obtained. The size of the tank may also be obtained from Tables 12.3 to 12.6.

Stays A suitable diagonal stay arrangement is provided to resist water pressure.

Water pressure diagram for a 1.25 m high tank is shown in Fig. 12.15.

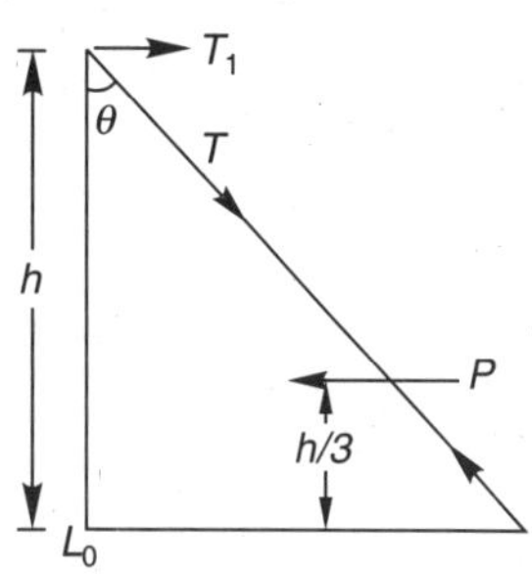

Fig. 12.15 *Water Pressure Diagram*

$$\text{Water pressure, } P = \frac{wh^2}{2} \times b$$

where b = width of the plate (1.25 m)

h = height of the tank (1.25 m)

θ = inclination of stay with vertical (=45°)

Taking moment about L_0

$$T_1 \times h = P \times \frac{h}{3} = \left(w \times \frac{h^2}{2}\right) \times b \times \frac{h}{3}$$

$$\Rightarrow \qquad T_1 = \frac{wh^2 b}{6} \tag{12.16}$$

Tension in stay, $T = T_1 \operatorname{cosec} \theta$

$$\text{The net area of stay} = \frac{T}{\sigma_{at}} \tag{12.17}$$

σ_{at} = permissible tensile stress = $0.8 \times 0.6 f_y$.

Tank bearers An arrangement of longitudinal and transverse beams similar to rectangular tanks is provided. The longitudinal beams are provided along the tank length below each joint. These rest over cross-beams, two or more in numbers depending on the size of tank.

The design load for longitudinal beams consists of self-weight of the joists, weight of the tank and weight of water as a uniform load over the joists. In addition, the reactions from stays also act on them as concentrated loads. Cross-beams are designed for their self-weight and reactions from longitudinal beams. Both longitudinal as well as cross-beams are designed as laterally unsupported beams.

12.10 WIND FORCE

For the calculation of wind load on towers the following procedure is followed.

1. Design wind speed is given by

$$V_z = V_b\, k_1\, k_2\, k_3$$

where, V_b = basic wind speed of the place in m/s
k_1 = probability factor (risk coefficient), given in Table 1.4
k_2 = terrain height and structure size factor, given in Table 1.7
k_3 = topography factor (1.0 for Plains).

The value of k_1 is obtained from Table 1.4, after deciding the age of the tower and the zone in which it is to be built.

The value of k_2 is fixed by deciding the class of the structure which itself depends upon the terrain category. For example, a tower located outside the city will fall in category 2 and class C. The topography factor k_3 is ascertained from the topography of the area and by the use of Appendix C of IS 875 (Part 3) (See Section 1.10).

2. Design wind pressure $p_z = 0.6\, V_z^2$ (at a height z).
3. The solidity ratio (ϕ) is calculated and corresponding to this the force coefficient (C_f) is obtained from Table 1.11.
4. Wind load on the tower, $F = C_f\, A_e\, p_z$

   where A_e = effective frontal area of the tower normal to the wind direction
   p_z = design wind pressure.
5. Wind load on the tank container is the product of the intensity of wind pressure and the longer side area exposed to the wind. This force will act at mid-height of container.

12.11 EARTHQUAKE FORCE

As per I.S: 1893–1984 (which is still not revised for tanks) following is the procedure for calculating the lateral force on tanks due to earthquakes.

1. The fundamental period of free oscillation T of the tank, $T = 2\pi\sqrt{\frac{\Delta}{g}}$ (12.18)

   where Δ = static horizontal deflection at the centre or gravity (C.G.) of the tank under the action of an imaginary static lateral equivalent to a load W acting at C.G. of the tank

   g = acceleration due to gravity

   W = (i) For empty condition = load of overhead water tank + $\frac{1}{3}$ weight of staging

   (ii) For tank-full condition = as in empty condition + weight of water.

2. The design seismic coefficent α_h is obtained by the response spectrum method, and is given by

$$\alpha_h = \beta\, I\, F_0 \frac{Sa}{g}$$

   where β = a factor depending upon the soil-foundation system (Table 12.8)

   I = importance factor, depending upon the life and function of the structure

   F_0 = Seismic zone factor for average acceleration spectra (Table 12.9)

   $\frac{Sa}{g}$ = average acceleration coefficient specified for various fundamental time periods and damping of structures (Appendix XI)

   The damping in steel tanks may be assumed to be 2%.

Table 12.8 *Soil Foundation Factor β*

Type of Soil Mainly Constituting the Foundation	*Value of Factor β for*			
	Bearing Piles Resting on soil Type I or Raft Foundation	*Friction Piles, Combined Footing, RCC Footing with the tie Beams*	*Isolated RCC Footings without tie Beams or Strip Foundations*	*Well Foundation*
Type I Rock or hard soils having $N^* > 30$	1.0	1.0	1.0	1.0
Type II medium soild having N from 10 to 30	1.0	1.0	1.0	1.0
Type III Soft soils having $N < 10$	1.0	1.2	1.5	1.5

N^* Standard penetration value according to I.S: 2131–1963.

Table 12.9 *Seismic Zone Factor for Average Acceleration Spectra*

Zone	*Seismic Zone Factor (F_0)*
I	0.05
II	0.10
III	0.20
IV	0.25
V	0.40

3. Using the fundamental period T, damping of 2%, the average acceleration coefficient is read from the curves shown in Appendix IX. Thereafter selecting suitable values of β, I, and F_0, the value of α_h is worked out.
4. The actual design lateral force will be equal to $\alpha_h W$ and is assumed to act at the centre of gravity of the tank. The member stresses in staging are obtained in the usual way for this force.

12.12 STAGING

An open tower structure is used to support a prefabricated elevated steel water storage tank. A tower may be regarded as four vertical cantilever lattice girders formed together in the shape of a square or rectangular box.

In the case of circular tanks an even number of properly braced columns are provided below the circular girder. For rectangular and pressed steel tanks a tower is most suited. The corner legs of the tower are placed so as to derive economy in the tank design. A maximum spread of the legs is desirable to resist overturning whereas a narrow spacing of the legs allows the tank supporting beams to cantilever over the sides of the tower. The bending moment induced in the cantilever portion of the beam offsets to some extent the bending moment in the central portion of the beams within the legs. Since about $\frac{1}{3}$rd of the weight of the tank is contained in the beams it would obviously be desirable to effect the economy by a compromise between the two. It could be argued that both the purposes could be served by giving a generous overhang to the beams and providing battered columns giving the required spread at the base of the tower. This would mean that pattern of side bracing does not repeat as in the case of vertical columns and the overall economy of the structure would be impaired. However, battered columns are justified for towers over 25 m height.

Generally, two channels with lacing on both the faces or with lacing on one face and a cover plate on the other face forms the most common section. For large capacity tanks, the leg section is built-up of two web plates, four angles and a cover plate on the inside face and lacing on all outer faces of the leg. This arrangement has an inherent advantage in proper detailing the upper end of the top panel tower leg.

For small capacities single angle, double angle or ISHB sections may be the choice. However, none of them permit a good column to tank connection or a combined column and bracing connection. Also, these are more expensive to erect.

The splices in the legs are made just above the strut connection but as near the centre line of the strut as possible.

The tower legs are braced in the horizontal as well as in the vertical plane. The purpose of bracing is two fold. It must be capable of resisting forces induced by wind pressure and it must be of such a form as to provide restraint at suitable points on the tower legs. A typical system of horizontal and diagonal bracing is shown in Fig. 12.16(c). For wind blowing in the direction shown in Fig. 12.16(a) a *N*-form of bracing is economical because the long diagonals are in tension. A sub-framing shown by dotted lines may be introduced to provide additional restraints to the tower legs. However, for a reversal of the wind direction a similar lattice girder 'opposite hand' in form is required (Fig. 12.16(b)). Since wind may blow in any direction the bracing system shown in Fig. 12.16(c), obtained by superimposition of the two cases is provided. This is known as a *duplicate tie system* in which the diagonals are designed as tie members. It is assumed that only one diagonal is in operation in the panel, depending upon the direction of wind and it is known as the *effective diagonal*. The other diagonal will be imperative. The horizontal braces preserve the square of the tower under conditions of diagonal wind.

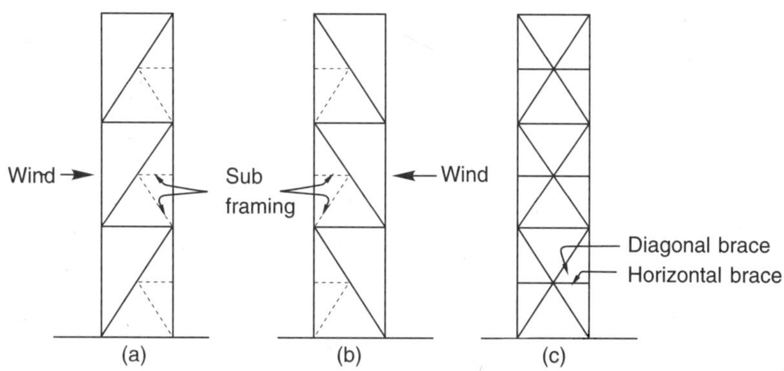

Fig. 12.16 *Development of Side Bracing System*

Conditions of Wind Loading to be Considered Consider a tank supported on four tower-legs *A, B, C* and *D*. Wind may blow normal to any of the sides *AB, BC, CD, DA* or may blow across the diagonals *AC* or *BD*.

Case I Wind normal to the sides AB and CD (Fig. 12.17(a)) It is assumed that wind pressure is resisted by the vertical lattice girders in sides *AD* and *BC* only. The effect of local bending of lattice girder *AB* and *CD*, and the effect of wind blowing parallel to sides *AD* and *BC* are ignored. Assuming the overturning of the tower about the centre line, the columns *A* and *B* would be subjected to uplift and the columns *C* and *D* would be subjected to thrust. Also, the bracings of sides *AD* and *BC* would be stressed. If M_w is wind moment and R the vertical reaction in each tower leg, then

$$2Rx = M_w$$

$$\Rightarrow \qquad R = \frac{M_w}{2x} \qquad \text{(For a square or rectangular tower)} \quad (12.20)$$

$$\text{and} \qquad R = \frac{M_w}{2\sqrt{2}r} \qquad \text{(For circular tanks)} \quad (12.21)$$

$$\text{Generalised formula: } R = \frac{M_w}{0.5\sqrt{2}\,Nr} \qquad \text{(For circular tanks)} \quad (12.22)$$

where N is the number of columns in the tower.

Case II Wind Blowing Across the Diagonal BD (Fig. 12.17(b)) Wind load is assumed to be resisted by the vertical lattice girders on all the four sides. The local bending on the sides of the tower is neglected. Column A is subjected to uplift and column C to thrust.

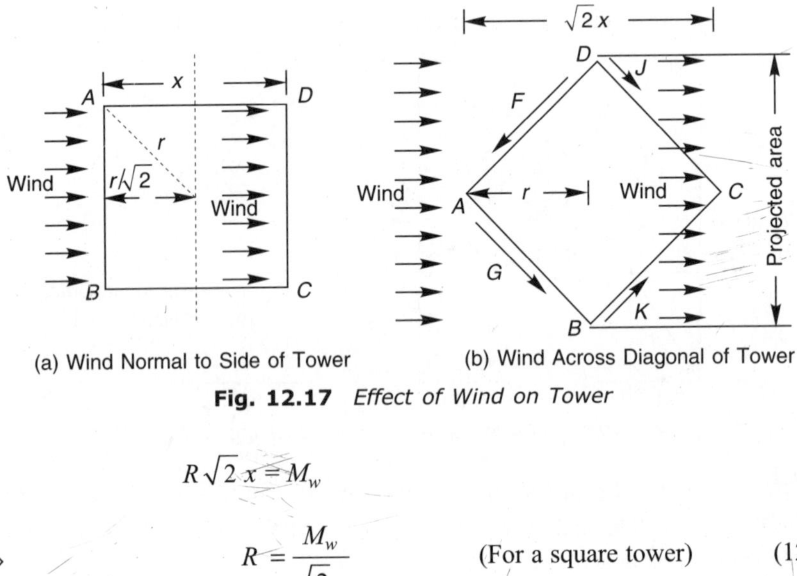

Fig. 12.17 *Effect of Wind on Tower*

$$R\sqrt{2}\,x = M_w$$

$$\Rightarrow \qquad R = \frac{M_w}{\sqrt{2}\,x} \qquad \text{(For a square tower)} \quad (12.23)$$

$$\text{and} \qquad R = \frac{M_w}{2r} \qquad \text{(For circular tanks)} \quad (12.24)$$

$$\text{Generalised formula: } R = \frac{M_w}{0.5\,Nr} \qquad \text{(For circular tanks)} \quad (12.25)$$

This gives the worst case of wind load on tower. Therefore, from the design consideration, bracing should be designed to suit Case I loading and the tower legs to suit Case II loading.

Solved Examples

Example 12.1 Design a circular elevated water tank for a capacity of 2,50,000 litres. The height of the tank bottom above the ground level is 8.7 m. The tank is supported over eight columns and is situated at the railway station in Allahabad.

Solution

Size of the tank

Capacity of tank = 2,50,000 litres = 250 m^3

Assume hemispherical bottom and the height of cylindrical shell to be 0.8 times the diameter of tank. Refer to Fig. Ex. 12.1(i).

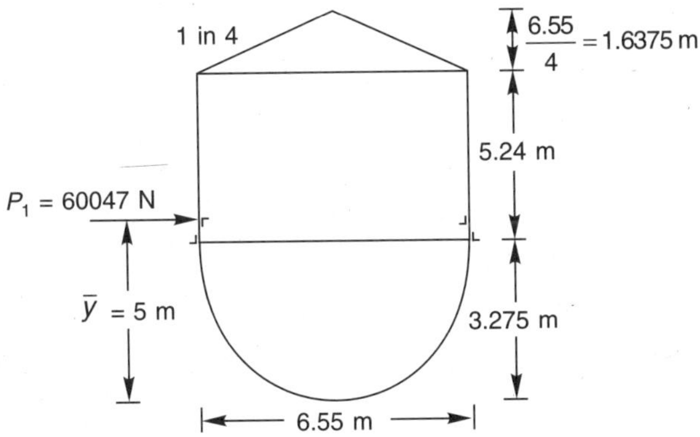

Fig. Ex. 12.1(i) *Cylindrical Tank with Hemispherical Bottom*

$$\frac{\pi}{4} \times D^2 \times 0.8\, D + \frac{1}{2} \times \frac{4}{3} \times \pi \times \left(\frac{D}{2}\right)^3 = 250$$

$\Rightarrow$ $D = 6.55$ m

Hence, $H = 0.8 \times D = 0.8 \times 6.55 = 5.24$ m

Thickness of the plates

Let us provide 16 mm ϕ power driven rivets for making the connections. Also let the efficiency of the joints be 75%.

Thickness of shell plates

$$t = \frac{wHD}{2\eta\sigma_{at}} = \frac{9.81 \times 10^{-6} \times 5.24 \times 10^3 \times 6.55 \times 10^3}{2 \times 0.75 \times (0.8 \times 0.6 \times 250)} = 1.87 \text{ mm}$$

Add 1.5 mm to account for corrosion.

$$t = 1.87 + 1.5 = 3.37 \text{ mm} \nless 6 \text{ mm}$$

Provide 6 mm thick plates in the cylindrical shell of the tank.

Thickness of suspended bottom plates

$$h = \left(5.24 + \frac{6.55}{2}\right) = 8.515 \text{ m}$$

$$t = \frac{whD}{4\eta\,\sigma_{at}} = \frac{9.81\times10^{-6}\times8.54\times10^{3}\times6.55\times10^{3}}{4\times0.75\times(0.8\times0.6\times250)} = 1.52 \text{ mm}$$

Add 1.5 mm to account for corrosion.

$$t = 1.52 + 1.5 = 3.02 \text{ mm} \not< 6 \text{ mm}$$

Provide 6 mm thick plates in the hemispherical bottom of the tank.

Conical roof

Provide 5 mm thick plates for the conical roof. The pitch of the roof may be kept 1 in 4.

Connections

Power driven 16 mm ϕ (gross diameter = 17.5 mm) field driven rivets and double riveted lap joints have been used all through.

Cylindrical shell plates Hoop stress per linear vertical height,

$$F_1 = \frac{wHD}{2} = \frac{9.81\times10^{-6}\times5.24\times10^{3}\times6.55\times10^{3}}{2} = 168.35 \text{ N/mm}$$

$$\text{Strength of the rivet in single shear} = \frac{\pi}{4}\times17.5^2\times0.8\times90 = 17318.03 \text{ N}$$

$$\text{Strength of the rivet in bearing} = 17.5\times6\times0.8\times270 = 22680 \text{ N.}$$

$$\text{Rivet value, } R_V = 17318.03 \text{ N}$$

$$\text{Pitch of the rivets} = 2\times\frac{17318.3}{168.35} = 205.74 \text{ mm}$$

$$\not> 60 \text{ mm} \qquad (10t = 10\times6 = 60).$$

Rivets for horizontal joints are provided at the same pitch as that for vertical joints.

Hemispherical bottom plates

Hoop stress per unit length in the radial joints,

$$F_2 = \frac{whD}{4} = \frac{9.81\times10^{-6}\times8.515\times10^{3}\times6.55\times10^{3}}{4} = 136.78 \text{ N/mm}$$

Since the force F_2 is less than F_1, rivets for making connections of hemispherical bottom plates are also provided at a pitch of 60 mm (same as calculated for the cylindrical portion of the tank) for radial as well as horizontal joints.

Cylindrical shell with hemispherical bottom shell

There are no inclined or compressive stresses on the hemispherical portion and therefore connection of the two shells need not be designed again. Hence, provide 16 mm ϕ power driven field rivets at a pitch of 60 mm and double riveted lap joint.

Circular girder

Weight of water, $W_1 = 10 \times 250 = 2500 \text{ kN} = 2500 \times 10^3 \text{ N}$

$$\text{Self weight of tank, } W_2 = \left[\pi \times 6.55 \times 10^3 \times 5.24 \times 10^3 \times 6 + \frac{1}{2} \times 4\pi \times \left(\frac{6.55 \times 10^3}{2}\right)^2 \times 6\right] \times 7.9 \times 10^{-5}$$

$$= 83052.77 \text{ N}$$

Self-weight of conical roof,

$$W_3 = \frac{\pi}{2} \times 6.55 \times 10^3 \sqrt{\left(\frac{6.55 \times 10^3}{2}\right)^2 + \left(\frac{6.55 \times 10^3}{4}\right)^2} \times 5 \times 7.9 \times 10^{-5}$$

$$= 14880.74 \text{ N}$$

The self-weight of the tank and that of the conical roof may be increased by 20 % to account for additional weight of overlap of plates and connectors.

$$W_4 = 1.2(W_2 + W_3)$$
$$= 1.2 \times (83052.77 + 14880.74) = 117520.21 \text{ N}$$

Assume, self-weight of girder = 1500 N/m.

Total self-weight of girder, $W = 1500 \times (\pi \times 6.55) = 30866.14 \text{ N}$

Total load on the girder, $W = 2500 \times 10^3 + 117520.21 + 30866.14$

$$= 2648386.35 \text{ N} \simeq 2650.0 \times 10^3 \text{ kN}$$

For circular girder supported over eight columns (Table 12.1):

Maximum bending moment (at supports) = $-0.00827\ WR$

$$= -0.00827 \times 2650 \times 10^3 \times \frac{6.55}{2}$$

$$= -71773.26 \text{ Nm}$$

Maximum torsion (at 9°33′ from support) = $0.00063\ WR$

$$= 0.00063 \times 2650 \times 10^3 \times \frac{6.55}{2}$$

$$= 5467.6 \text{ Nm}$$

$$\text{Maximum shear force (at support)} = \frac{2650 \times 10^3}{16} = 165625.00 \text{ N}$$

$$\text{Section modulus, } Z_{req} = \frac{71773.26 \times 10^3}{0.66 \times 250} = 434989.8 \text{ mm}^3$$

Let us try a girder section as shown in Fig. Ex. 12.1(ii)(a). The section has been built-up by using 1 m depth of cylindrical shell of the tank and 2 ISA 150 × 150 × 18 mm sections.

$$I_{xx} = \frac{6 \times 1000^3}{12} + 2 \times 1048.9 \times 10^4 + 2 \times 5079 \times (500 - 43.8)^2$$

$$= 26350.45 \times 10^4 \text{ mm}^4$$

$$\text{Section modulus,} \quad Z_{Prov} = \frac{I}{y} = \frac{263504.5 \times 10^4}{500} = 5270090 \text{ mm}^3$$

$$> 438310 \text{ mm}^3$$

which is all right.

$$\text{Torsional constant, } J = \sum \frac{1}{3} bt^3$$

$$= 2 \times \left[\frac{1}{3} \times 150 \times 18^3 + \frac{1}{3}(150 - 18) \times 18^3 \right] + \frac{1}{3} \times 1000 \times 6^3$$

$$= 1168416 \text{ mm}^4$$

$$\text{Shear stress due to torsion} = \frac{T}{J}\tau_{max} = \frac{5467.6 \times 10^3}{1168416} \times (18 + 6)$$

$$= 112.30 \text{ N/mm}^2$$

$$\text{Shear stress due to shear force} = \frac{165625.00}{2 \times 5079 + 1000 \times 6}$$

$$= 10.2503 \text{ N/mm}^2$$

$$\text{Total shear stress} = 112.30 + 10.2503 = 122.55 \text{ N/mm}^2$$

$$> 100 \text{ N/mm}^2 \qquad (0.4 \times 250)$$

Hence, the section needs a revision. The next trail section is shown in Fig. Ex. 12.1(ii)(b).

$$I_{xx} = \frac{6 \times 1800^3}{12} + 2 \times 2588.7 \times 10^4 + 2 \times 6881 \times (900 - 56.1)^2$$

$$= 1276861.9 \times 10^4 \text{ mm}^4$$

$$Z_{prov} = \frac{1276861.9 \times 10^4}{900} = 14187354 \text{ mm}^3 > 434989.8 \text{ mm}^3$$

which is all right.

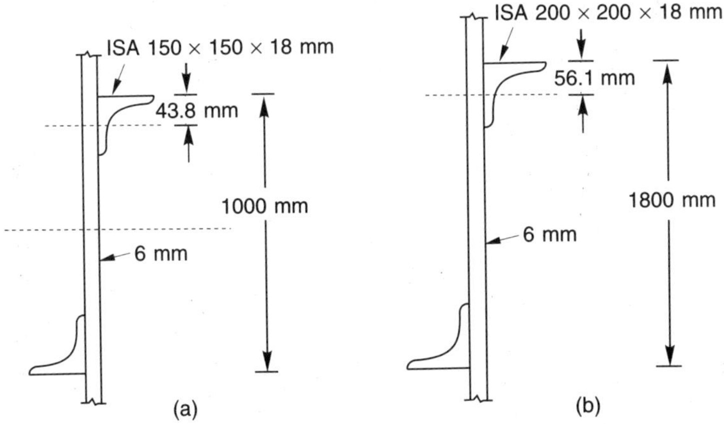

Fig. Ex. 12.1(ii) *Circular Girder*

Torsional constant,

$$J = 2 \times \left[\frac{1}{3} \times 200 \times 18^3 + (200 - 18) \times 18^3 \right] + \frac{1}{3} \times 1800 \times 6^3$$

$$= 3030048 \text{ mm}^4$$

$$\text{Shear stress due to torsion } = \frac{5467.6 \times 10^3}{3030048} \times (18 + 6) = 43.30 \text{ N/mm}^2$$

$$\text{Shear stress due to shear force } = \frac{165625.00}{2 \times 6881 + 1800 \times 6} = 6.743 \text{ N/mm}^2$$

$$\text{Total shear stress} = 43.30 + 6.743 = 50.04 \text{ N/mm}^2$$

$$< 100 \text{ N/mm}^2 \ (0.4 \times 250)$$

which is all right.

$$\text{Bending stress}, \sigma_{bct} = \frac{71773.26 \times 10^3}{14187354} = 5.059 \text{ N/mm}^2$$

$$\text{Hoop stress, } \sigma_{at_1} = \frac{wHD}{2\eta t} = \frac{9.81 \times 10^{-6} \times 5.24 \times 10^3 \times 6.55 \times 10^3}{2 \times 0.75 \times 6}$$

$$= 37.41 \text{ N/mm}^2$$

Maximum principal stress,

$$\sigma_1 = \frac{37.41 + 5.059}{2} + \sqrt{\left(\frac{37.41 + 5.059}{2}\right)^2 + (50.04)^2}$$

$$= 75.59 \text{ N/mm}^2 < 165 \text{ N/mm}^2 \ (0.6 \times 250)$$

which is all right.

Example 12.2 Design the staging for the cylindrical tank of Example 12.1.

Solution

Wind Load

For Allahabad $V_b = 47$ m/s

$k_1 = 1.07$ (100 years)

$k_2 = 1.0$ (category 2, class A)

$k_3 = 1.0$ (level ground)

$V_z = V_b\, k_1\, k_2\, k_3$

$= 47 \times 1.07 \times 1 \times 1 = 50.29$ m/s

Wind pressure, $p_z = 0.6\ V_z^2$

$= 0.6 \times 50.29^2 = 1517.45$ N/m^2

For circular plan shape and H/D ratio of 0.8, $C_f = 0.7$ (Table 1.8)

Wind load, $F_1 = C_f\, A_{e1}\, P_z$

$= 0.7 \times 1517.45 \times A_{e1} = 1062.215\ A_{e1}$ Newton

A_{e1} = projected area of tank

= areas of conical roof + cylindrical shell + hemispherical bottom

$$= \frac{1}{2} \times 6.55 \times \frac{6.55}{4} + 5.24 \times 6.55 + \frac{1}{2} \times \frac{\pi}{4} \times 6.55^2$$

$= 56.53$ m^2

$F_1 = 1062.215 \times 56.53 = 60047$ N

The wind load will act at the C.G. of the projected area of the tank.

Let the C.G. of the projected area be at a distance of $\bar{y}$ from the bottom of the tank (Fig. Ex. 12.1(i)).

$$\bar{y} = \frac{1}{56.53} \times \left[5.363 \times \left(\frac{1}{3} \times \frac{1}{4} \times 6.55 + 5.24 + 3.275\right) + 34.32 \times (2.62 + 3.275) + 16.85 \times \left(3.275 - \frac{4 \times 3.275}{3\pi}\right)\right] = 5.0 \text{ m}$$

Hence, the wind force F_1 will act at 1.725 m (5 – 3.275) above the junction of the cylindrical shell and bottom of tank as shown in Fig. Ex. 12.1(i).

For solidity ratio $\phi = 0.2$; $C_f = 3.3$ (Table 1.11)

Wind load, $F_2 = C_f A_{e2}\, P_d$

$= 3.3 \times 1517.45 \times 0.2 \times (12 \times 3.275) = 39359.62$ N

Height of columns = 8.7 + (6.6/2) = 12.0 m

This force F_2 acts at $\frac{12}{2}$ = 6 m above the ground level.

Overturning moment due to wind,

$$M_w = 39359.62 \times 6 + 60047 \times (12 + 1.725) = 1060302.79 \text{ Nm}$$

Addition axial thrust in the extreme column,

$$R_1 = \frac{M_w}{4r} = \frac{1060302.79}{4 \times 3.275} = 80939.14 \text{ N} \qquad \text{(8 columns)}$$

The additional axial thrust (80939.14 N) in the extreme column due to wind is less than $\frac{1}{3}$rd of the gravity loads $\left(\frac{2647427.02}{8} = 330928 \text{ N}\right)$ in the column. Hence, the column will be designed for vertical loads only and the permissible stresses will not be increased by $33\frac{1}{3}$% to account for wind.

Seismic force

$$\text{Weight of full tank} = 2647427.02 \text{ N}$$

$$\text{Weight of water} = 2500 \times 10^3 \text{ N}$$

$$\text{Weight of empty tank} = 2647427.02 - 2500 \times 10^3 = 147427.02 \text{ N}$$

$$\text{Weight of staging} = 0.20 \times 147427.02 = 29485.4 \text{ N}$$

(20% of weight of empty tank)

For empty condition, $W_e = 147427.02 + \frac{29485.4}{3} = 157255.48$ N

For full condition, $W_f = 157255.48 + 2500 \times 10^3 = 2657255.48$ N

The fundamental period of free oscillation,

$$T = 2\pi\sqrt{\frac{\Delta}{g}}$$

where Δ = static horizontal deflection (Fig. Ex. 12.2(c)

$$= Q\sum\frac{l^3}{a^2 AE} \quad \text{(for all panels)}$$

a = spacing of columns = 2.713 m, h = effective length of column 4 m, number of panels = 3

A = area of horizontal brace (say 200 mm^2)

$$l = \sqrt{2.713^2 + 4^2} = 4.833 \text{ m}$$

$$\Delta = \frac{3 \times Q \times 4833^3}{2713^2 \times 200 \times 2.1 \times 10^5} = 1.095 \times 10^{-3} Q$$

Empty condition

$$Q = \frac{W_e}{4} = \frac{157255.48}{4} = 39313.87 \text{ N}$$

$$\Delta = 39313.87 \times 1.095 \times 10^{-3} = 43.048 \text{ mm} = 4.30 \text{ cm}$$

$$T = 2\pi\sqrt{\frac{4.30}{981}} = 0.416 \text{ sec}$$

For $T = 0.416$ sec and damping = 2%, $\frac{Sa}{g} = 0.261$ (Appendix XI)

For hard soil and isolated footing, $\beta = 1.0$ (Table 12.8)
Seismic zone factor for Allahabad, $F = 0.1$
Importance factor for water tank, $I = 1.5$

$$\alpha_h = \frac{Sa}{g} F \beta I = 0.261 \times 0.1 \times 1 \times 1.5 = 0.03915$$

Earthquake force = 0.03915 × 157255.48 = 6156.55 N

Full condition

$$Q = \frac{w_f}{4} = \frac{2657255.48}{4} = 664313.87 \text{ N}$$

$$\Delta = 664313.87 \times 1.09 \times 10^{-3} = 724.102 \text{ mm} = 72.41 \text{ cm}$$

$$T = 2\pi\sqrt{\frac{72.41}{981}} = 1.707 \text{ sec}$$

For $T = 1.707$ sec and damping = 2%; $\alpha = 0.084$ (Appendix XI)

$$\alpha_h = 0.084 \times 0.1 \times 1 \times 1.5 = 0.0126$$

Earthquake force = 0.0126 × 2657255.48 = 33481.42 N

Since earthquake force in each of the two conditions is less than the wind load, the governing one for the design of the staging will be the wind load.

Columns

8 columns of height 12 m above the ground level are to be provided. These are divided into 3 panels as shown in Fig. Ex. 12.2(a).

$$\text{Height of each panel} = \frac{12}{3} = 4 \text{ m}$$

Total vertical load supported by staging = 2647427.02 N

$$\text{Weight supported by each column} = \frac{2647427.02}{8} = 330928.37 \text{ N}$$

Assuming self-weight of column (@ 500 N/m) = 12 × 500 = 6000 N

Design column load = 330928.37 + 6000 = 336928.37 N

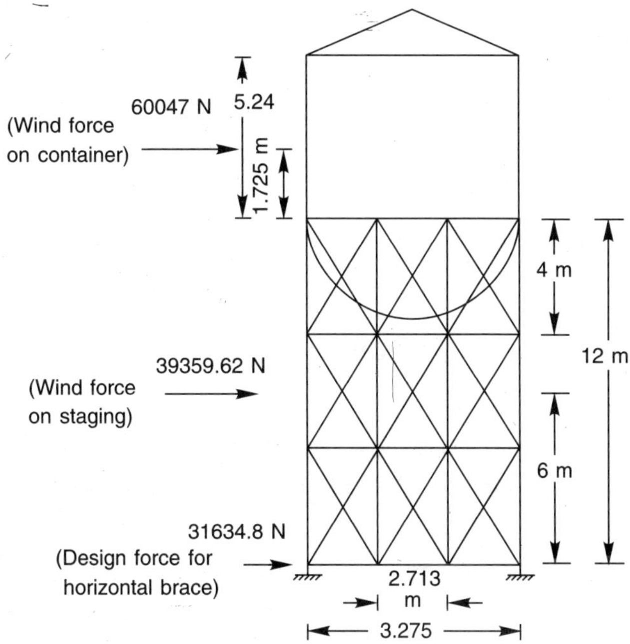

Fig. Ex. 12.2(a)

$$\text{Required area of column} = \frac{336928.37}{85} = 3963.86 \text{ mm}^2$$

From IS Handbook No. 1 select ISHB 200 @ 365.9 N/m.
The relevant properties of the section are,

$$A = 4754 \text{ mm}^2,\ r_{yy} = 45.1 \text{ mm}$$

$$\lambda = \frac{4000}{45.1} = 88.7$$

For $f_y = 250$ N/mm^2 and $\lambda = 88.7$; $\sigma_{ac} = 88.57$ N/mm^2.
Load carrying capacity = 4754 × 88.57= 421062 N > 336928.37 N
which is all right.

Bracing

Bracings are designed to resist combined shear because of wind and 2.5% of axial load.

Total lateral force in critical horizontal braces near the base of the staging is due to wind force on the container and that at the critical brace level (Fig. Ex. 12.2(a))

$$P = 60047 + \frac{5}{6} \times 39359.62 = 92846.68 \text{ N}$$

$$\text{Lateral force per column, } P_1 = \frac{92846.68}{8} = 11605.8 \text{ N}$$

Refer to Figs Ex. 12.2(b) and (c).

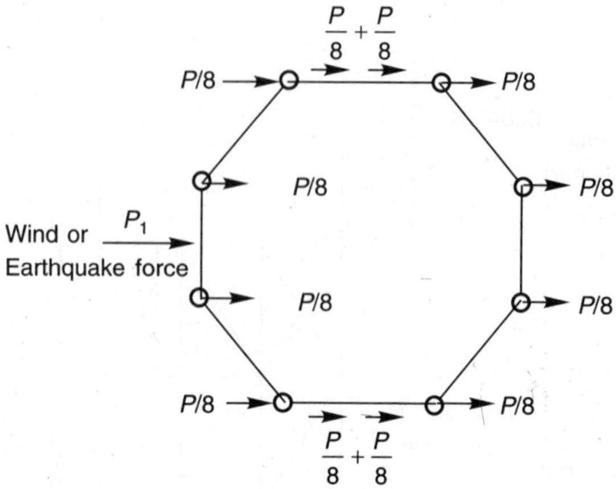

P = lateral force (maximum of wind or earthquake force)

Fig. Ex. 12.2(b) *Distribution of Shear*

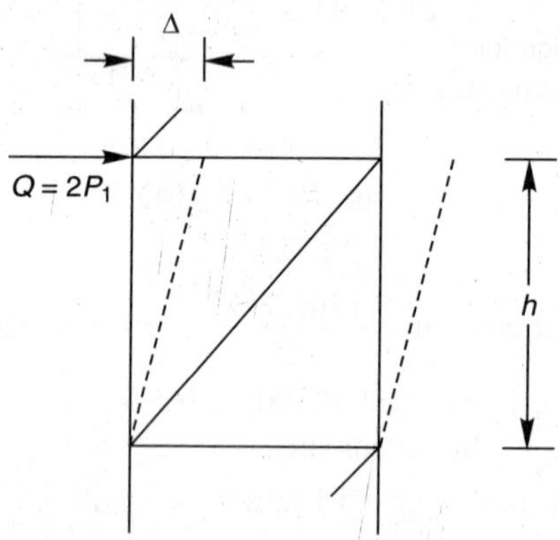

Fig. Ex. 12.2(c) *Deflection of Braced Panel*

Horizontal shear in bracing = Q + 2.5% of column load

$$= 2 \times 11605.8 + \frac{2.5}{100} \times 336928.37 = 31634.8 \text{ N}$$

Horizontal brace

The horizontal brace will be in compression.

$$\text{Design force} = 31634.8 \text{ N}$$

$$\text{Area} = \frac{31634.8}{60} = 527.24 \text{ mm}^2$$

Try ISA 65 × 65 × 6 mm section. From IS Handbook No. 1, the relevant properties of the section are:

$$A = 744 \text{ mm}^2,\ r_{min} = 12.6 \text{ mm}$$

Let us provide 16 mm ϕ power driven rivets. Assume more than one rivet for connecting the brace with the column.

$$\lambda = \frac{0.85 \times 2713}{12.6} = 183$$

For $\lambda = 183$ and $f_y = 250$ N/mm^2,

$$\sigma_{ac} = 32.1 \text{ N/mm}^2$$

Load carrying capacity = 1.33 × 32.1 × 744 = 31763.6 N > 31634.8 N
which is all right.
Strength of the rivet in single shear

$$= \frac{\pi}{4} \times 17.5^2 \times 1.25 \times 100 = 30066.02 \text{ N}$$

Strength of the rivet in bearing = 17.5 × 6 × 1.25 × 300 = 39375 N

Hence rivet value = 30066.02 N

$$\text{Number of rivets required} = \frac{31634.8}{30066.02} = 1.052 \simeq 2$$

Diagonal brace

The diagonal member of bracing will be in tension.

$$\tan\theta = \frac{4}{2.173} = 1.474,\ \theta = 55^\circ\ 51'$$

Force in diagonal bracing = 31634.8 sec 55° 51′ = 56353.71 N

Try ISA 125 × 75 × 6 mm. Connect this angle with its long leg to the column by 16 mm ϕ rivets. The relevant properties of the section are:

$$A = 1161 \text{ mm}^2,\ r_{min} = 16.2 \text{ mm}$$

$$\text{Area of connected leg, } A_1 = \left(125 - 17.5 - \frac{6}{2}\right) \times 6 = 627 \text{ mm}^2$$

$$\text{Area of outstanding leg, } A_2 = \left(75 - \frac{6}{2}\right) \times 6 = 432 \text{ mm}^2$$

$$k = \frac{3A_1}{3A_1 + A_2} = \frac{3 \times 627}{3 \times 627 + 432} = 0.813$$

$$A_{net} = A_1 + kA_2 = 627 + 0.813 \times 432 = 978.315$$

Load carrying capacity = 1.33 × 150 × 978.315 = 195173.87 N
> 56353.71 N

which is all right.

Number of rivets required = 56353.71/30066.02 = 1.874 ≃ 2

Base plate

$$\text{Load on each column} = \frac{2647427.02}{8} = 330928.37 \text{ N}$$

Self-weight of column = 373 × 12 = 4476 N

Load on base plate = 335404.37 N

(Assuming bearing pressure of concrete 4 N/mm^2)

$$\text{Area of base plate} = \frac{335404.37}{4} = 83851.09 \text{ N}$$

Provide square base plate.

$$L = B = \sqrt{83851.09} = 289.57 \text{ mm} \simeq 290 \text{ mm}$$

Depth of I.S.H.B. 200 = 200 mm

Flange width of I.S.H.B. 200 = 200 mm

$$\text{Projections of base plate, } a = b = \frac{(290 - 200)}{2} = 45 \text{ mm}$$

$$w = \frac{335404.37}{290 \times 290} = 3.988 \text{ N/mm}^2$$

$$t = \sqrt{\frac{3 \times 3.988}{185}\left(45^2 - \frac{45^2}{4}\right)} = 9.91 \text{ mm}$$

Provide 10 mm thick base plate.

Example 12.3 Design a rectangular tank of capacity 1,10,000 litres of water supported over a 12 m high staging. Columns are supported over concrete pedestal of M–15 concrete. The bearing capacity of soil is 100 kN/m^2. Design wind pressure may be assumed to be 1.05 kN/m^2. Plates of 1.25 width and 8.75 m length are available.

Solution

Tank

$$\text{Capacity of tank} = 1{,}10{,}000 \text{ litres} = 110 \text{ m}^3$$

Let us assume the height of the tank 2.4 m and free board 15 cm.

$$\text{Plan area of the tank} = \frac{110}{(2.4 - 0.15)} = 48.89 \text{ m}^2$$

The plates in the bottom of the tank will be placed with their width (1.25 m) parallel to the longer direction of the tank. These plates will be bent up (curves of 30 cm radius consuming $\frac{\pi \times 2 \times 30}{4} \simeq 47.1 \text{ cm}$ length of plate on each end) and made vertical and butt jointed to the side plates.

Hence, the width of tank = 8.75 – 2 × (0.85 + 0.471) + 2 × 0.3 = 6.708 $\simeq$ 6.7 m

Note Side of the water tank will be formed by one 1.25 m high plate and balance 1.15 m height will be provided by the length of the bottom plate. In calculation of the width 0.477 m has been subtracted instead of 0.3 m because 0.177 m extra plate is being used in forming the curve at each end.

$$\text{Length of the tank} = 48.89/6.7 = 7.29 \text{ m}$$

Provide 6 plates giving a length = 6 × 1.25 = 7.5 m. In addition to these provide one corner plate at each end. Corner end plates will also be bent at the same radius (30 cm) giving a height of side walls of 0.779 m (1.25 – 0.471 = 0.779 m) and then 0.071 m and 1.25 m wide plates will be butt jointed to it to give a height of 2.4 m.

Length of the tank available = 6 × 1.25 + 2 × 0.3 = 8.1 m. So provide a tank 8.1 m × 6.7 m × 2.4 m.

The joints of the bottom plates of the tank will be provided with Tee-sections from the inside and a cover plate from the outside (Fig. Ex. 12.3(a)). The Tee-sections will be placed at 1.25 m c/c with their webs upward forming vertical stiffeners, resting on longitudinal beam provided at 1.22 m c/c. These beams will rest over three cross beams supported over six columns.

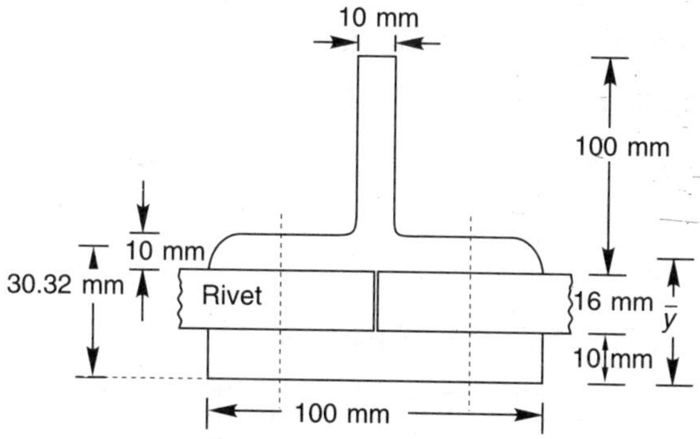

Fig. Ex. 12.3(a)

Design of the bottom plates

Span of bottom plate = c/c distance between Tee covers = 1.25 m

Consider unit width of plate.

Load of water/m^2 = wh = 9810 × 2.4 = 23544 N/m^2

Assume self-weight of plate = 1200 N/m^2

Total weight = 23544 + 1200 = 24744 N/m^2

Total weight = 24744 × 1 = 24744 N/m

$$\text{Maximum bending moment, } M = \frac{wl^2}{8} = \frac{24744 \times 1.25^2}{8} = 4832.8 \text{ Nm}$$

$$\text{Moment of resistance} = \frac{1}{6} bt^2 \sigma_{bct}$$

$$= \frac{1}{6} \times 1000 \times t^2 \times (0.8 \times 0.66 \times 250)$$

Hence, $\frac{1}{6} \times 1000 \times t^2 \times 132 = 4832.8 \times 10^3$

$\Rightarrow$ $t = 14.82 \simeq 16$ mm

Provide 16 mm thick plates for the tank bottom Thickness of the side plates will also be kept 16 mm.

Tee-Covers

Let us try ISNT 100 @ 150 N/m and 100 × 10 mm cover plate for the Tee-cover. The Tee-covers act as continuous beams as well as stiffeners. They transfer load from the plates to the longitudinal beams.

$$\text{Load from one panel} = 24744 \times 1.25 \times 1.22 = 37734.6 \text{ N}$$

$$\text{Assume self-weight} = 255.4 \text{ N}$$

$$\text{Total load} = 37734.6 + 255.4 = 37990 \text{ N}$$

Maximum bending moment,

$$M = \frac{WL}{10} = \frac{37990 \times 1.22}{10} = 4634.78 \text{ Nm}$$

Moment of resistance of the section Refer to Fig. Ex. 12.3(a). The relevant properties of ISNT 100 from IS Handbook No. 1 are,

$B = 100$ mm, $T = 10$ mm, $t = 10$ mm

$C_{xx} = 27.9$ mm, $A = 1910 \text{ mm}^2$ $I_{xx} = 173.8 \times 10^4 \text{ mm}^4$

$$\bar{y} = \frac{A_1 y_1 + A_2 y_2 + \ldots}{A_1 + A_2 + \ldots}$$

$$= \frac{1910 \times (10 + 16 + 27.9) + 100 \times 16 \times \left(10 + \frac{16}{2}\right) + 100 \times 10 \times \left(\frac{10}{2}\right)}{1910 + 100 \times 16 + 100 \times 10}$$

$$= 30.32 \text{ mm}$$

$$I = [173.8 \times 10^4 + 1910 \times (10 + 16 + 27.9 - 30.32)^2]$$

$$+ \left[\frac{100 \times (16 + 10)^3}{12} + 100 \times (16 + 10) \times \left(30.32 - \frac{26}{2}\right)^2\right]$$

$$= 372.64 \times 10^4 \text{ mm}^4$$

$$z_{xx} = \frac{372.64 \times 10^4}{(100 + 16 + 10 - 30.32)} = 38.946 \times 10^3 \text{ mm}^3$$

$$\text{Moment of resistance} = 38.946 \times 10^3 \times (0.8 \times 0.66 \times 250)$$

$$= 5140872 \text{ Nmm} = 5140.9 \text{ Nm}$$

$$> 4634.78 \text{ Nm}$$

Stays

Stays will be provided in both the directions at mid-height of the tank. The stays will be placed at 1.25 mc/c in the transverse direction and 1.22 mc/c in the longitudinal direction. Water pressure on the side walls for 1.25 m width $= \frac{9810 \times 2.4^2}{2} \times 1.25 = 35316$ N

Overhang of the tank from beam = 0.3 m.

Weight of water in the overhang portion (Fig. Ex. 12.3(b)) from a length of 1.25 m,

$$W_1 = 0.3 \times 1.25 \times 2.4 \times 9810 = 8829 \text{ N}$$

Weight of plate for a length of 1.25 m,

$$W_2 = 0.016 \times 1.25 \times 2.4 \times 79000 = 3792 \text{ N}$$

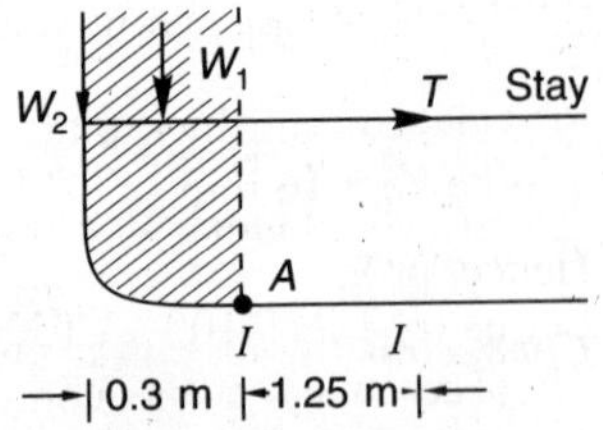

Fig. Ex. 12.3(b)

Let us assume the side plates to be hinged at the junction of the curved part and bottom plate. Taking moment about this hinge A,

$$T \times 1.20 = 35316 \times \frac{2.4}{3} + 8829 \times \frac{0.3}{2} + 3792 \times 0.3$$

$$\Rightarrow \qquad T = 25595.62 \text{ N}$$

$$A_{net} = \frac{25595.62}{0.8 \times 0.6 \times 250} = 213.30 \text{ mm}^2$$

Let us provide 50 ISF 10 mm section and connect it with stiffeners using 16 mm, power driven rivets.

$$\text{Net area provided} = (50 - 17.5) \times 10 = 325 \text{ mm}^2$$

$$> 213.29 \text{ mm}^2$$

which is all right.

The longitudinal stays are provided with the same cross section as that for transverse stays and will be placed above the transverse stays. The longitudinal stays will be subjected to bending stress due to their self-weight. The section provided is quite large and needs no check.

Check on the Tee-section as stiffener

Vertical Tee-stiffener is subjected to bending moment at the section where the stay is connected to it

$$B.M = \frac{9810 \times 1.20^3}{6} \times 1.25 = 3531.6 \text{ Nm} < 5140.9 \text{ Nm}$$

Hence, it is safe.

Connections

Connection of Tee with side plates

Shear per mm length

$$= \frac{FA\bar{y}}{I} = \frac{37990 \times \{(10+16) \times 100\} \times \left\{30.32 - \left(\frac{10+16}{2}\right)\right\}}{372.64 \times 10^4}$$

$$= 459.09 \text{ N/mm}$$

$$\text{Strength of the rivet in double shear} = 2 \times \frac{\pi}{4} \times 17.5^2 \times 0.8 \times 100 = 38484.51 \text{ N}$$

$$\text{Strength of rivet in bearing} = 17.5 \times 16 \times 0.8 \times 300 = 67200 \text{ N}$$

$$\text{Strength of rivet} = 38484.51 \text{ N}$$

Since 2 rivets are provided in one row,

$$\text{Pitch of rivets} = \frac{2 \times 38484.51}{459.09} = 167.65 \text{ mm}$$

$$> 100 \text{ mm} \qquad (10\,t)$$

Hence, provide rivets at a pitch of 100 mm.

Connection of stay with Tee-section

$$\text{Tension in stay} = 24571.8 \text{ N}$$

$$\text{Strength of rivet in single shear} = \frac{\pi}{4} \times 17.5^2 \times 0.8 \times 100 = 19242.25 \text{ N}$$

$$\text{No. of rivets} = \frac{24571.8}{19242.25} = 1.27 \simeq 2$$

Provide 2, 16 mm ϕ rivets to connect the stay with the Tee-section.

Beams

Refer to Fig. Ex. 12.3(c). The span of upper tier beams will be 7.5 m and that of the lower tier beams will be 6.1 m

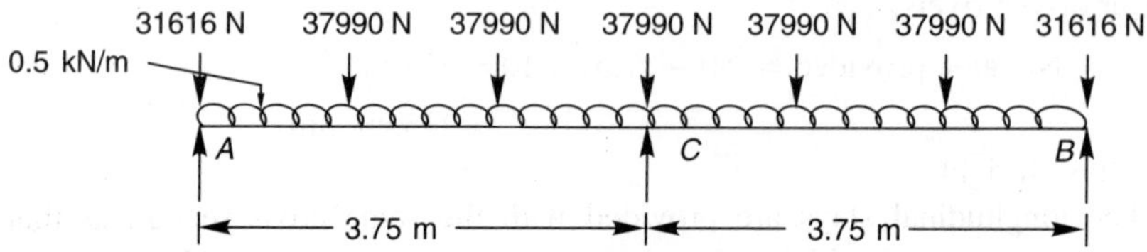

Fig. Ex. 12.3(c)

Upper tier beams

Intermediate beam

$$\text{Load from Tee-cover} = 37990 \text{ N}$$

$$\text{Load from end Tee-cover} = \frac{37990}{2} + 3792 + 8829 = 31616 \text{ N}$$

Assume self-weight of beam 0.5 kN/m. Considering each span *AC* and *CB* as a separate fixed beam, the fixed end moments may be calculated as follows

$$M_{F_{AC}} = -\frac{37990}{3.75^2} \times [1.25 \times 2.5^2 + 2.5 \times 1.25^2] - \frac{500 \times 3.75^2}{12} = -32244.27 \text{ Nm}$$

$$\simeq -32245 \text{ Nm}$$

$$M_{F_{CA}} = +\frac{37990}{3.75^2} \times [1.25^2 \times 2.5 + 2.5^2 \times 1.25] + \frac{500 \times 3.75^2}{12} = +32245 \text{ Nm}$$

Applying moment distribution method,

Joint	A		C			B
D.F.			0.5	0.5		
Member	AC		CA	CB		BC
FEM	−32245		+32245	−32245		+32245
	+32245	→	+16122.5	−16122.5	←	−32245
	0		+48367.5	−48367.5		0

Maximum bending moment = 48367.5 Nm

End reactions

$$= \frac{31616 \times 3.75 + 37990 \times 2.5 + 37990 \times 1.25 + 500 \times \frac{3.75^2}{2} - 48367.5}{3.75}$$

$= 57645.5$ N

$R_C = 2 \times 31616 + 5 \times 37990 + 500\ (2 \times 3.75) - 2 \times 57645.5 = 141641$ N

End beam

The loading on the end beam is shown in Fig. Ex. 12.3(d).

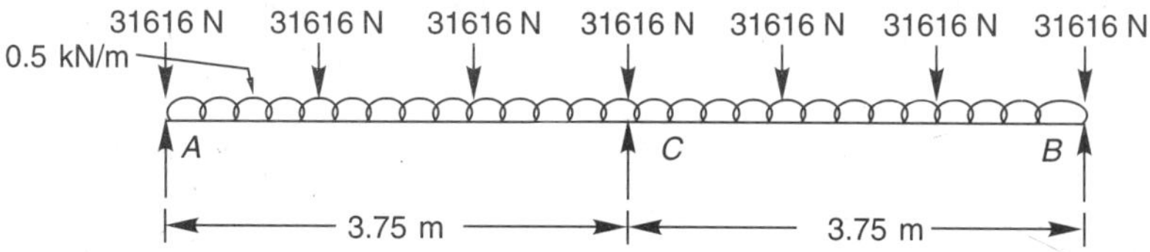

Fig. Ex. 12.3(d)

$$M_{F_{AC}} = -\frac{31616}{3.75^2} \times [1.25 \times 2.5^2 + 2.5 \times 1.25^2] - \frac{500 \times 3.75^2}{12}$$

$$= -25760.7 \text{ Nm}$$

$$M_{F_{CA}} = +25760.7 \text{ Nm}$$

Applying moment distribution method

$$\text{Maximum bending moment} = 25760.7 + \frac{25760.7}{2} = 38641.05 \text{ Nm}$$

End reaction, $R_A = R_B$

$$= \frac{31616 \times (3.75 + 2.5 + 1.25) + 500 \times \dfrac{3.75^2}{2} - 38641.05}{3.75} = 53865.22 \text{ N}$$

$R_C = 31616 \times 7 + 500 \times (2 \times 3.75) - 2 \times 53865.22 = 117331.56$ N

The design bending moment and shear force for the upper tier beam are 48367.5 Nm and 141641 N respectively.

$$Z = \frac{48367.5 \times 10^3}{120} = 403062.5 \text{ mm}^3$$

Let us try I.S.L.B. 325 @ 422.8 N/m. The relevant properties of the section from IS Handbook No. 1 are:

$Z_{xx} = 607.7 \times 10^3$ mm^3, $D = 325$ mm, $T = 9.8$ mm, $t = 7.0$, $r_y = 30.5$ mm

Since the beam is supported at the mid-point, the span, $l = 3.75$ m

The beam will be laterally unsupported for 3.75 m span.

$$\frac{D}{T} = \frac{325}{9.8} = 33.16$$

$$\frac{l}{r_y} = \frac{3750}{30.5} = 122.95$$

For $\dfrac{D}{T} = 33.16$, $\dfrac{l}{r_y} = 122.95$ and $f_y = 250$ N/mm^2,

$$\frac{T}{t} = \frac{9.8}{7.0} = 1.4 < 2, \quad \frac{d_1}{t} = 43.6,$$

$$\sigma_{bc} = 106.1 \text{ N/mm}^2$$

Moment carrying capacity of the beam

$$= 106.1 \times 607.7 \times 10^3 = 64476.96 \text{ Nm} > 48367.5 \text{ Nm}$$

which is all right

$$\text{Shear stress, } \tau_{va} = \frac{141641}{325 \times 7.0} = 62.26 \text{ N/mm}^2 < 100 \text{ N/mm}^2\ (0.4 \times f_y)$$

which is all right

Lower tier beams

The end reactions of the longitudinal beams spaced at 1.22 m is shown in Fig. Ex. 12.3(e). Assume self-weight of beam = 1 kN/m. The critical beam will be the intermediate beam,

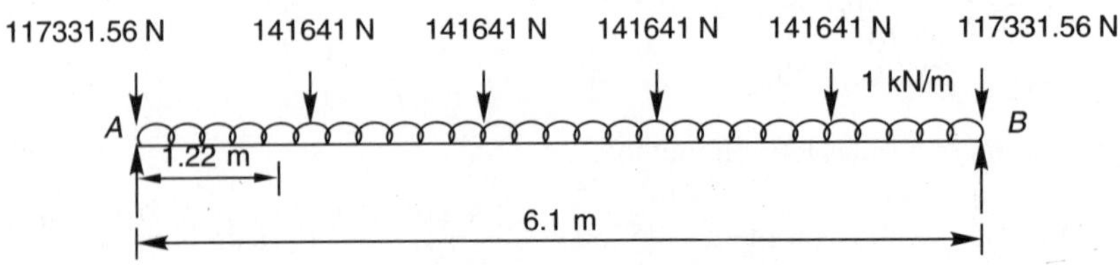

Fig. Ex. 12.3(e)

$$R_A = R_B = \frac{2 \times 117331.56 + 4 \times 141641 + 1000 \times 6.1}{2} = 403663.56 \text{ N}$$

$$\text{Maximum bending moment} = 403663.56 \times \frac{6.1}{2} - 117331.56 \times \frac{6.1}{2} - 141641 \times \left(\frac{6.1}{2} - 1.22\right) - 141641 \times \left(\frac{6.1}{2} - 2 \times 1.22\right) - 1000 \times \left(\frac{6.1}{2}\right)^2 \times \frac{1}{2}$$

$$= 523057.30 \text{ Nm}$$

The design bending moment and shear force are 523057.31 Nm and 403663.56 N respectively.

$$Z = \frac{523057.3 \times 10^3}{100} = 5230.57 \times 10^3 \text{ mm}^3$$

From IS Handbook No. 1 it is found that no rolled I-section provides this much section modulus. Let us try plate girder section from Table XXI of IS Handbook No. 1. The section selected is shown in Fig. Ex. 12.3(f). The relevant properties of the section are:

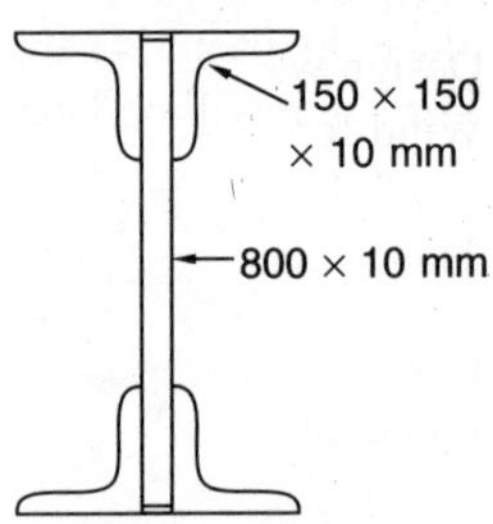

Fig. Ex. 12.3(f)

$z_{xx} = 4878.7 \text{ mm}^3$, $r_{yy} = 50$ mm

$T = 10$ mm, $t = 10$ mm, $T/t = 1$ (< 2)

Since $T/t < 2$, f_{cb} will be increased by 20%.

$$\frac{D}{T} = \frac{800}{10} = 80$$

$$\frac{l}{r_y} = \frac{6100}{50} = 122$$

$k_1 = 1$ (Table 7.3)

$k_2 = 0.5$ (Table 7.4)

$C_2 = C_1$

$Y = 26.5 \times 10^5/(122)^2 = 178.04 \text{ N/mm}^2$

$$X = 178.04\sqrt{1 + \frac{1}{20} \times \frac{122}{80}} = 184.70$$

$$f_{cb} = 1.2 \times k_1 (X + k_2 Y)\frac{C_2}{C_1}$$

$$= 1.2 \times 1 \times (184.70 + 0.5 \times 178.04) \times 1 = 328.46$$

$$\sigma_{bc} = \frac{0.66 \times 328.46 \times 250}{(328.46^{1.4} + 250^{1.4})^{1/1.4}} = 113.79 \text{ N/mm}^2$$

Moment carrying capacity $= 113.79 \times 4878.7 \times 10^3$

$= 555147.27 \times 10^3$ Nmm

$> 523057.31 \times 10^3$ Nmm

which is all right.

Shear stress $\tau_{Va} = \dfrac{403663.56}{800 \times 10} = 50.45 \text{ N/mm}^2$

$< 100 \text{ N/mm}^2$ $(0.4 f_y)$

which is all right.

Design of the supporting tower

Design wind pressure $= 1.05 \text{ kN/m}^2$

Wind force on tank, $P_1 = 8.1 \times (2.4 + 0.325) \times 1.05 \times 10^3 = 23176.125$ N

This acts the at a height of $12 + (0.8 + 0.325) + \dfrac{2.4}{2} = 14.325$ m from the ground level.

Force on Tower, $P_2 = 0.2 \times (12 + 0.8) \times 8.1 \times 1.05 \times 10^3 = 21772.8$ N

This acts at height of $\dfrac{12 + 0.8}{2} = 6.4$ m from the ground.

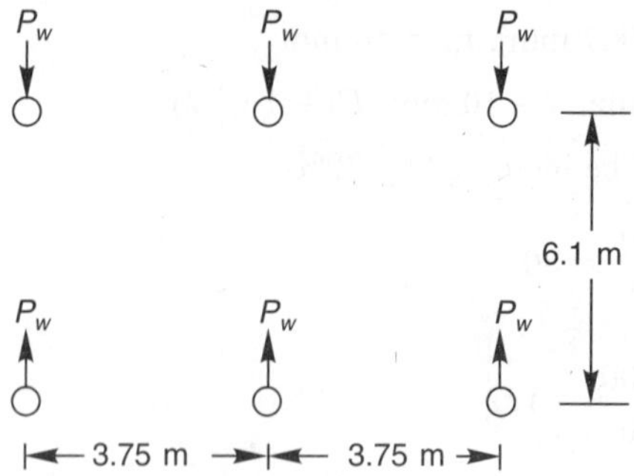

Fig. Ex. 12.3(g)

Bending moment due to wind,

$$M_w = 23176.125 \times 14.325 + 21772.8 \times 6.4 = 471343.91 \text{ Nm}$$

$$M_R = 3P_w \times 6.1$$

Since $M_R = M_w$

$$3P_w \times 6.1 = 471343.91$$

$\Rightarrow$ $P_w = 25756.49$ N

Force in each column due to the wind = 25756.50 N

Load from tank on each column = 403663.56 N

Since, wind load is less than $\dfrac{403663.56}{3} = 134554.52$ N, the columns will be designed for only gravity loads from tank and stresses will not be increased by $33\frac{1}{3}$%.

Load on column from the tank = 403663.56 N

Assuming self-weight (@ 400 N/m) = 400 × 12 = 4800 N

Total load = 403663.56 + 4800 = 408463.56 N

$$\text{Area} = \frac{408463.56}{85} = 4805.45 \text{ mm}^2$$

Let us try ISHB 200 @ 365.9 N/m. The relevant properties from IS Handbook No. 1 are:

$$A = 4754 \text{ mm}^2,\ r_y = 45.1 \text{ mm},\ D = 200 \text{ mm},\ B = 200 \text{ mm}$$

Let us divide the column height into 3 panels

$$\lambda = \frac{4.0 \times 10^3}{45.1} = 88.69$$

For $\lambda = 88.69$ and $f_y = 250$ N/mm²,

$$\sigma_{ac} = 91.44 \text{ N/mm}^2$$

Load carrying capacity = 91.44 × 4754 = 434710.5 N > 408463.56 N which is all right.

Bracing

The column height is divided into three panels by the horizontal bracing. The maximum shear will be in the lowest horizontal brace.

In the long direction three columns have been provided on one side. Assuming that the middle column shares half the wind load and that the two outer columns the remaining half,

Design load for the critical horizontal brace

$$= \frac{23176.125}{2} + \left(\frac{21772.8}{2}\right) \times \frac{5}{6} + \frac{2.5}{100} \times (408463.56 + 25756.49)$$

$$= 31515.56 \text{ N}$$

Horizontal brace

Let us provide 16 mm ϕ rivets, more than one in numbers, to connect the horizontal brace with the column.

$$A = \frac{31515.56}{1.33 \times 60} = 394.93 \text{ mm}^2$$

Try ISA 110 × 110 × 8 mm section. The relevant properties of the section from IS Handbook No. 1 are:

$$A = 1702 \text{ mm}^2,\ r_{\min} = 21.4 \text{ mm}$$

Note The length of the horizontal brace is quite large and therefore a section having large radius of gyration value should be selected.

$$\lambda = \frac{0.85 \times 6.1 \times 1000}{21.4} = 242.29 < 250$$

For $\lambda = 242.29$ and $f_y = 250 \text{ N/mm}^2$,

$$\sigma_{ac} = 19.542 \text{ N/mm}^2$$

Load carrying capacity of the strut

$$= 1.33 \times 19.542 \times 1702 = 44236 \text{ N} > 31515.56 \text{ N}$$

which is all right.

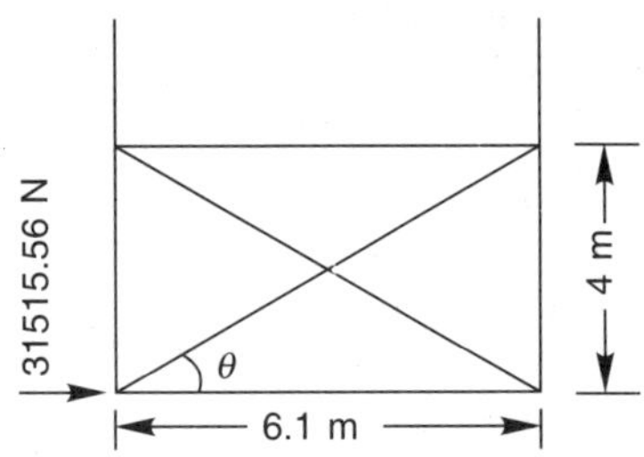

Fig. Ex. 12.3(h)

Diagonal brace

$$\text{Length, } L = \sqrt{6.1^2 + 4^2} = 7.29 \text{ m}$$

$$\sec\theta = \frac{7290}{6100} = 1.195 \qquad \text{(See Fig. Ex. 12.3(h))}$$

Force in the diagonal = 31515.56 × 1.195 = 37661.09 N

The diagonal braces will be connected at the point of intersection also by one 16 mm ϕ rivet to control the slenderness ratio.

$$A_{\text{net}} = \frac{37661.09}{1.33 \times 150} = 188.77 \text{ mm}^2$$

Try ISA 65 × 65 × 6 mm section. The relevant properties of the section from IS Handbook No. 1 are

$$A = 744 \text{ mm}^2, \; r_{min} = 12.6 \text{ mm}$$

$$\lambda = \frac{l}{r} = \frac{\frac{1}{2} \times 7290}{12.6} = 289.28 < 350$$

which is all right.

Let us provide 16 mm ϕ rivets for connecting the diagonal with the column.

$$A_1 = \left(65 - 17.5 - \frac{6}{2}\right) \times 6 = 267 \text{ mm}^2$$

$$A_2 = (65 - 17.5) \times 6 = 285 \text{ mm}^2$$

$$k = \frac{3 \times 267}{3 \times 267 + 285} = 0.7375$$

$$A_{net} = 267 + 0.7375 \times 285 = 477.2 \text{ mm}^2$$

Load carrying capacity = $1.33 \times 150 \times 477.2 = 95201.4 \text{ N} > 37661.09$ N

which is all right.

Connections

Strength of the rivet in single shear (16 mm ϕ)

$$= \frac{\pi}{4} \times 17.5^2 \times (1.25 \times 100) = 30066 \text{ N}$$

Strength of the rivet in bearing (6 mm) = $17.5 \times 6 \times (1.25 \times 300) = 39375$ N

Strength of the rivet in bearing (8 mm) = $17.5 \times 8 \times (1.25 \times 300) = 52800$ N

Horizontal brace

Rivet value = 30066 N

$$\text{Number of rivets} = \frac{31515.56}{30066} = 1.04 \simeq 2$$

Provide 2 rivets 16 mm ϕ for the connection.

Diagonal brace

Rivet value = 30066 N

$$\text{Number of rivets} = \frac{37661.09}{30066} = 1.252 \simeq 2$$

Provide 2 rivets 16 mm ϕ for the connection,

Base plate

Total load on the central column = 408463.56 N

(Assuming bearing pressure from concrete = 4 N/mm^2)

$$\text{Area of the base plate} = \frac{408463.56}{4} = 102115.89 \text{ mm}^2$$

$$\text{Provide a square base, } L = B = \sqrt{102115.89} = 319.55 \text{ mm}$$

$$\simeq 320 \text{ mm}$$

For I.S.H.B. 200 @ 365.9 N/m, $D = 200$ mm, $B = 200$ mm.

$$a = b = \frac{320 - 200}{2} = 60 \text{ mm} \qquad \text{(Refer to Fig. Ex. 12.3(i))}$$

$$w = \frac{408463.56}{320 \times 320} = 3.988 \text{ N/mm}^2$$

$$t = \sqrt{\frac{3 \times 3.988}{185}\left(60^2 - \frac{60^2}{4}\right)} = 13.24 \text{ mm} \simeq 16 \text{ mm}$$

Provide a 16 mm thick base plate.

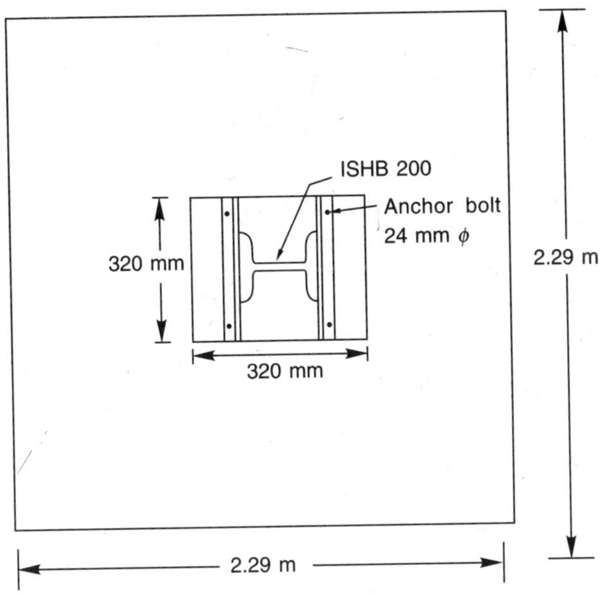

Fig. Ex. 12.3(i)

Anchor bolts

The critical condition in the middle windward column, for the design of the bolts, will be:

Tank is empty and wind is blowing.

Total load of water = 8.1 × 6.7 × 2.4 × 9810 = 1277732.88 N

Assuming central columns to share twice the load as that by the end columns. If the end column share load is W

Then, $\qquad 4W + 2W_1 = 1277732.88$

or $\qquad 4W + 2 \times (2W) = 1277732.88$

$\Rightarrow \qquad W = 159716.6$ N

Maximum load on the column when tank is full = 408463.56 N

Load on column when tank is empty = 408463.56 – 159716.6 = 248746.96 N

Load on the windward column when wind is acting

$\qquad = 248746.96 - 25756.49 = 222990.47$ N

Since the load on the windward column is thrust even in the empty tank condition the anchor bolts do not require any design. Provide 4,24 mm ϕ bolts on the four corners of the base plate.

Concrete pedestal

$$\text{Assume self-weight of pedestal} = 90{,}000 \text{ N}$$

$$\text{Load on central column when tank is full} = 408463.56 \text{ N}$$

$$\text{Thrust on central column due to wind} = 25756.49 \text{ N}$$

$$\text{Total load} = 408463.56 + 90{,}000 + 25756.49 = 524220.05 \text{ N}$$

$$\text{Area of base of pedestal} = \frac{524.22}{100} = 5.242 \text{ m}^2$$

Let us provide a square base.

$$L = B = \sqrt{5.242} = 2.289 \text{ m} \simeq 2.29 \text{ m}$$

Provide a pedestal 2.29 × 2.29 × 1 m (Fig. Ex. 12.3(i))

Example 12.4 Design a pressed steel tank for the following data:

Capacity = 85000 litres
Height of tank container = 2.5 m
Bottom of tank bearers above ground level = 9.0 m
Size of pressed plates = 1.25 m × 1.25 m
Horizontal acceleration = 6%
Design wind pressure = 930 N/m^2

Also, design the suppporting tower

Solution

Size

$$\text{Capacity} = 85000 \text{ litres} = 85 \text{ m}^3$$

Assume a free board of 15 cm

$$\text{Plan area of tank} = \frac{85}{2.5 - 0.15} = 36.17 \text{ m}^2$$

$$L = B = \sqrt{36.17} = 6.014 \text{ m}$$

Since the size of plate is 1.25 m × 1.25 m, provide length and width equal to 6.25 m.

Stays

The plan and elevation of the tank showing the stay arrangement is shown in Figs Ex. 12.4(a) and (b) (i) respectively.

The water pressure diagram on the side wall is shown in Fig. Ex. 12.4(b) (ii).

Pressure on the top 1.25 m height

$$p_1 = \frac{1}{2} \times wh_1 \times h_1 \times \text{length of the plate}$$

$$= \frac{1}{2} \times 9810 \times 1.25 \times 1.25 \times 1.25 = 9580 \text{ N}$$

Pressure on the bottom 1.25 height,

$$p_2 = \frac{1}{2} (wh_1 + wh_2) \times h_1 \times \text{length of the plate}$$

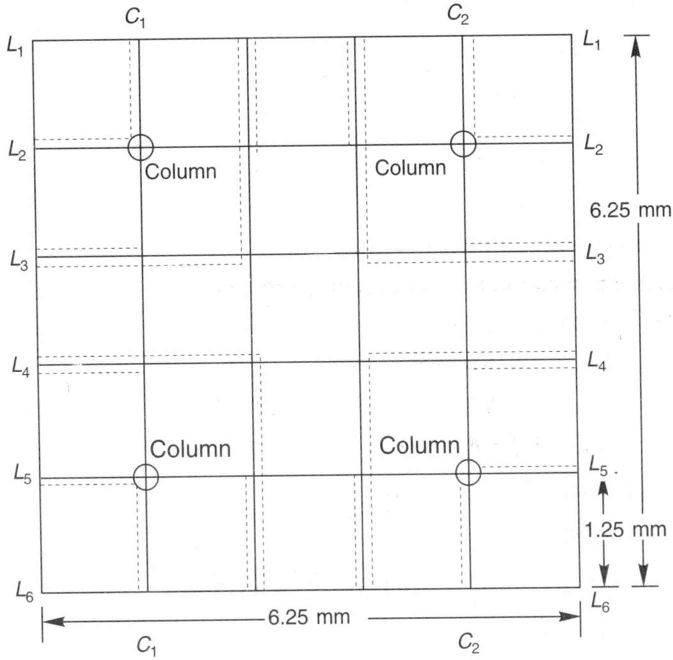

(a)

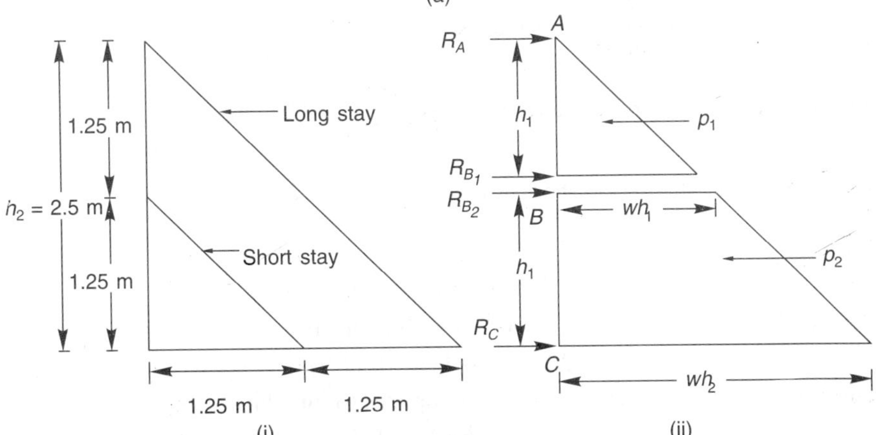

(b)

Fig. Ex. 12.4

$$= \frac{1}{2} \times 9810 \times (1.25 + 2.5) \times 1.25 \times 1.25 = 28740.23 \text{ N}$$

$$p_1 \text{ will act at } \frac{h_1}{3} = \frac{1.25}{3} = 0.4167 \text{ m from } B$$

$$p_2 \text{ will act at } \left[\frac{2 \times 1.25 \times 9810 + 2.5 \times 9810}{1.25 \times 9810 + 2.5 \times 9810}\right] \times \frac{1.25}{3} = 0.56 \text{ m from } C$$

Consider top 1.25 m height (AB)

Taking moment about joint B and equating it to zero.

$$R_A \times 1.25 - 9580 \times \frac{1.25}{3} = 0$$

$$\Rightarrow \qquad R_A = 3193.33 \text{ N} = 3194 \text{ N}$$

Hence, $\quad R_{B1} = 9580 - 3193.33 = 6386.6 \text{ N} \simeq 6387 \text{ N}$

Consider the bottom 1.25 m height (BC).

Taking moment about joint C and equating it to zero.

$$R_{B2} \times 1.25 - 28740.23 \times 0.56 = 0$$

$$\Rightarrow \qquad R_{B2} = 12875.623 \text{ N}$$

Hence, $\quad R_C = 28740.23 - 12875.623 = 15864.71 \text{ N} \simeq 15865 \text{ N}$

and $\quad R_B = R_{B1} + R_{B2} = 6387 + 12875.623 = 19262.623 \text{ N} \simeq 19263 \text{ N}$

Stays are inclined at 45°.

Tension in long stay,

$$T_1 = \frac{R_A}{\sin 45°} = 3194.00\sqrt{2} = 4517 \text{ N}$$

Tension in short stay,

$$T_2 = \frac{R_B}{\sin 45°} = 19263\sqrt{2} = 27242 \text{ N}$$

The required net area of the long stay $= \dfrac{4517}{0.8 \times 150} = 37.64 \text{ mm}^2$

The required net area of the short stay $= \dfrac{27242}{0.8 \times 150} = 227.0 \text{ mm}^2$

Weight of the tank

Weight of water = 6.25 × 6.25 × 2.5 × 9810 = 958007 N

Self weight of top tier plates (5 mm thick), 20 numbers

= 20 × 0.844 kN = 16880 N

Self-weight of lower tier plates and bottom plates (6 mm thick), 45 numbers

= 45 × 1.099 kN = 49455 N

Note weight of 5 mm plate= 0.844 kN, 6 mm plate = 1.099 kN, and 8 mm plate = 1.354 kN

Total weight of the plates = 16880 + 49455 = 66335 N

Neglecting the weight of the stays, weight of tank when full

= 958007 + 66335 = 1024342 N

Beams

Refer to Fig. Ex. 12.4(a). Since six longitudinal beams and two cross beams have been provided,

Load on intermediate beam $= \dfrac{1024342}{5} = 204868.4$ N

Uniformly distributed load on the intermediate beam

$$= \frac{204868.4}{6.25} = 32778.9 \text{ N/m} \simeq 33000 \text{ N/m}$$

Load on the end beam $L_1 L_1$ and $L_6 L_6 = \dfrac{204868.4}{2} = 102434.2$ N

Uniformly distributed load on the end beam

$$= \frac{102434.2}{6.25} = 16389.47 \text{ N/m} \simeq 17000 \text{ N/m}$$

Assume self-weight of each beam to be 1 kN/m.

Uniformly distributed load on intermediate beam = 33000 + 1000 = 34000 N/m

Uniformly distributed load on end beam = 17000 + 1000 = 18000 N/m

Bending moment and shear force

Longitudinal beams $L_1 L_1$ and $L_6 L_6$

Loads from stays and from the weight of water and tank are shown in Fig. Ex. 12.4(d).

$$R_A = R_B = \frac{2 \times 3194 + 4 \times 19263 + 18000 \times 6.25}{2} = 97970 \text{ N}$$

Cantilever bending moment (Span AE) $= \dfrac{18000 \times 1.25^2}{2} = 14062.5$ Nm

Simply supported bending moment (span AB)

$$= 97970 \times \frac{3.75}{2} - 19263 \times \frac{3.75}{2} - (19263 + 3194) \times \left(\frac{3.75}{2} - 1.25\right)$$

$$- 18000 \times \frac{3.75}{2} \times \frac{3.75}{4} = 101899.375 \text{ Nm}$$

Longitudinal beams $L_2 L_2$ and $L_5 L_5$

Load from stays and from the weight of the water and tank are shown in Fig. Ex. 12.4(e).

$$R_A = R_B = \frac{2 \times 19263 - 6 \times 19263 + 34000 \times 6.25}{2} = 67724 \text{ N}$$

Cantilever bending moment (span AE) $= \dfrac{34000 \times 1.25^2}{2} + 19263 \times 1.25$

$= 50641.25$ Nm

Simply supported bending moment (span AB)

$$= 67724 \times \frac{3.75}{2} + 2 \times 19263 \times \frac{3.75}{2} + 19263 \times \left(\frac{3.75}{2} - 1.25\right)$$

$$- 34000 \times \frac{3.75}{2} \times \frac{3.75}{4} = 151492.5 \text{ Nm}$$

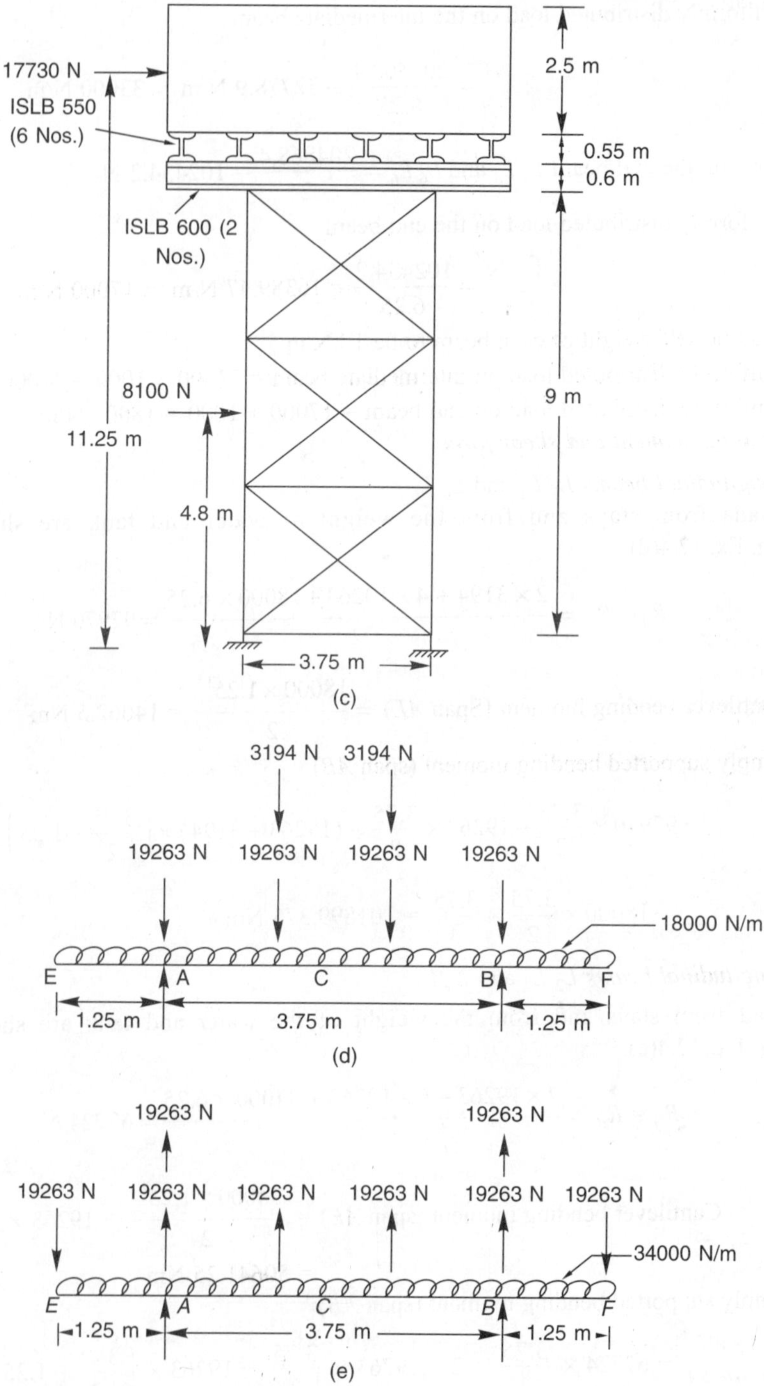

17730 N
ISLB 550
(6 Nos.)
2.5 m
0.55 m
0.6 m
ISLB 600 (2
Nos.)
8100 N
11.25 m
9 m
4.8 m
3.75 m
(c)
3194 N
3194 N
19263 N
19263 N
19263 N
19263 N
18000 N/m
E
A
C
B
F
1.25 m
3.75 m
1.25 m
(d)
19263 N
19263 N
19263 N
19263 N
19263 N
19263 N
19263 N
19263 N
34000 N/m
E
A
C
B
F
1.25 m
3.75 m
1.25 m
(e)

Fig. Ex. 12.4(c)-(e)

Longitudinal beams $L_3\,L_3$ and $L_4\,L_4$,

Load from stays and from weight of the water and tank are shown in Fig. Ex. 12.4(f).

$$R_A = R_B$$

$$= \frac{2 \times (19263 + 3194) - 2 \times 19263 - 4 \times 3194 + 34000 \times 6.25}{2}$$

$$= 103056 \text{ N}$$

Cantilever bending moment (span AE)

$$= \frac{34000 \times 1.25^2}{2} + (19263 + 3194) \times 1.25 = 54633.75 \text{ Nm}$$

Simply supported bending moment (span AB)

$$= (19263 + 103056) \times \frac{3.75}{2} + 2 \times 3194 \times \left(\frac{3.75}{2} - 1.25\right)$$

$$- 34000 \times \frac{3.75}{2} \times \frac{3.75}{4} = 173575 \text{ Nm}$$

Hence, the design bending moment and shear force for longitudinal beams are 173575 Nm and 103056 N respectively.

The longitudinal beams will be laterally unsupported. Hence, a reduced value of bending stress is assumed for calculating the required section modulus.

$$Z = \frac{173575 \times 10^3}{120} = 1446458 \text{ mm}^3$$

From IS Handbook No.1, let us try I.S.L.B. 550 @ 846.6 N/m. The relevant properties of the section are:

$$Z_{xx} = 1933.2 \times 10^3 \text{ mm}^3,\ D = 550 \text{ mm},\ T = 15 \text{ mm},\ t = 9.9 \text{ mm},\ r_y = 34.8 \text{ mm}$$

$$\frac{D}{T} = \frac{550}{15} = 36.666$$

$$\frac{l}{r_y} = \frac{3.75 \times 10^3}{34.8} = 107.758$$ (The effective length will be either equal to simple span or 3 times cantilever span whichever is more)

For $\frac{D}{T} = 36.666$, $\frac{l}{r_y} = 107.758$ and $f_y = 250$ N/mm^2,

$$\sigma_{bc} = 113.90 \text{ N/mm}^2$$

Moment carrying capacity $= 113.90 \times 1933.2 \times 10^3/10^3 = 220191.48$ Nm > 173575 Nm

which is all right.

Check for shear

$$\tau_{va} = \frac{103056}{550 \times 9.9} = 18.92 \text{ N/mm}^2 < 100 \text{ N/mm}^2\ (0.4 f_y)$$

which is all right.

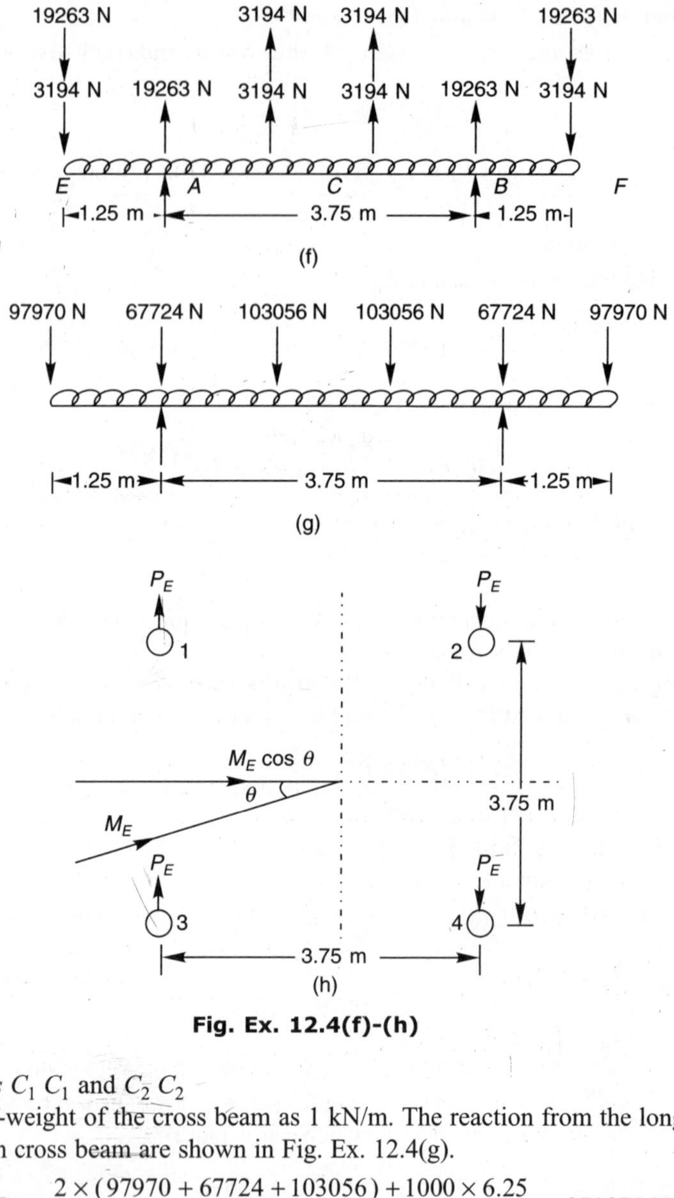

Fig. Ex. 12.4(f)-(h)

Cross beams $C_1\,C_1$ and $C_2\,C_2$

Assume self-weight of the cross beam as 1 kN/m. The reaction from the longitudinal beams on cross beam are shown in Fig. Ex. 12.4(g).

$$R_A = R_B = \frac{2\times(97970 + 67724 + 103056) + 1000\times 6.25}{2} = 271875 \text{ N}$$

$$\text{Cantilever bending moment (span } AE) = \frac{1000\times 1.25^2}{2} + 97970\times 1.25$$

$$= 123243.75 \text{ Nm}$$

Simply supported bending moment (span AB)

$$= (271875 - 67724)\times\frac{3.75}{2} - 103056\times\left(\frac{3.75}{2} - 1.25\right) - 1000\times\frac{3.75}{2}\times\frac{3.75}{4}$$

$$= 316615.3125 \text{ Nm}$$

Hence, the design bending moment and shear force for cross-beams are 316615.3125 Nm and 271875 N respectively.

$$Z = \frac{316615.3125}{120} \times 10^3 = 2638.46 \times 10^3 \text{ mm}^3$$

From IS Handbook No. 1, let us try I.S.M.B. 600 @ 1202.7 N/m. The relevant properties of the section are;

$Z_{xx} = 3060.4 \times 10^3$ mm^3, $D = 600$ mm, $T = 20.8$ mm,

$t = 12$ mm, $r_y = 41.2$ mm

$$\frac{D}{T} = \frac{600}{20.8} = 28.846$$

$$\frac{l}{r_y} = \frac{3.75 \times 10^3}{41.2} = 91.0194$$

For $D/T = 28.846$, $l/r_y = 91.0194$ and $f_y = 250$ N/mm^2;

$\sigma_{bc} = 130.39$ N/mm^2

Moment carrying capacity $= 130.39 \times 3060.4 \times 10^3/10^3 = 399045.56$ Nm > 316615.3125 Nm

which is all right.

Check for shear

$$\Sigma_{va} = \frac{271875}{600 \times 12} = 37.76 \text{ N/mm}^2 < 100 \text{ N/mm}^2 \qquad (0.4\,f_y)$$

which is all right.

Design of the supporting tower

Let us provide 4 columns constituting a tower. The total column height is divided in 3 panels as shown in Fig. Ex. 12.4(c).

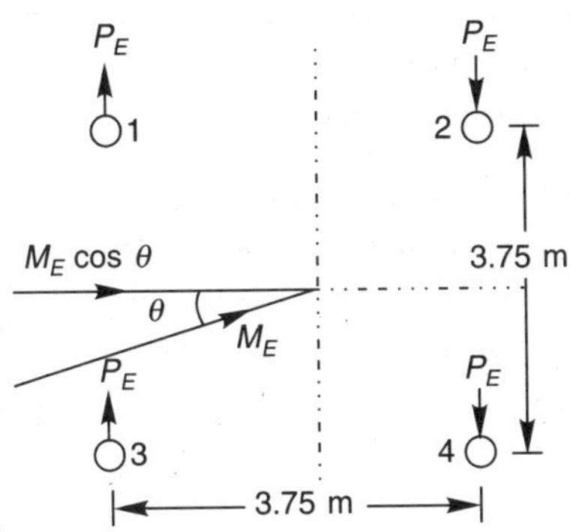

Fig. Ex. 12.4(i)

Earthquake forces

Horizontal acceleration = 6%

Horizontal earthquake force, $P_1 = 0.06 \times$ weight of full tank

$= 0.06 \times 1024342 = 61460.52$ N

Let horizontal force P_1 act at the C.G of the tank. The height of the C.G of the tank above the ground level = $9 + (0.6 + 0.55) + \frac{2.50}{2} = 11.40$ m. Let M_E be the overturning moment due to earthquake force P_1 P_E be the earthquake force in the columns.

$$P_E = \frac{M_E \cos\theta}{2 \times 3.75} + \frac{M_E \sin\theta}{2 \times 3.75}$$

For maximum value of θ, $\frac{\partial P_E}{\partial \theta} = 0$

$\Rightarrow$ $\theta = 45°$

$$P_E = \frac{61460.52 \times 11.4}{2 \times 3.75} (\cos 45° + \sin 45°) = 132115.82 \text{ N}$$

Wind load

Design wind pressure = 930 N/m^2

Wind force on tank, $P_1 = 6.25 \times (2.5 + 0.55) \times 930 = 17728.125$ N

≈ 17730 N

This acts from ground level at a height, $h_1 = 9 + (2.5 + 0.55)/2 + 0.6 = 11.125$ m.

Force on tower, $P_2 = 0.2 \times (9 + 0.6) \times 3.75 \times 930 = 6696 \text{ N} \approx 6700$ N

This will act at $h_2 = \frac{9 + 0.6}{2} = 4.8$ m

Bending moment due to wind,

$$M_w = 17730 \times 11.125 + 6700 \times 4.8 = 229406.25 \text{ Nm}$$

If P_w is the load in column because of wind as shown in Fig. Ex. 12.4(j), columns 1 and 3 will be subjected to uplift and columns 2 and 3 to thrust.

The moment of resistance,

$$M_R = 2P_w \times 3.75$$

Since $M_R = M_w$

$$2 \times P_w \times 3.75 = 229406.25$$

$\Rightarrow$ $P_w = 30587.5 \text{ N} < 132115.82$ N (Earthquake force)

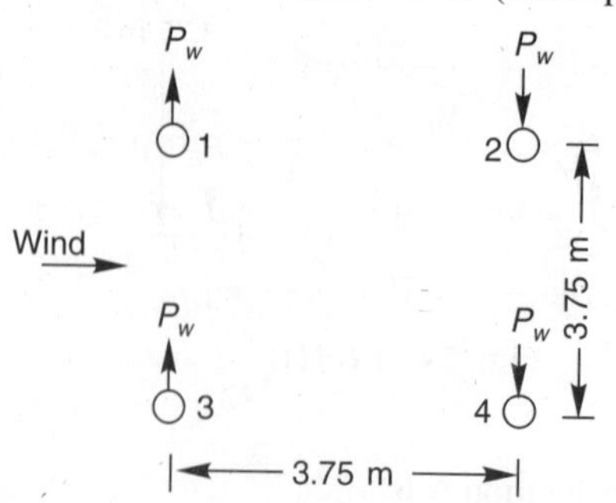

Fig. Ex. 12.4(j)

Hence the design force for the supporting tower will be the earthquake force.

Column Reaction from cross beam in the column = 271875 N

Design load for column = gravity load in column + earthquake load in column

$$= 271875 + 132115.82 = 403990.82 \simeq 403991 \text{ N}$$

$$\text{Area required by the column} = \frac{403991}{1.33 \times 80} = 3796.91 \text{ mm}^2$$

Let us try ISA 150 × 150 × 15 mm section for the column. The relevant properties from IS Handbook No. 1 are:

$$A = 4278 \text{ mm}^2, r_{\min} = 29.2 \text{ mm}$$

$$\lambda = \frac{l}{r} = \frac{3 \times 10^3}{29.2} = 102.74$$

For $\lambda = 102.75, f_y = 250 \text{ N/mm}^2,$

$$\sigma_{ac} = 77.80 \text{ N/mm}^2$$

Permissible average compressive stress = 1.33 × 77.80 N/mm^2.

Load carrying capacity = 1.33 × 77.80 × 4278 = 442661.77 N > 403400 N which is all right.

Bracing

Bracing will be designed for the earthquake force and 2.5% of axial loads in columns.

Earthquake force = 61460.52 N

This force will be divided in two parallel planes.

Hence, shear because of earthquake force in each plane

$$= \frac{61460.52}{2} = 30730.26 \text{ N}$$

2.5% of 2 column loads = 0.025 × 2 × 271875 = 13593.75 N

Horizontal brace

Design load = 30730.26 + 13593.75 = 44324.01 N

The horizontal brace will be in compression.

$$A = \frac{44324.01}{1.33 \times 60} = 555.43 \text{ mm}^2$$

Let us try ISA 90 × 90 × 8 mm. From IS Handbook No.1 the relevant properties of the section are:

$$A = 1379 \text{ mm}^2, r_{\min} = 17.5$$

Let us provide more than one rivet for connecting the brace.

$$\lambda = 0.85 \times \frac{3750}{17.5} = 182.142$$

For $\lambda = 182.142$ and $f_y = 250 \text{ N/mm}^2,$

$$\sigma_{ac} = 32.3574 \text{ N/mm}^2$$

Permissible average compressive stress = 1.33×32.3574 N/mm^2

Load carrying capacity = $1.33 \times 32.3574 \times 1379 = 59345.74$ N
> 44324.01 N

which is all right.

Diagonal brace

$$\text{Length of the diagonal} = \sqrt{3.75^2 + 3^2} = 4.802 \text{ m}$$

$$\text{Design force} = 44324.01 \times \frac{4.802}{3.75} = 56758.37 \text{ N}$$

The diagonal brace in tension is assumed to be effective.

$$\text{Net area required} = \frac{56758.37}{1.33 \times 150} = 284.5 \text{ mm}^2$$

Try ISA $65 \times 45 \times 6$ mm section with the long leg connected. From IS Handbook No. 1, the relevant properties of the section are,

$$A = 625 \text{ mm}^2,\ r_{min} = 9.5 \text{ mm}$$

Let us use 16 mm ϕ rivets for connection.

$$A_1 = \left(65 - 17.5 - \frac{6}{2}\right) \times 6 = 267 \text{ mm}^2$$

$$A_2 = \left(45 - \frac{6}{2}\right) \times 6 = 252 \text{ mm}^2$$

$$k = \frac{3A_1}{3A_1 + A_2} = \frac{3 \times 267}{3 \times 267 + 252} = 0.76$$

Net area provided = $A_1 + k A_2 = 267 + 0.76 \times 252 = 458.52$ mm^2
> 284.5 mm^2

which is all right.

Connections

16 mm ϕ power driven rivets are used. Increase permissible stresses by 25%.

$$\text{Strength of the rivet in single shear} = \frac{\pi}{4} \times 17.5^2 \times (100 \times 1.25) = 30066 \text{ N}$$

Strength of the rivet in bearing (6 mm) = $17.5 \times 6 \times (300 \times 1.25) = 39375$ N

Strength of the rivet in bearing (8 mm) = $17.5 \times 8 \times (300 \times 1.25) = 52500$ N

$$\text{Number of rivets for horizontal brace} = \frac{44324.01}{30066} = 1.474 \approx 2$$

$$\text{Number of rivets for diagonal brace} = \frac{56758.37}{30066}\ 1.88 \approx 2$$

Exercises

Note In the following problems, the statements are not intended to be complete. Any additional data reauired are to be assumed by the students.

12.1 Design a circular tank and its staging for the following data:

Capacity	120000 litres of water
Height of columns	12 m

The tank is situated at Kydganj, Allahabad. Plates and rolled sections of all sizes are available. The soil has a bearing capacity of 90 kN/m^2. Design the tank, ring girder, columns, bracings, base plate and footing.

12.2 Design the following for an elevated rectangular water tank to store 1×10^5 litres water.

1. Bottom plates
2. Side plates
3. Tee-covers
4. Stays
5. Upper tier beams
6. Lower tier beams

The tank is to be provided in Banglore.

12.3 A design for an overhead rectangular welded steel water tank is required for a residential colony having 200 flats in double storey buildings. Considering the water requirement as per the norms design the tank. Plates are available in widths of 1.5 m and lengths of 6 m. The depth of water in the tank is 2.25 m and the overhang on all the sides is 30 cm.

The staging consists of 4 columns suitably placed and the height of staging up to the bottom of the transverse beams is 8 m. Design the following:

(a) Tank walls and the supporting system of stays

(b) Bottom plates of tank and both the tiers of supporting beams

(c) Columns and bracings

The tank is to be designed for a locality in Delhi.

12.4 An overhead square riveted steel tank of capacity 60,000 litres is to be provided for a colony at Madras. Steel plates of width 1.25 m and length up to 6 m are available in addition to rolled sections of all sizes. The tank is to be supported on a staging consisting of four columns spaced suitably. The height of the staging to the bottom of the lower tier beams is 10 m. A free board of 10 cm is required. Design the tank supporting beams, columns and braces.

12.5 An overhead pressed steel tank is to be designed for a capacity of 75000 litres at Kerala railway station. Pressed steel plates of size 1.25×1.25 m are available. The tank is supported on beams suitably placed which in turn are supported on a staging consisting of 4 columns. The height of the tank from the ground level to the bottom of the supporting beams is 12 m. Design the tank, the supporting beams and the staging.

Chapter 13

Steel Stacks

13.1 INTRODUCTION

Steel chimneys, also known as steel stacks, are required for emission of the flue gases to a height so that exit velocity and dispersion of gases are within specified limits. The construction of tall stacks has been on the increase in the last couple of decades, due primarily to the increasing demand for air-pollution control.

Before designing a stack it is necessary to decide on the proper height and the diameter. The height must be such as will give the draft required and the cross-sectional area must be large enough to permit the passage of burnt gases. A draft is defined as the difference in pressures (absolute gas pressure—ambient atmospheric pressure) available for producing a flow of gases. It depends on the height of the stack, the height of the stack above mean sea level, the nature of fuel, the temperature of gases, and the furnace used. The cross-sectional area depends on the kind and quality of fuel to be burned in the plant, the draft available for carrying the burnt gases up in the stack, and the frictional losses within the stack.

Steel stacks are classified as self-supporting steel stacks and guyed steel stacks based on how the lateral forces (wind force, seismic force) are transmitted to the foundation. When the lateral forces are transmitted to the foundation by cantilever action, the stack is known as a *self-supporting steel stack* [Fig. 13.1(a)]. However,

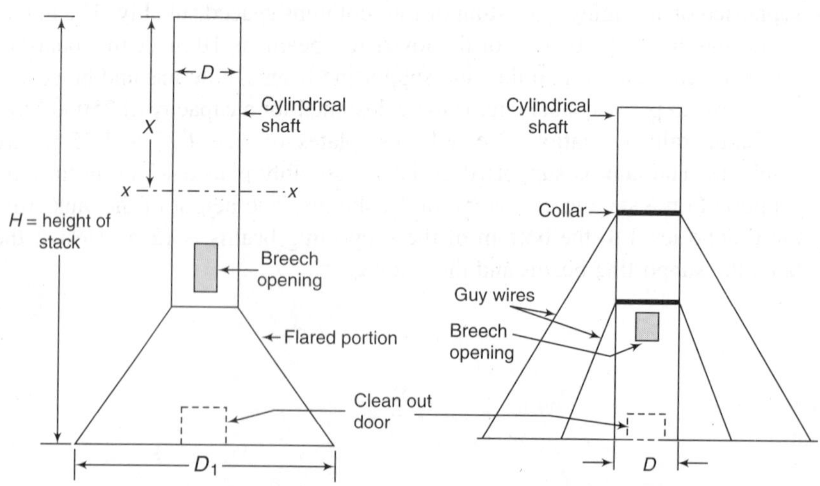

(a) Self-supporting steel stack (b) Guyed steel stack

Fig 13.1 *Types of Steel Stacks*

when the lateral loads are partly transmitted through the guy wires (one, two or three sets attached to the stack at different heights), the stack is known as a *guyed steel stack* [(Fig.13.1 (b))]. Guyed steel stacks are used for smaller heights, whereas self-supporting steel stacks may be used up to 275 m of height or even more.

Self-supporting steel stacks are usually cylindrical in shape with the lower portion widened or flared to provide a large base (bell shaped or conical) from stability consideration. A bell-mouthed base is costly, has no structural advantage and thus is seldom used. Guyed steel stacks are not provided with flare.

13.2 PROPORTIONING OF STACK

The inside diameter of the shaft is given by,

$$D = \sqrt{\frac{4Q}{\pi V}} \tag{13.1}$$

where Q = quantity of flue gas in m^3/s

V = velocity of flue gas in m/s

Note The diameter is so chosen that the velocity does not exceed 30 m/s.

As per the Central Board for Prevention and Control of Water Pollution, the following are the recommendations for the height of the stacks.

$$\left.\begin{aligned} H &= 74(Q_p)^{0.27} \quad \text{(for emission of particulate matters)} \\ H &= 14(Q_s)^{1/3} \quad \text{(for emission of } SO_2\text{)} \end{aligned}\right\} \tag{13.2}$$

where Q_p = particulate matter (tonnes/hour)

Q_s = SO_2 emission (kg/hour)

Note The height of the stack so obtained should not be less than 30 m for industries (except power plants), 220 m for thermal power plants (200 MW to 500 MW), and 275 m for thermal power plants of capacity more than 500 MW.

For self-supporting steel stacks, the height of flare from the ground level may generally be kept as 1/3 to 1/4 of total stack height and its bottom diameter not less than 1.25 times the shaft diameter.

13.3 CONSTRUCTIONAL DETAILS

The various elements of steel stacks are shown in Fig. 13.1. Steel stacks are made up of 0.9 m to 2.5 m wide steel plates. Since corrosion is likely to be more for the top two sections the plate thickness is generally kept not less than 6 mm. The thickness of plates in the flare should not be less than the thickness of plates used in the lowest portion of the cylindrical shaft. Also, the thickness of steel plates used for making the shaft should be not less than 6 mm in any case. The allowances to account for corrosion are given in Table 13.1.

Table 13.1 *Corrosion Allowance T_{ce} and T_{ci}*

Degree of Corrosion Expected	*Corrosion Allowance (mm)*			
	Copper Bearing Steel		*Non-copper Bearing Steel*	
	10 years	*20 years*	*10 years*	*20 years*
(a) External allowance T_{ce}				
(i) None (i.e. paint, insulation, cladding etc.)	Nil	Nil	Nil	Nil
(ii) Above average (i.e. unprotected)	1	2	1.5	3
(b) Internal allowance T_{ci}				
(i) None (i.e., non-corrosive flue gases)	Nil	Nil	Nil	Nil
(ii) Average (i.e., lined, insulated or natural gas fired)	1	2	1.5	3
(iii) Above average (i.e., unprotected coal fired	2	3	3	5

The steel plates forming the stacks are jointed with rivets. The recommended rivet diameters and types of joints are given in Table 13.2. The minimum pitch for rivets is 2.5ϕ. However, this spacing is increased somewhat (say about 50 per cent or so) for two or three rows of rivets. Some allowance in the rivet pitch is generally made to make the joint efficient and also to have an integral number of rivet spaces which is divisible by the number of plates. If one row of rivets at the minimum pitch is not enough, the size of the rivet should be increased or two rows of rivets used. Three rows are about the maximum number for the horizontal joints. The maximum pitch should not be more than 10 times the thickness of the plate, as it is difficult to secure a tight joint if this spacing is exceeded.

Table 13.2 *Type of Joint and Rivet Diameter Corresponding to Plate Thickness*

Thickness of Steel Plates (mm)	*Diameter of Rivet(mm)*	*Type of Riveted Joint*
6	16	Single riveted lap joint
8-10	18	Double riveted lap joint
12-16	22	Triple riveted lap joint
Thicker than 16	24	Butt joint

An opening known as *breech opening* or *flue opening* is provided in the cylindrical shaft of the stack, for entrance of flue gases coming from furnaces of boilers. The area of the breech opening is kept about 20 per cent larger than the cross-sectional area of the stack shaft. The width of the breech opening should not be more than two-thirds of the shaft diameter. Since an opening is created, it is reinforced vertically as well as horizontally. Both the reinforcements are kept the same. The vertical reinforcement provided should be 20 per cent larger than the area of steel removed in the ratio of diameter to the long chord perpendicular to the face of the opening. The reinforcing material should provide sufficient vertical stiffness. It should also extend a little above and below the opening so as to transfer and distribute the stress into the steel of the stack.

A clean out door of a minimum of 500 mm × 800 mm clear size is provided in the flare near the base for removal of ash. This also facilitates inspection of the inside of the stack.

Self-supporting steel stacks are usually lined to protect them from heat, corrosion, and to maintain temperatures of flue gases. The height of the lining should not be less than 10 times the diameter or one-third height above the breech opening. Generally, it is provided for the full height of the stack. The guyed steel stacks are however, unlined. The material used for lining should be capable of withstanding temperatures as high as 1500°C (such as fire bricks). For stacks up to 50 m of height and 100 mm of thick lining, the latter is supported over steel angles ISA 100 × 75 × 10 (or 12) mm riveted to the inside of the stack. For greater heights and lining thicknesses, ISA 150 × 75 × 10 (12) mm is used. A shorter leg of the angle is used for making connection with stack plates and is kept downwards. Besides supporting the lining and transfering the lining load to the steel stack, these angles stiffen the stack also.

A cap in the form of a ring made of a 300 mm wide and 25 mm thick cast-iron section is provided at the top of the stack to prevent collapse of the top of stack. In addition, it protects the lining also.

For maintaining the stack, a ladder is provided from the outside of the stack up to its full height.

13.4 CODAL PROVISIONS

Permissible Stresses To control buckling in steel stacks the compressive stress caused by the combination of extreme fibre stresses due to bending and direct load should not exceed the values given in Table 13.3 for steels conforming to IS: 226 and IS: 2062. For steels with yield stress higher than those of IS: 226 and IS: 2062 the maximum permissible stresses are obtained by multiplying the values in Table 13.3 by σ_y^* / σ_y where σ_y^* is the yield stress of steel used and σ_y is the yield stress of steel conforming to IS: 226 and IS: 2062. The values of maximum permissible stresses so obtained are further reduced for temperature by multiplying with a factor K_t given in Table 13.4.
The permissible stress in tension is taken as $0.6 f_y$.

Note The permissible stresses are increased by $33\frac{1}{3}$% when earthquake forces are considered. But no increase in permissible stress is allowed when wind is considered.

Table 13.3 *Maximum Permissible Compressive Stress, σ_c for Circular Stacks*

Ratio	*Maximum Permissible Compressive Stress in MPa, for D/t*													
h_e/D	≤140	150	160	170	180	190	200	225	250	300	350	400	450	500
upto20	126	124	123	120	118	115	112	105	99	87	78	70	64	58
30	108	107	105	103	101	99	96	90	85	75	67	60	55	50
40	89	88	86	85	83	81	79	74	70	62	55	50	45	41
50	72	71	70	69	68	66	64	60	57	50	45	40	37	34
60	59	58	57	56	55	54	52	49	46	41	36	33	30	27
70	48	48	47	46	45	44	43	40	38	33	30	27	24	22

(Contd.)

Table 13.3 (*Contd.*)

Ratio h_e/D	*Maximum Permissible Compressive Stress in Mpa, for D/t* ≤140	150	160	170	180	190	200	225	250	300	350	400	450	500
80	40	40	39	38	37	37	36	33	31	28	25	22	20	19
90	33	33	33	32	31	31	30	28	26	23	21	19	17	16
100	28	28	28	27	26	26	25	24	22	20	18	16	14	13
110	24	24	24	23	23	22	22	20	19	17	15	13	12	11
120	21	21	20	20	19	19	19	17	16	14	13	12	11	10
130	18	18	18	17	17	17	16	15	14	13	11	10	9	8
140	16	16	15	15	15	15	14	13	12	11	10	9	8	7
150	14	14	14	13	13	13	13	12	11	10	9	8	7	7

Note Intermediate values may be linearly interpolated.

Here, t = thickness of the plate at the level considered

D = mean diameter at the level considered

h_e = effective height from consideration of buckling

Table 13.4 K_t *values*

Temperature °C	*0-200*	*250*	*300*	*350*	*400*
K_t	1	0.75	0.67	0.6	0.5

Effective Height The effective height of steel stacks should be taken as that shown in Fig.13.2.

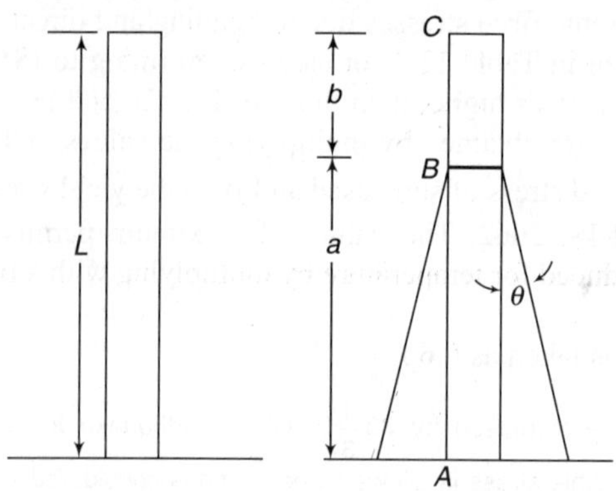

Self-supporting		Guyed with fixed base		Guyed with pinned base	
Stress		*BC*	*AB*	*BC*	*AB*
Axial	$h_e = 2L$	$h_e = 2b$	$h_e = 0.85a$	$h_e = 2b$	$h_e = a$
Bending	$h_e = L$	$h_e = b$	$h_e = 0.85a$	$h_e = b$	$h_e = a$

where h_e = effective height of stack.

Fig. 13.2 *Effective Height of Steel Stacks*

Deflection The maximum deflection of steel stacks is limited to 1/200 of the unsupported height. The thickness used in calculating the deflection should be the thickness provided minus the thickness allowed for corrosion.

13.5 FORCES ON A STACK

The forces acting on steel stacks are described as below.

Self Weight of steel stack W_s acts vertically downwards. Consider a horizontal section x–x as shown in Fig. 13.1. The thickness of steel plates of the stack above the section x–x may be assumed constant. The self-weight of the stack is given by

$$W_s = \rho\, \pi D\, t\, h$$

or

$$W_s = 79 \times (\pi Dth) \text{ kN} \quad (13.3)$$

where ρ = unit weight of steel (79 kN/m^3)
D = diameter of stack in meters
t = thickness of steel plates in meters
h = height of steel stack above the section x–x in meters

Weight of Lining in the steel stack W_L acts vertically downwards. The thickness of the brick lining may be assumed to be 100 mm. The weight of the lining

$$W_L = \rho_1 \pi D\,(0.1)h$$

or,

$$W_L = 2\pi Dh \text{ kN} \quad (13.4)$$

where ρ_1 = unit weight of brick lining ($\approx$ 20 kN/m^3)

Wind Pressure acting on a structure depends on the shape, the width, the height, the location of the structure, and the climatic conditions. The wind pressure per unit area increases with the height of the structure above the ground level. In order to simplify the design, the steel stack is divided into a number of segments of equal height. The height of each segment may be kept 10 m or so. Further, the intensity of wind pressure over each segment may be taken to be uniform and assumed to act at the mid-height of each segment. The intensity of wind pressure corresponding to the mid-height of each segment, which is at particular height, may be noted from IS: 875 (part 3) – 1987. The wind pressure on the flared portion may be found by using the average diameter of the flare. The wind force

$$P = k\, p_z \times \text{projected area of stack}$$
$$P = 0.7\, p_z\, Dh \text{ kN}$$

where k = shape factor. It accounts for the shape of the structure. The shape factor for cylindrical structures is 0.7.
p_d = design wind pressure
D = outer diameter of the stack
h = height of the segment of the stack over which the wind force is required.

Note In addition to the overturning effect due to wind pressure, the wind also has aerodynamic effect. However, the later has not been taken into consideration.

Seismic Forces The usual effect of ground motions is to cause the foundation to be moved in the horizontal direction. The damaging effect is proportional to the rate of acceleration—greater the effect the higher the acceleration. The stepwise procedure for calculating the earthquake forces is as follows.

1. The time period, T is computed by the following equation.

$$T = C_T \sqrt{\frac{W_T\, H}{E_S\, Ag}}$$

where C_T = a coefficient depending upon $k = \dfrac{H}{r_1}$ (Table 13.5)

H = height of stack above foundation pad
r_1 = radius of gyration at the base of stack
= 0.7 × radius of stack at the base
E_s = Young's modulus of elasticity of steel (2×10^5 N/mm^2)
A = area of cross-section of the stack at base ($\pi D_1 t$)
g = acceleration due to gravity
D_1 = diameter of stack at the base

Table 13.5 *Values of C_T and C_V*

Ratio $k = \frac{H}{r_1}$	5	10	15	20	25	30	35	40	45	50 & above
C_T	14.4	21.2	29.6	38.4	47.2	56	65	73.8	82.8	1.8k
C_V	1.02	1.12	1.19	1.25	1.30	1.35	1.39	1.43	1.47	1.50

2. The average spectral acceleration coefficient is determined from Appendix V, using the time period T, and damping as 5% of critical (for riveted steel stacks, 5% damping is permitted to account for frictional losses).
3. The horizontal seismic coefficient A_h is given by the following expression

$$A_h = \frac{Z}{2}\frac{I}{R}\frac{S_a}{g}$$

where all the above factors in the expression have already been defined in Section 1.11. However it should be noted that, IS: 1893 (Part 4)-2005 stipulates values of importance factors and response reduction factors as given in Tables 13.6 and 13.7, respectively.

Table 13.6 *Importance Factor Applicable to Stack-like Structures*

S. No.	Type of Structure	Importance Factor
1	Reinforced concrete ventilation stacks	1.5
2	Reinforced concrete chimneys	1.5
3	Reinforced brick masonry chimney for industry	1.5
4	Un-reinforced brick masonry chimney for industry	1
5	Reinforced concrete T.V. towers	1.5
6	Electric/traffic light poles	1
7	Steel stacks	1.5
8	Silos	1.5

Table 13.7 *Response Reduction Factor Applicable to Stack-like Structures*

S. No.	Type of Structure	Response Reduction Factor
1	Reinforced concrete T.V. towers	3.0
2	Reinforced concrete ventilation stacks	3.0
3	Reinforced concrete chimneys	3.0
4	Reinforced brick masonry	2.0
5	Steel chimney	2.0
6	Steel refinery vessels	2.0
7	Un-reinforced brick masonry chimney	1.0
8	Reinforced electric/traffic pole	2.0

4. The shear and moment at any section of the stack at a distance X from the top of the stack are calculated from Table 13.8 or by using expressions from Table 13.9.

$$V_X = C_v A_h W_T D_w$$
$$M_X = A_h W_T x D_m$$

where C_v = a coefficient (Table 13.5)
X = height of the centre of gravity of the stack above the base
D_v, D_m = distribution factors for shear and moment respectively at a distance X from the top as given in Table 13.8. The expres sions for these distributions for moment and shear along the height are given in Table 13.9.

Table 13.8 *Digitized Moment and Shear Distribution Factors D_m and D_v along the Height*

S. No.	X/H	Moment Distribution Factor (D_m)			Shear Distribution Factor (D_v)		
		Fixed Base	Raft on Soil	Pile Foundation	Fixed Base	Raft on Soil	Pile Foundation
1	0.00	0.00	0.00	0.00	0.00	0.00	0.00
2	0.05	0.09	0.13	0.11	0.28	0.21	0.14
3	0.10	0.13	0.19	0.16	0.42	0.27	0.19
4	0.20	0.18	0.27	0.22	0.64	0.34	0.26
5	0.30	0.22	0.33	0.28	0.83	0.38	0.31
6	0.40	0.27	0.39	0.33	1.00	0.41	0.35
7	0.50	0.32	0.45	0.38	1.00	0.44	0.40
8	0.60	0.39	0.52	0.45	1.00	0.49	0.46
9	0.70	0.48	0.60	0.54	1.00	0.55	0.54
10	0.80	0.60	0.70	0.65	1.00	0.65	0.65
11	0.90	0.77	0.83	0.80	1.00	0.80	0.80
12	0.95	0.88	0.91	0.89	1.00	0.89	0.89
13	1.00	1.00	1.00	1.00	1.00	1.00	1.00

Note X is the distance from top and H is the height of the stack above the base.

Table 13.9 *Expressions for Calculating D_m and D_v*

S. No.	Soil Foundation Condition	D_m	D_v
1	Fixed base or raft on hard soil (based on N values)	$0.4\left[\frac{x}{h}\right]^{1/2} + 0.6\left[\frac{x}{h}\right]^4$	$1.1\left[\frac{x}{h}\right]^{1/2} + 0.75\left[\frac{x}{h}\right] + 0.9\left[\frac{x}{h}\right]^4$
2	Raft on soil (based on N values)	$0.6\left[\frac{x}{h}\right]^{1/2} + 0.4\left[\frac{x}{h}\right]^4$	$1.1\left[\frac{x}{h}\right]^{1/2} - 0.75\left[\frac{x}{h}\right] + 0.65\left[\frac{x}{h}\right]^4$
3	Pile foundation	$0.5\left[\frac{x}{h}\right]^{1/2} + 0.5\left[\frac{x}{h}\right]^4$	$0.66\left[\frac{x}{h}\right]^{1/2} - 0.20\left[\frac{x}{h}\right] + 0.54\left[\frac{x}{h}\right]^4$

13.6 LOAD COMBINATIONS FOR SELF-SUPPORTING STEEL STACKS

The load combinations used in the design of a self-supporting steel stack are

(a) Dead load + Wind load
(b) Dead load + Earthquake load
(c) Dead load + Live load + Wind load + Temperature effect
(d) Dead load + Live load + Earthquake load + Temperature effect

13.7 STRESSES IN SELF-SUPPORTING STEEL STACKS

Stress Due to Self Weight

The compressive stress in the steel plates at section x–x due to the self-weight of the stack is given by

$$\sigma_s = \left(\frac{Ws}{\pi Dt}\right) = \left(\frac{79\,\pi Dth}{\pi Dt}\right)\text{kN/m}^2 \qquad (13.5)$$

or $$\sigma_s = 0.079 \text{ h N/mm}^2$$

where W_s = weight of steel plates above the section under consideration
D = diameter of stack

Stress Due to Weight of Lining

The compressive stress in the steel plates at the section x–x, due to the weight of lining is given by

$$\sigma_L = \left(\frac{W_L}{\pi Dt}\right) = \left(\frac{2\pi Dh}{\pi Dt}\right)\text{kN/m}^2$$

or $$\sigma_L = 0.002\left(\frac{h}{t}\right)\text{ N/mm}^2 \qquad (13.6)$$

where W_L = weight of lining above the section under consideration

Stress Due to Wind

$$M_W = P \times \frac{x}{2}$$

$$\sigma_m = \frac{M_W}{I} \times \frac{D}{2} = \frac{M_W}{Z} = \frac{4M_W}{\pi D^2 t} \qquad (13.7)$$

where M_W = moment due to wind
P = wind force
x = height of stack above the section under consideration
Z = section modulus ($\pi D^2 t/4$)

13.8 DESIGN PROCEDURE FOR SELF-SUPPORTING STEEL STACKS

In general, a self-supporting steel stack may be treated as a vertical cantilever beam subjected to loads caused by the horizontal wind pressure and the self-weight of the stack. The following are the steps involved in the design of a self-supporting steel stack.

1. The diameter and height of a steel stack are fixed in accordance with Eqs (13.1) and (13.2).

2. *Design of flare.* Once the height and diameter of the steel stack are known, the size of the flare is assumed in compliance with the following norms.

Height of flare, $H' = H/3$ to $H/4$

Diameter of flare at base $D_1 \nless 1.25D$

where H is height of the stack, D the diameter of the shaft and D_1, the diameter of the bottom of the flare.

3. *Design of thickness of steel plates forming the stack.* Since the wind pressure along the height of stack varies, increasing with height, the design for thickness of steel plates of stacks is simplified by dividing the cylindrical portion of the stack into segments of equal height.

The number of sections or courses including laps, on the stack, is chosen using a plate width of 0.9 to 2.5 m.

The thickness of the steel plates is designed for the various segments and sections of the stack shaft for windward and leeward side using the expressions given below. The thickness is also designed at the base of the flared portion and at the junction of the cylindrical and flared portion.

The windward side of the stack is subjected to tensile stress due to the combined effect of the wind and self-weight of the steel plates. The self-weight of the lining is not considered as tension will be maximum before lining.

$$\sigma_{t,\text{cal}} = \frac{4M_W}{\pi D^2 t} - \frac{W_S}{\pi D t} \text{ kN/m}^2 \qquad (13.8)$$

$$< \eta_1 \times 0.6 f_y$$

where η_1 = efficiency of the joint on tension side. (85%)

The leeward side of the stack is subjected to compressive stresses due to wind, and self-weights of stack and lining.

$$\sigma_{c,\text{cal}} = \frac{4M_W}{\pi D^2 t} + \frac{W_S}{\pi D t} + \frac{W_L}{\pi D t} \text{ kN/m}^2 \qquad (13.9)$$

$$< \eta_2 \sigma_c$$

where η_2 = efficiency of the joint on compression side (=1)

Note The thickness of steel plate that comes out to be maximum from the above two expressions is provided after making necessary correction to account for corrosion. The thickness so obtained should not be less than 6 mm.

4. *Design of the base plate.* The maximum compressive stress at the base plate is on the leeward side and is given by

$$\sigma_{c,\text{cal}} = \left[\frac{4M_W}{\pi D_1^2 t} + \frac{W_S}{\pi D_1 t} + \frac{W_L}{\pi D_1 t}\right] \text{kN/m}^2 \qquad (13.10)$$

The maximum compressive strength per unit circumferential length,

$$F_1 = \sigma_{c,\text{cal}} \times t \text{ kN/m}$$

The width of the base plate,

$$B = \frac{F_1}{\sigma_{p1}}$$

where, σ_{p1} = allowable bearing pressure on the foundation.

The base plate is kept symmetrical with respect to the stack as shown in Fig. 13.3.

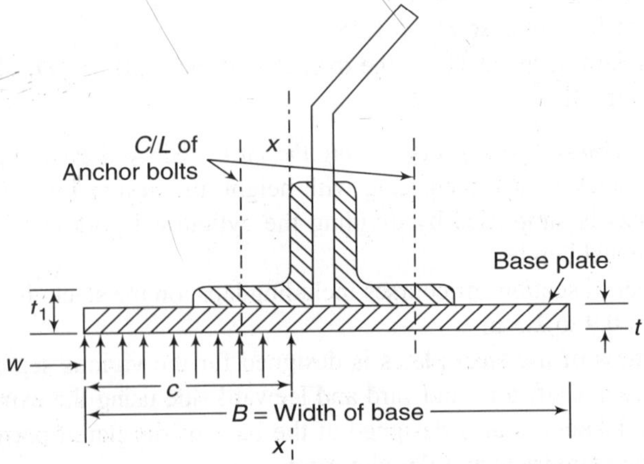

Fig. 13.3 *Details of Steel Stack Base*

Bending moment at the critical section x–x, $M = \dfrac{wc^2}{2}$

Moment of resistance at the critical section x–x,

$$M_R = \frac{1}{6} \times 1 \times t_1^2 \times \sigma_{bs} = \frac{t_1^2}{6}\sigma_{bs}$$

For equilibrium,

$$M = M_R$$

$$\frac{wc^2}{2} = \frac{t_1^2}{6}\sigma_{bs}$$

$$t_1 = c\sqrt{3w/\sigma_{bs}} \qquad (13.11)$$

Thickness of base plate, $t = t_1 - t_a$

where, t_1 = total thickness available at critical section

t_a = thickness of angle section

σ_{bs} = bending stress in slab base (185 N/mm^2)

5. *Design of the anchor bolts.* The self-supporting steel stack transmits the forces to the foundation by cantilever action. This causes uplift on the windward side and thrust on the leeward side of stack. Anchor bolts are provided to check the uplift. Anchor bolts are normally provided on the base plate from both the inside and outside of the stack as shown in Fig. 13.3.

Maximum tensile stress at the base plate on the windward side will be when the lining is absent.

$$\sigma_{t,\text{cal}} = \frac{4M_W}{\pi D_1^{\,2} t} - \frac{W_S}{\pi D_1 t} \text{ kN/m}^2 \tag{13.12}$$

Maximum force per unit circumferential length of the base plate on the windward side of stack

$$F_1 = \sigma_{t,\text{cal}} \times (t \times 1) = \sigma_{t,\text{cal}}\, t$$

$$F_1 = \left[\frac{4M_W}{\pi D_1^{\,2} t} - \frac{W_S}{\pi D_1 t}\right] t$$

Maximum uplift force to be resisted by anchor bolts

$$F_2 = F_1$$

Spacing S of anchor bolts $= \dfrac{R_t}{F_2}$

or

$$S = \frac{R_t}{\left[\dfrac{4M_W}{\pi D_1^2} - \dfrac{W_S}{\pi D_1}\right]} \tag{13.13}$$

where R_t = strength of anchor bolt.

6. *Design of plain concrete pedestal.* Steel stacks are subjected to large lateral forces. Usually, a massive plain concrete circular foundation in the shape of a frustum of a cone or an eight-sided pyramid, is provided. The depth of the foundation should not be less than 0.4 times the diameter of the foundation (D_2). Also, for no tension at the base, the maximum eccentricity (e) is limited to $D_2/8$. The diameter of the foundation is determined as follows.

$$e = \frac{D_2}{8} = \frac{M_W}{W_S + W_L + W_F}$$

or

$$M_W = \frac{D_2}{8}\,[W_S + W_L + W_F] \tag{13.14}$$

where W_F = weight of footing, and D_2 is the diameter of the base of pedestal.

Since the weight of lining and self-weight are negligible as compared to the weight of the footing, the former may be neglected.

Assuming unit weight of concrete to be 24 kN/m^3,

$$M_W = \frac{D_2}{8}\left[\frac{\pi}{4} D_2^2 \times 0.4\, D_2 \times 24\right]$$

$$D_2 \approx M_W^{1/4} \tag{13.15}$$

7. Check for the stability of the stack. The stability of the stack is checked by the following expression.

$$1.6(\sigma_w + \sigma_m) - 0.9\sigma_d < 1.8\sigma_a \qquad (13.16)$$

where σ_w = stress due to wind load
σ_m = stress by any other load causing an increase in the combined stress
σ_d = stress due to dead load
σ_a = permissible stress

13.9 GUYED STEEL STACKS

Guyed steel stacks are practically always unlined. These may be grouped into classes according to the number of sets of guys attached. Each set of guys is usually composed of generally three or four wires. Stacks built fairly close together in a continuous row are usually provided with a lattice bracing between them, and each of the end stacks has one or two sets of guys of three or four wires in each set, while each of the intermediate stacks has one or two sets of guys of two wires in each set. The classification is as follows:

1. Stacks with one set of guys (usually 3 or 4 and sometimes 6 wires) attached to a collar at one third or one quarter of the distance from the top.
2. Stacks with two sets of guys, usually 3 or 4 wires each, attached to collars [Fig.13.1(b)] at various heights.
3. Stacks with three sets of guys, usually 3 or 4 wires each, attached to collars at various heights.
4. Stacks in a continuous row with lattice bracing between them.

The forces for which the different components of guyed steel stacks are designed include self-weight of the steel stack (same as for self-supporting steel stack), vertical components of the pulls of the guy wires and bending and shear stresses due to wind. These are discussed in the sections to follow.

13.10 PULL ON GUY WIRES

The guy wires must be designed to take care of the entire wind reaction at a collar. The maximum pull in any guy wire will occur when the wind blows along that guy wire. This maximum pull P_g may be expressed by the equation

$$P_g = R_c/\sin\theta = R_c \operatorname{cosec}\theta \qquad (13.17)$$

where

R_c = the horizontal reaction at the collar (Fig.13.4) due to wind pressure, in kN

θ = angle that the guy makes with the vertical, in degrees.

The vertical component of the pull in the guy wire is $= P_g \cos\theta = \dfrac{R_c}{\sin\theta} \times \cos\theta = R_c \cot\theta$

The angle θ ranges between 30° and 75°, in most instances. The collar must be strong enough to withstand any tendency to buckle.

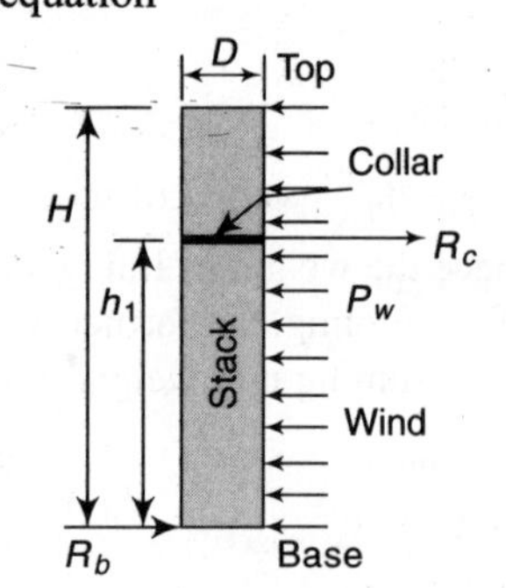

Fig. 13.4 *Horizontal Forces Acting on a Guyed Steel Stack with One Set of Guy*

Wind Moments and Reactions for Stacks with One Set of Guys

The simplest way is to consider the stack as a beam simply supported at the base and at the collar, where the guys are attached. If the base of the stack is securely bolted to a rigid foundation which is rare, the base of the stack could be considered as fixed and the moments, reactions, and shears computed accordingly. However, the base is rarely fixed though there are many instances where the base might be considered as partially fixed. The best and simplest method appears to be the one in which no allowance is made for even the partial fixity at the base. In the design of the base, the allowable unit-bearing and shearing stresses should not be exceeded.

The horizontal reaction R_c at the collar, as shown in Fig. 13.4, is found by taking moments about the base.

$$\Sigma M_{base} = R_c h_1 - \frac{P_W H}{2} = 0$$

$$R_c = \frac{P_W H}{2h_1}$$

The horizontal reaction at the base is

$$R_b = P_W - R_C = P_W - \frac{P_W H}{2h_1}$$

where R_c = horizontal reaction at the collar due to wind

P_w = total wind pressure on stack

R_b = horizontal reaction at base due to wind

H = total height of stack

h_1 = height of collar

The shear at the top of the collar $= -\dfrac{P_W}{H}(H - h_1)$

The shear just below the collar $= R_c - \dfrac{P_W}{H}(H - h_1) = \dfrac{P_W H}{2h_1} - P_W + \dfrac{P_W h_1}{H}$

The shear at the base, $R_b = P_W - \dfrac{P_W H}{2h_1}$

The bending moment at the collar $= -\dfrac{P_W}{H}(H - h_1)\dfrac{(H - h_1)}{2} = -\dfrac{P_W}{2H}(H - h_1)^2$

The maximum positive bending moment between collar and base $= \dfrac{P_W H}{2}\left(1 - \dfrac{H}{2h_1}\right)^2$

and occurs at a point of zero shear $= \left(H - \dfrac{H^2}{2h_1}\right)$ meters above the base.

Figures 13.5 and 13.6 show the variations of shear and bending moment from the top to the base of the guyed steel stack.

The values of the reactions and moments in a guyed steel stack with one collar are given in Table 10.1.

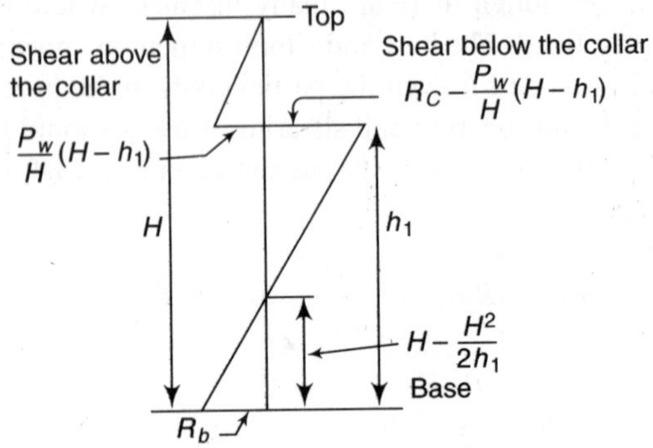

Fig 13.5 *Shear Diagram for a Guyed Steel Stack with One Set of Guys*

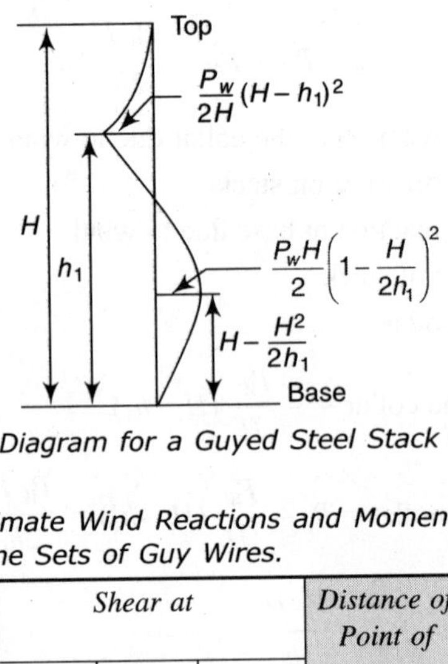

Fig.13.6 *Moment Diagram for a Guyed Steel Stack with One Set of Guys*

Table 13.10 *Approximate Wind Reactions and Moment in a Guyed Steel Stack With One Sets of Guy Wires.*

Height of Collar Above Base	*Reactions at*		*Shear at*			*Distance of Point of Zero Shear the Base*	*Moments to be used in Design*	
	R_c	R_b	*Top of Collar*	*Below the Collar*	*Base*		*At the Collar*	*At Point of Zero Shear*
$\frac{2H}{3}$	$\frac{3P_W}{4}$	$\frac{P_W}{4}$	$\frac{-P_W}{3}$	$\frac{5P_W}{12}$	$\frac{-P_W}{4}$	$\frac{H}{4}$	$\frac{P_W H}{18}$	$\frac{P_W H}{32}$
$\frac{3H}{4}$	$\frac{2P_W}{3}$	$\frac{P_W}{3}$	$\frac{-P_W}{4}$	$\frac{5P_W}{12}$	$\frac{-P_W}{3}$	$\frac{H}{3}$	$\frac{P_W H}{32}$	$\frac{P_W H}{18}$

Wind Moment and Reactions for Stacks With Two Sets of Guys

A guyed steel stack may be considered as a continuous beam in a vertical position with a number of simple rigid supports at the collars and base. The Three Moment Theorem can be applied to find the exact values of the reactions, shears, and moments caused by wind. However, since the supports, except the base, cannot be considered as rigid, the values determined this way will be erroneous. Further, the upper collar is rarely fastened at the top of the stack and the span between the upper and lower collars may be different than the span between the lower collar and the base. This makes the application of the Three Moment Theorem to be cumbersome and the computations to be tedious. To save analysis time and effort the shears and moments for stacks with two sets of guys may thus be determined by approximate method, as follows.

The method of determining the reactions, shears, and moments due to the wind considers that the upper collar carries the wind load from the top of the stack to a point at a distance $2x + x_1/2$, where x is the distance from the top of the stack to the collar and x_1 is half the distance between the top and bottom collars. The lower collar carries the wind load from this point to a point midway between the collar and the base, while the base carries the remainder of the load. This method gives reactions which agree quite closely with those found by the use of the Three Moment Theorem. Further, the sum of the moments of the reactions, found by this method, about the base also equals the moment of the wind pressure about the base. This approximate method thus may be used to determine the values of reactions, shears and moments.

Table 13.11 gives the values of the reactions and moments in a guyed steel stack with the collars placed at various heights by this method. From the table it will be

Table 13.11 *Approximate Wind Reactions and Moment for Guyed Steel Stacks with Two Sets of Guy Wires*

Height of Collar above Base		*Reactions at*			*Moments to be used in Design*			*Sum of Moment*
Upper	*Lower*	*Upper Collar*	*Lower Collar*	*Base*	*At Upper Collar*	*Between Upper and Lower Collars*	*Between Lower Collar and Base*	*of Reactions about Base should = $(0.5 P_w H)$*
$\frac{2H}{3}$	$\frac{H}{3}$	$\frac{2P_W}{3}$	$\frac{P_W}{6}$	$\frac{P_W}{6}$	$\frac{P_W H}{18}$	$\frac{P_W H}{18}$	$\frac{P_W H}{18}$	$\frac{P_W H}{2}$
$\frac{3H}{4}$	$\frac{H}{2}$	$\frac{P_W}{2}$	$\frac{P_W}{4}$	$\frac{P_W}{4}$	$\frac{P_W H}{32}$	$\frac{P_W H}{32}$	$\frac{P_W H}{32}$	$\frac{P_W H}{2}$
$\frac{8H}{10}$	$\frac{H}{2}$	$\frac{9P_W}{20}$	$\frac{3P_W}{10}$	$\frac{P_W}{4}$	$\frac{P_W H}{50}$	$\frac{P_W H}{50}$	$\frac{P_W H}{32}$	$\frac{51P_W H}{100}$
$\frac{85H}{100}$	$\frac{6H}{10}$	$\frac{7P_W}{20}$	$\frac{7P_W}{20}$	$\frac{3P_W}{10}$	$\frac{9P_W H}{800}$	$\frac{9P_W H}{800}$	$\frac{9P_W H}{200}$	$\frac{203P_W H}{400}$
$\frac{85H}{100}$	$\frac{5H}{10}$	$\frac{4P_W}{10}$	$\frac{7P_W}{20}$	$\frac{P_W}{4}$	$\frac{9P_W H}{800}$	$\frac{9P_W H}{800}$	$\frac{P_W H}{32}$	$\frac{103P_W H}{200}$

H = height of stack and P_w = total horizontal wind pressure.

noted that the reaction at the upper collar will be larger than that at the lower collar in most instances. The design of the upper collar and its guys should be based on the reaction value given in the table and that the sections of the lower collars and guys are kept same as that for the upper ones. This results in a stronger lower collar and guys than that required for tabulated reactions. However, this provision has an advantage. If the upper guys stretch a little, the upper collar reaction will be decreased and the lower collar reaction increased while the base reaction will remain about the same. Hence, it is advisable to make the lower collar and guys stronger than the ones required for tabulated reaction. Making both collars and sets of guys the same also simplifies the design and construction.

In designing for moment, it is practical and safe to design each of the middle and lower sections over their whole length for the maximum moment which occurs in that section. The upper section should be designed as a cantilever beam with the maximum moment at the upper collar.

Wind Moments and Reactions for Stacks with Three Sets of Guys

The discussion relating to the wind moments and reactions for a stack with two sets of guy applies to a stack with three sets of guys too. Table 13.12 presents the collar reactions in a stack with 3 sets of guys. The method used for finding the reactions is the same as that for stacks with two sets of guys.

Each collar and attached guys should be designed to carry the maximum tabulated collar reaction for safety and other practical reasons. The stack over the entire length of each section, except the top section, should be able to carry the tabulated bending moment for that section. The stack in the top section acts as a cantilever with the maximum moment at the upper collar.

13.11 DESIGN PROCEDURE FOR GUYED STEEL STACKS

Guyed steel stacks must be designed for the unit stresses caused by the self weight of the stack, horizontal wind pressure, and pull of the guy wires. Tension and compression will be caused by the moment due to the wind pressure, while the weight of the stack and the vertical components of the pulls of the guy wires will cause compression only. Hence, the unit compressive stress will be larger than the unit tensile stress. The wind pressure and the horizontal components of the pulls of the guys will cause horizontal shearing stress. The resulting unit shearing stresses will usually be quite small in the area on any cross section of a guyed stack. Provision must be made at the base of the stack so that the horizontal shear there will be transmitted to the supports.

In the design procedure outlined below it is assumed that a guyed steel stack is provided with guy wires at two levels along the height. Thus, the stack will be divided in three portions—the upper portion of the stack, i.e., from the top collar to the top of the stack, the portion between the top collar and the bottom collar, and the portion between the bottom collar and the base. No flare is required at the base of the stack.

A stepwise procedure is outlined for the design of a guyed steel stack with two sets of guys as follows:

Table 13.12 *Approximate Wind Reactions and Moments for Guyed Steel Stacks with Three Sets of Guy Wires*

Height of Collar Above Base			*Reactions at*				*Moments to be used in Design*				*Sum of Moment of Reactions about Base should = ($0.5\ P_w H$)*
Upper	*Middle*	*Lower*	*Upper Collar*	*Middle Collar*	*Lower Collar*	*Base*	*At Upper Collar*	*Between Upper and Middle Collars*	*Between Middle and Lower Collars*	*Between Lower Collar and Base*	
$\frac{3H}{4}$	$\frac{H}{2}$	$\frac{H}{4}$	$\frac{P_W}{2}$	$\frac{P_W}{8}$	$\frac{P_W}{4}$	$\frac{P_W}{8}$	$\frac{P_W H}{32}$	$\frac{P_W H}{32}$	$\frac{P_W H}{32}$	$\frac{P_W H}{32}$	$\frac{P_W H}{2}$
$\frac{8H}{10}$	$\frac{6H}{10}$	$\frac{3H}{10}$	$\frac{4P_W}{10}$	$\frac{3P_W}{20}$	$\frac{3P_W}{10}$	$\frac{3P_W}{20}$	$\frac{P_W H}{50}$	$\frac{P_W H}{50}$	$\frac{P_W H}{50}$	$\frac{P_W H}{50}$	$\frac{P_W H}{2}$
$\frac{85H}{100}$	$\frac{7H}{10}$	$\frac{4H}{10}$	$\frac{3P_W}{10}$	$\frac{3P_W}{20}$	$\frac{7P_W}{20}$	$\frac{P_W}{5}$	$\frac{9P_W H}{800}$	$\frac{9P_W H}{800}$	$\frac{9P_W H}{800}$	$\frac{P_W H}{50}$	$\frac{P_W H}{2}$

1. The number of collars and the number of guy wires to be attached with each collar are selected.
2. The angle (between 30° to 70°) with the vertical that the guy wires make is assumed and the horizontal wind reaction $R_{c,}$ at the upper collar is determined (Table 13.11). Maximum pull in the guy ($P_g = R_c$ cosec θ) is calculated, when the wind blows along it.
   Total pull in guy wire = P_g + pull due to initial tension in guy wire.
3. The total maximum vertical components of the pulls of the guys at the upper collar are calculated. The same force may be assumed to act at the lower collar. This assumption is safe.
4. The thickness of plates in the upper portion of the stack is designed using Eqs (13.18) and (13.19).

   Maximum tensile stress

$$\sigma_{t_1,\text{cal}} = \frac{4M_{C_1}}{\pi D^2 t} - \frac{W_{S_1}}{\pi D t} \qquad (13.18)$$

$$< \eta_1 \times 0.6 f_y$$

   where, M_{C_1} = moment at collar (Table 13.11)

   W_{S_1} = weight of stack above the top collar

   Maximum compressive stress

$$\sigma_{C_1,\text{cal}} = \frac{4M_{C_1}}{\pi D^2 t} + \frac{W_{S_1}}{\pi D t} \qquad (13.19)$$

$$< \eta_2\, \eta_c$$

   The thickness of steel plates that comes out to be maximum from the above two equations is provided after making necessary correction to account for corrosion and should not be less than 6 mm. The thickness so obtained will be provided for all the sections above it up to the top of the stack.
5. The section designed is checked in shear. The horizontal shears just above and below the upper collar are calculated, and the critical one is divided by the net cross-sectional area. This should be less than the permissible shear stress.
6. The thickness of the plates of the stack between the upper and lower collars is designed using Eqs (13.20) and (13.21). The maximum tensile stress occurs just below the upper collar.

$$\sigma_{t_2,\text{cal}} = \frac{4M_{C_2}}{\pi D^2 t} - \frac{W_{S_2}}{\pi D t} - \frac{P_g}{\pi D t} \qquad (13.20)$$

$$< \eta_1 \times 0.6 f_y$$

   where M_{C_2} = moment at bottom collar (Table 13.11)

   W_{S_2} = weight of stack above the bottom collar

The maximum compressive stress will occur on a section just above the lower collar and is

$$\sigma_{C_2,\text{cal}} = \frac{4M_{C_2}}{\pi D^2 t} + \frac{W_{S_2}}{\pi Dt} + \frac{P_g}{\pi Dt} \tag{13.21}$$

$$< \eta_2 \, \sigma_c$$

The thickness of steel plates that comes out to be maximum from the above two equations is provided after making necessary correction to account for corrosion and should not be less than 6 mm. The thickness so obtained will be provided for all the sections above it up to the top collar.

7. Maximum shear stress is calculated and checked in a way similar to as in the step 5.
8. The thickness of plates to be used in the lowest portion of the stack is designed using Eqs (13.22) and (13.23).
   The maximum tensile stress will occur just below the lower collar.

$$\sigma_{t_3,\text{cal}} = \frac{4M_{C_3}}{\pi D^2 t} - \frac{W_{S_2}}{\pi Dt} - \frac{P_g}{\pi Dt} \tag{13.22}$$

$$< \eta_1 \times 0.6 f_y$$

   The maximum compressive stress will occur just above the base.

$$\sigma_{C_3,\text{cal}} = \frac{4M_{C_3}}{\pi D^2 t} + \frac{W_{S_3}}{\pi Dt} + \frac{P_{g_1}}{\pi Dt} \tag{13.23}$$

$$< \eta_2 \times \sigma_c$$

   where

   M_{C_3} = maximum moment in this section (Table 13.11)
   W_{S_3} = weight of stack above the base
   P_{g_1} = total sum of vertical components of the pulls of the guys at the upper and lower collar.

9. The foundation for the stack is designed as a concrete pedestal.
   The design procedures of self-supporting and guyed steel stacks are supported with the Examples 13.1 and 13.2, respectively.

Solved Examples

Example.13.1 Design a welded self-supported steel stack located in the outskirts of Bhopal for the following data.

Terrain category	2
Topography	almost flat
Height of steel stack	80 m
Diameter of steel stack	3 m
Thickness of Brick lining	100 mm
Corrosion allowance	3 mm

Solution

Height of stack, $H = 80$ m

Diameter of stack, $d = 3$ m

Design of flare

Height of flare = $H/4$ = 20 m

Diameter of flare, $d_1 = (4/3) \times 3 = 4$ m

Wind load

Basic wind speed, V_b = 39 m/sec (Bhopal)

Assume height of each segment to be 10 m as shown in Fig. Ex. 13.1.

Segment 0–1

For height = 80 m, diameter = 3 m;

Class of structure = C

Risk factor, $k_1 = 1$

Topography factor, $k_3 = 1$

$$k_2 = 1.1 + \frac{0.07}{50} \times 30 = 1.142$$

Design velocity, $V_z = k_1 k_2 k_3 V_b = 1 \times 1.142 \times 1 \times 39 = 44.538$ m/s

Design wind pressure,

$$p_z = 0.6 V_z^2 = 0.6 \times (44.538)^2 = 1190 \text{ N/mm}^2 = 1.19 \text{ kN/m}^2$$

Wind force in this segment,

$$P_1 = kp_z \times \text{projected area} = 0.7 \times 1.19 \times (10 \times 3) = 24.99 \text{ kN.}$$

Segment 1–2

For height = 70 m, diameter = 3 m;

Class of structure = C

$$k_1 = k_3 = 1$$

$$k_2 = 1.1 + \frac{0.07}{50} \times 20 = 1.128$$

$$V_z = k_1\, k_2\, k_3\, V_b = 1.128 \times 39 = 43.992 \text{ m/sec}$$

$$p_z = 0.6 \times (43.992)^2 = 1161 \text{ N/mm}^2 = 1.161 \text{ kN/m}^2$$

$$P_2 = 0.7 \times 1.161 \times (10 \times 3) = 24.381 \text{ kN.}$$

Segment 2–3

For height = 60 m, diameter = 3 m;

Class of structure = C

$$k_1 = k_3 = 1$$

$$k_2 = 1.1 + \frac{0.07}{50} \times 10 = 1.114$$

$$p_z = 0.6 \times (1.114 \times 39)^2 \times 10^{-3} = 1.132 \text{ kN/m}^2$$

$$P_3 = 0.7 \times 1.132 \times (10 \times 3) = 23.772 \text{ kN.}$$

Segment 3–4

For height = 50 m, diameter = 3 m;

Class of structure = B

$$k_1 = k_3 = 1$$

$$k_2 = 1.15$$

$$p_z = 0.6 \times (1.15 \times 39)^2 \times 10^{-3} = 1.21 \text{ kN/m}^2$$

$$P_4 = 0.7 \times 1.21 \times (10 \times 3) = 25.41 \text{ kN}$$

Segment 4–5

For height = 40 m, diameter = 3 m;

Class of structure = *B*

$$k_1 = k_3 = 1$$

$$k_2 = 1.1 + \frac{0.05}{20} \times 10 = 1.125$$

$$p_z = 0.6 \times (1.125 \times 39)^2 \times 10^{-3} = 1.155 \text{ kN/m}^2$$

$$P_5 = 0.7 \times 1.155 \times (10 \times 3) = 24.255 \text{ kN.}$$

Segment 5–6

For height = 30 m, diameter = 3 m;

Class of structure = *B*

$$k_1 = k_3 = 1$$

$$k_2 = 1.1$$

$$p_z = 0.6 \times (1.1 \times 39)^2 \times 10^{-3} = 1.104 \text{ kN/m}^2$$

$$P_6 = 0.7 \times 1.104 \times (10 \times 3) = 23.184 \text{ kN}$$

Segment 6–7

For height = 20 m, diameter = 3.25 m;

Class of structure = *B*

$$k_1 = k_3 = 1$$

$$k_2 = 1.05$$

$$p_z = 0.6 \times (1.05 \times 39)^2 \times 10^{-3} = 1.006 \text{ kN/m}^2$$

$$P_7 = 0.7 \times 1.006 \times [20 \times (3 + 4)/2] = 49.294 \text{ kN}$$

Stress Calculations

Section modulus of steel stack, $Z = \dfrac{\pi D^2 t}{4}$

Stress due to wind moment, $\sigma_m = \dfrac{4M_w}{\pi D^2 t} = \dfrac{M_w}{250\pi D^2 t} \times 10^3$ N/mm²

Stress due to self weight of steel stack, $\sigma_s = 0.079\,h$ N/mm²

Stress due to lining (125mm thick brick), $\sigma_L = 0.002\left(\dfrac{h}{t}\right)$ N/mm².

Minimum thickness of shell from stability criteria, $t = \dfrac{D}{500} = \dfrac{3000}{500} = 6$ mm

Section 1–1

Height of steel stack above section 1–1, $h_e = 10$ m

Diameter of steel stack at the section, $D = 3$ m

Assuming efficiency of Butt weld to be 85% on the tension side and 100% on the compression side, and to start with thickness of steel plates as 6 mm.

Allowable tensile stress, $\sigma_{at} = \eta \times 0.6 f_y = 0.85 \times 0.6 \times 250 = 127.5$ N/mm²

$$\frac{D}{t} = \frac{3000}{6} = 500 \quad \text{and} \quad \frac{h_e}{D} = \frac{10}{3} = 3.33$$

From Table 13.3,

allowable stress in compression, $\sigma_c = 58$ N/mm^2

$$M_{w_1} = P_1 \times \frac{x}{2} = 24.99 \times 5 = 124.95 \text{ kNm}$$

$$\sigma_m = \frac{124.95 \times 10^3}{250 \times \pi \times 3^2 \times t} = \frac{17.68}{t} \text{ N/mm}^2$$

$$\sigma_s = 0.079 \times 10 = 0.79 \text{ N/mm}^2$$

$$\sigma_L = 0.002 \times \frac{10 \times 10^3}{t} = \frac{20}{t} \text{ N/mm}^2$$

Maximum tensile stress at the section,

$$\sigma_{\text{t, cal}} = \sigma_m - \sigma_s \text{ (wind load may occur before the lining)}$$

$$= \frac{17.68}{t} - 0.79 < 127.5 \text{ N/mm}^2$$

$\Rightarrow \quad t = 0.137$ mm

Maximum compressive stress at the section,

$$\sigma_{\text{c1, cal}} = \sigma_m + \sigma_s + \sigma_L$$

$$= \frac{17.68}{t} + 0.79 + \frac{20}{t} < 58 \text{ N/mm}^2$$

$\Rightarrow \quad t = 0.658$ mm

Add 3 mm to account for corrosion.

$$t_{\text{required}} = 0.658 + 3 = 3.658 \nless 6 \text{ mm}$$

Provide 6 mm thick plate.

Section 2–2

$$h_e = 20 \text{ m}, D = 3 \text{ m}$$

$$\frac{D}{t} = \frac{3000}{6} = 500 \quad \text{and} \quad \frac{h_e}{D} = \frac{20}{3} = 6.67$$

From Table 13.3 $\sigma_c = 58$ N/mm^2

$$M_{w_2} = 24.99 \times 15 + 24.381 \times 5 = 496.76 \text{ kNm.}$$

$$\sigma_m = \frac{496.7 \times 10^3}{250 \times \pi \times 3^2 \times t} = \frac{70.3}{t} \text{ N/mm}^2$$

$$\sigma_s = 0.079 \times 20 = 1.58 \text{ N/mm}^2$$

$$\sigma_L = 0.002 \times \frac{20 \times 10^3}{t} = \frac{40}{t} \text{ N/mm}^2$$

$$\sigma_{\text{t, cal}} = \frac{70.3}{t} - 1.58 < 127.5 \text{ N/mm}^2$$

$$\Rightarrow \quad t = 0.544 \text{ mm}$$

$$\sigma_{c1,\,cal} = \frac{70.3}{t} + 1.58 + \frac{40}{t} < 58 \text{ N/mm}^2$$

$$\Rightarrow \quad t = 1.954 \text{ mm}$$

Add 3 mm in account for corrosion.

$$t_{required} = 1.954 + 3 = 4.954 \nless 6 \text{ mm}$$

Provide 6 mm thick plate.

Section 3–3

$$h_e = 30 \text{ m}, \quad D = 3 \text{ m}$$

Assuming $t = 8$ mm

$$\frac{D}{t} = \frac{3000}{8} = 375 \quad \text{and} \quad \frac{h_e}{D} = \frac{30}{3} = 10$$

From Table 13.3 $\sigma_c = 74$ N/mm^2

$$M_{w_3} = 24.99 \times 25 + 24.381 \times 15 + 23.772 \times 5 = 1109.33 \text{ kN m}$$

$$\sigma_m = \frac{1109.33 \times 10^3}{250 \times \pi \times 3^2 \times t} = \frac{157}{t} \text{ N/mm}^2$$

$$\sigma_s = 0.079 \times 30 = 2.37 \text{ N/mm}^2$$

$$\sigma_L = 0.002 \times \frac{30 \times 10^3}{t} = \frac{60}{t} \text{ N/mm}^2$$

$$\sigma_{t,\,cal} = \frac{157}{t} - 2.37 < 127.5 \text{ N/mm}^2$$

$$\Rightarrow \quad t = 1.21 \text{ mm}$$

$$\sigma_{c1,\,cal} = \frac{157}{t} + 2.37 + \frac{60}{t} < 74 \text{ N/mm}^2$$

$$\Rightarrow \quad t = 3.029 \text{ mm}$$

Add 3 mm to account for corrosion.

$$t_{required} = 3.029 + 3 = 6.0294 \simeq 8 \text{ mm}$$

Provide 8 mm thick plate.

Section 4–4

$$h_e = 40 \text{ m}, D = 3 \text{ m}$$

Assuming $t = 8$ mm

$$\frac{D}{t} = \frac{3000}{8} = 375 \quad \text{and} \quad \frac{h_e}{D} = \frac{40}{3} = 13.3$$

From Table 13.3 $\sigma_c = 74$ N/mm^2

$$M_{w_4} = 24.99 \times 35 + 24.381 \times 25 + 23.772 \times 15 + 25.41 \times 5$$

$$= 1967.81 \text{kNm}$$

$$\sigma_m = \frac{1967.81 \times 10^3}{250 \times \pi \times 3^2 \times t} = \frac{278.4}{t} \text{ N/mm}^2$$

$$\sigma_s = 0.079 \times 40 = 3.16 \text{ N/mm}^2$$

$$\sigma_L = 2 \times \frac{40}{t} = \frac{80}{t} \text{ N/mm}^2$$

$$\sigma_{t,\text{ cal}} = \frac{278.4}{t} - 3.16 < 127.5 \text{ N/mm}^2$$

$$\Rightarrow \quad t = 2.13 \text{ mm}$$

$$\sigma_{c1,\text{ cal}} = \frac{278.4}{t} + 3.16 + \frac{80}{t} < 74 \text{ N/mm}^2$$

$$\Rightarrow \quad t = 5.06 \text{ mm}$$

Add 3 mm in account for corrosion.

$$t_{\text{required}} = 5.06 + 3 = 8.06 \approx 8 \text{ mm}$$

Provide 8 mm thick plate.

Section 5–5

$$h_e = 50 \text{ m}, \quad D = 3 \text{ m}$$

Assuming $t = 10$ mm

$$\frac{D}{t} = \frac{3000}{10} = 300 \quad \text{and} \quad \frac{h_e}{D} = \frac{50}{3} = 16.67$$

From Table 13.3 $\sigma_c = 87$ N/mm^2

$$M_{w_5} = 24.99 \times 45 + 24.381 \times 35 + 23.772 \times 25 + 25.41 \times 15 + 24.255 \times 5 = 3074.61 \text{ kNm}$$

$$\sigma_m = \frac{3074.61 \times 10^3}{250 \times \pi \times 3^2 \times t} = \frac{435}{t} \text{ N/mm}^2$$

$$\sigma_s = 0.079 \times 50 = 3.95 \text{ N/mm}^2$$

$$\sigma_L = 0.002 \times \frac{50 \times 10^3}{t} = \frac{100}{t} \text{ N/mm}^2$$

$$\sigma_{t,\text{ cal}} = \frac{435}{t} - 3.95 < 127.5 \text{ N/mm}^2$$

$$\Rightarrow \quad t = 3.27 \text{ mm}$$

$$\sigma_{c1,\text{ cal}} = \frac{435}{t} + 3.95 + \frac{100}{t} < 87 \text{ N/mm}^2$$

$$\Rightarrow \quad t = 6.44 \text{ mm}$$

Add 3 mm to account for corrosion.

$$t_{\text{required}} = 6.44 + 3 = 9.44 \simeq 10 \text{ mm}$$

Provide 10 mm thick plate.

Section 6–6

$$h_e = 60 \text{ m}, \quad D = 3 \text{ m}$$

Assuming $t = 12$ mm

$$\frac{D}{t} = \frac{3000}{12} = 250 \quad \text{and} \quad \frac{h_e}{D} = \frac{60}{3} = 20$$

From Table 13.3 $\sigma_c = 99 \text{ N/mm}^2$

$$M_6 = 24.99 \times 55 + 24.381 \times 45 + 23.772 \times 35 + 25.41 \times 25 + 24.255 \times 15 + 23.184 \times 5 = 4418.61 \text{ kN-m.}$$

$$\sigma_m = \frac{4418.61 \times 10^3}{250 \times \pi \times 3^2 \times t} = \frac{625.1}{t} \text{ N/mm}^2$$

$$\sigma_s = 0.079 \times 60 = 4.74 \text{N/mm}^2$$

$$\sigma_L = 0.002 \times \frac{60 \times 10^3}{t} = \frac{120}{t} \text{ N/mm}^2$$

$$\sigma_{t,\text{ cal}} = \frac{625.1}{t} - 4.74 < 127.5 \text{ N/mm}^2$$

$\Rightarrow$ $t = 4.73$ mm

$$\sigma_{cl,\text{cal}} = \frac{625.1}{t} + 4.74 + \frac{120}{t} < 99 \text{ N/mm}^2$$

$\Rightarrow$ $t = 7.90$ mm

Add 3 mm to account for corrosion.

$$t_{\text{required}} = 7.90 + 3 = 10.90 \simeq 12 \text{ mm}$$

Provide a 12 mm thick plate.

Section 7–7 (at the base)

$$h_e = 80 \text{m}, \quad D_1 = 4 \text{ m}$$

Assuming $t = 16$ mm

$$\frac{D_1}{t} = \frac{4000}{16} = 250 \quad \text{and} \quad \frac{h_e}{D_1} = \frac{80}{4} = 20$$

From Table 13.3 $\sigma_c = 99 \text{ N/mm}^2$

$$M_7 = 24.99 \times 75 + 24.381 \times 65 + 23.772 \times 55 + 25.41 \times 45 + 24.255 \times 35 + 23.184 \times 25 + 49.294 \times \left(\frac{4 + 2 \times 3}{4 + 3}\right) \times \frac{20}{3} = 7807.92 \text{ kN m}$$

$$\sigma_m = \frac{7807.92 \times 10^3}{250 \times \pi \times 4^2 \times t} = \frac{621.33}{t} \text{ N/mm}^2$$

$$\sigma_s = 0.079 \times 80 = 6.32 \text{ N/mm}^2$$

$$\sigma_L = 0.002 \times 60 \times 10^3 \times \frac{3}{4} = \frac{90}{t} \text{ N/mm}^2$$

$$\sigma_{t,\,cal} = \frac{621.33}{t} - 6.32 < 127.5 \text{ N/mm}^2$$

$\Rightarrow$ $$t = 4.58 \text{ mm}$$

$$\sigma_{c1,\,cal} = \frac{621.33}{t} + 6.32 + \frac{90}{t} < 99 \text{ N/mm}^2$$

$\Rightarrow$ $$t = 7.7 \text{ mm}$$

Add 3 mm in account for corrosion.

$$t_{required} = 7.7 + 3 = 10.7 < 12 \text{ mm}$$

Provide a 12 mm thick plate.

Check for Seismic Forces

Bhopal is in Zone II.

Equivalent thickness of the stack,

$$t_e = \frac{2 \times 6 + 2 \times 8 + 1 \times 10 + 1 \times 12 + 1 \times 12}{7} = 8.85 \text{ mm} \approx 9.0 \text{ mm}$$

Weight of steel, $W_s = \pi \times 3 \times 0.009 \times 80 \times 78.5 = 532.69$ kN

Weight of lining, $W_L = \pi \times 3 \times 0.125 \times 60 \times 20 = 1413.72$ kN

Increase the weight by 5% for stiffeners, platforms, ladder, etc.

Total weight = (532.69 + 1413.72) × 1.05 = 2043.73 kN

Area of cross section at base, $A = \pi D_1 t = \pi \times 4 \times 0.012 = 0.150 \text{ m}^2$

Radius of gyration, $r_1 = 0.7 \times 2 = 1.4$ m

$$K = \frac{H}{r_1} = \frac{80}{1.4} = 57.14$$

$$C_T = 1.8k = 1.8 \times 57.14 = 102.852$$

Time period, $$T = C_T \sqrt{\frac{W_t H}{E_s A g}}$$

$$= 102.852 \times \sqrt{\frac{2043.73 \times 80 \times 10^6}{2 \times 10^5 \times 0.15 \times 10^6 \times 9810}} = 2.42 \text{ sec}$$

$$A_h = \frac{ZI(S_a/g)}{2R}$$

$$Z = 0.1$$

$$I = 1.5 \text{ (Table 13.6)}$$

$$R = 2.0 \text{ (Table 13.7)}$$

$S_a/g = 0.562$ (for 5 % damping, medium soil) (Appendix V)

$$A_h = \frac{0.1 \times 1.5 \times 0.562}{2 \times 2} = 0.021$$

Shear at base, $$V = C_V A_h W_T D_v$$

$= 1.5 \times 0.021 \times 2043.73 \times 1$

$= 64.38$ kN (From Table 13.8, for $\frac{X}{H} = 1, D_v = 1$)

Since the shear at base of the steel stack is very small as compared to that due to the wind load, the design load will be the wind load.

Breech Opening

Maximum width of opening $= \frac{2}{3} \times 3000 = 2000$ mm

Let us provide a 1600 mm wide opening.

Area of steel removed is $1600 \times 12 = 19200$ mm^2

Reinforcement required for the area of steel removed $= 1.2 \times 19200 = 23040$ mm^2

Reinforcement required for each of the vertical side $= 23040/2 = 11520$ mm^2

Let us provide,

2 I.S.A $150 \times 150 \times 12$, A $= 3459 \times 2$	$= 6918$ mm^2
I.S.PL 400×12	$= 4800$ mm^2
Total area provided	$= 11718$ mm^2
	> 11520 mm^2

which is as should be.

Provide same reinforcement at the horizontal cut section also.

Base Plate

Maximum compressive stress at the base plate,

$$\sigma_{C_i} = \left[\frac{4M_W}{\pi D_1^2 t} + 0.079H + \frac{0.002H}{t}\right]$$

$$= \left[\frac{7807.92}{250 \times \pi \times 4^2 \times 0.012} + 0.079 \times 80 + \frac{0.002 \times 60}{0.012}\right]$$

At the *Section 7–7*

$$\sigma_{c1} = 68.10 \text{ N/mm}^2 = 68.10 \times 10^3 \text{ kN/m}^2$$

Maximum compressive strength per unit circumferential length

$$F_1 = 68.10 \times 12 = 817.2 \text{ kN/m} = 817.2 \text{ N/mm}$$

Width of base, $B = \frac{F_1}{\sigma_{p1}} = \frac{817.2 \times 10^3}{4 \times 10^3} = 204.3$ mm

σ_{p1} = allowable bearing pressure on foundation = 4 kN/m^2 (Assuming)

Provide width of base = 300 mm

Pressure under the base $= \frac{817.2}{300} = 2.72$ N/mm^2

Thickness of base plate required at critical section x–x (Fig. 13.3)

$$t_1 = c\sqrt{\frac{3w}{\sigma_{bs}}} = 150 \times \sqrt{\frac{3 \times 2.72}{185}} = 31.50 \text{ mm}$$

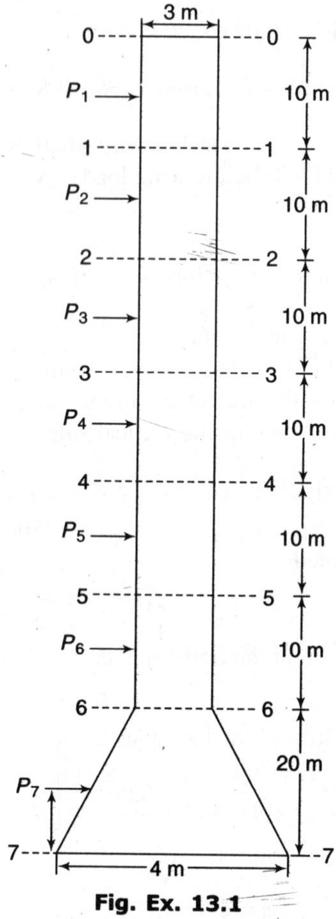

Fig. Ex. 13.1

Base plate thickness $t = t_1 - t_a$

$$= 31.50 - 12 = 19.5 \text{ mm} \approx 22 \text{ mm thick}$$

Use a base plate 300 mm wide and 22 mm thick.

Holding Down Bolts

Try 45 mm ϕ high tensile bolts with $f_y = 340$ N/mm^2

Allowable tensile stress of bolts, $\sigma_{tf} = 0.6 \times f_y = 204$ N/mm^2

$$\text{Strength of bolt, } R_t = \pi \times \frac{45^2}{4} \times 204 \times 10^{-3} \times 0.8 = 260 \text{ kN}$$

$$D_1 = 4000 + 300 = 4300 \text{ mm}$$

Increase moment by 50% for stability criteria and 25% for wind load.

$$\text{Spacing of anchor bolts, } S = \frac{R_t}{\left[\dfrac{4M_W}{\pi D_1^2} - \dfrac{W_S}{\pi D_1}\right]}$$

$$= \frac{260 \times 10^3}{\left[\dfrac{4 \times 7807.92 \times 1.25 \times 1.5}{\pi \times 4.3^2} - \dfrac{532.69}{\pi \times 4.3}\right]}$$

$$= 268.40 \simeq 250 \text{ mm}$$

Use 45 mm ϕ bolts at a spacing of 250 mm c/c.

Design of Foundation

Diameter of cement concrete foundation, $D_2 = [M_W]^{1/4}$

M_w = Moment due to wind at base in kN-m

$D_2 = [7807.2]^{1/4} = 9.4 \text{ m} \cong 9.5 \text{ m}$

Depth of foundation = $0.4 \times D_2 = 0.4 \times 9.5 = 3.76 \text{ m} \approx 3.8 \text{ m}$

Provide foundation of diameter and depth as 9.5 m and 3.8 m, respectively.

Example 13.2 Design a guyed steel stack for the following data:

Diameter = 1.5 m

Height = 30 m

Shape factor = 0.7

Average wind pressure, P_d = 1.2 kN/m^2

Assume initial tension in guy wire = 35 N/m^2

Total corrosion allowance = 4.5 mm

Solution Wind force on stack, $P_w = k\, p_z \times$ projected area

$= 0.7 \times 1.2 \times (30 \times 1.5) = 37.8$ kN

Total bending moment due to wind about the base

$$= 37.8 \times \frac{30}{2} = 567 \text{ kN-m}$$

Two collars each with 12 mm galvanized wire cables 4 in numbers should be sufficient for guying stack.

Let one collar be located at $\dfrac{8H}{10}$ or 24 m from the base of the stack and the other collar be at $\dfrac{H}{2}$ or 15 m above the base of the stack.

The guys connected at the upper collar will make an angle of approximately 60° with the vertical.

Horizontal wind reaction at the upper collar (Table 13.11)

$$R_c = \frac{9P_w}{20} = \frac{9 \times 37.8}{20} = 17.01 \text{ kN}$$

Maximum pull in one guy when wind blows along it,

$$P_g = R_c \operatorname{cosec} \theta = 17.01 \operatorname{cosec} 60° = 19.64 \text{ kN}$$

Total pull in guy wire = 19.64 + initial tension in the wire

$$= 19.64 + \left[35 \times \frac{\pi \times 12^2}{4}\right] \times 10^{-3} = 23.6 \text{ kN}$$

Total maximum vertical component of pulls of guys at the upper collar

$$= 17.01\cot 60° + 4 \times \left[\frac{35 \times \pi \times 12^2 \times 10^{-3}}{4}\right] \times \cos 60° = 9.82 + 7.91 = 17.73 \text{ kN.}$$

The lower collar will be considered to have same forces acting on it as that on the upper collar.

Upper Portion of the Stack

Top 6 m height of stack which is above the top collar

Assume thickness of steel stack, $t = 6$ mm

$$t_{\text{available}} = 6 - 4.5 = 1.5 \text{ mm}$$

Moment at the upper collar, $M_c = \dfrac{P_w H}{50}$

$$= \frac{37.8 \times 30}{50} = 22.68 \text{ kNm} \qquad \text{(From Table 13.11)}$$

Stress due to wind moment at the upper collar,

$$\sigma_m = \frac{4M_c}{\pi D^2 t} = \frac{4 \times 22.68}{\pi \times 1.5^2 \times 1.5 \times 10^{-3}} = 8556 \text{ kN/m}^2 = 8.55 \text{ N/mm}^2$$

Stress due to self-weight of the steel stack,

$$\sigma_s = \frac{W_S}{\pi D t} = 0.079\, h \text{ N/mm}^2 = 0.079 \times 6 = 0.474 \text{ N/mm}^2$$

Maximum compressive stress

$$\sigma_{c,\text{cal}} = \sigma_m + \sigma_s$$
$$= 8.55 + 0.474 = 9.02 \text{ N/mm}^2$$

Check for compressive stress

For $\dfrac{D}{t} = \dfrac{1500}{1.5} = 1000$, and $\dfrac{h_e}{D} = \dfrac{6}{1.5} = 4$;

From Table 13.3, allowable stress in compression, $\sigma_c = 58 \text{ N/mm}^2 > 9.02 \text{ N/mm}^2$ which is all right.

Maximum tensile stress,

$$\sigma_{t,\text{cal}} = \sigma_m - \sigma_s = 8.55 - 0.474 = 8.08 \text{ N/mm}^2$$

considering the efficiency of butt weld, $\eta = 85\%$

Allowable tensile stress $\sigma_t = 0.85 \times 150 = 127.5 \text{ N/mm}^2 > 8.08 \text{ N/mm}^2$.

which is all right.

Horizontal shear just above the upper collar (Table 13.11)

$$= -\frac{P_w}{H}(H - h_1) = -\frac{P_w\left(H - \frac{8}{10}H\right)}{H} = -\frac{1}{5}P_w = -\frac{37.8}{5} = -7.56 \text{ kN}$$

Horizontal shear just below the upper collar (Table 13.11)

$$= \frac{9}{20}P_w - \frac{1}{5}P_w = \frac{1}{4}P_w = \frac{1}{4} \times 37.8 = 9.45 \text{ kN}$$

$$\text{Shear stress at this section} = \frac{9.45\times10^3}{\pi Dt} = \frac{9.45\times10^3}{\pi\times1500\times1.5}$$

$$= 1.336 \text{ N/mm}^2 < 100 \text{ N/mm}^2$$

$$\delta_{va} = 0.4\times f_y = 0.4\times250 = 100 \text{ N/mm}^2$$

which is all right.

Central Portion of the Stack

The portion between the upper collar and lower collar

The maximum Bending moment (from Table 13.11) in this section,

$$M_c = \frac{P_w H}{50} = \frac{37.8\times30}{50} = 22.68 \text{ kNm}$$

Assume this moment to be acting anywhere in this section.

$$\sigma_m = \frac{4M_c}{\pi D^2 t} = \frac{4\times22.68}{\pi\times1.5^2\times1.5\times10^{-3}} = 8556 \text{ kN/m}^2 = 8.55 \text{ N/mm}^2$$

Stress due to self-weight of the steel stack,

$$\sigma_s = \frac{W_S}{\pi Dt} = 0.079\ h = 0.079\times15 = 1.185 \text{ N/mm}^2$$

Stress due to vertical component of pull of guys,

$$\sigma_{Pg} = \frac{p_g}{\pi Dt} = \frac{17.73\times10^3}{\pi\times1500\times1.5} = 2.5 \text{ N/mm}^2$$

Maximum compressive stress

$$\sigma_{c,\text{ cal}} = \sigma_m + \sigma_s + \sigma_{Pg}$$

$$= 8.55 + 1.185 + 2.5 = 12.24 \text{ N/mm}^2$$

Check for compressive stress

For $\frac{D}{t} = \frac{1500}{1.5} = 1000$ and $\frac{h_e}{D} = \frac{9}{1.5} = 6$;

From Table 13.3, allowable stress in compression, $\sigma_c = 58 \text{ N/mm}^2$
$> 12.24 \text{ N/mm}^2$

which is all right.

Maximum tensile stress

$$\sigma_{t,\text{ cal}} = \sigma_m - \sigma_s - \sigma_{Pg}$$

$$= 8.55 - 1.185 - 2.5 = 4.865 \text{ N/mm}^2$$

$$< 127.5 \text{ N/mm}^2$$

which is all right.

Lower Portion of the Stack

Maximum bending moment (from Table 13.11) in this section,

$$M_c = \frac{P_w H}{32} = \frac{37.8\times30}{32} = 35.44 \text{ kNm}$$

Stress due to wind load moment at upper collar

$$\sigma_m = \frac{4M_c}{\pi D^2 t} = \frac{4 \times 35.44}{\pi \times 1.5^2 \times 1.5 \times 10^{-3}} = 13370 \text{ kN/m}^2 = 13.370 \text{ N/mm}^2$$

Stress due to self-weight of the steel stack,

$$\sigma_s = \frac{W_S}{\pi D t} = 0.079\, h = 0.079 \times 30 = 2.37 \text{ N/mm}^2$$

Stress due to vertical component of pull of guys,

$$\sigma_{Pg} = 2 \times \frac{P_g}{\pi D t} = 2 \times \frac{17.73 \times 10^3}{\pi \times 1500 \times 1.5} = 5 \text{ N/mm}^2$$

Maximum compressive stress

$$\sigma_{c,\text{ cal}} = \sigma_m + \sigma_s + \sigma_{Pg} = 13.370 + 2.37 + 5 = 20.74 \text{ N/mm}^2$$

Check for compressive stress

For $\frac{D}{t} = \frac{1500}{1.5} = 1000$ and $\frac{h_e}{D} = \frac{15}{1.5} = 10$;

From Table 13.3, allowable stress in compression,

$$\sigma_c = 58 \text{ N/mm}^2$$
$$> 20.74 \text{ N/mm}^2$$

which is all right.

Maximum tensile stress

$$\sigma_{t,\text{ cal}} = \sigma_m - \sigma_s - \sigma_{Pg} = 13.37 - 2.37 - 5 = 6.0 \text{ N/mm}^2$$
$$< 127.5 \text{ N/mm}^2$$

which is all right.

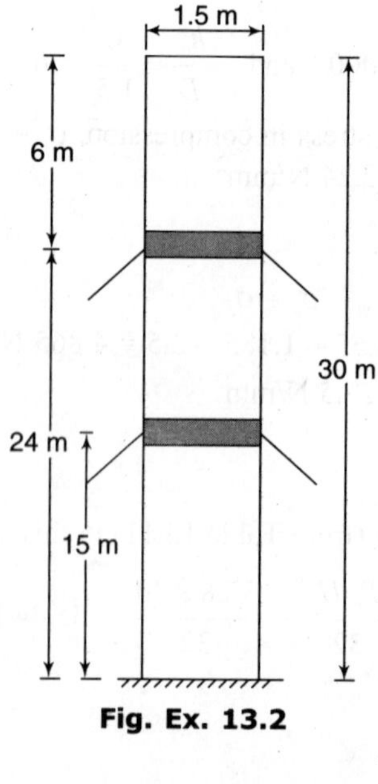

Fig. Ex. 13.2

$$\text{Horizontal shear force} = \frac{P_w}{4} = \frac{37.8}{4} = 9.45 \text{ kN}$$

$$\text{Shear stress in this section at base} = \frac{9.45 \times 10^3}{\pi Dt} = \frac{9.45 \times 10^3}{\pi \times 1500 \times 1.5}$$

$$= 1.336 \text{ N/mm}^2$$
$$< 100 \text{ N/mm}^2$$

which is all right.
Foundations and base plates may be designed similar to as in Example 13.1

Exercises

1. Design a self-supporting steel stack having a height of 75 m and 3 m in diameter. It is to be constructed in Gujarat. The bearing capacity of soil is 150 kN/m^2. The necessary relevant data may be assumed.
2. Design a self-supporting steel stack with a 100 m height to be constructed at Allahabad. The diameter of the cylindrical shaft is 4 m. The stack has a 110 mm thick lining. Design the plates of the stack, base plate and anchor bolts only. Assume the necessary relevant data.
3. Design a guyed steel stack of 21 m height with one set of guy wires attached at 7 m from the top. The diameter of the shaft is 1.2 m. Assume the necessary relevant data.

Bridges

14.1 INTRODUCTION

Bridges are the structures intended to carry rail-road traffic, highway traffic or both of these together over a crossing which may be a water body, canyon or another line of traffic. The first ever bridge having trusses with wrought iron as tension verticals and timber for all other members was built around 1840. The first all steel bridge was started across the Mississippi river in 1869. It consisted of three hingless steel arch spans. However, in India steel bridges came into existence around 1900.

A typical rail-road bridge is shown in Fig. 14.1. The Bridges constructed to carry highway traffic are known as highway bridges and those carrying railway traffic are known as railway bridges. Bridges carrying both of these are known as combined highway and railway bridges. Foot bridges are constructed to carry pedestrians only. Depending upon the span of the bridge, funds available and the type of traffic, various materials of construction like timber, masonry, reinforced cement concrete, prestressed concrete or steel may be used suitably and the bridge is classed accordingly, e.g. a bridge built up of steel sections is called steel bridge.

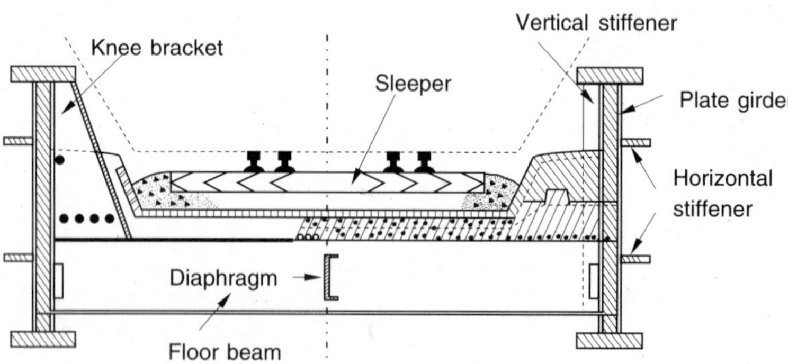

Fig. 14.1 *Rail-Road Bridge*

Besides serving its purpose safely and economically, a bridge should fit into and enhance the beauty of its surroundings. The factors that must be considered may be grouped as economics of the bridge, functional planning and physical planning.

Economics is not only concerned with the type of structure, the materials to be used, the number of spans and the method of erection, but factors like its expected life, the cost of maintenance and repairs, the effect of alignment and grade on the cost of traffic using it and the adaptability to future enlargement also effect it to a great extent.

Functional planning involves consideration of grade and alignment, number of tracks or traffic lanes, approach facilities and appearance. Moreover, the bridge should be in harmony with its setting, and in many cases this in itself can well determine the type of bridge that should be built.

Physical planning involves consideration of foundation material, the character of the stream, the type of bridge and number of spans. Evidently decisions taken have a direct bearing on both the economy and beauty of the structure.

This chapter cannot adequately acquaint readers with the broad aspects of bridge planning and engineering. It is intended rather to provide an introduction to the structural principles and procedures involved in the design of bridge framing, members and connections of simple plate girder and truss bridges. Specifications laid down by the Indian Railways, the Bridge Rules and the steel bridge code IS 1915 will be used to illustrate the bridge designs.

14.2 COMPONENTS

All bridges are composed of the superstructure or span and the substructure or foundation. This chapter deals mainly with the superstructure of railway bridges.

The four major parts of the superstructure are the floor system, the girders or the trusses, the end shoe and the bearings, and the lateral bracing system. A schematic arrangement of the components of a truss girder (open web girder) bridge is shown in Fig. 14.2.

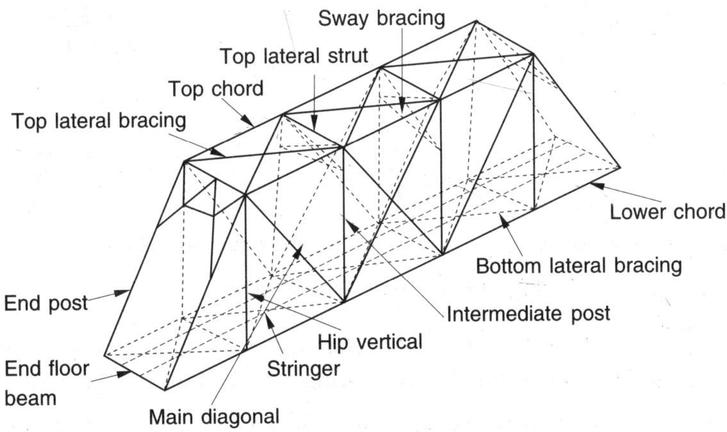

Fig. 14.2 *A Typical through Truss Highway Bridge*

The function of the floor system is to provide a flat surface over which the traffic moves. It is designed to transfer its own weight and the wheel reactions of the moving vehicles to the supporting girders or to the panel points of the trusses, which in turn transfer it to the piers or abutments. The end shoe spreads the reaction from the girder or truss over the required area of substructure. It also provides for the longitudinal movement caused by temperature change. The lateral bracing system imparts stiffness to the structure, reduces vibrations and holds the trusses upright preventing their collapse under the lateral force of a side wind.

14.3 TYPES

Steel bridges may be classified in a number of ways. The more important of the these are:

According to the Cross-section

Deck Bridge A deck bridge is one having its floor resting on top of all the main carrying members (Fig. 14.3).

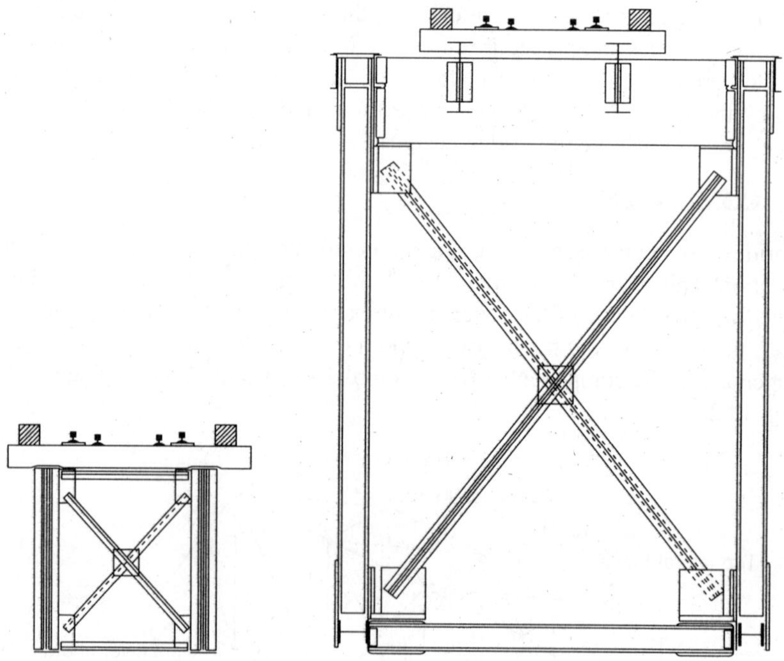

(a) Plate Girder Bridge

(b) Truss Girder Bridge

Fig. 14.3 *Deck Type Bridges*

Through Bridge A through bridge has its floor resting on the lower portion of the main carrying members, thus placing the bracing over the traffic (Fig. 14.4).

Semi-through Bridge A semi-through bridge has the main carrying members projected above the floor level and without overhead bracing. It is also known as a *pony bridge* (Fig. 14.5).

Double Deck Bridge A bridge having two decks supported over different levels both of which can be through decks or one can be through deck and the other an open one. A typical double deck bridge is shown in Plate 3.

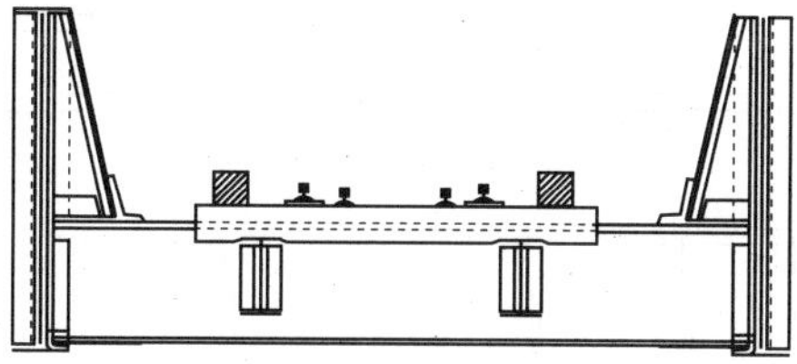

(a) Plate Girder Bridge

(b) Truss Girder

Fig. 14.4 *Through Type Bridge*

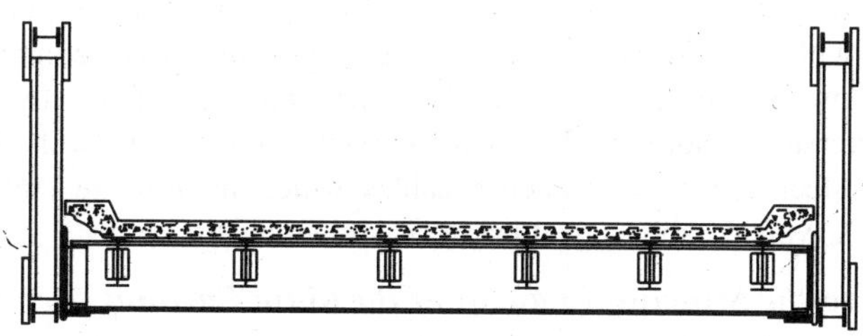

Fig. 14.5 *A Pony Truss Bridge*

According to the Make-up of the Main Carrying Members

I-beam Bridge It has rolled beams as the main carrying members (Fig. 14.6(a)).

Plate Girder Bridge It has plate girders (Fig. 14.6(b)) as the main carrying members. A typical railway plate girder bridge is shown in Plate 4.

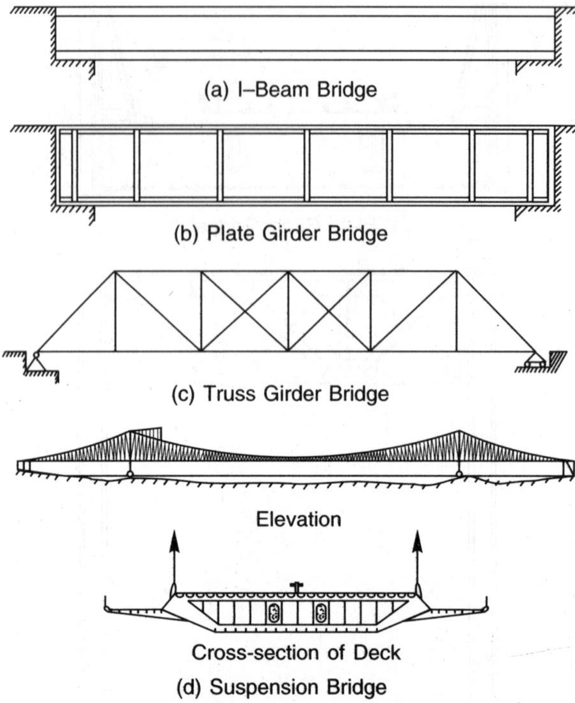

Fig. 14.6 *Types of Bridges*

Orthotropic Bridge Plate girders acting integrally with the steel deck is known as orthotropic bridge.

Truss Girder Bridge It has trusses (Fig. 14.6(c)) as the main carrying members. Usually triangular framed and occasionally quadrangular framed (vierendeel) trusses are provided.

Suspension Bridge It has high tensile strength steel cables (Fig. 14.6(d)) carrying the main load. These are usually stiffened with trusses to obtain rigidity and to keep deflections within limit. The cables, acting in tension, cause compression in the supporting towers and on inclined pull at each end which normally is resisted by massive concrete anchorages. The wires in each cable are splayed at the ends and attached to steel eye bars or anchor cables which in turn are embedded in the anchorage. A typical cable stayed bridge is shown in Plate 5.

According to the Structural Layout of the Main Carrying Members

Simple Span Bridge The main carrying member spans from one support to another, but is discontinuous over piers.

Continuous Bridge The main carrying member is continuous over supports.

Cantilever Bridge It consists of two simple spans, each carrying an overhang or cantilever portion as shown in Fig. 14.7. Almost all cantilever bridges have anchor arms in common and a suspended span (Fig 14.7(a)). The inserted hinges as shown

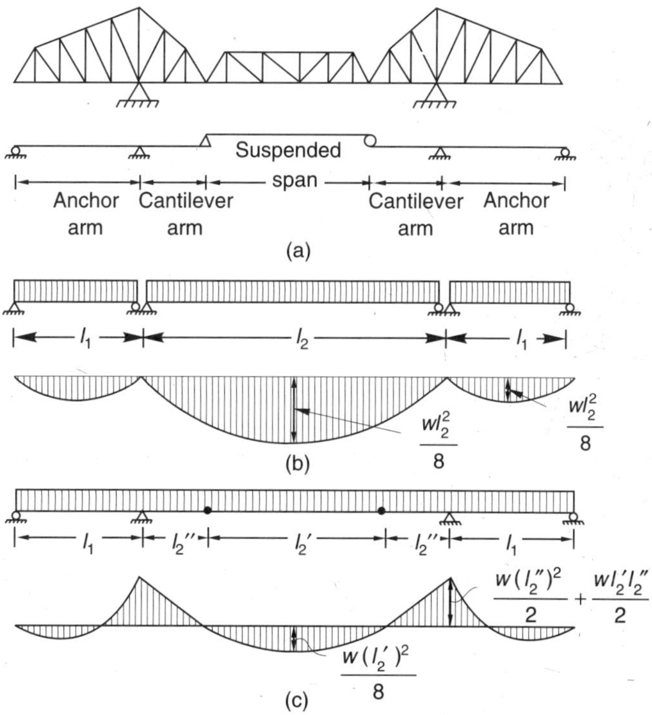

Fig. 14.7 *Principle of Cantilever Bridge*

in Fig. 14.7(c) reduce the maximum moment below that which would occur for the same load acting on a simple span between the two interior piers (Fig. 14.7(b)). The main advantage of a cantilever bridge over a continuous bridge of the same span arrangement is that it is statically determinate and hence easier to analyse and it is also not subjected to changes in stress due to support movements. However, continuous spans may permit a slightly greater moment reduction.

Arch Bridge Two-hinged arch bridges are very common three-hinge, one-hinge and hingeless arches are also sometimes used. An arch itself is made of girders or trusses (Fig. 14.8).

According to the Level of Crossing

Over Bridge When a highway is carried over a railway track it is called an over bridge.

Under Bridge When a highway is carried under a railway track it is known as under bridge.

According to the Nature of the Movement of the Bridge Girders

Fixed Bridge These bridges do not have any movement.

Moveable Bridge These can be moved in the horizontal or vertical direction. Examples are swing bridge, Bascule bridge, Rolling bridge, Vertical lift bridge, Transporting bridge and Floating swing bridge. The obvious reason for building a move-

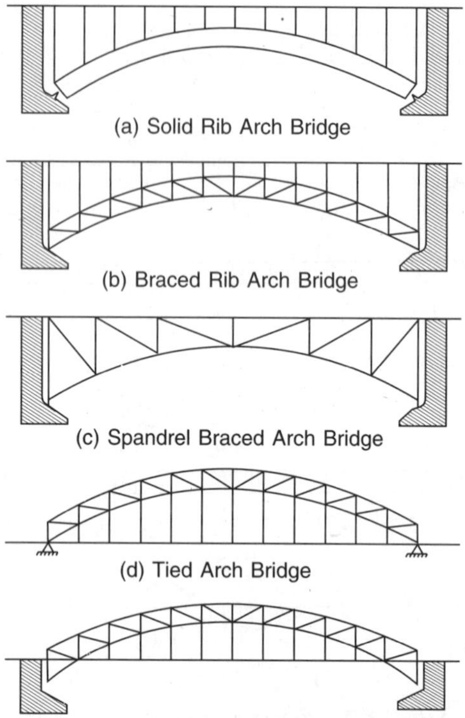

(a) Solid Rib Arch Bridge

(b) Braced Rib Arch Bridge

(c) Spandrel Braced Arch Bridge

(d) Tied Arch Bridge

(e) Tied Arch Bridge with Partial Hung Deck

Fig. 14.8 *Types of Arch Bridges*

able bridge is to provide clearance for shipping while avoiding the cost of high piers and costly approaches. Some of the moveable bridges are shown in Fig. 14.9.

According to the Type of Connection Pinned, riveted, bolted or welded connections may be used in fabrication and the bridge is named accordingly. The majority of existing bridges are riveted. The first all-welded plate girder railway bridge was constructed in 1955 at New York.

According to Span Length

Short Bridge Bridges spanning between 30 to 60 m.

Intermediate Bridges Bridges spanning between 60 to 120 m.

Long Bridges Bridges spanning between 120 to 180 m.

According to Degree of Redundancy Determinate and indeterminate bridges.

According to the Function

Vehicular Bridge Examples are highway and railway bridges.

Pedestrian Bridges Foot bridge is an example of a pedestrian bridge.

Material Handling Bridge These are used for carrying pipe lines, coal or ore unloading, etc.

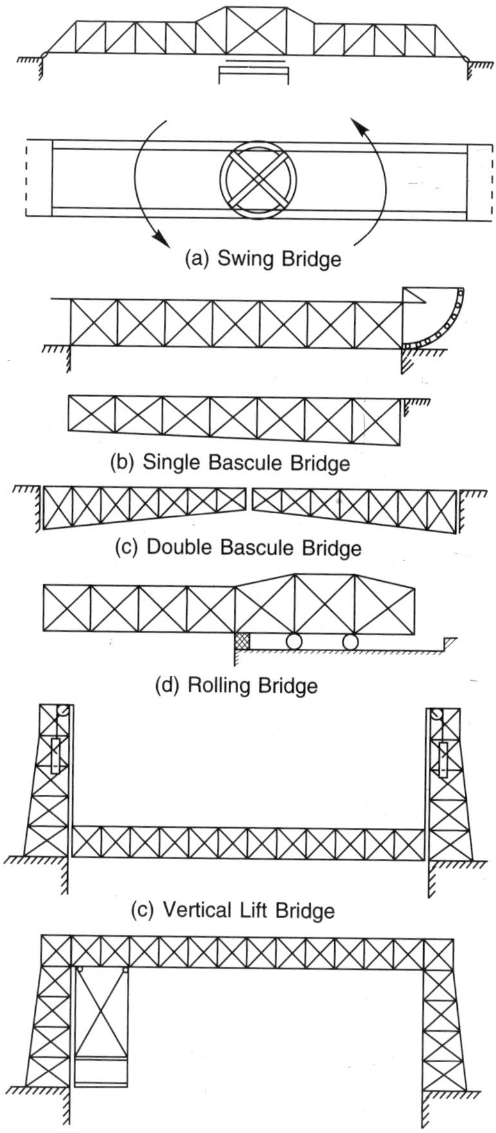

Fig. 14.9 *Type of Moveable Bridges*

14.4 CHOICE OF THE TYPE OF BRIDGE

The prime consideration in selecting the type of structure for a steel bridge is cost. The total cost of a bridge is minimum when the combined cost of its superstructure and substructure is minimum. In general economy will depend upon the type of construction and the number of spans. The following types of bridges are recommended for the length of spans against them.

Type of bridge	Span up to (m)
I-beam	10
Riveted plate girder	10–35
Welded plate girder	15–100
Truss girder	more than 30
Suspension	more than 150

Rolled beams (up to 10 m) and plate girders (up to 20 m) can be fabricated to a cost less than any other type of bridge span. Span lengths more than 20 m are difficult to transport in single length and are made in pieces and spliced in the field. The extra cost of material and labour offsets any other advantage that plate girders may have over truss girders of the same length.

The composition of plate girders has been dealt in Chapter 9. When plate girders are used for large spans, they have a relatively low depth to span ratio and usually result in more weight than a well designed truss. However, because of an optimum distance between floor beams and saving of labour in the manufacture of fewer structural pieces and fewer joints and consequently reduction in erection cost may result in a lower cost of girder. This is particularly true for spans up to 40 m.

In a truss girder bridges generally riveted trusses of Warren or Pratt type are used for spans between 30 to 50 m. The Warren trusses require somewhat less material than Pratt trusses of same span. Pin connected trusses are used, at present, for spans more than 50 m. For spans between 45 to 75 m riveted or pin connected curved chord Pratt type trusses are used. For spans 75 to 150 m pin connected or riveted Baltimore, Pennsylvania, or K-type trusses are used. For spans more than 150 m it has been a general practice to recommend an arch, cantilever or suspension bridge.

At present the general practice is to use riveted trusses up to 60 m only. Large span trusses lead to large joints and wide members resulting in large secondary stresses. To overcome this effect eye bars and pin connections are adopted. Deck trusses are somewhat cheaper than through trusses of the same span. In the former, the trusses may be placed closer together than in through bridges where clearance has to be kept for the passage of traffic. This somewhat reduces the length of the lateral members and floor beams. In general, the deciding factor between deck and through spans is the matter of clearance underneath the structure since the lowest steel must be placed well above the high water level. Where the surrounding country is low, the cost of fill or trestle approaches to a deck bridge may offset the difference in the cost between deck and through bridges and decide in favour of the latter.

14.5 THE FLOOR SYSTEM

Floors used in bridges may be classified in accordance with the type of traffic carried—highway or railway floors; structural action of floor—one way or two way slabs; orthotropic steel plate floors; principal material used—steel, timber or concrete floors. Selection of the appropriate floor system for a bridge is governed by quality of the roadway surface, proper drainage, weight of the floor system, construction time and overall cost including maintenance.

For railway bridges two types of floors are generally provided, the open deck type (Figs 14.10(b, d)) and the ballast deck type. A solid floor totally covers the main girders, while an open floor covers only the necessary part of it. Figs 14.10(a, c) depict through truss and through girder railway bridges. The most useful railway floor is the open deck type (Fig. 14.10(d)). It consists of timber sleepers resting on stringers framing into cross beams. Sometimes sleepers may be omitted and the rails in such a case rest directly on the stringers. The stringers rest on the cross girders and their bottom flanges are connected to the top flanges of the cross girders. However, the open floor can be used only over streams and secondary highway crossings. The ballast deck type floor (which is a solid one) may consist of a ballast supported on all metal–deck, or concrete or timber floor (Fig. 14.10(d)) supported on stringers or cross beams or on cross beams alone. In addition to the protection it offers, it also provides a continuous ballast road bed.

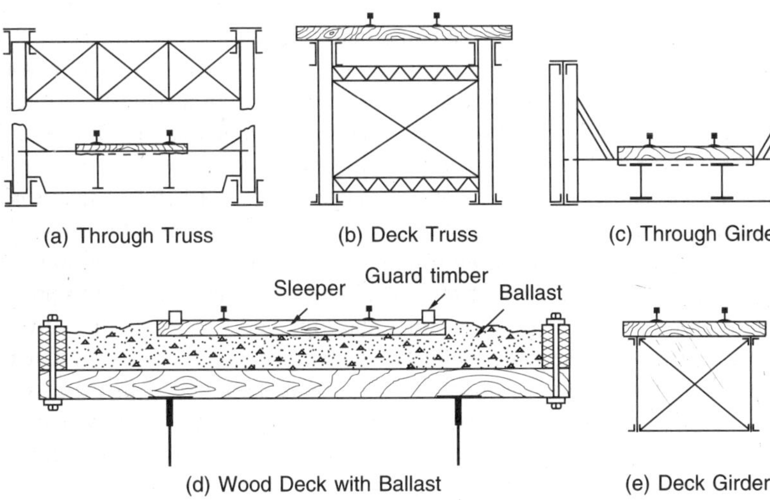

Fig. 14.10 *Cross-section of Railway Bridges and Floor Systems*

14.6 END SUPPORTS

Ordinary bridge trusses have a pin-connected shoe at each of their ends over the abutments. For small span bridges, a simple slotted bearing plate, (Fig. 14.11(a)) may be sufficient for distributing the end reaction over the required area of pier/abutment. Here, the girder rests on a sole plate which in turn rests on a wall plate, connected rigidly to the bed block, the sole plate being free to slide on it. To achieve this, the bolts connecting the sole plate and the wall plate pass through slotted holes in the sole plate. Thus the girder being rigidly connected to the sole plate, is also free to move in the longitudinal direction. This arrangement allows free longitudinal expansion and contraction due to temperature changes or live load stresses. At the other end of the truss a fixed connection is provided. However, for long bridges this arrangement has two disadvantages. First, the deflection of the bridge tends to lift the edge of the plate that is nearest to the end of the span making the bearing pressure non-uniform. Second, although the slotted holes at one end

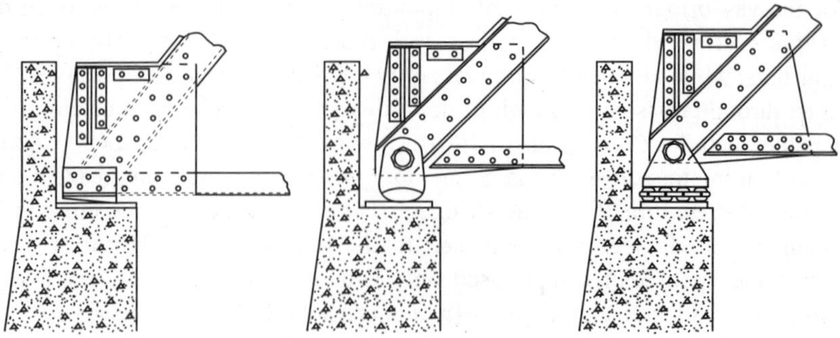

(a) Sliding Plate (b) Rocker and Pin (c) Rollers and Pin

Fig. 14.11 *Truss End Supports*

allow adjustment of temperature changes and for elongation of the bottom chord due to the stress friction must be overcome before necessary movement can take place. This would be objectionable, for it might produce crushing of the masonry and would certainly prevent the reaction from being concurrent with the forces in the bottom chord and end post. As a result a bending moment is introduced in the truss members at the end joint which is objectionable.

Figure 14.11(b) shows a bearing that makes use of a rocker between the bearing plate and the truss end. The main girder is connected to a top inverted shoe. This rests on a round cylindrical rocker pin resting on the bottom shoe. The bottom shoe may directly rest on the masonry or on a steel block over the masonry to provide more height and consequent rigidity. Since the top shoe is rigid to the girder, it rotates over the pin when the girder deflects. The bottom shoe is not affected by this movement and thus angular movement of the girder is permitted. This permits free deflection of the girder along the length. Although there will be only a line of contact between an unloaded rocker and its bearing plate, deformation under load distributes the reaction over a finite area. Evidently, at a given load this area increases with increase in the radius of the rocker.

Since the necessary radius and length of a given rocker depends upon the load to be supported, it is obvious that for a very large reaction the required radius may be too large, or else the rocker may be too long to keep down the radius. This problem is overcome by using a series (nest) of rollers as shown in Fig. 14.11(c).* The rollers may be coupled with the side bars to keep up their spacing and the entire nest is held by the tooth guides. A rocker-roller bearing allows free deflection as well as longitudinal movement of the girder. If the top shoe moves in the longitudinal direction, the pin and hence the bottom shoe also moves with it. The length of the rollers is kept 6 D to 10 D, where D is the diameter of rollers in mm. The gap between adjacent rollers is kept 5 mm.

The minimum size of the roller and pin should not be kept less than 100 mm. Segmental rollers are recommended for diameters greater than 250 mm.

---

* Rollers are segmented cylinders placed below the bottom shoe parallel to one another at a uniform spacing.

The metal bearings as discussed above are now being replaced by rubber bearings, the latter having advantages of being cheap, easily replaceable. They also permit rotation due to squeezing of the rubber. These bearings are made of natural rubber or synthetic rubber (neoprene) reinforced with steel plates. These can be replaced as and when desired (normally 25 years) for which provision of lifting one end of the bridge is made. These are of two types viz, plain bearing allowing rotation with little or no horizontal displacement used for concrete slab bridges of spans less than 20 m. For large span concrete bridges, heavily loaded girders and steel bridges, laminated neoprene bearings are used. For the design of neoprene bearings (also called elastomeric bearings) IRC:83 (Part II)-87 may be referred.

14.7 TRUSS GIRDER BRIDGES

There are three types of truss girder bridges: *the deck type, the through type* and the *semi-through type*. A deck bridge, in the first instance may appear to be economical because of smaller cross beams and consequently less costly floors and also shorter piers, but the additional cost of higher approaches usually offsets any savings. The general arrangement of different members in a through truss girder bridge is shown in Fig. 14.22.

A truss girder may be regarded as an *open web beam* with the top and bottom chords as flanges and the diagonal and vertical members as an open web. The end members of the truss form a part of the web and are called *end posts*. Panel points or *nodes* are the points of intersection of the web members and chord members. The lower panel points of the two trusses are joined by girders. These girders are known as *cross-girders* or *floor beams*. Beams supporting sleepers and running parallel to the trusses are known as *stringers*. These are connected to the cross beams. The lateral loads are transferred to the bearings by means of a bracing system. The spacing between the centres of the truss girder should be sufficient to resist overturning and should not be less than 1/20 of the effective span of the truss and also 1/3 of the height of the truss girder.

Types of Trusses Some of the common types of trusses used for truss girder bridges are shown in Fig. 14.12; the most common type is the parallel-chord high Pratt truss (Fig. 14.12(a)) for through spans of moderate length. Though it is economical only up to 50 m span it is recommended even up to 70 m. The Warren trusses for through and deck bridges are shown in Figs 14.12(b, c). A high Warren truss is most economical for a deck bridge and is seldom used for a through bridge. The curved chord Pratt truss (Figs 14.12(d)), also called Parker or camel back truss, is recommended for spans more than 50 m. For spans more than 90 m a K-bar truss (Fig. 14.12(e)) or a Pettit truss with subdivided panels (Fig. 14.12(f)) or with a broken end post (Fig. 14.12(g)) is used. The Baltimore truss, shown in Fig. 14.12(h)), may be looked upon as intermediate between Parker and Pettit trusses, but being uneconomical is not used widely.

The economic height-to-span ratio of a bridge is about one sixth to one eighth. The optimum inclination of the diagonal is 45°. To maintain the inclination for long span bridges, the Pratt or the Warren trusses result in long panel lengths. One way to shorten the panel length—in order to shorten the span of the stringers—is to subdivide these trusses. The examples are Pettit and Baltimore trusses. However, high secondary stresses in these trusses limit their use and a K-truss is preferred.

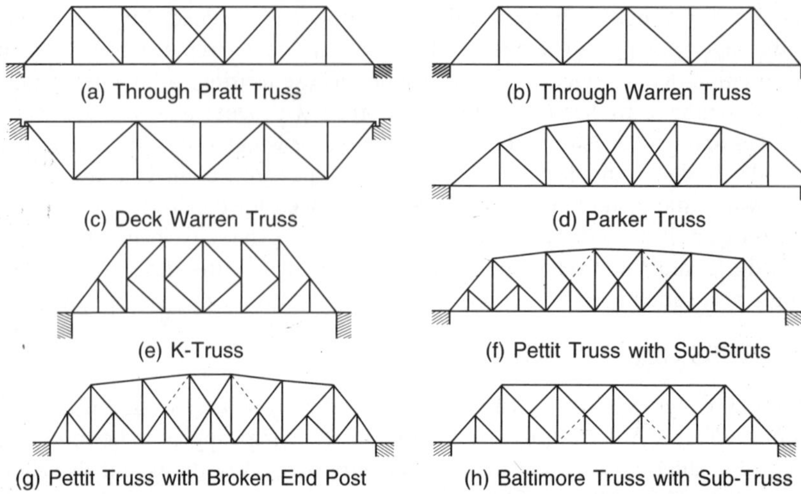

Fig. 14.12 *Bridge Trusses*

14.8 BRACING OF TRUSS GIRDER BRIDGES

A truss girder bridge, with properly designed trusses, under gravity loads will have adequate strength to perform its necessary functions if it remains upright. The trusses are kept, upright in their proper vertical planes, to resist either wind forces or the effect of lateral vibrations caused by live load with the help of lateral and portal bracing. Furthermore, the top chords of through bridges being in compression require more area for lateral stability if the two trusses are not tied together so as to provide shorter unsupported lengths. Thus a system of lateral bracing is a must to assure a stable and rigid structure. A typical bracing system for a through truss bridge is shown in Fig. 14.13.

Lateral Bracing It consists of a horizontal or lateral truss formed at the level of each chord of the main truss. The chords of the main trusses thus become the chords of the lateral trusses. The cross-beams act as the cross struts of the lower lateral truss and by merely adding diagonals, the lower lateral truss is formed. In the case of the upper lateral truss, the upper horizontal members of the sway frames or end portals form the cross-struts and thus by merely adding diagonals the upper lateral truss is formed.

There can be two possible ways in which the stresses can be transferred through the bracing system. One is that the end portal is designed to furnish the upper lateral wind forces. Then the stress path would be through the upper laterals to the end portals and down the portals to the abutments. However, an alternative path is possible through sway frames, provided at each panel point to stiffen laterally the internal cross-sections of the bridge and to hold the trusses in their vertical planes. A sway frame consists of two posts connected rigidity to the floor beam at the bottom and jointed by two struts and by diagonal bracing at the top. The diagonal reaches as far down the posts as is permitted by clearance requirements. Hence, as a result these become as stiff as an end portal.

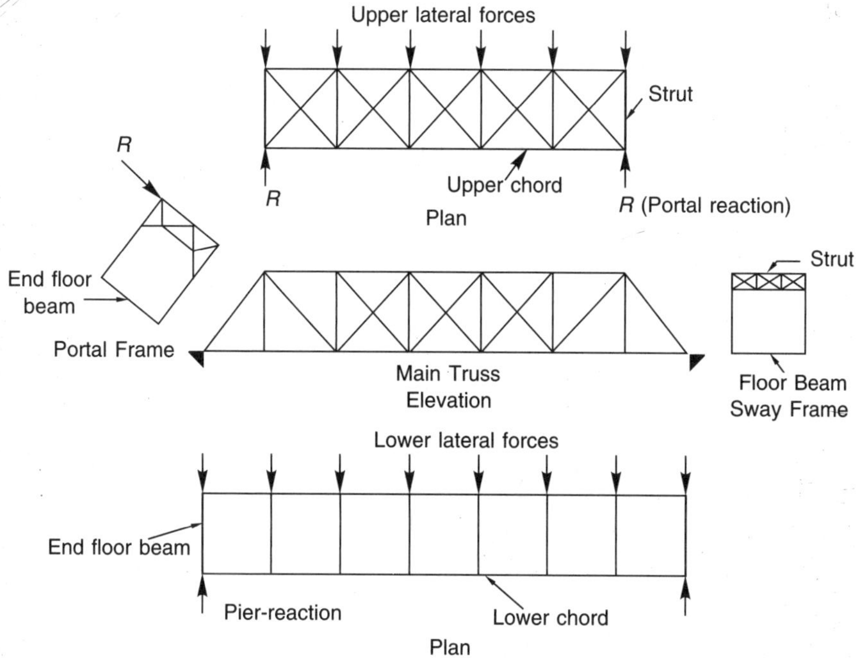

Fig. 14.13 *Lateral System for a Through Truss Bracing*

A sway frame or a portal can resist lateral forces at the top only. But the lateral forces act simultaneously on the upper and lower lateral trusses deflecting them sideways more or less equally. The posts of the sway frames tend to remain vertical and nearly straight. Then the upper lateral forces must be largely transferred down the end portals whose bases are on the fixed abutments. A reverse of this will be true when a concrete floor slab or a welded grid floor is used. In a through bridge the diagonals may be single or double angle sections connected to gusset plates on the bottom flanges of the chords.

Lateral bracing of a deck truss bridge usually consists of horizontal trusses in the planes of the two chords of the main truss girders. Since the lateral forces resulting from the live load are resisted almost entirely by the lateral truss at the floor level, the top lateral truss becomes more important than the bottom one. The top lateral truss helps also in the erection of bridge and to furnish the wind resistance until the floor is laid. For the diagonal members, (in tension), of the bottom lateral truss of deck bridge, usually a single angle section is sufficient. This is assumed to resist shear also. Obviously, the other diagonal is designed as a strut and a double angle section is provided.

The top lateral truss of both the deck and through bridges are usually four angles laced together (in the form of an I-Section).

Portal Bracing The lateral bracing system stabilizes the bridge in vertical and horizontal planes, but leaves it unstable in the transverse vertical planes.

For the deck truss bridge, one diagonal in the transverse plane containing the vertical members at each end of the bridge is enough, but usually two diagonals in each of these planes are provided. These diagonals along with the verticals, the floor beam and, the bottom strut form a statically indeterminate frame called *sway frame*. It transmits the end reaction of the top lateral truss to the abutment.

In a through truss bridge, diagonal bracing cannot be provided. The end posts are tied together to form a rigid frame called a *portal frame*. It transmits the end reaction of the top lateral truss to the abutments. The portal bracing should be as deep as the head room allows, so as to mini-mise bending stresses in the end posts and provide maximum rigidity. The frame shown in Fig. 14.14(a) is commonly used for single track railway and that shown in Fig. 14.14(b) for double track railway.

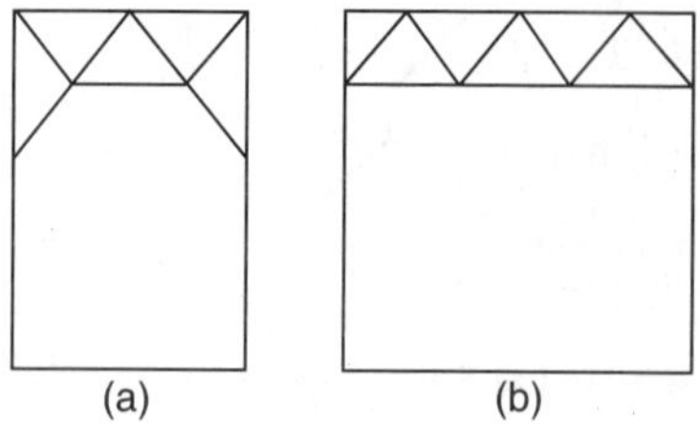

Fig. 14.14 *Portal Bracing*

A combination of lateral trusses and end portal frames is sufficient as bracing system. But to add rigidity, intermediate sway frames are provided in the planes of opposite verticals. In the deck bridge these are similar to ones at the ends, while in the through bridge they are similar to portal frames. Since it is not supposed to resist a specified force, the sizes of its members are a matter of judgement except for limitations such as the slenderness ratio.

14.9 PLATE GIRDER BRIDGES

Plate girder bridges are recommended for medium spans. However, their use for long spans has often been dictated by the fact that these can be fabricated at the site in much less time than the truss girder bridges. Also, the fabrication and erection cost for plate girder bridges is less than that for truss girder bridges. Plate girder bridges may be either of through or deck construction. The floor may span laterally from girder to girder or floor beams may be used, as in a truss girder bridge, with stringers to span from floor beam to floor beam.

14.10 BRACING OF PLATE GIRDER BRIDGES

Plate girder bridges are braced to provide lateral and torsional rigidity. It consists of lateral truss bracing and transverse bracing.

In the case of a deck bridge the upper flange is the loaded flange and the lateral truss bracing is provided at this level (Fig. 14.15(a)). But, usually a truss bracing is provided also at the bottom flange level. Transverse bracing in the form of cross

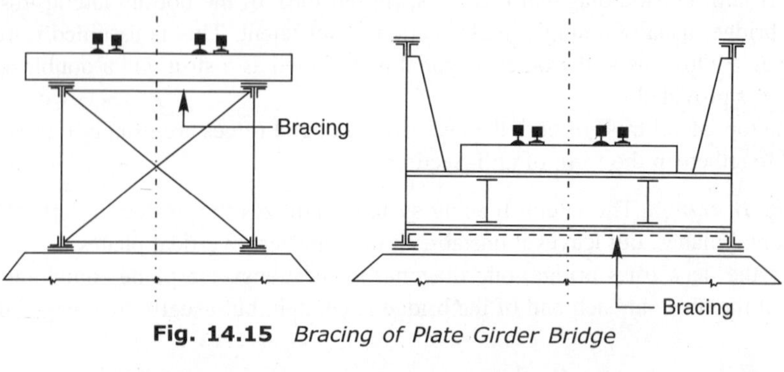

Fig. 14.15 *Bracing of Plate Girder Bridge*

frames or portal frames is provided in the vertical plane. For a through plate girder bridge the loaded flange is the bottom one and the lateral truss bracing is provided at this level (Fig. 14.15(b)). The upper flanges of the plate girders are braced by means of triangular gusset plates.

14.11 LOADS

The main loads to be considered for the design of superstructure are dead load, live load, impact load, seismic load and wind load. Snow load is not considered to be important because a heavy accumulation of snow would practically stop traffic.

14.12 DEAD LOAD

It is the weight of the structure itself together with the permanent loads carried thereon. The dead load of a bridge can be divided into two parts: that of the floor and that of the main structure itself.

The weight of the floor system consists of the floor slab, tracks, ties and the ballast on a concrete slab or steel deck, guard rails, etc. The weight of the main structure itself must be estimated before analysing or designing the structure. It consists of floor beams and the stringers, the weight of the main trusses or girders with the bracing system and the weight of the railings.

For a short span bridge the weight of the main structure may represent only a minor portion of the total dead load and live load. But for a large span the force in a member due to the dead load is often a large portion of total force in the member. Therefore, it is essential to have a good estimate of the dead weight. For short and medium span bridges routine formulae give fairly reasonable estimates of dead weights. However, for large span bridges or unusual bridges several revisions of the estimate may be needed. The estimated weight is compared with the actual weight of the bridge after the design and if the difference is significant, the design is revised.

The dead weight of a bridge depends on its span, depth of girder or truss, number of paneis, width of the railway or highway, design specifications and the individuality of the designer. Many rules and formulae have been suggested from time to time, some of which are as follows.

Hudson's Formula It estimates quite satisfactorily the dead weight of the truss girder bridges. In this the dead weight is assumed to be a function of bottom chord area.

$$w = 0.785\ A$$

where w = weight of the two trusses and bracings in N/m

A = net area of largest tension chord in mm^2.

In the above formula it is assumed that the average weight per meter of the truss is proportional to the net area of the largest tension chord as follows.

$$\begin{aligned} \text{Bottom chord} &= 1.00\ A \\ \text{Top chord} &= 1.25\ A \\ \text{Web system} &= 1.25\ A \\ \text{Connections} &= 1.00\ A \\ \text{Bracings} &= 0.50\ A \\ \hline \text{Total for one truss} &\quad 5.00\ A \end{aligned}$$

If the weight of steel is 78500 N/m^3, (i.e., 0.785 N/m length of one sq. mm of area), the weight of both the trusses and bracings in N/m will be

$$w = 2 \times 5A \times 0.0785 = 0.785\ A$$

Note The Hudson's formula may lead to quite erroneous estimates for trusses with a high depth to span ratio (having proportionately more weight in webs) and also for relatively shallow girders having proportionately more weight in flanges.

Wadell's Charts for the bridge structure give fairly conservative results for the plate girder bridges. These give weights about 10 to 15 per cent greater than the actual ones.

For a single track railway

$$w = kL\sqrt{W}$$

where w = weight of the two girders with bracing system (N/m length of bridge)

k = a constant (52.177 for deck bridge)

L = effective span of bridge (m)

W = heaviest axle load of engine (kN).

$w = 817\ L$ (for axle load = 245.2 kN)

14.13 LIVE LOAD

Theoretically it seems logical to design a bridge for the actual live loads they are expected to carry. In practice, this is difficult to accomplish since there are all kinds of vehicles and their combinations that may be carried by a bridge. Hence, the practice is to design bridges for standard loadings.

Live load may be specified as a line of wheel loads representing several heavy vehicles or a locomotive, or otherwise as an equivalent uniform load. The most common procedure is to design railway bridges for a conventional system of locomotive concentrations followed by a uniform load to represent the train.

Live loads due to train loadings have been specified in Bridge Rules for various types of tracks. It is cumbersome to calculate the member forces due to the moving train with concentrated wheel loads. Therefore, to simplify the analysis, the actual standard loadings have been expressed as equivalent uniformly distributed loads (EUDL) for computing the maximum bending moments and shear forces.

Railway bridges including combined rail and road traffic should be designed for the following standard of railway loading as specified by Bridge Rule.

For Broad Gauge (1676 mm) Modified BG loading of 1987 with a maximum axial load of 245.2 kN for the locomotive and a train load of 80.9 kN/m on both sides of the locomotive. Diagrams of standard loading are shown in Fig. 14.16. Equivalent uniformly distributed live loads on each track for calculating maximum bending moments and shear forces are given in Table 14.1. The tractive effort and braking force may be obtained from Table 14.2.

Table 14.1 *EUDL for Modified B.G. Loading and CDA*

(Metres)	Total Load for Bending Moment kN	Total Load for Bending Moment t	Total Load for Shear kN	Total Load for Shear t	CDA $\left(0.15+\frac{8}{6+L}\right)$
1.0	490	50.0	490	50.0	1.000
1.5	490	50.0	490	50.0	1.000
2.0	490	50.0	519	52.9	1.000
2.5	490	50.0	598	61.0	1.000
3.0	490	50.0	662	67.5	1.000
3.5	516	52.6	707	72.1	.992
4.0	596	60.8	778	79.3	.950
4.5	677	69.0	838	85.5	.912
5.0	741	75.6	888	90.5	.877
5.5	794	81.0	941	95.9	.846
6.0	838	85.5	985	100.4	.817
6.5	876	89.3	1024	104.4	.790
7.0	911	92.9	1068	108.9	.765
7.5	948	96.7	1111	113.3	.743
8.0	881	100.0	1154	117.7	.721
8.5	1010	102.9	1210	123.4	.702
9.0	1040	106.1	1265	129.0	.633
9.5	1070	109.1	1315	134.1	.666
10.0	1101	112.3	1377	140.4	.630
11.0	1282	130.7	1492	152.2	.621
12.0	1377	140.4	1589	162.0	.594
13.0	1475	150.4	1670	170.3	.571
14.0	1558	158.9	1740	177.4	.550
15.0	1631	166.3	1801	183.6	.531
16.0	1695	172.8	1853	189.0	.514
17.0	1751	178.5	1926	196.4	.498
18.0	1820	185.6	1999	203.9	.483
19.0	1886	192.4	2080	212.1	.470
20.0	1964	200.3	2168	221.1	.458
21.0	2039	207.9	2254	229.8	.446
22.0	2123	216.5	2337	238.3	.436
23.0	2203	224.7	2420	246.8	.426
24.0	2280	232.5	2503	255.2	.417
25.0	2356	240.2	2586	263.7	.408
26.0	2431	247.9	2668	272.1	.400
27.0	2506	255.5	2751	280.5	.392
28.0	2580	263.1	2833	288.9	.385
29.0	2654	270.6	2915	297.3	.379
30.0	2727	278.1	2997	305.7	.372
32.0	2874	293.0	3161	322.4	.361
34.0	3034	309.3	3325	339.1	.350
36.0	3191	325.3	3489	355.8	.340
38.0	3345	341.1	3652	372.4	.332
40.0	3498	356.7	3815	389.1	.324

(*Contd.*)

Table 14.1 (*Contd.*)

(Metres)	Total Load for Bending Moment kN	t	Total Load for Shear kN	t	CDA $\left(0.15+\frac{8}{6+L}\right)$
42.0	3649	372.1	3978	405.7	.317
44.0	3798	387.3	4141	422.3	.310
46.0	3947	402.4	4304	438.9	.304
48.0	4094	417.4	4467	455.5	.298
50.0	4253	433.7	4630	472.1	.293
55.0	4658	474.9	5036	513.6	.281
60.0	5051	515.1	5442	555.0	.271
65.0	5436	554.3	5848	596.4	.263
70.0	5831	594.6	6254	637.7	.255
75.0	6220	634.3	6660	679.1	.249
80.0	6603	673.3	7065	720.4	.243
85.0	6986	712.4	7470	761.8	.238
90.0	7391	753.7	7876	803.1	.233
95.0	7796	795.0	8281	844.4	.229
100.0	8201	836.2	8686	885.7	.225
105.0	8606	877.7	9091	927.0	.222
110.0	9011	918.8	9495	968.3	.219
115.0	9415	960.1	9901	1009.6	.216
120.0	9820	1001.4	10305	1050.9	.213
125.0	10225	1042.7	10711	1092.2	.211
130.0	10630	1083.9	11115	1133.5	.209

Table 14.2 *Longitudinal Loads for Modified B.G. loading-1987*

L (Loaded Length in metres)	Tractive Effort kN	t	Braking Force kN	t
1.0	81	8.3	62	6.3
1.5	81	8.3	62	6.3
2.0	164	16.7	123	12.5
2.5	164	16.7	123	12.5
3.0	164	16.7	123	12.5
3.5	245	25.0	166	16.9
4.0	245	25.0	184	18.8
4.5	245	25.0	184	18.8
5.0	245	25.0	184	18.8
5.5	245	25.0	184	18.8
6.0	245	25.0	185	18.9
6.5	327	33.3	221	22.5
7.0	327	33.3	221	22.5
7.5	327	33.3	221	22.5
8.0	409	41.7	276	28.1
8.5	409	41.7	276	28.1
9.0	409	41.7	279	28.1
9.5	409	41.7	276	28.1

(*Contd.*)

Table 14.2 (*Contd.*)

L (Loaded Length in metres)	*Tractive Effort*		*Braking Force*	
	kN	*t*	*kN*	*t*
10.0	490	50.0	331	33.8
11.0	490	50.0	331	33.8
12.0	490	50.0	331	33.8
13.0	490	50.0	368	37.5
14.0	490	50.0	368	37.5
15.0	490	50.0	368	37.5
16.0	572	58.3	386	39.4
17.0	572	58.3	386	39.4
18.0	654	66.7	441	45.0
19.0	654	66.7	441	45.0
20.0	735	75.0	496	50.6
21.0	735	75.0	500	51.0
22.0	735	75.0	516	52.6
23.0	735	75.0	533	54.3
24.0	735	75.0	552	56.3
25.0	735	75.0	565	57.6
26.0	817	83.3	581	59.2
27.0	817	83.3	597	60.9
28.0	899	91.7	613	62.5
29.0	981	100.0	662	67.5
30.0	981	100.0	662	67.5
32.0	981	100.0	686	70.0
34.0	981	100.0	755	75.0
36.0	981	100.0	751	76.6
38.0	981	100.0	784	79.0
40.0	981	100.0	816	83.2
42.0	981	100.0	848	86.5
44.0	981	100.0	881	89.8
46.0	981	100.0	913	93.1
48.0	981	100.0	945	96.4
50.0	981	100.0	978	98.7
55.0	981	100.0	1058	107.9
60.0	981	100.0	1140	116.2
65.0	981	100.0	1220	124.4
70.0	981	100.0	1301	132.7
75.0	981	100.0	1382	140.9
80.0	981	100.0	1463	149.2
85.0	981	100.0	1544	157.4
90.0	981	100.0	1625	165.7
95.0	981	100.0	1705	173.9
100.0	981	100.0	1787	182.2
105.0	981	100.0	1867	190.4
110.0	981	100.0	1949	198.7
115.0	981	100.0	2029	206.9
120.0	981	100.0	2110	215.2
125.0	981	100.0	2191	223.4
130.0	981	100.0	2272	231.7

For Meter Gauge (1000 mm)

(i) *Modified meter gauge loading*-1988 with maximum axle load of 156.9 kN for locomotive and a train load of 53.9 kN/m on both sides of locomotive with a maximum axle load of 137.29 kN for the train load.

(ii) *Standard main line* (M.L.) of 1929 for 129.4 kN axle load and a train load of 37.95 kN/m behind the engine.

(iii) *Standard branch line* (B.L.) of 1929 for 104.9 kN axle load and a train load of 37.95 kN/m behind the engine.

(iv) *Standard C* of 1929 for 79.4 kN axle load and a train load of 37.95 kN/m behind the engine.

Diagrams of modified loadings are shown in Fig. 14.17 to Fig. 14.20. Equivalent uniformly distributed loads for calculating maximum bending moments and shear forces are given in Table 14.3 and 14.4 respectively. Maximum tractive effort and braking load are given in Table 14.5.

Table 14.3 *EUDL for MG Loading and CDA*

L	*Total Load for Bending Moment*								
(m)	*MG Loading-1988*		*ML Standard of 1929*		*BL Standard of 1929*		*C Loading of 1929*		*CDA* $0.15 + \frac{8}{6+L}$
	kN	*t*	*kN*	*t*	*kN*	*t*	*kN*	*t*	
1.0	313.3	32.0	258.9	26.4	209.9	21.4	158.9	16.2	1.000
1.5	313.8	32.0	258.9	26.4	209.9	21.4	158.9	16.2	1.000
2.0	313.3	32.0	253.9	26.4	209.9	21.4	170.6	17.4	1.000
2.5	313.0	32.0	276.5	28.2	224.6	22.9	197.1	20.1	1.000
3.0	325.6	33.2	313.3	32.0	252.0	25.7	213.7	22.3	1.000
3.5	378.5	38.6	373.6	33.6	306.9	31.3	255.0	26.0	.992
4.0	428.6	43.7	428.6	43.7	347.2	35.4	283.4	23.9	.950
4.5	467.8	47.7	167.8	17.7	373.5	33.6	301.0	31.0	.912
5.0	501.1	51.1	498.2	50.3	106.1	41.1	331.5	33.3	.877
5.5	541.3	55.2	539.1	55.0	437.4	44.6	357.9	36.5	.846
6.0	581.5	59.8	581.5	59.3	470.7	48.0	381.5	38.9	.317
6.5	611.0	62.3	611.0	62.3	493.2	50.8	401.1	13.9	.790
7.0	644.3	65.7	611.3	65.7	531.5	51.2	423.6	43.7	.753
7.5	676.7	69.0	676.7	69.0	564.9	57.6	150.1	45.9	.713
8.0	713.9	72.8	713.9	72.3	593.3	60.5	472.7	81.2	.721
8.5	746.3	76.1	746.3	76.1	617.8	63.0	489.4	49.9	.702
9.0	773.7	78.9	773.7	78.9	610.7	65.3	506.0	51.6	.633
9.5	800.2	81.6	800.2	81.6	661.0	67.1	526.6	58.7	.666
10.0	827.7	84.4	827.7	84.4	683.5	69.7	545.2	55.6	.650
11.0	834.6	90.2	834.6	90.2	724.7	73.9	587.4	59.9	.621
12.0	953.2	97.2	953.2	97.2	734.5	80.0	636.5	61.9	.594
13.0	1015.0	103.5	1015.0	103.5	817.3	86.1	684.5	69.8	.571
14.0	1076.8	109.3	1076.8	109.3	903.2	92.1	733.5	74.3	.550
15.0	1137.6	116.0	1137.6	116.0	961.1	98.0	780.6	79.6	.531

(Contd.)

Table 14.3 (*Contd.*)

L (*m*)	*Total Load for Bending Moment*								CDA
	MG Loading-1988		*ML Standard of 1929*		*BL Standard of 1929*		*C Loading of 1929*		$0.15 + \frac{8}{6+L}$
	kN	*t*	*kN*	*t*	*kN*	*t*	*kN*	*t*	
16.0	1202.3	122.6	1202.3	122.6	1015.0	103.5	820.8	33.7	.514
17.0	1261.4	128.6	1261.4	128.6	1061.0	108.5	862.0	87.9	.493
18.0	1316.1	134.2	1316.1	134.2	1111.1	113.3	901.2	91.9	.483
19.0	1368.0	139.5	1368.0	139.5	1159.1	118.2	910.5	95.9	.480
20.0	1421.0	141.9	14214.0	141.9	1202.3	122.6	978.7	99.3	.453
21.0	1466.1	149.5	1466.1	149.5	1243.5	126.3	1920.9	101.1	.446
22.0	1518.1	154.8	1518.1	154.8	1233.6	131.4	1063.0	108.4	.430
23.0	1574.9	160.6	1574.9	160.6	1337.6	136.4	1105.2	112.7	.426
24.0	1625.9	165.8	1625.9	165.8	1380.8	140.3	1145.4	116.8	.417
25.0	1676.9	171.0	1676.9	171.0	1426.9	145.5	1189.5	121.3	.408
26.0	1743.6	177.8	1743.6	177.8	1482.8	151.2	1230.7	125.5	.408
27.0	1808.8	184.4	1808.8	184.4	1537.7	156.3	1275.8	130.1	.392
28.0	1869.1	190.6	1869.1	190.6	1588.7	162.0	1318.0	134.4	.386
29.0	1929.0	196.7	1929.0	196.7	1636.7	166.9	1357.2	133.4	.373
30.0	1998.7	203.0	1998.7	203.0	1635.8	171.9	1398.4	182.6	.373
32.0	2103.5	214.5	2103.5	214.5	1731.9	171.7	1490.6	152.0	.364
34.0	2221.2	226.5	2221.2	226.5	1891.7	192.9	1531.8	161.3	.352
36.0	2343.8	239.0	2343.8	239.0	2000.6	204.0	1666.1	169.9	.345
38.0	2466.4	251.5	2466.4	251.5	2102.5	214.4	1719.5	178.4	.332
40.0	2589.0	261.0	2589.0	261.0	2193.7	224.2	1830.9	186.7	.324
42.0	2691.9	274.5	2691.9	274.5	2290.8	233.6	1912.3	195.0	.317
44.0	2799.8	285.5	2799.8	285.5	2383.0	213.0	1994.7	203.4	.310
46.0	3902.8	296.0	3902.8	296.0	2471.3	252.0	2071.5	211.6	.304
48.0	3000.8	306.0	3000.8	306.0	2561.5	261.2	2154.5	219.7	.298
50.0	3093.9	316.0	3093.5	316.0	2650.7	270.3	2232.0	227.6	.293
55.0	3384.8	340.0	3384.3	310.0	2864.5	222.1	2121.3	246.9	.281
60.0	3624.7	369.6	3559.8	363.0	3068.5	312.9	2615.4	266.7	.271
65.0	3901.2	397.8	3775.6	335.0	3267.6	333.2	2811.6	236.7	.263
70.0	4177.8	426.0	3971.7	405.0	3172.5	354.1	3009.7	306.9	.255
75.0	4452.4	454.0	4182.5	426.5	3661.7	373.7	3205.3	326.9	.249
80.0	4727.0	482.0	4393.4	448.0	3854.0	393.0	3397.0	343.4	.243
85.0	5000.6	509.9	459.4	468.5	4059.0	413.9	3587.3	365.8	.233
90.0	5274.2	537.3	4786.6	488.1	4252.2	433.6	3780.5	335.5	.233
95.0	5547.3	565.7	4981.3	508.0	4440.0	452.3	3968.8	404.7	.229
100.0	5820.5	593.5	5177.9	528.0	4637.6	472.9	4150.2	423.2	.225
105.0	6092.1	621.2	5374.0	548.0	4823.9	491.9	4341.4	412.7	.222
110.0	6364.7	649.0	5565.3	567.5	5027.1	511.6	4524.3	461.4	.219
115.0	6636.4	676.7	5756.5	587.0	5202.4	530.5	4707.2	480.0	.216
120.0	6908.1	704.4	5952.6	607.0	5384.8	549.1	4903.2	500.5	.213
125.0	7179.7	732.1	6148.8	627.0	5571.2	563.1	5079.3	513.0	.211
130.0	7451.4	759.8	6335.1	646.0	5779.1	589.9	5281.8	538.9	.209

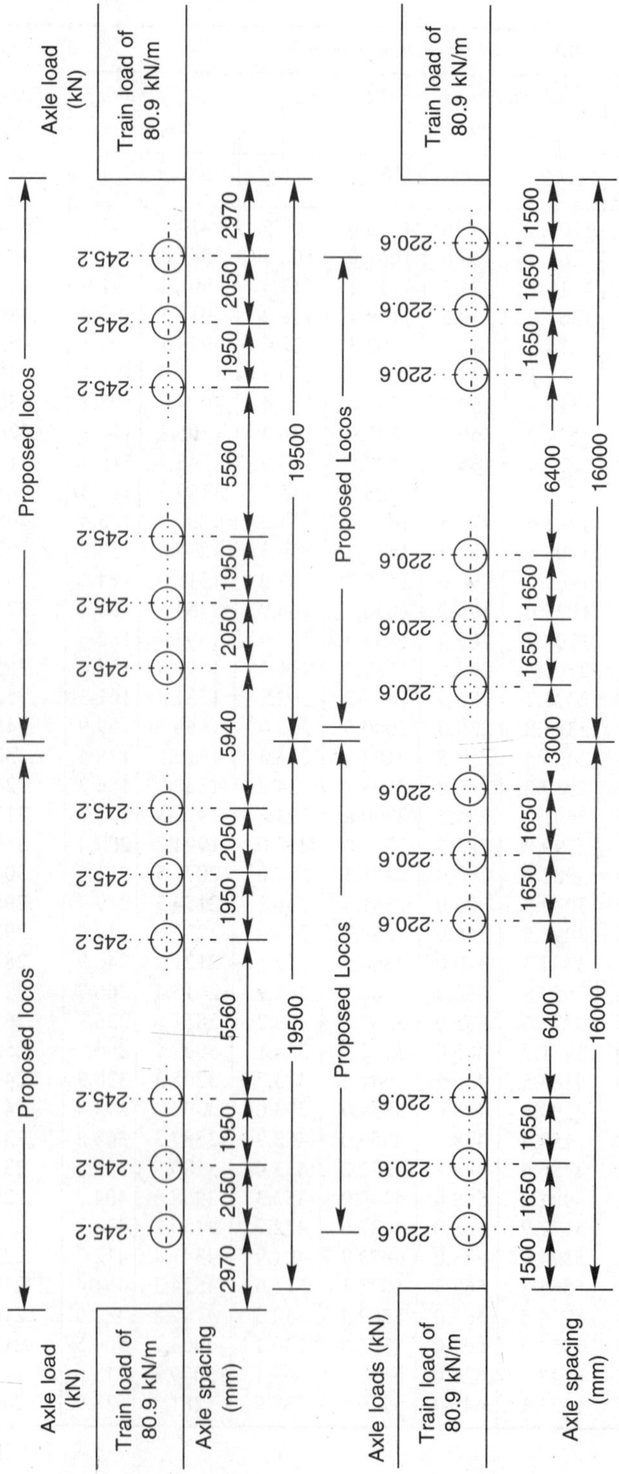

Longitudinal Forces

Tractive effort per loco	490.3 kN
Braking force per loco axle	25% of axle load
Braking force of train load	20% of train load

Fig. 14.16 *Modified Broad Gauge Loading-1987*

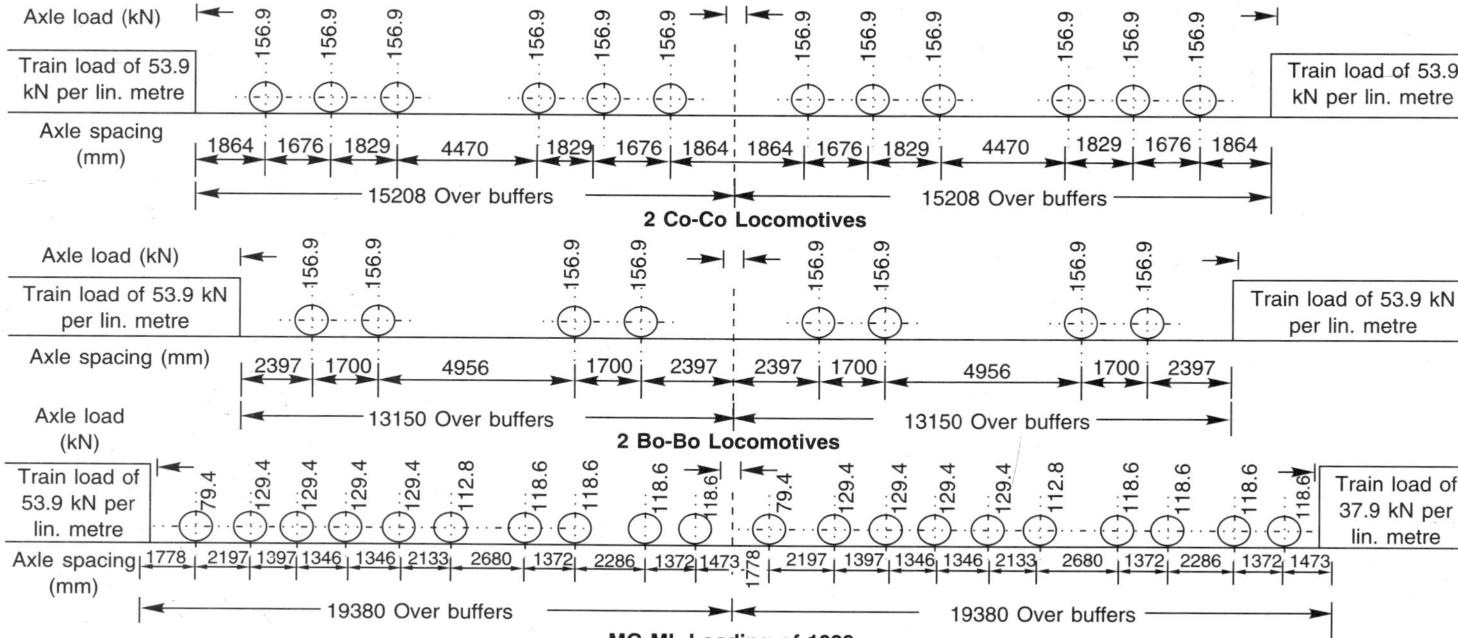

MG ML Loading of 1929

Intensity of Loading Per Locomotive = 61.13 kN/m

Longitudinal Forces	
Tractive effort per loco	313.7 kN for Co-Co Type and 235.3 kN for Bo-Bo Type
Braking force per loco axle	25% of axle load
Braking force of proposed MG train load	20% of train load

Fig. 14.17 *Modified Metre Gauge Loading 1988*

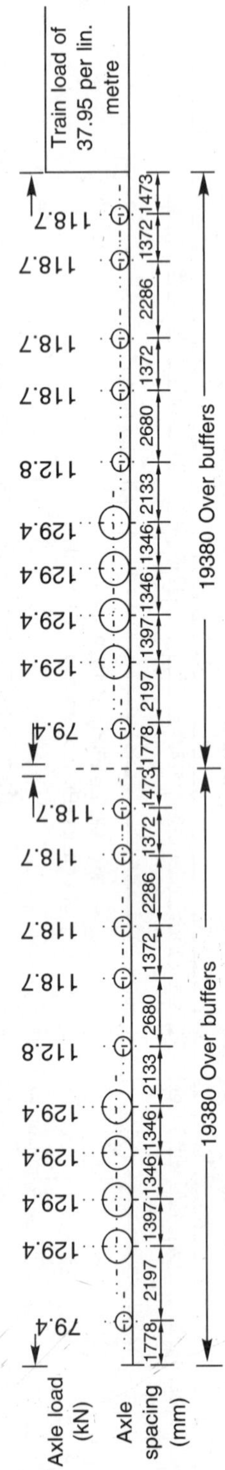

(Main Line)

Fig. 14.18 *MG Standard Loading of 1929*

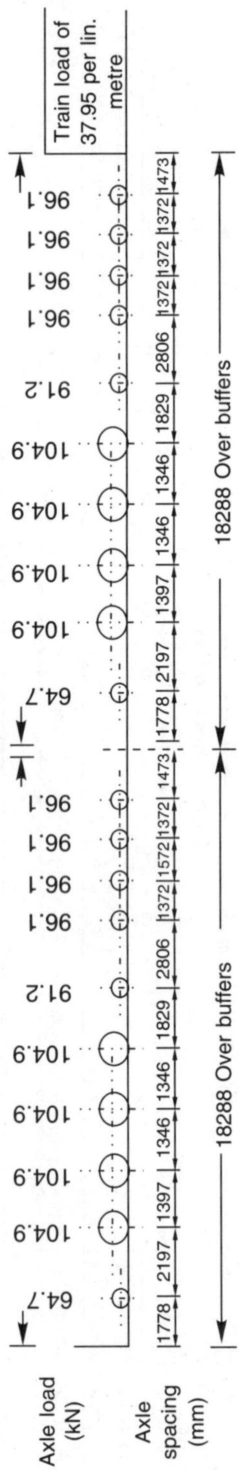

(Branch Line)

Fig. 14.19 *MG Standard Loading of 1929*

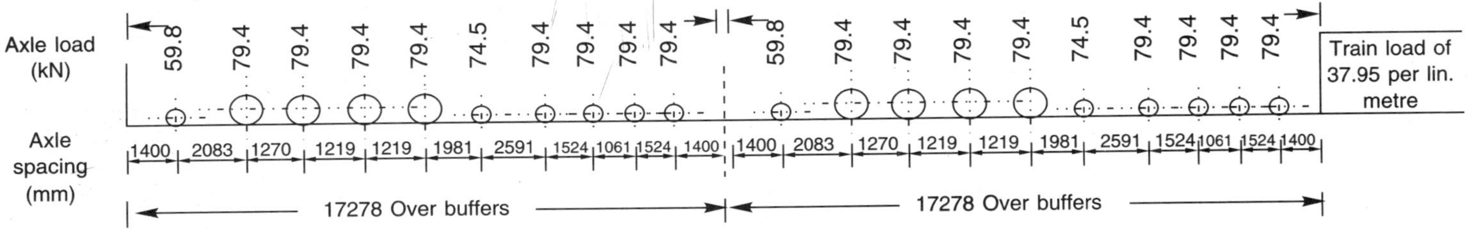

(C)

Fig. 14.20 *MG Standard Loading of 1929*

Table 14.4 *EUDL for M.G. Loading and CDA*

L (metres)	Total Load for Shear								CDA
	MG Loading-1988		ML Standard of 1929		BL Standard of 1929		C Loading of 1929		$0.15 + \frac{8}{6+L}$
	kN	t	kN	t	kN	t	kN	t	
1.0	313.8	32.0	258.9	26.4	209.9	21.4	158.9	16.2	1.000
1.5	313.8	32.0	285.4	29.1	231.4	23.6	138.3	19.2	1.000
2.0	364.8	37.2	313.2	35.0	278.5	28.4	233.4	23.3	1.000
2.5	416.8	42.5	378.5	38.6	306.9	31.3	250.1	25.5	1.000
3.0	452.1	46.1	428.6	48.7	347.2	35.4	283.4	28.9	1.000
3.5	478.6	48.8	478.6	48.8	386.4	39.4	310.9	31.7	.992
4.0	535.5	54.6	514.3	52.5	417.8	42.6	343.2	35.0	.950
4.5	580.6	59.2	567.8	57.9	460.9	47.0	375.6	33.3	.912
5.0	615.9	62.8	614.9	62.7	498.2	50.8	401.1	40.9	.377
5.5	653.1	66.6	653.1	66.6	529.6	54.0	422.7	43.1	.846
6.0	685.5	695.9	685.5	69.9	555.1	56.6	415.2	45.4	.817
6.5	716.9	73.1	716.9	73.1	589.4	60.1	472.7	48.2	.790
7.0	755.1	77.0	755.1	77.0	620.3	63.3	493.3	50.3	.765
7.5	790.4	80.6	690.4	80.6	647.2	66.0	512.9	52.3	.743
8.0	818.9	88.5	818.9	88.5	670.8	68.4	530.5	54.1	.721
8.5	845.4	86.2	845.3	86.2	692.3	70.6	549.2	56.0	.702
9.0	870.9	88.8	870.8	88.8	715.9	73.0	570.7	58.2	.633
9.5	904.2	92.2	904.2	92.2	741.4	75.6	590.4	60.2	.666
10.0	933.6	95.2	933.6	95.2	764.9	78.0	611.9	62.4	.650
11.0	1000.3	102.0	1000.3	102.0	321.8	83.8	658.0	67.1	.621
12.0	1061.1	108.2	1061.1	108.2	878.7	89.6	708.0	72.2	.594
13.0	1123.9	114.6	1123.8	114.6	936.5	95.5	758.1	77.3	.571
14.0	1182.7	120.6	1182.7	120.6	993.3	101.8	805.1	82.1	.550
15.0	1252.4	127.7	1252.3	127.7	1052.3	107.31	850.2	86.7	.531
16.0	1312.2	133.8	1312.1	133.8	1103.2	112.5	393.4	91.1	.514
17.0	1370.0	139.7	1370.0	139.7	1153.3	117.6	936.5	95.5	.493
18.0	1425.9	145.4	1425.9	115.4	1202.3	122.6	979.7	99.9	.483
19.0	1479.8	150.9	1479.8	110.9	1250.3	127.5	1020.9	104.1	.470
20.0	1531.8	156.2	1531.8	106.2	1297.4	132.3	1068.0	108.1	.453
21.0	1532.8	161.4	1582.8	161.4	1341.5	136.3	1104.2	112.0	.446
22.0	1642.7	167.5	1637.7	167.0	1887.6	141.5	1113.4	117.1	.436
23.0	1708.5	173.7	1700.5	173.4	1443.5	147.2	1193.5	121.7	.426
24.0	1767.2	130.2	1767.2	180.2	1483.5	152.3	1240.5	126.5	.417
25.0	1832.9	136.9	1832.9	186.9	1555.3	158.3	1236.7	130.9	.408
26.0	1896.6	193.4	1896.6	193.4	1608.3	164.0	1325.9	135.2	.400
27.0	1960.3	199.9	1960.3	199.9	1657.3	169.0	1369.0	139.6	.392
28.0	2020.2	206.0	2020.2	206.0	1710.3	174.4	1410.2	143.8	.385
29.0	2081.0	212.2	2031.0	212.2	1763.2	179.8	1461.2	149.0	.379
30.0	2143.7	213.6	2143.7	213.6	1817.2	135.3	1503.4	153.3	.372
32.0	2267.3	231.2	2267.3	231.2	1929.9	196.8	1599.5	163.1	.361
34.0	2394.8	244.2	2394.8	244.2	2035.9	207.6	1687.7	172.1	.350

(Contd.)

Table 14.4 *(Contd.)*

L (metres)	*Total Load for Shear*								*CDA* $0.15 + \frac{8}{6+L}$
	MG Loading-1988		*ML Standard of 1929*		*BL Standard of 1929*		*C Loading of 1929*		
	kN	*t*	*kN*	*t*	*kN*	*t*	*kN*	*t*	
36.0	2518.3	256.8	2518.3	256.8	2133.8	218.1	1775.0	131.0	.340
38.0	2635.0	268.7	2635.0	263.7	2239.8	228.4	1861.3	139.3	.332
40.0	2747.8	230.2	2747.3	230.2	2337.9	288.4	1946.6	193.5	.324
42.0	2857.7	291.4	2357.7	291.4	2433.0	213.1	2030.0	207.0	.317
44.0	2964.6	302.3	2964.6	302.3	2527.2	257.7	2114.3	215.6	.310
46.0	3069.5	313.0	3069.5	318.0	3019.4	267.1	2196.0	221.0	.304
48.0	3172.5	323.5	3172.5	323.5	2710.6	276.4	2279.1	232.4	.298
50.0	3269.5	333.4	3269.5	333.4	2830.2	283.6	2360.5	240.7	.293
55.0	3538.4	360.8	3509.8	357.9	3020.4	303.0	3562.5	261.3	.281
60.0	3817.9	389.3	3744.2	381.8	3235.2	329.9	2761.6	281.6	.271
65.0	4095.4	417.6	3968.8	404.7	3116.1	351.4	3961.6	302.0	.263
70.0	4372.0	145.8	4191.1	427.4	3654.9	372.7	3157.7	322.0	.255
75.0	4647.5	473.9	1114.0	150.1	3359.9	393.6	3355.3	312.2	.249
80.0	4922.1	501.9	4621.8	471.6	4062.9	414.3	3549.0	361.9	.243
85.0	5196.7	529.9	4831.7	492.7	4265.9	435.0	3742.2	331.6	.238
90.0	5470.3	557.8	5012.6	514.2	1166.9	155.5	3910.3	101.3	.233
95.0	5741.0	585.7	5323.0	543.3	4666.0	175.3	4132.5	121.4	.229
100.0	6026.6	613.5	5111.7	555.2	4365.1	496.1	4328.7	141.1	.225
105.0	6289.2	641.3	5653.5	567.5	5062.2	516.2	4532.6	162.2	.222
110.0	6561.9	669.2	5353.5	597.4	5259.3	536.3	4713.1	430.6	.219
115.0	6834.5	696.9	6057.6	617.7	5455.4	556.3	4995.3	500.2	.216
120.0	7106.2	724.6	6257.6	633.4	5651.6	576.3	5099.5	520.0	.213
125.0	7377.8	752.8	6155.7	658.3	5816.7	596.2	5291.7	539.6	.211
130.0	7619.5	780.0	6656.8	678.8	6041.9	616.1	5181.9	559.0	.209

Table 14.5 *Longitudinal Load (without deduction for MG (1000 mm) for dispersion)*

Loaded length in metres	*Tractive Effort kN (tonnes)*						*Braking Force kN (tonnes)*					
	ML		*BL*		*C*		*ML*		*BL*		*C*	
	t	*kN*	*t*	*kN*	*t*	*kN*	*t*	*kN*	*t*	*kN*	*t*	*kN*
1.0	9.1	89.2	7.4	72.5	5.6	54.9	5.8	56.9	4.7	46.1	3.5	34.3
1.5	8.8	86.3	7.2	70.6	5.4	52.9	5.7	55.9	4.6	45.1	3.5	34.3
2.0	8.6	84.3	7.0	68.6	5.7	55.9	5.6	54.9	4.6	45.1	3.7	36.3
2.5	8.9	87.3	7.3	71.6	6.0	62.8	6.0	58.0	4.9	48.8	4.3	42.2
3.0	9.9	97.1	7.9	77.5	6.9	67.7	6.7	65.7	5.4	52.9	4.7	46.1
3.5	11.6	113.8	9.4	92.2	7.8	76.5	8.0	78.5	6.5	63.7	5.4	52.9
4.0	12.3	125.5	10.4	101.9	8.5	83.4	9.0	88.3	7.3	71.6	6.0	53.8
4.5	13.7	134.4	11.1	108.8	8.9	87.3	9.7	95.1	7.9	77.4	6.3	61.8
5.0	14.2	139.3	11.5	112.8	9.5	93.2	10.3	101.1	8.3	81.4	6.9	67.7
5.5	15.1	148.1	12.2	119.6	10.0	98.1	11.1	108.8	9.0	88.3	7.3	71.6

(Contd.)

Table 14.5 *(Contd.)*

Loaded Length in metres	*Tractive Effort kN (tonnes)*						*Braking Force kN (tonnes)*					
	ML		*BL*		*C*		*ML*		*BL*		*C*	
	t	*kN*	*t*	*kN*	*t*	*kN*	*t*	*kN*	*t*	*kN*	*t*	*kN*
6.0	15.9	155.9	12.9	126.5	10.4	101.9	11.8	115.7	9.6	94.1	7.8	76.5
6.5	16.5	161.8	13.3	130.4	10.7	104.9	12.4	121.6	10.0	98.1	8.1	79.4
7.0	16.9	65.7	13.9	136.3	11.2	109.8	12.9	126.5	10.6	103.9	8.6	84.3
7.5	17.4	170.6	14.5	142.2	11.5	112.8	13.4	131.4	11.2	109.8	8.9	87.3
8.0	17.9	175.5	14.9	146.1	11.9	116.7	14.0	137.3	11.7	114.7	9.8	91.2
8.5	18.4	180.4	15.2	149.1	12.1	118.7	14.6	143.2	12.1	118.7	9.6	94.1
9.0	18.7	183.4	15.5	152.0	12.2	119.6	15.0	147.1	12.4	121.6	9.8	96.1
9.5	10.0	186.3	15.7	153.9	12.5	122.6	15.4	151.0	12.7	124.5	10.1	99.2
10.0	19.2	183.3	15.9	155.9	12.7	124.5	15.8	154.9	13.0	127.5	10.4	161.9
11.0	19.8	194.2	16.2	158.9	13.2	129.5	16.6	162.8	13.6	133.9	11.0	107.9
12.0	20.6	202.0	17.0	166.7	13.8	135.3	17.6	172.6	14.5	142.8	11.8	115.7
13.0	21.2	207.9	17.7	173.6	14.3	140.2	18.5	181.4	15.4	151.0	12.5	122.5
14.0	21.8	213.8	18.3	179.5	14.8	145.1	19.3	189.3	16.2	158.8	13.2	129.5
15.0	22.3	281.9	18.8	184.4	15.3	150.0	20.1	197.1	17.0	166.7	13.8	135.0
16.0	22.0	223.9	19.3	189.3	15.6	152.9	21.0	205.9	17.7	173.5	14.3	140.2
17.0	23.3	228.5	19.6	192.2	15.9	155.9	21.7	212.8	18.8	179.5	14.8	145.2
18.0	23.6	231.4	19.9	195.2	16.1	157.8	22.3	218.9	18.9	185.4	15.3	150.0
19.0	23.8	238.4	20.2	198.1	16.4	160.8	22.9	224.6	19.4	190.3	15.7	153.9
20.0	24.0	235.4	20.3	199.1	16.6	162.8	23.4	229.5	19.8	194.2	16.1	157.9
21.0	24.2	237.3	20.5	201.0	16.8	164.8	23.9	234.4	20.2	198.1	16.6	162.8
22.0	24.4	239.3	20.7	203.0	17.1	167.1	24.4	239.3	20.7	203.0	17.1	167.0
23.0	24.6	241.2	20.9	204.9	17.3	19.6	25.0	245.2	21.2	207.9	17.5	171.0
24.0	24.8	243.2	21.1	206.9	17.5	171.6	25.5	250.1	21.6	211.8	18.0	176.0
25.0	24.8	243.2	21.1	206.9	17.5	171.6	26.0	254.9	22.1	216.7	18.4	180.0
26.0	24.8	243.2	21.1	206.9	17.5	171.6	26.6	260.9	22.7	222.9	18.8	184.0
27.0	24.8	243.2	21.1	206.9	17.5	171.6	27.3	267.7	23.2	227.5	19.3	189.0
28.0	24.8	243.2	21.1	206.9	17.5	171.6	27.9	273.6	23.7	232.4	19.7	193.2
29.0	24.8	243.2	21.1	206.9	17.5	171.6	28.4	278.5	24.1	236.3	20.0	196.1
30.0	24.8	243.2	21.1	206.9	17.5	171.6	29.0	284.4	24.5	240.4	20.4	200.1
32.0	24.8	243.2	21.1	206.9	17.5	171.6	30.0	294.2	25.4	249.1	21.2	207.9
34.0	24.8	243.2	21.1	206.9	17.5	171.6	30.9	303.0	26.3	257.9	22.0	815.7
36.0	24.8	243.2	21.1	206.9	17.5	171.6	32.0	313.3	27.3	267.7	22.7	222.6
38.0	24.8	243.1	21.1	206.9	17.5	171.6	32.9	322.6	28.1	275.5	23.4	229.5
40.0	24.8	243.2	21.1	206.9	17.5	171.6	33.9	332.5	28.7	281.4	23.9	234.4
42.0	24.8	243.2	21.1	206.9	17.5	171.6	34.5	338.4	29.4	288.3	24.5	240.3
44.0	24.8	243.2	21.1	206.9	17.5	171.6	35.2	345.2	29.9	293.2	25.1	246.2
46.0	24.8	243.2	21.1	206.9	17.5	171.6	35.8	351.1	30.5	299.1	25.6	251.1
48.0	24.8	243.2	21.1	206.9	17.5	171.6	36.3	355.9	31.0	304.0	26.1	255.9
50.0	24.8	243.2	21.1	206.9	17.5	171.6	36.8	360.9	31.5	308.9	26.5	259.9
55.0	24.8	243.2	21.1	206.9	17.5	171.6	37.9	371.7	32.5	318.7	27.5	269.7
60.0	24.8	243.2	21.1	206.9	17.5	171.6	38.8	380.5	33.5	328.5	28.5	279.5
65.0	24.8	243.2	21.1	206.9	17.5	171.6	39.5	387.3	34.2	335.4	29.4	288.3
70.0	24.8	243.2	21.1	206.9	17.5	171.6	40.0	392.2	35.0	343.2	30.3	297.1

(Contd.)

Table 14.5 *(Contd.)*

Loaded Length in metres	Tractive Effort kN (tonnes)						Braking Force kN (tonnes)					
	ML		BL		C		ML		BL		C	
	t	kN	t	kN	t	kN	t	kN	t	kN	t	kN
75.0	24.8	243.2	21.1	206.9	17.5	171.6	40.6	398.2	35.6	349.1	31.2	305.9
80.0	24.8	243.2	21.1	206.9	17.5	171.6	41.2	404.0	36.2	355.0	31.9	312.8
85.0	24.8	243.2	21.1	206.9	17.5	171.6	41.7	408.9	36.8	360.8	32.6	319.7
90.0	24.8	243.2	21.1	206.9	17.5	171.6	42.1	412.8	87.4	366.9	33.2	325.6
95.0	24.8	243.2	21.1	206.9	17.5	171.6	42.4	415.8	37.8	370.7	33.8	331.5
100.0	24.8	243.2	21.1	206.9	17.5	171.6	42.8	419.7	38.3	375.6	34.3	336.4
105.0	24.3	243.2	21.1	206.9	17.5	171.6	43.2	428.6	38.8	380.5	34.9	342.3
110.0	24.8	243.2	21.1	206.9	17.5	171.6	43.5	423.7	39.2	334.4	35.4	347.2
115.0	24.8	243.2	21.1	206.9	17.5	171.6	43.8	429.5	39.6	388.8	35.9	352.1
120.0	24.8	243.2	21.1	206.0	17.5	171.6	44.2	433.5	40.0	392.2	36.4	356.9
125.0	24.8	243.2	21.1	206.9	17.5	171.6	44.5	436.4	40.3	395.2	36.8	360.9
130.0	24.8	243.2	21.1	206.9	17.5	171.6	14.8	439.3	40.8	400.1	37.4	366.8

For Narrow Gauge (762 mm)

(i) *Heavy* (*H*) class loading with a maximum axle load of 95.1 kN and a train load of 27.8 kN/m behind the engine.

(ii) *A-Class* main line loading with a maximum axle load of 79.4 kN and a train load of 27.8 kN/m behind the engine.

(iii) *B-class* branch line loading with a maximum axle load of 59.8 kN and a train load of 27.8 kN/m behind the engine.

Diagrams of standard loadings are shown in Figs 13.21 to 13.23. Equivalent uniformly distributed loads for calculating maximum bending moments and shear forces are given in Tables 13.6 and 13.7 respectively.

Table 14.6 *EUDL for N.G. (762 mm) and CDA*

L (Metres)	Total Load for Bending Moment						CDA
	Class Loading		Class Loading		Class Loading		91.5
	kN	t	kN	t	kN	t	91.5 + L
1.0	189.3	19.3	159.8	16.3	119.6	12.2	.989
1.5	189.3	19.3	159.8	16.3	119.6	12.2	.984
2.0	240.3	24.5	205.9	21.0	156.9	16.0	.979
2.5	30*.0	31.0	259.9	26.5	186.3	19.0	.973
3.0	398.8	34.5	279.5	28.5	215.7	22.0	.968
3.5	372.7	38.0	299.1	30.5	235.4	24.0	.963
4.0	392.3	40.0	343.2	35.0	250.1	25.5	.958
4.5	41.08	42.5	372.7	38.0	264.8	27.0	.953
5.0	447.3	45.0	397.2	40.5	289.3	29.5	.948
5.5	465.8	47.5	421.7	43.0	333.4	34.0	.943
6.0	425.4	49.5	441.3	45.0	372.7	38.0	.938
6.5	509.9	52.0	456.0	46.5	397.2	40.5	.934

(Contd.)

Table 14.6 *(Contd.)*

L (Metres)	Total Load for Bending Moment						CDA
	Class Loading		Class Loading		Class Loading		91.5
	kN	t	kN	t	kN	t	91.5 + L
7.0	544.3	55.5	470.7	48.0	416.8	42.5	.929
7.5	568.8	58.0	485.4	49.5	436.4	44.5	.924
8.0	588.4	60.0	505.0	51.5	446.2	45.5	.920
8.5	603.1	62.5	524.7	53.5	456.0	46.5	.915
9.0	617.8	63.0	549.2	56.0	465.8	47.5	.910
9.5	887.6	64.0	578.6	59.0	470.7	48.0	.906
10.0	642.3	65.5	608.0	62.0	475.6	48.5	.901
11.0	681.6	69.5	632.5	64.5	490.3	50.0	.893
12.0	750.2	76.5	657.0	67.0	509.9	52.0	.884
13.0	863.0	88.0	976.7	69.0	539.4	55.0	.876
14.0	931.6	95.0	711.0	72.5	588.4	60.0	.867
15.0	970.9	99.0	784.5	80.0	686.5	70.0	.859
16.0	1029.7	105.0	843.4	86.0	755.1	77.0	.851
17.0	1078.7	110.0	882.6	90.0	784.5	80.0	.843
18.0	1127.8	115.0	921.8	94.0	823.8	84.0	.836
19.0	1176.8	120.0	961.1	98.0	853.2	87.0	.828
20.0	1225.8	125.0	1000.3	102.0	882.6	90.0	.821
21.0	1274.9	130.0	1039.5	106.0	921.8	94.0	.813
22.0	1323.9	135.0	1088.5	111.0	951.2	97.0	.806
23.0	1372.9	140.0	1127.8	115.0	980.7	100.0	.799
24.0	1422.0	145.0	1176.8	120.0	1019.9	104.0	.792
25.0	1471.0	150.0	1225.8	125.0	1049.3	107.0	.785
26.0	1520.0	155.0	1274.9	130.0	1088.5	111.0	.779
27.0	1559.3	159.0	1323.9	135.0	1118.0	114.0	.772
28.0	1598.5	163.0	1363.1	139.0	1157.2	118.0	.766
29.0	1637.7	167.0	1412.2	144.0	1186.6	121.0	.759
30.0	1676.9	171.0	1451.4	148.0	1216.0	124.0	.753
32.0	1755.4	179.0	1539.9	157.0	1284.7	131.0	.741
34.0	1833.8	187.0	1618.1	165.0	1353.3	138.0	.729
36.0	1902.5	194.0	1696.6	173.0	1422.0	145.0	.718
38.0	1971.1	201.0	1775.0	181.0	1490.6	152.0	.707
40.0	2039.8	208.0	1853.5	189.0	1549.5	158.0	.696
42.0	2098.6	214.0	1922.1	196.0	1618.1	165.0	.685
44.0	2167.3	221.0	1980.9	202.0	1686.7	172.0	.675
46.0	2235.9	228.0	2049.6	209.0	1745.6	173.0	.665
48.0	2394.8	234.0	2108.4	215.0	1814.2	185.0	.656
50.0	2353.6	240.0	2177.1	222.0	1873.1	191.0	.647
55.0	2510.5	256.0	2343.8	239.0	2020.2	206.0	.625
60.0	2667.4	272.0	2490.9	254.0	2147.7	219.0	.604
65.0	2775.3	283.0	2638.0	269.0	2275.1	232.0	.585
70.0	2824.3	283.0	2785.1	284.0	2373.2	242.0	.567
75.0	2553.7	291.0	2932.2	299.0	2441.9	249.0	.550

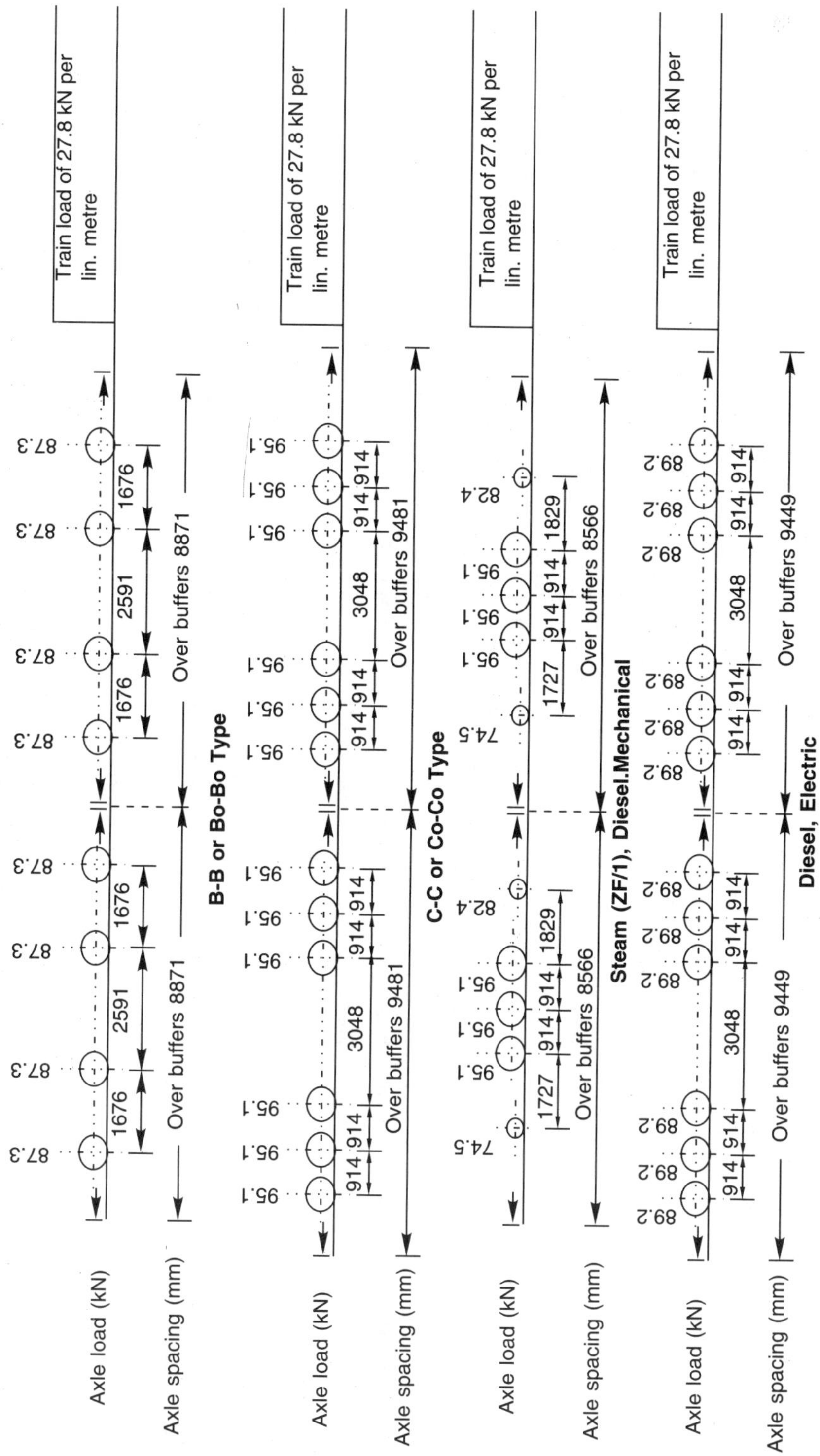

Fig. 14.21 *762 mm Gauge – H Class Loading*

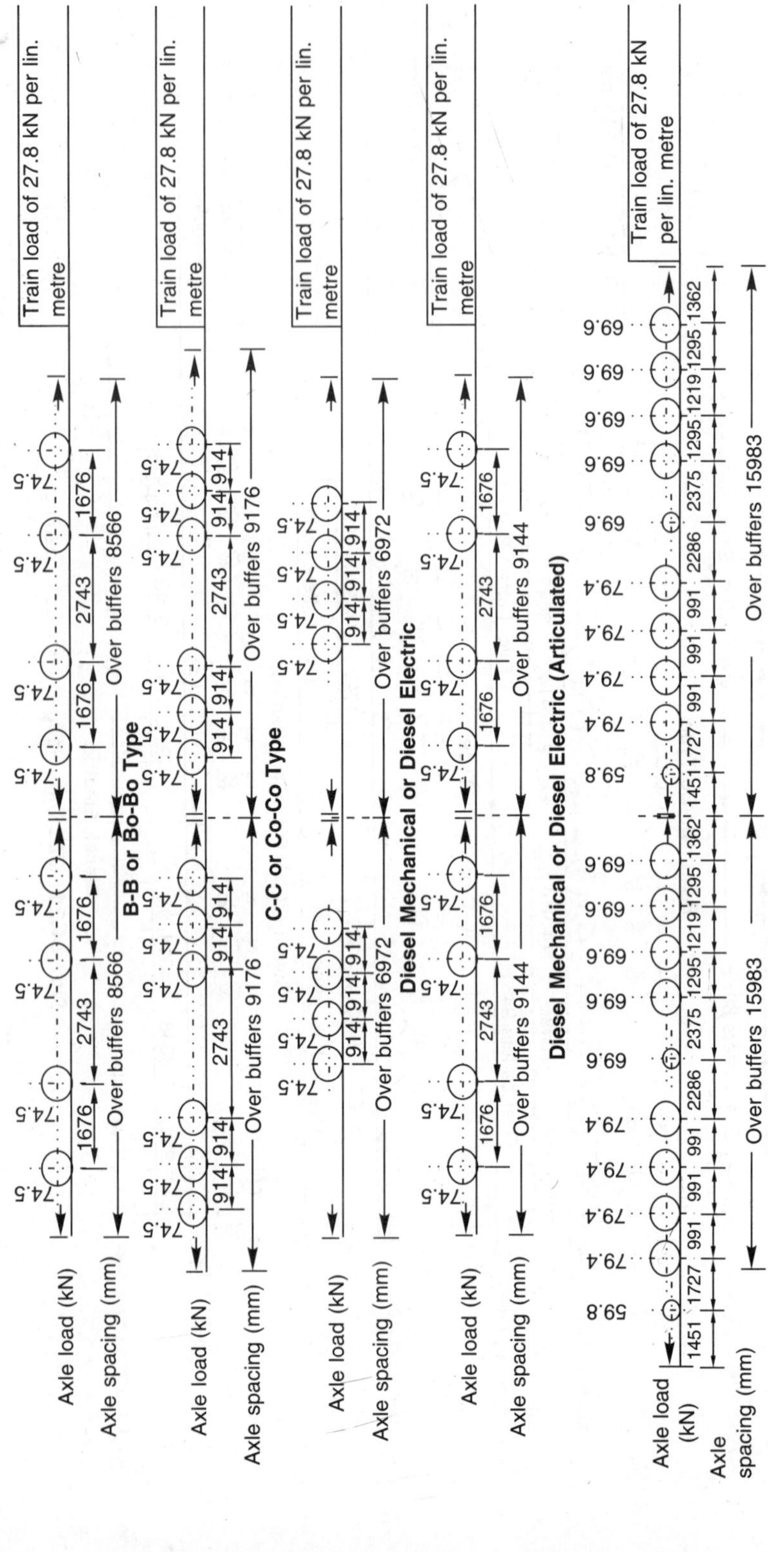

Engine as Per Drg. No. CSO/C-873

Fig. 14.22 *762 mm Gauge – A Class Loading*

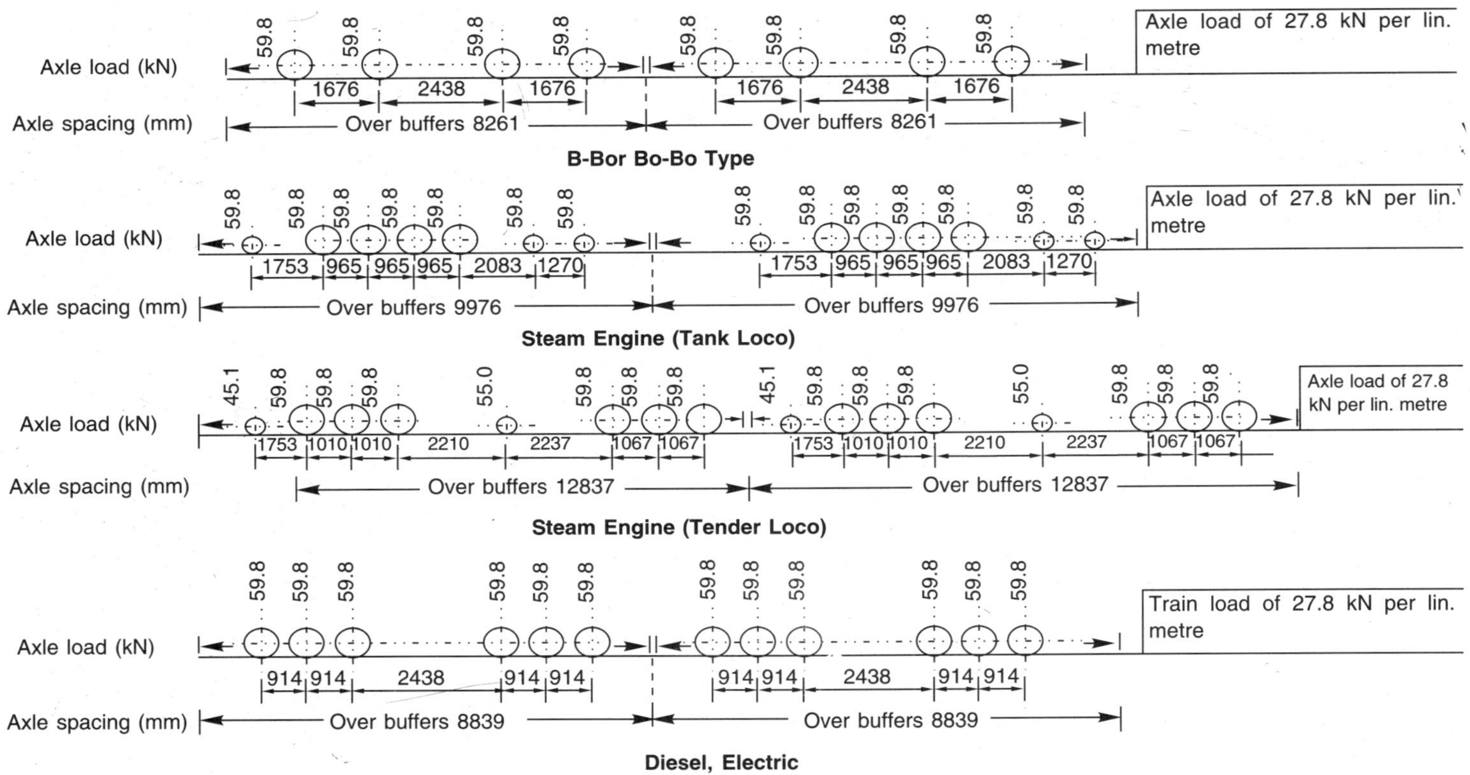

Fig. 14.23 *762 mm Gauge – B Class Loading*

Table 14.7 *EUDL for NG (762 mm) and CDA*

L (metres)	Total Load for Shear Class Loading kN	t	Class Loading kN	t	Class Loading kN	t	CDA 91.5 / 91.5 + L
1.0	237.3	24.2	189.3	19.3	149.1	15.2	.989
1.5	264.8	27.0	215.7	22.0	166.7	17.0	.984
2.0	308.9	31.5	250.1	25.5	196.1	20.0	.979
2.5	362.8	37.0	294.2	30.0	225.6	23.0	.973
3.0	397.2	40.5	323.6	33.0	350.1	25.5	.968
3.5	421.7	43.0	362.8	37.0	274.6	28.0	.963
4.0	451.1	46.0	392.3	40.0	304.0	31.0	.958
4.5	485.4	49.5	421.7	43.0	323.6	33.0	.953
5.0	505.0	51.5	451.1	46.0	343.2	35.0	.948
5.5	524.7	53.5	475.6	48.5	362.8	37.0	.943
6.0	544.3	55.5	500.1	51.1	382.5	39.0	.938
6.5	568.8	58.0	519.8	53.0	402.1	41.0	.934
7.0	598.2	61.0	539.4	55.0	421.7	43.0	.929
7.5	627.6	64.0	559.0	57.0	441.3	45.0	.924
8.0	657.0	67.0	578.6	59.0	460.9	47.0	.920
8.5	681.6	69.5	603.1	61.5	475.6	48.5	.915
9.0	711.0	72.5	617.8	63.0	495.2	60.5	.910
9.5	740.4	75.5	637.4	65.0	514.8	52.5	.906
10.0	769.8	78.5	657.0	67.0	534.5	54.5	.910
11.0	833.6	85.0	706.1	72.0	573.7	58.5	.893
12.0	897.3	91.5	750.2	76.5	612.9	62.5	.884
13.0	951.2	97.0	794.3	81.0	652.1	66.5	.876
14.0	1010.1	103.0	838.5	85.5	696.3	71.0	.867
15.0	1068.9	109.0	882.6	90.0	735.5	75.0	.859
16.0	1118.0	114.0	941.4	96.0	774.7	79.0	.851
17.0	1176.8	120.0	990.5	101.0	823.8	84.0	.843
18.0	1345.4	127.0	1039.5	106.0	863.0	88.0	.836
19.0	1314.1	134.0	1078.7	110.0	902.2	92.0	.828
20.0	1382.7	141.0	1127.8	115.0	941.4	96.0	.821
21.0	1441.6	147.0	1167.0	119.0	980.7	100.0	.813
22.0	1490.6	152.0	1206.2	123.0	1010.1	103.0	.806
23.0	1529.8	156.0	1255.3	128.0	1039.5	106.0	.799
24.0	1569.1	160.0	1294.5	132.0	1078.7	110.0	.792
25.0	1608.3	164.0	1343.5	137.0	1108.2	113.0	.785
26.0	1657.3	169.0	1382.7	141.0	1147.4	117.0	.779
27.0	1696.6	173.0	1422.0	145.0	1176.8	120.0	.772
28.0	1735.8	177.0	1471.0	150.0	1206.2	123.0	.766
29.0	1784.8	182.0	1510.2	154.0	1235.6	126.0	.759
30.0	1824.0	186.0	1559.3	159.0	1274.9	130.0	.753
32.0	1902.5	194.0	1637.7	167.0	1333.7	136.0	.741
34.0	1971.1	201.0	1726.0	176.0	1402.4	143.0	.729
36.0	1971.1	210.0	1726.0	176.0	1402.4	143.0	.789
36.0	2049.6	209.0	1804.4	184.0	1461.2	149.0	.718

(Contd.)

Table 14.7 (*Contd.*)

L	*Total Load for Shear*						*CDA*
(*metres*)	*Class Loading*		*Class Loading*		*Class Loading*		*91.5*
	kN	*t*	*kN*	*t*	*kN*	*t*	*91.5 + L*
38.0	2128.0	217.0	1882.9	192.0	1529.8	153.0	.707
40.0	2196.7	224.0	1951.5	199.0	1588.7	162.0	.696
42.0	2265.3	231.0	2030.0	207.0	1657.3	169.0	.685
44.0	2334.0	238.0	2098.6	214.0	1716.2	175.0	.675
46.0	2402.6	245.0	2167.3	221.0	1765.2	180.0	.665
48.0	2461.5	251.0	2235.9	228.0	1824.0	196.0	.565
50.0	2530.1	258.0	2304.6	235.0	1882.9	192.0	.647
55.0	2687.0	274.0	2461.5	251.0	2030.0	207.0	.625
60.0	2853.7	291.0	2628.2	268.0	2177.1	222.0	.604
65.0	3020.4	308.0	2785.1	284.0	2314.4	236.0	.585
70.0	3177.4	324.0	2942.0	300.0	2451.7	250.0	.567
75.0	3334.3	340.0	3089.1	315.0	2579.1	263.0	.550

14.14 IMPACT LOAD

The principal cause of impact under locomotives is the effect of unbalanced rotating weights in drivers. These weights are present in the form of counterweights used to balance the reciprocating parts of the locomotive mechanism. Figure 14.24 shows a locomotive driver with a counter weight at *B*. In the forward motion when the weight is at *A* or *C*, the centrifugal force due to the rotation of the wheel is directed downwards and when at *B* it is directed upwards. Thus, the drivers exert an alternating upward and downward force on the structure tending to set the structure into vibration. When the period of rotation of the moving weight and period of vibration of the structure are equal, the amplitude of vibration of the structure is cumulative and may cause large impact stresses. Figure 14.25(a) shows the deflection curve of a structure due to slowly moving loads and Fig. 14.25(b) shows it for a rapidly moving load. The latter shows the effect of alternating upward and downward forces on the deflection of the structure. As shown in Fig. 14.25(b), one half of the amplitude of the cumulative vibration represents the impact effect. The other causes of impact are the effect of roughness of track such as rail joints and play between wheels and rails, and eccentric wheels and defective springs under coaches which tend to produce a vibratory effect on bridge. But the effect of these secondary forces are not measured since their effect is no greater than those due to rotating counter weights.

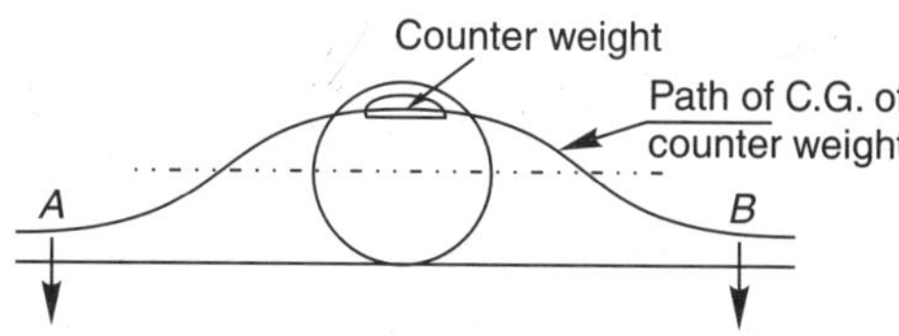

Fig. 14.24 *Locomotive Driver with Counter Weight*

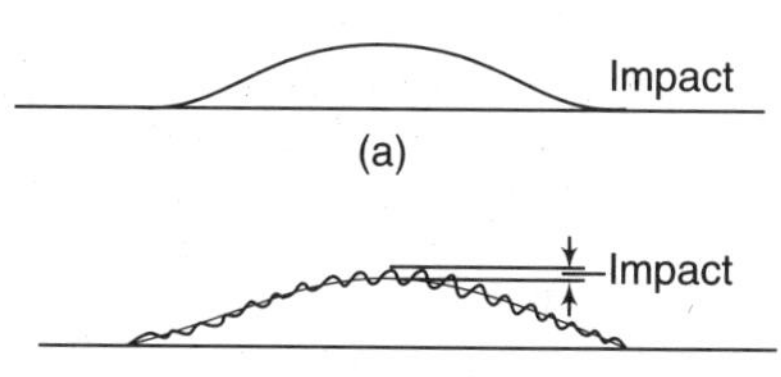

Fig. 14.25 *Deflection Curves due to Moving Load*

To account for impact, the stresses due to live loads, considered as static loads, are increased by an arbitrary factor known as the *impact coefficient* or *coefficient of dynamic augment* (CDA). CDA depends upon several factors like the type of loading, speed of the locomotive, loaded span length, type of structure, type of material, etc. It is not possible to incorporate all the parameters in one formula and it is usual to specify the impact factor involving only the loaded length. CDA should be obtained as follows.

For BG and MG CDA applicable up to 160 kmph for BG and 100 kmph for MG should be obtained by using the following relations.

(a) *For single track spans*

$$\text{CDA} = 0.15 + \frac{8}{6 + L} \quad \text{subject to maximum of one.}$$

where L is given by

(i) The loaded length of span in metres for the position of the train giving the maximum stress in the member under consideration.

(ii) 1.5 times the cross girder spacing in the case of stringers.

(iii) 2.5 times the cross girder spacing in the case of cross-girders.

(b) *For main girders of double track spans with two girders* CDA as calculated above may be multiplied by a factor of 0.72 and subjected to a maximum of 0.72.

(d) *For intermediate main girders of multiple track spans* CDA should be as specified in (a) multiplied by 0.6 and subjected to a maximum of 0.6.

(c) *For the outside main girder of multiple track spans with intermediate girders*, CDA should be as specified in (a) or (b) whichever is applicable.

(e) *For cross-girders carrying two or more tracks*, CDA as calculated in (a) or (b) may be multiplied by a factor of 0.72 and subjected to a maximum of 0.72.

(f) Where rails with ordinary fish plated joints are supported directly on transverse steel troughing or steel sleepers, the dynamic augment for calculating stresses in such troughing or sleepers should be taken as $\frac{7.32}{B + 5.49}$ for BG and $\frac{5.49}{B + 4.27}$ for MG

where B = the spacing of the main girders in metres.

For Narrow Gauge (NG) Railways (762 mm and 610 mm)

$$\text{CDA} = \frac{91.5}{91.5 + L}$$

where L = loaded length of the span as defined in (a).

14.15 WIND FORCE

The concept of wind load has undergone considerable change in the past few years. The code of practise now treats wind pressure as a dynamic force which in the past was expressed in terms of basic wind pressure (static pressure in direction of wind).

The intensity of wind pressure depends upon the wind velocity. The wind velocity depends upon the geographical location, terrain condition, topography, height, shape and size of structure, orientation of the elements and solidity ratio. The net wind pressure on a structure may be calculated by considering all the above factors. However, for the purpose of design, the basic wind pressure at various heights are adopted from maps and Tables given in IS 875, the total wind load on a bridge under consideration equals the wind pressure multiplied by the area obstructed in elevation with a minimum wind force per metre fixed by specifications to assure that the design is of a reasonable stiff structure. The obstructed area for the bridge is calculated as below.

(i) *For unloaded spans* The net exposed area should be considered as one and half times the horizontal projected area of the span, except for plate girders. For a plate girder, the area of the leeward girder should be multiplied by the factors listed below and added to the area of the windward girder.

(a) When the spacing of the leeward girder does not exceed half its depth	0.0
(b) For spacing exceeding half the depth and up to full depth	0.25
(c) For spacing exceeding full depth and up to one and half times depth	0.5
(d) For spacing exceeding one and half times depth and up to twice its depth or more	1.0

(ii) *For loaded spans* The net exposed area should be sum of

(a) One and half times that portion of the horizontal projected area of the span not covered by the moving load, except for plate girders for which the area of leeward girders not covered by the moving loads should be multiplied by the factors listed under (i) above and added to the area of the windward girder above or below the moving load.

(b) The horizontal projected area of the moving load.

Notes

1. Wind on moving loads is assumed to act at 1.5 m above the deck and may be assumed as 3 kN/m for ordinary high bridges.
2. No live load should be assumed if the wind velocity is taken more than 130 km/hr (wind pressure of 1.08 kN/m^2).
3. Wind load should not be less than 4.5 kN/m for a loaded chord and 2.25 kN/m for an unloaded chord for through and semi-through bridges.
4. Wind load should not be less than 4.5 kN/m for deck bridges.
5. The area of the moving load should be taken as from 600 mm above the rail level to the top of the highest stock for which the bridge is designed.
6. The bridge should not be considered to carry any live load when the wind pressure at deck level exceeds

1.47 kN/m^2	for BG bridge
0.98 kN/m^2	for MG bridge
0.74 kN/m^2	for NG bridge

7. A more rational analysis should be made for suspension and cable stayed bridges.

According to the bridge code, the entire country is divided in two zones—the coastal zone and the normal wind zone. However, IS 875 (Part 3) divides the country in six zones (Appendix IV).

Coastal Zone (Zone 1) These are along Orissa, Andhra Pradesh, West Bengal, Tripura and on the west coast of Kathiawar Peninsula. This is a high wind zone.

Normal Zone (Zone 2) All the zones not included in coastal zones.
The normal wind pressures on bridges should be taken from Table 14.8.

Table 14.8 *Normal Wind Pressure on Bridge (Zone 2)*

Height (m)	0	10	25	50	80	100
Pressure (kN/m^2)	0.8	+ 1.18	+ 1.42	+ 1.62	+ 1.77	+ 1.83

Note Double the values specified in Table 14.8 should be used for Zone 1.

14.16 SEISMIC FORCES

A bridge situated in a region subjected to earthquakes should be designed as per the specifications of IS 1893–1984 (which is still not revised for bridges). Seismic load and wind load should not be considered to act simultaneously.

For the purpose of determining seismic forces, the country is divided into five zones. The basic horizontal seismic coefficient α_0 for different zones is listed in Table 14.9. The design horizontal seismic coefficient, $\alpha_h = \beta I \alpha_0$, where β depends upon soil foundation system (Table 12.8) and I is a coefficient depending upon the importance of the structure, being 1.5 for important bridges and 1.0 for other bridges. An important bridge is one having a linear water way of 300 m or total waterway of 1000 sq m or more.

Table 14.9 *Basic Seismic Coefficient*

Zone No.	*Horizontal Seismic Coeefficient (a0)*
I	0.01
II	0.02
III	0.04
IV	0.05
V	0.08

The design vertical seismic coefficient may be taken as 0.5 times the horizontal seismic coefficient.

Horizontal seismic force due to live load on the bridge should be ignored when acting in the direction of the traffic, but when acting in the direction perpendicular to the traffic this is to be considered for 50 per cent of the design live load without impact.

$$\text{Seismic force to be resisted, } F = W_m\ \alpha_h \text{ (or } W_m\ \alpha_v)$$

where W_m = weight of the mass under consideration ignoring reduction due to buoyancy.

Notes Modal analysis is performed for the following bridges situated in Zone IV and V.

1. Cable stayed bridges, horizontally curved girder bridge, steel arch bridge, and
2. When the height of substructure from the base of the foundation to the top of pier is more than 30 m or when the bridge span is more than 120 m.

14.17 LONGITUDINAL FORCES

These may be caused due to one or more of the following.

1. Tractive effort caused through coupled or driving wheels.
2. Braking effect from the application of brakes to wheels.
3. Frictional resistance offered to the movement of free bearings due to change of temperature and other causes.

Notes

1. No increase for dynamic effects should be made for longitudinal loads.
2. For railway bridges the value of the longitudinal load due to either the tractive effort or the braking force for the loaded length (L) should be obtained from Tables 14.2 and 14.5.

The loaded length L should be taken equal to

(a) The length of one span when considering the effect of the longitudinal loads
   (i) on the girders
   (ii) on the stability of abutments
   (iii) on the stability of piers under the condition of one span loaded
   (iv) on the stability of piers carrying one fixed and one roller bearing

(b) The length of two spans when considering the condition of both the spans loaded, the stability of piers carrying fixed or sliding bearings. In this case the total longitudinal load should be divided between the two spans in proportion to their lengths.

(c) For determining the value of tractive effort, L should not exceed 34 m for BG bridge and 24 m for MG bridge.

14.18 RACKING FORCES

These are caused by the side swing of train and locomotive, play between the rails and wheels, sidesway of the rolling stock due to the action of car and engine springs and, due to unbalanced forces in the locomotive. The effect of the racking force is not considered in the design of main chord members, instead lateral bracing is designed for it. As per specifications it is taken to be 5.88 kN/m span. For spans less than 20 m a combined moving lateral load of 8.82 kN/m due to wind and racking force may be considered.

14.19 CENTRIFUGAL FORCE

For a bridge on curve, all of its members will be affected by the centrifugal action of the moving vehicles. The centrifugal force is given by

$$C = \frac{WV^2}{12.7\,R}$$

where C = centrifugal force in kN/m span
W = EUDL in kN/m run
V = maximum speed m kmph
R = radius of curve in m.

The centrifugal force can be assumed to act at a height of 1830 mm for BG and 1470 mm for MG above the rail level. The bridge members should be proportioned so as to carry safely the stresses induced by this action in addition to all other stresses to which they may be subjected.

14.20 TEMPERATURE EFFECTS

The variation in temperature results in expansion and contraction of the structural material. Usually a bridge is a determinate structure and as such will deform freely and thus no stresses are generated due to temperature changes. However, support or joint conditions may not permit free movement resulting in secondary stresses. Moreover bridge bearings do offer some resistance against movement. Indeterminate bridges do have temperature forces. The coefficient of expansion for steel is 12×10^{-6}/°C, for RCC is 11.7×10^{-6}/°C and that for concrete is 10.8×10^{-6}/°C. The recommended temperature variation of metal structures is –18°C to +50°C (moderate climate) and –35°C to +50°C (extreme climate).

14.21 SECONDARY STRESSES

In the analysis of trusses the following assumptions are made leading to primary stresses.

1. All members are straight and free to rotate at panel points.
2. All panel points lie at the intersection of gravity axes of members.
3. All loads including the weight of members lie at the panel points.

However, in practice these assumptions are not realised and the members are subjected to bending and shear stresses in addition to direct stresses. These stresses are known as *secondary stresses*. Other causes of secondary stresses are the elastic deformation of the structure combined with the rigidity of joints.

Notes

1. Secondary stresses generated due to unknown eccentricities arising out of inaccuracy in fabrication are already accounted for in the factor of safety.
2. The bridge should be designed, fabricated and erected in a way so as to minimise, as far as possible, secondary stresses.

14.22 DESIGN PROCEDURE OF BRIDGES

A bridge is actually a space structure. The analysis and design of a bridge, however, is simplified by assuming plannar and linear components, such as main trusses or main girders, floor beams, stringers, bracing frames, etc. The sequence of bridge design, as in building construction, arranges itself automatically into the order dictated by the necessary loading information. For instance the floor is carried by the stringers, therefore the design of the former must proceed before the stringer design.

The design procedure for truss girder and plate girder bridges is illustrated with the help of solved examples. Some of the specifications to be observed strictly are as below.

1. The thickness of the flat, plate, angle or, Tee-bar should not be less than 8 mm in the main members of the bridge when both sides are accessible for painting. However, if only one side is accessible the thickness should not be less than 10 mm. In floor plates and parapets a minimum thickness of 6 mm may be used.
2. In a riveted construction, angles for the main members of the girders should not be less than 75 × 50 mm. Also, angles less than 65 × 45 mm and flats less than 50 mm width should not be used in any part of bridge except for hand railing.
3. The distance between centres of trusses or girders should be sufficient to prevent overturning by the specified lateral forces. In no case it should be less than 1/20 of span for truss girders and 1/16 of span for plate girders.

4. The depth between gravity axes of the top and bottom chords should not be greater than three times the width between the centres of the main girders.
5. Depth of truss girder should preferably be not less than 1/10 of span
6. Depth fo plate girder and rolled beams should not be less than 1/12 of span.
7. Where provision for expansion and contraction due to change of temperature is necessary it should be provided to the extent of not less than 25 mm for every 30 m span.

Solved Examples

Example 14.1 An open web girder bridge consists of two pratt trusses of the type shown in Fig. Ex. 14.1. The span of the truss is 24 m c/c of bearings. The bridge supports an equivalent uniformly distributed live load of 175 kN per metre run. The dead load transmitted to each truss inclusive of self-weight is 15 kN/m. Design the members U_2U_3, U_4L_4, U_2L_3 and L_2L_3. Assume the impact factor to be 15%.

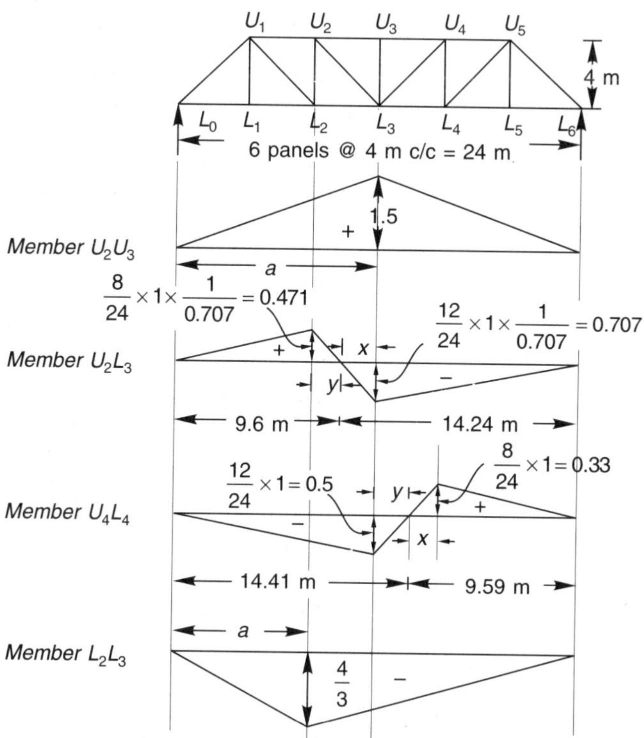

Fig. Ex. 14.1(a) *Influence Line Diagram for the Truss*

Solution

Member Forces

Member U_2U_3

It is a compression member.

$$\frac{a \times (l-a)}{l \times h} = \frac{12 \times (24-12)}{24 \times 4} = 1.5$$

$$\text{Area of I.L.D} = \frac{24}{2} \times 1.5 = 18 \text{ units}$$

$$\text{Dead load} = 15 \times 18 = 270 \text{ kN}$$

$$\text{Live load} = 175 \times 18 = 3150 \text{ kN}$$

$$\text{Impact load} = \frac{15}{100} \times 3150 = 472.5 \text{ kN}$$

$$\text{Design force} = 270 + 3150 + 472.5 = 3892.5 \text{ kN}$$

Member U_2L_3

$$\frac{0.707}{x} = \frac{0.471}{y}$$

and $$x + y = 4 \text{ m}$$

$\Rightarrow$ $x = 2.4$ m and $y = 1.6$ m

$$\text{Area of positive zone of I.L.D.} = \frac{1}{2} \times 0.471 \times 9.6 = 2.261 \text{ units}$$

$$\text{Area of negative zone of I.L.D.} = \frac{1}{2} \times 0.707 \times 14.4 = -5.09 \text{ units}$$

$$\text{Dead load} = 15 \times 2.261 + 15 \times (-5.09) = -42.44$$
$$= 42.44 \text{ kN (T)}$$

$$\text{Live load} = 175 \times 2.261 = 395.67 \text{ kN (C)}$$

$$\text{Corresponding impact load} = \frac{15}{100} \times 395.67 = 59.35 \text{ kN (C)}$$

$$\text{Live load} = 175 \times (-5.09) = -890.75 \text{ kN}$$
$$= 890.75 \text{ kN (T)}$$

$$\text{Corresponding impact load} = \frac{15}{100} \times 890.75 = 133.61 \text{ kN (T)}$$

$$\text{Design force} = -42.44 - 890.75 - 133.61 = 1066.8 \text{ kN (T)}$$

and, also $$= -42.44 + 59.35 + 395.67$$
$$= 412.58 \text{ kN (C)}$$

Member U_4L_4

$$\frac{0.33}{x} = \frac{1}{2y}$$

also $$x + y = 4$$

$\Rightarrow$ $x = 1.59$ m and $y = 2.41$ m

$$\text{Area of positive zone of I.L.D.} = \frac{1}{2} \times 0.33 \times 9.59 = 1.582 \text{ units (C)}$$

$$\text{Area of negative zone of I.L.D.} = \frac{1}{2} \times 0.5 \times 14.41 = -3.603 \text{ units}$$
$$= 3.603 \text{ units (T)}$$

$$\text{Dead load} = 15 \times 1.582 + 15 \times (-3.603)$$
$$= -30.315 \text{ kN} = 30.315 \text{ kN (T)}$$

$$\text{Live load} = 175 \times 1.582 = 276.85 \text{ kN (C)}$$

$$\text{Corresponding impact load} = \frac{15}{100} \times 276.85 = 41.53 \text{ (C)}$$

$$\text{Live load} = 175 \times (-3.603) = -630.525 \text{ kN}$$
$$= 630.525 \text{ kN (T)}$$

$$\text{Corresponding impact load} = \frac{15}{100} \times 630.525 = 94.579 \text{ kN (T)}$$

$$\text{Design force} = -30.315 - 630.525 - 94.579$$
$$= -755.42 \text{ kN} = 755.42 \text{ kN (T)}$$

and also
$$= -30.315 + 276.85 + 41.53$$
$$= 288.07 \text{ kN (C)}$$

Member L_2L_3

It is a tension member.

$$\frac{a \times (l - a)}{l \times h} = \frac{8 \times (24 - 8)}{24 \times 4} = \frac{4}{3}$$

$$\text{Area of ILD} = -\frac{1}{2} \times \frac{4}{3} \times 24 = -16.0 \text{ units}$$

$$\text{Dead load} = 15 \times (-16) = -240 \text{ kN} = 240.0 \text{ kN (T)}$$

$$\text{Live load} = 175\ (-16) = -2800 \text{ kN} = 2800.0 \text{ kN (T)}$$

$$\text{Impact load} = \frac{15}{100} \times 2800.0 = 420.0 \text{ kN (T)}$$

$$\text{Design force} = 240 + 2800 + 420 = 3460 \text{ kN (T)}$$

Design of members

Top chord U_2U_3

$$\text{Design force} = 3892.5 \text{ kN (C)}$$

$$\text{Length of member} = 4000 \text{ mm}$$

Assuming the member to have section as shown in Fig. Ex. 14.1(b).

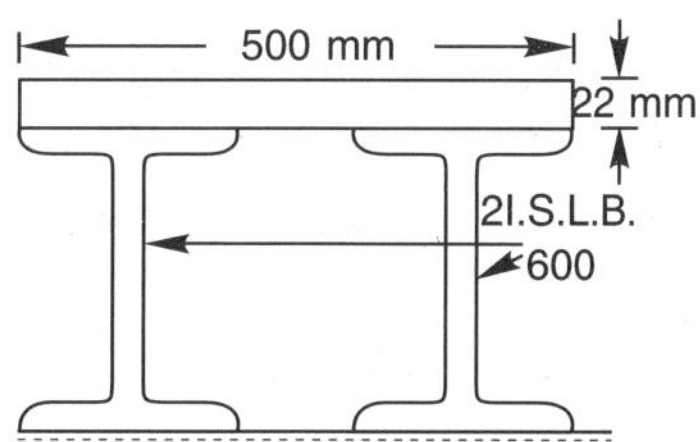

Fig. Ex. 14.1(b)

$$\text{Area} = \frac{3892.5 \times 10^3}{110} = 35386.36 \text{ mm}^2$$

$$I_{yy} = 2 \times \left[1821.9 \times 10^4 + 12669 \times \left(\frac{500}{2} - \frac{210}{2} \right)^2 \right] + \frac{22 \times 500^3}{12}$$

$$= 79833.6 \times 10^4 \text{ mm}^4$$

$$A = 2 \times 12669 + 500 \times 22 = 36338 \text{ mm}^2$$

$$r_y = \sqrt{\frac{79833.6 \times 10^4}{36338}} = 148.22 \text{ mm}$$

Let us provide rakers at every 8 m. The effective length of the member about the *yy* axis, l_y = 8000 mm.

$$\lambda = \frac{l_y}{r_y} = \frac{8000}{148.22} = 53.974$$

for $\lambda = 53.97$ and $f_y = 250$ N/mm^2,

$$\sigma_{ac} = 128.026 \text{ N/mm}^2$$

$$\text{Load carrying capacity} = 128.026 \times 36338 = 4652208.78 \text{ N}$$
$$= 4652.2 \text{ kN} > 3892.5 \text{ kN}$$

Member U_2L_3

$$\text{Tensile force} = 1066.80 \text{ kN}$$
$$\text{Compressive force} = 412.58 \text{ kN}$$
$$\text{Length of member} = \sqrt{4^2 + 4^2} = 5.656 \text{ m}$$

Let us try 4 angles laced together as shown in Fig. Ex. 14.1(c).

$$\text{Net area required} = \frac{1066.8 \times 10^3}{150} = 7112 \text{ mm}^2$$

$$\text{Gross area} = 4 \times 2102 = 8408 \text{ mm}^2$$

The 4 angles have been laced. The section will be weakened by 6 rivet holes.
Let us provide 20 mm ϕ rivets for connection.

$$\text{Net area provided} = 8408 - 6 \times 21.5 \times 10$$
$$= 7118 \text{ mm}^2$$

which as all right.

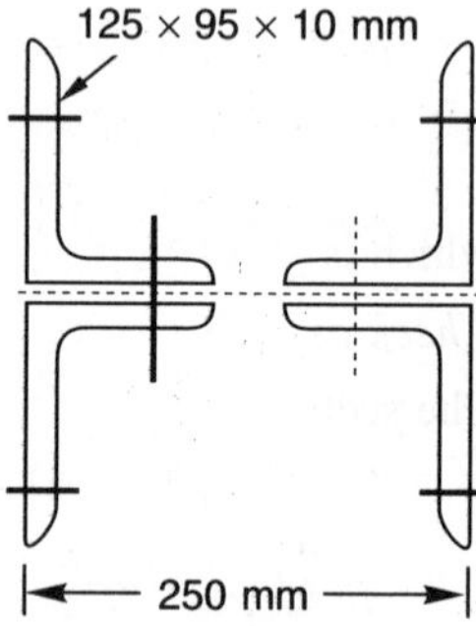

Fig. Ex. 14.1(c)

Check as a compression member

$$I_{xx} = 4 \times 325.8 \times 10^4 + 4 \times 2102 \times 38.8^2$$
$$= 2568.97 \times 10^4 \text{ mm}^4$$

$$I_{yy} = 4 \times 162.7 \times 10^4 + 4 \times 2102 \times \left(\frac{250}{2} - 23.9\right)^2 = 9244.79 \times 10^4 \text{ mm}^4$$

$$r_{xx} = \sqrt{\frac{2568.97 \times 10^4}{8408}} = 55.27 \text{ mm}$$

$$\lambda = \frac{0.85 \times 5.656 \times 10^3}{55.27} = 86.98$$

For $\lambda = 86.98$ and $f_y = 250$ N/mm^2,

$$\sigma_{ac} = 93.32 \text{ N/mm}^2$$

$$\text{Load carrying capacity} = 93.3 \times 8408$$
$$= 784634.56 = 784.64 \text{ kN}$$
$$> 412.58 \text{ kN}$$

which is all right.

Member U_4L_3

Compressive force = 288.07 kN

Tensile force = 755.42 kN

Length = 4000 mm

Let us try 4 angles as shown in Fig. Ex. 14.1(d)

$$\text{Area} = 4 \times 1538 = 6152 \text{ mm}^2$$

$$I_{xx} = 4 \times 245.5 \times 10^4 + 4 \times 1538 \times 41.5^2$$
$$= 2041.52 \times 10^4 \text{ mm}^4$$

$$r_{xx} = \sqrt{\frac{2041.52 \times 10^4}{6152}} = 57.60 \text{ mm}$$

$$\lambda = \frac{0.85 \times 4000}{57.60} = 59.03$$

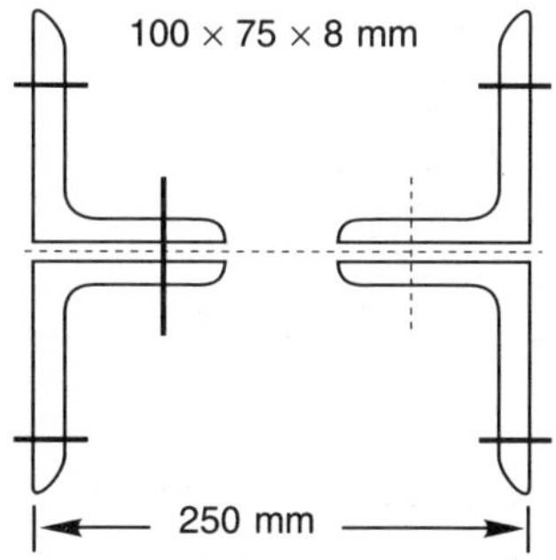

Fig. Ex. 14.1(d)

For $\lambda = 59.03$ and $f_y = 250$ N/mm^2,

$$\sigma_{ac} = 122.07 \text{ N/mm}^2$$

Load carrying capacity = 122.07 × 6152 = 750974.64 N = 750.97 kN

> 288.07 kN

which is all right.

Check as tension member

The section will be weakened by 6 rivet holes. Assuming 20 mm ϕ rivets.

Net area provided = 6152 – 4 × 21.5 × 18 = 5120 mm^2

Load carrying capacity = 5120 × 150 = 768 N = 768 kN

> 755.42 kN

which is all right.

Member L_2L_3

Tensile force = 3460 kN

Length = 4000 mm

$$\text{Net area required} = \frac{3460 \times 10^3}{150} = 23066.66 \text{ mm}^2$$

Let us try a section as shown in Fig. Ex. 14.1(e)

Net area provided

= 2 × 8314 + 500 × 20 – 2 × 23.5 × 20

= 25688 mm^2

Load carrying capacity = 25688 × 150

= 3853200 N

= 3853.2 kN

> 3460 kN

which is all right.

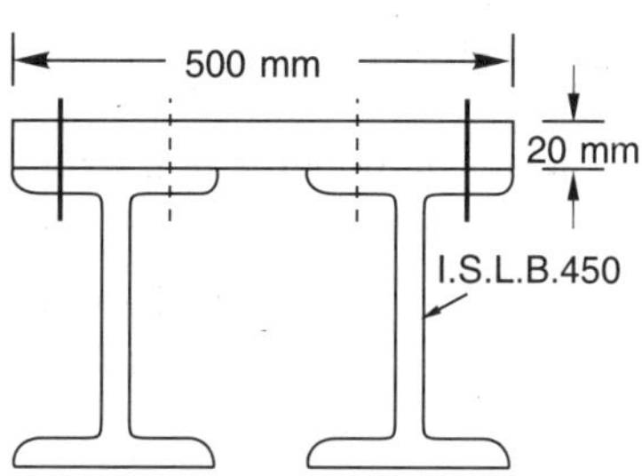

Fig. Ex. 14.1(e)

Example 14.2 Design a foot bridge for the following particulars.

Type of girder	N-type trusses
Span of girders	18 m c/c
Spacing of cross girders	2.25 m c/c
Clear walking width between main girders	3 m
Live load	4.0 kN/m^2

Flooring: Timber planks on cross girders.

Design the timber planks, cross girders, rakers, central top chord and bottom chord members.

Solution Let us provide N-type truss with eight panels 2.25 m each, height of the truss 1.8 m and cross-girders at the panel points.

Timber planks

$$\text{Span} = 2.25 \text{ m } (= \text{spacing of cross-girders})$$

$$\text{Live load} = 4.0 \text{ kN/m}^2 = 4000 \text{ N/m}^2$$

Assume thickness of the planks as 65 mm @ 8000 N/m^3.

$$\text{Self weight of planks} = 8000 \times .065 = 520 \text{ N/m}^2$$

$$\text{Total load} = 4000 + 520 = 4520 \text{ N/m}^2$$

Consider 1.0 m width of the planks.

$$\text{Maximum bending moment} = \frac{wl^2}{8} = \frac{4520 \times 2.25^2}{8} = 2860.3 \text{ Nm}$$

$$\text{Moment of resistance} = \sigma_{bct} \times \frac{1}{6}\, bd^2$$

Equating the moment of resistance to maximum bending moment,

$$\frac{1}{6} \times \sigma_{bct} \times 1000 \times 65^2 = 2860.3 \times 10^3$$

$$\Rightarrow \qquad \sigma_{bct} = 4.06 \text{ N/mm}^2 < 10.0 \text{ N/mm}^2$$

which is all right.

$$\text{Maximum shear force} = \frac{wl}{2} = \frac{4520 \times 2.25}{2} = 5085 \text{ N}$$

$$\text{Average shear stress, } \tau_{va} = \frac{5085}{1000 \times 65} = 0.0782 \text{ N/mm}^2$$

$$\text{Maximum shear stress, } \tau_{vm} = 1.5 \times 0.0782 = 0.1173 \text{ N/mm}^2 < 0.8 \text{ N/mm}^2$$

which is all right.

$$\text{Moment of inertia of the plank} = \frac{1000 \times 65^3}{12} = 2288.5 \times 10^4 \text{ mm}^4$$

$$\text{Maximum deflection} = \frac{5}{384}\frac{wl^4}{EI}$$

$$= \frac{5}{384} \times \frac{4520 \times 2.25 \times 2250^3}{1.0 \times 10^4 \times 2288.5 \times 10^4} = 6.59 \text{ mm}$$

$$\text{Permissible deflection} = \frac{\text{span}}{325} = \frac{2250}{325} = 6.92 \text{ mm} > 6.59 \text{ mm}$$

which is as it should be.

Cross-beams

Clear walking width between main girders = 3.0 m

Assume the length of the cross-girder to be a little more, say by 0.3 m. This will also be equal to the c/c distance between the lattice girders as the cross-girders will be supported over panel points of trusses.

$$\text{Length of cross girder} = 3.0 + 0.3 = 3.3 \text{ m}$$

Load transmitted to the cross-girder from the floor,

$$W_1 = 3 \times 2.25 \times 4520 = 30510 \text{ N}$$

Assume self-weight of cross beam to be 150 N/m.

$$\text{Total load, } W = 30510 + 3.3 \times 150 = 31005 \text{ N}$$

$$\text{Maximum bending moment} = \frac{WL}{8} = \frac{31005 \times 3.3}{8} = 12789.56 \text{ Nm}$$

$$\text{Required section modulus, } Z = \frac{12789.56 \times 10^3}{165} = 77512.5 \text{ mm}^3$$

Let us try ISLC 150 @ 141.2 N/m. From IS Handbook No 1, the relevant properties of the section are:

$$Z = 93000 \text{ mm}^3,\ t = 4.8 \text{ mm},\ I_{xx} = 697.2 \times 10^4 \text{ mm}^4$$

Checks

(i) *Shear force* $= \dfrac{31005}{2} = 15502.5$ N

$$\text{Shear stress, } \tau_{va} = \frac{15502.5}{150 \times 4.8} = 21.53 \text{ N/mm}^2$$

$$< 100 \text{ N/mm}^2 \qquad (0.4 f_y)$$

which is all right.

$$\text{Permissible deflection} = \frac{\text{span}}{325} = \frac{3.3 \times 10^3}{325} = 10.15 \text{ mm}$$

(ii) *Maximum deflection* $= \dfrac{5}{384} \times \dfrac{wl^4}{EI}$

$$= \frac{5}{384} \times \frac{31005 \times (3.3 \times 10^3)^3}{2 \times 10^5 \times 697.2 \times 10^4}$$

$$= 10.40 \text{ mm} > 10.15 \text{ mm}$$

Note Since the cross-beam is failing marginally (by 0.25 mm) in deflection I.S.L.C. 150 may be provided. If the designer feels, a camber of 3 mm or so may be provided at midspan so as to bring down the maximum deflection to 7.4 mm (10.4 – 3.0 = 7.4). As an other alternative I.S.M.C. 150 may be provided.

Forces in the truss members

The N-type truss used is shown in Fig. Ex. 14.2(a).

$$\text{Dead load from the cross-beam} = \frac{31005 - 4000 \times 2.25 \times 3.0}{2 \times 2.25} = 890 \text{ N/m.}$$

Note Only 3 m span of cross girder is loaded.

Assume weight of each truss = 350 N/m.

Total dead load on each truss = 890 + 350 = 1240 N/m

$$\text{Live load on each truss} = \frac{4000 \times 3}{2} = 6000 \text{ N/m}$$

Total load on each truss = 1240 + 6000 = 7240 N/m

To find forces in the members of the truss let us draw influence line diagrams.

Top chord

It is a compression member. Maximum force in this member will occur when the whole span is loaded. Because of symmetry forces in only left half of the members are computed.

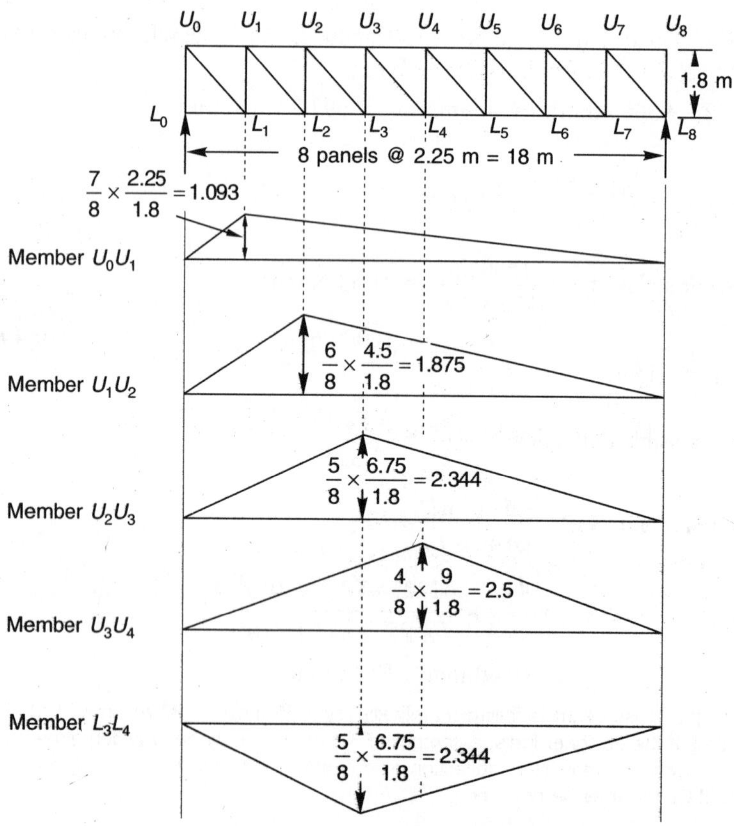

Fig. Ex. 13.2(a) *Influence Line Diagram for Member Forces*

Member U_0U_1

$$\text{Maximum force} = 7240 \times \frac{1}{2} \times 18 \times 1.093 = 71219.88 \text{ N (C)}$$

Member U_1U_2

$$\text{Maximum force} = 7240 \times \frac{1}{2} \times 18 \times 1.875 = 122175 \text{ N (C)}$$

Member U_2U_3

$$\text{Maximum force} = 7240 \times \frac{1}{2} \times 18 \times 2.344 = 152735.04 \text{ N (C)}$$

Member U_3U_4

$$\text{Maximum force} = 7240 \times \frac{1}{2} \times 18 \times 2.5 = 162900 \text{ N (C)}$$

Central bottom chord

It is tension member. Members L_3L_4 and L_4L_5 will be stressed the most.

Member L_3L_4 or L_4L_5

$$\text{Maximum force} = 7240 \times \frac{1}{2} \times 18 \times 2.344 = 152735.04 \text{ N (T)}$$

Design of central top chord members U_3U_4 and U_4U_5

Since members U_3U_4 and U_4U_5 are subjected to maximum compressive force, these are designed and same section will be provided for all the other members of top chord.

$$\text{Maximum compressive force} = 162900 \text{ N}$$

$$\text{Area required} = \frac{162900}{80} = 2036.25 \text{ mm}^2$$

Try 2 ISA 70 × 70 × 8 mm. From IS Handbook No. 1, the relevant properties are:

$$A = 2116 \text{ mm}^2, r_{\min} = 21.2 \text{ mm}$$

$$\lambda = \frac{0.85 \times 2250}{21.2} = 90.21$$

For $\lambda = 90.21$ and $f_y = 250$ N/mm^2,

$$\sigma_{ac} = 89.79 \text{ N/mm}^2$$

$$\text{Load carrying capacity} = 89.79 \times 2116 = 189995.64 \text{ N} > 162900 \text{ N}$$

which is all right.

Design of central bottom chord members L_3L_4 and L_4L_5

Since members L_3L_4 and L_4L_5 are subjected to maximum tensile force, these are designed and same section will be provided for all the other members of lower chord.

$$\text{Maximum tensile force} = 152735.04 \text{ N}$$

$$\text{Net area required} = \frac{152735.04}{150} = 1018.23 \text{ mm}^2$$

Try 2ISA 60 × 60 × 6 mm. From IS Handbook No. 1 the relevant properties of the section are:

$$A = 1368 \text{ mm}^2, r_{min} = 18.2 \text{ mm}$$

Let us provide 16 mm ϕ rivets for connection. Assuming the two angles to be provided on opposite side of the gusset plate and that the section is weakened by one rivet hole,

$$\text{Net area provided} = 1368 - 17.5 \times 6 \times 2 = 1158 \text{ mm}^2$$

$$\text{Load carrying capacity} = 150 \times 1158 = 173700 \text{ N} > 152735.04 \text{ N}$$

which is all right.

Design of the raker

Rakers are provided at an angle θ with the vertical and at 0.4 m from the trusses as shown in Fig. Ex. 14.2(b).

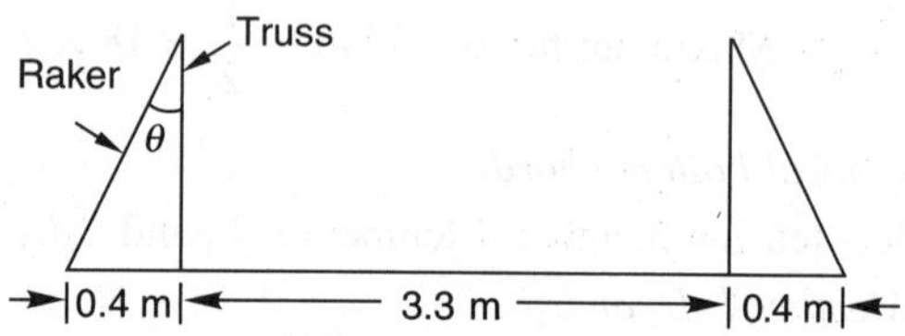

Fig. Ex. 13.2(b)

$$\tan \theta = \frac{0.4}{1.8}$$

$\Rightarrow$ $\theta = 12° \, 31' 43''$

The horizontal component of thrust in the raker is assumed to be 2.5% of the maximum compressive force in the top chord.

If T is the thrust in the raker, then

$$T \sin \theta = \frac{2.5}{100} \times 162900$$

or $T \times \sin 12° \, 31' \, 43'' = 4072.5$

$\Rightarrow$ $T = 18773.58 \text{ N} \simeq 18774 \text{ N}$

$$\text{Area} = \frac{18774}{60} = 312.89 \text{ mm}^2$$

Try ISA 60 × 60 × 8 mm. From IS Handbook No. 1 the relevant properties of the section are:

$$A = 896 \text{ mm}^2, r_{min} = 11.5 \text{ mm}$$

Let us provide more than one rivet for connection of the raker with the top chord and bottom.

$$l = 0.85 \times L$$

$$L = 1.8 \sec 12° \, 31' \, 43''$$

$$= 1.843 \text{ m} = 1843 \text{ mm}$$

$$\lambda = \frac{0.85 \times 1843}{11.5} = 136.22 < 140$$

which is as it should be.

For $\lambda = 136.22$ and $f_y = 250 \text{ N/mm}^2$,

$$\sigma_{ac} = 53.268 \text{ N/mm}^2$$

$$\text{Load carrying capacity} = 53.268 \times 896 = 47728.12 \text{ N} > 18774 \text{ N}$$

which is all right.

Example 14.3 A deck type plate girder railway bridge is to be constructed for a broad gauge single line track on the main line. The following data is available.

Effective span = 20 m
c/c distance between plate girders = 2 m
Dead load on each girder = 220L + 600 N/m
Dead load of track with sleepers = 6800 N/m
Lateral load = 9000 N/m

Design the superstructure of the bridge with welded plate girders.

Solution Two plate girders at a c/c distance of 2 m will be provided to support the track.

$$\text{Dead load of each girder} = 220 \times 20 + 600 = 5000 \text{ N/m}$$

$$\text{Dead load from track per girder} = \frac{6800}{2} = 3400 \text{ N/m}$$

$$\text{Total dead load} = 5000 + 3400 = 8400 \text{ N/m}$$

$$\text{Total dead load} = 8400 \times 20 = 168{,}000 \text{ N} = 168 \text{ kN}$$

Live load

Refer to Table 14.1. (For 20 m span and B.G.)

$$\text{Live load for bending moment} = 1964 \text{ kN}$$

$$\text{Live load for bending moment per girder} = \frac{1964}{2} = 982 \text{ kN}$$

$$\text{Live load for shear force} = 2168 \text{ kN}$$

$$\text{Live load for shear force per girder} = \frac{2168}{2} = 1084 \text{ kN}$$

$$\text{Impact factor} = 0.458$$

$$\text{Maximum bending moment due to dead load} = \frac{168 \times 20}{8} = 420 \text{ kNm}$$

$$\text{Maximum bending moment due to live load} = \frac{982 \times 20}{8} = 2455 \text{ kNm}$$

$$\text{Bending moment due to impact} = 0.458 \times 2455 = 1124.39 \text{ kNm}$$

$$\text{Design bending moment} = 420 + 2455 + 1124.39$$
$$= 3999.39 \text{ kNm} \simeq 4000 \text{ kNm}$$

$$\text{Maximum shear force due to dead} = \frac{168}{2} = 84 \text{ kN}$$

$$\text{Maximum shear force due to live load} = \frac{1084}{2} = 542 \text{ kN}$$

$$\text{Shear force due to impact} = 0.458 \times 542 = 248.23 \text{ kN}$$

$$\text{Design shear force} = 84 + 542 + 248.23 = 874.23 \text{ kN}$$
$$\simeq 880 \text{ kN}$$

Notes

1. Plate girders will be designed for a bending moment of 4000 kNm and shear force of 880 kN.
2. Lateral bracings connecting the main girders and also the end cross frames are designed to transmit the racking forces, wind pressure effects and the centrifugal forces, if any, to the bearings. These are specified in the Bridge Rule. However, in the present example these lateral forces are given to be 9000 N/m. In addition to this the lateral bracing between compression flanges of all spans should also be able to resist a transverse shear of 2.5% of the total compressive force carried by both the flanges at the section under consideration. However, if the sleepers rest directly on the compression flanges and thus prevent their lateral buckling, this additional transverse shear may be neglected.

Plate girder

$$\text{Span} = 20 \text{ m}$$
$$\text{B.M} = 4000 \text{ kNm}$$
$$\text{S.F} = 880 \text{ kN}$$

Web

Assuming that the flanges of the plate girder will be curtailed, the economical depth is given by

$$d = 1.1\sqrt{\frac{M}{\sigma_{bc,t}\; t}}$$

Let the thickness of web be 10 mm.

$$d = 1.1\sqrt{\frac{4000 \times 10^6}{165 \times 10}} = 1712.69 \text{ mm}$$

Let us provide 1800 × 10 mm web.
Minimum web thickness from shear consideration

$$t = \frac{880 \times 10^3}{100 \times 1800} = 4.88 \text{ mm} < 10 \text{ mm}$$

which is as it should be.

Flanges

$$A_f = \frac{M}{\sigma_{bc,t}\, d_1} = \frac{4000 \times 10^6}{165 \times 1800} = 13468 \text{ mm}^2$$

The required flange area will be provided by the plate sections. Let us try a 30 mm thick flange plate.

$$\text{Width of the flange plate} = \frac{13468}{30} = 448.9 \text{ mm} \simeq 450 \text{ mm}$$

$$\text{Flange outstand} = \frac{450 - 10}{2} = 220 \text{ mm}$$

$$\not> 360 \text{ mm } (12t = 12 \times 30)$$

which is all right.

Flange area provided by the plate = 450 × 30 = 13500 mm^2

> 13468 mm^2

Check Moment of inertia of the plate girder (Fig. Ex. 14.3(a)).

$$I_{xx} = \frac{10 \times 1800^3}{12} + 2 \times \left[\frac{450 \times 30^3}{12} + 450 \times 30 \times \left(\frac{1800}{2} + \frac{30}{2}\right)^2\right]$$

$$= 2746710 \times 10^4 \text{ mm}^4$$

Maximum bending tensile stress

$$= \frac{4000 \times 10^6}{2746710 \times 10^4} \times \left(\frac{1800}{2} + 30\right)$$

$$= 135.43 \text{ N/mm}^2 < 165 \text{ N/mm}^2$$

which is all right (safe in tension).

Lateral bracing will be provided at compression flange levels. Since the lateral struts are provided at every 2 m, the effective length of compression flange will be 2 m. The compression flange of the plate girder for this 2 m length will be laterally unsupported and the maximum permissible bending compressive stress will be less than $0.6 f_y$. To find the permissible stress in bending compression a ratio ψ is required which needs determination of curtailment of flange plate. Therefore, first curtailment is designed and then the permissible bending compressive stress is found.

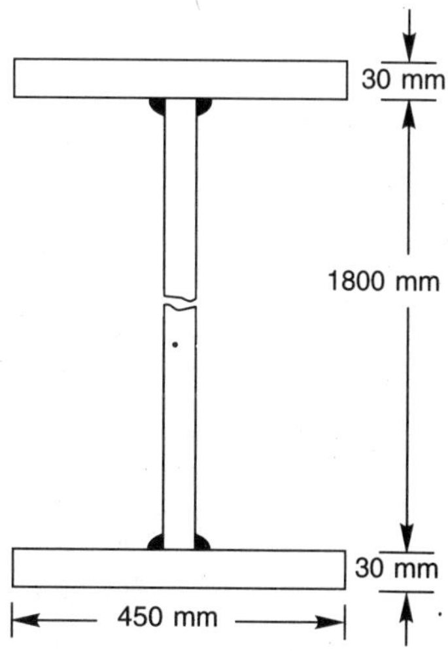

Fig. Ex. 14.3(a)

Curtailment

Flange outstand = 220 mm

If we curtail the flange plate from 30 mm to 20 mm.

The permissible maximum outstand will be

$$12 \times t = 12 \times 20 = 240 \text{ mm} > 220 \text{ mm}$$

which is all right.

We cannot make any appreciable reduction further since it will violate the permissible outstand condition.

Moment of inertia of the plate girder with 20 mm thick plate,

$$I_{xx} = \frac{10 \times 1800^3}{12} + 2 \times \left[\frac{450 \times 20^3}{12} + 450 \times 20 \times \left(\frac{1800}{2} + \frac{20}{2}\right)^2\right]$$

$$= 1976640 \times 10^4 \text{ mm}^4$$

$$M_R = \frac{165 \times 1976640 \times 10^4}{\left(\frac{1800}{2} + 20\right)} = 3545.06 \times 10^6 \text{ Nmm}$$

The bending moment diagram is parabolic.

$$4000 = K \times \frac{20}{2}\left(20 - \frac{20}{2}\right)$$

$\Rightarrow$ $$K = 40$$

$$x \text{ @ } y = 3545.06 \text{ kNm}$$

$$3545.06 = 40\,x\,(20 - x)$$

or $$3545.06 = 800x - 40x^2$$

or $$x^2 - 20x + 88.6265 = 0$$

$\Rightarrow$ $$x = 6.6275 \text{ m}$$

Reduce the plate thickness from 30 mm to 20 mm up to 6.6 m from the supports (Fig. Ex. 14.3(b)).

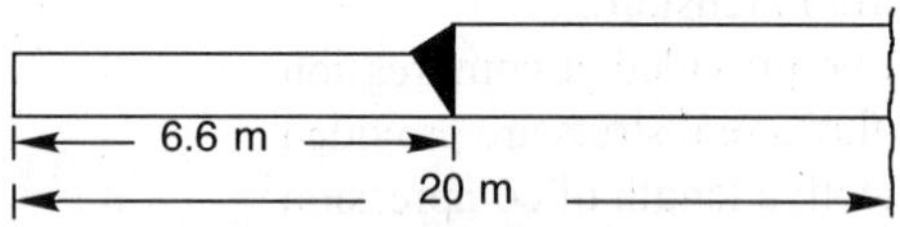

Fig. Ex. 14.3(b)

$$\psi = \frac{\text{area of both flanges at the point of least B.M}}{\text{area of both flanges at the point of maximum B.M}}$$

$$= \frac{2 \times 450 \times 20}{2 \times 450 \times 30} = 0.666$$

for $$\psi = 0.666,\ k_1 = 0.866 \quad \text{(Table 7.3)}$$

$$\omega = \frac{\text{moment of inertia of compression flange about } yy\text{-axis}}{\text{moment of inertia of both the flanges about } yy\text{-axis}} = 0.5$$

for $$\omega = 0.5,\ k_2 = 0 \quad \text{(Table 7.4)}$$

Moment of inertia of compression flange of plate girder about yy-axis

$$= \frac{1}{2} \times \text{moment of inertia of plate girder about } yy\text{-axis}$$

$$= \frac{1}{2} \times \left[\frac{1800 \times 10^3}{12} + 2 \times \frac{30 \times 450^3}{12}\right] = 22788.75 \times 10^4 \text{ mm}^4$$

$$\text{Area} = 1800 \times 10 + 2 \times (30 \times 450) = 45000 \text{ mm}^2$$

$$r_y = \sqrt{\frac{22788.75 \times 10^4}{45000}} = 71.16 \text{ mm}$$

$$Y = \frac{26.5 \times 10^5}{\left(\frac{2000}{71.16}\right)^2} = 3354.73 \text{ N/mm}^2$$

$$X = 3354.73 \sqrt{\left[1 + \frac{1}{20} \times \left(\frac{2000 \times 30}{71.16 \times (1800 + 60)}\right)^2\right]} = 3371.9 \text{ N/mm}^2$$

$$f_{cb} = 0.866 \times (3371.9 + 0 \times 3354.73) \times 1 = 2920$$

$$\sigma_{bc} = 0.66 \times \frac{2920 \times 250}{\left[2920^{1.4} + 250^{1.4}\right]^{1/1.4}}$$

$$= 161.3 \text{ N/mm}^2 > 135.45 \text{ N/mm}^2$$

which is all right (safe in compression).

Connection

Flange to web

$$\text{Horizontal shear/mm} = \frac{VA\bar{y}}{I_{xx}}$$

$$V = 880 \text{ kN}$$

$$\overline{AY} = (450 \times 20) \times \left(\frac{1800}{2} + \frac{20}{2}\right) = 819 \times 10^4 \text{ mm}^3$$

$$I_{xx} = 1976640 \times 10^4 \text{ mm}^4$$

$$\text{Horizontal shear/mm} = \frac{880 \times 10^3 \times 819 \times 10^4}{1976640 \times 10^4} = 364.62 \text{ N/mm}$$

Continuous welding is done on both sides of the web.

$$364.62 = 2 \times (0.7 \times S \times 1 \times 108)$$

$$\Rightarrow \quad S = 2.41 \text{ mm}$$

$$\simeq (5 \text{ mm for 20 mm thick plate})$$

Provide a 5 mm continuous fillet weld.

Bearing stiffener

Maximum shear force = 880 kN

$$\text{Bearing area required} = \frac{880 \times 10^3}{187.5} = 4693.33 \text{ mm}^2$$

Let us try 2 plates 160 mm wide as stiffener.

$$\text{Thickness of the plate} = \frac{4693.33}{2 \times 160} = 14.66 \text{ mm} \simeq 15 \text{ mm}$$

Outstand (160 mm) should not be more than 12 × 15 = 180 mm (12t) which is all right.

Bearing area provided = 2 × 160 × 15 = 4800 mm^2

> 4693.33 mm^2

which is sufficient.

Refer to Fig. Ex. 14.3(c)

Area of stiffener = 2 × (160 × 15) + 2 × 20 × 10 × 10 = 8800 mm^2

$$I_{xx} = 2\times\left[\frac{15\times160^3}{12}+15\times160\times\left(\frac{160}{2}+\frac{10}{2}\right)^2\right] = 4492\times10^4\ \text{mm}^4$$

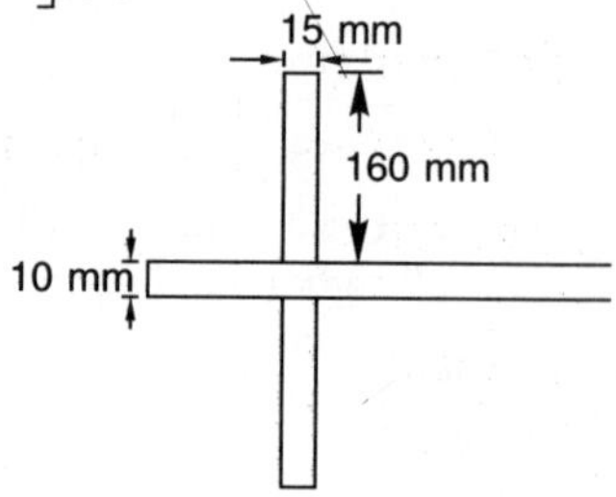

Fig. Ex. 14.4(c)

$$r = \sqrt{\frac{4492\times10^4}{8800}} = 71.44\ \text{mm}$$

$$\lambda = \frac{0.7\times1800}{71.44} = 17.64$$

For $\lambda = 17.64$ and $f_y = 250$ N/mm^2

$$\sigma_{ac} = 148.47\ \text{N/mm}^2$$

$$\text{Safe load} = 148.47\times8800$$
$$= 1306.53\times10^3\ \text{N} > 880\times10^3\ \text{N}$$

which is sufficient.

Hence, provide 160 × 15 mm plates as stiffener.

Connection

$$\text{Load} = 880\ \text{kN} = 880\times10^3\ \text{N}$$
$$880\times10^3 = 2\times(0.7\times1800\times S\times108)$$
$$\Rightarrow \quad S = 3.23\ \text{mm}$$

Provide 5 mm size fillet weld on both sides of the stiffener

Intermediate stiffeners

$$\tau_{va,\ \text{cal}} = \frac{880\times10^3}{1800\times10} = 48.88\ \text{N/mm}^2$$

For unstiffened web

(i) $t_{w,\ \text{min}} = \dfrac{1800\sqrt{48.88}}{816} = 15.42$ mm

(ii) $t_{w,\ \text{min}} = \dfrac{1800\sqrt{250}}{1344} = 21.17$ mm

(iii) $t_{w,\ \text{min}} = \dfrac{1800}{85} = 21.17$ mm

The web thickness provided is 10 mm (less than 21.17 mm). So intermediate stiffeners are required.

For vertical stiffeners

(i) $\dfrac{1}{180}\times1800 = 10$ mm

(ii) $t_{w,\ \text{min}} = \dfrac{1800\sqrt{250}}{3200} = 8.89$ mm

(iii) $t_{w,\,min} = \dfrac{1800}{200} = 9$ mm

The web thickness provided is 10 mm which is same as 10 mm (greater of 10 mm or 8.89 mm but not less than 9 mm) so provide a vertical stiffener.

The smallest clear panel dimension $\not>$ 180t i.e. 1800 mm. Provide vertical stiffeners at a spacing of 1800 mm.

$$\text{Number of stiffeners} = \frac{20 \times 10^3}{1800} = 11.11 \simeq 12$$

$$\text{Actual smaller clear panel dimension} = \frac{20 \times 10^3}{12} = 1666.66 \text{ mm}$$

$$\simeq 1660 \text{ mm}$$

$$\text{Spacing of the stiffener} = \frac{1660}{1800} d = 0.922 \text{ d}$$

$$\frac{d}{t} = \frac{1800}{10} = 180$$

For $\dfrac{d}{t}$ = 180, and c = 0.922 d

$$\tau_{va} = 83.56 \text{ N/mm}^2 \text{ (Table 6.6A, IS 800)} > 48.88 \text{ N/mm}^2$$

which is all right.

$$\text{Outstand of vertical stiffener} \not> 12\,t$$

$$\text{Maximum outstand} = 12 \times 10 = 120 \text{ mm}$$

Minimum required web thickness from shear consideration,

$$t = \frac{880 \times 10^3}{1800 \times 83.56} = 5.85 \text{ mm}$$

Try a flat section 80 × 10 mm.

$$\text{Required moment of inertia} = \frac{1.5 \times 1800^3 \times 5.85^3}{1660^2}$$

$$= 63.56 \times 10^4 \text{ mm}^4$$

Moment of inertia about the face of web

$$= \frac{10 \times 80^3}{12} + (10 \times 80) \times 40^2$$

$$= 170.66 \times 10^4 \text{ mm}^4$$

$$> 63.56 \times 10^4 \text{ mm}^4$$

which is all right.

Connection

$$\text{Shear force} = \frac{125 \times 10^2}{80} = 156.25 \text{ kN/m}$$

$$= 156.25 \text{ N/mm}$$

$$\text{Strength of weld/mm} = 0.7 \times S \times 1 \times 108$$

$$2 \times (0.7 \times S \times 1 \times 108) = 156.25$$

$$\Rightarrow \quad S = 1.033 \text{ mm}$$

Provide a 3 mm fillet weld on both sides of the stiffener.

Lateral bracing

Lateral bracing is provided at the compression flange level as shown in Fig. 14.3(d). The diagonals in tension will be effective only. The influence line diagram for the bracing system with the diagonals in tension is shown in Fig. 14.3(e). The force in the members L_0U_1 and $L_{10}U_9$ will be maximum and also equal. Therefore, L_0U_1 (or $L_{10}U_9$) is designed and the same section is provided for the other diagonals.

$$\text{Force in } L_0U_1 = 0.9 \times \sqrt{2} \times \frac{20}{2} \times 9000 = 114551.3 \text{ N}$$

$$A_{req} = \frac{114551.3}{150} = 763.67 \text{ mm}^2$$

Since the angle will not be accessible for clearing and painting (thickness $\nless$ 8 mm), ISA 100 × 65 × 8 mm is tried from IS Handbook No. 1.

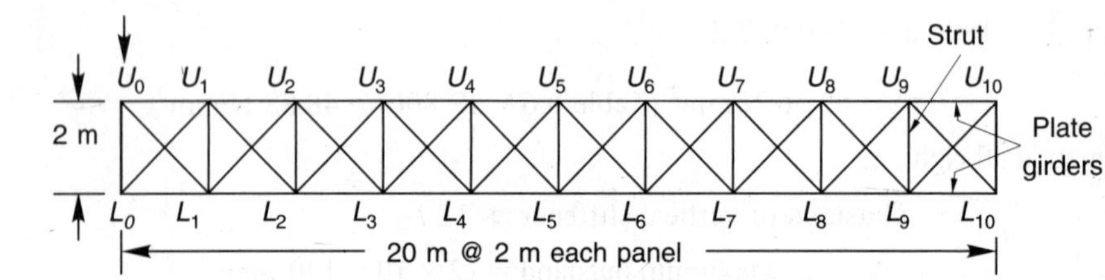

Fig. Ex. 14.3(d)

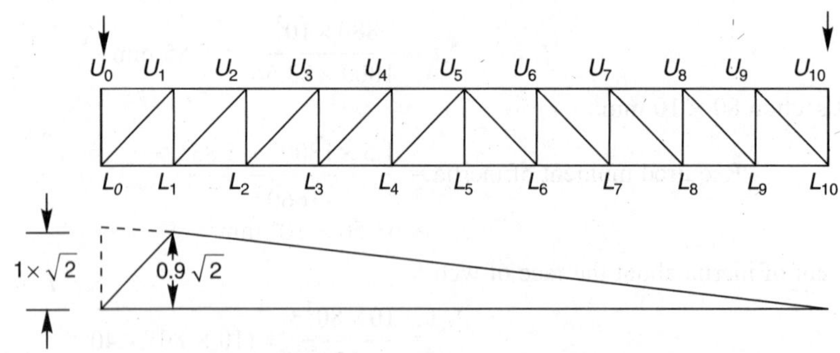

Fig. Ex. 14.3(e) *I.L.D. for L_0U_1*

The relevant properties of the section, $C_{xx} = 32.8$ mm.

$$A_1 = \left(100 - \frac{8}{2}\right) \times 8 = 768 \text{ mm}^2$$

$$A_2 = \left(65 - \frac{8}{2}\right) \times 8 = 488 \text{ mm}^2$$

$$k = \frac{3 \times 768}{3 \times 768 + 488} = 0.825$$

$$\text{Net area provided} = 768 + 0.825 \times 488 = 1170.6 \text{ mm}^2$$

$$\text{Load carrying capacity} = 1170.6 \times 150 = 175590 \text{ N} > 114551.3 \text{ N}$$

which is all right.

Connection

Let the force resisted by the weld near the outer edge of flange (i.e. at B) be P_B. Taking moment about A (Fig. Ex. 14.3(f))

$$114551.3 \times 32.8 = P_B \times 100$$

$$\Rightarrow \quad P_B = 37572.8 \text{ N}$$

Force resisted by weld at A,

$$P_A = 114551.3 - 37572.8$$

$$= 76978.5 \text{ N}$$

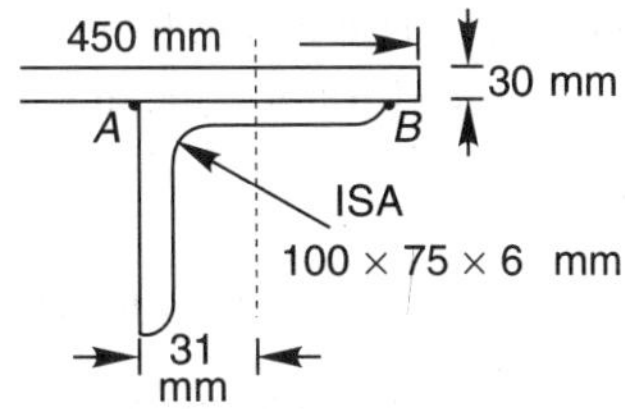

Fig. Ex. 14.3(f)

Let us provide a 5 mm weld for making the connection.

Length of welds

$$37572.8 = 0.7\, l_B \times S \times 108$$

$$\Rightarrow \quad l_B = 99.39 \text{ mm} \simeq 100 \text{ mm}$$

also

$$76978.5 = 0.7 \times l_A \times S \times 108$$

$$\Rightarrow \quad l_A = 203.6 \text{ mm} \simeq 204 \text{ mm}$$

End strut

$$\text{Force in the end strut, } L_0U_0 = 9000 \times \frac{20}{2} = 90000 \text{ N}$$

$$\text{Area} = \frac{90000}{70} = 1285.7 \text{ mm}^2$$

Let us try ISA 90 × 90 × 8 mm from IS Handbook No. 1. The relevant properties are:

$$A = 1379 \text{ mm}^2, r = 17.5 \text{ mm}$$

The strut will be welded to the flange of the plate girder (Fig. Ex. 14.3(g))

$$\lambda = \frac{0.85 \times 2000}{17.5} = 97.14$$

For $\lambda = 97.14$ and $f_y = 250$ N/mm^2,

$$\sigma_{ac} = 82.86 \text{ N/mm}^2$$

Load carrying capacity $= 82.86 \times 1379$

$= 114263.94$ N

> 90000 N

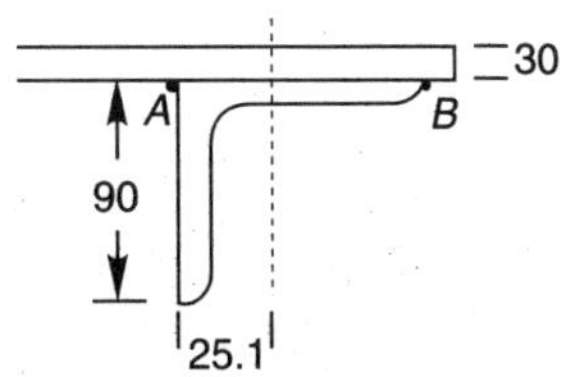

Fig. Ex. 14.3(g)

which is all right.

Connection

Taking moment about A,

$$90000 \times 25.1 = P_B \times 90$$

$$\Rightarrow \qquad P_B = 25100 \text{ N}$$
$$P_A = 90000 - 25100 = 64900 \text{ N}$$

Length of welds

$$25100 = 0.7 \times l_B \times 5 \times 108$$
$$\Rightarrow \qquad l_B = 66.40 \text{ mm} \simeq 70 \text{ mm}$$
$$64900 = 0.7 \times l_A \times S \times 108$$
$$\Rightarrow \qquad l_A = 171.69 \text{ mm} \simeq 175 \text{ mm}$$

End Cross-frame

Double diagonals as shown in Fig. Ex. 14.3(h) are provided of which the one in tension is assumed to be effective.

Two end frames one at each end of the plate girder are provided.

$$\text{Lateral force} = 9000 \times \frac{20}{2} = 90000 \text{ N}$$

$$\text{Force in the diagonal} = 90000 \sec \theta = 121082.61 \text{ N}$$

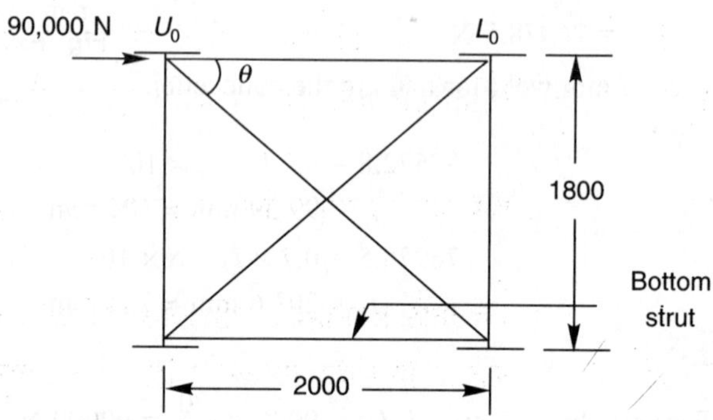

Fig. Ex. 13.3(h)

Provide ISA 100 × 65 × 8 mm (load carrying capacity = 175590 N) same as designed in lateral bracing (member L_0V_1). The connection will also be same as designed in for the section in lateral bracing system. Provide a nominal section say 90 × 90 × 8 mm for the bottom strut.

Let us provide intermediate cross frames at every 4 m. Since the lateral force will be much less than that on the end cross frame, ISA 75 × 75 × 8 mm sections for diagonals as well as for struts will be sufficient.

Exercises

Note In the following problems, the statements are not intended to be complete. Any additional data required are to be assumed by the students. Unless specified use riveting or welding for the connections as described.

14.1 A Pratt truss for a bridge is to carry 2 numbers of 700 mm diameter M.S. water mains and a walkway 1.5 m wide over a span of 20 metres as shown in Fig. Prob. 14.1. The pipes have a self-weight of 2.0 kN/m excluding the weight of water. The walkway may be assumed to carry a live load of 4 kN/m^2. The cross-beams are spaced 2.5 m c/c and height of the bridge is 2.5 m above the wooden platform laid over the cross-beams, in the walkway space only.

Design the wooden platform section, the cross girders and the truss.

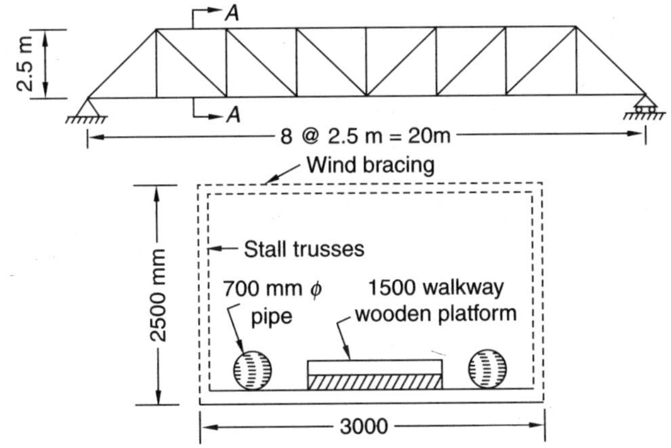

Fig. Prob. 14.1

14.2 A pratt truss shown in Fig. Prob. 14.2. is to be used as one of the two main girders of a single track bridge carrying rolling loads on stringers which in turn are carried by cross-girders connected to the nodes of the trusses. The truss centre lines are 6 m apart while a pair of stringer divide this distance into three equal parts of 2 m each. The following additional data is given.

Span of the main girder	36 m
Dead load on each truss	20 kN/m
Equivalent uniformly distributed live load per girder	60 kN/m
Impact factor	20%

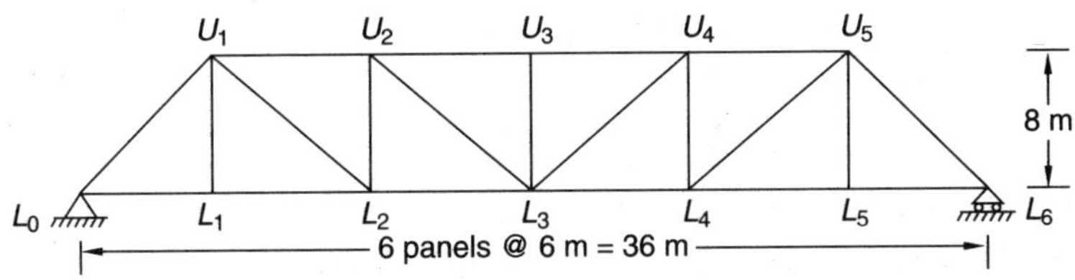

Fig. Prob. 14.2

Draw influence lines for forces in members L_2L_3, U_2L_2, U_1L_2 and U_3L_3.

Design sections for members L_2L_3, U_2L_2, U_1L_2 and U_3U_3. Design also the stringers and cross-girders.

14.3 In a through type riveted plate girder bridge, the effective span of the plate girders is 18 m. Each girder supports a live load inclusive of impact of 150 kN/m run in addition to its own weight. Design the mid-span section, the bearing stiffener and the intermediate stiffener.

14.4 A deck type plate girder railway bridge is required for a broad guage single line track on the main line for the following data.

Effective span	16 m
Distance between c/c of plate girders	2 m
Dead load of each girder	4 kN/m
Dead load of track and timber sleepers	7 kN/m
Lateral load	9 kN/m

Design the components of the bridge.

Chapter 15

Tubular Structures

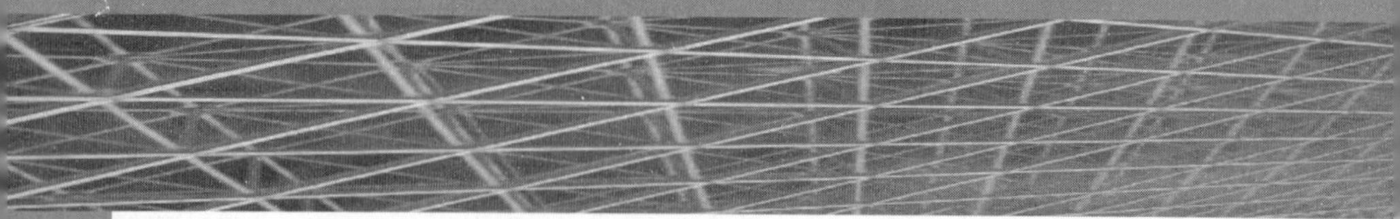

15.1 INTRODUCTION

Tubular sections form the most efficient sections for some of the structural elements. However, in the past few decades their use was restricted for the following reasons.

1. It was difficult to have a smooth cut surface.
2. It was difficult to cut the tube end to a correct profile, for making joints between steel tubes at angles.
3. Connections were a problem.

The first two problems were overcome with the development of the oxyacetylene cutting machine, which cuts the surface to the correct profile leaving a bevelled edge and facilitates welding operation. The joints represent the intersections of curved surfaces, and therefore extra care and time is involved in cutting the pipe to prepare the joints. Connections, however, are now not a problem with the development of welding techniques.

The economy of steel tube construction is uncomparable. The large span roof trusses with tube sections have smaller self-weight and many a times the supporting R.C.C. columns may even be replaced by masonry columns deriving considerable economy. These are most suitable for roofing systems of industrial buildings and warehouses and for large space frame lattice structures for stadia and arenas. Since tubular sections subjected to wind or water current have a low drag coefficient, they provide an ideal section for transmission line towers, masts and offshore drilling installations.

IS: 1161 specifies the use of the following types of tubes for structural purposes.

HFW: Hot finished welded

HFS: Hot finished seamless

ERW: Electric resistance welded or induction welded

15.2 CLASSIFICATION

Steel tubes are classified as light, medium and heavy depending upon the wall thickness. On the basis of yield stress, these are classed as Y_{st} 210—tensile strength 3300 kg/cm^2, yield stress 210 N/mm^2; Y_{st} 240—tensile strength 4100 kg/cm^2, yield stress 240 N/mm^2 and; Y_{st} 310—tensile strength 4500 kg/cm^2, yield stress 310 N/mm^2.

Note The values of the stresses in IS:806 1968 and 1161 are given in MKS units. To convert the stress values in SI units, these may be multiplied with a factor of 0.0981.

15.3 ADVANTAGES AND DISADVANTAGES

1. These have small self-weights. Also because of direct connections, gusset plates are eliminated further reducing dead loads.
2. Tubes have uniform radius of gyration and for the same weight their torsional strength is more than any other rolled section.
3. For the same load the surface area of a tube is about 60 to 70% of that for other rolled sections. Consequently wind forces are small. Also, because of less surface area considerable economy is achieved in maintenance, painting and fire proofing.
4. For dynamic loads tubes have higher frequency of vibration than other rolled sections.
5. Due to a smooth finished surface, dirt and moisture do not collect over the surface, reducing the possibility of corrosion.
6. The change in load with the floor levels can be accommodated by varying the tube thickness and the external tube dimension may be maintained.
7. The internal hollow space of tubular columns may be used for carrying drain pipes, wires, cables, etc. Also these spaces may be filled with concrete to increase the load carrying capacity and to improve fire resistance.

The only disadvantage of the tubular section is its high manufacturing cost.

15.4 BEHAVIOUR OF TUBULAR SECTIONS

Compression Members In a tubular section the material is disposed of symmetrically about its centre of gravity. Because of this it has a uniform radius of gyration. This makes a tubular section the most ideal section for compression members.

The diameter of the tube should be as large as possible with the ratio d/t as small as possible to avoid local buckling. Here, d is the mean diameter of the tube and t is the thickness of the tube. The specifications for effective length and slenderness ratio are same as that for other rolled sections (Chapter 4). The effective length of truss members is same as that of continuous struts. The permissible stress in compression is given in Table 15.1.

Table 15.1 *Permissible Axial Stress in Compression,* σ_{ac} *(kg/cm²)*

Slenderness Ratio $\lambda = l/r$	*Grade* $(Y_{St}\ 210)$	*I.S: 1161 Grade* $(Y_{St}\ 240)$	*Grade* $(Y_{St}\ 310)$
0	1250	1500	1900
10	1217	1448	1821
20	1175	1400	1750
30	1131	1352	1679
40	1088	1303	1610
50	1046	1255	1539
60	1002	1207	1468
70	970	1155	1375
80	929	1088	1263
90	876	1003	1128

(Contd.)

Table 15.1 *(Contd.)*

Slenderness Ratio $\lambda = l/r$	*Grade* $(Y_{St}$ *210)*	*I.S: 1161 Grade* $(Y_{St}$ *240)*	*Grade* $(Y_{St}$ *310)*
100	814	910	989
110	745	813	869
120	674	721	758
130	603	638	665
140	540	565	584
150	490	503	517
160	432	443	450
170	381	392	398
180	339	348	353
190	304	311	315
200	271	278	280
210	243	249	250
220	219	225	227
230	198	204	205
240	180	185	187
250	162	167	167
300	106	106	106
350	71	71	72

Notes

1. Intermediate values may be obtained by linear interpolation.
2. The formulae used to derive the above tabulated values are as below,
   (i) For $\lambda < 60$, σ_{ac} should not exceed the value obtained by linear interpolation between 0.6 f_y for $\lambda = 0$ and σ_{ac} for $\lambda = 60$ as obtained by (ii).
   (ii) For $\lambda = 60 - 150$, σ_{ac} should not exceed F_c

$$F_c = \frac{\sigma_y / m}{1 + 0.15 \sec \frac{l}{r} \times \sqrt{\frac{m F_c}{4E}}}$$

   (iii) For $\lambda > 150$, σ_{ac} should not exceed a value $F_c \left(1.2 - \frac{l}{750\, r}\right)$

   where F_c = permissible average axial compressive stress as obtained in (ii)
   m = factor of safety (1.67)

Tension Members The design criteria for tension members is the net cross-sectional area. For the same cross-sectional area tubes cost more than the other rolled sections and are hence uneconomical. The design procedure is same as that for other rolled sections (Chapter 6). Permissible stresses for tubes in tension are given in Table 15.2

Table 15.2 *Permissible Stress for Steel Tubes as per IS 806:1968*

Grade (IS:1161-1979)	*Axial Stress in Tension (kg/cm²)*	*Bending Stress (kg/cm²)*	*Minimum Shear Stress (kg/cm²)*	*Minimum Bearing Stress (kg/cm²)*
Y_{St} 210	1250	1400	900	1700
Y_{St} 240	1500	1650	1100	1900
Y_{St} 310	1900	2050	1350	2500

Beams Tubular sections for beams are usually found in industrial buildings as well as in roof truss members subjected to transverse loads at points other than panel points. The torsional strength of tubular sections is 200 to 300 times that of an open section. Also, their deformation is about 1000 times smaller than that of open sections of the same size. When used as a beam, the risk of lateral instability is the least. The guide lines for the design of beams and purlins are same as for other rolled sections (Chapter 7). The permissible bending stresses in tubes are given in Table 15.2. The maximum bending stress and shear stress (total shear divided by half the net cross-sectional area of the tube) should not exceed the permissible stresses as specified in Table 15.2.

Tubular beams are provided with shoes at supports. These stiffen the tube ends and transmit the load adequately to the support. For beams subjected to point loads stiffeners will be required to check local stresses.

The maximum deflection of the tubular beams should not exceed 1/325 of the span. This requirement may be deemed to be satisfied, if the bending stress in compression or in tension does not exceed $31500\frac{D}{l}$ kg/cm^2, where l is the effective length of the beam. In the case of purlins the deflection limitations are waved out provided the requirements given in Table 15.3 are satisfied.

Table 15.3 *Minimum Section Modulus and Outside Diameter of Tubes for Waving Deflection Limitations*

Grade (IS:1161-1979)	*End Condition*	
	Simply Supported	*Continuous*
1. Section Modulus cm^3		
Y_{St} 210	*WL*/11200	*WL*/16800
Y_{St} 240	*WL*/13230	*WL*/19840
Y_{St} 310	*WL*/16400	*WL*/24600
2. Maximum outside diameter (D) cm		
Y_{St} 210, Y_{St} 240, Y_{St} 310	*L*/40	*L*/70

where W = total distributed load in Newtons on the purlins from the dead loads and snow loads excluding wind

L = distance in mm between trusses supporting purlins or other supports (in mm).

15.5 MINIMUM THICKNESS

For tubes painted with one prime coat of red oxide and then painted periodically, the thickness should not be less than

(i) for construction exposed to weather — 4 mm
(ii) for construction not exposed to weather — 3.2 mm
(iii) for members not readily accessible for maintenance — 5 mm

For tubes painted with one coat of zinc primer followed by two coats of paint, the thickness should not be less than

(i) for construction exposed to weather — 3.2 mm
(ii) for construction not exposed to weather — 2.6 mm

15.6 COMBINED STRESSES

1. When a tubular member is subjected to bending moments about both the axes, the combined bending stress is given by

$$\sigma_{b,\text{cal}} = \sqrt{\sigma^2_{bx,\text{cal}} + \sigma^2_{by,\text{cal}}}$$

2. When a tubular member is subjected to combined axial and bending stresses it should be so proportioned that

$$\frac{\sigma_{ac,t,\text{cal}}}{\sigma_{ac,t}} + \frac{\sigma_{bc,t,\text{cal}}}{\sigma_{bc,t}} \le 1$$

where $\sigma_{ac,t\,\text{cal}}$ = calculated axial compressive or tensile stress

$\sigma_{ac,t}$ = permissible axial compressive or tensile stress

$\sigma_{bc,t\,\text{cal}}$ = calculated bending compressive or tensile stress

$\sigma_{bc,t}$ = maximum permissible bending compressive or tensile stress

15.7 CONNECTIONS

Tubular sections in trusses may be jointed either by means of a gusset plate or member-to-member by welding.

Gusset plates are used in tubes connections for three reasons:

1. They provide additional length of fillet welding to the tube. Since tubes have thin sections, it is easier to use fillet welds. Trying complete penetration butt weld requires extra care in cutting, beveling and fitting of thin wall tubes as shown in Fig. 15.1.
2. They allow the intersecting tube members to be cut short and the gusset plate carries the entire load back to the main member.
3. They provide a direct transfer of force through the main tube member when other members are connected on opposite sides of the member.

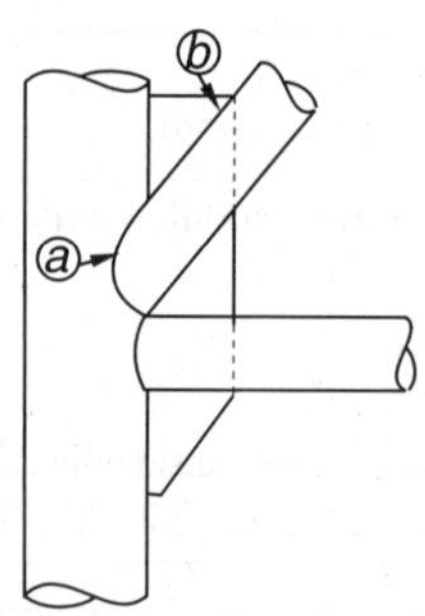

Weld ⓐ does not have to be made as carefully because fillet weld ⓑ provides additional strength

Fig. 15.1

Although gusset plates are used in tube connections, they tend to stiffen the tube and as a result, concentrate the stress in the pipe at the end of the gusset plate for the shape of gusset plate as shown in Fig. 15.2(a). However, this stiffening effect can be reduced by tapering the gusset plates at their ends as shown in Fig. 15.2(b) and a more even stress distribution within the tube at this connection can be obtained. Under static loads, any reasonable stress concentration in the tube near the end of the gusset plate would probably be reduced by some localised plastic yielding; so this would not be a problem. However, gusset plates should be avoided for connections subjected to fatigue loading.

A more uniform stress distribution can be had by welding member-to-member as shown in Fig. 15.3. Rather than to follow that customary practice of having all axes

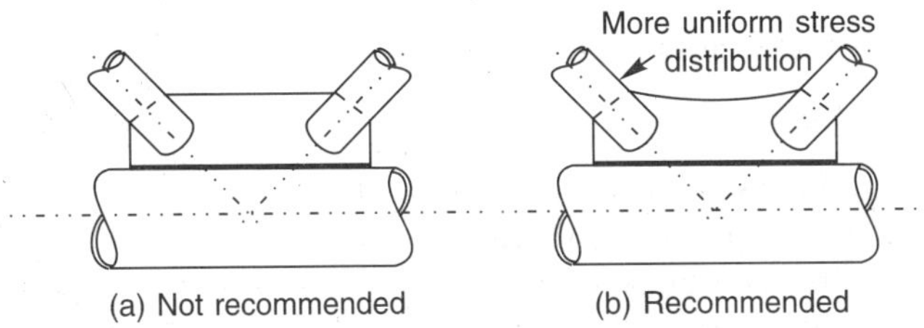

Fig. 15.2 *Gusset Plate Connections for Tubes*

intersecting at a point (Fig. 15.4(a)) it is better to have branch tubes overlap as shown in Fig. 15.3(a). In the latter arrangement the normal components of the branch tube forces are brought closer together, the moment causing local longitudinal bending is reduced and further, at least part of the normal component may be transferred directly from branch to branch without affecting the chord wall at all. Also, when members overlap it is preferable to continue the tension branch tube to the main tube (chord) and to cut the compression branch tube to fit around it rather than vice-versa.

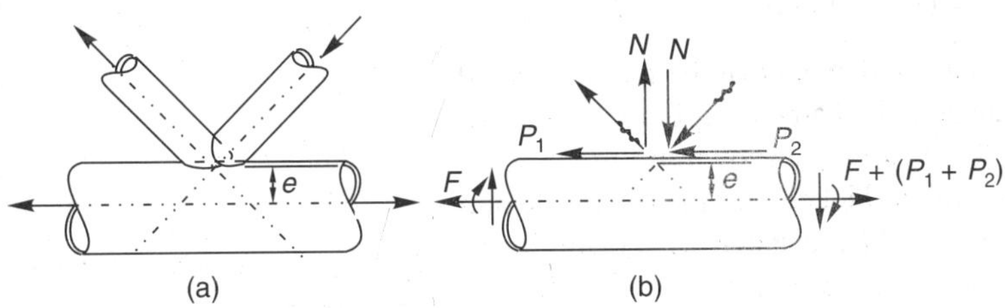

Fig. 15.3 *Welding of Tubes Member to Member*

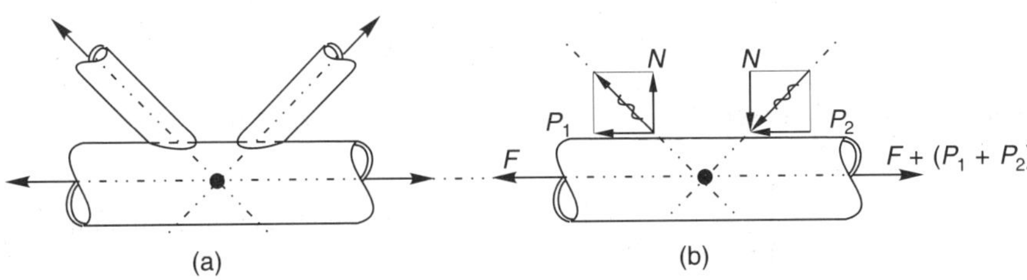

Fig. 15.4 *Effect of Two Branch Tubes*

Welded connections provide rigid joints and result in saving in material and greater overall economy. However, it introduces secondary stresses which may be neglected until the axes of the members do not meet in a point. If eccentricity exists in the connection, the secondary stresses must be given due consideration.

Specifications permit welding of the tubular members only when their axes intersect at an angle not less than 30°. For intersection angles less than 30°, the joint is permitted only when adequate efficiency of the joint is ensured. The members may be joined by complete penetration butt weld, fillet weld or fillet-butt weld as shown in Fig. 15.5.

A butt weld is provided throughout and the eccentricity is limited to $\frac{1}{2}(D - d)$. The throat thickness of butt weld is taken as the thickness of the thinner part jointed. When the ratio of diameters of the branch pipe and main pipe is less than 1/2, a fillet weld is provided. The maximum eccentricity is limited to $\frac{1}{2}\left(\frac{D}{3} - d\right)$. In case the ratio of diameters of the branch pipe to the main pipe is more than 1/3, a fillet-butt weld is provided. Fillet weld is provided in a portion of the joint and butt weld for the remainder with a continuous change from one form to another. The maximum eccentricity lies between $\frac{1}{2}(D - d)$ and $\frac{1}{2}\left(\frac{D}{3} - d\right)$. Here, D is the outside diameter of the main tube and d is the outside diameter of branch tube. All these three cases are shown in Fig. 15.5.

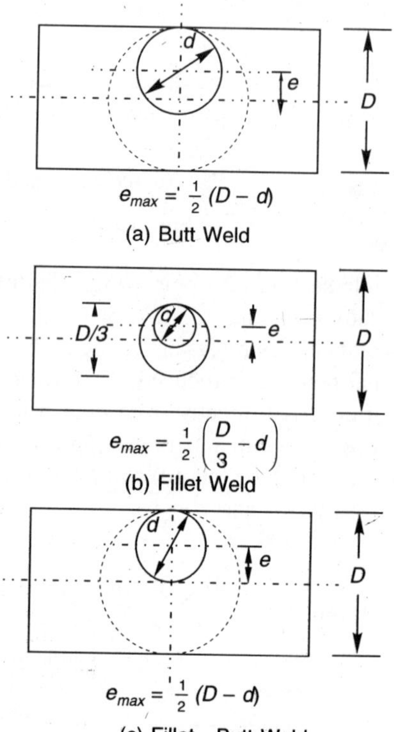

Fig. 15.5 *Maximum Eccentricity in Various Welds*

The throat thickness of the fillet weld and fillet-butt weld are taken as the minimum effective throat thickness.

The length of the weld to join the two tubes is given by

$$L = a + b + 3\sqrt{a^2 + b^2}$$

where $a = \frac{d}{2}\ \text{cosec}\ \theta$ (Fig. 15.6)

$b = \frac{d}{3} \times \left[\frac{3 - (d/D)^2}{2 - (d/D)^2}\right]$ (Intersection with a tube)

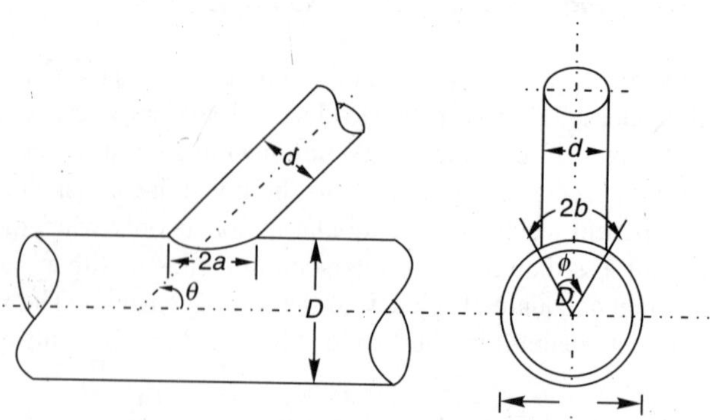

Fig. 15.6 *Intersection of Two Tubes*

also, $b = \frac{d}{2}$ (Intersection with plate)

θ = angle of intersection of the branch tube and the main tube.

Solved Examples

Example 15.1 Design members *AB*, *AC* and joint *A* of a roof truss as shown in Fig. Ex. 15.1, for the following data:

Member	Length	Compressive force	Tensile force
AB	2.3 m	60 kN	55 kN
AC	1.8 m	55 kN	80 kN

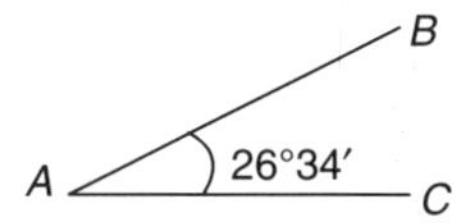

Fig. Ex. 15.1

Solution

Member AB

Let us try a 65 mm nominal bore light tube with an outside diameter of 76.1 mm. From Appendix XII, the relevant properties of the section are,

$$A = 733 \text{ mm}^2, r = 25.8 \text{ mm}$$

$$\lambda = \frac{l}{r} = \frac{0.7 \times 2.3 \times 10^3}{25.8} = 62.40$$

From Table 15.2 and for steel grade Y_{St} 210,

$$\sigma_{ac} = 0.0981 \times 994.3 = 97.54 \text{ N/mm}^2$$

$$\text{Load carrying capacity} = 97.54 \times 733 \times 10^{-3} = 71.49 > 60 \text{ kN}$$

which is all right.

Check as a tension member

$$\text{Load carrying capacity} = 733 \times 0.0981 \times 1250 \times 10^{-3} = 89.86 \text{ kN} > 50 \text{ kN}$$

which is all right.

Member AC

$$A_{\text{required}} = \frac{80 \times 10^3}{0.0981 \times 1250} = 652.40 \text{ mm}^2$$

Let us try 50 mm nominal bore medium tube with an outside diameter of 60.3 mm. From Appendix XII the relevant properties of the section are,

$$A = 641 \text{ mm}^2, \quad r = 20.1 \text{ mm}$$

Since the area provided (641 mm^2) is more than the area required (640 mm^2) the section is safe in tension.

Check as compression member

$$\lambda = \frac{0.7 \times 1.8 \times 10^3}{20.1} = 62.68$$

For $\lambda = 62.68$ and grade of steel Y_{St} 210

$$\sigma_{ac} = 0.0981 \times 993.4 = 97.4 \text{ N/mm}^2$$

$$\text{Load carrying capacity} = 97.4 \times 641 \times 10^{-3} = 62.47 \text{ N}$$
$$> 55 \text{ kN}$$

which is all right.

Joint A

$$\text{Outside diameter of the rafter, } AB = 76.1 \text{ mm}$$
$$\text{Outside diameter of the principal tie, } AC = 60.3 \text{ mm}$$
$$\text{Angle of intersection} = 26° \, 34'$$

$$a = \frac{d}{2} \text{ cosec } \theta = \frac{60.3}{2} \text{ cosec } 26°34' = 67.413 \text{ mm}$$

$$b = \frac{d}{3} \times \left[\frac{3-(d/D)^2}{2-(d/D)^2}\right] = \frac{60.3}{3} \times \left[\frac{3-(60.3/76.1)^2}{2-(60.3/76.1)^2}\right] = 34.746 \text{ mm}$$

The length of the curve of intersection,

$$L = a + b + 3\sqrt{a^2+b^2}$$

$$= 67.413 + 34.746 + 3 \times \sqrt{67.413^2 + 34.746} = 329.68 \text{ mm}$$

Let S be the size of the weld.

$$\text{Strength of the weld} = 0.7 \, lS \, \tau_{vf}$$
$$80 \times 10^3 = 0.7 \times 329.68 \times S \times 108$$
$$\Rightarrow \quad S = 3.20 \text{ mm} \simeq 4$$

Provide a 4 mm fillet weld for connection.

Example 15.2 Design a purlin section for the following data:

Spacing of roof trusses c/c	5 m
Dead load of roofing	0.5 kN/m
Live load on purlin	1.1 kN/m
Wind load on purlin	–1.5 kN/m

Solution

Loads are assumed to act normal to the roof truss.

Assume self-weight of the purlin = 0.1 kN/m.

$$\text{Total dead load} = 0.5 + 0.1 = 0.6 \text{ kN/m}$$
$$\text{Dead load + live load} = 0.6 + 1.1 = 1.7 \text{ kN/m}$$
$$\text{Dead load + wind load} = 0.6 - 1.5 = -0.9 \text{ kN/m}$$
$$\text{Design load} = 1.7 \text{ kN/m} = 1.7 \times 10^3 \text{ N/m}$$

$$M = \frac{1.7 \times 10^3 \times 5 \times (5 \times 10^3)}{10} = 42.5 \times 10^5 \text{ Nmm}$$

$$Z = \frac{42.5 \times 10^5}{0.0981 \times 1650} = 26256.45 \text{ mm}^3$$

If the deflection of the purlin is not to be checked the following criteria should be satisfied

(i) Z_{req} from deflection criteria $= \dfrac{0.6 \times 10^3 \times 5 \times (5 \times 10^3)}{16800} = 892.86 \text{ mm}^3$

$< 26256.45 \text{ mm}^3$

which is all right.

(ii) Provided the nominal bore of tube is more than

$$\frac{L}{700} = \frac{5000}{70} = 71.42 \text{ mm}$$

Select a medium tube 90 mm nominal bore (> 71.42 mm) and of section modulus 28800 mm^3 (> 26256.45 mm^3) from Appendix XII.

Exercises

Note In the following problems, the statements are not intended to be complete. Any additional data required are to be assumed by the students.

15.1 Design joint *A* of a tubular roof truss and the members meeting at the joint *A*, the line diagram of which is shown in Fig. Prob. 15.1.

Member	*Length*	*Force*	
		(*C*)	(*T*)
AB	2.2 m	110 kN	38 kN
AC	2.8 m	32 kN	87 kN

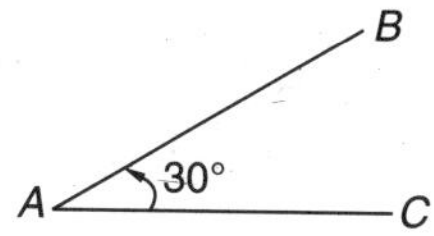

Fig. Prob. 15.1

15.2 A tie member in a roof truss is connected to the principal rafter at an angle of 90°. Design the members and the connections for the following data. Use grade Y_{st} 240 tubes.

S. No.	*Member*	*Length*	*Force*
1.	Principal rafter panel	2.6 m	85 kN
2.	Tie member	2.2 m	35 kN

15.3 Design the columns and beams of the framed structure shown in Fig. Prob. 15.3. Use grade Y_{st} 310 tubes. Live load over the roof may be assumed to be 1.5 kN/m^2.

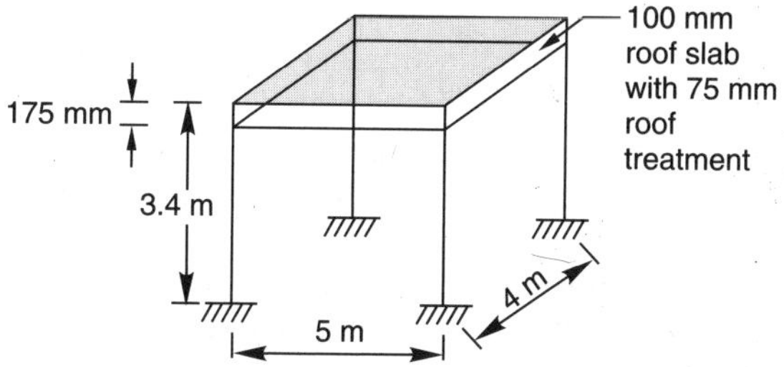

Fig. Prob. 15.3

Light-Gauge Steel Construction

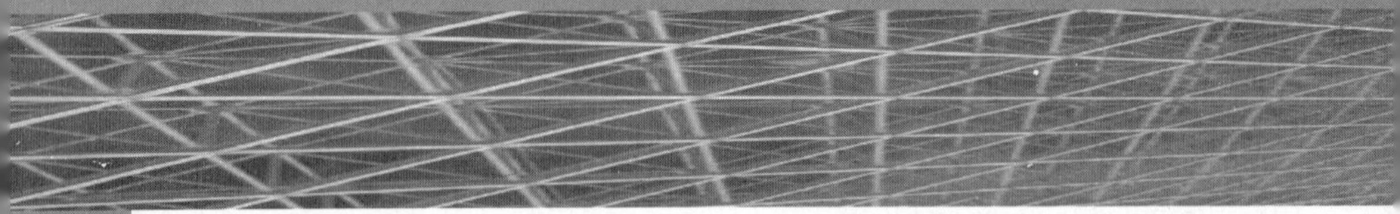

16.1 INTRODUCTION

Light-gauge steel evolved as a building material in the 1930's and reached large-scale usage only after the Second World War. In comparison with conventional steel construction, where standard hot-rolled shapes are used, the cold-formed light-gauge steel structures are a relatively new development. Light-gauge steel sections are cold-formed in rolls by rolling the material in cold condition or by bending the steel sheets or strips in press brakes; cold-rolling being used for mass production while press brakes are used for economical production of small quantities of special shapes. These are used widely in structures subjected to light or moderate loads or for members of short span lengths. For such structures, the use of conventional hot-rolled shapes is often uneconomical because the stresses developed in the smallest available shape may be very low. Further, a variety of light-gauge members can be formed in the cold state with ease and the material can be used in the most effective manner. Earlier, use of light gauge steel was limited to building construction only, but now it finds application in trucks and trailer bodies, rail coaches, etc. Light gauge members are connected by spot, fillet, plug or slot welds or by screws, rivets, bolts, etc. For specifications of light-gauge steel sections IS 811 and for analysis and design, IS 801 should be referred.

16.2 SHAPES

The shapes which can be cold-formed are many and varied. Generally, the shape varies with its applications. Engineers have learned to adopt this versatility to advantage in design of compression members, studs, joists, beams, roof and floor panels and other industrial structural members. In the design of structural sections for framing members the main aim is to develop shapes which combine economy of material (that is a favourable strength–weight ratio) with versatility, ease of mass production, and provision for effective and simple connection to other structural members or to non-structural collateral material or both of them.

The usual shapes of framing members are channels, zees, angles, hat sections, tees, I-Sections and tubular sections and are shown in Fig. 16.1. These sections, 50 to 300 mm in depth can carry substantial loads and are used as primary framing members in residential, commercial and industrial buildings up to two stories in height and roof trusses up to 15 m span. Framing members formed with light-gauge structural steel are cold-formed from steel sheets or strips. Thickness of the framing members ranges from 1.2 to 4.0 mm.

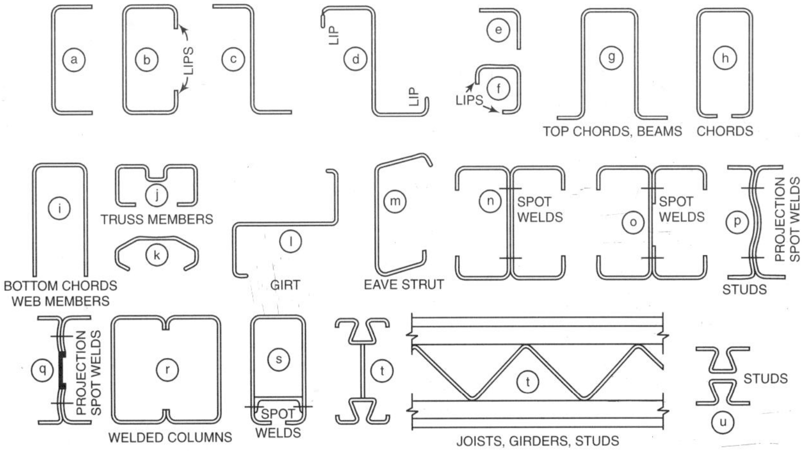

Fig. 16.1 *Common Shapes of Cold-Formed Light-Gauge Sections*

Shapes *a* to *u* of Fig. 16.1 are similar to hot-rolled shapes, except that in shapes *b*, *d* and *f*, lips are used to stiffen the thin flanges. These shapes are easily produced but have the disadvantage of being unsymmetrical. Shapes *g* to *k* of Fig. 16.1 are to be found only in cold-formed construction; they have the advantage of being symmetrical. Shapes *g*, *h*, *j* and *k* are adapted for use in trusses and latticed girders; these sections are compact, well stiffened and have large radii of gyration in both principal directions. Shape *i* of Fig. 16.1, lacking edge stiffeners on the vertical sides, is better adapted for use as a tension member. Shapes *l* and *m* of Fig. 16.1 are used specifically as girts and eave struts, respectively. In all-metal buildings, shape *l* being the same as shape *d* of Fig. 16.1 which is also, used for purlins. The above members are all one-piece shapes produced merely by cold-forming.

When automatic welding is combined with cold rolling, it is possible to obtain additional shapes. Shapes *n* and *o* of Fig. 16.1 are two varieties of I shapes, the former better adapted for use as studs or columns, the latter for joists or beams. Two of the most successful shapes, namely, shapes *p* and *q* of Fig. 16.1 are further adaptations of shapes *n* and *o*. By deforming the webs and by using projection spot welding, curved slots are formed which provide nailing grooves for connecting collateral material, such as wall boards and wood floors. Shapes *r* and *s* of Fig. 16.1 represent closed sections particularly favourable in compression the former primarily for columns, the latter for compression chords of trusses. Shape *t* of Fig. 16.1 shows one of a variety of open web joists, with chords shaped for nailing, and shape *u* shows a section similar to the chords of shape *t* connected directly to form a nailable stud.

Figure 16.2 shows another category of cold-formed sections, manufactured in panel sizes, used for roof and floor decks, sidings, and walls. Standard roof decks are usually 58 mm deep, with a rib spacing of 130 mm and are used on spans between purlins up to 5 m. Floor and roof panels are made to cover spans from 3 to 10 m. They are usually cellular in shape and permit a wide variety of ancillary uses. The thickness used generally range from 1.2 to 2.5 mm and for standard roof deck and wall cladding from 0.8 to 1.2 mm. The corrugated sheets are the most common use of this category and are in use for decades.

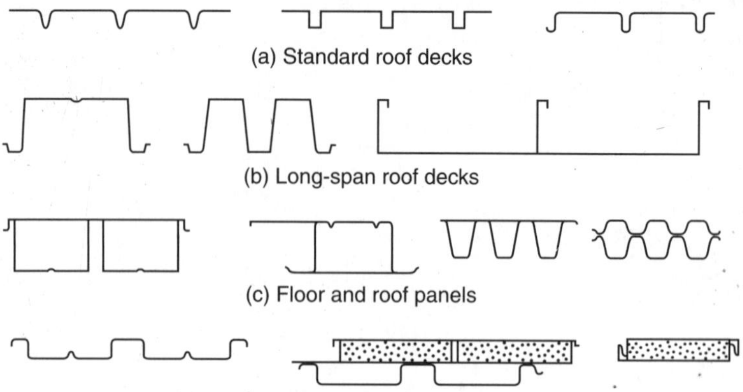

Fig. 16.2 *Floor and Roof Decks, and Wall Panels*

In the design of decks and panels, structural efficiency is only one of the many criteria since the shape should also be selected to minimize deflections, provide maximum coverage, permit adequate insulation, and accessibility of cells for housing conduits, etc. Optimum strength, that is, optimum strength–weight ratio, therefore, is desired only conditionally, that is, in so far as it is compatible with the other enumerated features. It is evident that the shapes used in light-gauge construction are quite different from, and considerably more varied than, those employed in hot-rolled framing.

Some of the structural sections used for light-gauge construction shown in Figs 16.1 and 16.2 are but a few examples and many more shapes and configurations are possible.

16.3 DEFINITIONS

Stiffened Compression Element is a flat compression element of which both edges parallel to the direction of stress are stiffened by another element (Fig. 16.3).

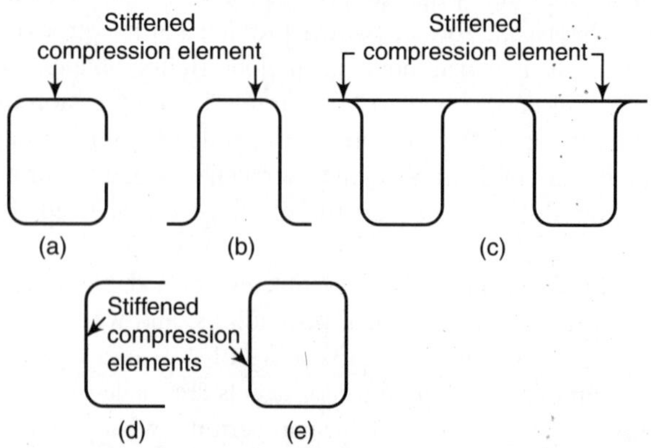

Fig. 16.3 *Stiffened Compression Elements*

Unstiffened Compression Element is a flat compression element stiffened at only one edge parallel to the direction of stress (Fig. 16.4).

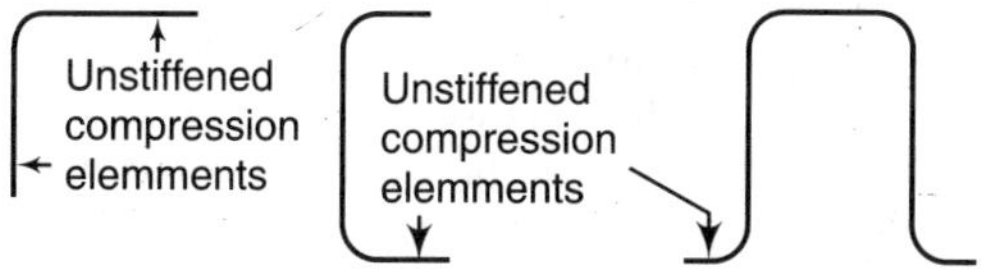

Fig. 16.4 *Unstiffenened Compression Elements*

Multiple Stiffened Element is an element that is stiffened between webs, or between a web and a stiffened edge (Fig. 16.5) by means of intermediate stiffeners parallel to the direction of stress.

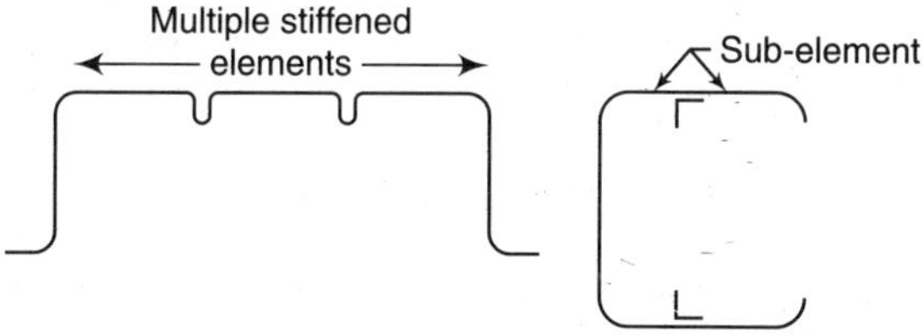

Fig. 16.5 *Multiple Stiffened Element and Sub-Element*

Flat Width Ratio of a single flat element is the ratio of the flat width *w*, exclusive of edge fillets, to the thickness *t* (Fig. 16.6).

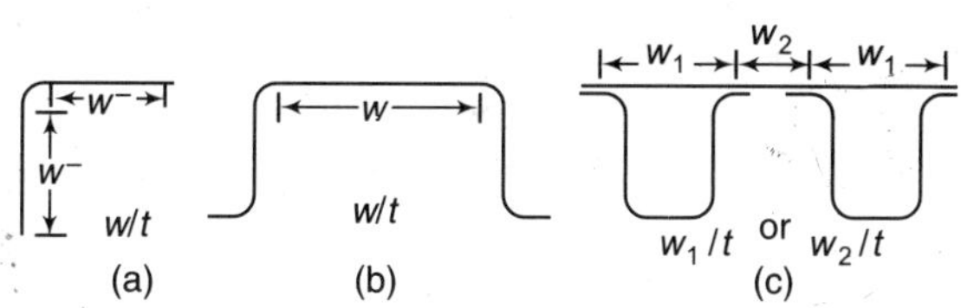

Fig. 16.6 *Flat-Width Ratio*

Effective Design Width of a flat element is its reduced width from design consideration (Fig. 16.7). The portion of the total width which is considered removed to arrive at the effective design width is located symmetrically about the centre line of the element.

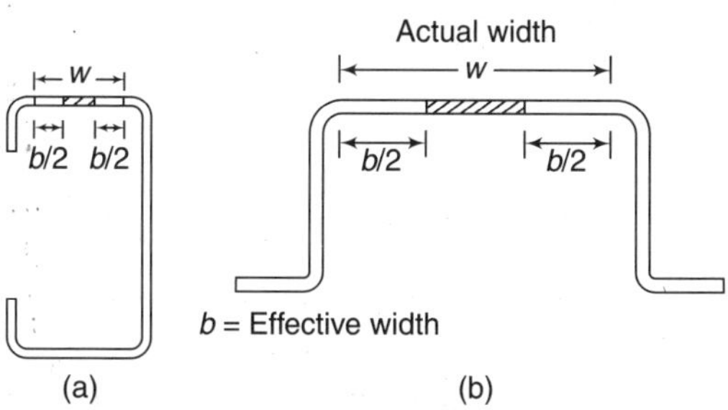

Fig. 16.7 *Effective Design Width*

16.4 PROPERTIES OF SECTIONS

The determination of section properties of light-gauge sections are based on reduced effective area for load determination. The sectional properties may be based on linear or mid-line method. In this method, material of the section is assumed to be concentrated along the centre line of the steel sheet and therefore considered to be a line, and area elements are considered as line elements. The thickness t is introduced after all linear computations are made.

The formulae for line elements are exact, since the line as such has no thickness dimension. However, in computing the properties of an actual section, where the line element represents an actual element with a thickness dimension, the results will be approximate for the following reasons.

1. The moment of inertia of a straight (actual) element about its longitudinal axis is considered negligible.
2. The moment of inertia of a straight (actual) element inclined to the axis of reference is slightly larger than that of the corresponding line element.
3. Small errors are involved in using the properties of a linear arc to find those of an actual curved corner. However, the error is negligible or disappears when the mean radius of a circular element is over four times its thickness.

$$\text{Moment of inertia } I \text{ of a light-gauge section} = I' \times t$$

$$\text{Total area of a light-gauge section} = L_t \times t$$

$$\text{Section modulus } Z \text{ of the section} = \frac{I}{y}$$

where L_t = total length of the elements

t = thickness of the element

I' = moment of inertia about the centre line of the section

y = distance from neutral axis to the extreme fibre of the extreme element

Notes

1. First power dimensions, for example, the radius of gyration, are obtained directly by linear method and do not involve the thickness dimension.
2. When the flat width w of a stiffened compression element is reduced for design purposes, the effective design width is used directly to compute the total effective length L_e of the line elements.

Figure 16.8 shows properties of some of the line elements.

16.5 LOCAL BUCKLING OF PLATE ELEMENTS

Most structural members, either hot or cold rolled, are composed of flat plate elements. These elements under compression may buckle locally out of their original planes. The effect of local-plate buckling on the strength of the entire member depends upon (1) the location of the buckled element, (2) its buckling and post-buckling strength, and (3) the type of member. The design methods used in handling these problems rely in varying degrees upon the results of a few elementary cases in the theory of elastic buckling of ideal flat plates.

The three most important cases are

- uniform compression, as in the flange of beams and the webs and flanges of columns,
- pure bending, as in the web of a beam in a region of zero shear, and
- pure shear, as in the web of a beam in a region of zero bending.

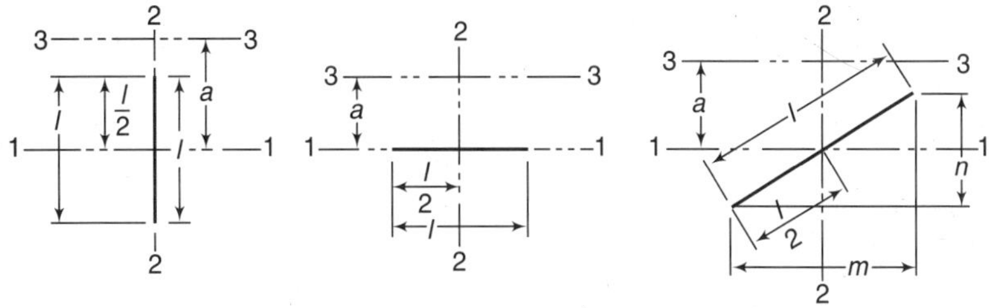

(a) (b) (c)

(a)

$$I_1 = \frac{l^3}{12}$$

$$I_2 = 0$$

$$I_3 = la^2 + \frac{l^3}{12} = l\left(a^2 + \frac{l^2}{12}\right)$$

(b)

$$I_1 = 0$$

$$I_2 = \frac{l^3}{12}$$

$$I_3 = la^2$$

(c)

$$I_1 = \frac{ln^2}{12}$$

$$I_2 = \frac{lm^2}{12}$$

$$I_3 = la^2 + \frac{ln^2}{12} = l\left(a^2 + \frac{n^2}{12}\right)$$

(a) Moment of Inertias of Common Line Elements

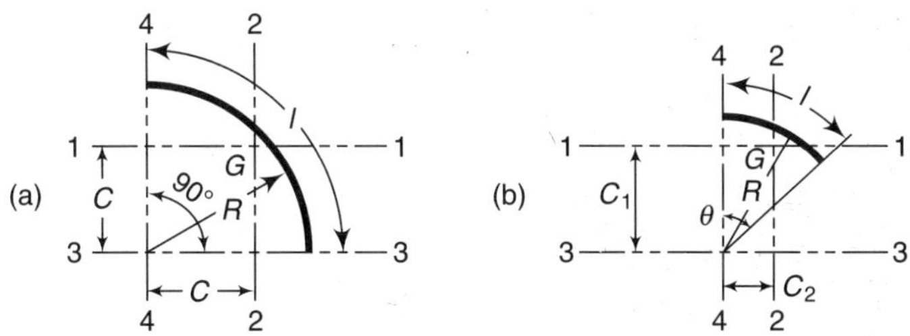

(a)

$l = 1.57\ R$; $C = 0.637\ R$

$I_1 = I_2 = 0.149\ R^3$

$I_3 = I_4 = 0.785\ R^3$

G = centre of gravity

(b)

θ (radians) $= 0.1745 \times \theta$ (degrees)

$$l_1 = \theta R$$

$$C_1 = \frac{R \sin\theta}{\theta}; \quad C_2 = \frac{R(1-\cos\theta)}{\theta}$$

$$I_1 = \left[\frac{\theta + (\sin\theta)(\cos\theta)}{2} - \frac{(\sin\theta)^2}{2}\right]R^3$$

$$I_2 = \left[\frac{\theta - (\sin\theta)(\cos\theta)}{2} - \frac{(1-\cos\theta)^2}{2}\right]R^3$$

$$I_3 = \left[\frac{\theta + (\sin\theta)(\cos\theta)}{2}\right]R^3$$

$$I_4 = \left[\frac{\theta - (\sin\theta)(\cos\theta)}{2}\right]R^3$$

(b) Moment of Inertias of Circular Line Elements

Fig. 16.8 *Properties of Light-Gauge Elements*

While uniform compression is common, the other two are rather rare in practice, the usual conditions being a combination of bending and shear with both flexural and shearing stresses varying continuously along the member. In design it is generally satisfactory to isolate bending and shear and to develop design guides for each as if the other did not exist. In most regions one or the other effect is small enough to neglect, and the guides may be used directly. Approximate interaction formulae may be developed for cases where both shear and bending are substantial, as at the interior supports of continuous beams.

In the first two elementary cases, the critical stress may be written as

$$f_{cr} = \frac{K\pi^2 E}{12(1-\mu^2)(b/t)^2} \quad (16.1)$$

and, in the third,

$$\tau_{cr} = \frac{K\pi^2 E}{12(1-\mu^2)(b/t)^2} \quad (16.2)$$

where K is a buckling coefficient which depends upon the support conditions and the ratio of length to width of the plate in each case. b, t are the width and thickness of the plate element, respectively, and E, μ are elastic constants. There is, of course, no fundamental difference between the buckling phenomena represented by the two equations. In pure shear, buckling is the result of the inclined compressive stresses which tend to cause diagonal waves in the plate.

Flat plates with different boundary conditions and corresponding values of K are shown in Table 16.1. Simple edge supports (those which provide no restraint against rotation about the edge and no restraint against deformation along the edge) are shown by simple cross-hatching. Fixed-edge supports (full restraint against rotation about the edge but no restraint against deformation along the edge) are indicated by double cross-hatching. Completely free supports are unmarked. The ratio of length to width, a/b, is designated α, which is called the *aspect ratio*. Deflections of typical sections are also indicated.

For light-gauge steel members the ratio of width to thickness of the plate element is quite large causing the plate element to fail invariably in buckling rather than yielding as is usually for hot-rolled sections; Limits are placed on width to thickness ratio to check that the plate elements do not fail in buckling.

For light-gauge members the critical buckling is generally of a local nature and precedes general buckling of the member. Light-gauge design criteria are based on post-buckling strength of the member after local buckling has occurred. Most design formulae for local-plate buckling boil down to some function of b/t with, where necessary, recognition of the effect of the aspect ratio, the conditions of edge restraint, and the post buckling strength. Due to the post-buckling strength a stiffened light-gauge wide plate element has ultimate strength many times more than the critical stresses. A wide light-gauge plate may be considered to be a grid of longitudinal and transverse bars as shown in Fig. 16.9. Under the loading shown in the figure, longitudinal bars are loaded in compression and hence buckle while the transverse bars stretch like ties. This stretching action of transverse bars restrains the lateral deflection of longitudinal bars and is responsible for post-buckling strength of plates.

Table 16.1 *Approximate Buckling Coefficients for Flat Plates*

Uniform compression 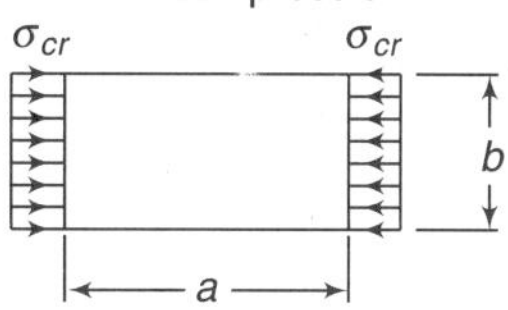	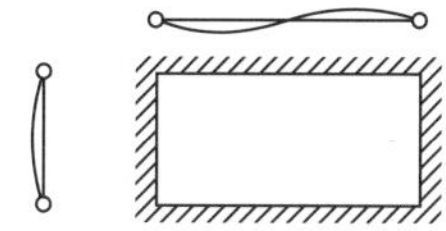	$K = 4.00$
	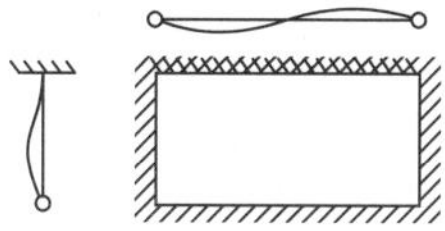	$K = 5.42$
	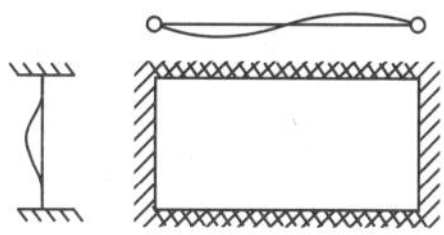	$K = 6.97$
	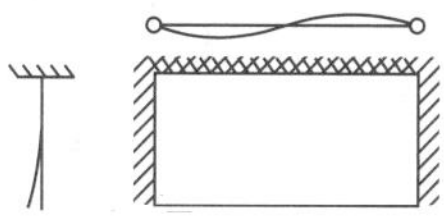	$K = 1.277$
	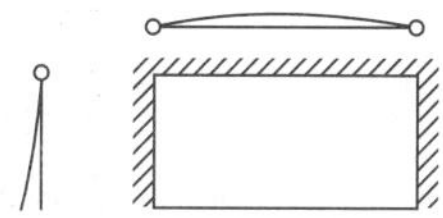	$K = 0.425$
Pure bending 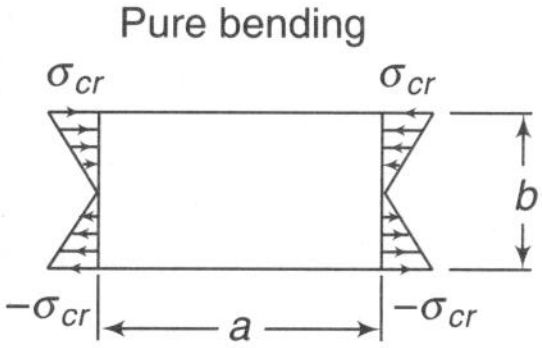	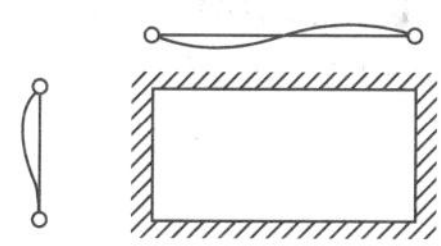	$\alpha \geq 0.67$: $K = 24$
	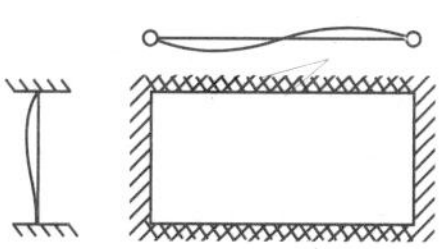	$\alpha \geq 0.40$: $K = 40$
Pure shear 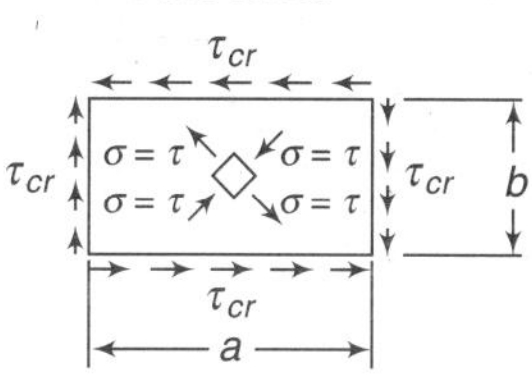	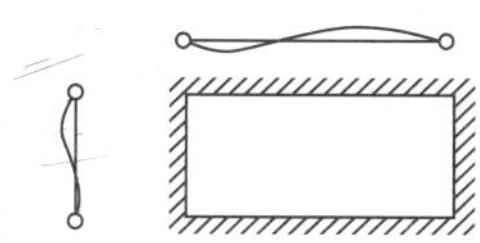	$\alpha \geq 1.0$: $K = 5.34 + \frac{4.0}{\alpha^2}$
	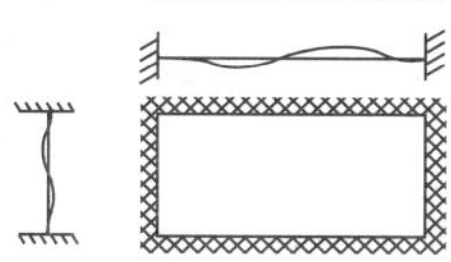	$\alpha \geq 1.0$: $K = 8.98 + \frac{5.6}{\alpha^2}$

Note All k values for uniform compression are for long plates only.

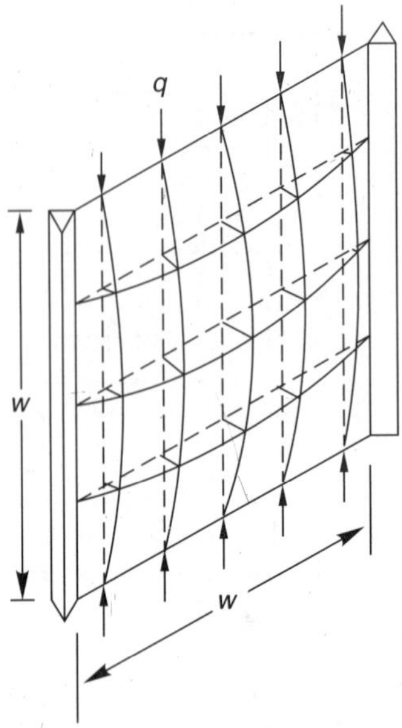

Fig. 16.9 *Model for Post Buckling Strength of Plate*

Note Plates with large width-to-thickness ratios have high post-buckling strengths than those with low *w/t* ratios. That is why for narrow thick plates, failure load never exceeds the critical load.

The development of the design formulae and the necessary definitions and concepts required for understanding the analysis of light-gauge sections are presented in the sections to follow.

16.6 EFFECTIVE DESIGN WIDTH

Consider a plate simply supported on two edges and loaded with a uniformly distributed load p as shown in Fig. 16.10(a). The stress in the plate will be uniform (f_1) for its values less than or equal to critical stress, f_{cr} [Fig. 16.10(b)]. On any further increase of stress beyond f_{cr}, the plate at the centre buckles and the stress distribution will be as shown in Fig. 16.10(c). This reveals that the unbuckled

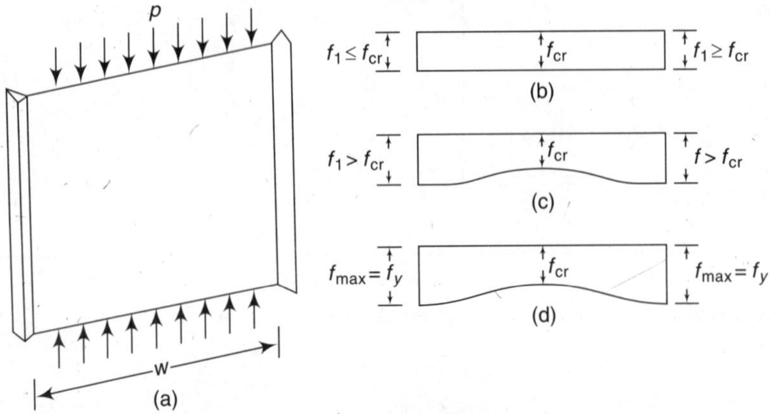

Fig. 16.10 *Compressive Stress Distributionof Thin Plate Elements*

portion of plate now resists the increased load. On further increase of load, failure occurs when the stress at the supported edge reaches yield stress f_y [Fig. 16.10(d)]. Therefore, the ultimate strength of the plate is reached when the maximum stress in the fibres of the plate near the supported edge reaches the yield stress. The load-carrying capacity of the plate will be the area of the stress curve of Fig. 16.10(d) multiplied by the thickness of the plate. It is difficult to account for the non-uniform stress distribution in the design and as such the total force is assumed to be distributed over lesser width (approximated as that width of plate which just buckles when compressive stress reaches the value f_{max}) with uniform stress. This reduced width b is called the *effective design width* of plate (Fig. 16.11) and is given by,

$$\int_0^w f\,dx = f_{max} \cdot b \tag{16.3}$$

where b is the effective design width.

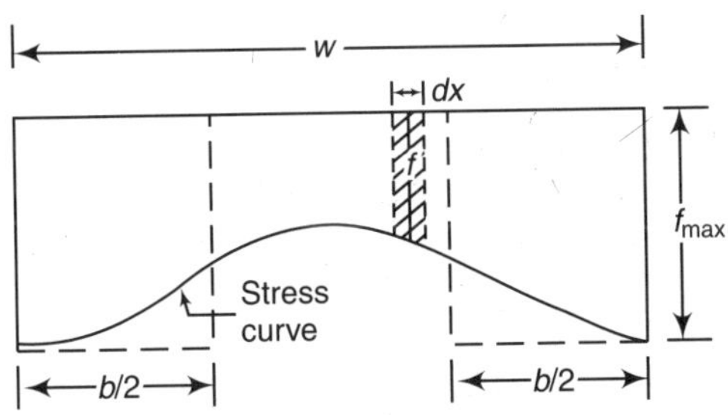

Fig. 16.11 *Compressive Stress Distribution and Effective Width*

The concept of effective width is based on replacing the plate width w, subjected to a varying stress distribution across the width, by an equivalent width b, in which the total load is carried by a uniformly distributed stress equal to the edge stress. The expressions for determining effective design width of light-gauge steel plate elements given in IS 801 are detailed below. For determination of loads and deflections of plate elements, the expressions to be used are different. Also, expressions for determination of loads and deflection for box and tubular sections are different than those for plates.

Eq. (16.1) for simply supported plate may be rewritten as

$$\left(\frac{b}{t}\right)^2 = \frac{\pi^2 E}{3(1-\mu^2) f_y} \qquad \text{(Taking } k = 4\text{, Table 16.1)}$$

or

$$\frac{b}{t} = 1.9\sqrt{\frac{E}{f_y}} \qquad \text{(Taking } \mu = 0.3\text{)} \tag{16.4}$$

The above expression based on experimental results is known as the Von–Karman equation. This expression modified by Winter is as

$$\frac{b}{t} = 1.9\sqrt{\frac{E}{f_{max}}}\left[1 - 0.475\left(\frac{t}{w}\right)\sqrt{\frac{E}{f_{max}}}\right] \tag{16.5}$$

IS 801 uses still a modified expression as

$$\frac{b}{t} = 1.9\sqrt{\frac{E}{f_{max}}}\left[1 - 0.415\left(\frac{t}{w}\right)\sqrt{\frac{E}{f_{max}}}\right] \tag{16.6}$$

Taking $E = 2074000$ kg/cm^2 (IS 800: 1975), the Eq. (16.6) reduces to

$$\frac{b}{t} = \frac{2736}{\sqrt{f}}\left(1 - \frac{598}{(w/t)\sqrt{f}}\right) \tag{16.7}$$

Equation (16.7) is modified for determination of load-carrying capacity and deflection of the light-gauge sections. The reason for this is as follows.

The ultimate moment of the section,

or $$M_{ult} = Z_1 \times 1.67 \times f_b \qquad \text{(factor of safety = 1.67)}$$

or, $$\frac{M_{ult}}{1.67} = Z_1 \times f_b$$

or, $$M_{allowable} = Z_1 \times f_b \tag{16.8}$$

where f_y = yield stress

f_b = basic design stress

Z_1 = section modulus at yield stress

It may be confusing for a designer to calculate $M_{allowable}$ for section modulus at yield stress. To avoid this, Eq. (16.7) is modified for load determination by replacing f by 1.67 f.

For load determination, Eq. (16.7) would change to

$$\frac{b}{t} = \frac{2117}{\sqrt{f}}\left(1 - \frac{462}{(w/t)\sqrt{f}}\right)$$

The above expression after rounding off (in IS 801: 1975) is as follows:

$$\frac{b}{t} = \frac{2120}{\sqrt{f}}\left(1 - \frac{465}{(w/t)\sqrt{f}}\right) \tag{16.9}$$

In case of small moments ($M < M_{allowable}$)

If effective width (b) = full width (w), then Eq. (16.9) can be written as

$$\left(\frac{w}{t}\right)^2 - \frac{2120}{\sqrt{f}}\left(\frac{w}{t}\right) + \frac{95800}{f} = 0$$

$$\frac{w}{t} = \frac{1431}{\sqrt{f}} \tag{16.10}$$

IS 801: 1975 rounds off Eq. (16.10) as under.

$$\left(\frac{w}{t}\right)_{\lim} = \frac{1435}{\sqrt{f}} \tag{16.11}$$

Thus flanges are fully effective ($b = w$) up to $(w/t)_{\lim} = 1435/\sqrt{f}$. For flanges with (w/t) larger than $(w/t)_{\lim}$, effective width is given by Eq. (16.9).

For deflection determination

Equation (16.7) has been rounded off (in IS 801: 1975) as

$$\frac{b}{t} = \frac{2710}{\sqrt{f}}\left(1 - \frac{600}{(w/t)\sqrt{f}}\right) \tag{16.12}$$

The value of $(w/t)_{\lim}$ for deflection determination can be obtained as has been done in the case for load determination, and is

$$\frac{w}{t} = \frac{1813}{\sqrt{f}} \tag{16.13}$$

IS 801: 1975 rounds off Eq. (16.13) as under.

$$\left(\frac{w}{t}\right)_{\lim} = \frac{1850}{\sqrt{f}} \tag{16.14}$$

Thus flanges are fully effective ($b = w$) up to $(w/t)_{\lim} = 1850/\sqrt{f}$. For flanges with (w/t) larger than $(w/t)_{\lim}$, effective width is given by Eq. (16.12).

All the equations from Eq. (16.5) to Eq. (16.14) are for plate elements without intermediate stiffeners. The elements with large flat width ratios become uneconomical because of small effective widths and require stiffening (multiple stiffened elements as shown in Fig. 16.5). When flat width ratio of a sub-element of a multiple stiffened element exceeds 60, the effective design width is reduced and is given by

$$\frac{b_e}{t} = \frac{b}{t} - 0.10\left(\frac{w}{t} - 60\right) \tag{16.15}$$

The area of stiffeners is reduced to an effective area given by

$$A_{ef} = \alpha A_{st} \tag{16.16}$$

where $\alpha = \left(3 - 2 \times \dfrac{b_e}{w}\right) - \dfrac{1}{30}\left[1 - \dfrac{b_e}{w}\right]\dfrac{w}{t} \qquad \left(\dfrac{w}{t} = 60 \text{ to } 90\right)$

where A_{st} = area of stiffener

and

$$A_{ef} = \left(\frac{b}{w}\right) A_{st} \tag{16.17}$$

The effective width expressions arrived at as above are not applicable for square and tubular sections. Since square and tubular sections are rolled under strict quality control, a higher value of effective widths are permitted by IS 801 as follows.

For load determination
$$\frac{b}{t} = \frac{2120}{\sqrt{f}}\left[1 - \frac{420}{(w/t)\sqrt{f}}\right] \tag{16.18}$$

For deflection determination
$$\frac{b}{t} = \frac{2710}{\sqrt{f}}\left[1 - \frac{545}{(w/t)\sqrt{f}}\right] \tag{16.19}$$

IS 801: 1975 is still in MKS units and is not revised. To maintain SI units throughout the book the formulae given in the code are converted to SI units and are presented in Table 16.2.

Table 16.2 *Formulae for Effective Width of Stiffened Compression Members Without Intermediate Stiffener*

MKS Units	*SI Units*
1. For compression elements other than tubes	1. For compression elements other than tubes
(a) *For load determination*	(a) *For load determination*
For (w/t) up to $(w/t)_{lim} = 1435/\sqrt{f}$, $b = w$	For (w/t) up to $(w/t)_{lim} = 446/\sqrt{f}$, $b = w$
For $(w/t) > (w/t)_{lim}$, $\frac{b}{t} = \frac{2120}{\sqrt{f}}\left(1 - \frac{465}{(w/t)\sqrt{f}}\right)$	For $(w/t) > (w/t)_{lim}$, $\frac{b}{t} = \frac{658}{\sqrt{f}}\left(1 - \frac{145}{(w/t)\sqrt{f}}\right)$
(b) *For deflection determination*	(b) *For deflection determination*
For (w/t) up to $(w/t)_{lim} = 1850/\sqrt{f}$, $b = w$	For (w/t) up to $(w/t)_{lim} = 574/\sqrt{f}$, $b = w$
For $(w/t) > (w/t)_{lim}$, $\frac{b}{t} = \frac{2710}{\sqrt{f}}\left(1 - \frac{600}{(w/t)\sqrt{f}}\right)$	For $(w/t) > (w/t)_{lim}$, $\frac{b}{t} = \frac{842}{\sqrt{f}}\left(1 - \frac{186}{(w/t)\sqrt{f}}\right)$
2. For closed square and rectangular tubes	2. For closed square and rectangular tubes
(a) *For load determination*	(a) *For load determination*
For (w/t) up to $(w/t)_{lim} = 1540/\sqrt{f}$, $b = w$	For (w/t) up to $(w/t)_{lim} = 478/\sqrt{f}$, $b = w$
For $(w/t) > (w/t)_{lim}$, $\frac{b}{t} = \frac{2120}{\sqrt{f}}\left(1 - \frac{420}{(w/t)\sqrt{f}}\right)$	For $(w/t) > (w/t)_{lim}$, $\frac{b}{t} = \frac{658}{\sqrt{f}}\left(1 - \frac{130}{(w/t)\sqrt{f}}\right)$
(b) *For deflection determination*	(b) *For deflection determination*
For (w/t) up to $(w/t)_{lim} = 1990/\sqrt{f}$, $b = w$	For (w/t) up to $(w/t)_{lim} = 618/\sqrt{f}$, $b = w$
For (w/t) up to $(w/t)_{lim}$, $\frac{b}{t} = \frac{2710}{\sqrt{f}}\left(1 - \frac{545}{(w/t)\sqrt{f}}\right)$	For (w/t) up to $(w/t)_{lim}$, $\frac{b}{t} = \frac{842}{\sqrt{f}}\left(1 - \frac{169}{(w/t)\sqrt{f}}\right)$

The formulae in SI units can be obtained by dividing the constants of equation, given in IS 801: 1975 in MKS units, by a factor $\sqrt{2074000/200000} = 3.22$, where 2074000 and 2×10^5 are values of modulus of elasticity in MKS and SI units respectively. Also, the multiplying factor for stress conversion from MKS to SI is taken as 9.81/100 = 0.0981.

For example, Eq. (16.7) in SI units may be rewritten as

$$\frac{b}{t} = \frac{850}{\sqrt{f}}\left(1 - \frac{185.6}{(w/t)\sqrt{f}}\right) \tag{16.20}$$

where

μ = Poisson's ratio of steel (0.3)

f_{max} = actual maximum stress in compression element computed on the basis of effective design width

f = actual unit stress in the compression element computed on the basis of effective design width

16.7 SPECIFICATIONS

Stiffeners

Edge stiffeners in the form of web or lip and intermediate stiffeners between webs, or between web and a stiffened edge, are used to stiffen the compression elements. The minimum moment of inertia of an edge stiffener is given by

$$I_{min} = 1.83t^4 \sqrt{\left(\frac{w}{t}\right)^2 - \frac{281200}{f_y}}, \quad \text{but not less than } 9.2t^4 \tag{16.21}$$

where the stiffener consists of a simple lip, the required overall depth of the lip

$$d_{min} = 2.8t\sqrt[6]{\left(\frac{w}{t}\right)^2 - \frac{281200}{f_y}}, \quad \text{but not less than } 4.8t \tag{16.22}$$

Notes

1. Equations (16.21) and (16.22) are in MKS units. To change them to SI units the coefficient 281200 may be replaced by 27590.
2. The minimum moment of inertia required by an intermediate stiffener is double of that of the edge stiffener, since it has to stiffen the two compression elements on either side of the intermediate stiffener.
3. For the conditions to be satisfied by an intermediate stiffener, IS 801 may be referred.

Maximum Allowable Overall Flat-Width Ratios

The maximum allowable overall flat-width ratios *w/t*, disregarding intermediate stiffeners and taking *t* as the actual thickness of the element are as given in Table 16.3.

Table 16.3 *Maximum Overall Flat-Width Ratio*

(a) Stiffened compression element having one longitudinal edge connected to a web or flange element, the other stiffened by:	
(i) *Simple lip*	60
(ii) *Any other kind of stiffener*	90
(b) Stiffened compression element with both longitudinal edges connected to other stiffened elements	500
(c) Unstiffened compression element	60

Notes

1. Unstiffened compression elements that have *w/t* exceeding approximately 30 and stiffened compression elements that have *w/t* ratios exceeding approximately 250 are likely to develop noticeable deformation at the full allowable working stresses without affecting the ability of the member to carry design loads.
2. Stiffened elements having *w/t* ratios larger than 500 may be used with safety to support loads, but substantial deformation of such elements under load may occur and may render the design formulae given in the code inapplicable.

For unusually wide flanges the width is limited by

$$w_f = \sqrt{\frac{126500\,td}{f_{av}}} \times \sqrt[4]{\frac{100\,c_f}{d}} \tag{16.23}$$

where w_f = the width of flange projection beyond the web for I-beams or similar sections, or half of the distance between webs for box- or U-type beams (cm)

t = flange thickness (cm)

d = depth of beam (cm)

c_f = amount of curling (< 5% not important)

f_{av} = the average stress in the full, unreduced flange width in kg/cm^2

Effective Design Width of Short Span Beams

Where the span of the beam is less than $30w_f$ and it carries one concentrated load or several loads spaced farther apart than $2w_f$, the effective design width of any flange, whether in tension or compression, should be limited to as given in Table 16.4.

Table 16.4 *Maximum Allowable Ratio of Effective Design Width to Actual Width*

L/w_t (1)	Ratio (2)	L/w_t (1)	Ratio (2)
30	1.00	14	0.82
25	0.96	12	0.78
20	0.91	10	0.73
18	0.89	8	0.67
16	0.86	6	0.55

where L = full span for simple spans; or the distance between inflection points for continuous beams; or twice the length of cantilever beams in cm.

Note For flanges of I-beams and similar sections stiffened by lips at the outer edges, w_f should be taken as the sum of the flange projections beyond the web plus the depth of the lip.

Maximum Allowable Web Depth

The ratio h/t of the webs of flexural members should not exceed the following limitations:

(a) *For members with unstiffened webs:*

$$(h/t)_{max} = 150$$

(b) *For members which are provided with adequate means of transmitting concentrated loads or reactions or both into the web.*

$$(h/t)_{max} = 200$$

where h = clear distance between flanges measured along the plane of web
t = web thickness

Note Where the web consists of two or more sheets, the h/t ratio should be computed for individual sheet.

Deflection of Beams

The maximum deflection of the beam is limited to span/325.

Notes

1. Since the effective width of stiffened compression element changes with the stress, the moment of inertia I also changes with the B.M. at that section. Thus, in computing the deflection, a variable moment of inertia must be considered. However, such accuracy is generally not required and one may use the effective section at maximum bending moment for computing the deflection.
2. For beams with unstiffened flanges only gross moment of inertia is used for computing deflection.

16.8 BASIC ALLOWABLE DESIGN STRESSES

Stresses in tension, compression and flexural members should not exceed $0.6f_y$, where f_y is the yield stress. Some of the grades of steel with their yield stress and basic allowable design stress as specified in IS: 801: 1975 are given in Table 16.5.

Table 16.5 *Basic Allowable Design Stress*

Minimum Yield Strength of Steel *(Kg/cm²)*	*Basic Allowable Design Stress (f)* *(Kg/cm²)*
2100	1250
2400	1450
3000	1800
3600	2160

Note The allowable stresses may be increased by $33\frac{1}{3}$% for members and assemblies subjected to wind or earthquake forces.

16.9 ALLOWABLE COMPRESSIVE STRESSES IN UNSTIFFENED ELEMENTS

An unstiffened compression element may fail in yielding if it is short and its w/t ratio is less than a certain value. If, however, the w/t ratio is large, it will fail by buckling. For intermediate values of w/t ratios, the failure will be by inelastic buckling. Figure 16.13 shows the failure stresses and allowable stresses for limits 0 < w/t < 60. Compression f_c (in kg/cm^2) on flat unstiffened element is given by:

$$\left.\begin{array}{l}\text{(a) For } w/t \text{ ratio} \ngtr 530/\sqrt{f_y}: \\ \qquad f_c = 0.60\, f_y \\ \text{(b) For } w/t \text{ ratio} > 530/\sqrt{f_y}: \\ \qquad \text{but} \ngtr 1210/\sqrt{f_y^*}: \\ \qquad f_c = f_y[0.767\ (3.15/10^4)\ (w/t)\sqrt{f_y}\ ] \\ \text{(c) For } w/t \text{ ratio} > 1210/\sqrt{f_y} \\ \qquad \text{but} \ngtr 25^*: \\ \qquad f_c = 562\ 000/(w/t)^2 \end{array}\right\} \qquad (16.24)$$

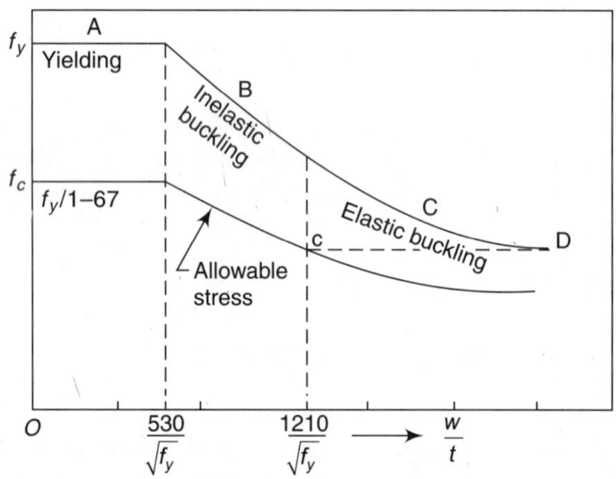

Fig. 16.12 *Unstiffened Element Failure Stresses and Allowable Stresses for 0 < (w/t) <60*

(d) For w/t ratio from 25 to 60[†]

For angle struts: $f_c = 562\ 000/\ (w/t)^2$ (16.26)

For all other sections: $f_c = 1390 - 20\ (w/t)$ (16.27)

---

* When the yield point of steel is less than 2320 kg/cm^2 then for w/t ratios between $530/\sqrt{f_y}$ and 25:

$$f_c = 0.6 f_y - \frac{\left[\dfrac{w}{t} - \dfrac{530}{\sqrt{f_y}}\right](0.6\, f_y - 900)}{25\left[1 - 21.2/\sqrt{f_y}\right]} \qquad (16.25)$$

[†] Unstiffened compression elements having ratios of w/t exceeding approximately 30 may show noticeable distortion of the free edges under allowable compressive stress without detriment to the ability of the member to support load. For ratios of w/t exceeding approximately 60 distortion of the flanges is likely to be so pronounced as to render the section structurally undesirable unless load and stress are limited to such a degree as to render such use uneconomical.

Notes

1. Unstiffened compression elements having $(w/t) > 60$ should not be used.
2. The expressions for compression f_c (in N/mm^2) on flat unstiffened elements are given in Table 16.6.

Table 16.6 *Allowable Stresses in Unstiffened Compression Elements (IS: 801–1975)*

MKS Units	*SI Units*
For $w/t \le 530/\sqrt{f_y}$ $f_c = 0.60\sqrt{f_y}$	For $w/t \le 165/\sqrt{f_y}$ $f_c = 0.60\sqrt{f_y}$
For $1210/\sqrt{f_y} > w/t > 530/\sqrt{f_y}$ $f_c = f_y[0.767 - (3.15/10^4)(w/t)\sqrt{f_y}\,]$	For $375/\sqrt{f_y} > w/t > 165/\sqrt{f_y}$ $f_c = f_y[0.767 - 10^{-3}(w/t)\sqrt{f_y}\,]$
$25 > w/t > 1210/\sqrt{f_y}$ $f_c = 562\ 000/(w/t)^2$	$25 > w/t > 375/\sqrt{f_y}$ $f_c = 54200/(w/t)^2$
For $60 > w/t > 25$ (a) *Angle struts:* $f_c = 562\ 000/(w/t)^2$ (b) *Other sections:* $f_c = 1390 - 20\ (w/t)$	For $60 > w/t > 25$ (a) *Angle struts:* $f_c = 542\ 000/(w/t)^2$ (b) *Other sections:* $f_c = 134 - 1.93\ (w/t)$

16.10 COMPRESSION MEMBERS

The column strength of thin plates is defined as the load-carrying capacity of the member controlled by one or a combination of the following four types of failure: (a) crushing, (b) local buckling of thin-plate elements of the section, (c) overall or primary column buckling by lateral bending over the unsupported length of the member, and (d) torsional buckling or twisting of the section about a longitudinal axis.

In cold-formed light-gauge sections, the width–thickness ratios of component elements are large enough to cause local buckling. This necessitates consideration of the local buckling effects in the allowable stress computations. This is done by incorporating a factor Q, called *form factor*, in the allowable stress expressions. The form factor is less than one and by substituting Qf_y for f_y in the axial stress expressions, the expressions for allowable compressive stress can be arrived at.

Determination of Form Factor

For Members Composed Entirely of Stiffened Elements The cross sections of two typical light-guage steel columns consisting entirely of stiffened elements are shown in Fig. 16.13. It is easy to see that, when axially compressed, each side of the box section will act as a uniformly compressed plate having a buckling coefficient somewhere between 4.00 and 6.97 and probably much closer to the lower value. The same is true for the web of the C section, but there is a little more uncertainty regarding its flanges. If there were no turned-down lips to support the outer edges of the flanges, they would be free and the buckling coefficient would be somewhere between 1.277 and 0.425, probably much closer to the latter. The lips give some support to these edges, and, if they are sufficiently stiff, their resistance is equivalent to that of an idealized simple support.

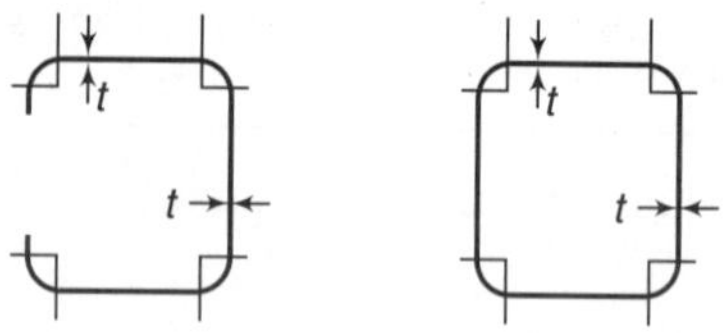

Fig. 16.13 *Sections with all Stiffened Elements*

We should assume that the stiffening lips on the channel in the figure are adequate and that all major plate elements—the web and the two flanges—may be treated as plates supported along both longitudinal edges. The Q factor for columns of this type is a function of the effective area.

$$P_{ult} = A_{eff} \times f_y$$

where P_{ult} = yield load
A_{eff} = effective area of all stiffened elements computed for basic design stress

$$\text{Yield stress} = \frac{P_{ult}}{A} = \frac{A_{eff}}{A} f_y$$

But,
$$\text{yield stress} = Qf_y$$

Hence,
$$Q = \frac{A_{eff}}{A} \tag{16.28}$$

That is for a section with entirely stiffened elements, Q is a factor equal to the effective design area of the section divided by the gross area. The (A_{eff}) used in the Eq. (16.28) is the effective area at the yield point.

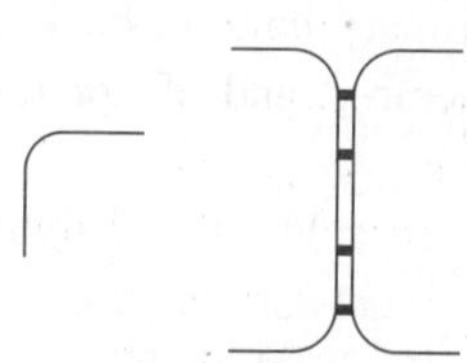

(a) (b)

Fig. 16.14 *Sections with Unstiffened Flanges*

For Members Composed Entirely of Unstiffened Elements Aside from the angle section [Fig. 16.14(a)], it is rather difficult to think of columns made up entirely of unstiffened elements. However, a double channel in which the webs are interconnected or sufficiently thick so that the full area will be effective up to the point of failure is equivalent to one in so far as the computation of Q is concerned [Fig. 16.14(b)]. The Q factor for columns of this type is a function of the ultimate strength of the unstiffened plate.

An unstiffened plate having a small width-to-thickness ratio will fail by yielding prior to buckling. The upper limit to this may be generally set at $b/t = 530/\sqrt{f_y}$. Between $b/t = 530/\sqrt{f_y}$ and $b/t = 1210/\sqrt{f_y}$, failure will occur by inelastic buckling with negligible post-buckling reserve.

$$\therefore \quad P_{ult} = f_{cr} \times A$$

or
$$\frac{P_{ult}}{A} = \frac{f_{cr}}{f_y} \times f_y = \frac{1.67\, f_c}{1.67\, f_b} f_y = \frac{f_c}{f_b} f_y$$

where f_c and f_b are the allowable compressive and bending stresses, respectively.

That is for a member composed entirely of unstiffened elements, Q is a factor equal to the allowable compressive stress for the weakest plate of the section, divided by the basic design stress ($0.6f_y$). The weakest plate is, of course, the one with the largest b/t ratio.

But, yield stress = Qf_y

Hence,
$$Q = \frac{f_c}{f_b} = \frac{f_c}{f} \tag{16.29}$$

where $f = f_b$ = basic design stress = $0.6f_y$

The permissible compressive stress f_c is computed by equations given in Table16.6.

For Members Consisting of both Stiffened and Unstiffened Elements The cross sections of two light-gauge steel columns consisting of both stiffened and unstiffened elements are shown in Fig. 16.15. The capacity of a member of this type is reached when the weakest unstiffened plate element fails at the stress $\sigma_{cr} = 1.65f_c$, where f_c is computed from the applicable Eq. 16.24. At failure, the effective area of the section is equal to the full area of all unstiffened elements plus the effective area of all stiffened elements at the stress f_{cr} (not f_y as in the case of columns with all stiffened elements). Still calling this total effective area A_{eff}, the ultimate capacity of the plate may be written as

$$P_{ult} = A_{eff} f_{cr} \tag{16.30}$$

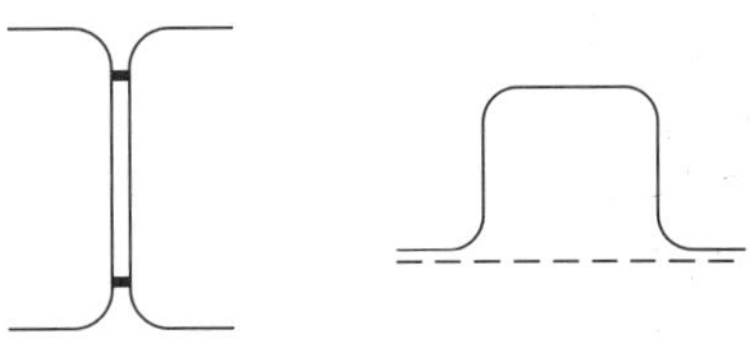

Fig. 16.15 *Sections with Stiffened and Unstiffened Plates*

The member consisting of both stiffened and unstiffened elements will attain its failure load when the weaker of the unstiffened element buckles at the critical stress (f_{cr}). At this stress, A_{eff} will consist of unreduced area of unstiffened elements and effective area of stiffened elements computed for f_{cr} (or value of f_c from Eq. (16.24) is used).

Therefore, $P_{ult} = f_{cr} \times A_{eff}$

This can be arranged as

$$\frac{P_{ult}}{A} = \frac{f_{cr}}{f_y} \times \frac{A_{eff}}{A} f_y = \left(\frac{f_c}{f_b} \times \frac{A_{eff}}{A}\right) f_y$$

But, since yield stress = Qf_y

Hence,
$$Q = \frac{f_c}{f_b} \times \frac{A_{eff}}{A} = \frac{A_{eff}}{A} \times \frac{f_c}{f}$$

Thus, $$Q = Q_{\text{unstiffened}} \times Q_{\text{stiffened}} \tag{16.31}$$

Allowable Stresses for Shapes Not Subjected to Torsional-Flexural Buckling

For doubly symmetric shapes, closed cross-sectional shapes or cylindrical sections, and any other shapes which can be shown not to be subject to torsional-flexural buckling, and for members braced against twisting, The average axial stress P/A, in compression members should not exceed the following values of f_{a1}.

For $$kL/r < \frac{C_c}{\sqrt{Q}}$$

$$f_{a1} = \frac{12}{23} Qf_y - \frac{3}{23} \times \frac{(Qf_y)^2}{\pi^2 E}\left(\frac{kL}{r}\right)^2 \tag{16.32}$$

or $$f_{a1} = 0.522 Qf_y - \left(\frac{Qf_y kL/r}{12500}\right)^2$$

for $$kL/r \geq \frac{C_c}{\sqrt{Q}}$$

$$f_{a1} = \frac{12\pi^2 E}{23(kL/r)^2}$$

or $$f_{a1} = \frac{10680000}{(kL/r)^2} \tag{16.33}$$

where

$C_c = \sqrt{2\pi^2 E/f_y}$
P = total load;
A = full unreduced cross-sectional area of the member
f_{a1} = allowable average compression stress under concentric loading
E = modulus of elasticity = 2074000 kgf/cm^2
k = effective length factor*
L = unbraced length of member
r = radius of gyration of full, unreduced cross section
f_y = yield point of steel
Q = form factor

---

* In frames where lateral stability is provided by diagonal bracing, shear walls, attachment to an adjacent structure having adequate lateral stability, or by floor slabs or roof decks secured horizontally by walls or bracing systems parallel to the plane of the frame, and in trusses the effective length factor k for the compression members should be taken as unity, unless analysis shows that a smaller value may be used. The effective length kL of compression members in a frame which depends upon its own bending stiffness for lateral stability, should be determined by a rational method and should not be less than the actual unbraced length.

Note When the factor Q is equal to unity, the steel is 2.29 mm or more in thickness, and kL/r is less than C_c:

$$f_{a_1} = \frac{\left[1 - \dfrac{(kL/r)^2}{2(C_c)^2}\right] f_y}{\dfrac{5}{3} + \dfrac{3}{8}\dfrac{(kL/r)}{(C_c)} - \dfrac{(kL/r)^3}{8(C_c)^3}}$$

The factor of safety in the expressions above is 1.92 (i.e. 23/12).

The expressions for allowable stresses for shapes subjected to torsional flexural buckling may be referred from IS 801.

16.11 FLEXURAL MEMBERS

The strength of a light-gauge-formed beam is limited by the smallest shear or bending moment that will produce yielding, buckling, or excessive distortion of any of its elements. Where the proper function of the beam depends on its deflection, the useful strength of the beam may be less than the value obtained by shear or moment limitation.

Flanges of Beams

The maximum bending stress occurs in the most distant fibres of flanges. When the centroid of the effective section is at or below mid-depth of the beam, the upper (compression) fibre will be most highly stressed. This will always be the case in a section symmetrical about the xx-axis or in which tension flanges are heavier than the compression flanges; the effective width of the compression flange will be computed on the basis of the basic design stress f_b (= 0.6 f_y). If, however, the centroid of the effective section is above mid-depth, as is often the case in the deck type of sections, or for a beam having compression flange heavier than tension flange, the stress at the instance of failure, will be less than f_y, the tension flange thus will yield prior to compression flange. Since the actual stress will depend upon the position of neutral axis which will depend on the effective area of compression flange which in turn varies with stress in compression flange, a trial assumption must be made either as to stress or effective width and, possibly, several trial calculations made before convergence is obtained.

Laterally Unbraced Beams To prevent lateral buckling, the maximum compressive stress f_b on extreme fibers of laterally unsupported straight flexural members* should not exceed the allowable stress as specified in Section 16.8 or 16.9 nor the following maximum stresses:

(a) When bending is about the centroidal axis perpendicular to the web for either I-shaped sections symmetrical about an axis in the plane of web or symmetrical channel shaped sections:

---

* These provisions apply to *I*-, *Z*-, or channel-shaped flexural members (not including multiple web deck, *U*- and closed-box type members and curved or arch members). These provisions do not apply to laterally unbraced compression flanges of otherwise laterally stable sections.

For $\dfrac{L^2 S_{xc}}{dI_{yc}}$

$$> \frac{0.36\pi^2 EC_b}{f_y}$$

but $$< \frac{1.8\pi^2 EC_b}{f_y}$$

$$f_b = \frac{2}{3} f_y - \frac{f_y^2}{5.4\pi^2 EC_b}\left(\frac{L^2 S_{xc}}{dI_{yc}}\right) \tag{1.34}$$

For $\dfrac{L^2 S_{xc}}{dI_{yc}} \geq \dfrac{1.8\pi^2 EC_b}{f_y}$

$$f_b = 0.6\pi^2 EC_b \frac{dI_{yc}}{L^2 S_{xc}} \tag{1.635}$$

(b) For point-symmetrical Z-shaped sections bent about the centroidal axis perpendicular to the web:

For $\dfrac{L^2 S_{xc}}{dI_{yc}}$

$$> \frac{1.8\pi^2 EC_b}{f_y}$$

but $$< \frac{0.9\pi^2 EC_b}{f_y}$$

$$f_b = \frac{2}{3} f_y - \frac{f_y^2}{2.7\pi^2 EC_b}\left(\frac{L^2 S_{xc}}{dI_{yc}}\right) \tag{16.36}$$

for $\dfrac{L^2 S_{xc}}{dI_{yc}} \geq \dfrac{0.9\pi^2 EC_b}{f_y}$

$$f_b = 0.3\pi^2 EC_b \frac{dI_{yc}}{L^2 S_{xc}} \tag{16.37}$$

where

L = the unbraced length of the member

I_{yc} = the moment of inertia of the compression portion of a section about the gravity axis of the entire section parallel to the web

S_{sc} = compression section modulus of entire section about major axis I_x divided by distance to extreme compression fibre

C_b = bending coefficient which can conservatively be taken as unity, or calculated from:

$$C_b = 1.75 + 1.05\left(\frac{M_1}{M_2}\right) + 0.3\left(\frac{M_1}{M_2}\right)^2, \text{ but not more than } 2.3$$

where M_1 is the smaller and M_2, the larger bending moment at the ends of the unbraced length, taken about the strong axis of the members, and where M_1/M_2, the ratio of end moments is positive when M_1 and M_2 have the same sign (reverse curvature bending) and negative when they are of opposite sign (single curvature bending).When the bending moment at any point within an unbraced length is larger than that at both ends of this length, the ratio M_1/M_2 should be taken as unity.

For members subjected to combined axial and bending stress, C_b should be 1.

E = modulus of elasticity = 2074000 kg/cm^2

d = depth of section

Web of Beams

Shear Stresses in Webs The maximum average shear stress $f_{v,cal}$ in kg/cm^2, on the gross area of a flat web should not exceed:

(a) For $h/t \ngtr 4590/\sqrt{f_y}$:

$$f_v = \frac{1275\sqrt{f_y}}{h/t} \quad \text{with a maximum of } 0.40 f_y \tag{16.38}$$

(b) For $h/t > 4590/\sqrt{f_y}$:

$$f_v = \frac{5850000}{(h/t)^2} \tag{16.39}$$

where

t = web thickness

h = clear distance between flanges measured along the plane of the web

f_y = yield point in kg/cm^2

Note Where the web consists of two or more sheets, these should be considered as separate members carrying their share of the shear.

Bending Stresses in Webs The compressive stress $f_{bw,\text{cal}}$ in kg/cm^2, in the flat web of a beam due to bending in its plane, should not exceed f_{bw} nor should it exceed

$$f_{bw} = \frac{36560000}{(h/t)^2} \text{ kg/cm}^2 \tag{16.40}$$

Combined Bending and Shear Stresses in Webs For webs subject to both bending and shear stresses, the member should be so proportioned that such stresses do not exceed the allowable values specified for shear and bending stresses in webs as above and that the quantity $\sqrt{(f_{bw,\text{cal}}/f_{bw})^2 + (f_{v,\text{cal}}/f_v)^2}$ does not exceed unity.

where

$f_{bw,\text{cal}}$ = actual compression stress at junction of flange and web

$$f_{bw} = \frac{36560000}{(h/t)^2} \text{ kg/cm}^2$$

$f_{v,\text{cal}}$ = actual average shear stress, that is, shear force per web divided by web area

f_v = allowable shear stress except that the limit of 0.4 f_y should not apply

Web Crippling of Beams

Web crippling, as described in hot rolled beams (Section 7.9), is also a problem in light-gauge steel beams, and, in these, a new dimension is added. In hot-rolled beams the load is usually applied in the plane of the web, but normally this is not feasible for cold-formed sections because of their make-up. Typical cold-formed members having loads on their flanges are shown in Fig. 16.12. In the two cases, which represent extreme conditions, the radius of bend between flange and web causes the load to be eccentric to the webs. In the first case, the eccentricity has little effect on the web capacity since, because of the symmetry of loading and configuration, there is some end restraint on the webs and since, if they do buckle, they must buckle together. The web which is forced to buckle toward its flange tips furnishes some support for the other. In the second case, however, there is no such restraint, eccentricity aggregates the buckling tendency in this case, and the critical stress becomes a function of the bend radius as well as the other parameters important in hot-rolled members.

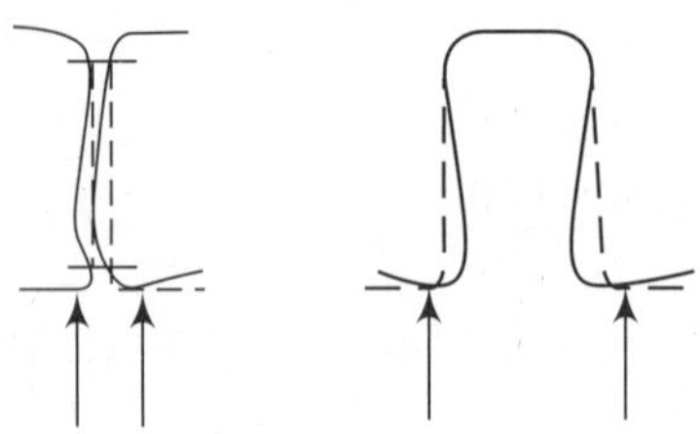

Fig. 16.16 *Web Crippling of Light-Gage Beams*

To avoid crippling of unreinforced beam webs having a flat-width ratio h/t equal to or less than 150, concentrated loads and reactions should not exceed the values of P_{max} given below, where P_{max} is the allowable concentrated load or reaction. Webs of beams for which the ratio h/t greater than 150 should be provided with adequate means of transmitting concentrated loads and reactions directly into the web.

(a) Beams having Single Unreinforced Webs

1. For end Reactions or for Concentrated Loads on Outer Ends of Cantilevers

For inside corner radius equal to or less than the thickness of sheet

$$P_{\max} = 70t^2[98 + 4.20(N/t) - 0.022\,(N/t)\,(h/t) - 0.011\,(h/t)] \times [1.33 - 0.33\,(f_y/2320)]\,(f_y/2320) \quad (16.41)$$

For other corner radii up to 4t, the value P_{max} given by the above formula should be multiplied by (1.15 – 0.15 R/t)

2. For Reactions of Interior Supports or for Concentrated Loads Located Anywhere on the Span

For inside corner radius equal to or less than the thickness of sheet

$$P_{\max} = 70t^2[305 + 2.30(N/t) - 0.009(N/t)(h/t) - 0.5(h/t)] \times [1.22 - 0.22(f_y/2320)]\,(f_y/2320) \quad (16.42)$$

For other corner radii up to 4t, the value P_{max} given by the above formula is to be multiplied by $(1.06 - 0.06\ R/t)$.

3. *For Corner Radii Larger than 4t, Tests shall be Made in Accordance with IS: 801.*

(b) *For I-beams Made of Two Channels Connected Back to Back or for Similar Sections which Provide a High Degree of Restraint Against Rotation of the Web, such as I-Sections Made by Welding Two Angles to a Channel*

1. *For End Reactions or for Concentrated Loads on the Outer Ends of Cantilevers*

$$P_{\max} = t^2 f_y (4.44 + 0.558\sqrt{N/t}) \tag{16.43}$$

2. *For Reactions of Interior Supports or for Concentrated Loads Located Anywhere on the Span*

$$P_{\max} = t^2 f_y (6.66 + 1.146\sqrt{N/t}) \tag{16.44}$$

where t = web thickness
N = actual length of bearing, expect that in the above formulae the value of N should not be taken greater than h
h = clear distance between flanges measured along the plane of web
f_y = yield point
R = inside bend radius

Notes

1. In all of the above expressions, P_{max} represents the load or reaction for one solid web sheet connecting top and bottom flanges. For webs consisting of two or more such sheets, P_{max} should be computed for each individual sheet and the results added to obtain the allowable load or reaction for the composite web.
2. For loads located close to ends of beams, provisions of (a)(2) and (b)(2) apply, provided that for cantilevers the distance from the free end to the nearest edge of bearing, and for a load close to an end support the clear distance from edge of end bearing to nearest edge of load bearing is larger than 1.5h. Otherwise provisions of (a)(1) and (b)(2) apply.

Solved Examples

Example 16.1 Find the column section properties and allowable load for the column section shown in Fig. Ex. 16.1 (a). The effective length of column is 3.0 m. Take f_y = 235 N/mm^2.

Solution

Linear properties

Radius of corner, $R = 2.4 + 0.8 = 3.2$ mm

Length of corner $= 1.57R = 1.57 \times 3.2 = 5.024$ mm

$$C_{xx} = 0.637R = 0.637 \times 3.2 = 2.038 \text{ mm}$$

$$L = 2 \times 112 + 2 \times 112 + 4 \times 5.024 = 468.1 \text{ mm}$$

$$A = 468.1 \times 1.6 = 748.96 \text{ mm}^2$$

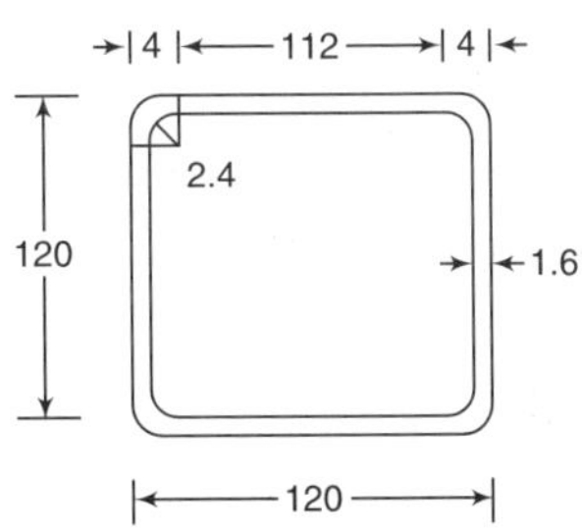

Fig. Ex. 16.1(a)

Linear $I_{xx} = 2 \times 112 \times 59.2^2 + 2 \times \dfrac{112^3}{12} + 4 \times 5.024 \times 58.038^2 = 1086885.58 \text{ mm}^3$

Actual $I_{xx} = 1086885.58 \times 1.6 = 1739016.92 \text{ mm}^4$

$$r = \sqrt{\frac{I_{xx}}{A}} = \sqrt{\frac{1739016.92}{748.96}} = 48.18 \text{ mm} = 4.8 \text{ cm}$$

$$f = 0.6 f_y = 0.6 \times 235 = 141 \text{ N/mm}^2$$

To calculate the form factor, effective width and effective area of the section will be required.

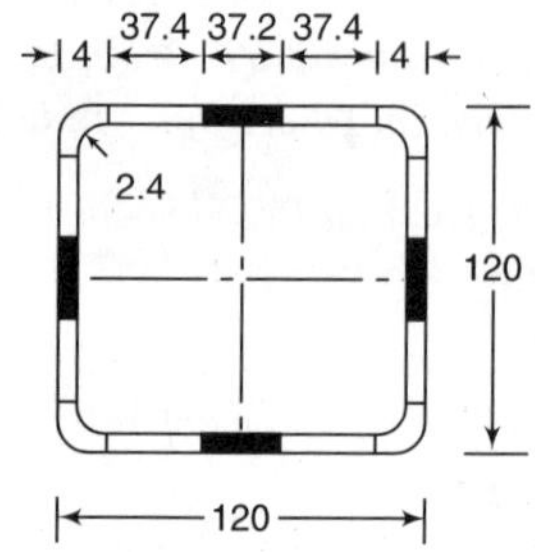

Fig. Ex. 16.1(b)

$$\left(\frac{w}{t}\right)_{\text{lim}} = \frac{478}{\sqrt{f}} = \frac{478}{\sqrt{141}} = 40.25$$

$$\frac{w}{t} = \frac{112}{1.6} = 70 > 40.25$$

$$\therefore \quad \frac{b}{t} = \frac{658}{\sqrt{f}}\left(1 - \frac{130}{(w/t)\sqrt{f}}\right) \quad \text{(From Table 16.2)}$$

$$= \frac{658}{\sqrt{141}}\left(1 - \frac{130}{70\sqrt{141}}\right) = 46.75$$

$$\therefore \quad b = 46.75 \times 1.6 = 74.8$$

$$A_{eff} = 749 - 4 \times 37.2 \times 1.6 = 510.92 \text{ mm}^2$$ [refer Fig. Ex. 16.1(b))]

Form factor, $Q = \dfrac{A_{eff}}{A} = \dfrac{510.92}{749} = 0.682$

$$C_c = \sqrt{\frac{2\pi^2 E}{f_y}} = \sqrt{\frac{2\pi^2 \times 2 \times 10^5}{235}} = 129.61$$

$$\left(\frac{l}{r}\right)_{\text{lim}} = \frac{C_c}{\sqrt{Q}} = \frac{129.61}{\sqrt{0.682}} = 157$$

Actual $\dfrac{l}{r} = \dfrac{3.0 \times 100}{4.8} = 62.5 < 157$

Hence,

$$f_{a1} = \frac{12}{23} Q f_y - \frac{3}{23} \times \frac{(Q f_y)^2}{\pi^2 E}\left(\frac{kL}{r}\right)^2 \quad \text{(From Eq. (16.32))}$$

$$= \frac{12}{23} \times 0.682 \times 235 - \frac{3}{23} \times \frac{(0.682 \times 235)^2}{\pi^2 \times 2 \times 10^5} \times 62.5^2$$

$$= 76.99 \text{ N/mm}^2$$

Allowable load $= f_a \times A = 76.99 \times 748.96 = 57662 \text{ N} = 57.66 \text{ kN}$

Example 16.2 A hat of 100 mm × 80 mm × 4 mm section with a 25 mm lip as shown in Fig. Ex. 16.2 is to be used as a concentrically loaded column of 3.1 m effective length. Determine the allowable load. Take f_y = 235 N/mm^2.

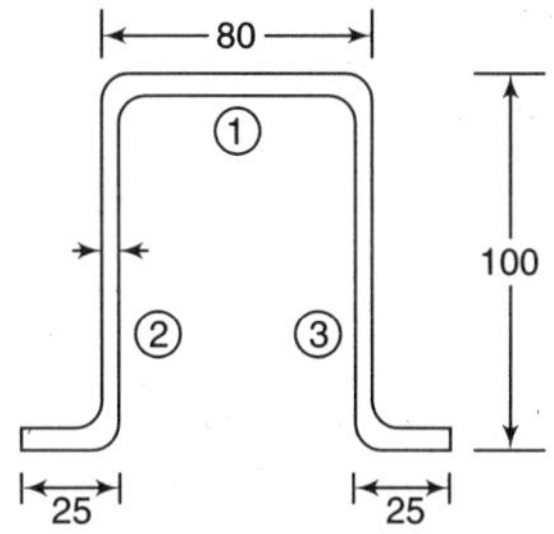

Fig. Ex. 16.2

Solution

$$f = 0.6 f_y = 0.6 \times 235 = 141 \text{ N/mm}^2$$

$$\left(\frac{w}{t}\right)_{\text{lim}} = \frac{446}{\sqrt{f}} = \frac{446}{\sqrt{141}} = 37.6$$

$$\frac{w_1}{t} = \frac{72}{4} = 18 \qquad < 37.6$$

$$\frac{w_2}{t} = \frac{92}{4} = 23 \qquad < 37.6$$

Since w/t ratio for both flange and web is less than $(w/t)_{\text{lim}}$, full width will be effective.

i.e., $b = w$ and $Q = 1$.

From IS: 811, the relevant properties of the section are

$$A = 1180 \text{ mm}^2, C_y = 45.1 \text{ mm}$$

$$I_{xx} = 152 \times 10^4 \text{ mm}^4, I_{yy} = 161 \times 10^4 \text{ mm}^4$$

$$r_{xx} = 35.8 \text{ mm}, r_{yy} = 36.9 \text{ mm}$$

$$\frac{l}{r} = \frac{3100}{35.8} = 86.6 \qquad (kL = 3.1 \text{ m})$$

$$C_c = \sqrt{\frac{2\pi^2 E}{f_y}} = \sqrt{\frac{2\pi^2 \times 2 \times 10^5}{235}} = 129.61$$

$$\left(\frac{l}{r}\right)_{\text{lim}} = \frac{C_c}{\sqrt{Q}} = \frac{129.61}{\sqrt{1}} = 129.61 \qquad < 86.6$$

Since $\frac{l}{r} < \left(\frac{l}{r}\right)_{\text{lim}}$

Hence

$$f_{a1} = \frac{12}{23} Q f_y - \frac{3}{23} \times \frac{(Q f_y)^2}{\pi^2 E} \left(\frac{kL}{r}\right)^2 \qquad \text{(From Eq. (16.32))}$$

$$= \frac{12}{23} \times 1 \times 235 - \frac{3}{23} \times \frac{(1 \times 235)^2}{\pi^2 \times 2 \times 10^5} \times 86.6^2 = 95.24 \text{ N/mm}^2$$

Allowable load $= f_a \times A$

$= 95.24 \times 1180 = 112383 \text{ N} = 112.4 \text{ kN}$

Example 16.3 A square box of 180 mm × 180 mm × 2 mm section as shown in Fig. Ex. 16.3 is to be used as a column of 4 m effective length. It is stiffened on all four sides. Find the maximum load it can carry. Design also the stiffener. Take $f_y = 235$ N/mm^2.

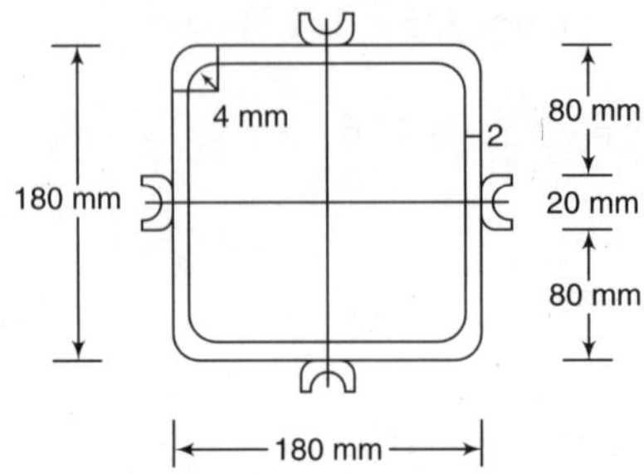

Fig. Ex. 16.3

Solution

Linear properties

Radius of corner, $R = 4 + 1 = 5$ mm

Length of corner = $1.57R = 1.57 \times 5 = 7.85$ mm

$$C_{xx} = 0.637R = 0.637 \times 5 = 3.185 \text{ mm}$$

$$L = 2 \times 168 + 2 \times 168 + 4 \times 7.85 = 703.4 \text{ mm}$$

$$A = 703.4 \times 2 = 1406.8 \text{ mm}^2$$

Linear $\quad I_{xx} = 2 \times 168 \times 89^2 + 2 \times \frac{168^3}{12} + 4 \times 7.85 \times 87.185^2 = 3511397.6$

Actual $\quad I_{xx} = 3511397.6 \times 2 = 7022795.22 \text{ mm}^4$

$$r = \sqrt{\frac{I_{xx}}{A}} = \sqrt{\frac{7022795.22}{1406.8}} = 70.65 \text{ mm} = 7.1 \text{ cm}$$

$$\left(\frac{w}{t}\right)_{\lim} = \frac{478}{\sqrt{f}} = \frac{478}{\sqrt{141}} = 40.25$$

$$\frac{w}{t} = \frac{172}{2} = 86 > 40.25$$

$$\frac{b}{t} = \frac{658}{\sqrt{f}}\left[1 - \frac{130}{\left(\frac{w}{t}\right)\sqrt{f}}\right] \qquad \text{(From Table 16.2)}$$

$$= \frac{658}{\sqrt{141}}\left[1 - \frac{130}{86\sqrt{141}}\right] = 48.36$$

When flat width ratio of the sub-element of a multiple stiffened element exceeds 60, the effective design width is reduced and is given by

$$\frac{b_e}{t} = \frac{b}{t} - 0.10\left(\frac{w}{t} - 60\right) \qquad \text{(From Eq. (16.16))}$$

$$= 48.36 - 0.10 \times (86 - 60) = 45.76$$

or, $$b_e = 2 \times 45.76 = 91.52 \text{ mm}$$

$$A_{eff} = 1406.8 - 4 \times (168 - 91.52) \times 2 = 794.96 \text{ mm}^2$$

$$Q = \frac{A_{eff}}{A} = \frac{794.96}{1406.8} = 0.565$$

$$C_c = \sqrt{\frac{2\pi^2 E}{f_y}} = \sqrt{\frac{2\pi^2 \times 2 \times 10^5}{235}} = 129.61$$

$$\left(\frac{l}{r}\right)_{\text{lim}} = \frac{C_c}{\sqrt{Q}} = \frac{129.61}{\sqrt{0.565}} = 172.43 \qquad (kL = 4.0 \text{ m})$$

$$\text{Actual } \frac{l}{r} = \frac{4 \times 100}{7.1} = 56.34 < 172.43$$

$$f_{a1} = \frac{12}{23} Q f_y - \frac{3}{23} \times \frac{(Q f_y)^2}{\pi^2 E}\left(\frac{kl}{r}\right)^2 \qquad \text{(From Eq. (16.32))}$$

$$= \frac{12}{23} \times 0.565 \times 235 - \frac{3}{23} \times \frac{(0.565 \times 235)^2}{\pi^2 (2 \times 10^5)}(56.34)^2 = 65.57 \text{ N/mm}^2$$

Allowable load $= f_a \times A$

$$= 65.57 \times 1406.8 = 92243.8 \text{ N} = 92.24 \text{ kN}$$

For stiffener,

$$I_{\min} = 1.83 t^4 \sqrt{\left(\frac{w}{t}\right)^2 - \frac{27590}{f_y}}$$

$$= 1.83 \times 2^4 \times \sqrt{86^2 - \frac{27590}{235}} = 2498 \text{ mm}^4$$

Use 20 mm × 20 mm × 1.60 mm channels as stiffeners. (From Table 3 of IS: 811) $I_{xx} = 5540$ mm^4, $I_{yy} = 3450$ mm^4 > 2498 mm^4

Example 16.4 Design a 4.0 m long column to carry a load of 170 kN. Take $f_y = 235$ N/mm^2.

Solution Use two channels of 140 mm × 70 mm × 4 mm section having 25 mm lips with a back-to-back distance of 50 mm as shown in Fig. Ex. 16.4.

$$\text{Area, } A = 2 \times [4 \times 140 + 2 \times (4 \times 62) + 2 \times (4 \times 25)] = 2512 \text{ mm}^2$$

$$I_{xx} = 2\times\left[\frac{(140)^3}{12}\times 4 + 2(4\times 62)\times 68^2 + \left\{4\times\frac{(140)^3}{12} - 4\times\frac{(140-2\times 5)^3}{12}\right\}\right]$$

$$= 7759674.67 \text{ mm}^4$$

$$r_x = \sqrt{\frac{I_{xx}}{A}} = 55.58 \text{ mm}$$

$$\frac{l}{r_x} = \frac{4000}{55.58} = 71.97$$

$$\left(\frac{w}{t}\right)_{\text{lim}} = \frac{446}{\sqrt{f}} = \frac{446}{\sqrt{141}} = 37.6$$

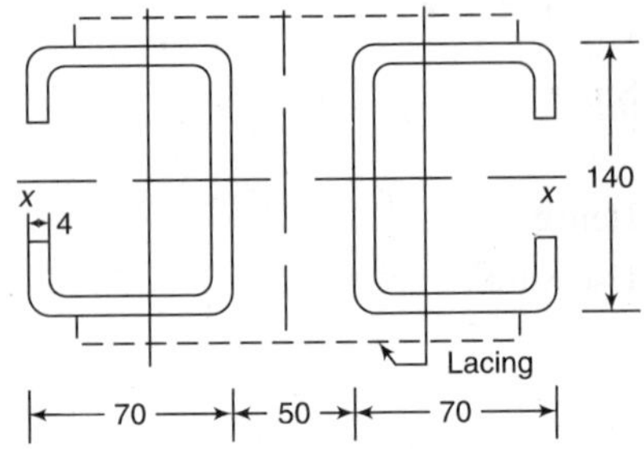

Fig. Ex. 16.4

For web

$$\frac{w}{t} = \frac{132}{4} = 33 < 37.6$$

Hence, $b = w$ and $Q = 1$.

$$C_c = \sqrt{\frac{2\pi^2 E}{f_y}} = \sqrt{\frac{2\pi^2\times(2\times 10^5)}{235}} = 129.1$$

$$\left(\frac{l}{r}\right)_{\text{lim}} = \frac{C_c}{\sqrt{Q}} = \frac{129.61}{\sqrt{1}} = 129.61$$

Since $\frac{l}{r_x} < \left(\frac{l}{r}\right)_{\text{lim}}$

Hence $f_{a1} = \frac{12}{23}Qf_y - \frac{3}{23}\times\frac{(Qf_y)^2}{\pi^2 E}\left(\frac{kL}{r}\right)^2$ (From Eqn. 16.32)

$$= \frac{12}{23}\times 1\times 235 - \frac{3}{23}\times\frac{(1\times 235)^2}{\pi^2\times 2\times 10^5}\times 71.97^2 = 103.71 \text{ N/mm}^2$$

Allowable load = $f_a \times A$

= 103.71 × 2512 = 260519.52 N = 260.52 kN > 170 kN

Hence safe.

Lacing

Transverse shear per lacing = $\frac{1}{2}\times\frac{2.5}{100}\times 170\times 10^3$ = 2125 N = 2.125 kN

Compressive force in lacing bars = 2.125 × cosec 45° = 3.005 kN

Let us consider a weld of 3 mm size and 45 mm length.

Assuming a 5 mm margin on both sides.

Effective spacing = 190 – (2 × 5) = 180 mm

Length of lacing flat, l = 180 cosec 45°

$\frac{l}{r}$ should not exceed 145.

$$r \geq \frac{l}{145} = \frac{180 \text{ cosec } 45°}{145} = 1.756 \text{ mm}$$

Since $r = \frac{t}{\sqrt{12}}$

Hence $t \geq 1.756 \times \sqrt{12} = 6.08$ mm

Use 8 mm thick flat of width 20 mm.

$$r = \frac{t}{\sqrt{12}} = \frac{8}{\sqrt{12}} = 2.31 \text{ mm}$$

$$\frac{l}{r} = \frac{254.6}{2.31} = 110.2$$

$$\frac{w}{t} = \frac{20}{5} = 4 < 37.6$$

Hence $b = w$ and $Q = 1$.

$$\therefore \quad f_{a1} = \frac{12}{23} Q f_y - \frac{3}{23} \times \frac{(Q f_y)^2}{\pi^2 E} \left(\frac{kl}{r}\right)^2 \qquad \text{(From Eq. (16.32))}$$

$$= \frac{12}{23} \times 1 \times 235 - \frac{3}{23} \times \frac{(1 \times 235)^2}{\pi^2 \times 2 \times 10^5} \times 110.2^2 = 78.29 \text{ N/mm}^2$$

Allowable load = 78.29 × (8 × 20) = 12526.8 N = 12.53 kN > 3.005 kN

Example 16.5 Two channels of 180 mm × 80 mm section with bent lips as shown in Fig. Ex. 16.5 are connected with webs to act as beam. The thickness of the plate is 2.5 mm and the depth of the lip is 25 mm. The beam has an an effective span of 4.1 m. Determine the allowable load per m run on the beam. Take f_y = 235 N/mm².

Solution Basic design stress, $f = 0.6 f_y = 0.6 \times 235$
$= 141$ N/mm²

$$\left(\frac{w}{t}\right)_{\text{lim}} = \frac{446}{\sqrt{f}} = \frac{446}{\sqrt{141}} = 37.6$$

$$\frac{w}{t} = \frac{64}{2.5} = 25.6 < 37.6$$

$$\frac{b}{t} = \frac{658}{\sqrt{f}}\left(1 - \frac{145}{(w/t)\sqrt{f}}\right)$$

(From Table 16.2)

$$= \frac{658}{\sqrt{141}}\left(1 - \frac{145}{25.6\sqrt{141}}\right) = 28.98$$

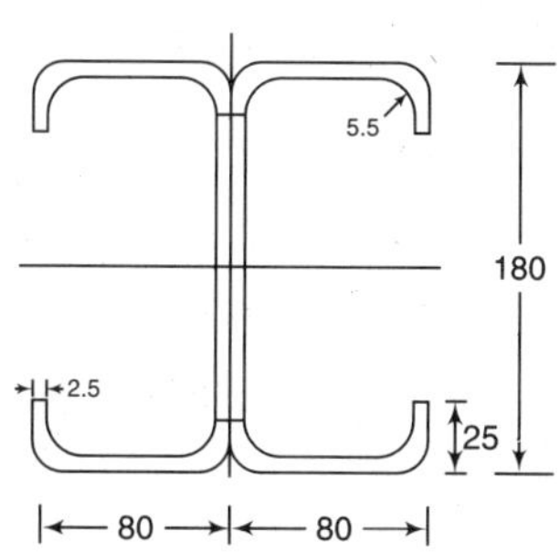

Fig. Ex. 16.5

$$b = 28.98 \times 2.5 = 72.4 > 64$$

Hence, $b = w = 64$ mm

$$I_{xx} = 4 \times 80 \times 2.5 \times (90 - 1.25)^2 + \frac{1}{12} \times (2.5 \times 2) \times 175^3 + 4 \times 2.5 \times 22.5 \times (90 - 25 + 11.25)^2$$

$$= 9842486.98 \text{ mm}^4$$

$$Z = \frac{I_{xx}}{y} = \frac{9842486.98}{90} \simeq 109.36 \times 10^3 \text{ mm}^3$$

$$M = fZ$$

$$= 141 \times 109.36 \times 10^3 = 15.42 \times 10^6 \text{ Nmm} = 15.4 \text{ kNm}$$

Let, w be the load in kN/m.

$$\frac{w \times 4.1^2}{8} = 15.4 \text{ kN/m}$$

$$\Rightarrow \quad w = 7.33 \text{ kN/m}$$

Example 16.6 A top chord member of a roof truss is of hat section as shown in Fig. Ex. 16.6. It is subjected to a compression of 132.5 kN and a moment of 1636 kNm. Check the safety of the section if $f_y = 210$ N/mm^2 and the length of the member is 1.70 m.

Solution For flange

$$\frac{w_1}{t} = \frac{92}{4} = 23 \quad (w_1 = 108 - 16 = 92)$$

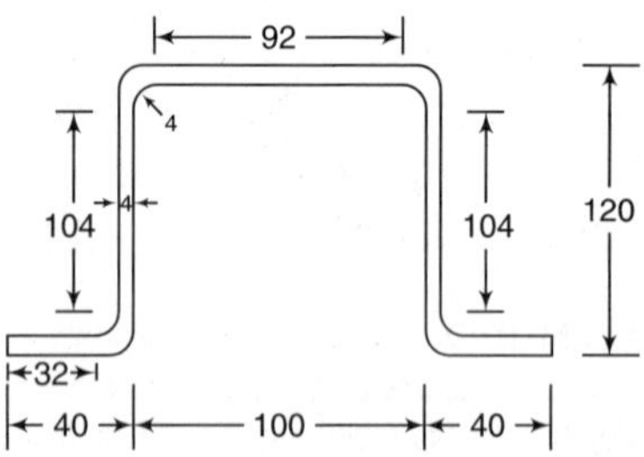

Fig. Ex. 16.6

For web

$$\frac{w_2}{t} = \frac{104}{4} = 26 \quad (w_2 = 120 - 16 = 104)$$

$$\left(\frac{w}{t}\right)_{lim} = \frac{446}{\sqrt{f}} = \frac{446}{\sqrt{125}} = 40.58$$

Hence, all the elements are fully effective (i.e., $b = w$) and $Q = 1$.

Linear properties

$$L = 2 \times 32 + 2 \times 104 + 92 + 4 \times 9.42 = 401.68 \text{ mm}$$

$$A = 401.68 \times 4 = 1607 \text{ mm}^2$$

$$C_{xx} = 0.637 \times (4 + 2) = 3.822$$

Weight per metre length $= 1607 \times 0.0785 = 126$ N

$$C_y = \frac{92 \times 2 + 2 \times 104 \times 60 + 2 \times 32 \times 118 + 2 \times 9.42 \times 4.178 + 2 \times 9.42 \times 115.822}{401.68}$$

$$= 55.95 \text{ mm}$$

Linear $$I_{xx} = 92 \times 54^2 + 2 \times \frac{104^3}{12} + 2 \times 104 \times 4^2 + 2 \times 32 \times 62^2 + 2 \times 9.42 \times 51.822^2 + 2 \times 9.42 \times 59.822^2$$

$$= 823.11 \text{ mm}^3$$

Actual $\quad I_{xx} = 823.11 \times 10^3 \times 4 = 329.24 \times 10^4 \text{ mm}^4$

Linear $\quad I_{yy} = \dfrac{92^3}{12} + 2 \times \dfrac{32^3}{12} + 2 \times 32 \times 74^2 + 2 \times 104 \times 52^2 + 2 \times 9.42 \times 49.822^2 + 2 \times 9.42 \times 54.178^2 = 1085.31 \times 10^3 \text{ mm}^3$

Actual $\quad I_{yy} = 1085.31 \times 10^3 \times 4 \text{ mm}^4 = 434.1 \times 10^4 \text{ mm}^4$

$$r_{xx} = \sqrt{\frac{I_{xx}}{A}} = \sqrt{\frac{329.24 \times 10^4}{1607}} = 45.26 \text{ mm}$$

$$\frac{l}{r_{xx}} = \frac{1700}{45.26} = 37.6$$

$$f_{a1} = \frac{12}{23} Q f_y - \frac{3}{23} \times \frac{(Q f_y)^2}{\pi^2 E}\left(\frac{kl}{r}\right)^2 \qquad \text{(From Eq. (16.32))}$$

$$= \frac{12}{23} \times 1 \times 235 - \frac{3}{23} \times \frac{(1 \times 235)^2}{\pi^2 \times 2 \times 10^5} \times 37.6^2 = 117.45 \text{ N/mm}^2$$

$$f_{a,cal} = \frac{132.5 \times 10^3}{1607} = 82.45 \text{ N/mm}^2$$

$$f_{b,cal} = \frac{1636 \times 10^3 \times 56}{329.24 \times 10^4} = 27.83 \text{ N/mm}^2$$

$$f_e' = \frac{12}{23} \times \frac{\pi^2 E}{(kL_b / r_b)^2} = \frac{12 \times \pi^2 \times 2 \times 10^5}{23 \times 37.6^2} = 728.46 \text{ N/mm}^2$$

$$C_m = 1.00$$

From the interaction formula (IS 801: 1975)

$$\frac{f_{a,cal}}{f_{a1}} + \frac{f_{b,cal} C_m}{f_{b,cal}\left[1 - \dfrac{f_{a,cal}}{f_e'}\right]} \le 1$$

$$= \frac{82.45}{117.45} + \frac{27.83 \times 1}{125 \times \left[1 - \dfrac{82.45}{728.46}\right]} = 0.95$$

which is less than 1.
Hence, this section is safe.

Exercises

16.1 Compute the elastic critical stresses, for the plate shown in Fig. Prob. 16.1, for different edge conditions. Thickness of plate = 2 mm.
Take $E = 2 \times 10^5$ N/mm^2, $\mu = 0.3$

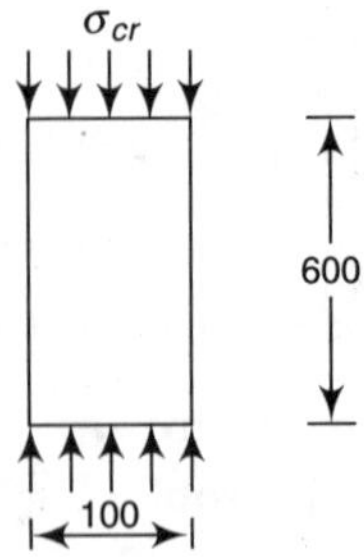

Fig. Prob. 16.1

Ans. 14.46 kN/mm^2 (when the edges parallel to the compressive stress are simply supported.)

16.2 Determine the sectional properties (area, moment of inertia) of light-gauge steel shown in Fig. Prob. 16.2 by linear method. Also find the percentage error.

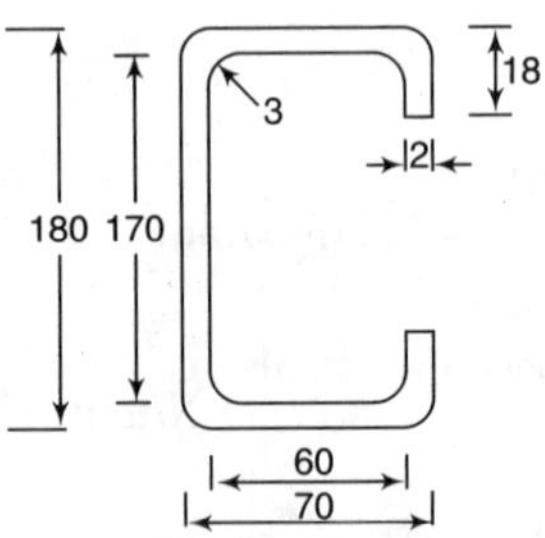

Fig. Prob. 16.2

Ans. 682.2 mm^2, error 2.05%; 342.54 × 10^4 mm^4, error 3.24%

16.3 Figure Prob. 16.3 shows a light-gauge column section with all elements stiffened. Compute the allowable load for the following data.

Length of column = 3.2 m

Yield stress, f_y = 235 N/mm^2

Both ends of column: pinned.

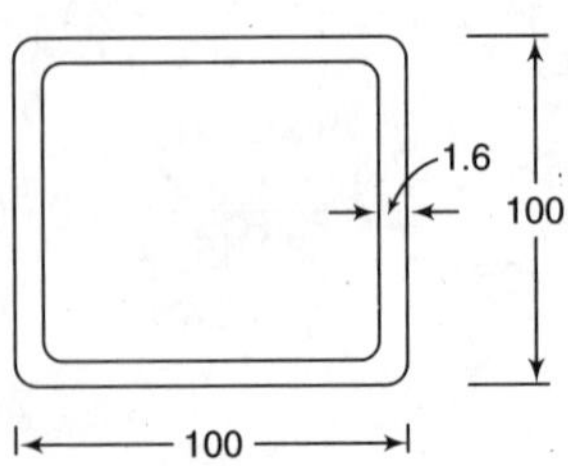

Fig. Prob. 16.3

Ans. 51.2 kN

16.4 Compute the allowable load on the pinned-end light gauge steel strut of 2.4 m length (f_y = 206 N/mm^2).The section of the column is composed of all unstiffened elements as shown in Fig. Prob. 16.4.

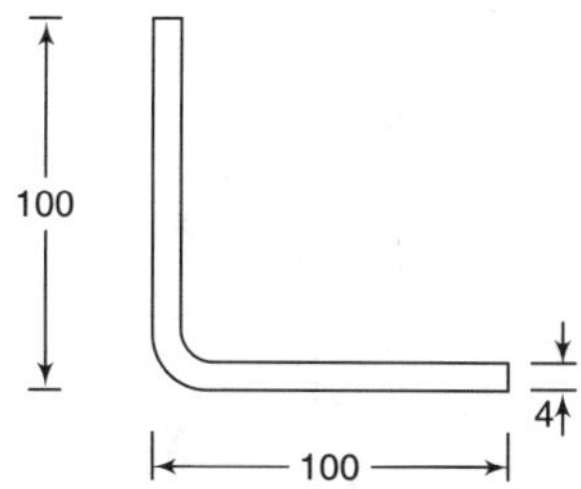

Fig. Prob. 16.4

Ans. 51.9 kN

16.5 Compute the allowable load on the pinned-end light-gauge steel column shown in Fig. Prob. 16.5. The length of column is 3.1 m and f_y = 235 N/mm^2. The two sections are jointed together by spot welding.

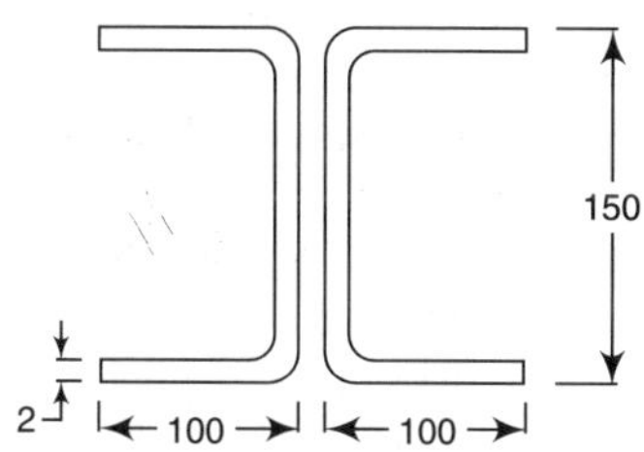

Fig. Prob. 16.5

Ans. 121.53 kN

16.6 Compute the allowable load on the light-gauge steel beams shown in Fig. Prob. 16.6. f_y = 235 N/mm^2. Span of beam is 3.0 m.

Ans. 8.75 kN, 150 kN

Fig. Prob. 16.6

chapter 17

Elementary Plastic Analysis and Design

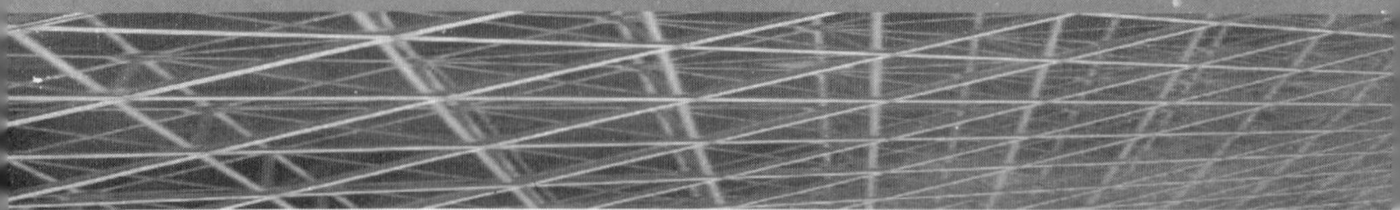

17.1 INTRODUCTION

In conventional design methods steel structures are designed according to elastic theory, wherein a structure is designed for the strength of steel till its yield. In the elastic design, working stresses are specified by using factor of safety with respect to the yield stress. These working stresses are used to find the section of members for the working loads. It has been observed that steel structures behave in a manner quite different from that assumed in conventional design methods.

Steel has a unique physical property, ductility, because of which it is able to absorb large deformations beyond the elastic limit without fracture. Due to this property, steel possesses a reserve of strength beyond its yield, as is evident from stress-strain curve (Fig. 17.1) of mild steel, which engineers have tried to utilise in the plastic method of design.

Plastic design is an aspect of the limit design* that extends the structural usefulness up to the plastic strength or ultimate load carrying capacity. The concept of design based on ultimate load was first utilized by a Hungarian, Dr. Gabor Kazinczy. The other terms referred to the plastic method of design are limit design, collapse method of design and ultimate design.

In the plastic method, the design criterion is the ultimate strength and hence the behaviour of members beyond the yield stress in the inelastic (or) plastic range are considered. In this method, the working loads are multiplied with specified factors known as load factors to obtain ultimate loads under which a structure collapses. Since with the help of actual stress-strain curve the ultimate strength of actual structures can easily be worked out, a review of the stress-strain curve is as follows.

17.2 IDEALIZED STRESS-STRAIN CURVE FOR MILD STEEL

The stress-strain curve for an annealed mild steel specimen in tension has the typical form as shown in Fig. 17.1. Various elements of the curve are explained below:

OAB represents a straight line curve. The strain produced is proportional to the stress producing it, i.e. it obeys *Hooke's law*. Point *A* represents the *limit of proportionality*, beyond which linear variation ceases. *B* represents the *elastic limit* or the maximum stress up to which the specimen regains its original length on removal of the applied load. *C* represents the *upper yield point* at which there

*A limit design method is based on any chosen limit of structural usefulness. The limit can be the attainment of the yield point stress (conventional design), brittle fracture, fatigue, instability, deflection or maximum plastic strength.

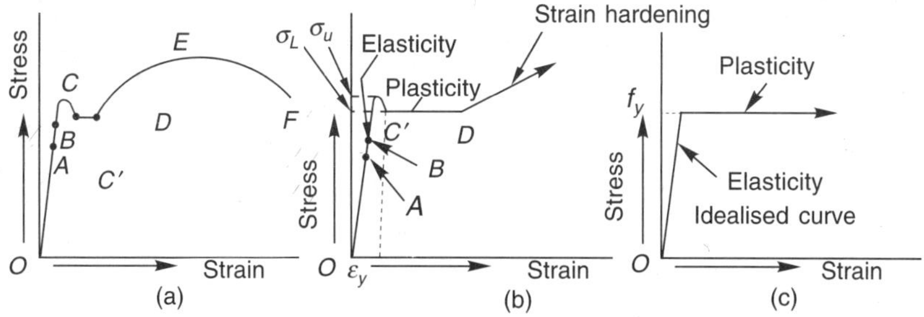

Fig. 17.1 *Complete Stress Strain Curve of Structural Steel*

is a definite increase in strain without any further increase in stress. Actually, it is difficult to determine the actual stress-strain relation in the elastic range near the yield point; proportional and elastic limit are considered to be coincident. The stress drops abruptly from upper yield to the lower yield stress and strain increases at constant stress up to the point *D*. This portion represents a purely plastic flow. Beyond *D* strain increases further with the increase of stress and the material is said to be in a *strain hardening range*. *E* represents the point of maximum stress corresponding to the ultimate load, beyond which an increase of strain occurs with a decrease of stress until the specimen breaks at *F*. The yield range *OABCD* is most important from the view of the plastic theory. The yield range can be studied more conveniently by enlarging considerably the strain scale, (Fig. 17.1(b)). σ_u and σ_L are the upper and lower yield stress respectively. *E* is the Young's modulus, the slope of the initial elastic line *OC* and E_s is the slope of the initial portion of the strain hardening line beyond *D*.

The upper yield effect, though very real, is variable since it is dependent upon the method and the rate of load application. Also, it does not affect the value of the fully plastic moment. For the purpose of plastic design of steel structures, the elastic limit and the lower yield point may be assumed to be numerically identical. The strain at the first yield point is 0.00125. The strain at the initiation of strain hardening is 0.015 and that the maximum strain is 0.25. From these values it may be seen that the elastic deformation is only about 1/12 of strain at the initiation of strain hardening and about 1/200 of maximum strain. Because of the large reserve of ductility in mild steel, fracture of steel members is non-existent except in the case of tension members. The plastic range is sufficiently long and it seems reasonable to extend it without limit, that is to ignore the effect of strain hardening. Thus in Fig. 17.1(c) the extension is supposed to be unlimited at the constant yield stress f_y. Also in the idealized elasto-plastic stress-strain curve the strain hardening range is neglected. It is difficult to justify the negligence of the strain hardening range as it is certain that strain will enter this range. However, neglecting the increase of stress during strain hardening adds to the margin of safety.

The ultimate strength of simple structural members is worked out with the help of stress-strain curve of Fig. 17.1(c) in the sections to follow.

17.3 ULTIMATE LOAD-CARRYING CAPACITY OF TENSION MEMBERS

Consider a member of cross-sectional area, *A*, subjected to axial tension *P*. Figure 17.2 shows a load vs deflection curve. The deflection at the elastic limit is given by

$$\delta_y = \frac{P_y l}{A_s E} \qquad (17.1)$$

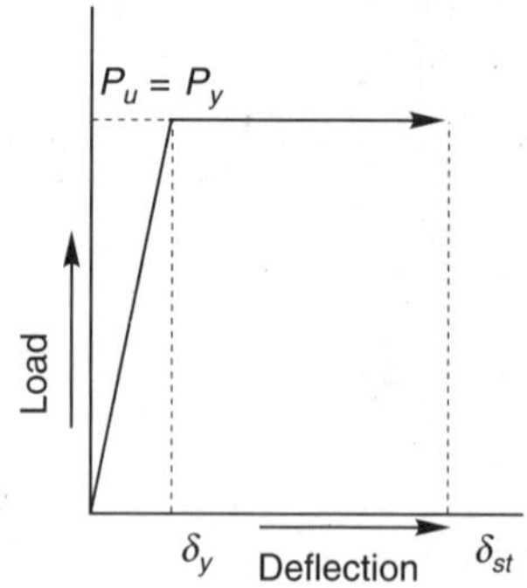

Fig. 17.2 *Load-Deflection Curve for a Tension Member*

where P_y = yield load.

The plastic flow in the member sets in as soon as the load reaches the value $P_y = f_y A_s$. As soon as P equals P_y, the deflection increases instantaneously and unobstructed to δ_{st} (deflection at the start of strain hardening). Since the deformation at this stage of δ_{st} is 1.5% of the length of member, which is too large for most structures, the usefulness of the tension members is practically finished at the start of strain hardening. Hence, for all practical purposes the yield load P_y is considered to be the ultimate load P_u. Hence, the value of P_u remains constant and is termed as the ultimate load for a tension member.

The ultimate load concept is of little use in the design of tension members, as is obvious from the curve that due to a load producing an average direct stress slightly above the elastic limit, the section ruptures. The average rupture stress in tension may be nearly twice the elastic limit based upon the original cross-sectional area. This reflects that there should be a large reserve capacity in a tension member. Therefore, in a tension member the eccentricity of load, if any, can be ignored as it is taken care of by the margin.

As per the specification of I.S: 800–1984 the ultimate load carrying capacity is calculated by the following expression,

$$P_{at} = 0.85 A_s f_y \qquad (17.2)$$

where P_{at} = the calculated maximum load capacity
A_s = effective cross-section area of the member
f_y = yield stress of steel.

17.4 ULTIMATE LOAD-CARRYING CAPACITY OF COMPRESSION MEMBERS

A steel compression member of normal stiffness as used in structures buckles at an average stress somewhere near the elastic limit (or the yield point). Hence, there may be no reserve capacity beyond the yield load in a column. If there is no buckling, the ultimate load carrying capacity is the product of the cross-sectional area and the maximum stress attained before the onset of strain hardening.

$$P_u = f_y A_s \qquad (17.3)$$

As per the specification of I.S: 800–1984, the maximum load carrying capacity of a compression member is given by the following expression:

$$P_{ac} = 1.7 A_s \sigma_{ac} \qquad (17.4)$$

where P_{ac} = calculated maximum load carrying capacity
A_s = effective cross-sectional area of the member,
σ_{ac} = maximum permissible stress in axial compression, from Table 4.4 of the text using the effective length l equal to the actual length L.

In practice, the compression members are slender and may buckle at an average stress far below the elastic limit. The maximum moment capacity of a beam-column member is reduced and should be checked by the following expression, depending upon the ratio of the axial force P to the yield strength of the member P_y.

(i) if $$P/P_y > 0.15,$$

then
$$\frac{P}{P_y} + \frac{M_{pc}}{1.18 M_p} \le 1.0 \tag{17.5}$$

(ii) if
$$\frac{P}{P_y} > 0.15 \text{ and} > \frac{1+\beta-\lambda_0}{1+\beta+\lambda_0},$$

then
$$\frac{P}{P_{ac}} + \frac{M_{pc}\, C_m}{M_0\left(1-\dfrac{P}{P_e}\right)} \le 1.0 \tag{17.6}$$

(iii) Any other case not covered by above cases:
$$\frac{M_{pc}}{M_p} \le 1.0 \tag{17.7}$$

where P = axial compressive force

P_y = yield strength of the axially loaded section = $A_s f_y$

P_e = Euler load = $\dfrac{\pi^2 E A_s}{(l/r)^2}$

M_p = plastic moment capacity of the section

M_{pc} = maximum plastic moment capacity of the beam column

M_0 = lateral buckling strength in the absence of axial load. If the column is laterally braced, M_0 is equal to M_p

β = ratio of end moments

λ_0 = characteristic slenderness ratio = $\sqrt{P_y/P_e} = \dfrac{L}{\pi r}\sqrt{f_y/E}$,

l = effective length and is taken equal to actual length. The effective length factor in plastic analysis is taken as 1.0

C_m = coefficient defined in article 7.1.3 of I.S: 800–1984

r = radius of gyration about the same axis about which the bending moment acts.

Note The above Eqs (17.5) to (17.7) are also applicable to tension members subjected to bending moment.

17.5 FLEXURAL MEMBERS

In treating the subject of the plastic design of structures, resistance to bending action is the primary means by which the loads are supported. The application of this method is best suited to the analyses of continuos beams and rigid frames. The concepts used in the plastic analysis of flexural members are explained as follows.

Bending of Beams

Let us consider a beam with O as the centroid and whose cross-section has an axis of symmetry (yy), subjected to a steadily increasing bending moment M. i.e. from the elastic limit (stage 1) up to the plastic limit (stage 4). The corresponding distributions of stresses for stages 1 to 4, are plotted in Fig. 17.3.

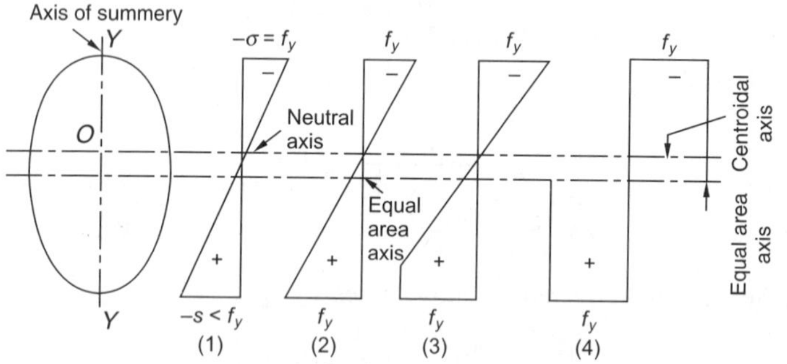

Fig. 17.3 *Development (in Stages) of Full Plasticity of a Beam Section*

It is assumed that the stress-strain relation is an ideal elasto-plastic stress-strain curve (Fig. 17.1(c)). The beam is assumed to be bent by a pure couple and the effect of axial and shear forces are neglected. The deformations and strains are assumed to be small so that stresses other than longitudinal normal stresses are neglected. Further, it is assumed that a plane section before bending remains plane after bending and the cross-section is assumed to be symmetrical with respect to an axis in the plane of bending.

Let the bending moment be increased steadily from zero value. The longitudinal normal stress σ, and longitudinal strain ε both vary linearly across the section with distance y from the neutral axis, so that the behaviour is entirely elastic (stage 1). In the second stage the moment is increased further. If a value is reached beyond the yield moment, the strain in the outermost fibres reaches yield value ε_y and the yield stress is just attained on the lower surface of the beam. The neutral axis no longer passes through the centroid and is shifted to maintain equilibrium. The yield spreads inwards from the upper surface of the beam.

A further increase in the bending moment causes the yield to spread inwards from the lower surface of the beam as well, in addition to spreading further inwards from the upper surface as shown in stage 3. This stage of partial plasticity occurs because of the yielding of the outer fibres without increase of stress as predicted by the horizontal line of the idealized stress-strain diagram (Fig. 17.1(c)). On increasing the bending moment further, the whole section yields as shown in stage 4. The two zones of yield grow in depth from the extreme fibre towards the neutral axis and the corresponding bending moment is the fully plastic moment. Any further increase in the bending moment results only in rotation, since no greater resisting moment than the fully plastic moment can possibly be developed until strain hardening occurs.

Plastic Hinge

A plastic hinge is a zone of yielding due to flexure in a structural member. Although hinges do not actually form, it can be seen that large changes of slope occur over small lengths of the member at positions of maximum moments. A strain hardening action usually occurs at these hinges so that large deflections are accompanied by a slight increase in loads.

A structure can support the computed ultimate load due to the formation of plastic hinges at certain critical sections. The members remain elastic until the moment reaches a value M_p, the maximum moment of resistance of a fully yielded cross-section. Any additional moment will cause the beam to rotate with little increase in stress. The rotation occurs at a constant moment, M_p. The zone acts as if it was hinged except with a constant restraining moment M_p. The plastic hinge, therefore, can be defined as *a yielded zone due to flexure in a structure in which infinite rotation can take place at a constant restraining moment* M_p of the section. A plastic hinge is represented by a black dot. The value of the moment at the adjacent sections of the yield zone for a certain length, is more than the yield moment. This length depends upon the loading and geometry of the section. To simplify the analysis, this small length is neglected and the plastic hinge is assumed to be formed at discrete points of zero length. But, it cannot be neglected for the calculation of deflections and the design of bracings as the length over which yielding extends is quite important.

The plastic hinges are reached first at the sections subjected to the greatest deformation (curvature). The possible places for plastic hinges in a structure with prismatic members are points of concentrated loads, at the ends of member meeting at a connection involving a change in geometry and at the point of zero shear in a span under distributed load.

Hinge Length Consider a simply supported rectangular beam subjected to a gradually increasing concentrated load P, at the centre. A plastic hinge will be formed at the centre. Here, reference may be made to Fig. 17.4.

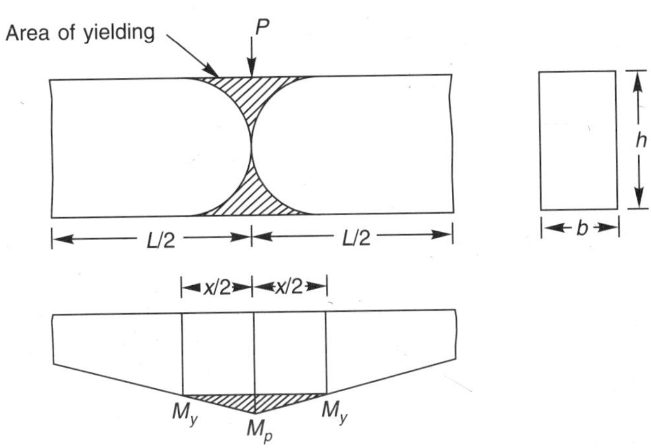

Fig. 17.4 *Development of Plastic Hinge and Hinge Length*

$$M_p = \frac{PL}{4}$$

$$M_y = f_y Z$$

$$= f_y \times \frac{bh^2}{6}$$

$$= f_y \times \frac{1}{6} \times 4 \times \frac{bh^2}{4}$$

$$= \frac{2}{3} \times f_y \times \frac{bh^2}{4}$$

$$= \frac{2}{3} \times f_y \times Z_p$$

$$= \frac{2}{3} M_p \tag{17.8}$$

From the bending moment diagram (Fig. 17.4),

$$\frac{M_p}{\frac{L}{2}} = \frac{M_y}{\frac{L}{2} - \frac{x}{2}}$$

or $$\frac{(L-x)}{2} M_p = \frac{L}{2} M_y$$

or $$(L-x)\, M_p = LM_y$$

or $$(L-x)\, M_p = L \times \frac{2}{3} \times M_p$$

or $$(L-x) = \frac{2}{3} L$$

or $$x = \frac{L}{3} \tag{17.9}$$

Therefore the hinge length of the plasticity zone is equal to 1/3rd of the span.

Note The hinge length of the plasticity zone for a simple beam subjected to uniformly distributed load is $L/\sqrt{3}$.

Fully Plastic Moment of a Section

The fully plastic moment is the maximum moment of resistance of a fully yielded cross-section. The following assumptions are made to evaluate its value in addition to the assumptions made earlier while discussing bending of beams.

1. Instability of the structure will not occur prior to the attainment of the ultimate load.
2. The connections provide full continuity such that the fully plastic moment can be transmitted.

3. The loading is proportional, i.e. all the loads increase in a fixed proportion to one another.

Consider a fully yielded beam section. The bending stress distribution will be rectangular, as in Fig. 17.5.

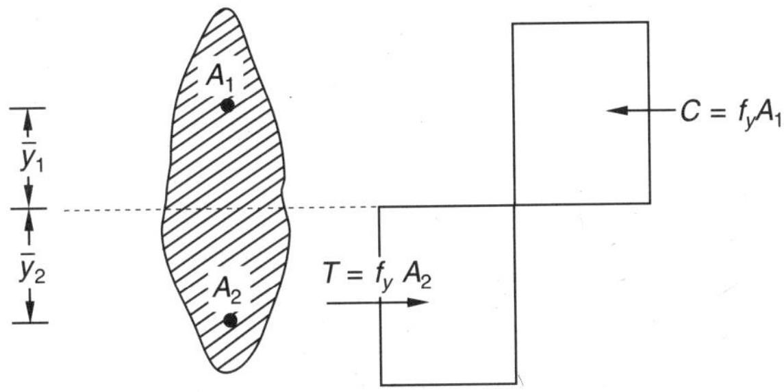

Fig. 17.5 *Plastic Moment of a Beam Section*

Let the nature of the bending moment be sagging such that the fibres above N.A. are in compression and those below N.A. are in tension.

$$\text{Force in compression } C = f_y A_1$$

$$\text{Force in tension } T = f_y A_2$$

If an equilibrium condition exists, the force in compression must be equal to the force in tension, i.e.,

$$C = T$$

$$f_y A_1 = f_y A_2$$

or
$$A_1 = A_2 = \frac{A}{2}$$

where A_1 is the area of the section above N.A., A_2 is the area of the section below N.A., and A is the area of cross-section of the beam. Therefore, the neutral axis of the plastified section is sometimes called *equal area axis*.

The two forces, tensile and compressive, form a couple which resist the plastic moment.

$$M_p = f_y A_1 \bar{y}_1 + f_y A_2 \bar{y}_2$$

or
$$M_p = f_y \frac{A}{2} \bar{y}_1 + f_y \frac{A}{2} \bar{y}_2$$

or
$$M_p = f_y \frac{A}{2} (\bar{y}_1 + \bar{y}_2)$$

or
$$M_p = f_y Z_p \tag{17.10}$$

where
$$Z_p = \text{plastic modulus of section} = \frac{A}{2} (\bar{y}_1 + \bar{y}_2)$$

$\bar{y}_1, \bar{y}_2$ = distance of centre of gravity of the area above and below the neutral axis, respectively.

Redistribution of Moment and Reserve of Strength

The redistribution of moment can be explained on the basis of the action and successive formation of the plastic hinges. Under the gradually increasing load on a structure, the plastic moment is reached at a section that is most highly stressed. On further increase of load, the value of the plastic moment is maintained and the section rotates. A plastic hinge is formed at the section. Other less highly stressed sections maintain an equilibrium with the increased load by a proportionate increase in the moment and successive formation of plastic hinges occurs at other sections, where the moment reaches the value of the plastic moment. This process of formation of successive plastic hinges continues till the ultimate load is reached. Therefore, it can be said that flexural members can sustain the ultimate loads due to the redistribution of moments.

Let us take an example of a fixed ended beam with a concentrated load near one of its support (Fig. 17.6(a)). As the load P is increased, the beam reaches its elastic limit at end A, and a plastic hinge is formed. The moments at sections B and C are less than the plastic moment. On a further increase of load, the plastic section at A rotates without absorbing any more moment. The resisting moment at C increases until the plastic moment value is reached and a plastic hinge is formed at C. On still further increase of load, a plastic hinge is formed at B and the ultimate load is reached. Therefore, it can be said that the formation of plastic hinges allow a subsequent redistribution of moment until the fully plastic moment is reached at each critical section. The redistribution of the moment is the main contributing factor to the reserve of strength. An idea of the margin provided by the reserve of strength for a particular problem is illustrated with an example of a fixed ended beam of span L, loaded uniformly over the entire span (Fig. 17.6(b)).

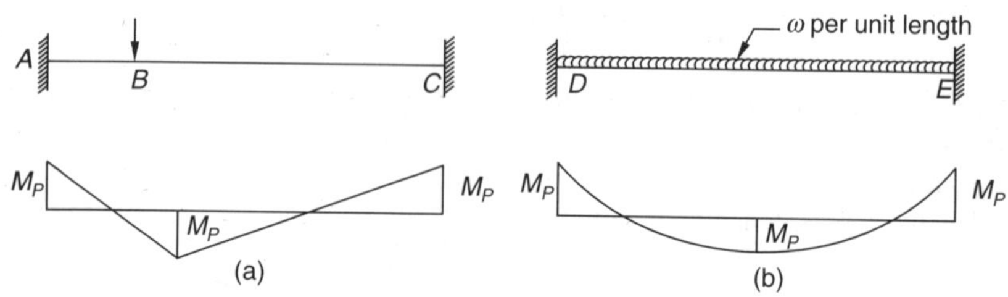

Fig. 17.6

By equilibrium

$$\frac{(w_u L)\times L}{8} \quad \frac{P_u L}{8} = 2M_p$$

or
$$P_u = 16\frac{M_P}{L} \tag{17.11}$$

At the elastic limit, the centre moment is one-half the end moment.

$$\frac{P_y L}{8} = M_y + \frac{M_y}{2}$$

or
$$\frac{P_y L}{8} = \frac{3}{2}M_y$$

or $$P_y = \frac{12 M_y}{L} \tag{17.12}$$

From Eqs 17.11 and 17.12,

$$\frac{P_u}{P_y} = \frac{16\dfrac{M_p}{L}}{12\dfrac{M_y}{L}} = \frac{4 M_p}{3 M_y}$$

Assuming a shape factor $f = 1.12$ for an I-Section,

$$\frac{P_u}{P_y} = \frac{4}{3} \times 1.12 = 1.493$$

For this particular problem the ultimate load is 49.3% greater than the load at the first yield. This is the reserve of strength which is actually disregarded in conventional design.

17.6 SHAPE FACTOR

Shape factor is defined as the ratio of the plastic moment and the yield moment of the section. It is a function of the cross-section form or shape and is represented by f.

$$f = \frac{M_P}{M_y} = \frac{f_y Z_p}{f_y Z} = \frac{Z_p}{Z}$$

where Z and Z_p are the elastic and plastic section moduli respectively.

Rectangular Section For a rectangular section (Fig. 17.7), the elastic modulus of section

$$Z = \frac{bh^3}{12} \Big/ \frac{h}{2} = \frac{bh^2}{6} \tag{17.13}$$

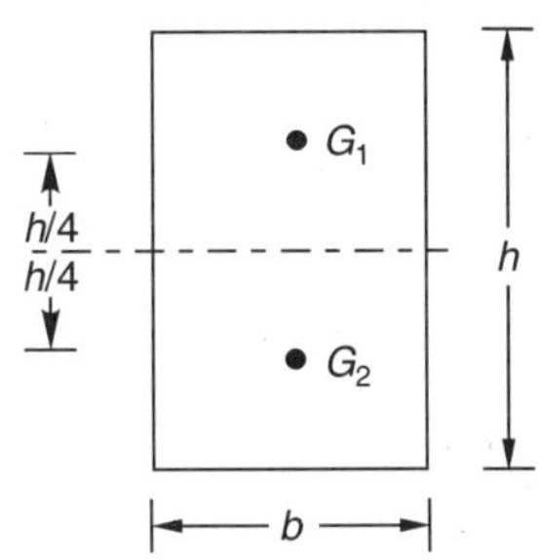

Fig. 17.7

and the plastic modulus of section

$$Z_p = \frac{A}{2}(\bar{Y}_1 + \bar{Y}_2)$$

$$= \frac{bh}{2} \times \left(\frac{h}{4} + \frac{h}{4}\right)$$

$$= \frac{bh^2}{4} \tag{17.14}$$

$$f = \frac{Z_p}{Z}$$

$$= \frac{\dfrac{bh^2}{4}}{\dfrac{bh^2}{6}} = 1.5$$

The values of shape factor for some of the other sections are given in Appendix XIII.

17.7 LOAD FACTOR

Load factor is defined as the ratio of the collapse load to the working load. It is represented by F.

$$F = \frac{P_u}{P_w} \tag{17.15}$$

where P_u and P_w are the collapse and working loads respectively.
For a simple beam the variation of the bending moment with the load is linear.

$$F = \frac{P_u}{P_w} = \frac{M_p}{M_w} = \frac{f_y Z_p}{\sigma Z} = \frac{f_y}{\sigma} f = (\text{F.O.S})f \tag{17.16}$$

where $\dfrac{f_y}{\sigma}$ = factor of safety as in elastic design

f_y = yield stress

σ = permissible or working stress

f = shape factor.

Therefore, the load factor may also be defined as the product of the factor of safety and shape factor. Let us take the example of an I-Section beam whose shape factor f = 1.12.

$$\text{F.O.S} = \frac{f_y}{\sigma} = \frac{f_y}{0.66 f_y} = 1.515$$

$$\text{Load factor, } F = 1.515 \times 1.12$$

$$= 1.70$$

In actual practice a load factor varying from 1.70 to 2.0 is assumed depending upon the designer's judgement. The load factors for various types of loads are listed in Appendix XIV. When the structures are subjected to wind the corresponding load factor for plastic design is reduced by 25%. The prime function of the load factor is to ensure that the structure will be safe under service conditions. Therefore, it may be regarded as a factor of safety based upon the collapse load. It depends upon the nature of loading, the support conditions, and the geometrical shape of structural members. Uncertainty of the loads, imperfection in workmanship and errors in fabrication are some of the other factors which influence the choice of load factor.

17.8 MECHANISM

When a structure is subjected to a system of loads it is stable and hence functional until a sufficient number of plastic hinges have been formed to render the structure unstable. As soon as the structure reaches an unstable condition, it is considered to have been failed. The segments of the beam between the plastic hinges are able to move without an increase of load. This condition in a member is called *mechanism*. The concepts of mechanism formation in a structure due to loading beyond the elastic limit and virtual work are used in the plastic analysis and design of steel structures. If an indeterminate structure has redundancy r, the insertion of r plastic hinges makes it statically determinate. Any further hinge converts this statically determinate structure into mechanism. Hence, for collapse the number of plastic hinges required are $(r + 1)$.

Types of Mechanism

Various possible mechanisms, as listed below, are shown in Fig. 17.8.

1. Beam mechanism
2. Panel mechanism
3. Joint mechanism
4. Gable mechanism
5. Composite (combined) mechanism

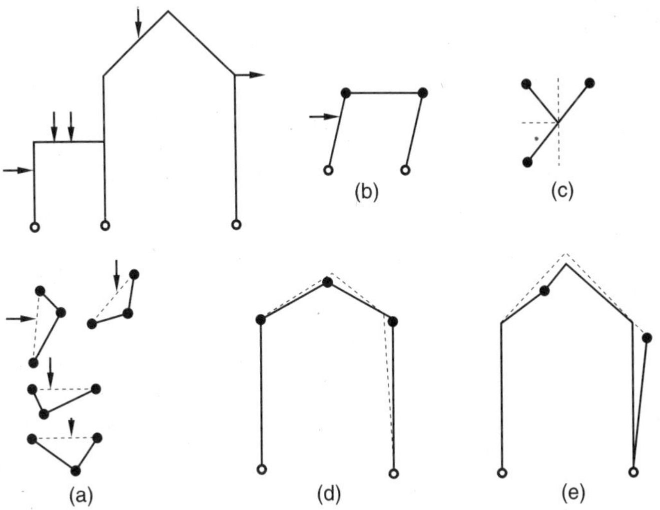

(a) Beam Mechanisms—All the loaded spans behave as beam mechanism.
(b) Sway Mechanism—It is formed due to lateral loads
(c) Joint Mechanism—It is formed due to action of moment. The number of members at the joint should be three or more.
(d) Gable Mechanism—It is exhibited in gable frames. Columns spread more at the top than that at the base.
(e) Composite Mechanism—Any of the two independent mechanisms may be combined to form composite mechanism. A beam and gable mechanism are joined to form a gable mechanism in this figure.

Fig. 17.8 *Types of Mechanisms*

Number of Independent Mechanisms

To avoid any possible omission of a combined mechanism it is essential to have in advance a clear idea of the number of independent mechanisms.

Let

N = number of possible plastic hinges
r = number of redundancies
n = possible independent mechanisms.

Then,
$$n = N - r$$

After finding out the number of independent mechanisms all the possible combinations are made in such a ways so as to make the external work maximum or the internal work a minimum. This is done to obtain the lowest possible load.

17.9 PLASTIC COLLAPSE

The plastic collapse of a structure depends upon its redundancy. When a sufficient number of plastic hinges are formed to convert a structure into a mechanism, the structure collapses. At such a stage the deflection increases very fast at constant load. The collapse of a structure can be partial, complete or over complete. These terms can be explained on the basis of indeterminacy r, and the number of plastic hinges developed N. If the number of plastic hinges in the collapse mechanism are less than $(r + 1)$ the collapse is called *partial collapse*. In such a case, part of the structure may fail making it useless as a whole. If the number of plastic hinges in the collapse mechanism are $(r + 1)$ the collapse is called *complete collapse*. Such a mechanism has only one degree of freedom. If the number of plastic hinges developed are more than $(r + 1)$, the collapse is called *overcomplete collapse*. In such a case there are two or more mechanisms for which the corresponding value of the load is the same, this load value being the actual collapse load.

17.10 CONDITIONS IN PLASTIC ANALYSIS

In the elastic method of design the conditions to be satisfied are the equilibrium condition, the compatibility (or continuity) condition and the moment curvature relation (initial yield or limiting stress condition). The conditions to be satisfied for the plastic methods of analysis are as follows:

1. *Equilibrium Condition* All the equilibrium conditions, i.e. summation of all the forces and moments should be equal to zero.

2. *Mechanism Condition* The structure at collapse must be capable of deforming as a mechanism due to the formation of plastic hinges, i.e. the ultimate load is reached when a mechanism forms. This is also called the *continuity condition*.

3. *Yield Condition* This is also called *plastic moment condition*. The bending moment at any section must not exceed the full plastic moment of the section $(M \leq M_p)$.

If all the three conditions are satisfied, a unique value, the lowest plastic limit load, is obtained. However, if the equilibrium condition and any of the above two, i.e. mechanism or yield conditions is satisfied an approximate value above or below the true plastic limit load is obtained.

17.11 PRINCIPLE OF VIRTUAL WORK

The principle of virtual work may be stated as follows:

If a system of forces in equilibrium is subjected to a virtual displacement, the work done by the external forces equals the work done by the internal forces, i.e.

$$W_E = W_i$$

where W_E and W_i are the external and internal work done.

This is simply a method to express an equilibrium condition. Therefore, if the work equation is formulated correctly the equilibrium condition is automatically satisfied. This principle can be used to determine the collapse load. The structure is assumed to deflect through a small additional displacement after the ultimate load is reached. The work done by external forces is equated to the work absorbed by the plastic hinges and the collapse load is worked out. As the collapse equation is derived from work considerations, the bending moment diagram must be drawn to check that the yield condition is not violated.

17.12 THEOREMS OF PLASTIC ANALYSIS

Usually it is not possible to satisfy all the three conditions of plastic analysis in one operation. A solution arrived at on the basis of assumed mechanism will yield a load carrying capacity that is either correct or too high, whereas a solution arrived at on the basis of a statical moment diagram not violating the plastic moment condition, will either be correct or too low. The equilibrium condition however has to be satisfied in both the cases. Thus, depending upon the method chosen to obtain a solution to the problem, one obtains either an 'upper limit' (upper bound) below which the correct answer must lie, or a 'lower limit' (lower bound), above which the correct answer lies.

Static or Lower Bound Theorem

For a given frame and loading if there exists any distribution of bending moments throughout the frame which is both safe and statically admissible with a set of loads P, then the value of the load P must be less than or equal to the collapse load P_u ($P \leq P_u$). The static theorem was first suggested by Kist and its proof was given by Gvozder, Greenberg and Prager, and also by Horne.

A load computed on the basis of an assumed equilibrium moment diagram in which the moments are not greater than M_p is less than or at best equal to the true ultimate load. Hence, the static method represents the lower limit to the true ultimate load and has a maximum factor of safety. The static theorem satisfies the equilibrium and yield conditions.

Kinematic or Upper Bound Theorem

For a given frame subjected to a set of loads P, the value of P which is found to correspond to any assumed mechanism, must be either greater than or equal to the collapse load P_u ($P \geq P_u$). A proof of this theorem was established by Gvozder and also by Greenberg and Prager.

A load computed on the basis of an assumed mechanism will always be greater than or at least equal to the true ultimate load. Hence, the kinematic method represents an upper limit to the true ultimate load and has a smaller factor of safety. The kinematic theorem satisfies the equilibrium and continuity conditions.

Uniqueness Theorem

For a given frame and loading if at least one safe and statically admissible bending moment distribution can be found, and if in this distribution the bending moment is equal to the fully plastic moment at sufficient cross-sections to cause failure of the frame as a mechanism due to rotation of plastic hinges, then the corresponding load will be equal to the collapse load ($P = P_u$). All the three conditions, i.e., the equilibrium, continuity and yield conditions are satisfied. This theorem was proved by Horne.

A structure can be analysed by using either the Static or the Kinematic theorem. A solution arrived at by drawing a statical moment diagram (not violating the plastic moment condition) will either be correct or too low. On the other hand a solution arrived at by assuming the mechanism will give a load carrying capacity, either correct or too high. Thus, depending upon the theorem used, we can obtain a lower limit (or bound) which is less than the true load capacity, or the upper limit (or bound) which is more than the true load capacity. When both the theorems are satisfied in a given problem then the solution is said to be the correct (unique) one.

17.13 METHODS OF ANALYSIS

With the principle of virtual work and the upper and lower bound theorems, a structure can be analysed for its ultimate load by any of the following methods:

1. Static method
2. Kinematic method

For complicated frames, the static method of analysis is more difficult, and finding the correct equilibrium equation becomes illusive. In these cases, the kinematic method is more practical.

Static Method of Plastic Analysis

This consists of selecting the redundant forces. The free and redundant bending moment diagrams are drawn for the structure. A combined bending moment diagram is drawn in such a way that a mechanism is formed. The collapse load is found by working out the equilibrium equation. It is checked that the bending moment is not more than the fully plastic moment at any section.

Kinematic Method of Plastic Analysis

This consists of locating the possible places of plastic hinges. The possible independent and combined mechanisms are ascertained. The collapse load is found by applying the principle of virtual work. A bending moment diagram corresponding to the collapse mechanism is drawn and it is checked that the bending moment is not more than the fully plastic moment at any section.

A more rigorous formulation of the mechanism method is also possible. With this approach, known as the *method of inequalities*, the correct mechanism can be determined directly by linear programming techniques. For most of the structures for which plastic design would ordinarily be used, however, the mechanism method as presented in this chapter usually will be adequate.

17.14 CANCELLATION OF HINGE IN THE COMBINED MECHANISM (BEAM + SWAY)

When the beam and sway mechanism are combined one hinge is cancelled. Each of the two mechanisms has one degree of freedom. For the equilibrium condition to be satisfied the hinge rotation in the two mechanisms α, β should be equal, which results in the cancellation of one hinge. For instance, let us assume that the beam and sway mechanisms have different hinge rotations α and β. The resultant combined mechanism will be as shown in Figs 17.9(b) (c) depending upon whether $\alpha \lesseqgtr \beta$. None of the two combined mechanisms in Fig. 17.9(b) and (c) represent a state of equilibrium, as a frame with three redundancies specifies completely the bending moment diagram with the development of four plastic hinges. Therefore, it is essential that one hinge must be eliminated. The solution can be $\alpha = 0$, $\beta = 0$ or $\alpha = \beta$.

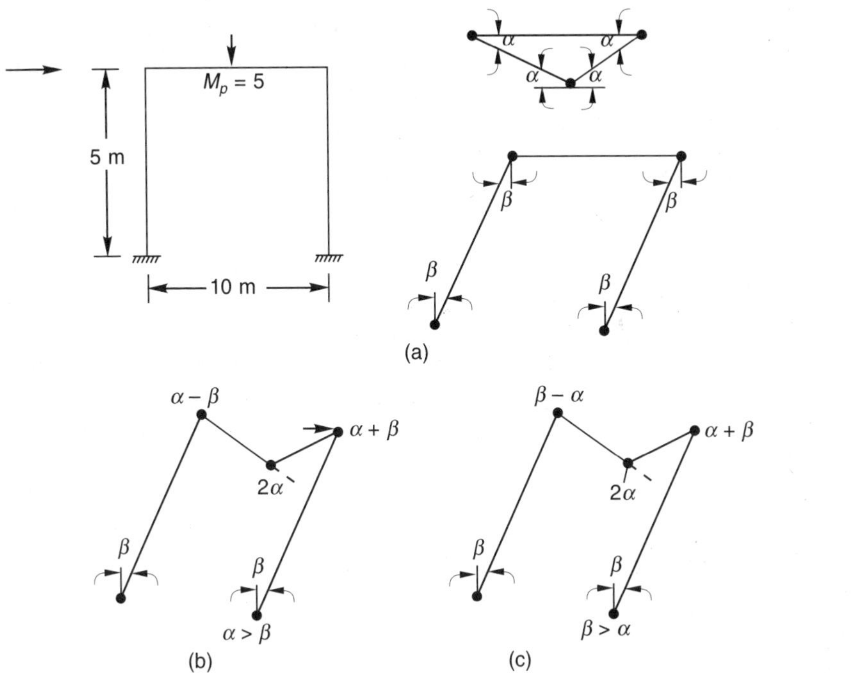

Fig. 17.9 *Cancellation of Plastic Hinge in Combined Mechanism*

Case I (Fig. 17.9(b): By the virtual work equation it can be shown that $P_u = \dfrac{4\alpha + 2\beta}{\alpha + \beta}$, to obtain the least collapse load (as it is upper bound). Let α be constant and β varied. β is increased to the maximum limiting value which is α, ($\alpha > \beta$). Thus, the plastic hinge at the left corner disappears.

Case II (Fig. 17.9(c)): By the virtual work equation it can be shown that $P_u = \dfrac{2\alpha + 4\beta}{\alpha + \beta}$.

In this case it can also be shown that the plastic hinge at the left hand corner disappears when α is made equal to β.

Therefore, it can be said that the combined mechanism can form only when the rotation in the different mechanisms of one degree freedom to be combined are equal, resulting in a single degree freedom mechanism.

17.15 DESIGN

The important functional requirements to be satisfied during design are that the structure should be strong enough to resist external loading without collapsing, and stiff enough not to deflect unduly under that loading. Simple plastic theory in common with the elastic theory assumes in the first instance that deflections of a structure are not the critical design criteria, thus the structure as a whole and its components, can be proportioned on the basis of the collapse load. In case if the designer feels that the deflections might be large enough to cause inconvenience and may influence the design, then he must apply the usual checks. In fact, both elastic and plastic methods of analysis assume that deflections are small compared to the span of beam. One should not get confused by the distorted shapes usually plotted since these are exaggerated to illustrate the 'kinking' at the plastic hinge. Plastic design methods consist of proportioning a structure under factored loads at the point of collapse. The estimated working loads are multiplied by the load factor to work out the factored loads. Thereafter the maximum plastic moment M_p is found. The plastic modulus of the section is found by dividing the maximum plastic moment with the yield stress. A section is selected which furnishes the required elastic modulus ($Z = Z_p/f_y$).

In case of continuous beams the plastic moment value for various spans may be different. To keep a uniform cross-section the largest of the plastic moment values for various spans is chosen and a suitable beam section is selected. But to achieve economy, the section is selected for the least plastic moment found for all the spans and cover plates are provided for the excess plastic moments in other spans. Finally, it should be kept in mind that overall economy is not necessarily associated with the most efficient choice of the section for each span. It is necessary to consider fabrication conditions, which may dictate the uniform section where, theoretically, sections of different weight might be used.

17.16 SCOPE OF PLASTIC ANALYSIS

The plastic methods of structural analysis find their principal application in the design of redundant framed mild steel structures, continuous or restrained beams and girders which carry load by virtue of flexural resistance. Plastic design for steel structures is neither suggested for statically determinate beams and girders nor for structures with pin connected members. The reason can be appreciated by the following examples.

Consider a simply supported beam with a point load at the centre, where the moment is maximum. As the load is increased, the moment increases proportionately until the extreme fibre of material is stressed to the yield. On further increase of load, the material deformations and deflections increase rapidly. Therefore, the load causing the first yield is assumed to be the critical load for a simple beam where large unacceptable deflections occur. On the contrary, a fixed ended beam, behaves in a different manner. Under the load system, it will have three peak moments, two at the ends (will be equal) and one at the centre. As load is applied to the structure, the cross-section at the greatest of these peaks will reach the yield point. As the load is increased a zone of yielding develops there but elsewhere the structure is elastic and controls the deflections. Due to the ductility of steel the moment at the first zone of yielding remains constant and the structure calls upon its less heavily stressed portions to carry a further increase of load. Subsequently the zones of yielding are developed at other points of peak moments. The corresponding load deflection curve is shown in Fig. 17.10. It depicts that failure does not correspond to the attainment of the elastic limit at the ends. There is considerable reserve of load carrying capacity beyond the yield load P_y. The ultimate load P_u is reached only when the yield zones have developed at the centre also. During this development there is no possibility of growth of excessive deflections as these are prevented by the continuity of the central portion of the beam till a yield zone develops at the centre also. Thus, it can be concluded that the redundant structures can be designed more economically due to the plastic reserve of strength.

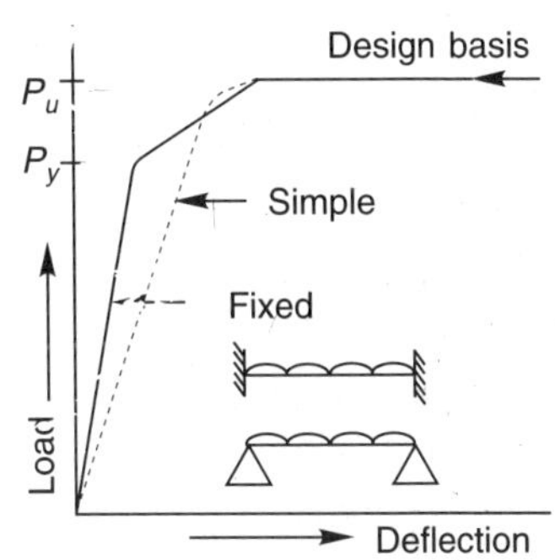

Fig. 17.10 *Load-deflection Curve for Simple and Fixed End Beams*

Plastic methods of analysis are basically intended for plane frames in which the joints can transmit bending moments and the members can resist bending, as the hypothesis of plastic analysis (bending moment and curvature relation) is in close accordance with the actual properties of rolled steel members used in the construction of frames. The concept of plastic analysis would be quite elementary in the case of truss members (tension and compression members) as the basic hypotheses are not obeyed at all. The plastic method of analysis developed for tension and compression members in Sections 17.3 and 17.4 are of purely theoretical interest only.

17.17 LIMITATIONS OF PLASTIC ANALYSIS

Simple plastic theory makes the following assumptions:

1. The fabrication is done in ductile steel.
2. The loads are carried by bending and the effects of axial load and shear force on a member are neglected. The theory of simple plastic analysis cannot be applied to trusses, as the members are subjected to axial forces rather than bending. Special methods are employed for such cases.
3. Strength is assumed to be the main criteria. Deflections are checked only when expected to be large.

17.18 PLASTIC DESIGN VS. ELASTIC DESIGN

The advantages of plasticity of metals are consciously or unconsciously made use of even in elastic design methods, e.g., in the elastic method of design if a design is too conservative for a given permissible working stress, then the stress value is changed, indicating that plasticity is made use of.

There are two distinct advantages of plastic design methods over the conventional or elastic design methods. Firstly, they are more economical as they make full use of the material's strength beyond the elastic limit. Secondly, the design procedures are much simpler and rational. One very important point to favour plastic method of design is that it gives some idea of the collapse mode and strength of the structure.

Some of the metals, especially steel has considerable reserve of strength beyond its elastic limit. Also ultimate load for steel can be computed more precisely and accurately. Taking advantage of the above the plastic design methods permit use of much smaller structural sections to safely support the working loads. As regards simplicity of procedures the plastic methods are inherently simple as they do not take into consideration the elastic conditions of continuity which involve tedious and complicated calculations. It is for these reasons that plastic design methods are advocated.

Following are some of the important aspects of the two design methods:

1. The structure designed by the plastic method is economical. The saving in steel may be about 10–15%.
2. For indeterminate and complicated structures the plastic method is more conveniently applicable and avoids analytical approximations.
3. In the elastic method of design, the design process is repeated several times to obtain an optimum section, which consumes more time. The plastic method of design produces a balanced section in a single attempt and thereby, saving the computational time.
4. The factor of safety is the same in both plastic and elastic design of indeterminate structures. The concept of this factor in the plastic design is more realistic.
5. Stresses produced by settlement, erection, etc. can be determined more correctly by the plastic method as the calculations are based on plastic deformations.

Offsetting the above advantages, there are several disadvantages of using plastic design method as follows:

1. It is difficult to obtain the collapse mode if the structure is reasonably complicated.
2. There is little savings in column design.
3. It is difficult to design for fatigue.
4. Lateral bracing requirements are more stringent than the elastic design.
5. Most designers have more familiarity with elastic design methods.

Collapse load of standard cases

Concentrated Load at Centre	
Simply Supported Beam	*Fixed Beam*
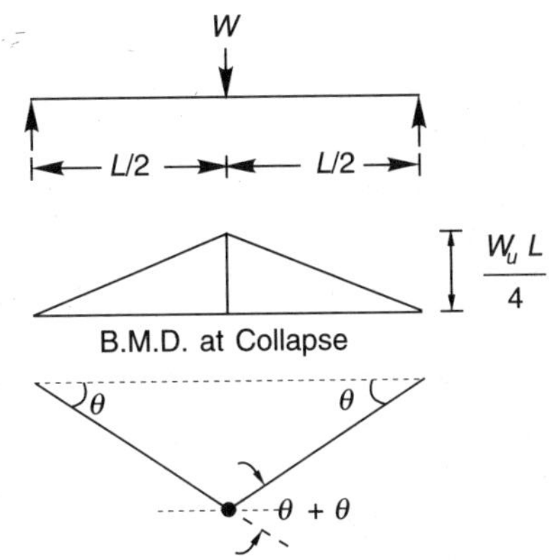	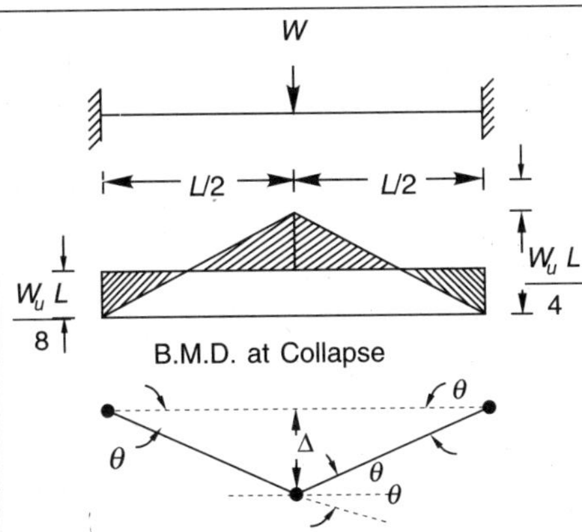
Fig. 17.11	**Fig. 17.12**
Beam Mechanism	Beam Mechanism
Static method	*Static method*
$\frac{W_u L}{4} = M_p$	$\frac{W_u L}{4} = 2M_p$
or $W_u = \frac{4M_p}{L}$	or $W_u = \frac{8M_p}{L}$
Kinematic method	*Kinematic method*
$W_u \left(\frac{L}{2}\right)\theta = M_p\,(\theta + \theta)$	$W_u \left(\frac{L}{2}\right)\theta = M_p\,\theta + M_p\,(\theta + \theta) + M_p\theta$
or $W_u = \frac{4M_p}{L}$	or $W_u = \frac{8M_p}{L}$

Eccentric Load

Simply Supported Beam

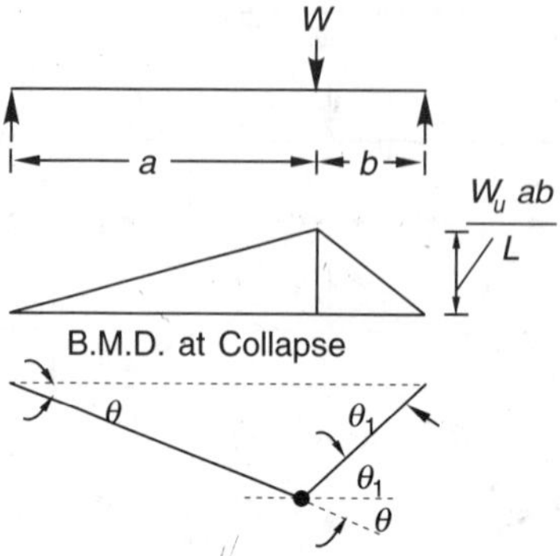

Fig. 17.13

Static method

$$\frac{W_u ab}{L} = M_p$$

or $$W_u = \frac{M_p L}{ab}$$

Kinematic method

$$W_u a\theta = M_p(\theta + \theta_1)$$

$$= M_p\left(\theta + \frac{a}{b}\theta\right)$$

or $$W_u a\theta = M_p\,\theta\left(1 + \frac{a}{b}\right)$$

$$= M_p\,\theta\left(\frac{b+a}{b}\right)$$

$$W_u a\theta = M_p\,\theta\left(\frac{L}{b}\right)$$

or $$W_u = M_p\,\frac{L}{ab}$$

Fixed Beam

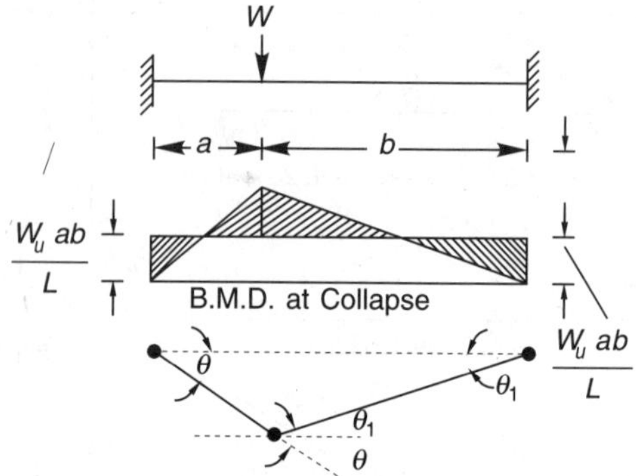

Fig. 17.14

Static method

$$\frac{W_u ab}{L} = 2M_p$$

or $$W_u = \frac{2M_p L}{ab}$$

Kinematic method

$$W_u a\theta = M_p\theta + M_p(\theta + \theta_1) + M_p\theta_1$$

or $$W_u a\theta = M_p\,\theta + M_p\left(\theta + \frac{a}{b}\theta\right) + M_p\,\frac{a}{b}\theta$$

$$= 2M_p\,\theta\left(\frac{a+b}{b}\right)$$

$$= 2M_p\,\theta\frac{L}{b}$$

or $$W_u = \frac{2L}{ab}M_p$$

or $$W_u = \frac{2M_p L}{ab}$$

Uniformly Load Distributed

Simply Supported Beam

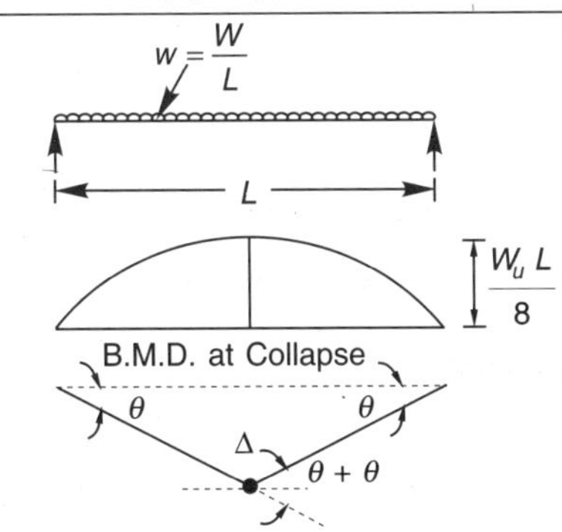

Fig. 17.15

Static method

$$\frac{W_u L}{8} = M_p$$

$$W_u = \frac{8M_p}{L}$$

Kinematic method

$$W_u \frac{1}{2}\left(\frac{L}{2}\theta\right) = M_p(\theta + \theta)$$

or $$W_u = \frac{8M_p}{L}$$

Fixed Beam

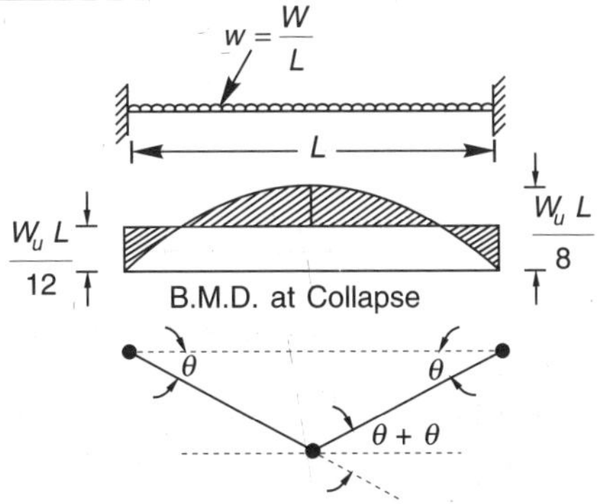

Fig. 17.16

Static method

$$\frac{W_u L}{8} = 2M_p$$

$$W_u = \frac{16M_p}{L}$$

Kinematic method

$$W_u \frac{1}{2}\left(\frac{L}{2}\theta\right) = M_p\theta + M_p(\theta + \theta) + M_p\theta$$

or $$W_u = \frac{16M_p}{L}$$

Concentrated Load

Propped Cantilever

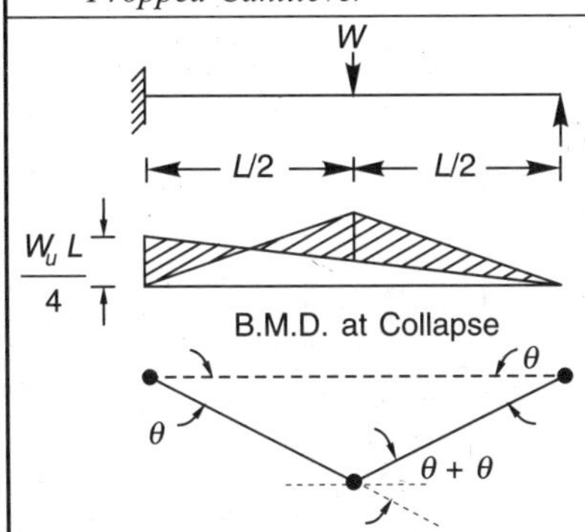

Fig. 17.17

Static method

$$\frac{W_u L}{4} = M_p + M_p \frac{L/2}{L}$$

or $$W_u = \frac{6M_p}{L}$$

Kinematic method

$$W_u \frac{L}{2}\theta = M_p\theta + M_p(\theta + \theta)$$

Propped Cantilever

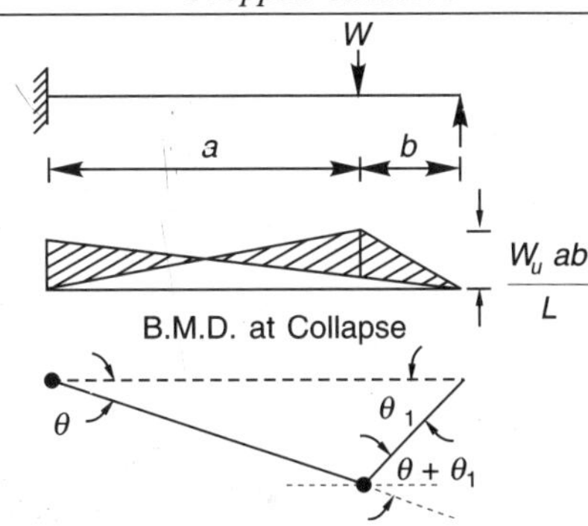

Fig. 17.18

Static method

$$\frac{W_u ab}{L} = M_p + M_p \frac{b}{L}$$

or $$W_u = \frac{L+b}{ab} M_p$$

Kinematic method

$$W_u a\theta = M_p\theta + M_p(\theta + \theta_1)$$

(Contd.)

or $\quad W_u = \dfrac{6M_p}{L}$	or $\quad = M_p\theta + M_p\left(\theta + \dfrac{a}{b}\theta\right)$ or $\quad = \dfrac{b+(a+b)}{b} M_p\ \theta$ or $\quad W_u = \dfrac{L+b}{ab} M_p$

Solved Examples

Example 17.1 Find the shape factors for the following sections:

(a) Square of side *a* with its diagonal parallel to the *xx*-axis.

(b) A diamond section with unequal diagonals, the shorter being *b* and longer *d*; the shorter diagonal placed parallel to the *xx*-axis.

(c) Hollow tube section of external diameter *D* and internal diameter *d*.

(d) Triangular section of base *b* and height *h*.

Solution

(a) (Refer Fig. Ex. 17.1(a))

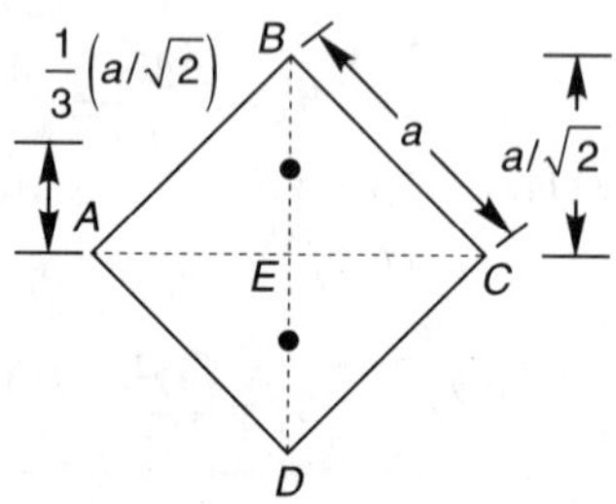

Fig. Ex. 17.1(a)

$$AC = \sqrt{a^2 + a^2} = \sqrt{2}\, a$$

$$BE = \frac{BD}{2} = \frac{AC}{2} = \frac{\sqrt{2}a}{2} = \frac{a}{\sqrt{2}}$$

$$I_{xx} = \frac{2\sqrt{2}}{12} a \left(\frac{a}{\sqrt{2}}\right)^3 = \frac{a^4}{12}$$

$$Z_{xx} = \frac{a^4/12}{a/\sqrt{2}} = \frac{a^3}{6\sqrt{2}}$$

Plastic section modulus, $Z_p = \dfrac{A}{2}(\bar{y}_1 + \bar{y}_2)$

$$= \frac{a^2}{2} \times \left(\frac{a}{3\sqrt{2}} + \frac{a}{3\sqrt{2}}\right) = \frac{a^3}{3\sqrt{2}}$$

$$\text{Shape factor} = \frac{Z_p}{Z} = \frac{a^3/3\sqrt{2}}{a^3/6\sqrt{2}} = 2$$

(b) Moment of inertia about *xx*-axis, (Ref. Fig. Ex. 17.1(b))

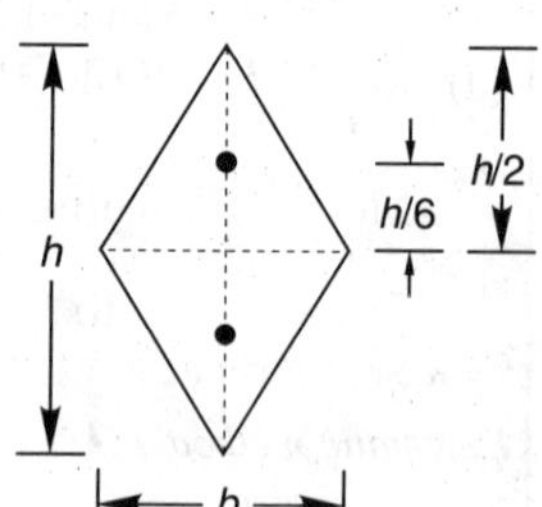

Fig. Ex. 17.1(b)

$$I_{xx} = 2 \times \frac{1}{12} b \left(\frac{h}{2}\right)^3 = \frac{bh^3}{48}$$

Section modulus,

$$Z = \frac{I_{xx}}{y} = \frac{bh^3/48}{h/2} = \frac{bh^2}{24}$$

Plastic section modulus, $Z_p = \frac{A}{2}(\bar{y}_1 + \bar{y}_2) = \frac{1}{2} \times \left(\frac{bh}{2}\right) \times \left(\frac{h}{6} + \frac{h}{6}\right) = \frac{bh^2}{12}$

$$\text{Shape factor} = \frac{bh^2/12}{bh^2/24} = 2$$

(c) Refer Fig. Ex. 17.1(c)

$$I_{xx} = \frac{\pi}{64}(D^4 - d^4)$$

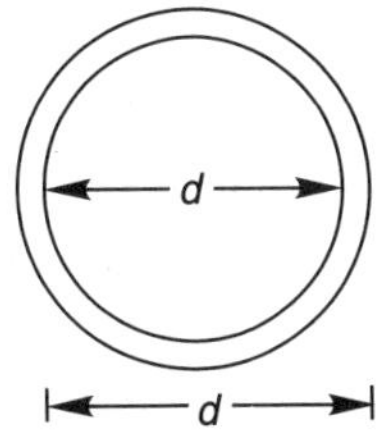

Fig. Ex. 17.1(c)

Section modulus, $Z = \frac{\pi}{64} \times \frac{(D^4 - d^4)}{D/2} = \frac{\pi}{32} \times \frac{D^4 - d^4}{D}$

Plastic section modulus, $Z_p = \frac{A}{2}(\bar{y}_1 + \bar{y}_2)$

$$A = \frac{\pi}{4}(D^2 - d^2)$$

$$\bar{y}_1 = \bar{y}_2 = \frac{\frac{1}{2} \times \left[\frac{\pi D^2}{4}\left(\frac{2}{3} \times \frac{D}{\pi}\right) - \frac{\pi d^2}{4}\left(\frac{2}{3} \times \frac{d}{\pi}\right)\right]}{\frac{1}{2} \times \left[\frac{\pi}{4}(D^2 - d^2)\right]} = \frac{2}{3\pi} \times \frac{(D^3 - d^3)}{(D^2 - d^2)}$$

$$Z_p = \frac{1}{2} \times \frac{\pi}{4}(D^2 - d^2)\left[2 \times \frac{2}{3\pi} \times \frac{(D^3 - d^3)}{(D^2 - d^2)}\right] = \frac{1}{6}(D^3 - d^3)$$

$$\text{Shape factor} = \frac{Z_p}{Z} = \frac{\frac{1}{6}(D^3 - d^3)}{\frac{\pi}{32}\left(\frac{D^4 - d^4}{D}\right)} = \frac{32}{6\pi} \times \frac{D^3\left(1 - \frac{d^3}{D^3}\right)}{D^3\left(1 - \frac{d^4}{D^4}\right)} = \frac{32}{6\pi} \times \frac{\left(1 - \frac{d^3}{D^3}\right)}{\left(1 - \frac{d^4}{D^4}\right)}$$

$$\text{Let } k = \frac{d}{D}$$

$$\text{Shape factor} = \frac{32}{6\pi} \times \frac{(1 - k^3)}{(1 - k^4)} = 1.7\,\frac{(1 - k^3)}{(1 - k^4)}$$

when $d \approx D, k = 3/4$

Shape factor $= 1.27 = 4/\pi$

(d) Refer Fig. Ex. 17.1(d)

Moment of inertia, $I_{xx} = \frac{bh^3}{36}$

Section modulus, $Z = \frac{bh^3}{36} \Big/ \frac{2}{3}h$

$$= \frac{bh^2}{24}$$

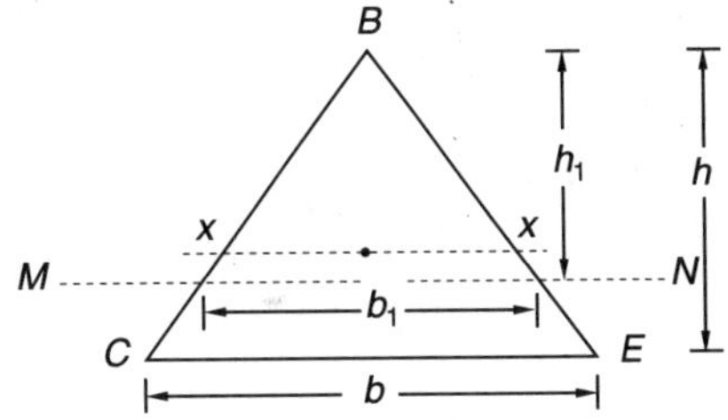

Fig. Ex. 17.1(d)

Plastic section modulus, $= \frac{A}{2}(\bar{y}_1 + \bar{y}_2)$

Let *MN* be the equal area axis, then

$$\frac{1}{2}\,b_1h_1 = \frac{1}{2}\left(\frac{bh}{2}\right) \quad \Rightarrow \quad b_1 h_1 = \frac{bh}{2} \qquad (1)$$

$$\frac{h_1}{b_1} = \frac{h}{b} \qquad (2) \qquad \text{(From similar } \Delta^s\ BMN \text{ and } BCE\text{)}$$

From Eqs (1) and (2)

$$h_1 = h/\sqrt{2}$$

and

$$b_1 = b/\sqrt{2}$$

$$A = \frac{bh}{2}$$

$$\bar{y}_1 = \frac{h_1}{3} = \frac{h}{3\sqrt{2}}$$

$$\bar{y}_2 = \left(\frac{h-h_1}{3}\right)\times\left(\frac{b_1+2b}{b_1+b}\right)$$

$$= \left(\frac{h-\frac{h}{\sqrt{2}}}{3}\right)\times\left(\frac{2b+\frac{b}{\sqrt{2}}}{b+\frac{b}{\sqrt{2}}}\right) = \frac{h(\sqrt{2}-1)}{3\sqrt{2}}\times\frac{(2\sqrt{2}+1)}{(\sqrt{2}+1)} = \frac{0.4643\,h}{3}$$

$$Z_p = \frac{A}{2}(\bar{y}_1+\bar{y}_2) = \frac{1}{2}\times\frac{bh}{2}\left[\frac{h}{3}\times\frac{1}{\sqrt{2}}+\frac{h}{3}\times 0.4643\right]$$

$$= \frac{bh^2}{12}[0.7072+0.4643] = 0.0976\,bh^2$$

$$\text{Shape factor} = \frac{Z_p}{Z} = \frac{0.0976\,bh^2}{bh^2/24} = 2.343$$

Example 17.2 Find the shape factor of a Tee-section with a flange 100 × 10 mm and web 150 × 10 mm (see Fig. Ex. 17.2).

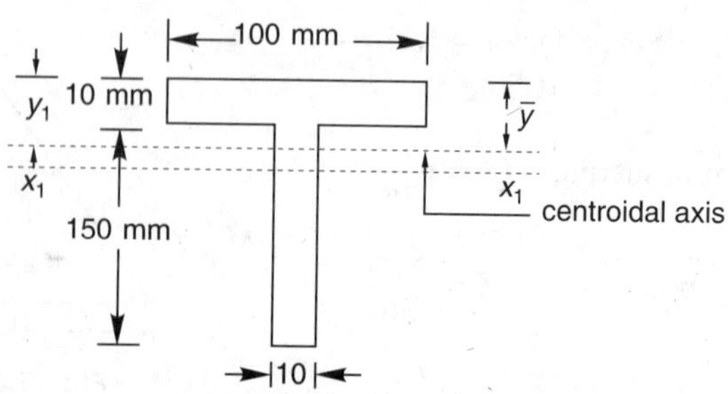

Fig. Ex. 17.2

Solution Let the distance of the neutral axis from the top be $\bar{y}$.

$$\bar{y} = \frac{100 \times 10 \times 10/2 + 150 \times 10 \times (75 + 10)}{100 \times 10 + 150 \times 10} = 53 \text{ mm}$$

Moment of inertia about the *xx*-axis,

$$I_{xx} = \frac{100 \times 10^3}{12} + 100 \times 10 \times (53 - 5)^2 + \frac{10 \times 150^3}{12} + 10 \times 150 \times (85 - 53)^2$$
$$= 666.0833 \times 10^4 \text{ mm}^4$$

$$\text{Section modulus, } Z_{xx} = \frac{I_{xx}}{y_{max}} = \frac{666.0833 \times 10^4}{(160 - 53)} = 62.250 \times 10^3 \text{ mm}^3$$

Let x_1x_1 be the equal area axis at a distance y_1 from top.

Total area of section, $A = 100 \times 10 + 150 \times 10 = 2500 \text{ mm}^2$

Then,

$$100 \times 10 + (x_1 - 10) \times 10 = 1250$$

$$\Rightarrow \quad x_1 = 35 \text{ mm}$$

$$Z_p = \frac{A}{2}(\bar{y}_1 + \bar{y}_2)$$

$\bar{y}_1$ = the distance of C.G of the area above the equal area axis

$$= \frac{100 \times 10 \times (35 - 10/2) + 10 \times (35 - 10) \times (35 - 10)/2}{100 \times 10 + 10 \times (35 - 10)} = 26.5 \text{ mm}$$

$\bar{y}_2$ = the distance of CG of the area below the equal area axis

$$= \frac{(150 + 10) - 35}{2} = 62.5 \text{ mm}$$

$$\text{Plastic section modulus, } Z_p = \frac{2500}{2} \times (26.5 + 62.5) = 111250 \text{ mm}^3$$

$$\text{Shape factor} = \frac{111250}{62.250 \times 10^3} = 1.7871$$

Example 17.3 Find out the collapse load for the cantilever shown in Fig. Ex. 17.3.

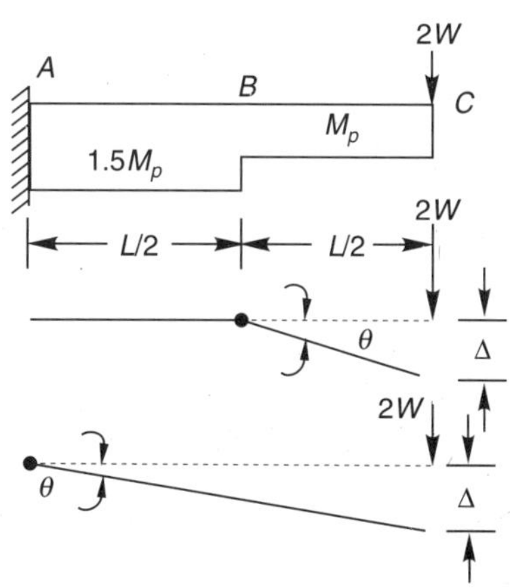

Fig. Ex. 17.3

Solution The beam as shown is of a non-uniform cross-section. A plastic hinge can form at the point where the section changes. A plastic hinge can also form at the fixed support.

Case I Plastic hinge at the point, where the cross-section changes.

External work done = load × deflection

$$= 2W_u \times \Delta$$

$$= 2W_u \frac{L}{2}\theta$$

but lies off the centre and towards the propped end. Therefore, it becomes essential to first locate the position of the plastic hinge. Let us assume that it forms at a distance x from the propped end.

External work done in the case of uniformly distributed load is computed by multiplying the load and the average deflection or by multiplying the load/unit length and the area of the deflected shape (mechanism).

From the mechanism,

$$\Delta = (L - x)\,\theta = x\,\theta_1$$

or

$$\theta_1 = \frac{L - x}{x}\,\theta$$

$$\text{External work done} = \text{load} \times \text{average deflection}$$

$$= (w_u L) \times \frac{1}{2}(L - x)\theta$$

$$= \frac{1}{2}\,w_u L(L - x)\theta$$

$$\text{Internal work done} = M_p\theta + M_p(\theta + \theta_1)$$

$$= M_p\theta + M_p\left(\theta + \frac{L - x}{x}\,\theta\right)$$

$$= M_p\left(1 + 1 + \frac{L - x}{x}\right)\theta$$

$$= M_p\left(\frac{L + x}{x}\right)\theta$$

By the principle of virtual work,

$$\text{External work done} = \text{Internal work done}$$

$$\frac{1}{2}\,w_u L(L - x)\,\theta = M_p\left(\frac{L + x}{x}\right)\theta$$

or

$$M_p = \frac{w_u L}{2} \times \frac{x(L - x)}{(L + x)} \qquad (1)$$

For the maximum value of M_p,

$$\frac{dM_p}{dx} = 0$$

or

$$\frac{dM_p}{dx} = 0 = \frac{\dfrac{(w_u L)}{2}(L + x)(1 \times (L - x) + x \times (-1)) + x(L - x) \times 1}{(L + x)^2}$$

or

$$(L + x)(L - 2x) - x(L - x) = 0$$

or $$x^2 + 2xL - L^2 = 0$$

$\Rightarrow$ $$x = 0.414L \quad (2)$$

Put the value of x in Eq. (1)

$$M_p = \frac{w_u L}{2} \times \frac{x(L-x)}{(L+x)}$$

$$= \frac{w_u L}{2} \times \frac{0.414L \times (L - 0.414L)}{(L + 0.414L)}$$

$$= \frac{w_u L}{2} \times \frac{0.414 \times 0.586 \times L^2}{0.414 L}$$

or $$w_u = \frac{M_p}{L^2} \times 2 \times \frac{1.414}{0.414 \times 0.586}$$

$$= 11.656 \frac{M_p}{L^2}$$

If W_u = total uniformly distributed load = $w_u L$, then

$$W_u = 11.656 \frac{M_p}{L}$$

Example 17.6 A beam fixed at both the ends is subjected to uniformly distributed load W on its right half portion as shown in Fig. Ex. 17.6. Determine the collapse load if the beam has uniform cross-section.

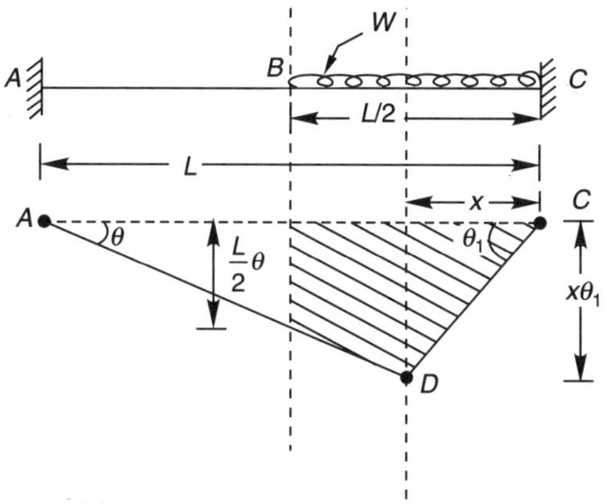

Fig. Ex. 17.6

Solution For a fixed ended beam and loaded partly for its span as shown in Fig. Ex. 17.6, the possible plastic hinges will be at A and C, and at a point D at a distance x from support C.

External work done = load × deflection

$$= \frac{W}{8} \times \frac{L}{3}\theta = \frac{WL}{24}\theta$$

Internal work done = $M_p\theta$

By principle of virtual work.

External work done = Internal work done

$$\frac{WL}{24}\theta = M_p\theta$$

$$\Rightarrow \qquad W_u = \frac{24M_p}{L}$$

Span AC

Since the end C is propped, the end D will lift up when a mechanism is formed and a negative work is done by the load acting at D. Plastic hinges will form at A and B only. At C no hinge will be formed since it is propped.

External work done = work done by load W

+ negative work done by load $W/8$

$$= W\frac{L}{2}\theta + \left(-\frac{W}{8} \times \frac{L}{3}\theta\right) = \frac{11}{24}WL\theta$$

Internal work done = $M_p\theta + M_p(\theta + \theta) = 3M_p\theta$

By principle of virtual work.

External work done = Internal work done

$$\frac{11}{24}WL\theta = 3M_p\,\theta$$

$$\Rightarrow \qquad W_{u_2} = 6.545\,\frac{M_p}{L}$$

Hence, the collapse load for the beam AD is $\dfrac{6.545\,M_p}{L}$

Example 17.8 A beam of uniform cross-section and span L is built-in at one end and simply supported at the other. It carries a concentrated load at a distance x from the built-in end. Show that the collapse load has the value $\dfrac{2L-x}{x(L-x)}M_p$, where M_p is the fully plastic moment.

If both the ends had been built-in, show that the load at collapse would have increased in the ratio $\dfrac{2L}{2L-x}$.

Solution Since end A of the beam is fixed, a plastic hinge will develop at A. A plastic hinge will also be formed below the concentrated load and a mechanism is formed as shown in Fig. Ex. 17.8(a).

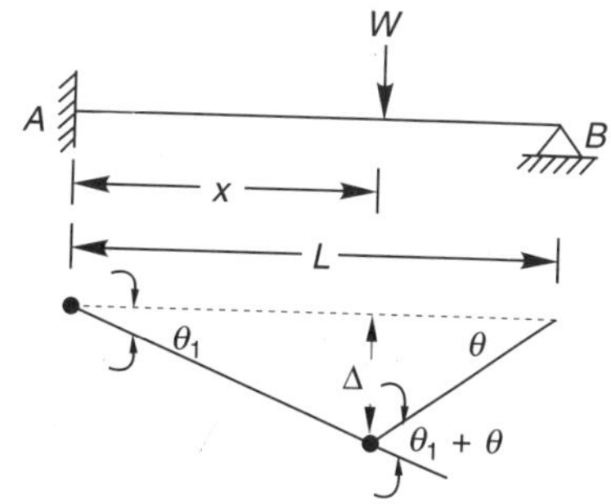

Fig. Ex. 17.8(a)

$$\Delta = x\theta_1 = (L - x)\theta$$

or
$$\theta_1 = \frac{L-x}{x}\theta$$

External work done = load × deflection

$$= W_u \times (L - x)\theta$$

Internal work done = moment × rotation

$$= M_p\theta_1 + M_p(\theta_1 + \theta)$$

$$= M_p\left(\frac{L-x}{x}\right)\theta + M_p\left(\frac{L-x}{x}\theta + \theta\right)$$

$$= M_p\left(\frac{L-x}{x} + \frac{L-x}{x} + 1\right)\theta$$

$$= M_p \frac{2L-x}{x}\theta$$

By the principle of virtual work,

External work done = Internal work done

$$W_u(L - x)\theta = M_p \frac{2L-x}{x}\theta$$

or
$$W_u = \frac{2L-x}{x(L-x)}M_p$$

If both the ends are fixed than one more plastic hinge will develop at B (Fig. Ex. 17.8(b)

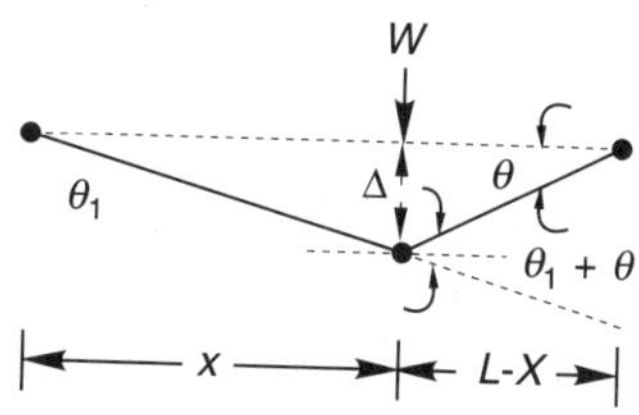

Fig. Ex. 17.8(b)

External work done = $W_u(L - x)\theta$

Internal work done = $M_p\theta_1 + M_p(\theta_1 + \theta) + M_p\theta$

$$= M_p\frac{(L-x)}{x}\theta + M_p\left(\frac{L-x}{x}\theta + \theta\right) + M_p\theta$$

$$= M_p\left(\frac{L-x}{x}\theta + \frac{L-x}{x}\theta + \theta + \theta\right)$$

$$= M_p\left(\frac{L-x+L-x+x+x}{x}\right)\theta$$

$$= M_p\frac{2L}{x}\theta$$

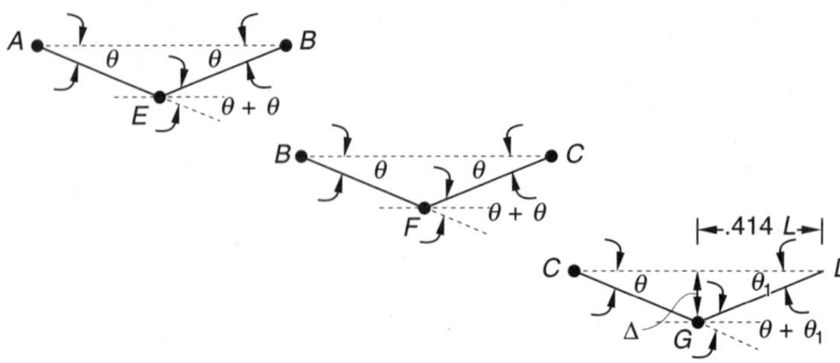

Fig. Ex. 17.10(b)

$$\text{External work done} = \text{load} \times \text{deflection} = W_u \times \frac{L}{2}\theta$$

$$\text{Internal work done} = \text{moment} \times \text{rotation}$$

$$= M_p\theta + M_p(\theta + \theta) + M_p\theta = 4M_p\theta$$

By principle of virtual work,

$$\text{External work done} = \text{Internal work done}$$

$$W_u \frac{L}{2}\theta = 4M_p\theta$$

or
$$W_u = \frac{8M_p}{L}$$

Span BC The ends *B* and *C* are intermediate supports. Therefore, the beam *BC*, acts as a fixed ended beam. The possible locations of plastic hinges are at supports *B*, *C* and *F* (the centre of the beam).

$$\text{External work done} = \text{load/unit length} \times \text{area of the mechanism}$$

$$= \frac{W_u}{L} \times \frac{1}{2} L\left(\frac{L}{2}\theta\right) = \frac{1}{4} W_u L\theta$$

$$\text{Internal work done} = \text{moment} \times \text{rotation}$$

$$= M_p\theta + M_p(\theta + \theta) + M_p\theta = 4M_p\theta$$

By the principle of virtual work,

$$\text{External work done} = \text{Internal work done}$$

$$\frac{1}{4} W_u L\theta = 4M_p\theta$$

or
$$W_u = 16\frac{M_p}{L}$$

Span CD The end *C* is an intermediate support and the end *D* is a simple support. Therefore, the beam *CD* acts as a propped cantilever. The possible locations of the plastic hinges are support *C* and point *G* (0.414 *L* from support *D*).

$$\Delta = 0.414L\theta_1 = 0.586L\theta \qquad \text{or} \qquad \theta_1 = \frac{0.586}{0.414}\theta$$

External work done = load × average deflection

$$= \left(\frac{W_u}{L} \times L\right) \times \frac{1}{2} \times 0.586\, L\theta$$

$$= \frac{1}{2} \times 0.586 W_u L\theta$$

Internal work done = moment × rotation

$$= M_p\theta + M_p(\theta + \theta_1)$$

$$= M_p\theta + M_p\left(\theta + \frac{0.586}{0.414}\theta\right) = \frac{1.414}{0.414} M_p\theta$$

By principle of virtual work,

External work done = Internal work done

$$\frac{1}{2} \times 0.586 W_u L\theta = \frac{1.414}{0.414} M_p\theta$$

or

$$W_u = \frac{2 \times 1.414}{0.586 \times 0.414} \times \frac{M_p}{L} = 11.656 \frac{M_p}{L}$$

The collapse load is the least of the three ultimate loads. Hence, the collapse load $W_u = 8M_p/L$. The plastic moment diagram is shown in Fig. Ex. 17.10(c).

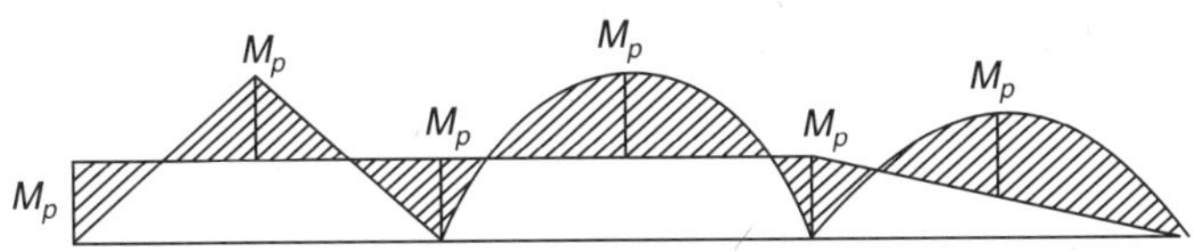

Fig. Ex. 17.10(c)

Example 17.11 Find out the collapse load for a portal frame shown in Fig. Ex. 17.11(a). The beams and columns are of same cross-section. Also draw the plastic moment diagram.

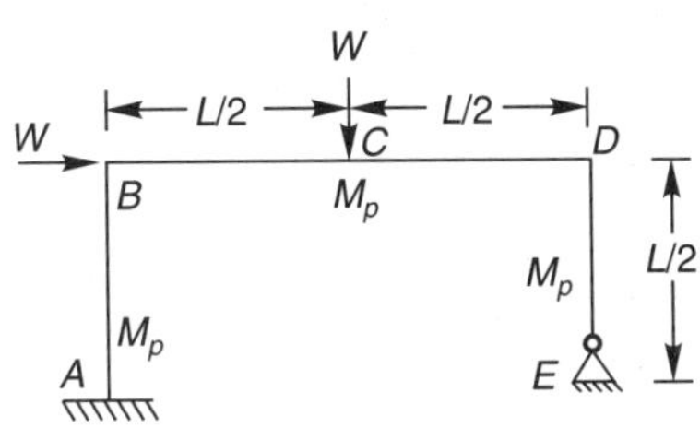

Fig. Ex. 17.11(a)

Solution In the case of portal frames, the first step is to find out the number of independent mechanisms and then to form all the possible combined mechanisms.

Possible location for plastic hinges are A, B, C and D.

Number of possible plastic hinges, $N = 4$

Degree of redundancy, $r = 5 - 3 = 2$

Plastic moment diagram The moment diagram is drawn on the tension side. (See Fig. 17.11(e)). E is a mechanical hinge so the moment value is zero. D and C are the points where fully plasuic moments are developed so ordinates, equal to M_p, can be drawn. At B, the value of moment is less than the plastic moment and is not known. To calculate the value of the moment at B, the value of the elastic moment in terms of the plastic moment at C is calculated. The moment is assumed to vary linearly and by this the value of moment at B is found. The process is illustrated below.

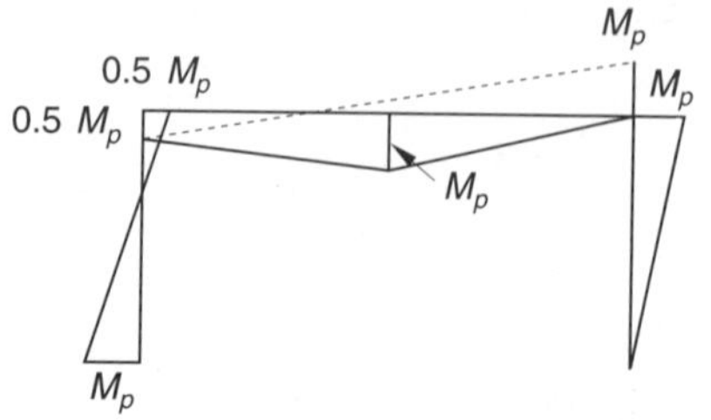

Fig. Ex. 17.11(e)

The value of moment at A is also fully plastic moment M_p.

For the beam BD,

$$M = \frac{WL}{4}$$

$$= 5\frac{M_p}{L} \times \frac{L}{4}$$

$$= \frac{5}{4}M_p = 1.25\ M_p$$

Thus, the ordinate is plotted as measured from the bottom of the plastic moment ordinate at C. From D to C (its span) the fall in plastic moment is of 0.75 M_p. For next half span, i.e. from C to B, it falls further by 0.75 M_p. Therefore, the net ordinate of the plastic moment for the beam BD is 0.5 M_p. For equilibrium, the plastic moment in column at B will also be 0.5 M_p.

Example 17.12 Compute the collapse load for the portal frame as shown in Fig. Ex. 17.12(a).

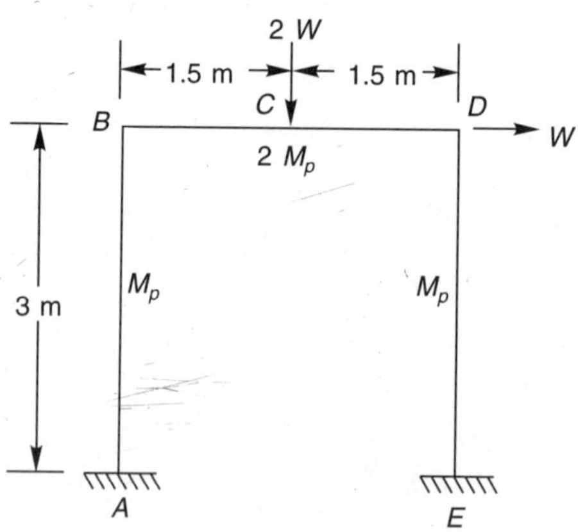

Fig. Ex. 17.12(a)

Solution The possible locations of the plastic hinges are at A, B, C, D and E.

$$\text{Number of possible plastic hinges, } N = 5$$

$$\text{Degree of redundancy, } r = 6 - 3 = 3$$

$$\text{Number of possible independent mechanisms, } n = N - r = 5 - 3 = 2$$

The two independent mechanisms are:

1. Beam mechanism
2. Sway mechanism

These two independent mechanism can form a combined mechanism.

Beam mechanism The ends B and D of the frame $ABDE$ are fixed joints. Therefore, the beam BD acts as a fixed end beam. The possible locations of plastic hinges are B, D and C (below the concentrated load) as shown in Fig. Ex. 17.12(b). Since the plastic moments for columns is M_p and that of beam is $2M_p$, the plastic hinges will form in the columns.

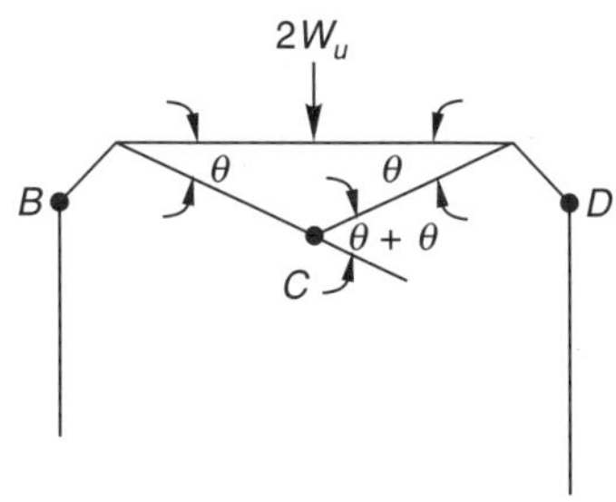

Fig. Ex. 17.12(b)

$$\text{External work done} = \text{load} \times \text{deflection} = 2W_u \times 1.5\theta = 3W_u\theta$$

$$\text{Internal work done} = \text{moment} \times \text{rotation} = M_p\theta + 2M_p(\theta + \theta) + M_p\theta = 6M_p\theta$$

By the principle of virtual work,

$$\text{External work done} = \text{internal work done}$$

$$3W_u\theta = 6M_p\theta$$

or

$$W_u = 2M_p$$

Sway mechanism The column AB and ED are of the same length. Since they deflect by the same amount Δ, hence the rotations at the ends are same as θ. Possible locations of plastic hinges are at A, B, D and E (Fig. Ex. 17.12(c)).

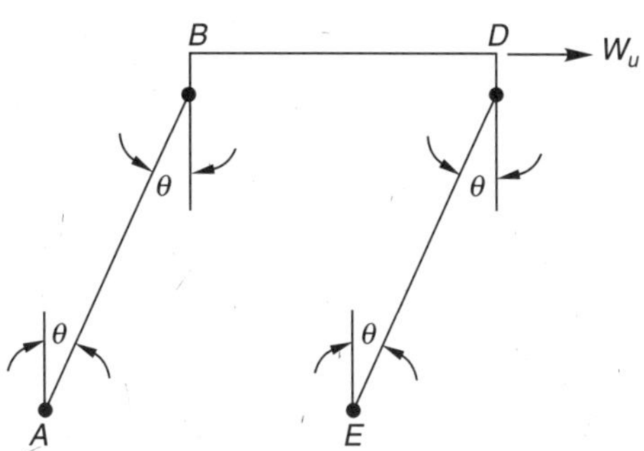

Fig. Ex. 17.12(c)

$$\text{External work done} = W_uL\theta = W_u \times 3 \times \theta = 3W_u\theta$$

$$\text{Internal work done} = M_p\theta + M_p\theta + M_p\theta + M_p\theta = 4\,M_p\theta$$

Sway mechanism The columns *AB* and *ED* are of different lengths. These deflect by the same amount. Therefore, the rotations at the ends of the columns will be different. From Fig. Ex. 17.13(c),

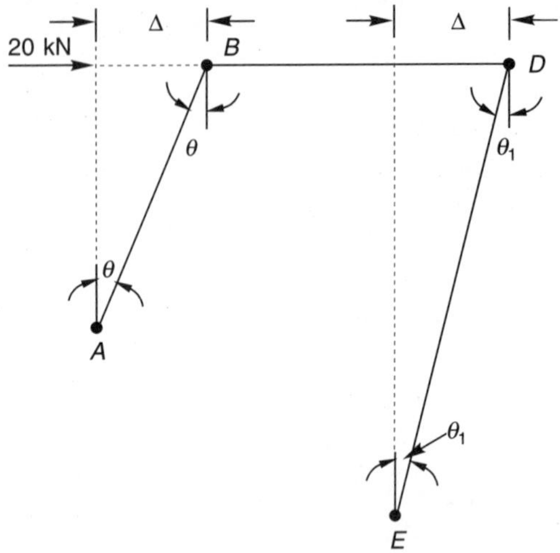

Fig. Ex. 17.13(c)

$$\Delta = 3\theta = 6\theta_1$$

or $$\theta_1 = \theta/2$$

$$\text{External work done} = \text{load} \times \text{deflection} = 20 \times 3 \times \theta = 60\theta$$

$$\text{Internal work done} = \text{moment} \times \text{rotation}$$
$$= M_p\theta + M_p\theta + M_p\theta_1 + M_p\theta_1$$
$$= M_p\theta + M_p\theta + M_p\theta/2 + M_p\theta/2 = 3M_p\theta$$

By the principle of virtual work,

$$\text{Internal work done} = \text{External work done}$$

$$3M_p\theta = 60\theta$$

or $$M_p = 20 \text{ kNm}$$

Combined mechanism The possible locations of plastic hinges are *A*, *C*, *D* and *E* only. Refer to Fig. Ex. 17.13(d).

$$\text{External work done} = \Sigma\,(\text{load} \times \text{deflection})$$
$$= 40 \times 2 \times \theta + 20 \times 3 \times \theta = 140\theta$$

$$\text{Internal work done} = \Sigma\,(\text{plastic moment} \times \text{rotation})$$
$$= M_p\theta + M_p(\theta + \theta) + M_p\theta + M_p\theta_1 + M_p\theta_1$$
$$= M_p\theta + 2M_p\theta + M_p\theta + M_p\theta/2 + M_p\theta/2 = 5M_p\theta$$

By the principle of virtual work,

$$\text{Internal work done} = \text{External work done}$$

$$5M_p\theta = 140\theta$$

or $$M_p = \frac{140}{5} = 28 \text{ kNm}$$

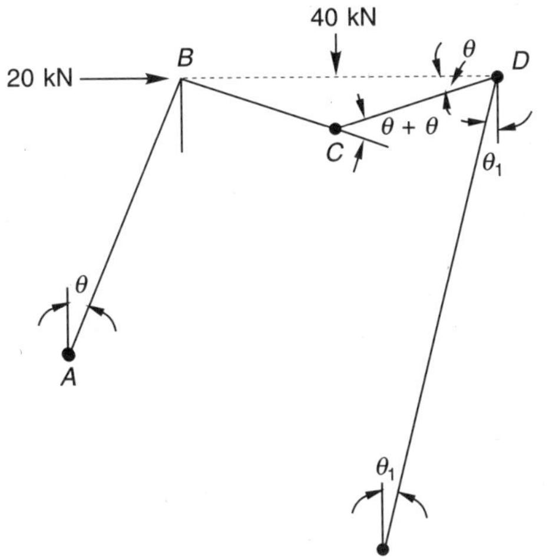

Fig. Ex. 17.13(d)

The plastic moment for the frame is the maximum of the three plastic moments found. Therefore, the value of the plastic moment is 28 kNm.

Example 17.14 Design a continuous beam *ABCD* as shown in Fig. Ex. 17.14(a). The load factor and the shape factor may be assumed to be 1.70 and 1.12 respectively.

(a) Provide a uniform cross-section throughout the beam.
(b) Provide the most economical section.

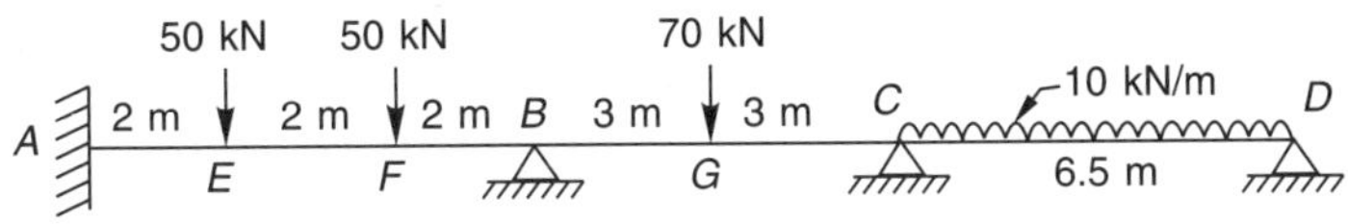

Fig. Ex. 17.14(a)

Solution The factored loads are

$$50 \times 1.70 = 85.0 \text{ kN}$$
$$70 \times 1.70 = 119 \text{ kN}$$
$$10 \times 1.70 = 17 \text{ kN/m}$$

Let us apply these factored loads on the beam and analyse it using the Kinematic theorem.

Span AB The possible places for the plastic hinges will be supports *A* and *B*, and below the concentrated loads at either *E* or *F* or under both *E* and *F* as shown in Fig. Ex. 17.14(b).

$$\Delta = 2\theta_1 = 4\theta$$

or

$$\theta_1 = 2\theta$$

(i) The plastic hinges can form at *A*, *B* and *E*.

$$\text{External work done} = 85 \times 2 \times \theta_1 + 85 \times 2 \times \theta$$
$$= 85 \times 2 \times 2\theta + 85 \times 2 \times \theta = 510\ \theta$$

$$\text{Internal work done} = M_p\theta_1 + M_p(\theta_1 + \theta) + M_p\theta$$
$$= M_p 2\theta + M_p(2\theta + \theta) + M_p\theta = 6M_p\theta$$

$$\text{Plastic section modulus required, } Z_p = \frac{M_p}{f_y}$$

$$= \frac{61.62 \times 10^3 \times 10^3}{250} = 246.48 \times 10^3 \text{ mm}^3$$

$$\text{Elastic section modulus required, } Z = \frac{Z_p}{f}$$

$$= \frac{246.48 \times 10^3}{1.12} = 220.07 \times 10^3 \text{ mm}^3$$

From I.S. Handbook No. 1, select an I.S.L.B. 225 @ 230.5 N/m,

Section modulus provided = 222.4×10^3 mm^3

Plastic section modulus provided = $222.4 \times 10^3 \times 1.12$

= 249.088×10^3 mm^3

Plastic moment = $Z_p f_y$

= $249.088 \times 10^3 \times 250 = 62.272$ kNm

Design of cover plates

Span AB Additional plastic moment to be resisted by the cover plate

= 85 – 62.272 = 22.728 kNm

Required plastic section modulus of cover plate

$$= \frac{22.728 \times 10^3 \times 10^3}{250} = 90.912 \times 10^3 \text{ mm}^3$$

$$\text{Required area of cover plate } = \frac{Z_p}{D}$$

$$= \frac{90.912 \times 10^3}{225} = 404.05 \text{ mm}^2$$

Provide one cover plate 85 mm × 5 mm.

Plastic section modulus provided = (85 × 5) × (225 + 5)

= 97.75×10^3 mm^3 > 90.912×10^3 mm^3

which is as it should be.

Span BC Addition plastic moment to be resisted by the cover plate

= 89.25 – 62.272 = 26.978 kNm

Required plastic section modulus of cover plate

$$= \frac{26.978 \times 10^3 \times 10^3}{250} = 107.912 \times 10^3 \text{ mm}^3$$

$$\text{Required plastic area of cover plates } = \frac{Z_p}{D} = \frac{107.912 \times 10^3}{225} = 479.608 \text{ mm}^2$$

Provide a cover plate 100 mm × 5 mm.

Plastic section modulus provided = (100 × 5) × (225 + 5)

= 115×10^3 mm^3 > 106.912×10^3 mm^3

which is all right.

Example 17.15 Find out the plastic moment for a given frame of uniform cross-section under the applied factored loads as shown in Fig. Ex. 17.15(a).

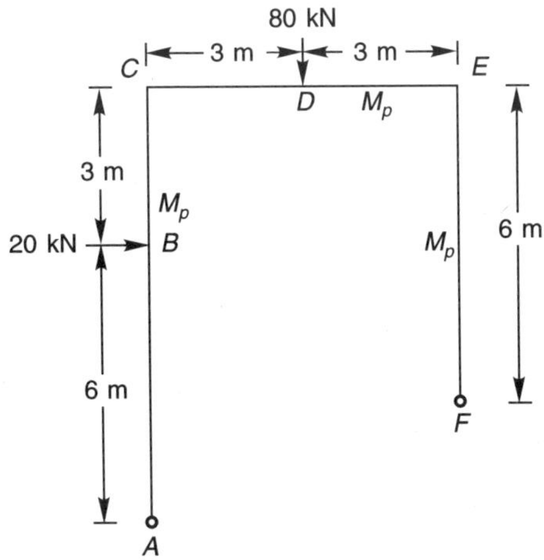

Fig. Ex. 17.15(a)

Solution The possible locations of plastic hinges are B, C, D and E.

Number of possible plastic hinges; $N = 4$

Degree of redundancy, $r = 4 - 3 = 1$

Number of possible independent mechanisms, $n = N - r = 4 - 1 = 3$

The three independent mechanisms are:

1. Beam mechanism, span CE
2. Beam mechanism, span AC
3. Sway mechanism

Two possible combinations of the independent mechanisms can be made to obtain combined mechanisms, i.e. the sway mechanism can be combined with each of the beam mechanisms.

Beam mechanism (span CE) The ends C and E of the beam CE are fixed joints. Therefore, the plastic hinges will form at C and E and also at D, i.e. at the point of concentrated load as shown in Fig. Ex. 17.15(b).

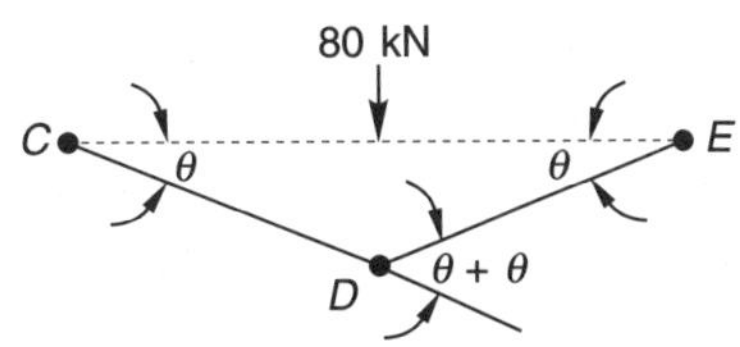

Fig. Ex. 17.15(b)

External work done = load × deflection

$$= 80 \times 3 \times \theta = 240\ \theta$$

Internal work done $= M_p\theta + M_p(\theta + \theta) + M_p\theta = 4M_p\theta$

By the principle of virtual work,

Internal work done = External work done

$$4M_p\theta = 240\ \theta$$

or
$$M_p = 60 \text{ kNm}$$

Beam mechanism (span AC) The end A is a mechanical hinge; so no plastic hinge will develop there. The end C is a joint so a plastic hinge will develop there.

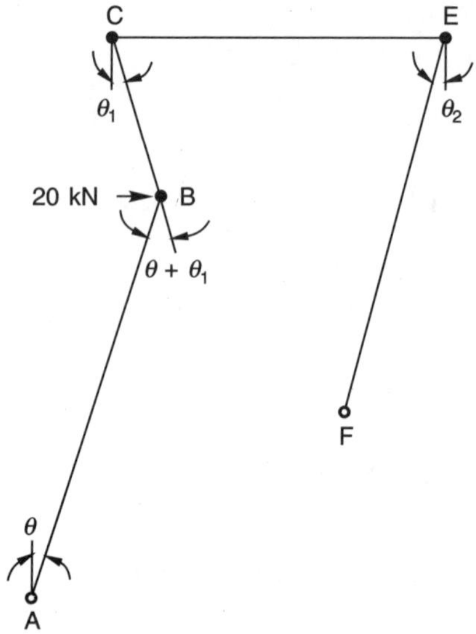

Fig. Ex. 17.15(f)

By the principle of virtual work,

$$\text{Internal work done} = \text{External work done}$$

$$6.5M_p\theta = 240\ \theta$$

$$M_p = 36.9 \text{ kNm}$$

Therefore, the plastic moment M_p for the section is 80 kNm.

Exercises

17.1 Obtain the length and profile of a plastic hinge for a simply supported beam subjected to a uniformly distributed load.

17.2 Find out the value of a plastic modulus for the following sections:

(a) A square of side x, bent about a diagonal $(Z_p = x^3/3\sqrt{2})$

(b) An isosceles triangle of height h and base b, bent about an axis parallel to the base. $(Z_p = bh^2(\sqrt{2}-1)/3\sqrt{2})$

(c) An I-Section of overall depth 300 mm and flanges 200 × 20 mm, web 10 mm thick, bent about its strong axis. $(Z_p = 1289 \times 10^3 \text{ mm}^3)$

(d) A thin walled circular tube, mean diameter d, wall thickness t.

$$\left[\frac{1}{6}(d+t)^3 - \frac{1}{6}(d-t)^3 \approx td^2\right]$$

17.3 Deduce the shape factors for the following sections:

(a) triangular

(b) diamond

(c) hollow tube (2.34, 2, $4/\pi$)

17.4 A beam of uniform cross-section and span $2L$ is subjected to equal concentrated loads at a distance a, from each end of the beam. The beam is propped at the centre. Show that the collapse load is equal to $\dfrac{(L+a)}{(L-a)}\dfrac{M_p}{a}$.

17.5 A simply supported beam of uniform cross-section and span $2L$ is propped at the centre. Find out the collapse load if,

(i) a concentrated load is applied at the centre of the left hand span,

(ii) equal concentrated loads are applied at the centre of each span

$(6M_p/L, 6M_p/L)$

17.6 A beam of uniform cross-section, fully plastic moment M_p, of length $2l$ rests on simple supports at its ends and on a central prop. Equal concentrated loads are applied at a distance al from each end of the beam, what would be the value of the collapse load?

$$\left[\frac{1+a}{a(1-a)}\times\frac{M_p}{l}\right]$$

17.7 A fixed ended beam of uniform cross-section, full plastic moment M_p, and length $2l$ rests on a central prop. Equal concentrated loads are applied at a distance al from each of the fixed ends. What would be their value at collapse?

$$\left[\frac{2}{a(1-a)}\times\frac{M_p}{l}\right]$$

17.8 A steel I-beam is designed to carry, as a simply supported beam, a uniformly distributed load of intensity w over a span $2l$ with a load factor of 2. When a load of intensity $2w$ has been applied, so that the beam is about to collapse, a rigid prop is inserted at the centre. Show that more load can be added until yield occurs at section $(2/5)l$ from each end. Show further that complete collapse of the propped beam does not take place until the total intensity of load carried is $2(3+2\sqrt{2})w$.

17.9 A continuous beam ABC rests on simple supports at A, B and C, the span AB being l and the span BC being $2l$. The beam is composed of two uniform members AB and BC rigidly jointed at B, the member AB having a fully plastic moment M_p and the member BC having a fully plastic moment kM_p. The span AB carries a concentrated load W, and the span BC carries a uniformly distributed load $(5/6)\,W$.

The beam is to be designed according to the plastic theory, so that for any given value of k the value of M_p is to be such that one or other of the spans fails by plastic collapse. Show that the collapse will occur in the span AB rather than in the span BC if k exceeds a value of about 0.8.

17.10 Determine the collapse load for the continuous beam section shown in Fig. Prob. 2.9. The beam has a non-uniform cross-section.

$$\left(\frac{10}{3}\frac{M_p}{L}\right)$$

Chapter 18

Introduction to Limit State Design

18.1 INTRODUCTION

The relevant Code of Practice, IS: 800–1984, applicable to the structural use of hot-rolled steel is largely based on Working Stress Method and results in uneconomic designs. The Limit State Design approach is technically sound and results in significant economy and uniform reliability in completed structures. This method of design also known as load and resistance factor method is not a recent concept. Since 1974 it has been used in Canada and Europe as Limit States design and as Load and Resistance Factor method in America. IS: 800 is in the process of revision and recommends Limit State method for the design of steel elements and structures. Also, it recommends Working Stress method in situations where Limit State method cannot be conveniently adopted.

18.2 BASIS FOR DESIGN

For a structural design to be satisfactory, generally four major objectives—utility, safety, economy and elegance must be fulfilled. Also, the structure must fulfill its intended function for the stipulated period for which it is designed. Therefore, steel structures should be designed and constructed to satisfy the design requirements for stability, strength, serviceability, brittle fracture, fatigue, fire, and durability with due regards to economy, in such a way that they

(a) behave as a three dimensional identity. Its constituent parts and other components should constitute a stable and robust structure under normal loading
(b) remain fit with adequate reliability and are able to sustain all actions (loads) and other influences experienced during construction and use
(c) have adequate durability under normal maintenance
(d) are not seriously damaged or do not collapse under accidental events like explosions, impact or due to consequences of human error to an extent not originally expected to occur.

The potential for catastrophic damage is limited or avoided by appropriate choice of one or more of the following:

(i) Avoiding, eliminating or reducing exposure to hazards, which the structure is likely to sustain
(ii) Choosing structural forms, layouts, details and design such that the structure has low sensitivity to hazardous conditions and survives with only local damage even after serious damage to any individual element by the hazard.

(iii) by appropriate choice of suitable material, design and detailing procedure, construction specifications, and control procedures for shop fabrication and field construction, as relevant to the particular structure.

In the Limit State Design method, the structure is designed to withstand safely all loads likely to act on it throughout its life. It is expected that the structure will satisfy the serviceability requirements, such as limitations of deflection and vibration and will not collapse under accidental loads such as from explosions or impact or due to consequences of human error to an extent not originally expected to occur. The acceptable limit for the safety and serviceability requirements before failure is called a *limit state.*

The objective of the Limit State design is to achieve a structure that will not become unfit for use with an acceptable target reliability. In other words, the probability of a limit state being reached during its lifetime should be very low. In general, the structure should be designed on the basis of the most critical limit state (on the basis of strength and stability at ultimate load) and then checked for other limit states (deflections, etc., at serviceability loading).

Note Structures designed for unusual or special functions should comply with relevant additional limit states considered appropriate to that structure.

18.3 LIMIT STATES

Limit states are the states beyond which the structure no longer satisfies the specified performance requirements. The limit states are normally grouped under two headings:

(a) Limit state of strength or safety also known as ultimate limit state
(b) Limit state of serviceability

Limit state of strength are those associated with failures (or imminent failure), under the action of probable and most unfavorable combination of loads on the structure using the appropriate partial safety factors, which may endanger the safety of life and property. The limit state of strength include

(a) loss of stability/equilibrium of the structure (including the effect of sway where appropriate and overturning) or any of its parts including supports and foundations
(b) strength limits (general yielding, formation of mechanism, rupture of the structure or any of its parts or components)
(c) fatigue (leading to fracture)
(d) brittle fracture

Limit state of serviceability are limit states beyond which specified service criteria are no longer met. These include

(a) deformation and deflections, which may adversely affect the appearance or effective use of the structure, or may cause improper functioning of equipment or services, or may cause damages to finishes and non-structural members
(b) vibrations in the structure or any of its components causing discomfort to people, damages to the structure, its contents or which may limit its functional effectiveness.

Note Special consideration should be given to floor systems susceptible to vibration, such as large open-floor areas free of partitions to ensure that such vibrations are acceptable for the intended use and occupancy.

(c) Repairable damage due to fatigue
(d) Corrosion and durability
(e) Ponding of structures

Note The attainment of one or more limit state of strength may be regarded as an inability to sustain any increase in load whereas the limit state of serviceability provides a check against the need for remedial action or some other loss of utility. Thus limit state of strength are conditions to be avoided whilst limit state of serviceability could be considered as merely undesirable.

Thus the following limit states may be identified for design purposes and are provided for in terms of partial factors reflecting the severity of the risks.

Ultimate Limit State is related to the maximum design load capacity under extreme conditions. The partial load factors are chosen to reflect the probability of extreme conditions, when loads act alone or in combination.

Serviceability Limit State is related to the criteria governing normal use. Unfactored loads are used to check the adequacy of the structure.

Fatigue Limit State is important where distress to the structure by repeated loading is a possibility.

18.4 ACTIONS (LOADS)

The actions to be considered in design include direct actions (loads) experienced by the structure due to self weight, external actions, etc., and imposed deformations such as those due to temperature and settlement. Actions are classified by their variation with time as given below:

Permanent Actions Actions due to self-weight of structural and non-structural components, fittings, ancillaries, and fixed equipment, etc.

Variable Actions Actions due to construction and service stage loads such as imposed (live) loads (crane loads and snow loads), wind loads, earthquake loads etc.

Accidental Actions Actions due to explosions, impact of vehicles, fires, etc.

Characteristic Actions The characteristic values of loads generally correspond to the value specified by relevant code, standard or client. The characteristic actions are the values of the different actions that are not expected to be exceeded, with more than 5% probability, during the life of the structure and they are taken as

(a) the self-weight (in most cases may be calculated on the basis of nominal dimensions and unit weights (IS:875, Part 1)).
(b) the variable loads (the value specified in relevant code or standard (IS: 875, IS:1893))

For the purpose of designing any element or a structure as a whole, the following loads (actions) and their effects should be taken into account, where applicable, with partial safety factors and combinations.

(a) Dead loads; IS 875 ; (Part I)
(b) Imposed loads; (live load, crane load, snow load, and dust load etc.; IS 875 : (Part II/IV)
(c) Wind loads; IS 875 : (Part III)
(d) Earthquake loads; IS 1893
(e) Erection loads;
(f) Accidental loads such as those due to blast (This load was not covered in IS:800–1984); and
(g) Secondary effects due to contraction or expansion resulting from temperature changes, differential settlement of the structure as a whole or of its components, eccentric connections, rigidity of joints differing from design specifications.

Load Combinations Load combinations for design purposes should be those that produce maximum forces and effects and consequently maximum stresses and deformations. The following combination of loads with appropriate load factors specified in the code may be considered.

(a) Dead load + imposed load,
(b) Dead load + imposed load + wind or earthquake load,
(c) Dead load + wind or earthquake load, and
(d) Dead load + erection load.

Note In the case of structures supporting cranes, imposed loads should include the crane effects.

18.5 DESIGN CRITERIA

A good design of a structure consists of so proportioning the structure that the probability of its failure is the least. In traditional allowable stress design this is achieved by using the concept of factor of safety (considering lesser strength in the design equation). In limit states design, for ensuring the design objectives, the design is based on characteristic values for material strengths and applied loads, which take into account the probability of variations in the material strengths and loads. The characteristic values are based on statistical data, if available. Where such data is not available they are based on experience. Structure is designed considering the limit states at which they would become unfit for their intended purpose. For verifying the adequacy of the structure, appropriate partial safety factors, based on semi-probabilistic methods are used. Two separate partial safety factors, one applied to forces due to loading (γ_{fk}) and another to the material strength (γ_m) are employed to cater for different amounts of variability associated with these.

γ_{fk} considers variability of loading and allows for the

(i) possibility of unfavourable deviation of the load from the characteristic value,
(ii) possibility of inaccurate assessment of the load,
(iii) uncertainty in the assessment of the effects of the load, and
(iv) uncertainty in the assessment of the limit states being considered.

γ_m considers variability of material strength and takes account of the following:

(i) The possibility of unfavourable deviation of material strength from the characteristic value. Material strengths may initially vary appreciably from their assumed values and may vary more with time due to creep, corrosion, and fatigue.

(ii) The possibility of unfavourable variation of member sizes.

(iii) The possibility of unfavourable reduction in member strength due to fabrication and tolerances. This includes reckless handling of steel shapes in shop, forcing of members in position to line up the bolt holes, etc.

(iv) Uncertainty in the calculation of strength of the members.

(v) Mode of failure (ductile/brittle).

The design values are derived from the characteristic values through the use of both of these partial safety factors. The reliability of design is expressed as

$$\text{Design Action } (Q_d) \leq \text{Design Strength } (S_d) \qquad (18.1)$$

The left-hand side of Eq (18.1) refers to the load effects on the structure, while the right-hand side refers to the resistance or capacity of the structure.

The design action, Q_d, is expressed as

$$Q_d = \sum \gamma_{fk} Q_{ck} \qquad (18.2)$$

and, the design strength, S_d is obtained as given below from ultimate strength, S_u and partial safety factors for materials, γ_m.

$$S_d = S_u/\gamma_m \qquad (18.3)$$

where

Q_{ck} = characteristic load

γ_{fk} = partial safety factor for loads, given in Table 18.1.

γ_m = partial safety factor for materials, given in Table 18.2.

Table 18.1 *Partial Safety Factors For Loads, γ_{fk} For Limit States*

Combination	*Limit State of Strength*					*Limit State of Serviceability*			
	DL	*LL*		*WL/EL*	*AL*	*DL*	*LL*		*WL/EL*
		Leading	*Accompanying (CL, SL etc.)*				*Leading*	*Accompanying (CL etc.)*	
DL+LL+CL	1.5	1.5	1.05	—	—	1.0	1.0	1.0	—
DL+LL+CL+	1.2	1.2	1.05	0.6	—	1.0	0.8	0.8	0.8
WL/EL	1.2	1.2	0.53	1.2					
DL+WL/EL	1.5 (0.9)*	—	—	1.5	—	1.0	—	—	1.0
DL+ER	1.2 (0.9)*	1.2	—	—	—	—	—	—	—
DL+LL+AL	1.0	0.35	0.35	—	1.0	—	—	—	—

*This value is to be considered when stability against overturning or stress reversal is critical .

Abbreviations: DL = Dead Load, LL = Imposed Load (Live Loads), WL = Wind Load, SL = Snow Load, CL = Crane Load (Vertical/horizontal), AL = Accidental Load, ER = Erection Load, EL = Earthquake Load.

Note The effects of loads in terms of stresses or stress resultants may be obtained from an Appropriate method of analysis.

Table 18.2 *Partial Safety Factor for Materials, γ_m*

S. No.	*Definition*	*Partial Safety Factor*	
1	Resistance, governed by yielding γ_{mo}	1.10	
2	Resistance of member to buckling γ_{mo}	1.10	
3	Resistance, governed by ultimate stress γ_{ml}	1.25	
4	Resistance of connection	Shop Fabrication	Field Fabrication
	(i) Bolts-Friction Type, γ_{mf}	1.25	1.25
	(ii) Bolts-Bearing Type, γ_{mb}	1.25	1.25
	(iii) Rivets γ_{mr}	1.25	1.25
	(iv) Welds γ_{mw}	1.25	1.50

18.6 LIMIT STATES OF STRENGTH

For checking the strength and stability of a structure the loads are multiplied by the relevant γ_{fk} factors given in Table 18.1. The factored loads are then applied in the most unfavorable realistic combination for the part or effect under consideration. The load capacity of each member and its connections, as determined by the relevant provisions of the Code, should be such that the factored loads would not cause failure.

The ultimate strength calculation may require consideration of

(a) loss of equilibrium of the structure or any part of it, considered as a rigid body, and

(b) failure by excessive deformation, rupture or loss of stability of the structure or any part of it including support and foundation.

Also, limit state of fatigue and of strength corresponding to formation of mechanism may be investigated.

Stability

Stability should be ensured for the structure as a whole and for each of its elements. This includes, overall frame stability against overturning and sway. The loads considered should be multiplied by the relevant γ_{fk} factors, given in Table 18.1, to get the factored loads.

Stability Against Over-turning The structure as a whole or any part of it should be designed to prevent instability due to overturning, uplift or sliding under factored load as given below:

(a) The actions should be divided into components aiding instability and components resisting instability.

(b) The actions and their effects causing instability should be combined using appropriate load factors as per the Limit States requirements.

(c) The permanent loads and effects causing resistance should be multiplied with a partial safety factor 0.9 and added together with design resistance (after multiplying with appropriate partial safety factor).

(d) The resistance effect should be greater than or equal to the destabilizing effect.

Note

(1) Combination of imposed and dead loads should be such as to cause most severe effect on overall stability.
(2) In assessing the stability against overturning, the probable variations in dead load during construction or other temporary conditions must be accounted for.

Stability Against Sway The whole structure, including portions between expansion joints, should be adequately stiff against sway. To ensure this, in addition to designing for applied horizontal loads, a separate check should be carried out for notional horizontal loads. The notional horizontal forces may arise from practical imperfections and may be considered as (a) 1% of factored dead load from that level, applied horizontally; (b) 0.5% of factored total gravity load (dead plus vertical imposed loads) from that level, applied horizontally. These notional forces should be assumed to act on structure in any one orthogonal direction at a time and should be applied at each roof and floor level or their equivalent. They should be taken as acting simultaneously with vertical loads.

Note The notional force should not be: applied when considering overturning; combined with horizontal loads; combined with temperature effects; taken to contribute to net shear on the foundations.

Sway stability may be provided, for example, by braced frames, joint rigidity or by utilizing staircases, lift cores and shear walls. Whatever system is used, reversal of loading should be considered. The cladding, floors and roof should have adequate strength and be so secured to the structural framework as to transmit all horizontal forces to the points of sway resistance. Where such sway stability is provided by construction other than the steel framework, the steelwork designer should state clearly the need for such construction and the forces acting upon it.

Fatigue

Fatigue is unlikely to be a problem for ordinary building structures; it is more significant for bridges. Since fatigue is largely dependent upon stress range (The difference between maximum and minimum stresses expressed in service), most bridges even do not receive sufficient applications of load heavy enough to produce the necessary large changes in stresses and subsequently fatigue. Generally fatigue need not be considered unless a structure or element is subjected to numerous significant fluctuations of stress. However, in structures subjected to a very large number of cycles of fluctuating (say, at least 1,00,000) load applications, failure may occur by the continued growth of cracks in the material at stresses well below those necessary to cause ordinary static yielding or collapse. Stress changes due to fluctuations in wind loading need not be considered but care should be taken of wind induced oscillations.

Crane girders, railway bridges and offshore structures subjected to wave loading are but a few examples where fatigue needs a consideration and check. When designing for fatigue, the load factor for action, γ_{fk} equal to unity should be used for the load causing stress fluctuation and stress range.

Plastic Collapse

Plastic analysis and design may be used if the conditions under the plastic method of analysis are satisfied.

18.7 LIMIT STATES OF SERVICEABILITY

Serviceability Limit State is related to the criteria governing normal use. Serviceability limit state is the limit state(s) beyond which service criteria specified for deflection, vibration, durability and fire resistance are no longer met:

Load factor, γ_{fk} of value equal to unity should be used for all loads to check the adequacy of the structure under serviceability limit states, unless specified otherwise.

Deflection

Deflection under service loads of a building or a building component should not impair the strength of the structure or components or cause damage to finishings. Deflections are the main serviceability limit state that must be considered in design. Deflections are to be checked for the most adverse but realistic combination of service loads and their arrangement, by elastic analysis, using a load factor of 1.0. Recommended limits of deflections for certain structural member systems are given in Table 18.3 (taken from draft code IS:800). Circumstances may arise where greater or lesser values than those given in the code would be more appropriate depending upon the nature of material in element to be supported (vulnerable to cracking or not) and intended use of the structure.

Where the deflection due to dead load plus live load combination is likely to be excessive, consideration should be given to pre-camber the beams, trusses, and girders. The values of desired camber should be specified in design drawing. Generally for spans greater than 25 m camber approximately equal to the deflection due to dead loads plus half the live load, may be used.

Dynamic Effects

Suitable provisions in the design should be made for the dynamic effects of live loads, impact loads and vibration due to machinery operating loads. In severe cases possibility of resonance, fatigue or unacceptable vibrations should be investigated. Unusually flexible structures (generally the height to effective width of lateral load resistance system exceeding 5:1) should be investigated for lateral vibration under dynamic wind loads. Structures subjected to large number of cycles of loading should be designed against fatigue failure. Table 18.4 (taken from draft code IS:800) gives limiting values of the natural frequency or the alternative total deflection to avoid resonance.

The design must make suitable provision for

(a) The effects of imposed loads, which can induce impact, vibration, etc.
(b) Vibration caused by machines and oscillation caused by harmonic resonance

To avoid resonance, the natural frequencies of structures or parts of structures must be sufficiently different from those of the excitation source.

Notes

(1) The deflection of a member should be calculated without considering the impact factor or dynamic effect of the loads.
(2) Roofs, which are very flexible, should be designed to withstand any additional load that is likely to occur as a result of ponding of water or accumulation of snow.

Table 18.3 *Deflection Limits Other Than for Pitched Roof Portal Frame*

Type of Building		*Design Load*	*Member*	*Supporting*	*Maximum Deflection*
Industrial building	Vertical	Live load	Purlin	Roof cladding	Span/150
		Live load	Simple span	Brittle cladding	Span/240
		Live load	Cantilever	Brittle cladding	Span/120
		Live load	Simple span	Elastic cladding	Span/180
		Live load	Cantilever	Elastic cladding	Span/90
		Live load	Simple span	Floor	Span/300
		Live load	Cantilever	Floor	Span/150
		Crane load (Manual operation)	Gantry	Crane	Span/500
		Crane load (Electric operation up to 50 t)	Gantry	Crane	Span/750
		Crane load (Electric operation over 50 t)	Gantry	Crane	Span/1000
		Crane (Vertical) + Roof load	Gantry	Crane	Inward-12 mm Outward-25 mm
		Moving load (Charge cars, etc.)	Gantry	Crane	Span/600
	Lateral Crane + Wind	No cranes	Column	Elastic cladding	Height/150
		No cranes	Column	Masonry/brittle cladding	Height/240
		Crane	Gantry (lateral)	Crane	Span/400
				Relative between rails	10 mm
		Crane	Column/frame	Gantry (pendant operated)	Height/100
			Column/frame	Gantry (cab operated)	Height/240
Other Buildings	Vertical	Live load	Floors & roofs	Not susceptible to cracking	Span/300
		Live load	Floors & roofs	Susceptible to cracking	Span/360
	Lateral	Wind	Building	……	Height/300
		Wind	Inter storey drift		Storey height/ 300

Table 18.4 *Limiting Values of Frequency or Deflection to Avoid Resonance*

Conditions	*Minimum natural Frequency (Hz)*	*Maximum Deflection $\delta_1 + \delta_2$ (mm)*
Floors over which people walk frequently	3	28
Floors which can be loaded in a rhythmical manner e.g. gymnasium or dance hall	5	10

The maximum deflection (Fig. 18.1) may be expressed as

$\delta_{max} = \delta_1 + \delta_2 - \delta_0$

δ_{max} = sagging in the final state relative to the straight line joining the supports.

δ_0 = pre-camber (hogging) of the beam in the unloaded state, (State 0)

δ_1 = variation of the deflection of the beam due to permanent loads immediately after loading, (State 1)

δ_2 = variation of the deflection of the beam due to the variable loading plus any time dependant deformations due to the permanent load, (State 2)

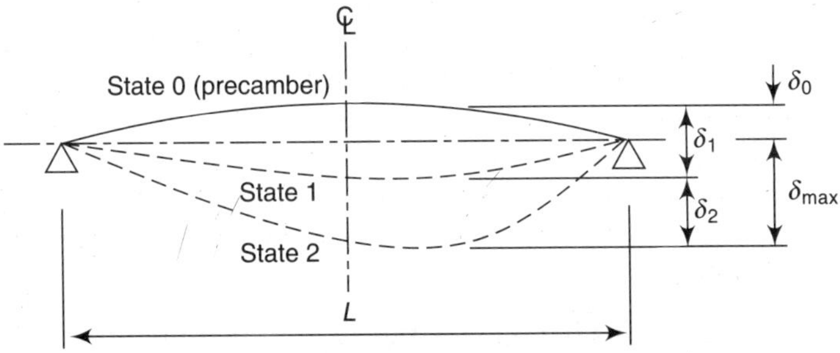

Fig. 18.1 *Vertical Deflections to be Considered*

Durability

The factors affecting the durability of the structures, under conditions relevant to their intended life are

(a) the environment,
(b) the degree of exposure,
(c) the shape of the member and the structural detailing,
(d) the protective measure, and
(e) ease of maintenance.

The durability of steel structures should be ensured by the recommendations of the code. Specialist literature may be referred to for more detailed additional information while designing for durability.

Fire Resistance

Although steel is an incombustible material, its strength may reduce substantially by the action of high temperatures, for example by a prolonged fire in a building. Fire resistance of a steel member is a function of its mass, its geometry, the actions to which it is subjected, its structural support condition, fire protection measures adopted and the fire to which it is exposed. Therefore in steel buildings the steel members must be provided with some form of fire protection. Such protection is afforded normally by encasing the steelwork in a suitable fire-resistant medium, the most suitable being encasing the member in concrete, fire board or cementitious fibre materials. Specialist literature may be referred to for detailed information in design of fire resistance of structures.

Ponding

Ponding will occur on almost all flat roofs to a certain degree, even though roof drains are present. The roof drains may be inadequate during heavy rains or may become partially or completely clogged. During strong winds this water may be pushed on one end of roof creating excessive unaccounted load in the analyses. Some of the recommendations are as follows:

(a) All roofs with a slope of less than 5% must be checked to ensure that rainwater is not collected in pools. Allowance must be made for possible construction inaccuracies, settlements of foundations, deflections of roofing materials and structural members and the effects of pre-camber. This also applies to floors of car parks and other open-sided structures.
(b) Pre-cambering of beams can be used to reduce the likelihood of rainwater collecting in pools, provided that rainwater outlets are appropriately located.
(c) Where the roof slope is less than 3%, it must be checked against collapse due to the weight of water (or snow, if applicable) collected in pools, which might be formed due to the deflection of structural members or roofing material.

18.8 METHOD OF DETERMINING ACTION EFFECTS

For the purpose of complying the requirements of the limit states of stability, strength and serviceability the effects of design actions on a structure, its members and connections should be determined either by elastic analysis or plastic analysis or by advance structural analysis.

The structure should be analysed in its entirety except as follows:

(a) Regular structures may be analysed as a series of parallel two -dimensional parts of structure, the analysis being carried out in each of the two directions, at right angles to each other, except when there is significant load redistribution between the parts of a structure.
(b) For vertical loading in a multi-storey structure, provided with bracing or shear walls to resist all lateral forces, each level thereof, together with the columns immediately above and below, may be considered as a part of structure, the columns being assumed fixed at the ends remote from the level under consideration.
(c) Where beams at a floor level in a multi-storey structure are considered as a part of the structure, the bending moment at the support of the beam may be determined based on the assumption that the beam is fixed at the far end support, one span away from the span under consideration, provided that the floor beam is continuous beyond that support point.

The effects of design action in the members and connections of a structure should be determined by assuming singly or in combination, the following forms of construction:

Rigid Construction In rigid construction, the connections between members (beam and column) at their junction are assumed to have sufficient rigidity to hold the original angles between the members connected at a joint unchanged under loading.

Semi-rigid Construction In semi-rigid construction, the connections between beams and column at their junction may not have sufficient rigidity to hold the original angles between the members at a joint unchanged, but are assumed to have the capacity to furnish a dependable and known degree of flexural restraint. The relationship between the degree of flexural restraint and the level of the load effects is established by any rational method or based on test results.

Simple Construction In simple construction, the connections between beams and column at their junction will not resist any appreciable moment and is assumed to be hinged.

Notes

1. In practice, structures are designed as simple or rigid construction; semi-rigid construction has never found a favour from designers.
2. The design of connections should be consistent with the form of construction, and the behaviour of the connections should not adversely affect any other part of the structure, beyond what is allowed for in design.

Summary

The Limit State method uses different partial safety factors based on theory of probability, involving the separate condition of different kinds of failure, type of materials and types of loads. Assessment of dead loads can be made more accurately than live loads. Allowable stress design assumes same factor of safety for dead loads and live loads, while for Limit State Design method the partial safety factors are different for dead and live loads. As a result, the weight of steel that will result by the two approaches will be different and will necessarily depend upon the ratios of live loads to dead loads. Generally, for structures, this ratio varies from 0.25 to as high as 4.0; low-rise steel buildings generally fall in the upper range of these ratios. Since in allowable stress design method safety factors for dead load and live load is same regardless of their ratio, heavier members will result and the factor of safety will increase as the live load to dead load ratio decreases. Savings in weight of steel of the order of 15 per cent for tension and compression members and 10 percent for beams may be achieved by Limit State method. On the other hand with high live load to dead load ratios, the increase in weight of steel is quite low. However, it should be noted and emphasized that the objective of recommending Limit State method is not the economy but to have a rational design based on uniform reliability for steel structures.

Appendices

I ABBREVIATIONS

I.S.J.B.	Indian Standard Junior Beam
I.S.L.B.	Indian Standard Light Beam
I.S.M.B.	Indian Standard Medium Beam
I.S.H.B.	Indian Standard Column Section
I.S.W.B.	Indian Standard Wide Flange Beam
I.S.J.C.	Indian Standard Junior Channel
I.S.L.C.	Indian Standard Light Channel
I.S.M.C.	Indian Standard Medium Channel
I.S.J.T.	Indian Standard Junior Tee Bar
I.S.S.C.	Indian Standard Special Channel
I.S.N.T.	Indian Standard Normal Tee Bar
I.S.H.T.	Indian Standard Wide Flange Tee Bar
I.S.S.T.	Indian Standard Long Legged Tee Bar
I.S.L.T.	Indian Standard Light Tee Bar
I.S.A.	Indian Standard Equal/Unequal Angle
I.S.B.A.	Indian Standard Bulb Angle
I.S.RO.	Indian Stand Round Bar
I.S.SQ.	Indian Standard Square Bar
I.S.F.	Indian Standard Flat
I.S.PL	Indian Standard Plate
I.S.SH	Indian Standard Sheet
I.S.ST	Indian Standard Strip

II UNIT WEIGHT OF BUILDING MATERIALS

1.	Asbestos cement sheets	120 to 156 N/m^2
2.	Asphalt	2200 N/m^3
3.	Brick	21600 N/m^3
4.	Brick work in cement mortar	19000 N/m^3
5.	Brick work in lime mortar	18000 N/m^3
6.	Cement	14400 N/m^3
7.	Cement mortar	20800 N/m^3
8.	Earth	18000 N/m^3
9.	Galvanised steel sheet	53.2 to 138.1 N/m^2
10.	Gravel	18000 N/m^3

11. Lime mortar — 16000 to 18400 N/m^3
12. Plain cement concrete — 23000 N/m^3
13. Reinforced cement concrete — 24000 N/m^3
14. Rolled steel — 79000 N/m^3
15. Roofing tiles — 350 to 850 N/m^2
16. Sand — 14000 N/m^3
17. Surkhi — 10100 N/m^3
18. Water — 10000 N/m^3

III LIVE LOADS

Live Loads on Floors

S. No.	*Loading Class Number*	*Type of Floor*	*Minimum Live Load in kN/m^2*	*Alternative Minimum Live Load*
1	2	Floors in dwelling houses, hospital, bed rooms, dormitories	2.0	Subject to a minimum total load of 2.5 times the values in column (4) for any slab and 6 times the value for any beam
2.	$2\frac{1}{2}$	Office floors	2.5	
3.	3	Floors of banking halls, entrance halls, reading rooms	3.0	
4.	4	Shop floors, class-room floors, assembly with fixed seating, restaurants	4.0	
5.	5	Floors of warehouses, workshops, factories, dance halls, waiting halls, for light weight loads	5.0	
6.	$7\frac{1}{2}$	Floors of warehouses, workshops, other buildings for medium weight loads	7.5	
7.	10	Floors of warehouses, workshops, other structures for heavy weight loads	10.0	
8.	Stairs	Stairs, landings, and corridors for class 2 loading but not liable to overcrowding.	3.0	Subject to a minimum of 1.3 kN concentrated load at the unsupported end of cantilever-step
9.	Balcony	Not liable to over-crowding		
		For class 2 loading	3.0	
		For all other classes	5.0	
		Liable to overcrowding	5.0	

(*Contd.*)

(Contd.)

S. No.	Loading Class Number	Type of Floor	Minimum Live Load in kN/m^2	Alternative Minimum Live Load
10.	Garage light	Floors using vehicles not exceeding 25 kN		
		Slabs	4.0	Worst combination of
		Beams	2.5	wheel loads
11.	Garage heavy	Vehicles not exceeding 40 kN	7.5	Subject to a minimum of one and half times wheel load but not less than 9 kN over 750 mm square

Live Loads on Roofs

S. No.	Type of Floor	Live Load Measured on Plan (kN/m^2)	Minimum Live Load Measured on Plan
1.	Flat sloping or curved roof with slopes up to 10°. (i) access provided	1.50	3.75 kN uniformly distributed over any span of 1 m width of roof slab and 9 kN uniformly distributed over the span in case of beams.
	(ii) access not provided	0.75	Half the load above, subject to a minimum of 400 N/m^2.
2.	Sloping roofs with slopes > 10°	(i) For roofing and purlins- 0.75 kN/m^2 less 20 N/m^2 for each one degree over 10° (ii) For beams, girders and trusses supporting roofing, 2/3 of load in (i). (iii) Loads in (i) and (ii) do not include loads due to rain, snow and dust.	

Note: All roof coverings (other than glass) should be capable of carrying an incidental load of 900 N concentrated over an area of 1250 mm square.

IV WIND MAP OF INDIA

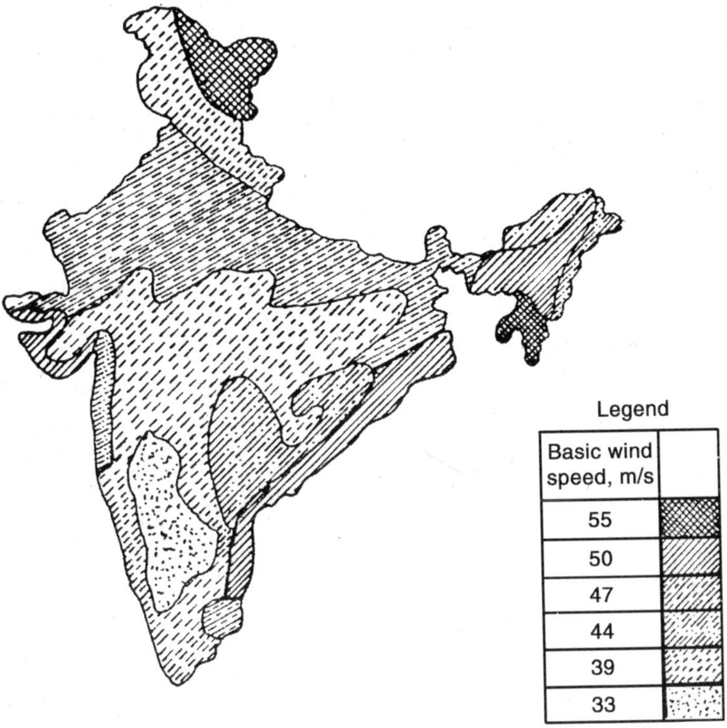

V RESPONSE SPECTRA FOR ROCKS AND SOIL SITES FOR 5% DAMPING (IS: 1893–2002)

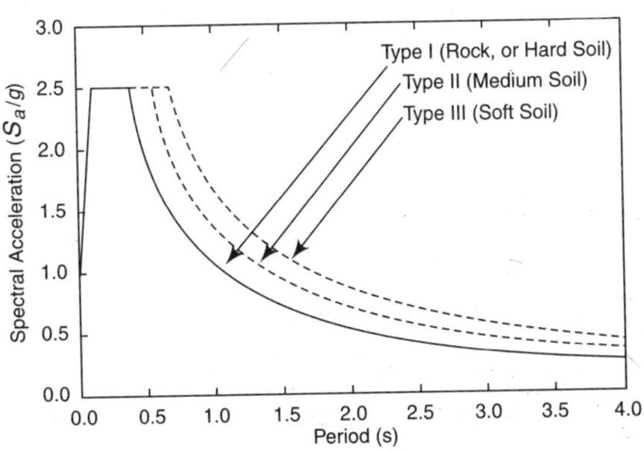

VI PERMISSIBLE BEARING PRESSURES ON SUBSOILS

Type	*kN/m²*
Alluvial soil, made ground, very wet sand	50
Soft clay, wet or loose sand	100
Ordinary fairly dry clay, sandy clay and fairly dry fine sand	200
Firm dry clay, firm ballast	300
Compact sand or gravel, hard compact clay	425

VII STRESS AREA OF BOLTS

Designation	*Size (mm)*	*Stress (net) Area (mm^2)*
M 2.5	2.5	3.39
M 3	3	5.03
M 4	4	8.78
M 5	5	14.2
M 6	6	20.1
M 8	8	36.6
M 10	10	58.0
M 12	12	84.3
M 16	16	157
M 20	20	245
M 24	24	353
M 30	30	561
M 33	33	694
M 36	36	817
M 39	39	976

VIII SHEAR CENTRE

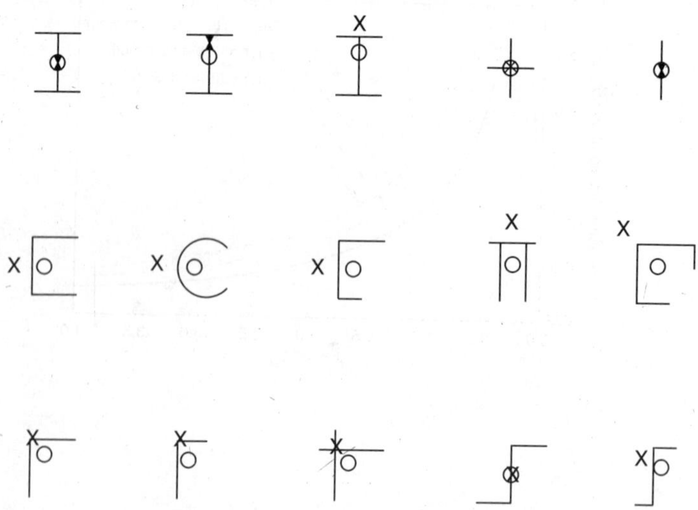

O Centroid
X Shear centre

IX EXTERNAL PRESSURE COEFFICIENTS (C_{PE}) FOR ROOFS OF RECTANGULAR CLAD BUILDINGS

Building Height Ratio	*Roof Angle* α	*Wind Angle* θ *0°* *EF*		*Wind Angle* θ *90°* *GH*		*Local Coefficients*			
							EG	FH	
	degrees								
$\frac{h}{w} < \frac{1}{2}$ 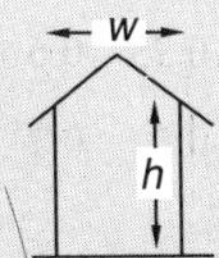	0	−0.8	−0.4	−0.8	−0.4	−2.0	−2.0	−2.0	—
	5	−0.9	−0.4	−0.8	−0.4	−1.4	−1.2	−1.2	−1.0
	10	−1.2	−0.4	−0.8	−0.6	−1.4	−1.4		−1.2
	20	−0.4	−0.4	−0.7	−0.6	−1.0			−1.2
	30	0	−0.4	−0.7	−0.6	−0.8			−1.1
	45	+0.3	−0.5	−0.7	−0.6				−1.1
	60	+0.7	−0.6	−0.7	−0.6				−1.1
$\frac{1}{2} < \frac{h}{w} \le \frac{3}{2}$ 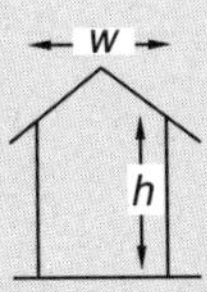	0	−0.8	−0.6	−1.0	−0.6	−2.0	−2.0	−2.0	—
	5	−0.9	−0.6	−0.9	−0.6	−2.0	−2.0	−1.5	−1.0
	10	−1.1	−0.6	−0.8	−0.6	−2.0	−2.0	−1.5	−1.2
	20	−0.7	−0.5	−0.8	−0.6	−1.5	−1.5	−1.5	−1.0
	30	−0.2	−0.5	−0.8	−0.8	−1.0			−1.0
	45	+0.2	−0.5	−0.8	−0.8				
	60	+0.6	−0.5	−0.8	−0.8				
$\frac{3}{2} < \frac{h}{w} < 6$ 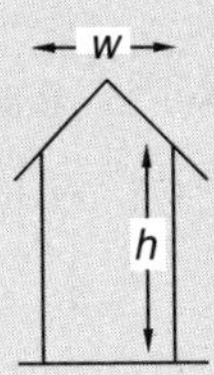	0	−0.7	−0.6	−0.9	−0.7	−2.0	−2.0	−2.0	—
	5	−0.7	−0.6	−0.8	−0.8	−2.0	−2.0	−1.5	−1.0
	10	−0.7	−0.6	−0.8	−0.8	−2.0	−2.0	−1.5	−1.2
	20	−0.8	−0.6	−0.8	−0.8	−1.5	−1.5	−1.5	−1.2
	30	−1.0	−0.5	−0.8	−0.7	−1.5			
	40	−0.2	−0.5	−0.8	−0.7	−1.0			
	50	+0.2	−0.5	−0.8	−0.7				
	60	+0.5	−0.5	−0.8	−0.7				

Notes

1. *h* is the height to eaves or parapet and *w* is the lesser horizontal dimension of a building.
2. Where no local coefficients are given, the overall coefficients apply.
3. For hipped roofs the local coefficient for the hip ridge may be conservatively taken as the appropriate ridge value.

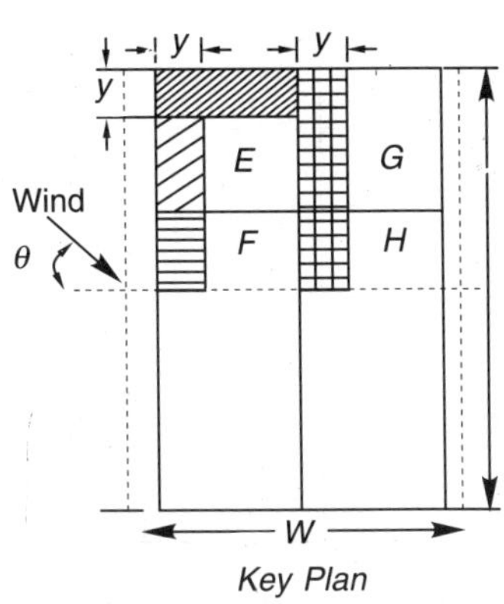

Key Plan

y = h or 0.15 w, whichever is the lesser

X EXTERNAL PRESSURE COEFFICIENTS (C_{PE}) FOR WALLS OF RECTANGULAR BUILDINGS

Building Height Ratio	Building Plan Ratio	Elevation / Plan	Wind Angle θ (degrees)	C_{pe} for Surface A	B	C	D	Local C_{pe}
$\frac{h}{w} < \frac{1}{2}$	$1 < \frac{l}{w} < \frac{3}{2}$	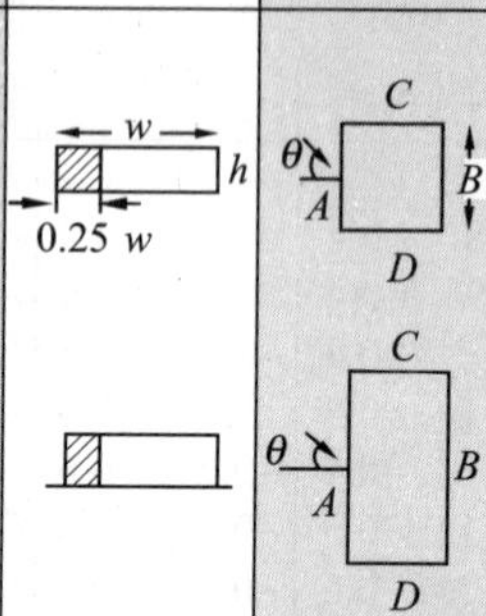	0	+0.7	–0.2	–0.5	–0.5	–0.8
			90	–0.5	–0.5	+0.7	–0.2	
	$\frac{3}{2} < \frac{l}{w} < 4$		0	+0.7	–0.23	–0.6	–0.6	–1.0
			90	–0.5	–0.5	+0.7	–0.1	
$\frac{1}{2} < \frac{h}{w} < \frac{3}{2}$	$1 < \frac{l}{w} < \frac{3}{2}$	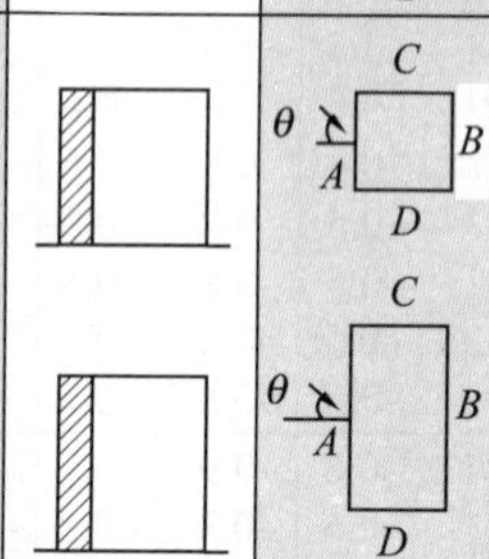	0	+0.7	–0.25	–0.6	–0.6	–1.1
			90	0.5	–0.5	+0.7	–0.25	
	$\frac{3}{2} < \frac{l}{w} < 4$		0	+0.7	–0.3	–0.7	–0.7	–1.1
			90	–0.5	–0.5	+0.7	–0.1	
$\frac{3}{2} < \frac{h}{w} < 6$	$1 < \frac{l}{w} < \frac{3}{2}$	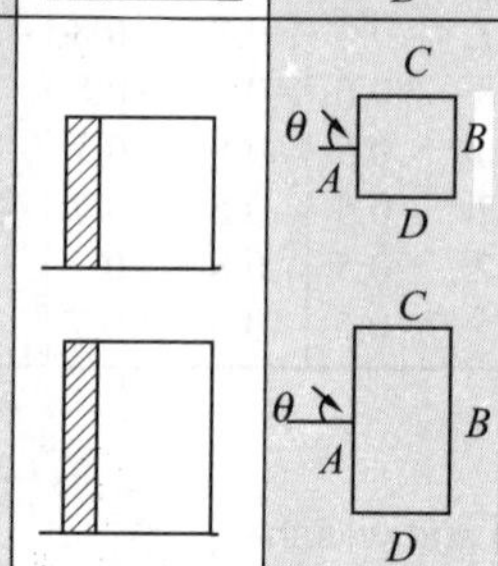	0	+0.8	–0.25	–0.8	–0.8	–1.2
			90	–0.8	–0.8	+0.8	–0.25	
	$\frac{3}{2} < \frac{l}{w} < 4$		0	+0.7	–0.4	–0.7	–0.7	–1.2
			90	–0.5	–0.5	+0.8	–0.1	
$\frac{h}{w} > \infty$	$\frac{l}{w} = \frac{3}{2}$	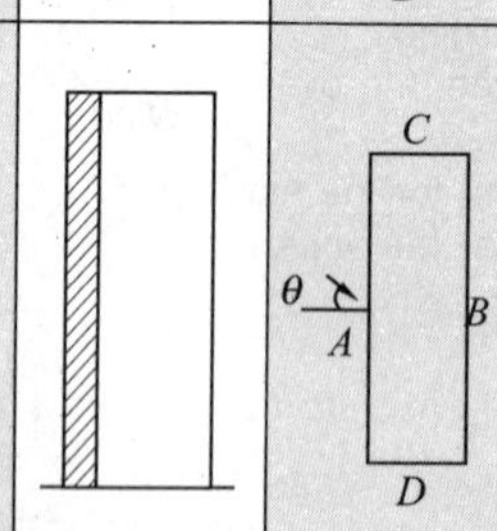	0	+0.95	–1.85	–0.9	–0.9	–1.25
			90	–0.8	–0.8	+0.9	–0.85	
	$\frac{l}{w} = 1.0$		0	+0.95	–1.25	–0.7	–0.7	–1.25
			90	–0.7	–0.7	+0.95	–1.25	
	$\frac{l}{w} = 2$		0	+0.85	–0.75	–0.75	–0.75	–1.25
			90	–0.75	–0.75	+0.85	–0.75	

Note h is the height to eaves or parapet, l is the greater horizontal dimension of a building and w is the lesser horizontal dimension of a building.

XI AVERAGE ACCELERATION SPECTRA (IS: 1893–1984)

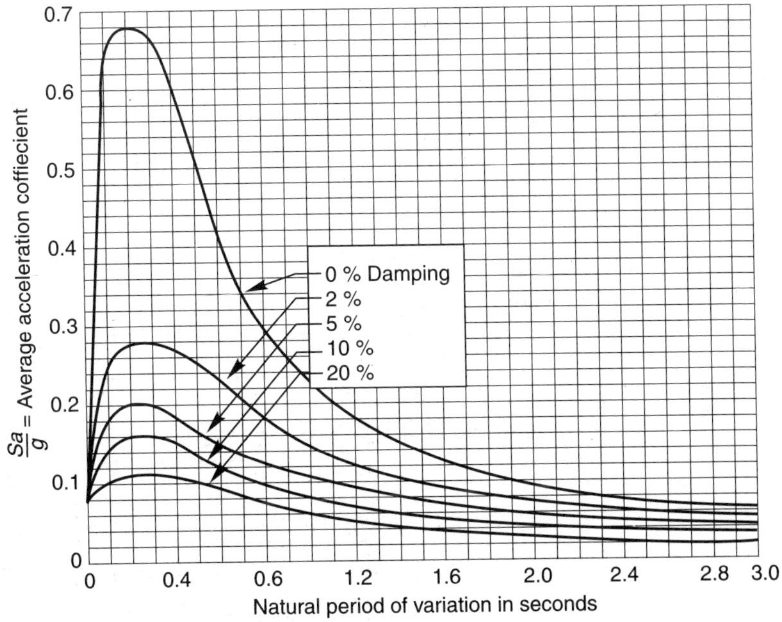

XII SIZES AND PROPERTIES OF STEEL TUBES FOR STRUCTURAL PURPOSES

Nominal Bore	Outside Diameter	Class	Thickness	Weight	Area of Cross-Section	Internal Volume	Surface Area		Moment of Inertia	Section Modulus Z	Radius of Gyration r	r^2
							External	Internal				
(mm)	(mm)		(mm)	(kg/m)	(cm^2)	(cm^3/m)	(cm^2/m)	(cm^2/m)	(cm^4)	(cm^3)	(cm)	(cm^2)
15	21.3	Light	2.0	0.962	1.21	235	669	543	0.57	0.54	0.69	0.47
		Medium	2.6	1.210	1.53	304		506	0.68	0.64	0.67	0.45
		Heavy	3.2	1.440	1.82	174		468	0.77	0.72	0.65	0.42
20	26.9	Light	2.3	1.410	1.78	391	845	701	1.36	1.01	0.87	0.76
		Medium	2.6	1.570	1.98	370		682	1.48	1.10	0.86	0.75
		Heavy	3.2	1.890	2.38	330		644	1.70	0.27	0.85	0.71
25	33.7	Light	2.6	2.010	2.54	638	1059	895	3.09	1.84	1.10	1.22
		Medium	3.2	2.420	3.07	585		858	3.60	2.14	1.08	1.18
		Heavy	4.0	2.950	3.73	519		807	4.19	2.49	1.06	1.12
32	42.4	Light	2.6	2.570	3.25	1087	1332	1169	6.46	3.05	1.41	1.99
		Medium	3.2	3.110	3.94	1018		1131	7.62	3.59	1.39	1.93
		Heavy	4.0	3.810	4.83	929		1081	8.99	4.24	1.36	1.86
40	48.3	Light	2.9	3.270	4.14	1419	1517	1335	10.70	4.43	1.61	2.59
		Medium	3.2	3.590	4.53	1379		1316	11.59	4.80	1.60	2.56
		Heavy	4.0	4.410	5.57	1276		1266	13.77	5.70	1.57	2.47
50	60.3	Light	2.9	4.140	5.23	2333	1894	1712	21.59	7.16	2.03	4.13
		Medium	3.6	5.070	6.41	2215		1668	25.87	8.58	2.01	4.03
		Heavy	4.5	6.170	7.89	2067		1612	30.90	10.20	1.98	3.92

(Contd.)

(Contd.)

Nominal Bore	Outside Diameter	Class	Thickness	Weight	Area of Cross-Section	Internal Volume	Surface Area External	Surface Area Internal	Moment of Inertia	Section Modulus Z	Radius of Gyration r	r^2
(mm)	(mm)		(mm)	(kg/m)	(cm^2)	(cm^3/m)	(cm^2/m)	(cm^2/m)	(cm^4)	(cm^3)	(cm)	(cm^2)
65	76.1	Light	3.2	5.800	7.33	3816	2391	2190	48.78	12.80	2.58	6.66
		Medium	3.6	6.490	8.20	3728		2165	54.01	14.20	2.57	6.59
		Heavy	4.5	7.920	10.10	3536		2108	65.12	17.10	2.54	6.48
80	88.9	Light	3.2	6.810	8.62	5346	2793	2592	79.21	17.80	3.03	9.19
		Medium	4.0	8.430	10.70	5140		2542	96.34	21.70	3.00	9.00
		Heavy	5.0	10.300	13.20	4889		2479	116.04	26.20	2.97	8.82
90	101.6	Light	3.6	8.760	11.10	6999	3192	2966	133.2	26.2	3.47	12.02
		Medium	4.0	9.700	12.30	6881		2941	146.3	28.8	3.45	11.98
		Heavy	5.0	11.900	15.20	6590		2878	177.5	34.9	3.42	11.70
100	114.3	Light	3.6	9.900	12.50	9009	3591	3365	192.0	33.6	3.92	15.33
		Medium	4.5	12.100	15.50	8707		3308	234.3	41.0	3.89	15.10
		Heavy	5.4	14.500	18.50	8413		3252	274.5	48.0	3.85	14.86
110	127.0	Light	4.5	12.200	17.30	10933	3990	3707	325.3	51.2	4.33	18.78
		Medium	5.0	15.000	19.20	10751		3676	357.1	56.2	4.32	18.64
		Heavy	5.4	16.200	20.60	10605		3651	382.0	60.2	4.30	18.52
125	139.7	Light	4.5	14.900	19.10	13417	4389	4106	437.2	62.6	4.78	22.87
		Medium	5.0	16.600	21.20	13212		4075	480.5	68.8	4.77	22.71
		Heavy	5.4	17.900	22.80	13050		4050	514.5	73.7	4.75	22.58
		Light	4.5	16.400	20.90	16151	4788	4505	572.2	75.1	5.23	27.37
135	152.4	Medium	5.0	18.200	23.20	15926		4474	629.5	82.6	5.21	27.19
		Heavy	5.4	19.500	24.90	15748		4479	674.5	88.5	5.20	27.05

(Contd.)

(Contd.)

Nominal Bore	Outside Diameter	Class	Thick-ness	Weight	Area of Cross-Section	Inter-nal Volume	Surface Area		Moment of Inertia	Section Modulus Z	Radius of Gyration r	r^2
							Exter-nal	Inter-nal				
(mm)	(mm)		(mm)	(kg/m)	(cm^2)	(cm^3/m)	(cm^2/m)	(cm^2/m)	(cm^4)	(cm^3)	(cm)	(cm^2)
		Light	4.5	17.800	22.70	19138	5187	4904	732.6	88.7	5.68	32.27
150	165.3	Medium	5.0	19.700	25.10	18894		4873	806.6	97.7	5.66	32.07
		Heavy	5.4	21.200	27.10	18699		4847	864.7	105.0	5.65	31.92
		Light	4.5	18.100	23.20	19931	5287	5005	777.2	92.4	5.79	33.56
160	168.3	Medium	5.0	20.100	25.70	19681		4973	855.8	102.0	5.78	33.36
		Heavy	5.4	21.700	27.60	19483		4948	917.7	109.0	5.76	33.27
		Light	5.0	23.300	29.60	26504	6085	5771	132.0	136.0	6.67	44.54
175	193.7	Medium	5.4	25.300	31.90	26274		5746	141.7	146.0	6.66	44.36
		Heavy	5.9	27.300	34.80	25987		5715	153.6	159.0	6.64	44.13
200	219.1	Light	5.0	26.400	33.6	34340	6883	6500	1928	176	7.57	57.33
		Medium	5.6	29.400	37.6	33947		6531	2142	195	7.55	57.02
		Heavy	5.9	31.000	39.5	33751		6513	2247	205	7.54	56.86
225	244.5	Heavy	5.9	34.200	44.2	42529	7681	7311	3149	258	8.44	71.21
250	273.0	–	5.9	38.800	49.5	53592	8578	8207	4412	323	9.45	89.30
300	323.9	–	6.3	49.500	63.4	76121	10177	9781	7992	493	11.20	125.44
350	355.6	–	8.0	68.800	86.5	90588	11173	10670	13111	737	12.30	151.29

XIII SHAPE FACTOR f

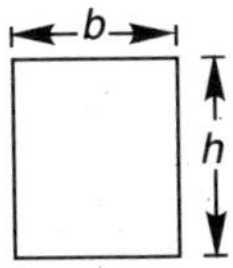	1.5
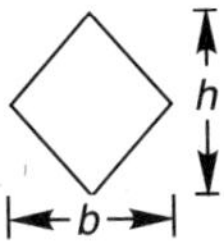	2
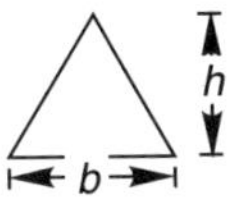	2.34
	1.7
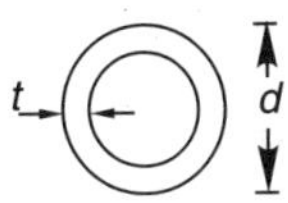	$4/\pi$
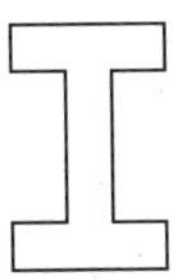 Wide flange (strong axis)	1.14
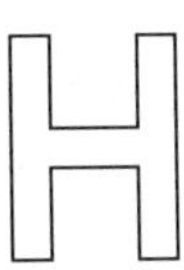 Wide flange (weak axis)	1.50

XIV LOAD FACTOR F

Dead load	1.7
Dead load + imposed load	1.7
Dead load + wind/seismic load	1.7
Dead load + imposed load + wind/seismic load	1.3

List of Symbols

A	Cross-sectional area
A_{net}	Net cross-sectional area
A_{gross}	Gross cross-sectional area
A_f	Flange area
A_w	Web area
a, b	Respectively the greater and the lesser projections of base plate beyond column
b_o	Width of steel flange in encased member
C	The c/c distance of battens, compressive force
c	Distance between vertical stiffeners
C_b	Bending coefficient
C_m	Coefficient (Reduction factor)
C_1, C_2	Respectively the lesser and greater distances from the section's neutral axis to the extreme fibres
D	Overall depth of beam
d	Gross diameter of rivet, clear distance between flange angles or between flanges ignoring fillets
d_1	Beam web without horizontal stiffeners—the clear distance between the flanges neglecting fillets or the clear distance between the inner toes of flange angles; beam web with horizontal stiffeners—the clear distance between horizontal stiffener and the tension flange neglecting fillets or the inner toes of tension flange
d_2	Twice the clear distance from the neutral axis of a beam to the compression flange neglecting fillets for the inner toes of flange angles
d_3	Clear distance between the two flanges
d_s	Depth of splice plate
E	Modulus of elasticity for steel = 2×10^5 MPa
e	Eccentricity
F	Load factor
f	Shape factor
f_b	Basic design stress
f_{cr}	Critical stress
f_y	Yield stress
f_{eb}	Elastic critical stress in bending
f_{cc}	Elastic critical stress in compression

g	Gauge distance
h	Outstand of stiffener
I	Moment of inertia, Importance factor
I_{net}	Net moment of inertia
I_{gross}	Gross moment of inertia
K	Buckling coefficient
k_1, k_2	Coefficients
L	Span, length of the member
l	Effective length of the member
M	Elastic bending moment
M_p	Plastic moment
M_w	Moment due to wind
N	Number of parallel planes of battens, number of possible plastic hinges
n	Number of rivets, number of possible independent mechanisms
P_{ac}	Maximum load carrying capacity of a strut
P_{at}	Maximum load carrying capacity of a tension member
P_E	Euler load
P_y	Yield strength of axially loaded section
p_d	Design wind pressure
Q	Form factor
R	Response reduction factor
R_v	Rivet value
r	Distance of rivet from C.G. of rivet group, radius of gyration of section
S	size of fillet weld
s	Pitch, staggered pitch
Sa/g	Average spectral acceleration coefficient
T	Mean thickness of compression flange, Time period
t	Thickness of web
V	Transverse shear
W	Concentrated load
w	Uniformly distributed load, pressure on the underside of the base
W_u	Ultimate load
Z	Elastic modulus of section, Zoning factor
Z_p	Plastic modulus of section
β	Ratio of smaller to larger moment
λ	Slenderness ratio
λ_0	Characteristic slenderness ratio = $\sqrt{P_y / P_e}$
σ_{ac}	Maximum permissible compressive stress in axially loaded strut.
$\sigma_{ac,\text{ cal}}$	Calculated average stress due to axial compressive stress
σ_{at}	Maximum permissible tensile stress in axially loaded tension member
$\alpha_{at,\text{ cal}}$	Calculated average stress due to axial tensile force
σ_{bs}	Maximum permissible bending stress in slab base

σ_{bc}	Maximum permissible compressive stress due to bending
$\sigma_{bc,\ cal}$	Calculated compressive stress due to bending
σ_{bt}	Maximum permissible tensile stress due to bending
$\sigma_{bt,\ cal}$	Calculated tensile stress due to bending
σ_e	Maximum permissible equivalent stress
σ_p	Maximum permissible bearing stress
σ_{pf}	Maximum permissible bearing stress in fastener
σ_{tf}	Maximum permissible stress in axial tension in fastener
τ_{cr}	Critical shear stress
τ_{va}	Maximum permissible average shear stress in a member
τ_{vm}	Maximum permissible shear stress in a member
τ_{vf}	Maximum permissible shear stress in fastener
τ_v	Resultant shear stress
θ	Ratio of the rotation at the hinge point to the relative elastic rotation at the far end of the beam segment containing plastic hinge
ψ	Ratio of total area of both the flanges at the point of least bending moment to the corresponding area at the point of greatest bending moment
ω	Ratio of moment of inertia of the compression flange alone to that of the sum of the moments of inertia of the flanges, each calculated about its own axis parallel to the *Y-Y* axis of the girder at the point of maximum bending moment
ρ	Unit weight of steel
μ	Poisson's ratio
ε	Strain
ϕ	Curvature, nominal diameter of rivet

Index

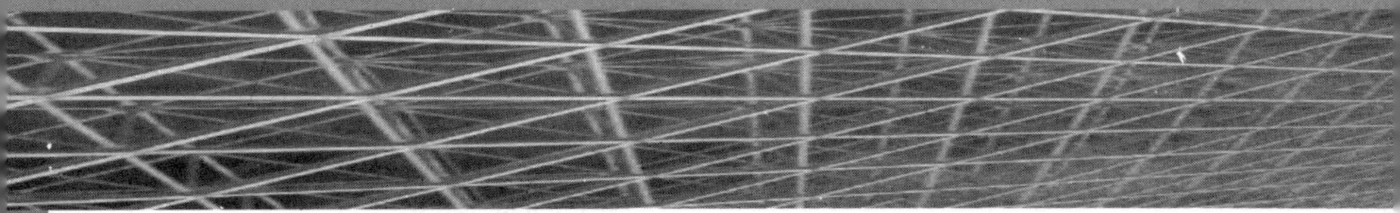